SUFFOLK UNIVERSITY
MILDRED F. SAWYER LIBRARY
8 ASHBURTON PLACE
BOSTON, MA 02108

D1523148

INTRACELLULAR CALCIUM

MONOGRAPHS IN MOLECULAR BIOPHYSICS AND BIOCHEMISTRY

Edited by
H. Gutfreund,
*Department of Biochemistry,
University of Bristol*

The Physical Behaviour of Macromolecules with biological functions
S. P. Spragg

Intracellular Calcium: its Universal Role as Regulator
A. K. Campbell

**Biothermodynamics
The Study of Biochemical Processes at Equilibrium**
J. T. Edsall and H. Gutfreund

Lewis Victor Heilbrunn
(1892–1959)
University of Pennsylvania, Philadelphia
A pioneer of intracellular Ca^{2+} as a universal regulator

INTRACELLULAR CALCIUM,
its Universal Role as Regulator

Anthony K. Campbell
Department of Medical Biochemistry,
Welsh National School of Medicine

JOHN WILEY & SONS LIMITED
Chichester · New York · Brisbane · Toronto · Singapore

Copyright © 1983 by John Wiley & Sons Ltd.

All rights reserved.

No part of this book may be reproduced by any means, nor transmitted, nor translated into a machine language without the written permission of the publisher.

Library of Congress Cataloging in Publication Data:

Campbell, Anthony K.
　Intracellular calcium.

　(Monographs in molecular biophysics and biochemistry)
　Includes index.
　1. Calcium—Physiological effect.　2. Cellular
control systems.　I. Title.　II. Series. [DNLM:
1. Calcium—Metabolism.　2. Calcium—Physiology. QV276
C187i]
QP535.C2C268　1983　　　599.01′9214　　　82-8656
ISBN 0 471 10488 4　　　　　　AACR2

British Library Cataloguing in Publication Data:

Campbell, Anthony K.
　Intracellular calcium.—(Monographs in molecular
　biophysics and biochemistry)
　1. Calcium—Physiological effect
　2. Cellular control mechanisms
　I. Title　　II. Series
　574.87′61　　QP535.C2
ISBN 0 471 10488 4

Phototypeset by Macmillan India Ltd., Bangalore, India and Printed in Great Britain by Page Bros. (Norwich) Ltd.

This book is dedicated to Professor C. Nicholas Hales and to Dr. Leslie H. N. Cooper, FRS, who introduced me to the problem of intracellular Ca^{2+} and whose encouragement, critical and stimulating advice over many years have greatly widened my horizons, providing me with the enthusiasm to write this book.

'Ja calcium, das ist alles!' Otto Loewi, 1959

Calcium (Ca)
Atomic weight   40.08
Atomic number   20
Melting point   842–848°C
Boiling point   1487°C
Specific gravity 1.55 (20°C)
Valence   2

Contents

Preface	xv
Chapter 1 The natural history of calcium	1
1.1. The natural occurrence of calcium	1
1.1.1 The discovery of calcium	1
1.1.2 The geology of calcium	4
1.1.3 Biomineralization	6
1.2. The requirement of cells for calcium	8
1.2.1 Calcium in water and extracellular fluids	8
1.2.2 Effect of removal of calcium on animal cells	10
1.2.3 Requirement of plants for calcium	12
1.2.4 Calcium and bacteria	12
1.2.5 Tissue content of calcium and other cations	13
1.3 The four main biological roles of calcium	13
1.3.1 Structural role	15
1.3.2 Electrical role	15
1.3.3 Cofactor for extracellular enzymes and proteins	16
1.3.4 Calcium as an intracellular regulator	17
1.4. An historical perspective	17
1.5. Some questions	21
Chapter 2 The investigation of intracellular Ca^{2+} as a regulator	23
2.1. What is meant by the term regulator?	23
2.2. The concept of 'threshold phenomena'	25
2.3. Experimental approaches	27
2.3.1 The basic questions	27
2.3.2 Does intracellular Ca^{2+} play a role in the phenomenon?	28
2.3.3 Is the role of intracellular Ca^{2+} 'active' or 'passive'?	37

2.3.4 What is the molecular basis of the regulatory role of intracellular Ca^{2+}?	38
2.3.5 Pathology and pharmacology	39
2.4. The need to measure cell Ca^{2+}	39
2.4.1 The need to measure total Ca^{2+}	40
2.4.2 The need to measure free Ca^{2+}	43
2.5. Measurement of Ca^{2+}	44
2.5.1 Total Ca^{2+}	44
2.5.2 Free Ca^{2+}	49
2.5.3 Localization of Ca^{2+} within the cell	66
2.6. The measurement of Ca^{2+} fluxes	72
2.6.1 Ca^{2+} pools within the cell	72
2.6.2 Methods of measuring Ca^{2+} flux	73
2.6.3 Analysis of data	73
2.6.4 Problems in interpretation	75
2.7. Ca^{2+} buffers	76
2.8. Other ions	82
2.9. Conclusions	84

Chapter 3 The chemistry of biological calcium — 85

3.1. The value of the chemical approach	85
3.2. The ligands	86
3.2.1 What is a ligand?	86
3.2.2 Inorganic Ca^{2+} ligands	88
3.2.3 Naturally occurring small organic Ca^{2+} ligands	89
3.2.4 Polymeric and polyvalent electrolytes as Ca^{2+} ligands	90
3.3 Thermodynamics and coordination chemistry of Ca^{2+} binding	96
3.3.1 Thermodynamic principles	96
3.3.2 The problem of activity coefficients	100
3.3.3 How much Ca^{2+} is bound to ligands and how fast does it equilibrate?	104
3.3.4 Solubility of calcium salts	114
3.3.5 Coordination chemistry of calcium	117
3.4. Is Ca^{2+} binding active?	120
3.4.1 Criteria	120
3.4.2 Regulatory Ca^{2+}-binding proteins	123
3.5. What are the chemical conditions for a 'threshold'?	130
3.6. Diffusion	132
3.7. Conclusions	133

Chapter 4 Intracellular Ca^{2+} and the electrical activity of cells	135
4.1. Some questions	135
4.2. The electrical properties of membranes	136
4.2.1 Historical	136
4.2.2 The cell membrane	137
4.2.3 Ca^{2+} and the resting membrane potential	139
4.2.4 How to establish an 'active' relationship between intracellular Ca^{2+} and the electrical properties of cells	144
4.3. Ca^{2+} and excitable cells	145
4.3.1 What is an excitable membrane?	145
4.3.2 Ca^{2+} as a current carrier across membranes	149
4.3.3 Can opening of Ca^{2+} channels lead directly to an increase in cytosolic free Ca^{2+}?	155
4.3.4 How does the Ca^{2+} movement through Ca^{2+} channels stop?	160
4.4. Effects of intracellular Ca^{2+} on the electrical properties of membranes	160
4.4.1 Possible physiological significance	160
4.4.2 Intracellular Ca^{2+} and K^+ conductance	162
4.4.3 Intracellular Ca^{2+} and communication between adjacent cells	165
4.4.4 Intracellular Ca^{2+} and the response of photoreceptors to light	171
4.5. Intracellular Ca^{2+} and bioluminescence	178
4.6. The regulation of intracellular Ca^{2+} through internal stores and pumps	185
4.6.1 The function of intracellular Ca^{2+} stores and pumps	185
4.6.2 Intracellular Ca^{2+} stores and the regulation of cytoplasmic free Ca^{2+}	186
4.6.3 Ca^{2+} efflux from cells	198
4.7. Transcellular Ca^{2+} transport	205
4.8. Conclusions	205
Chapter 5 Calcium and cell movement	206
5.1. Types of cell movement	206
5.1.1 Prokaryote motility	206
5.1.2 Eukaryote motility	209
5.2. The molecular basis of eukaryotic cell movements	210
5.2.1 Microtubules and tubulin	212
5.2.2 Microfilaments and actin	219
5.2.3 Actomyosin and troponin	225

5.3. Muscle	231
5.3.1 The problems of muscle contraction	231
5.3.2 Evidence for the role of intracellular Ca^{2+}	234
5.3.3 Regulation of intracellular free Ca^{2+} in muscle	238
5.3.4 The mechanism by which Ca^{2+} stimulates contraction	245
5.3.5 Secondary regulators	249
5.4. Chemotaxis in eukaryotes	252
5.5. The special case of the spasmoneme	254
5.6. Conclusions	255

Chapter 6 Intracellular Ca^{2+} and intermediary metabolism   257

6.1. How and why is intermediary metabolism regulated?	257
6.1.1 Intercellular pathways	257
6.1.2 Mechanisms of regulation	258
6.1.3 The need for 'second messengers'	261
6.2. Cyclic nucleotides and intracellular Ca^{2+}	268
6.2.1 Regulation of intracellular cyclic nucleotide concentration by Ca^{2+}	269
6.2.2 Regulation of intracellular Ca^{2+} by cyclic nucleotides	274
6.2.3 Coregulation of enzymes by Ca^{2+} and cyclic nucleotides	275
6.3. Glucose metabolism	277
6.3.1 Glucose transport across the cell membrane	278
6.3.2 Glycogen breakdown and synthesis	280
6.3.3 Glycolysis and glucose oxidation	284
6.3.4 Gluconeogenesis	286
6.4. Lipid metabolism	291
6.4.1 Lipolysis	291
6.4.2 Lipogenesis	296
6.4.3 Phospholipid metabolism	296
6.4.4 Prostaglandins, endoperoxides and thromboxanes	301
6.5. Conclusions	303

Chapter 7 Endocytosis and exocytosis: The uptake and release of substances from cells   305

7.1. What are endocytosis and exocytosis?	305
7.2. Exocytosis—vesicular secretion	307
7.2.1 What is secretion?	307
7.2.2 The stimuli of exocytosis	315

7.2.3 Is intracellular Ca^{2+} the trigger for exocytosis?	319
7.2.4 Ca^{2+} and the electrophysiology of secretory cells	329
7.2.5 The maintenance of cytosolic free Ca^{2+} and the source of Ca^{2+} for exocytosis	333
7.2.6 How does cytoplasmic Ca^{2+} initiate secretion?	335
7.2.7 Secondary regulators of exocytosis	344
7.3. Endocytosis	347
7.3.1 The phenomena of phagocytosis and pinocytosis	347
7.3.2 Is intracellular Ca^{2+} the trigger for endocytosis?	348
7.3.3 Intracellular Ca^{2+} and other phenomena associated with endocytosis	349
7.4. Non-vesicular secretions	353
7.4.1 Fluid secretions	356
7.4.2 Steroid hormone secretion	358
7.5. Overall conclusions	361

Chapter 8 The reproduction and development of cells — 362

8.1. Ca^{2+} and cell growth	362
8.2. The biology of cellular reproduction and development	364
8.2.1 The life cycle of cells	364
8.2.2 Sexual reproduction	368
8.2.3 Cell transformation	370
8.2.4 Embryonic growth and tissue development	371
8.2.5 Dormancy	372
8.3. Ca^{2+} and mitosis	373
8.3.1 The cell biology of mitosis	373
8.3.2 The potential role for intracellular Ca^{2+} in mitosis	374
8.3.3 Lymphocyte activation	376
8.3.4 Egg fertilization	380
8.3.5 Conclusions	386
8.4. Ca^{2+} and meiosis	386
8.5. Ca^{2+} and cell transformation	390
8.5.1 *Naegleria gruberii*	390
8.5.2 Capacitation of sperm	391
8.6. Overall conclusions	392

Chapter 9 The pathology and pharmacology of intracellular Ca^{2+} — 393

9.1. Tissue calcium and the medical problems	393
9.1.1 Types of tissue injury	395

9.1.2	Evidence for a role for intracellular ca^{2+} in cell injury	396
9.1.3	Infectious diseases	400
9.1.4	Toxins	405
9.1.5	Anoxia	408
9.1.6	Immune injury	408
9.1.7	Membrane abnormalities and changes in cell shape	413
9.1.8	Muscle diseases	414

9.2. Possible mechanisms of cell injury induced by a change in intracellular Ca^{2+} — 416
  9.2.1 How can a pathological change in cell Ca^{2+} occur? — 416
  9.2.2 Abnormalities in intracellular Ca^{2+} stores — 421
  9.2.3 Abnormalities in the regulation of intracellular free Ca^{2+} — 423
  9.2.4 Ca^{2+}-dependent regulatory proteins — 424
  9.2.5 Normally 'passive' or 'latent' Ca^{2+} ligands — 424
  9.2.6 Precipitation — 425

9.3. Intracellular Ca^{2+} and necrosis — 426

9.4. The pharmacology of intracellular Ca^{2+} — 429
  9.4.1 The problems — 429
  9.4.2 Anaesthetics — 430
  9.4.3 Phenothiazines — 437
  9.4.4 Cardiovascular agents — 439
  9.4.5 Disodium cromoglycate and antiallergics — 446
  9.4.6 Xanthines — 447
  9.4.7 Cations — 449
  9.4.8 Miscellaneous compounds — 450
  9.4.9 Conclusions — 450

9.5. Overall conclusions — 452

Chapter 10 Synthesis and perspectives — 455

10.1. Synthesis—what is so special about calcium? — 455
  10.1.1 Resumé — 455
  10.1.2 Unitary hypothesis — 459

10.2. Problems — 469
  10.2.1 General mechanisms — 469
  10.2.2 Plants — 471
  10.2.3 Prokaryotes — 473

10.3. Guidelines for establishing intracellular Ca^{2+} as a regulator — 474

10.4. Perspectives — 476
  10.4.1 Techniques for chemical studies on intact and single cells — 476

10.4.2 The evolutionary significance of intracellular Ca^{2+}	477
10.4.3 Coda	482
References	483
Species index	539
Subject index	541

Preface

Most of us are used to being asked to explain the nature of our work to family and friends who often have had little or no formal scientific training. 'Oh, calcium, isn't that something to do with teeth and bones?' The more informed layman might even have heard of vitamin D, parathyroid disease and the kidney in relation to whole-body calcium metabolism. Yet very few people are aware of the vital function calcium plays inside cells in controlling their behaviour, the subject of this book. Even within the scientific community the idea that intracellular calcium plays a wide-ranging role in cell activation has been surprizingly slow to gain acceptance. The significance of Sutherland's discovery of cyclic AMP was quickly grasped, and yet the concept of calcium as an intracellular messenger had been proposed some 30 or 40 years before.

During the last ten years or so there has been an enormous increase in literature relating to intracellular calcium, including thousands of original papers and reviews published each year. In 1980 a new journal, *Cell Calcium*, was started by Churchill-Livingstone. Most major biological meetings now include several papers concerned with this topic.

The aim of this book is to argue that calcium is the mediator inside cells of a wide range of biological phenomena involving cell activation. I have tried to highlight phenomena where there is good evidence for this assertion. In other cases, where little or no evidence exists, an experimental approach is suggested to answer the question, 'Is intracellular calcium the mediator of cell activation?' An attempt has been made to provide an historical perspective throughout the book, not simply for the historically minded but because close examination of classical experiments provides illuminating insights into how the scientific method works in practice.

Modern biochemistry has sometimes been open to the criticism that phenomena are generated in the test-tube which have little or no relevance to events occurring in the intact cell. It was Frederick Gowland Hopkins (Fig. 1.5), in the early years of this century, who argued that it was justifiable to study reactions in broken cells provided that the 'event' which occurred in the living cell also occurred in the *in vitro* system. The philosophy behind this book is that, if one is interested in understanding how life works at a chemical level, it is essential first to define the problem in an appropriate form using the living system. It is then possible to ask what the chemical mechanisms are which underlie the physiological or pathological phenomenon in question. The correct definition of the biological problem enables the biochemist to know what to look for, and characterize, when isolating and purifying structures and molecules from living tissues. Some may be encouraged to think that chemical physiologist, rather than biochemist, is a more appropriate title for themselves.

It has been my intention to provide a book which will be of interest to university teachers and researchers, clinicians, postgraduates and interested undergraduates. The emphasis on intracellular calcium might, in some cases, be criticised on the grounds that other intracellular mechanisms, which are independent of intracellular calcium, have been ignored. In these examples I have tried not to obscure these alternative hypotheses but rather to stimulate the reader to ask, 'Is this the complete explanation?' 'Could calcium also have a role in regulating this cell?'

In the first chapter attention is drawn to familiar examples of calcium salts in Nature. The theme is developed that there are four main biological roles for calcium and that, through an historical perspective, it has emerged that one, the particular topic for the remainder of the book, is crucial if animals and plants are to control their behaviour and relationship with the environment. Although less is known about calcium in bacteria, a few special examples are discussed. Chapter 2 deals with how one investigates intracellular calcium as a regulator. It deals not only with methodology but also attempts to lay down criteria which must be satisfied to justify the claim that intracellular calcium is the mediator of cell stimuli. Chapter 2 also establishes the philosophy of defining problems at a physiological level before examining the chemistry of the reactions involved, which is the subject of Chapter 3. Chapters 4–8 contain a detailed discussion of the evidence for intracellular calcium being the mediator of certain electrical changes in cells, cell movement, activation of intermediary metabolism, secretion and cell division. Many drugs and agents of cell injury cause alterations in intracellular calcium. Chapter 9 examines the question, 'Are these changes directly involved in the change in cell structure or function caused by the drug or injurious agent, or are they secondary to other events in the cell induced by a calcium-independent mechanism?' Chapter 10 tries to provide a resumé of the preceding chapters and to formulate a unitary hypothesis which embraces all of the phenomena discussed in the book. It is here, and in some other chapters, that I have proposed an hypothesis of my own. Throughout the book I have tried to make it clear when a particular conclusion is my own and when it is in agreement with the consensus of scientific opinion.

Although this book contains quotations from more than 1500 references it is inevitable that some have had to be omitted. It is my sincere hope that no critical experimental papers have been forgotten and that any other omissions cause no offence to the many researchers working in this field.

Over the past ten years since I finished my Ph.D., my research has taken me to many exciting laboratories, and in particular marine laboratories such as the Marine Biological Association, Plymouth, Station Biologique, Roscoff, and RRS *Discovery*. The exposure I have had to the enormous range of animals as a result of these visits has been a great joy and has greatly widened my horizons. I have tried to capture some of the enthusiasm I have for animals and plants, in an aesthetic sense, by drawing the reader's attention to the considerable range of organisms from various phyla where intracellular calcium seems to play a role as a cell regulator.

In addition to the pleasure I have had from visiting other laboratories I have been fortunate enough to discuss my ideas with many outstanding scientists. The preparation of this book has given me an immense amount of stimulation and I am greatly indebted to Professor H. Gutfreund FRS for encouraging me to write it, and for giving much invaluable advice. I should like to thank Sir Eric Smith FRS, past Director of the Marine Biological Association Laboratory, for allowing me to work there, and Dr. L. H. N. Cooper FRS for introducing me to the laboratory, where I have spent so many happy hours. I am also particularly grateful to the present Director, Professor E. J. Denton FRS, for much encouragement and for the use of his equipment, including that for my first experiments with obelin. I should also like to thank the staff of this laboratory for much help over the past eight years, and in particular Dr. Hans Meves, Dr. Eric Corner, Roger Swinfen, Dr. Geoff Potts, Bob Maddock, and Skipper Chris Knott and the crew of the Gammarus. I am also grateful to Plymouth Ocean Projects and members of the UWIST subaqua club for helping me with my annual collection of obelin.

Thanks go to Professor Nick Hales and his research group with whom I worked for ten years. I particularly thank Dr. J. Paul Luzio, Dr. Ken Siddle, Dr. Gerry V. Brenchley, and Dr. Chris C. Ashley (University of Oxford) for many enjoyable times experimenting together and for countless searching and provocative discussions. I should like to thank Dr. R. J. Thompson and the various people with whom I have collaborated, Dr. D. George Moisescu, Dr. David W. Yates, Dr. J. S. Woodhead, Dr. Peter J. Herring, Professor Frank McCapra and his research group, and Dr. Jean-Marie Bassot.

I am grateful to Professor George H. Elder, everyone in the Department of Medical Biochemistry, and many senior members of the Welsh National School of Medicine for much support, and the members of my research group over the past few years, Dr. Bob Dormer, Dr. Maurice Hallett, Richard Daw, Chris Davies, Ashok Patel, Stephanie Matthews, Ian Weeks, Mary E. Holt, Alan Davies, Paul Morgan, Steve Edwards, several of whom have given valuable advice about the manuscript, and Chris Hullin, Dot Thomas and Dilys Marks, who have persevered so magnificently with my handwriting and whims over the years. I thank also Malcolm E. T. Ryall for developing and constructing the equipment which has enabled me to carry out experiments using chemiluminescence.

I am greatly indebted to the Science and Medical Research Councils, the Arthritis and Rheumatism Council, the Department of Health and Social Security, the Welsh Office Scheme for the Development of Health and Social Research, and The Royal Society for financial support.

Finally, I thank my family, Sue, David and Neil, for much moral support and for putting up with my cyclothymic moods, particularly Sue for typing the manuscript of this book and for encouraging my singing career from which I derive such pleasure and which has helped to retain my sanity during difficult times with science. I also thank my mother and late father who encouraged me to study science from an early age, and my sister Dr. Caroline Sewry, particularly for many illuminating discussions and material for this book.

It is my hope that the deliberations in this book will help those who read it to be excited by the question 'What is so special about calcium?'

Anthony K. Campbell
Cardiff
July 1982

CHAPTER 1

The natural history of calcium

'All true biologists deserve the coveted name of naturalist. The touch stone of the naturalist is his abiding interest in living Nature in all its aspects' Frederick Gowland Hopkins (1936).

1.1. The natural occurrence of calcium

A glance into any rockpool on the seashore will reveal many beautiful animals and plants in which calcium plays a vital part in their skeletal structures. This occurrence in Nature of calcium, in the form of stable complexes with anions, has aroused the interest of scientists from many disciplines. Since these complexes occur commonly in both living organisms and in others long extinct, it is clear that calcium has played an important part in biological evolution. Life probably began on earth about 3500 million years ago, 1000 million years after the earth was first formed (Harling, 1967; Schopf, 1970, 1975), and calcareous remains of extinct organisms from as far back as 2000–3000 million years ago have been found. These have been of great value to the geologist in studying the fossil record, and in enabling him to unravel the course of evolution through the main geological eras.

Although the importance of the element calcium in the skeletal structures of living organisms has been recognized for more than a century, only recently has the attention of biologists focussed on other biological roles for calcium, and in particular its role as a regulator inside cells. This awareness raises three questions. Firstly, in what state does calcium occur in Nature? Secondly, what are the requirements of living organisms for calcium? Thirdly, what function does calcium play in them?

1.1.1. The discovery of calcium

Early in June, 1808, Humphry Davy (Fig. 1.1) carried out an experiment at the Royal Institution in London which involved connecting a large battery he had made across a moist mixture of lime and red oxide of mercury (one third by weight). Using an idea he developed from experiments carried out by Berzelius

Fig. 1.1. Humphry Davy (1778–1829) aged 23. Portrait by Howard. By courtesy of the Royal Institution

and Pontin at the negative electrode he added a small globule of mercury. An amalgam was formed which, on removal of the mercury by distillation, yielded a tiny amount of a greyish-white metal with the lustre of silver. This metal burnt avidly with a yellow-red flame when exposed to the air. This forms mainly the nitride but also quicklime. Davy called the metal calcium, after the Latin *calx* meaning lime. His discovery of the element calcium, together with that of the other alkaline earths, was reported to the Royal Society on June 30th, 1808 (Davy, 1808b). The following year, in his second Bakerian lecture, Davy reported another method for preparing these metals involving the preparation of a mercury amalgam and the use of potassium vapour (Davy, 1810). The previous year, 1807, Davy had observed 'small globules having a high metallic lustre' which appeared after electrolysis of molten soda or potash. Davy argued that electricity was to be of great value in discovery of 'the true elements' (Davy, 1808a).

Davy and his protégé Michael Faraday (Williams, 1965; Agassi, 1971) thus recognized that salts consisted of positively and negatively charged components combined in fixed proportions. It was they who provided biologists with the chemistry enabling them to investigate the role of cations such as Ca^{2+} in living organisms.

Six naturally occurring, stable isotopes of calcium exist (Table 1.1A), nearly 97% of the calcium in nature being ^{40}Ca. Six radioactive isotopes of calcium have been produced (Table 1.1B), but have such short half-lives that they do not occur naturally. One naturally occurring isotope, ^{48}Ca, is also radioactive, but has a very long half-life.

The element calcium is a greyish-white metal, highly reactive with nitrogen, oxygen and water. It therefore occurs in Nature only as salts formed between Ca^{2+} and various anions.

$$3Ca + N_2 \rightarrow Ca_3N_2$$
$$2Ca + O_2 \rightarrow 2CaO$$
$$Ca_3N_2 + 6H_2O \rightarrow 2NH_3 + 3Ca(OH)_2$$
$$CaO + H_2O \rightarrow Ca(OH)_2$$

$\rightarrow$ Salts

Table 1.1. The isotopes of calcium

Mass number (atomic mass)	Natural abundance (%)	Radioactive decay
A Naturally occurring isotopes		
40 (39.96259)	96.94	—
42 (41.95863)	0.64	—
43 (42.95878)	0.145	—
44 (43.95549)	2.1	—
46 (45.9537)	0.0033	—
48 (47.9524)	0.18	β^- (0.12 MeV), $t_{1/2} > 2 \times 10^{16}$ years
B Radioactive isotopes		
38	—	γ (3.5 MeV), $t_{1/2} = 0.7$ sec
39	—	β^+ (6 MeV), $t_{1/2} = 0.9$ sec
41	—	Nuclear transformation K, $t_{1/2} = 10^5$ years
45	—	β^- (0.26 MeV), $t_{1/2} = 165$ days
47	—	β^-, γ (2.0, 0.7 MeV), $t_{1/2} = 4.7$ days
49	—	β^-, γ (2.0 MeV), $t_{1/2} = 8.8$ months

Data from Kaye and Layby (1959) and Handbook of Chemistry.

1.1.2. The geology of calcium

Calcium is one of the commonest elements, making up about 3% of the earth's crust (Day, 1963; Fyfe, 1974). On a molar basis aluminium is the commonest metallic element (Fig. 1.2). This is followed by four metals of comparable abundance—sodium, calcium, magnesium and iron. Potassium is only about one-half as abundant as either sodium or calcium.

Calcium-containing minerals are found in the three main types of rock—igneous, sedimentary and metamorphic. Natural waters also contain a relatively high concentration of ionised calcium. In the sea, for example, the calcium concentration is about 10 mM. Igneous rocks, formed originally by fusion of molten rock, contain a variety of calcium silicates, phosphates and fluorides (Table 1.2). Calcium carbonate is also found in some igneous rocks. The ubiquity of calcium minerals, of various compositions, in igneous rocks is probably due to the ability of calcium to undergo isomorphous replacement with other cations, such as sodium (Day, 1963). In contrast to the variety of calcium minerals found in igneous rocks, in sedimentary rocks the major calcium salt is calcium carbonate (Table 1.2).

The oldest sedimentary rocks were formed about 3750 million years ago (Table 1.3). Since that time large sedimentary deposits have been laid down in rocks formed during all the main geological eras. At one time there was some argument concerning the proportion of sedimentary rocks that were of biogenic origin compared to those due simply to inorganic deposition. Until relatively recently

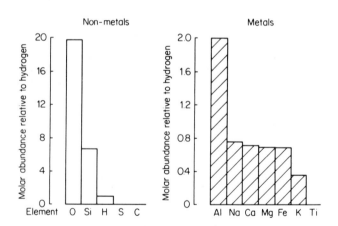

Fig. 1.2. Occurrence of the major elements in the earth's crust. Data converted from Day (1963), Fairbridge (1972) and Fyfe (1974). Hydrogen = 1460 moles per million gram. Relative abundance in many geological publications is calculated on a weight basis as parts per million. Molar ratios relative to hydrogen provide data more relevant to the chemical reactivity of the elements. The figures here only apply to the earth's crust. In the earth's mantle magnesium is at least thirty times as abundant as calcium

Table 1.2. Some calcium minerals

Igneous rocks	
Anorthite (a feldspar)	$CaAl_2SiO_8$
Diopside (a pyroxene)	$CaMgSi_2O_6$
Actinolite (an amphibole)	$Ca_2Mg_5Si_8O_{12}(OH)_2$
Wollastonite	$CaSiO_3$
Sphene	$CaTiSiO_5$
Apatite	$Ca_5(PO_4)_3F$
Sedimentary rocks	
Calcite	$CaCO_3$
Aragonite	$CaCO_3$
Vaterite	$CaCO_3$
Dolomite	$CaMg(CO_3)_2$
Gypsum	$CaSO_4 \cdot 2H_2O$

Precambrian (greater than 600 million years' old) fossils were almost unknown. However, during the last 15 years the discovery of microfossils, in a form known as microbiotas and stromatolites, has provided evidence that life began at least 3000 million years ago (Harling, 1967; Schopf, 1970, 1975) (Table 1.3). It is therefore likely that a considerable proportion of the sedimentary rocks were of biological origin.

The fossil record abounds with the calcareous remains of extinct organisms. Strictly the term 'fossil' should be reserved for the actual remains of a dead organism, whether complete or fragmentary. However, during the time between the death of the organism and the embedding of its remains in rock, the action of

Table 1.3. The time scale of evolution

Event	Geological era (period)	Millions of years ago (approx.)
Formation of the earth		4500
Oldest sedimentary rocks	Precambrian (Archaen or Eozoic)	3750
Origin of life	Precambrian (Archaen)	3500
Oldest stromatolites	Precambrian (Archaen)	3100
Oxygen-producing microbes	Precambrian (Proterozoic or Archaeozoic)	2300
Oldest calcareous blue-green algae	Precambrian (Proterozoic)	1900
Origin of the eukaryotic cell	Precambrian (Proterozoic)	1500
Development of meiosis	Precambrian (Proterozoic)	1000
Origin of metazoans	Precambrian (Proterozoic)	700
Explosion in metazoan populations	Palaeozoic (Cambrian)	600
Origin of vertebrates—primitive fishes	Paleozoic (Ordovician)	400
Appearance of mammals	Mesozoic (Triassic)	200
Appearance of man	Cainozoic (Pleistocene)	1.5

water may cause complete or partial replacement of the remains by minerals other than those originally present (Swinnerton, 1970). Whether a fossil consists of the actual remains or not, it is clear from both the Precambrian and Phanerozoic periods that calcium salts have played a vital part in the formation of skeletal structures of living organisms from earliest times. Although most of present day algae and protozoa do not have calcareous skeletons, deposits of calcareous algae and protozoa are found in rocks of the Precambrian era (Chilinger and Bissel, 1963; Glaessner, 1962; Harling, 1967). The remains of calcareous blue-green algae have been discovered which are at least 1900 million years old (Table 1.3). Massive deposits of the calcified skeletons of metazoans occur from the beginning of the Paleozoic era (Table 1.3), about 600 million years ago. These fossils of the Paleozoic, Mesozoic and Cainozoic eras include microscopic foraminifera, as well as the shells of molluscs and crustaceans, coral formations, and the bones of vertebrates.

1.1.3. Biomineralization

Whilst the calcareous remains of extinct organisms are common, the great variety of functions for precipitated calcium salts is even more striking when one examines living organisms (Wilbur and Simkiss, 1968; Pautard, 1970; Dawes, 1975; Aaron, 1976; Copp, 1969, 1970, 1976; Goreau, 1977). Calcium phosphate is the major constituent of bone and teeth, whereas the shells of molluscs and arthropods consist mainly of calcium carbonate. These calcium salts, as well as sometimes calcium sulphate, also have important functions in a number of other phyla (Table 1.4). For example, calcium carbonate is the main constituent of the calcareous skeletons of many Anthozoans, giving rise to the formation of coral

Table 1.4. Some examples of biomineralization

Organism group (example)	Structure	Major calcium salt
Unicellular		
Some protozoa	Shell	$CaCO_3$
Some algae	Shell	$CaCO_3$
Animals		
Coelenterata (sea corals)	Coral	$CaCO_3$
Coelenterata (jelly-fish)	Statocyst	$CaSO_4$
Porifera (sponges)	Spicules	$CaCO_3$
Platyhelmintha (tapeworms)	Calcareous corpuscles	$Ca_3(PO_4)_2$
Echinodermata (sea urchins)	Shell	$CaCO_3$
Molluscs (bivalves)	Shell	$CaCO_3$
Arthropoda (barnacles)	Shell	$CaCO_3$
Chordata (vertebrates)	Bone, teeth	$Ca_{10}(PO_4)_6(OH)_2$
Plants		
Higher plants	Intracellular granules	Calcium oxalate
Multicellular algae (*Corallina*)	Stem skeleton	$CaCO_3$

reefs. In contrast, the statocysts, which are the marginal sense organs in jelly-fish responsible for orientation of the animal, contain granules made up mainly of calcium sulphate (Russell, 1953, 1969; Spangeberg and Beck, 1968). It has been known for 80 years that the spicules of calcareous sponges, which are an essential part of their skeletal structure, consist mainly of calcium carbonate (Minchin, 1898; Brien, 1968). Extracellular calcium precipitates are also to be found in some unicellular and multicellular algae. The latter are marine, for example the calcareous seaweed *Corallina officinalis* (Turvey and Simpson, 1966), which is the ubiquitous pink seaweed found growing on the edge of so many rock pools.

In higher plants intracellular granules of calcium oxalate are sometimes found. In animals oxalic acid (COOH-COOH) can be formed by oxidation of glycine or ascorbic acid. Plants and microorganisms, however, contain a glyoxylic acid (CHO-COOH) pathway (the glyoxylate cycle) not found in animal tissues, and which is important in photosynthesis. Glyoxylic acid is a major source of oxalic acid in plants. Large quantities of calcium can also accumulate in spore-forming bacteria. In this case the main salt is calcium dipicolinate (Halverson, 1963; Kornberg *et al.*, 1975) (Fig. 3.2).

Table 1.5. Some pathological examples of calcification

Cause	Site of deposition
Dystrophic	
Tuberculosis	Caseating lesions of lung and lymph nodes
Old infarcts	Heart, kidney, spleen
Rheumatic heart disease	Heart valve
Atheroma	Intima of vessel (e.g. aorta or coronary arteries)
Mönckeberg's sclerosis	Media of artery wall
Haematomata	Anywhere, but calcification especially associated with bone
Senility	Cartilage, aortic valve, pineal gland, dura mater
Old thrombi (phleboliths)	Veins (usually pelvic)
Dead parasites—hydatid cysts	Usually liver
cysticercosis	Soft tissue
Trichinella spiralis	Muscle
Fat necrosis	Peritoneal (acute pancreatitis)
Degenerating tumour	Especially uterus (fibroids), ovary and breast
Mercury poisoning	Kidney tubules
Calcinosis circumscripta (very rare)	Skin
Metastatic	
Hyperparathyroidism	
Hypervitaminosis D	Many possible sites in all cases, particularly kidney
Extensive destructive bone lesions	(nephrocalcinosis), lung, atheromatous sites,
Multiple myeloma	cornea and stomach (around fundal glands)
Renal tubular acidosis	

These calcifications consist mainly of calcium phosphate, apart from in the kidney where a substantial amount of calcium oxalate and calcium carbonate can be found.

In man deposition of calcium salts in tissues occurs in several pathological conditions (Table 1.5). Two types of deposition have been defined (Rees and Coles, 1969; Walter and Israel, 1970; Smith and Williams, 1971; La Ganga, 1974). Dystrophic calcification occurs in dead or degenerating tissue; the serum calcium concentration is usually within the normal range. In contrast, the other type of calcification, metastatic, is the result of a disturbance in whole body calcium metabolism; in this case the serum calcium concentration is usually elevated above the normal range, leading to deposition in many otherwise healthy soft tissues. It is not always clear whether either of these two types of tissue calcification is initiated from within or outside cells. Another example of calcification, usually considered as a special case, is the formation of stones, known as calculi, in the kidney, gall-bladder and bladder.

From this discussion of the occurrence of calcium precipitates in Nature, it is clear that insoluble complexes of calcium with certain anions play an important part in the life of organisms from many different phyla. The question arises whether this is the only function of calcium in these organisms. Furthermore, does calcium have any function in non-calcareous organisms, or in the soft tissues of metazoans? In order to answer this question it is first necessary to discover whether calcium is an essential constituent of the fluid surrounding cells not involved in biomineralization processes.

1.2. The requirement of cells for calcium

1.2.1. Calcium in water and extracellular fluids

There is a substantial body of evidence that the majority of uni- and multicellular organisms require calcium for normal growth and function. In fresh water the number of species seems to be directly related to the concentration of calcium in the water; as the calcium concentration decreases so does the number of species (Macan, 1963; Macan and Worthington, 1974). Butcher in 1933 even classified rivers into distinct classes on the basis of the concentration of calcium in the water. The critical concentration for snails, for example, is 20 ppm (approx. 0.5 mM), and the freshwater shrimp *Gammarus pulex*, which is one of the most adaptable freshwater animals, is not found in very soft water.

The concentration of calcium in sea water is usually about 10 mM, being approximately one-fifth that of magnesium, one-fiftieth that of sodium, and about the same as that of potassium (Table 1.6). In fresh water, on the other hand, the concentration of calcium is considerably less than that of sea water (Reid, 1961) (Table 1.6). Everyone is familiar with the differences between hard and soft water observable on washing one's hands with soap when in different parts of the country. The waterworks engineer considers water to be hard at 60 parts of calcium per million (approx. 1.5 mM), whereas the housewife becomes aware of calcium at about 30 ppm (approx. 0.75 mM). The concentration of calcium varies considerably depending on the source of the water. In soft water the concentration may be as low as 20 μM, whereas in hard water concentrations

Table 1.6. Concentration of the major cations in extracellular fluids

Fluid	Total cation concentration (mM)			
	Na^+	K^+	Mg^{2+}	Ca^{2+}
Water				
Sea water	475	10	55	10 (9.3–11.8)
Fresh water ($Na^+ + K^+$)	0.1–80		0.02–0.4	0.02–2
Rain water ($Na^+ + K^+$)	0.01		0.004	0.002–0.02
Balanced salt solutions				
Locke's saline	156	6	—	2
Krebs-Ringer	145	6	1.2	2
Eagle's medium	143	5.4	0.8	1.8
Nematode (*Ascaris*) saline	130–168	3–24	0–16	2–7
Marine invertebrate saline	513	12.9	23.6	11.8
Serum and body fluids				
Adult human serum	140	4	1	2.5
Vertebrate serum	87–544	4–12	1–10	1.5–5
Pig intestinal fluid	124	27	6	14
Coelenterate (*Physalia*) gastrovasicular fluid	350	33	24	6
Nematode (*Ascaris*) body fluid	129	25	49	6
Molluscan serum—marine	475	10–22	55	9–15
fresh water	16–86	0.4–5	0.1–2.4	1.5–7.8
land	47–75	2.4–10	1–20	3.3–12.3

The figures shown here are only approximate; they are intended to give an indication of the range, or order of magnitude, of cation concentrations bathing cells. In most cases the major anion balancing all the cations is chloride (oceanic chlorinity is approx. 19 $^o/_{oo}$. Chlorinity is the total weight (g/kg) of chlorine, bromine and iodine in sea water, where bromine and iodide have been converted to their equivalent of chlorine).

References: Robertson (1941, 1949, 1953); Potts (1954); Shaw (1955); Eagle (1956); Reid (1961); Manery (1961); Macan (1963); Lenhoff & Loomis (1961); Burton (1973a, b); Henry (1964); Copp (1976).

of up to 2 mM are well known, and in some areas, famous for their beers, values as high as 4 mM have been reported.

The concentration of calcium in the body fluids of various animals also varies over a wide range (Table 1.6), though within any one species it has to be maintained within much closer limits than is necessary for the survival of animals in fresh water. In man, for example, the plasma concentration of calcium (total) is between 2.1 and 2.6 mM in normal individuals, being maintained by the hormones vitamin D and parathyroid hormone (Woodhead, 1982). The concentration of total calcium in the body fluids of most other animals is in the range 1–15 mM.

The precise concentration of *free* ionized calcium is less easy to determine, since extracellular fluids contain calcium-binding proteins which reduce the free Ca^{2+}, often to less than 50% of the total. Attempts have been made to derive a mathematical relationship between the concentrations of the four major cations,

K^+, Na^+, Mg^{2+} and Ca^{2+}, in the serum or haemolymph of vertebrates and invertebrates (Conway, 1943, 1945; Burton, 1973a, b). Such an equation has been derived for vertebrates (Burton, 1973a, b) and depends on the necessity of maintaining constant electrical gradients across cell membranes

$$([Ca^{2+}]_{free} + 0.0005)/([K^+]_{free} + 0.034[Na^+])_{free} = \text{constant} \quad (1.1)$$

where $[Ca^{2+}]_{free}$, $[K^+]_{free}$ and $[Na^+]_{free}$ refer to free ionized concentrations of these cations. Concentration of Mg^{2+} do not correlate well with other cations.

1.2.2. Effect of removal of calcium on animal cells

Sidney Ringer (Fig. 1.3), working at University College Hospital in London, was the first person to investigate systematically the requirement of inorganic

Fig. 1.3. Sidney Ringer, FRS (1835–1910). Reproduced by permission of University College Hospital Medical School, London

compounds for the normal function and survival of animal cells (Ringer, 1882, 1883a, b, c, 1886, 1890; Ringer and Sainsbury, 1894). He showed that calcium was required for normal contraction of frog heart, that it was necessary for the development of fertilized eggs and tadpoles, and that calcium was important in the adhesion of cells to each other, the latter also being demonstrated by Herbst (1900).

Ringer was also the first to realise how difficult it is to get rid of calcium contamination from a solution. In 1882 he had apparently shown that calcium was *not* necessary for the beating frog heart. However, in the following year he wrote, 'After the publication of a paper in the *Journal of Physiology*, Vol. III, No. 5, I discovered that the saline which I had used had not been prepared with distilled water but with pipe water supplied by the New River Water Company. As this water contains minute traces of various inorganic substances, I at once tested the action of saline solution made with distilled water and found that I did not get the effects described in the paper referred to. It is obvious therefore that the effects I had obtained are due to some of the inorganic constituents of the pipe water.' He found that this missing constituent was calcium. More recently it has been realized that even distilled water and Analar reagents contain significant quantities of calcium (Levitzki and Reuben, 1973; Shimomura and Johnson, 1973; Campbell and Siddle, 1976). Distilled water often contains $1-10\,\mu M\,Ca^{2+}$, and Ca^{2+} leaches off glassware which can raise the Ca^{2+} concentration to as much as $30\,\mu M$. In order to lower the free Ca^{2+} concentration to less than $0.1\,\mu M$ it is necessary to use spectroscopically pure reagents and Ca^{2+} chelators such as EGTA (see Chapter 2).

Ringer's experiments were followed by those of several other physiologists, who included Locke (1894), Loeb (1906), Mines (1910, 1913) and Loewi (1917, 1918) [See Berliner (1933) for review of Ca^{2+} and the heart]. They not only confirmed that calcium was necessary for maintaining the morphology and function of animal tissues, but also that calcium seemed to play a unique role in the action of cell stimuli such as adrenaline, and drugs such as digitalis. Further evidence for the physiological importance of calcium came as a result of the development of tissue culture (Eagle, 1956; Paul, 1961; Willmer, 1965). Calcium, together with sodium, potassium and magnesium, is an essential component of physiological salines (Table 1.6). Removal of calcium from tissue culture media, which include serum, using ion-exchange resins results in changes in cell structure (Shooter and Grey, 1952) and a decrease in growth rate of cells (Shooter and Grey, 1952; Owens *et al.*, 1956, 1958; Willmer, 1965, 1970, 1974, 1977), as well as reducing the strength of adhesion between cells (Coman, 1954; Curtis, 1962; Gingell *et al.*, 1970). In fact, removal of extracellular calcium often results in dissociation of tissues into isolated cells (Herbst, 1900; Gray, 1922; Heilbrunn, 1937, 1943; Willmer, 1965). Other tissues, such as rabbit lens, may swell and burst under these conditions (Rubino, 1936).

These observations made *in vitro* are consistent with those made in man regarding the importance of extracellular calcium. Pathologically low serum calcium concentrations can cause both acute and long-term changes in the behaviour of tissues. For example, it has been known for nearly one hundred years that serum free calcium concentrations below about 0.9 mM result in uncontrollable muscle spasms known as tetany (Thompson and Collip, 1932). The reason for this is that electrically excitable cells like nerves and muscle cells are particularly sensitive to a reduction in the Ca^{2+} concentration surrounding the cell (Brink, 1954; Frankenhauser, 1957; Frankenhauser and Hodgkin, 1957; Shanes, 1958). A reduction in extracellular Ca^{2+} concentration causes nerve axons to become hyperexcitable. Alkalosis or hypoparathyroidism can result in a decrease in the free Ca^{2+} concentration in serum to a level at which spontaneous firing of nerves may occur. These nerves then activate the muscles which undergo tetanic contractions.

The evidence that calcium is required for the growth and function of other animals besides the chordates is less well documented. Nevertheless, Lane (1968) has shown that growth of the coelenterate *Hydra littoralis* is retarded if the medium Ca^{2+} concentration is less than 0.1 mM. Calcium is also required for the development of the sting cells, the cnidocytes, in animals from this phylum (Loomis, 1954). In other phyla, for example *Nematoda*, the survival of organisms depends on the presence of calcium (Arthur and Sanbom, 1969).

1.2.3. Requirement of plants for calcium

Like animals, plants also must have calcium (Wyn-Jones and Lunt 1967; Burström, 1968; Hewitt and Smith, 1975), concentrations in the range 0.1–1 mM being normally required to maintain cell structure and function. The growth of plant roots is particularly sensitive to calcium deprivation. Lack of calcium may also result in changes in the structure of intracellular organelles, a decrease in cell elongation, as well as affecting cell walls and the permeability of cell membranes to solutes and other ions. Calcium may also act to antagonize the effects of toxic trace metals in the soil. Examples of special requirements for calcium in plants are at the onset of flowering (Hewitt and Smith, 1974) and the nodulation of legumes following infection by *Rhizobia* (Dixon, 1969).

1.2.4. Calcium and bacteria

Whilst calcium is required by many marine bacteria (MacLeod and Matula, 1961; MacLeod, 1965; Hutner, 1972), and by some, if not all, blue-green algae (Hölm-Hansen, 1968), a clear-cut requirement for calcium in microorganisms has rarely been demonstrated (Wyatt, 1961, 1964; Wyatt *et al.*, 1962; Hutner, 1972). The reason for this is that, as discussed earlier, it is only comparatively recently that it has been realized how difficult it is to reduce the concentration of ionized calcium in incubation fluids to less than 1 μM.

In spite of this problem, it has been shown that some bacteria, other than the marine forms, do require calcium for growth (Skankar and Bard, 1952; Shooter and Wyatt, 1955). Calcium is also of vital importance in the formation of bacterial spores (Halverson, 1963; Kornberg *et al.*, 1975) (see Chapter 8).

1.2.5. Tissue content of calcium and other cations

The potassium content of most cells is considerably greater than that of sodium, and the magnesium content greater than that of calcium (Albritton, 1951; Manery, 1954, 1961, 1966, 1969). There appears to be a relationship between the concentration ratios of Mg^{2+}/Ca^{2+} and K^+/Na^+ inside a number of different cells, this being related to the phosphorus content of the cells (Tempest et al., 1966; Williams and Wacker, 1967; Wacker and Williams, 1968). The intracellular concentration of potassium is usually greater than 100 mmoles/l cell water, whereas that of total cell calcium is usually in the range 1–10 mmoles/l cell water. This generalization does not, however, apply to all cells. For example, muscle cells, with large intracellular stores of calcium, can have up to 20 mmoles/l cell water in the cells, whereas cells without much bound intracellular calcium, such as bacteria and erythrocytes, may contain less than 0.1 mmole/l cell water. Nevertheless, all cells from soft tissues that have so far been examined contain detectable quantities of calcium, as well as sodium, potassium and magnesium. None of these cations are evenly distributed within the cell. For instance, much of the calcium and magnesium in the cell is bound to inorganic and organic ligands as well as being sequestered in organelles such as the endoplasmic reticulum (Ebashi, 1960; Endo, 1977), mitochondria (Manery, 1969; Carafoli and Crompton, 1976; Bygrave, 1977), and the nucleus (Williamson and Gulick, 1944; Thiers and Vallee, 1957; Williams and Wacker, 1967). All of these organelles contain higher total calcium concentrations than the cell cytoplasm. Cytoplasmic free Ca^{2+} in the resting cell is about 0.1 μM. In contrast nearly all of the K^+ and about 76% of the Na^+ is free.

It can therefore be concluded that, whilst definitive evidence is lacking for some organisms, calcium is required for the growth and normal function of the majority of unicellular and metazoan organisms, and that this requirement is independent of the role of calcium in biomineralization. Furthermore, all organisms exist in the presence of some free calcium in the solution which bathes their cells.

The weathering of rocks or the action of bacteria on calcium minerals maintains the level of Ca^{2+} in fresh water (Fig. 1.4). Rivers carry this calcium to the sea. The Ca^{2+} concentration is maintained in the sea at about 10 mM as a result of a balance between removal by sedimentation and addition of Ca^{2+} from the rivers.

Having established this universal requirement for Ca^{2+} in living organisms, the question arises whether it is possible to define more than one biological role for calcium?

1.3. The four main biological roles of calcium

There are three main groups of cations in living organisms—the alkali metals, the alkaline earths and the transition metals. Their biological roles can be classified into six categories (Table 1.7).

Table 1.7. The major biological roles of cations

Metal group	Cation	Function
Group I (alkali metals)	Na^+, K^+	Osmotic balance Electrical activity across membranes Activation of a few enzymes
Group II (alkaline earths)	Mg^{2+}	ATP Mg^{2-}, the substrate for all kinases and some phosphatases Activator of a few enzymes Chlorophyll
	Ca^{2+}	Biomineralization Structure of soft tissues (cell adhesion, membrane permeability) Electrical activity across some membranes Wide-ranging regulator of protein structure and enzyme activity
Transition metals	Fe^{2+}/Fe^{3+} Cu^+/Cu^{2+}	Redox catalysts (respiratory chain)
	Fe^{2+}, Cu^{2+}	Oxygen-carrying pigments (haemoglobin, haemocyanin)
	Co^{2+}	Vitamin B_{12}
	Cu^{2+}, Co^{2+}, Mn^{2+}, Mo^{2+}, V^{2+}, Zn^{2+}	Catalytic function at the active centre of some enzymes

Biological roles: (1) Structural; (2) electrical; (3) requirement for maximum activity and regulation of enzymes and proteins (i.e. cofactor); (4) intracellular regulator; (5) osmotic balance; (6) oxidoreduction reactions.

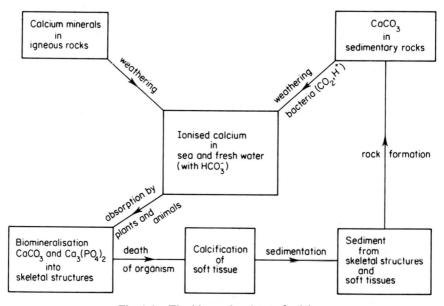

Fig. 1.4. The biogeochemistry of calcium

The first four of these categories apply to Ca^{2+}, the third being divided into extracellular and intracellular enzymes and proteins. Ca^{2+} does play a role indirectly in maintaining the osmotic balance of cells through its role in the integrity and permeability properties of biological membranes (Lucke and McCutcheon, 1932; Berntsson et al., 1965). Ca^{2+} plays no direct part in redox reactions since, unlike transition metals such as iron and copper, calcium has only one stable valency state.

The four biological roles of Ca^{2+} therefore are: (1) structural; (2) electrical; (3) cofactor for extracellular enzymes and proteins; (4) intracellular regulator.

1.3.1. Structural role

As we have already seen, precipitates of calcium phosphate and carbonate are the major inorganic constituents of the skeletal structures of most animals (Table 1.4). On their own these precipitates would be brittle, as can be seen if the organic components of bone are removed by ashing. However, in combination with collagen a rigid and strong structure is formed. This, however, is not the only structural role of calcium in tissues (Table 1.8). Ca^{2+} bound to phospholipids, other membrane components (Manery, 1969; Harrison and Harrison, 1974), protein and nucleic acid (Harris and Milne, 1975; Li et al., 1977) play a role in maintaining the structure and function of all eukaryotic cells and tissues.

1.3.2 Electrical role

Bernstein in 1902 first suggested that potassium ions were responsible for the resting membrane potential of cells, and that the action potential of electrically excitable cells might be due to a loss in selective permeability of the cell membrane to potassium. Some 30–40 years later the importance of the permeability of the cell membrane to sodium ions in the development of action potentials was fully realised (Cole and Curtis, 1939; Hodgkin and Huxley, 1945; Cole, 1949; Hodgkin, 1951), which lead to the so called 'sodium theory' of the

Table 1.8. Some structural roles of calcium

Form of calcium	Example
Skeletal	
Calcium carbonate	Coral, shells
Calcium phosphate	Bone, teeth
Soft tissue	
Calcium phospholipid	Membrane fluidity and integrity
Calcium proteinate	Electrically excitable cell stability
	Cell communication via gap junctions
	Cell adhesion
	Bacteriophages and virus particles
Calcium nucleate	Chromatin structure

action potential (Hodgkin and Katz, 1949b). More recently the importance of both Cl^- and Ca^{2+} in the electrical activity of some cells has been discovered. In particular, Ca^{2+} can carry a significant quantity of current during an action potential in cells from both vertebrates and invertebrates (see Chapter 4).

1.3.3. Cofactor for extracellular enzymes and proteins

Several extracellular degradative enzymes in both eukaryotes and prokaryotes require Ca^{2+} for stability or maximal activity (Table 1.9). These include proteases and certain enzymes involved in blood clotting. In addition, some regulatory proteins found outside the cell or on the cell surface require Ca^{2+}. The criterion for a functional role in these cases is that removal of Ca^{2+} with a chelating agent such as EDTA or EGTA inhibits the activity, or function, of the protein. Thus, for example, blood clotting is inhibited by adding citrate or EDTA to whole blood. Similarly, complement activation via the classical pathway, that is from antibody–antigen complex, is also inhibited if Ca^{2+} is removed (Fearon and Austen, 1976).

Vital as these functions of Ca^{2+} are they are not regulatory in a 'physiological' sense since changes in extracellular Ca^{2+} concentration do not play a significant part in either initiating or regulating the process concerned. Under physiological conditions in the animal a reaction such as blood clotting or complement activation is activated by a process not dependent on a change in extracellular Ca^{2+}. In contrast, in the fourth and final biological role of Ca^{2+} it is the change in Ca^{2+} within the cell which triggers the response of the cell to a stimulus.

Table 1.9. Ca^{2+} requirement for extracellular proteins

Protein	Source
Prokaryotes	
Collagenase	*Clostridium histolyticum*
α-Amylase	*Bacillus subtilis*
Haemolysin	*Bacillus thermoproteolyticus*
Eukaryotes	
Some hormone and transmitter receptors	Mammalian cells
α-Amylase	Mammalian saliva
Trypsin(ogen)	Mammalian pancreas and duodenum
DNAase I	Mammalian pancreas
Phospholipase A	Snake venom
Transglutaminase and four serine proteases in blood clotting (factor II—prothrombin; factor VII; factor IX—Christmas factor; factor X—Stuart factor)	Mammalian serum
Formation of C1q,r,s complex on antibody–antigen complex	Mammalian serum
Haemocyanin	Mollusc or arthropod haemolymph

For references see Chapter 3 and Kretsinger (1976a, b).

Table 1.10 Some phenomena regulated by a change in intracellular Ca^{2+}

Phenomenon	Example	Type of primary stimulus
Cell movement	Muscle contraction	Neurotransmitter
	Chemotaxis	Chemical substance
Secretion	Neurotransmitter release	Action potential
	Endocrine hormone release	Hormone
	Exocrine secretion	Neurotransmitter
Cell division	Cell transformation	Chemical
	Egg fertilization	Sperm
Intermediary metabolism	Glycogen or lipid degradation	Hormone
Membrane permeability	Substrate uptake	Hormone
	Cation permeability	Action potential
	Cell–cell communication	Unknown
Vision	Retinal rod	Light

1.3.4. Calcium as an intracellular regulator

For the remainder of this book attention will be focused on when and how changes in Ca^{2+} inside cells mediate the response of cells to a stimulus. The stimulus may be physical, electrical, or chemical. The number of phenomena where intracellular Ca^{2+} appears to be the trigger for cell activation is large (Table 1.10), and includes cells from all of the eukaryotic phyla as well as some prokaryotes. As we shall see (Chapter 9), disturbances in intracellular Ca^{2+} also play an important role in cell injury in disease as well as in the action of many drugs.

What is the essence of this biological role of calcium? As a result of an external or internal stimulus there is a change in the behaviour of a cell or group of cells which affects the functioning of the whole organism. It will be argued that the physiological phenomena (Table 1.10) considered in this book occur as a result of a change in the concentration of free Ca^{2+} at a specific site inside the cell. This change in free Ca^{2+} is brought about by a primary stimulus either releasing a store of Ca^{2+} from within or causing Ca^{2+} to enter the cell from outside. How has this concept developed which has attracted the attention of so many biologists in recent years?

1.4. An historical perspective

When Ringer began his investigations into the ionic requirements for heart muscle contraction, the 'cell' theory of Schwann was less than 50 years old and the laws of thermodynamics, which thanks to Kelvin were having a great impact on physics and chemistry as well as on geology and paleontology (Burchfield, 1975), had not yet penetrated into biology. The crystallization of the first enzyme, urease, did not occur until 1926 (Sumner, 1926), and ATP was not isolated for a further three years (Lohmann, 1929; Fiske and Subbarow, 1929), the structure of ATP being confirmed by synthesis in 1948 (Baddiley et al., 1948). It was therefore hardly surprising that physiologists, attempting to explain the effects of ions on living tissues and cells, were forced to resort to the mysterious 'protoplasm', which at one time seemed as difficult to investigate as the elusive 'aether' of the nineteenth-century physicist. By the turn of the century both the chemo-vitalist and the organo-vitalist schools begun by Leibig and Barthez and Bichat, respectively (Florkin, 1975a, b) were dying out. In their place came a new school of chemical physiologists led by men like Frederick Gowland Hopkins (Fig. 1.5) (Needham and Baldwin, 1949). This resulted not only in the isolation of proteins, enzymes and vitamins, but also in the discovery of two major groups of cell stimuli, namely hormones (Bayliss and Starling, 1902) and neurotransmitters (Dale, 1934).

During the 20–30 years that followed the pioneering experiments of Ringer, a number of other physiologists were able to show that the absence of extracellular calcium affected a wide variety of phenomena, quite apart from further studies

Fig. 1.5. Frederick Gowland Hopkins, FRS, University of Cambridge

on the requirement for calcium in heart muscle contraction (Straub, 1912; Mines, 1913). Loeb (1906, 1922) found that calcium was needed for the development of fertilized eggs of the marine teleost *Fundulus*, whilst the removal of extracellular calcium from blastomeres of dividing sea-urchin eggs resulted in a pulling apart of the cells (Herbst, 1900; Heilbrunn, 1943). The transmission of nerve impulses to muscle was shown to require calcium (Locke, 1894; Overton, 1904). A number of reports were also published concerning the effects of calcium on the cilia of the common mussel *Mytilus edulis* (Gray, 1922, 1924), and on protoplasmic streaming in *Amoeba proteus* (Chambers and Reznikoff, 1926; Pantin, 1926; Heilbrunn, 1927). Intermediary metabolism was also found to be dependent on the ionic composition of external media (Hastings *et al.*, 1952).

The discovery of hormones and neurotransmitters stimulated some workers to investigate the ionic dependence of the effects of these substances. It was proposed that the effect of adrenaline to increase motility and protoplasmic streaming in *Paramecium* was due to adrenaline increasing the concentration of intracellular calcium through an increase in membrane permeability to calcium (von der Wense, 1933), whilst others suggested that the action of adrenaline on the heart could be explained by a release of bound calcium inside the cell (Lawaczek, 1928; Hermann, 1932; Weise, 1934).

The puzzle was how to explain the effects of manipulation of external Ca^{2+}. Ringer developed an hypothesis, which was extended by Rubenstein (1928) and Höber (1945), that K^+ and Na^+ antagonized each other, Ca^{2+} and Mg^{2+} could both antagonize Na^+, and in some cases Ca^{2+} and Mg^{2+} could antagonize each other. Furthermore, there were other cases where cations like Ca^{2+} and Mg^{2+} could act synergistically. These cation interactions were explained in rather vague terms of colloid chemistry and cation binding groups in the 'protoplasm' (Heilbrunn, 1923; 1937, 1943, 1956, 1958), or through effects of Ca^{2+} on the permeability of the plasma membrane (Lillie, 1936).

It was Pollack in 1928 who first realized that in order to clarify the regulatory role of intracellular Ca^{2+} it was necessary to measure its concentration in the living cell. Using the microinjection techniques pioneered by Chambers and the Needhams (Chambers and Reznikoff, 1926; Chambers, 1928; Needham and Needham, 1925/26), Pollack injected the dye alizarin sulphonate into an amoeba. Calcium alizarin sulphonate is only sparingly soluble. Adjacent to the site of pseudopod formation in the amoeba Pollack observed a shower of red crystals. He concluded that this was due to an increase in Ca^{2+} in the cell. Although this conclusion can be criticised because of doubts as to the specificity of this technique for Ca^{2+}, it seems to have stimulated Heilbrunn and others to consider more closely the importance of intracellular Ca^{2+} in cell regulation.

Heilbrunn (frontispiece) extended the role of intracellular calcium to a much wider range of phenomena. The physiological processes he included were not only contraction of smooth and striated muscle and the action of adrenaline on the heart, but also protoplasmic flow and amoeboid movement (Gray, 1922, 1924; Kamada, 1940), parthenogenesis in the developing eggs of worms, molluscs and echinoderms (Hollingsworth, 1941; Tyler, 1941), and the action of anaesthetics. Furthermore, since there was evidence that calcium was not evenly distributed within the cell, the source of the calcium used in these stimuli could be either external or an internal pool of bound calcium (Heilbrunn, 1937).

In spite of these arguments the majority of biologists remained unconvinced about the general importance of calcium as an intracellular regulator. As recently as 1962 there was still a considerable amount of doubt about the 'calcium hypothesis' (Ebashi, 1980). Progress was hampered by the lack of convincing evidence that a physiological stimulus caused an increase in intracellular free Ca^{2+} and by the lack of a molecular explanation for the effects of Ca^{2+}. These two problems were first resolved in muscle, which provides a physiological process in which a role for intracellular Ca^{2+} has been best characterized.

Two major problems were faced by the early muscle physiologists. First, how was the energy from the metabolism of substrates such as glucose converted into mechanical energy, and second, what was the internal trigger which allowed this to take place? A number of hypotheses were proposed to overcome these problems (Höber, 1945; Needham, 1971). Hill and Meyerhoff proposed that the lactic acid from glycolysis provided the energy and stimulus for contraction. This idea was soon rejected on energetic grounds. Other physicochemical mechanisms proposed were the 'surface tension' and 'colloid-chemical' theories, neither of which was adequate to explain the extent of muscle shortening during a contraction. Szent-Györgi, on the other hand, recognized the importance of myosin, and suggested that the folding of this protein was stimulated by potassium. This seemed unlikely to be able to explain contraction since the free K^+ was already known to be high in the muscle cell.

An important breakthrough was the discovery that the contractile apparatus was an enzyme complex and that the energy for contraction came from the hydrolysis of ATP (Englehardt and Ljabimowa, 1939). Furthermore, the fact that the myosin ATPase in this complex could be stimulated specifically by calcium (Bailey, 1942; Needham, 1942) was consistent with intracellular calcium being the trigger for contraction. This was confirmed directly by an experiment in which muscle contraction was stimulated by the injection of calcium into the cell (Heilbrunn and Wiercinski, 1947).

It was pointed out by Hill (1948) that the diffusion of substances like calcium from the outer membrane into the interior of the cell would be too slow to account for the rapid contraction of vertebrate muscle. This problem was resolved by the discovery of the sarcoplasmic reticulum, an intracellular reservoir of calcium which could take up and release calcium under the appropriate conditions (Bennet and Porter, 1953; Kumagai *et al.*, 1955; Hasselbach and Makinose, 1963; Ebashi, 1960, 1961). The advent of calcium buffers, using chelators such as EGTA (Portzehl *et al.*, 1964), enabled biochemists to study the isolated contractile apparatus and calcium-accumulating system in a low, but defined ionised calcium concentration. These calcium buffers were also injected into intact cells in order to define the concentration range of calcium inside the cell over which contraction took place (Portzehl *et al.*, 1964). Finally, the direct measurement of intracellular free calcium during contraction showed that the concentration of free calcium inside the cell varied over the range 0.1–10 μM (Ridgway and Ashley, 1967; Ashley and Ridgway, 1968). The pioneering experiments of Ashley and Ridgway were carried out using the calcium-activated photoprotein aequorin (Shimomura et al., 1962) as an intracellular indicator of free Ca^{2+} (see Chapter 2).

These studies on muscle clarified two vital points concerning the role of intracellular calcium as a regulator. First, the release of a relatively low concentration of calcium inside the cell did not provide the energy for contraction but rather acted as a trigger to a system which had already been precharged by the metabolism of the cell. Second, calcium acted as a cofactor of an enzyme complex, rather than acting through some elusive effect on the protoplasm of the cell. This was in contrast to the role of magnesium whose major involvement in contraction was as a complex, $ATPMg^{2-}$ with the substrate for the ATPases.

In summary, by the mid 1960s it was clear that monovalent cations were mainly concerned with the osmotic and electrical behaviour of cells, whereas the divalent cations, calcium and magnesium, were involved in cell activation, calcium acting as the actual intracellular stimulus and magnesium being involved mainly with the substrate of the reactions concerned (Table 1.7). Transition metals, on the other hand, were either involved in redox reactions, or were tightly bound cofactors of enzymes with binding characteristics making them unsuitable mediators of cell stimuli.

1.5. Some questions

Natural Science is concerned with describing and understanding natural phenomena. As biologists we try to understand not only the problems of animals within our own phylum but those in the 29 others which contain the million or so animal species which inhabit the earth (Table 1.11). In addition, there are problems that are unique to plants and prokaryotes. How are muscles able to respond within a few milliseconds to a stimulus from outside the animal? What controls the rhythm of a beating heart or pulsating jelly-fish? How are various organisms able to adapt to different environments? How are the various tissues in an organism coordinated and what controls their growth and development? What were the stages in the evolution of these processes?

This book is concerned with identifying biological phenomena and with discovering what role intracellular calcium plays in regulating them. It will be concerned with five main questions:

(1) How do we show that intracellular Ca^{2+} does play a role in a particular phenomenon?

(2) How can we distinguish the role of intracellular Ca^{2+} as a regulator from the other biological roles of Ca^{2+}?

(3) What is so special about the chemistry of Ca^{2+} which makes it so suitable as an intracellular regulator?

Table 1.11. Approximate number of known eukaryote species

Phylum	Number of species	Percentage of all animal species
Animals	1 000 000	100
Arthropoda–Mandibulata–Insecta	750 000	75
Crustacea	20 000	2
Myriapoda	10 500	1.1
Chelicerata	32 000	3.2
Mollusca	80 000	8
Chordata–Vertebrate–Fishes	25 000	2.5
Birds	8 700	0.9
Reptiles	6 000	0.6
Mammals	4 500	0.5
Amphibians	3 000	0.3
Porifera	10 000	1
Nematoda	10 000	1
Coelenterata (Cnidaria and Ctenophora)	9 100	0.9
Platyhelmintha	9 000	0.9
Annelida	8 700	0.9
Echinodermata	5 300	0.5
Rotifera	1 500	0.2
Other small phyla (approx. 20)	3 300	0.3
Plants	250 000	
Protozoa	50 000	

(4) What is the molecular basis for the action of intracellular Ca^{2+} and the control of its concentration in the cell?

(5) Is it possible to rationalize what is known of intracellular Ca^{2+} as a regulator and to formulate a unitary hypothesis regarding its physiological and evolutionary significance?

We begin our search for the answers to these questions by examining the experimental approaches to the investigation of intracellular Ca^{2+} as a regulator.

CHAPTER 2

The investigation of intracellular Ca^{2+} as a regulator

2.1. What is meant by the term 'regulator'?

Imagine a child standing by the side of the road who, seeing its parents on the other side, suddenly runs across the road to join them. Unfortunately in doing so the child trips and falls, badly grazing its knee. The child need have no fear since within a few minutes the blood vessels near to the site of injury will have constricted, the blood will have clotted, and the bleeding stopped. During these events many biochemical changes will have taken place in the child, primarily as a result of the action of various nerves and the release of chemical stimuli into the blood. Calcium ions will have been involved in several of these changes both inside and outside the cells. Let us consider two of them, first the role of Ca^{2+} in the contraction of the muscles which enable the child to move and the blood vessels to constrict, and secondly, its role in the formation of the fibrin clot.

In the first of these examples Ca^{2+} can truly be considered as a physiological regulator since it is the increase in the concentration of Ca^{2+} in the cytoplasm of the muscle cells which provides the link between the external trigger of the cell and the contractile proteins inside the cell. In the second case, as was discussed in Chapter 1, extracellular Ca^{2+} is essential for the activity of four serine proteinases and a transglutaminase which form part of the cascade resulting in the formation of the fibrin clot (Esmon and Jackson, 1974; Suttie and Jackson, 1977). However, calcium is not itself responsible for initiating the response of this pathway to the child's injury, nor does one consider small diurnal changes in the concentration of plasma Ca^{2+} to have any significant control on whether the clot forms or not. Clotting can only be prevented by the addition of an inhibitor or a strong Ca^{2+}-chelator such as EDTA or citrate, which will lower the ionized calcium in the plasma to submicromolar concentrations. Hence, in the case of the prothrombin to thrombin conversion Ca^{2+} can be considered to play a 'passive', albeit vital, role in this process, whereas in muscle contraction intracellular Ca^{2+} can be considered to play an 'active' role in the regulation of this phenomenon.

The distinction between 'active' and 'passive' Ca^{2+} depends on whether a change in Ca^{2+}, as opposed to one of the other components in the system concerned, is involved in mediating the effect of the primary signal (Table 2.1).

Table 2.1. Some examples of 'active' and 'passive' Ca^{2+}–protein binding

Protein	Location	Approx. $K_d^{Ca^{2+}}$	Reference
Active Ca^{2+}			
Phosphorylase b kinase	Blowfly flight muscle	0.1 μM	Sacktor et al. (1974)
	Rabbit skeletal muscle	1 μM	Cohen (1974)
Troponin C	Muscle	0.2 μM	Potter and Gergely (1975)
Calmodulin	All eukaryotic cells	1 μM	Cheung (1980); Means and Dedman (1980)
Leiotonin	Smooth muscle	1 μM	Ebashi (1980)
Sarcoplasmic (Ca^{2+})-ATPase	Sarcoplasmic reticulum	0.3 μM	MacLennan et al. (1972)
(Ca^{2+})-ATPase	Erythrocyte cell membrane	1 μM	Knauf et al. (1974)
Passive Ca^{2+}			
α-Amylase	Saliva	1 μM	Vallee et al. (1959); Hsiu et al. (1974); Vallee and Wacker (1970)
Prothombin	Plasma	0.5 mM	Esmon and Jackson (1974); Suttie and Jackson (1977)
DNAase	Duodenal juice	16 μM	Price (1974)
Trypsinogen	Duodenal juice	0.6 mM	Abbot et al. (1975)
Acetylcholine receptor	Synapse	ca 1 mM	Lukas et al. (1979)
ACTH receptor	Adrenal cortex	ca 1 mM	Tait et al. (1980)
Coupling to cyclase	Adipose tissue		
Ca^{2+}-binding sites on outer surface	Squid giant axon	0.3 μM and > 20 mM	Baker and McNaughton (1978)

The concept of 'active' and 'passive' Ca^{2+} binding can also be applied to nucleic acids, phospholipids and membrane proteins. For example, Ca^{2+} plays an 'active' role in cells where a Ca^{2+} current is necessary for the generation of action potentials (Hagiwara, 1973; Reuter, 1973) (see Chapter 4). Ca^{2+}, bound to membranes, is also necessary for maintaining the structural integrity and normal electrical activity of many cells (Frankenhauser and Hodgkin, 1957; Shanes, 1958). Removal of external Ca^{2+} causes vertebrate nerves to become hyperexcitable and to generate spontaneous action potentials, whereas in certain invertebrate excitable cells a lowering of intracellular free Ca^{2+} enables action potentials to be generated more easily than in the normal cell. The question arises, is the removal of the Ca^{2+} under these non-physiological conditions exposing Ca^{2+}-binding sites which are normally 'passive', or could they play an active role by releasing or binding Ca^{2+} as part of the mechanism by which channels to specific ions are opened and closed in excitable cells?

Most of the examples of active Ca^{2+} are intracellular, whereas those of passive Ca^{2+} are extracellular (Table 2.1). There are several reasons for expecting this to be generally the case. For example, since the plasma Ca^{2+} concentration is maintained within very narrow limits in healthy organisms, it might be expected that only the mechanism responsible for this maintenance, i.e. parathyroid hormone/vitamin D (Woodhead, 1982), would have evolved the sensitivity necessary to respond to small diurnal changes in plasma Ca^{2+} concentration. In contrast, inside the cell, where the concentration of free Ca^{2+} is thought to be about $0.1 \mu M$ in the resting cell (Ashley and Ridgway, 1970; Baker, 1972; Duncan, 1976; Ashley and Campbell, 1979), if Ca^{2+} binding is to have any function at all and yet be passive, either the affinity for Ca^{2+} must be so high that the binding sites are virtually saturated (i.e. $>95\%$) at $0.1 \mu M$ Ca^{2+} or the association and dissociation rates must be slow enough so as not to allow any significant changes in the amount of Ca^{2+} binding during the time when the concentration of free Ca^{2+} in the cytoplasm of the cell changes. As we shall see in this and the following chapters, it is these thermodynamic restrictions imposed on Ca^{2+}-binding sites which make it so important that the concentration of intracellular free Ca^{2+} and the thermodynamic constants for Ca^{2+} binding be measured. An important consequence of the distinction between active and passive Ca^{2+} binding is the effect it has on the interpretation of experimental procedures which are designed to disturb cell calcium, and on the way the possible role of Ca^{2+} in the action of a pharmacological agent or a pathological phenomenon is investigated. These conditions may convert normally passive or latent Ca^{2+}-binding sites into active ones.

2.2. The concept of 'threshold phenomena'

Let us return to the two examples discussed in the opening section of this chapter—muscle contraction and the formation of fibrin clots. Both of these reactions belong to a class of phenomena which can be considered as 'threshold phenomena' in that before being exposed to the appropriate stimulus the muscles were relaxed and the blood was flowing freely in the blood vessels, no clots being present. This simple type of all-or-none response is characteristic of many biological phenomena, and forms an important part of the discussions in several of the chapters of this book. 'Threshold phenomena' are defined as cellular events involving the transition of the cell from one state to another under physiological conditions. Examples can be found in many types of cellular activity.

(1) Sensory mechanisms: including nerve conduction; the electrical activity of many cells (particularly those that can generate action potentials); visual processes; the flashing of many luminous organisms; sensory mechanisms in unicellular eukaryotes; and chemotaxis in prokaryotes.

(2) Movement: including the contraction of various types of muscle; flagellate and amoeboid cell movement.

(3) Secretion and the converse process, endocytosis: including the release of hormones, neurotransmitters and substances from granule-containing cells; the vesicular uptake of soluble material by pinocytosis; and the uptake of particulate material by phagocytosis.

(4) Cell and tissue development: including cell fertilization; cell division; amoeboid to flagellate cell transformation; cell differentiation; the formation of tissues; and the control of cell populations.

The generation of an action potential by a nerve, the response of the rods and cones in the eye to a photon hitting them, the flash of light emitted by a luminous animal, the contraction of a muscle cell or the division of one cell into two, represent clear examples of threshold phenomena. Consider, however, the release of a neurotransmitter at a synapse. The fact that miniature end-plate potentials can be detected in the adjoining cell, together with the fact that by adjusting the experimental conditions it is possible to observe the release of different quantities of transmitter from the nerve ending, might argue against this secretory mechanism being classified as a threshold phenomenon. However, the primary physiological stimulus, i.e. the action potential reaching the nerve terminal from the axon, is a threshold phenomenon and does transform the cell within a few milliseconds from one state to another.

Once a cell has been stimulated above the threshold it may be possible to regulate the magnitude of the response, as well as the time taken to complete the response. Furthermore, physiologically occurring substances and pharmacological agents may determine the conditions under which the threshold occurs. For example, in heart muscle cells the all-or-none phenomenon of beating in each cell is under regulation by adrenaline and acetylcholine which regulate the strength of the contraction and rate of beating.

Some phenomena exist which apparently are not threshold phenomena in that the tissue concerned appears to exhibit a continuous, graded response from the basal level in the absence of a stimulus to a maximum in the presence of a saturating concentration of the stimulus. It is usually assumed that the regulation of intermediary metabolism by hormones such as insulin and adrenaline falls into this category (Chapter 6). An alternative hypothesis would be that such graded tissue responses are the result of switching on different numbers of cells.

One of the objectives of this book is to highlight the importance of threshold phenomena in cell regulation and the role of intracellular Ca^{2+} in mediating the threshold response. Although non-threshold cell responses do exist it should be borne in mind that the measured response of a whole tissue is the integral of the responses of hundreds of millions of cells. The possibility that a graded tissue response to different concentrations of a stimulus might be explained by different cells in the population breaking through a threshold, in other words switching on or off, should also be considered.

The enormous gradient in Ca^{2+} concentration which exists across the cell membrane means that Ca^{2+} is uniquely situated to act as the intracellular mediator of a wide variety of threshold phenomena. Furthermore, the action of other intracellular regulators such as cyclic AMP which interact with Ca^{2+}-mediated mechanisms may function either to determine the conditions necessary for the threshold, for example the concentration of the primary stimulus necessary, or may alter the strength and time course of the Ca^{2+}-induced response.

2.3. Experimental approaches

2.3.1. The basic questions

The activation of a cell response is initiated by a primary stimulus and can be modified by a secondary regulator. These secondary regulators may enhance or inhibit the response of the cell to a primary stimulus (Table 2.2). Most of the examples of secondary regulators considered in this book act acutely. However, others, like thyroid and steroid hormones, may act as longer-term regulators controlling the sensitivity of tissues to primary stimuli.

The concept of primary and secondary regulators applies to many types of cell activation, particularly when a threshold response of the cell is involved. It is not always clear whether it also applies to phenomena such as intermediary metabolism (Chapter 6). This is activated during many threshold cell responses, for example muscle contraction and phagocytosis, to provide the energy required. However, cells such as those in adipose tissue and liver, responsible for regulating the energy metabolism of the whole body, appear to exhibit a graded response to hormones like adrenaline and insulin. For example, in adipose tissue, adrenaline-stimulated lipolysis is inhibited by insulin. In the same tissue insulin activates glucose uptake and its metabolism to glycogen and through glycolysis, and stimulates lipid synthesis. The balance between these effects results in a change in the overall direction of glycogen metabolism, and similarly for triglyceride, when the animal eats a meal. A similar argument can be applied to glycolysis versus gluconeogenesis in the liver. The point at which the overall net direction of these pathways changes could be considered to be a threshold. In order to resolve these problems completely it may be necessary to examine the effects of hormones on the metabolism of individual cells.

Thus, before examining the possible role of intracellular Ca^{2+} in the initiation of a cell or tissue response it is first necessary: (1) to identify the primary stimuli and secondary regulators, be they electrical, physical or chemical (Table 2.2); (2) to identify any physiological conditions that provide a secondary regulation of the primary stimulus, thereby altering either the strength or duration of the cell response; and (3) to define whether the primary stimuli provoke a threshold or continuously graded response in the cell and whether the secondary regulators act by altering the amount of primary stimulus necessary to provoke the threshold or alter the responsiveness above the threshold level. It is the answers to these physiological questions that provide the necessary framework for the chemist wishing to explain the role of Ca^{2+} as a regulator in molecular terms.

Table 2.2. Some examples of primary stimuli and secondary regulators

Phenomenon	Primary stimulus	Secondary regulator*
Cell movement—chemotaxis of polymorphs	Complement fragment C5a	Prostaglandin E_1 (i) Prostaglandin $F_{2\alpha}$ (a)
Heart muscle contraction	Action potential	Acetylcholine (i) β-Adrenaline (a)
Neurotransmitter release at nerve terminal	Action potential	Adenosine (i)
Insulin secretion from β-cell	Glucose	Adrenaline (i) Glucagon (a)
Excitation of retinal rods	Light	Dark adaptation (a)
Cell fertilization	Sperm	?
Lymphocyte transformation	Antigen	?

See also Tables 5.2, 5.14, 5.20, 6.3, 7.1, 7.12 and 8.3.
* i = inhibitor; a = activator.

2.3.2. Does intracellular Ca^{2+} play a role in the phenomenon?

There are two initial experimental approaches to investigate whether intracellular Ca^{2+} is required for the action of primary or secondary regulator: (1) manipulation of extracellular Ca^{2+}; and (2) manipulation of intracellular Ca^{2+}.

Manipulation of extracellular Ca^{2+}

As we saw in Chapter 1, this approach was first used by Ringer at the end of the last century to show that Ca^{2+} was required for the normal contraction of the heart. Since then many phenomena have been shown to be inhibited or abolished by removal of extracellular Ca^{2+} (Table 2.3). There are three main problems in carrying out these experiments. (1) Ca^{2+} contamination in the water and reagents (Mast and Pace, 1939; Campbell and Siddle, 1976) may produce up to 30 μM free Ca^{2+} in the external medium which may be sufficient for the cell to respond. (2) Removal of extracellular Ca^{2+} may result in loss of Ca^{2+} from external or internal structures with resulting loss in cell responsiveness unrelated to the primary mechanism being studied. (3) Removal of external Ca^{2+} results in the death of many cells, particularly after long periods.

The first problem can be solved by using chelating agents such as EGTA to lower free Ca^{2+} to less than 1 nM (EDTA also binds Mg^{2+}), by using spectroscopically pure reagents, by using highly purified water and by washing all glassware in acid just before use and never storing any solutions for long periods in a glass vessel.

Incubation times of cells with 'Ca^{2+}-free' media have varied from only a few minutes to several hours. A rapid loss of cell response on removal of Ca^{2+} is consistent with a requirement for external Ca^{2+} in the action of the cell stimulus. If longer incubations are required to inhibit the effect of the stimulus then either

Table 2.3. The effect of removal of extracellular calcium

Phenomenon	Example	Reference
(A) *Removal of, or a decrease in, extracellular Ca^{2+} inhibits the physiological response*		
1. Cell movement	Muscle contraction (heart, smooth, invertebrate)	Ringer (1883a,b); Bülbring and Tomita (1970, 1977); Ashley and Campbell (1978)
	Chemotaxis (leukocytes, fibroblasts, bacteria)	Becker and Showell (1972); Gail et al. (1973); Ordal (1977)
	Sensory movement (protozoa)	Hauser et al. (1978); Brehm and Eckert (1978)
	End-plate of skeletal muscle	Brecht and Gebert (1966)
2. Cell aggregation	Slime mould	Bonner (1971)
	Sea urchin blastula	Herbst (1900)
	Gap junction formation	Loewenstein (1966; 1975)
3. Membrane fusion	Erythrocyte fusion induced by Sendai virus	Hart et al. (1976); Volsky and Loyter (1978)
	Muscle myoblasts	Shainberg et al. (1969)
4. Cell transformation	Lymphocyte activation	Diamantstein and Odenwald (1974)
	Amoeboid → flagellate in *Naegleria*	Fulton (1977)
5. Vision	Vertebrate rods	Hagins and Yoshikami (1974)
6. Chemiluminescence	Coelenterates	Cormier et al. (1978); Campbell et al. (1979)
	Echinoderms	Brehm (1973)
	Macrophages and polymorphs	Hallett et al. (1981)
7. Phagocytosis and pinocytosis	Macrophages	See Table 7.12
8. Secretion	Many vesicular types	See Table 7.3; Rubin (1970)
9. Hormone action	ACTH	Farese (1971a,b); Tait et al. (1980)
	α-Adrenaline, glucagon, vasopressin (liver)	Keppens et al. (1977); Whitton et al. (1977); Blackmore et al. (1978)
	5-Hydroxytryptamine on brown adipose tissue lipolysis	Itaya (1978)
	Glucocorticoids on thymocytes	Kaiser and Edelmen (1977)
10. Membrane permeability	K^+ and other ions	Gilbert and Ehrenstein (1969); Gilbert (1972)
11. Extracellular pathways	Complement, blood clotting	Suttie and Jackson (1977)
(B) *Little or no acute effect of removing extracellular Ca^{2+}*		
1. Cell movement	Muscle contraction (mammalian skeletal)	Brink (1954)
	Contractions in *Spirostomum* (a protozoan)	Ettienne (1970)
2. Cell aggregation	Coupling of heart cells	Gilula and Epstein (1976)

Table 2.3. (*contd*)

Phenomenon	Example	Reference
3. Cell transformation	Maturation of starfish oocytes	Moreau and Guerrier (1979)
	Sperm fertilization of fish eggs	Gilkey *et al.* (1978)
4. Vision	Some invertebrate photoreceptors	Brown and Blinks (1974)
5. Chemiluminescence	Fertilized eggs (non-functional)	Foerder *et al.* (1978)
6. Secretion	Parotid (β-adrenaline-stimulated)	See Table 7.3
7. Hormone action	Insulin and adrenaline on adipose tissue	Siddle and Hales (1980)
(C) *Spontaneous activation or enhancement of cell response by removal of extracellular Ca^{2+}*		
1. Transmission of impulse	Mammalian nerve axons	Brink (1954); Shanes (1958)
2. Vision	*Limulus* photoreceptor hyperpolarization	Brown and Blinks (1974)

Note: (1) No attempt has been made to distinguish the time of incubation without Ca^{2+} before addition of the stimulus, but in most cases it was between 30 min and 2 h.
(2) Some of the Ca^{2+} effects are reflections of a passive role.

an intracellular Ca^{2+} store is the source of Ca^{2+} for triggering the cell, or the conditions have caused injury to the cells.

Additional information can be obtained by observing the effect of varying the concentration of extracellular Ca^{2+}, usually over the range 0–10 mM. The relationship between the response of the cell and Ca^{2+} concentration can be linear in the case of some secretory processes (see Chapter 7), or may be related to the fourth power of the Ca^{2+} concentration in the case of neurotransmitter release from nerve terminals (see Chapters 4 and 7), whilst in some phenomena a Ca^{2+} concentration > 2 mM is inhibitory. These power-law relationships may allow some predictions to be made regarding the stoichiometries of Ca^{2+} channels and Ca^{2+}-binding proteins. It is important to show whether both the maximal response of the cells to a saturating level of stimulus (viz. V_{max}) and their sensitivity to the stimulus (viz. K_m) are affected by the Ca^{2+} concentration, since Ca^{2+} is necessary for the coupling of some hormone and neurotransmitter receptors to the activated component in the cell membrane, for example adenylate cyclase.

Manipulation of intracellular Ca^{2+}

The aim of this experimental approach is to see whether changes in cytoplasmic Ca^{2+}, or Ca^{2+} in intracellular organelles, either mimic or modify the physiological response of the cells. Two typical ways of manipulating intracellular Ca^{2+} are either to inject Ca^{2+}, Ca^{2+}/EGTA or EGTA into the cell, or to add a substance to the extracellular fluid which increases or decreases the concentration of intracellular Ca^{2+}.

The first approach was adopted by Heilbrunn and Wiercinski, who in 1947 showed that injection of Ca^{2+} into frog muscle stimulated contraction; Mg^{2+}, Na^+ and K^+ did not. Similarly injection of Ca^{2+} into the presynaptic terminal of the giant synapse of the squid stimulates neurotransmitter release (Miledi and Katz, 1966; Miledi, 1973), its injection into mast cells stimulates histamine secretion (Kanno et al., 1973), and its injection into the salivary gland cells of *Chironomus* switches off the electrical conductivity between neighbouring cells normally linked through the gap junctions (Rose and Loewenstein, 1975, 1976), whilst injection of Ca^{2+} into frog oocytes induces furrow formation (Timourian et al., 1972), a phenomenon that can be inhibited by injecting EGTA into the cell. Simple as this approach might seem it is fraught with several difficulties. Firstly, the micropipette, or microelectrode if the injection is done by ionophoresis rather than pressure injection, can cause irreversible damage to the electrical and biochemical response of the cell. Secondly, since the concentration of intracellular free Ca^{2+} has only been measured in a few cell types, the quantity of Ca^{2+} that must be microinjected to raise the concentration of Ca^{2+} in the cytoplasm significantly, and yet not be too high to cause damage, is often unknown. Thirdly, it has been shown that Ca^{2+}-buffering systems exist inside cells which rapidly remove any Ca^{2+} injected into them (Baker, 1972; Rose and Loewenstein, 1976; Brinley et al., 1979). In order to maintain an elevated free Ca^{2+} concentration in the cytoplasm of the cell for more than a few seconds it is necessary either to add an inhibitor of the uptake system, or to inject a Ca/EGTA buffer (Portzehl et al., 1964; see also section 2.6 of this chapter).

Two other novel methods have been developed for manipulating cytoplasmic Ca^{2+} using chelating agents. One involves the production of a 'hole' in the cell membrane by a high-voltage discharge (Baker and Knight, 1978); the other utilizes an ester of a Ca^{2+} chelator which allows it to enter the cell where hydrolysis causes the compound to revert to the acid form (Tsien, 1980, 1981). In the first of these methods an electric field strength of several $kV\,cm^{-1}$ is applied to a suspension of cells for a few μsec. This causes a breakdown in the structure of parts of the cell membrane leaving the cell permeant to ions, metabolites and proteins. The number and size of the holes can be controlled by adjusting the field strength and time of the discharge (Riemann et al., 1975). By adding Ca/EGTA buffers outside the permeant cells it is possible to define the concentration of free Ca^{2+} inside the cell necessary to stimulate secretion, for example. In the other technique applicable to small cells, an acetoxymethyl tetraester of a Ca^{2+} chelator is allowed to penetrate the cells by diffusion across the lipid bilayer. Once inside an esterase hydrolyses off the ester thereby regenerating the four CO_2^- groups masked by the acetoxymethyl groups, producing a high-affinity Ca^{2+} chelator which is now in theory impermeable to the cell membrane. It is necessary to establish that the compound is in a defined compartment of the cell and that it does not leak out of the cell when the cell responds to a physiological stimulus.

An alternative approach is to add to the cells a substance which affects either the permeability of the cell membrane to Ca^{2+} or intracellular Ca^{2+} stores. Several substances are now known which, at least under some conditions, can increase or decrease the intracellular concentration of Ca^{2+} in this way (Table 2.4; Fig. 2.1). With many of these compounds there can be doubt about the specificity of their effects with respect to intracellular Ca^{2+}. It is therefore essential when using them to show that any effects are dependent on Ca^{2+} and that they do not cause any impairment of the electrical or biochemical properties of the cell. The major limitations to the interpretation of experiments using these compounds (Table 2.4) are their lack of specificity with respect to Ca^{2+} and an imprecise knowledge of their mechanism of action. Nevertheless, they have been widely used as a means of obtaining evidence for a role of Ca^{2+} in cell regulation.

Table 2.4. Some conditions which may affect intracellular free Ca^{2+}

Effect	Effector	Reference
Increase		
1. Electrically excitable cells with voltage-dependent Ca^{2+} channels	K^+-replaced saline or other means of depolarization	Baker (1972)
2. Cells with a Na^+-dependent Ca^{2+}-efflux	Replacement of extracellular Na^+ by Li^+ or choline	Baker (1972); Blaustein (1974)
3. Inhibitors of oxidative phosphorylation or the respiratory chain	Oligomycin, dinitrophenol, CN^-	Baker (1972)
4. Increased influx of Ca^{2+} across the cell membrane	Ionophores A23187, X-537A, and ionomycin	Pressman (1976); Campbell and Dormer (1978); Hainaut and Desmedt (1979)
5. Release of Ca^{2+} from the sarcoplasmic reticulum	Caffeine	Ashley et al. (1974, 1977); Ebashi (1976; Bianchi (1961); Holland and Porter (1969)
	Ryanodine	Jenden and Fairhurst (1969)
Decrease or inhibition of increase		
1. Inhibitors of the slow Ca^{2+} channel in electrically excitable cells	Iproveratril and D 600, La^{3+}, Mn^{2+}, Co^{2+}, Ni^{2+}	Baker (1972)
2. Inhibition of Ca^{2+} permeability to mitochondria	Ruthenium red	Moore (1971); Denton et al. (1980)
3. Local anaesthetics and other membrane 'stabilizers'	Tetracaine, procaine	Siddle and Hales (1974b)
4. Inhibition of Ca^{2+} release from sarcoplasmic reticulum	Dantralene	Hainaut and Desmedt (1979)

See Figs. 2.1 and 2.2 for molecular structures.

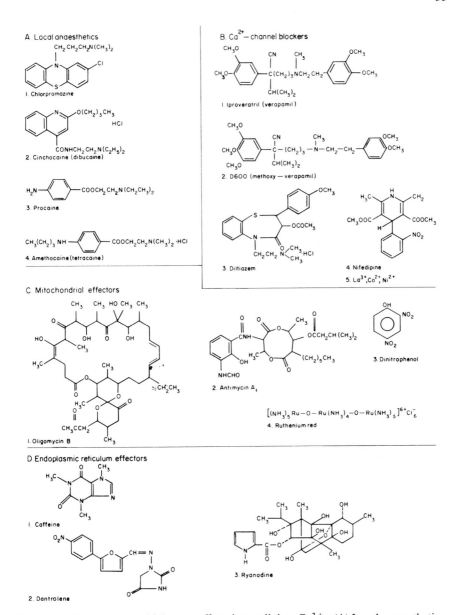

Fig. 2.1. Compounds which can affect intracellular Ca^{2+}. (A) Local anaesthetics. (B) Ca^{2+}-channel blockers. (C) Mitochondrial effectors (see also Fletcher et al., 1961). (D) Endoplasmic reticulum effectors. (see Chapters 4, 5 and 9 for mechanisms)

Ever since the discovery of lipophilic compounds which were able to make biological membranes permeable to monovalent cations (Moore and Pressman, 1964; Pressman, 1976), the search for naturally occurring and synthetic substances that could do the same for divalent cations has attracted the attention of many biologists and chemists (Trutter, 1975, 1976; Umen and Scarpa, 1978). Ca^{2+} ionophores, the name given to such compounds, have been found in mammalian tissues (Gomez-Puyou and Gomez-Lojero, 1977; Shamoo and Goldstein, 1977). However, the two most widely used Ca^{2+} ionophores, X-537A (Berger *et al.*, 1951), and A23187 (Reed and Lardy, 1972) (see Fig. 2.2 for structures), are both extracted from bacteria. These substances, and in particular A23187 extracted from the bacterium *Streptomyces chartreusensis*, which is thought to be more specific for Ca^{2+} relative to Mg^{2+} (Pressman, 1976), have been extensively used and shown to mimic, at least qualitatively, the effect of many physiological cell stimuli (Table 2.5). Other Ca^{2+} ionophores include ionomycin and specially synthesized compounds (Fig 2.2; Umen and Scarpa, 1978). Ionomycin is a polyether antibiotic from another species of *Streptomyces* (*Conglobatus*). It is more selective for Ca^{2+} than A23187 and binds with a 1:1 stoichiometry ($C_{41}H_{70}O_9Ca$, molecular weight 746).

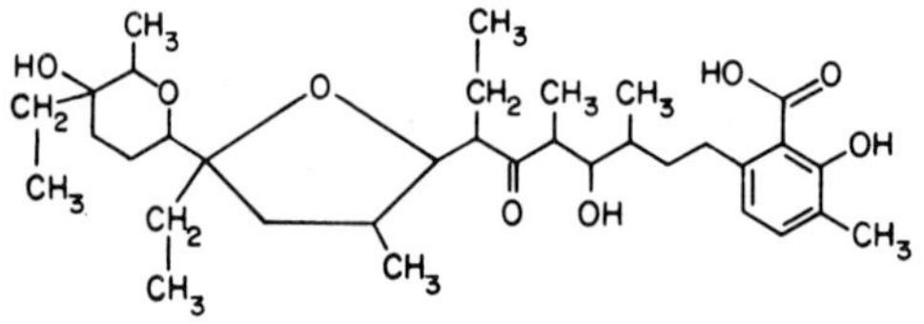

X-537A
(Lasalocid A)

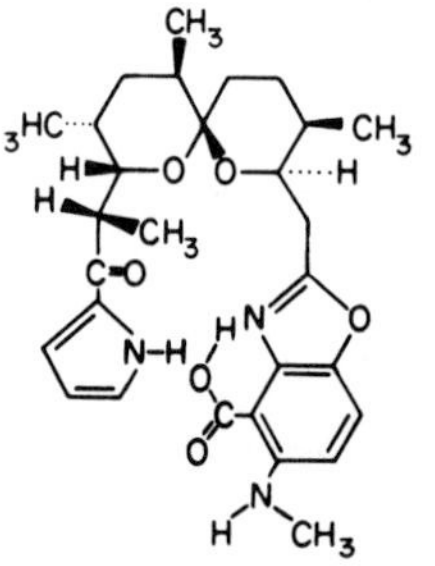

A23187

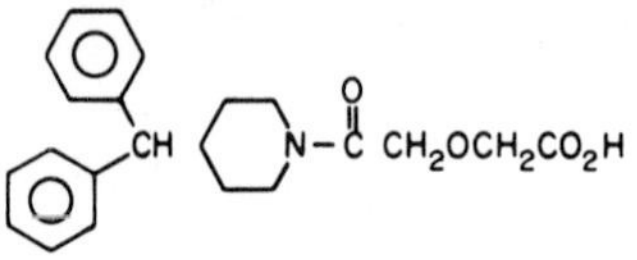

McN4308

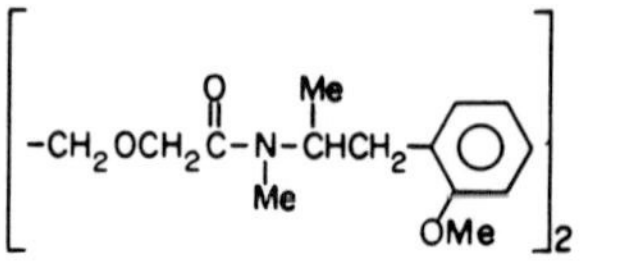

McN3849

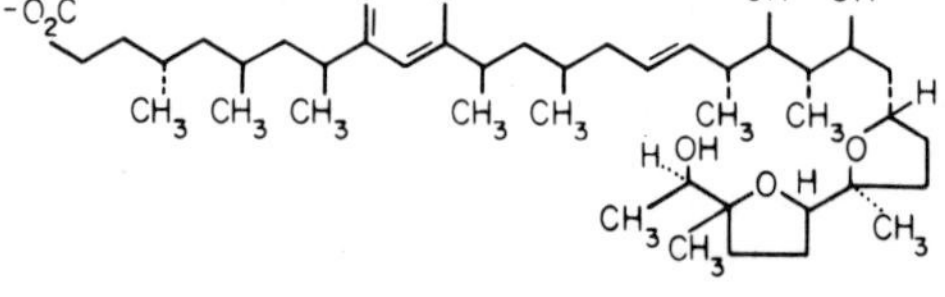

Ionomycin (E_m94) SQ23377
$C_{41}H_{70}O_9Ca$
Molwt 709·02 (747·09)

Fig. 2.2. Some Ca^{2+} ionophores. See Trutter (1976) and Umen and Scarpa (1978) for references

Table 2.5. Attempts to mimic physiological stimuli using ionophore A23187

Phenomenon	Example to be mimicked	Reference
(A) *Mimicking effects*		
1. Cell movement	Frog muscle contraction	Devore and Nastuk (1977)
	Barnacle muscle contraction	Hainaut and Desmedt (1979)
	Tumbling of chemotactic bacteria	Ordal (1977)
2. Cell aggregation	Sealing of gap junctions	Rose and Loewenstein (1976)
	Platelet aggregation	Massini and Luscher (1974)
3. Membrane fusion	Erythrocytes by fusogens	Lucy *et al.* (1970, 1975)
	Erythrocytes by Sendai virus	Volsky and Loyter (1978)
	Myoblast fusion	Schadt and Pelks (1975)
4. Cell transformation	Yeast cell division	Duffus and Patterson (1974)
	Lymphocyte proliferation	Maino *et al.* (1974)
	Egg maturation and meiosis	Steinhardt *et al.* (1974)
	Egg fertilization by sperm	Steinhardt *et al.* (1974)
5. Shape changes in cells	Sendai virus and other fusogens on erythrocytes	Volsky and Loyter (1978)
6. Response of cells to light	Rat retinal rods (hyperpolarize)	Yoshikami and Hagins (1977)
	Chloroplasts	Telfer *et al.* (1975)
7. Luminescence	Coelenterate	Campbell *et al.* (1979)
	Phagocytes	Wilson *et al.* (1978)
	Fertilized eggs	Foerder *et al.* (1978)
8. Secretion	Many vesicular secretions (e.g. mast cells)	Foreman *et al.* (1976)
9. Hormone action	Liver glycogenolytic hormones (adrenaline, glucagon, antidiuretic hormone)	Keppens *et al.* (1977); Blackmore *et al.* (1978)
	Glucocorticoids on thymocytes	Kaiser and Edelman (1977)
	Hormonal elevation of cyclic GMP	Goldberg and Haddox (1977)
10. Membrane ion conductance	K^+ in *Paramecium*	Eckert and Brehm (1978); Brehm *et al.* (1978)
(B) *Negative effects of A23187*		
1. Hormone action	Insulin and adrenaline on adipocytes	Siddle and Hales (1980)
2. Neurotransmitter action	Muscarinic acetylcholine receptor and phosphatidylinositol turnover	Michell *et al.* (1977)

In spite of their popularity a number of problems should be borne in mind when critically evaluating the results of experiments using Ca^{2+} ionophores.

(A) How closely does the ionophore response mimic the natural one particularly with respect to time course and magnitude? Since A23187 has been shown to increase the intracellular concentration of Ca^{2+} within a few seconds

(Campbell and Dormer, 1978), a difference of many minutes between the onset of the response induced by the physiological stimulus and that induced by the ionophore, e.g. cell fusion, is not consistent with Ca^{2+} being the mediator of the physiological effector. Furthermore, the possible effect of the ionophore on some pool of Ca^{2+} inside the cell which is normally passive should be considered.

(B) Specificity for Ca^{2+}. The stoichiometry of ionophore A23187 is thought to be $(A23187^-)_2 Ca^{2+}$, and it is therefore a neutral charge carrier without direct effect on membrane potential. Nor would the effect of A23187 be expected to be influenced by the membrane potential. However, since this ionophore probably exchanges $2H^+$ for Ca^{2+} it may influence the intracellular pH. The dissociation constants of various ions for A23187 (Table 2.6) illustrate that its affinity for Ca^{2+} is only three times that for Mg^{2+}. It is not surprising, therefore, that ionophore A23187 can also have profound effects on cell Mg^{2+} (Campbell and Siddle, 1976). Furthermore, this ionophore can also transport amines and amino acids across hydrophobic layers (Pfeiffer et al., 1974; Hovi et al., 1975).

The requirement for extracellular Ca^{2+} should be determined. Metabolic effects of A23187 have been found in the absence of external Ca^{2+} (Campbell and Siddle, 1976) and in some cases this may be caused by the ionophore penetrating intracellular membranes and stimulating Ca^{2+} release from intracellular stores. It is also important to remember that when μM concentrations of ionophore are added to cells much of this will dissolve in the membrane phospholipid where its concentration will be equivalent to several mM. Only a few turnovers of the ionophore are therefore required to cause large changes in cytoplasmic free Ca^{2+}.

(C) By how much does A23187 elevate the intracellular Ca^{2+} concentration? Ionophore A23187 causes a large uptake of Ca^{2+} by cells, except when EGTA is added to the external medium. In the latter case A23187 causes depletion of intracellular Ca^{2+} stores by more than 50% within a few minutes. Using red cells

Table 2.6. Dissociation constants for A23187 and certain cations

Cation	K_d
1. *Divalent*	$A_2 M$
Ca^{2+}	2.7×10^6
Mg^{2+}	7.7×10^6
Sr^{2+}	9.1×10^8
2. *Monovalent*	$A_2 HM$
Na^+	1.7×10^3 M
K^+	1.4×10^4 M

Data taken from Pfeiffer et al. (1978).

1. $(A_2 Ca)_{org} + 2H^+_{aq} \rightleftharpoons 2AH_{org} + Ca^{2+}_{aq}$ $K_d = [AH]^2 [Ca^{2+}]/[A_2 Ca][H^+]^2$ (2.1)

2. $(A_2 NaH)_{org} + H^+_{aq} \rightleftharpoons 2AH_{org} + Na^+_{aq}$ $K_d = [AH]^2 [Na^+]/[A_2 NaH][H^+]$ (2.2)

where A = A23187; org = organic phase; aq = aqueous phase, $K_d^H = 0.2\ \mu$M for $HA \rightleftharpoons H^+ + A^-$

it is possible to adjust the concentration of intracellular free Ca^{2+} within the μM range with appropriate ionophore and extracellular Ca^{2+} and Mg^{2+} concentrations (Campbell and Siddle, 1976; Flatman and Lew, 1977). However, in cells which contain efficient Ca^{2+}-buffering systems it is impossible to estimate the effect of A23187 on the absolute free Ca^{2+} concentration in the cell unless it is measured directly.

(D) Potency. The concentrations of ionophore A23187 usually employed range between 1 and 20 μM. Its potency in raising the concentration of intracellular Ca^{2+} can be markedly reduced by increasing the cell concentration—which presumably decreases the concentration of A23187 in the lipid bilayer of each individual cell—or by the presence of Mg^{2+} or albumin in the incubation medium.

(E) Irreversible effects in cells. Ionophore A23187 causes isolated cells in culture to round up (Rasmussen and Goodman, 1977), and membrane vesiculation (Allan and Michell, 1977). Since the effects of A23187 are very difficult to reverse and since A23187 can cause large decreases in cell ATP (Campbell and Siddle, 1976), and hence increases in ADP and AMP, care should be taken when using ionophores that the experimental conditions do not cause irreversible cell damage.

Inhibitors

Inhibitors of Ca^{2+} permeability such as ruthenium red, La^{3+} (Weiss, 1974), and the verapamil derivative D600 (Fig. 9.16), as well as 'membrane stabilizers' such as local anaesthetics (Table 2.3 and Fig. 2.1) have been used to provide evidence of a role for Ca^{2+} in several systems. However, lack of knowledge regarding both their specificity for Ca^{2+} and their mechanism of action cause problems in the interpretation of such experiments.

Replacement of extracellular Ca^{2+} by Sr^{2+} or Ba^{2+}

In cells where Ca^{2+} plays either an active or passive role it is often possible to replace Ca^{2+} partially by two other group II cations, Sr^{2+} and Ba^{2+} (Skoryna, 1981). Whilst these experiments may shed some interesting light on the chemistry of the Ca^{2+}-binding sites, they seldom help to provide the experimental evidence required to demonstrate a role for intracellular Ca^{2+} in a particular phenomenon.

2.3.3. Is the role of intracellular Ca^{2+} 'active' or 'passive'?

In Chapter 1 we saw that there are four main biological roles for Ca^{2+}. If Ca^{2+} plays a role in the activation of a cell by a primary stimulus it is necessary to establish that this role is intracellular and is an 'active' one. Four questions need to be answered in order to do this. (1) Does the primary stimulus cause a change in the concentration of free Ca^{2+} in the cell compartment where the reactions concerned with cell activation are found? (2) Does the primary stimulus alter the affinity of regulatory Ca^{2+}-binding sites with or without a change in free Ca^{2+}? (3) If the primary stimulus causes a change in intracellular free Ca^{2+} how does it do it? Is the source of Ca^{2+} extracellular or intracellular stores? (4) What is the site of action of the Ca^{2+} and how does it work?

Secondary regulators which act through a Ca^{2+}-dependent mechanism will either modify the response of intracellular free Ca^{2+} to the primary stimulus or will interact with the effect of Ca^{2+} on the process concerned. They may require another intracellular messenger such as cyclic AMP for their action (see Chapter 6).

As we shall see in the chapters which follow, these questions have yet to be answered completely for any type of cell activation, although in several examples of muscle contraction the experimental evidence has gone a long way towards answering the four questions posed above.

The establishment of a correlation between a change in intracellular free Ca^{2+} and the time scale of cell activation is crucial. The identification of intracellular Ca^{2+} stores, specific Ca^{2+} channels in the cell membrane and Ca^{2+} pumps enables the mechanism of regulation of intracellular free Ca^{2+} to be defined. Thus a primary stimulus (Table 2.2) may alter the uptake and release of Ca^{2+} across the cell membrane or across the membrane of intracellular organelles. In order to understand how Ca^{2+} actually causes the physiological response of the cell it is necessary to isolate and characterize the Ca^{2+}-binding proteins responsible (see Chapter 3). The thermodynamics of Ca^{2+} binding to, and release from, these proteins must be defined. It is also necessary to show what structural change Ca^{2+} causes in the ligand to which it is bound and how this change causes a change in the structures ultimately responsible for the cell response, for example, actomyosin in muscle or granule–cell membrane fusion in secretory cells.

The search for answers to the four main questions posed above requires investigations using a combination of intact cell preparations, isolated organelles and purified proteins. It is essential to remember, however, that if the primary intracellular event caused by a cell stimulus is an increase in intracellular free Ca^{2+} then this event can only be demonstrated whilst the cell remains intact.

2.3.4. What is the molecular basis of the regulatory role of intracellular Ca^{2+}?

In order to discover the molecular basis of effects of Ca^{2+} inside cells it is necessary to extract, purify and reconstitute a Ca^{2+}-activated 'event' in the test tube which mimics that in the intact cell. A full characterization requires six experimental procedures:

(1) Identification of the Ca^{2+}-binding site. Is it the protein or intracellular structure itself responsible for the phenomenon which binds the Ca^{2+}? Is it a calmodulin? in which case the criteria laid down in Chapter 3 section 3.4.2 must be satisfied. Is the effect of Ca^{2+} mediated indirectly through effects of Ca^{2+} on

phospholipids and membrane proteins? Is Ca^{2+} interacting with the substrate (e.g. CaATP) of an enzyme, or with a cofactor (e.g. NADH), or regulatory metabolite (e.g. citrate)?

(2) Extraction and purification of the Ca^{2+}-binding site.

(3) Reconstitution of the Ca^{2+}-activated process *in vitro* using purified components. The concentration of Ca^{2+} required and the time course should correlate with that defined in the intact cell using Ca^{2+} indicators.

(4) Characterization of the Ca^{2+} sites. How many Ca^{2+} sites are there per unit structure? What is the K_d^{Ca} for each site? K_d^{Ca} should be approximately 1 μM if it is to be significant as a regulatory site. What is the specificity for Ca^{2+} of these sites? They should be at least 1000 times higher in affinity for Ca^{2+} relative to Mg^{2+} if they are to bind significant quantities of Ca^{2+} in the cell. What are the 'on' and 'off' rates (i.e. k_1 and k_{-1}) for Ca^{2+} binding and for the activation of the phenomenon *in vitro*? Are these fast enough to explain the 'event' in the intact cell?

(5) The chemistry of the Ca^{2+} sites. What is the amino acid sequence of the Ca^{2+}-binding protein and the amino acid composition of the Ca^{2+}-binding site? What is the coordination number for Ca^{2+}? Is it possible to use X-ray crystallography to understand the mechanism of any protein conformational change induced by Ca^{2+}, and how can this explain activation of the process in the cell?

(6) Test of Ca^{2+} site in the cell. Is it possible to design experiments using the intact cell to test the identification and characterization of the proposed Ca^{2+} site, which have previously been studied in broken cells and purified systems? Some pharmacological compounds can be used to inhibit Ca^{2+}-dependent processes in cells. Furthermore, antibodies to Ca^{2+}-binding proteins, such as calmodulin, can be used to localize the Ca^{2+} sites in the cell and to inhibit Ca^{2+}-dependent cell activation.

2.3.5. Pathology and pharmacology

There are a considerable number of pathological conditions where abnormalities in regulation of intracellular Ca^{2+} could explain many of the clinical manifestations of the disease (see Chapter 9). These abnormalities could involve Ca^{2+}-binding sites which are either active or passive. Alternatively it may be necessary to invoke latent Ca^{2+} sites not normally involved in physiological cell responses. Similarly the action of many pharmacological substances depends on their ability to disturb intracellular Ca^{2+} or its interaction with primary or secondary stimuli (see Chapter 9).

2.4. The need to measure cell Ca^{2+}

In order to study the biological role of Ca^{2+} it is essential to be able to measure the Ca^{2+} concentration, or more correctly activity, in tissues and body fluids. There are four main questions which can be asked. (1) What is the total Ca^{2+}

concentration in the cells and external fluid and does this change in response to a primary or secondary stimulus (Table 2.2), or under pathological, pharmacological or experimental conditions? (2) What is the distribution of Ca^{2+} within the cell? (3) How much of the Ca^{2+} in any compartment is free in solution compared with bound Ca^{2+}? (4) Does the proportion of bound to free Ca^{2+} change at a particular location in the cell under the influence of a stimulus?

2.4.1. The need to measure total Ca^{2+}

Much has been published on the content of Ca^{2+} and other cations in the tissues of organisms from many phyla (Robertson 1941, 1949, 1953; Manery, 1969; Burton, 1973 a, b). However, the overall conclusions that can be drawn from this information are somewhat disappointing. The two main reasons for this are, first, the lack of any appropriate conceptual framework with respect to the possible biological roles of the various cations studied, and secondly, the fact that the majority of these measurements were made using whole tissue or organism extracts, hence it is impossible to estimate how much of the cation was cellular and how much was extracellular. With regard to this latter problem it is of little use having a highly sensitive and specific method for measuring Ca^{2+} unless careful thought is put towards deciding how the results will be expressed.

It is first necessary to decide how the extracellular volume can be estimated. This can be done, for example, using a radioactively labelled extracellular marker such as inulin (a polymer of the sugar fructose, molecular weight approx. 3000) which does not penetrate cells. Alternatively, another physiologically occurring substance, known to be predominantly extracellular, can be measured. For example, the Na^+ content often enables an estimate of the extracellular volume to be made since in most tissues the intracellular concentration of Na^+ is small (approx. 1/10 of that outside the cell). Once the extracellular volume and concentration of the external cations are known the extracellular content of the cations can be estimated and subtracted from the total tissue content to obtain the cellular content of the cation concerned. Second, it is necessary to measure other parameters so that the cellular cation content can be expressed per cell, per unit weight, and as a molar concentration relative to the cell water. In order to do this the following tissue parameters must be measured: cell number, DNA content, wet and dry weight, intracellular water.

In spite of these problems measurement of total Ca^{2+} has provided some important information about the biological role of Ca^{2+}. It is possible to draw up a Ca^{2+} balance for a hypothetical cell (Table 2.7, Fig. 2.3). It has been known for nearly a century that the nucleus contains Ca^{2+}, mainly as chromatin and in the nucleolus, and this Ca^{2+} could be physiologically significant (Loew, 1892). Although precise figures are not available for intact cells it appears that between 25% and 50% of the Ca^{2+} may be in the nucleus, the mean nuclear Ca^{2+} being 2–6 mmoles/kg nuclear water. Of course cells like skeletal muscle, platelets and other granule-containing cells (Clemente and Meldolesi, 1975) with particu-

Table 2.7. Estimated cation content for a hypothetical cell

Intracellular compartment		Percentage of total cell ion content			
		Ca^{2+}	K^+	Na^+	Mg^{2+}
Nucleus		50	28	25	48
Mitochondria		30	8	3	17
Microsomes		14	3	3	14
Cell membrane (mainly extracellular)		5	1	1	0.05
Cytoplasm	bound	0.5	1	1	10
	free	0.005	59	67	10
Total		99.5	100	100	99.5

These figures are only approximate and are intended to illustrate the sort of distribution of Ca^{2+} that might exist in a cell with total cell Ca^{2+} concentration of 2 mmoles/kg cell water. Obviously a cell with a particularly large intracellular store of Ca^{2+} will alter these figures somewhat (e.g. secretory cells with Ca^{2+} in the granules, or muscle). In some cells nuclear Ca^{2+} may be less than 50% of the total. References: Thiers and Vallee (1957); Claret-Berthon *et al.* (1977); Baker (1972); Brinley (1978); Klein *et al.* (1970).

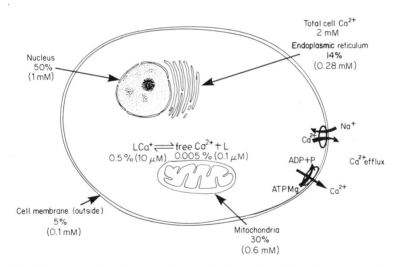

Fig. 2.3. The calcium balance of a typical eukaryotic cell. The figures are only approximate and are intended to indicate where Ca^{2+} is to be found in the cell

larly large intracellular Ca^{2+} stores will have a much higher total cell Ca^{2+} content than other cells (Manery, 1969) (Table 2.8).

Some attempts have been made to rationalize the general observations that most cells are high in magnesium and potassium whilst being low in calcium and sodium (Williams and Wacker, 1967; Wacker and Williams, 1968). This can partially be explained by the existence of active calcium and sodium pumps in the cell membrane, responsible for maintaining the low concentration of free sodium

Table 2.8. The cation content of some cells and tissues

Cell type	Example	Cation content (mg atom per kg wet weight)				Reference
		Ca^{2+}	K^+	Na^+	Mg^{2+}	
Prokaryote	*Bacillus*	2	350	70	130	1, 2
	Bacillus spores	40	140	5	70	1, 2
Isolated eukaryotic cells	Yeast	0.3	33	4	4	3
	Euglena	0.3	103	5	5	3
	Erythrocytes (human)	0.02	100	10	3	3, 4, 5
	Parotid acinar cells	6				6
Invertebrate nerve	Squid axoplasm	0.4	400	50	10	3, 7
Vertebrate soft tissues	Heart muscle (human)	4	65	50	10	8
	Rat liver	1.6	65	22	5	9, 10
	Human skin	4	35	90	3	8
	Brain	1.5	90	55	6	8

When the original data were expressed per gram dry weight, it was assumed that dry weight = 0.3 wet weight. Isolated cells were usually washed in Ca^{2+} medium before estimation of ion content but not in the case of whole tissues, hence the high Na^+ values.
References: 1, Rouf (1964); 2, Dally and Gray (1974); 3, Williams and Wacker (1967); 4, Long and Mouat (1973); 5, Lichtman *et al.* (1972); 6, Kanagasuntheram and Randle (1976); 7, Keynes and Lewis (1956); 8, Widdowson and Dickerson (1964); 9, Thiers and Vallee (1957); 10, Bresciani and Auricchio (1962).

and calcium ions inside the cell. However, this does not explain why the cell needs to maintain these intracellular conditions. The necessity for this lies in the evolutionary significance of the regulation, by these cations, of the electrical and mechanical properties of cells. These in turn will be dependent on the chemistry of the cations, and on the chemical composition of the fluids surrounding the precursors of living organisms during the evolution of the two principle cell types—pro- and eukaryotes.

Changes in total cell Ca^{2+} have been reported under several experimental conditions; for example, the activation of some cells by K^+ (van Breeman *et al.*, 1973) or the effect of prolonged cholinergic stimulation on exocrine cells (Dormer *et al.*, 1981). However, it seems unlikely that a physiological stimulus will severely change the total Ca^{2+} of the cell *in situ*, except when the cell becomes fatigued. In which case it will take some time to recover its responsiveness and its normal Ca^{2+} balance.

The Ca^{2+} content of some tissues does vary in pathological states. For example, the Ca^{2+} content of some rat liver hepatomas is approximately twice that of normal liver (Hickie and Kalant, 1968; Everett *et al.*, 1964). Necrotic liver induced by anoxia or toxic chemicals contains twice the normal amount of Ca^{2+} within 2 hours of injury and up to ten times the normal amount after 1–2 days (Majno, 1964). Lack of oxygen is a common cause of increased Ca^{2+} content in pathological conditions in man and can be demonstrated experimentally in animals. For example, transient ischaemia in dog myocardium, a model for heart attack in man, results in an increase in Ca^{2+} content of the tissue, associated with a large increase in mitochondrial Ca^{2+} (Shen and Jennings, 1972). In view of the 10 000–100 000-fold concentration gradient of Ca^{2+} which exists across the membrane of most cells it is not surprising that damage to the cell results in an increase in the Ca^{2+} content of the cell.

Although it could be argued that the measurement of tissue Ca^{2+} has not contributed greatly to our understanding of the biological role of this cation, except perhaps its structural role, there are several experimental situations in the investigation of intracellular Ca^{2+} as a regulator where it is imperative to be able to measure the total concentration of Ca^{2+}.

(1) Effect of experimental conditions, pharmacological agents, and pathological states on cell Ca^{2+} balance. For example, it is important to know whether substances such as ionophore A23187 (Fig. 2.2) significantly change the Ca^{2+} content of intracellular Ca^{2+} stores (e.g. mitochondria) and thereby affect the physiological significance of the interpretation of results.

(2) Estimation of the specific activity of Ca^{2+} in cells and organelles in experiments using radioactive Ca^{2+}. The interpretation of experiments using radioactive Ca^{2+} is crucially dependent on knowing whether changes in the specific activity (i.e. the ratio of radioactive to total Ca^{2+}) occur during the experiment (see this chapter, section 2.6).

(3) Estimation of Ca^{2+} concentrations in the fluids bathing cells *in vivo* and *in vitro*.

(4) Estimation of Ca^{2+} contamination of reagents and solutions so that the concentration of free and bound Ca^{2+} can be precisely defined.

2.4.2. The need to measure free Ca^{2+}

It can be seen from many of the preceding arguments that measurement of ionized (free or unbound) Ca^{2+} plays a central part in obtaining definitive evidence for the regulatory role of intracellular Ca^{2+}. If an extracellular signal or an intracellular regulator, be it chemical or electrical, is to regulate a physiological process through Ca^{2+} then it must be possible to detect a change in the concentration of intracellular free Ca^{2+} after addition of the stimulus. In most cases the initial change will be in cytoplasmic Ca^{2+}, but even if the proteins or other molecules regulated by Ca^{2+} are inside an intracellular organelle one would still expect changes in free Ca^{2+} inside the organelle to result in detectable changes in cytoplasmic Ca^{2+}. There are several reasons why methods for measuring free Ca^{2+} in solution and in cells are necessary:

(1) Detection of changes in intracellular free Ca^{2+} on addition of the stimulus to the cell provides direct evidence of a link between the primary signal (usually

extracellular; Table 2.2) and the intracellular physiological event, provided that the time course of the change can be correlated with cell activation.

(2) The effect of experimental procedures to mimic, enhance or inhibit the normal response of the cell through manipulation of intracellular Ca^{2+} can only be completely interpreted if the procedure, e.g. ionophore A23187 or the injection of EGTA into the cell, can be shown to cause a change in intracellular free Ca^{2+} which correlates with the time course of the effect of the agent concerned.

(3) Absolute measurement of intracellular free Ca^{2+} provides the range of Ca^{2+} concentrations, together with response times, which are necessary in Ca^{2+}-binding proteins if they are to be involved in the physiological response.

(4) Concomitant measurement of intracellular free Ca^{2+} with radioactive Ca^{2+} fluxes in and out of cells plays a crucial part in the interpretation of these experiments (see this chapter, section 2.6). In particular it may be possible to interpret some of the changes in measured $^{45}Ca^{2+}$ flux through changes in the specific activity of the Ca^{2+} rather than through an effect on the permeability of the cell to Ca^{2+}.

(5) Measurement of free Ca^{2+} is required for the standardization of solutions, the estimation of the free Ca^{2+} concentration bathing cells and the estimation of absolute Ca^{2+} uptake and release from organelles and isolated molecules.

Although the first attempt to detect changes in intracellular Ca^{2+} was carried out about 50 years ago (Pollack, 1928), even today changes in intracellular Ca^{2+} have only been detected directly in a few cell types (Ashley and Campbell, 1979) (Table 2.13). The technical problems when using small cells such as those from mammals have been considerable.

Successful quantitative studies have highlighted two important features of cell Ca^{2+}. First, the concentration of free cytoplasmic Ca^{2+} in resting cells is of the order of 0.1 μM (Ashley and Ridgway, 1970; Baker, 1972; Ashley and Campbell, 1979). This means that a concentration gradient of Ca^{2+} exists across the cell membrane of some 10 000–100 000-fold, which is several orders of magnitude greater than for any of the other univalent or divalent ions. This suggests why Ca^{2+} has been utilized in so many regulated phenomena since one would expect only a small change in the permeability of the cell membrane to Ca^{2+} to cause a large fractional change in free Ca^{2+} in the cytoplasm of the cell. Secondly, absolute measurements of intracellular free Ca^{2+} have emphasized the importance of relating studies with isolated proteins to the conditions to which they are exposed in the cell. Many effects of Ca^{2+} that have been demonstrated on proteins *in vitro* are now thought not to be physiologically significant (see Table 2.1 and Chapter 3).

2.5. Measurement of Ca^{2+}

2.5.1. Total Ca^{2+}

The concentration of Ca^{2+} bathing most cells is in the range 0.1–10 mM (Table 1.6). The Ca^{2+} content of tissues varies from approx. 20 μmoles/kg water in non-

Table 2.9. Methods of measuring total calcium in tissue extracts

Methods	Useful range (detection limit* in 1 ml)	Reference
1. *Precipitation*		
Oxalate	0.1–10 mM (0.1 μg atoms)	Kramer and Tisdall (1921); Clark and Collip (1925)
Molybdate	mM	Harrison and Raymond (1953)
Carbonate	mM	Weir and Hastings (1936)
2. *Titration with Ca^{2+} ligand + Ca^{2+} indicator*		
EDTA—cal-red as indicator	10 μM–mM	Weissmann and Pileggi (1974)
EGTA—pH	10 μM–mM	Moisescu and Pusch (1975)
3. *Metallochromic indicators*		
Eriochrome black	mM	Henry (1964)
Murexide (ammonium purpurate)	1 μM–20 mM (0.8 ng atom)	Scarpa (1972)
Arsenazo III	0.05–200 μM (10 pg atom)	Kendrick (1976)
4. *Photoproteins*		
Aequorin and obelin	0.1–100 μM (10 pg atom)	Campbell (1974); Blinks (1978)
5. *Photometry*		
Emission flame photometry	10 μM–10 mM (10 ng atom)	Raminez–Munoz (1968)
Atomic absorption spectrophotometry	0.1 μM–1 mM (100 pg atom)	Haljamae and Wood (1971); Willis (1963, 1965); Kuntziger *et al.* (1974)

* The detection limit can be reduced by decreasing the volume used in the assay. In some cases as little as 10 nl are required.

nucleated erythrocytes to 4 mmoles/kg cell water in muscle to 10 mmoles/kg water in cartilage (Table 2.8). In damaged cells the Ca^{2+} content may be many times that of the normal cell (Majno, 1964). In order to estimate the Ca^{2+} content of small pieces (e.g. 1 mg) of tissue, to estimate the total Ca^{2+} concentration in extracellular fluids, and to estimate Ca^{2+} contamination of solutions, it is necessary to have a method which can detect Ca^{2+} down to at least 0.1 μM (Table 2.9) in the solution in which the final measurement is carried out. First the Ca^{2+} must be extracted from the biological sample and presented to the assaying device in a suitable form.

Extraction

The tissue is weighed wet and then dried in an oven at up to 150 °C. Ashing of the tissue can be used to remove all the organic matter from the sample. The dry weight of the tissue is then determined. To ensure that no calcium remains bound

A. Absorbing dyes

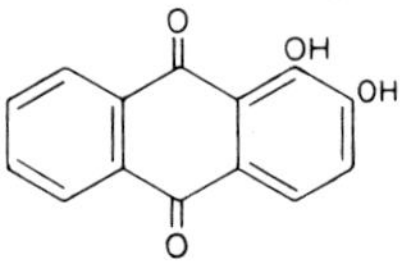

Alizarin

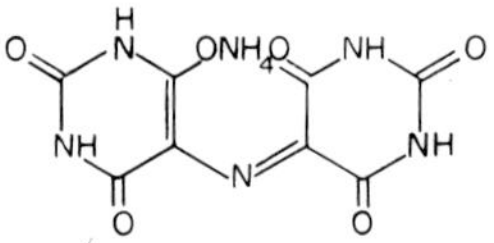

Murexide (ammonium purpurate)

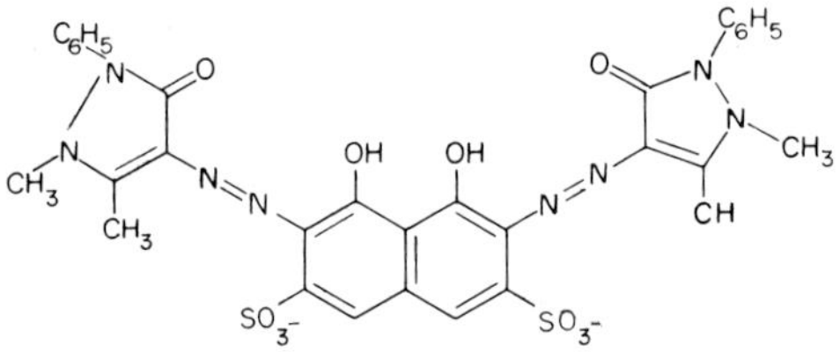

Antipyrylazo III

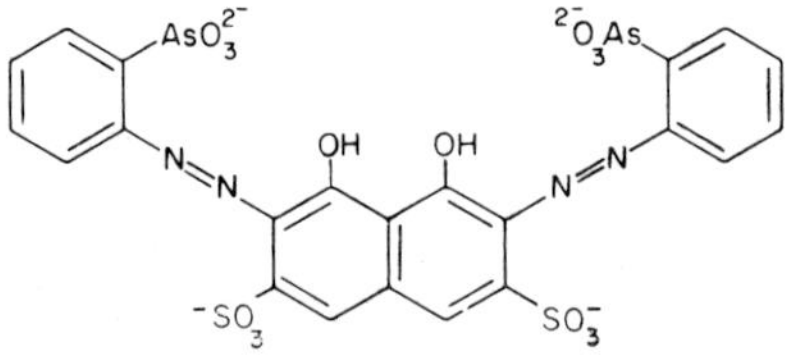

Arsenazo III

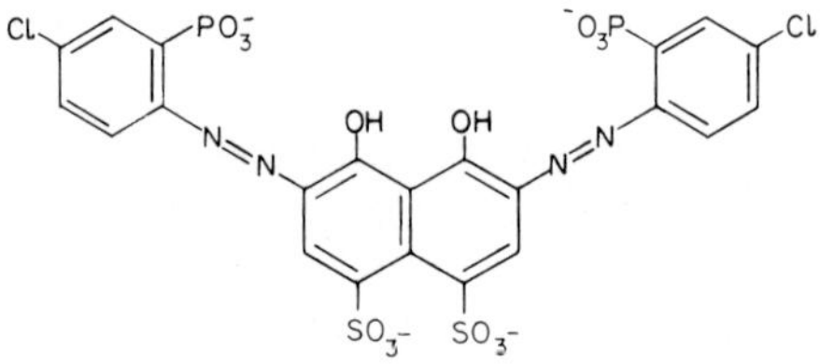

Dichlorophonazo III

B. Fluorescent dyes

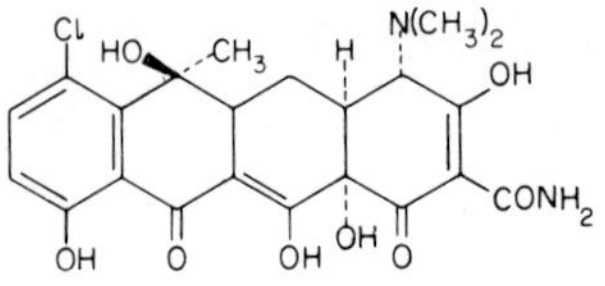

Chlortetracycline (CTC)

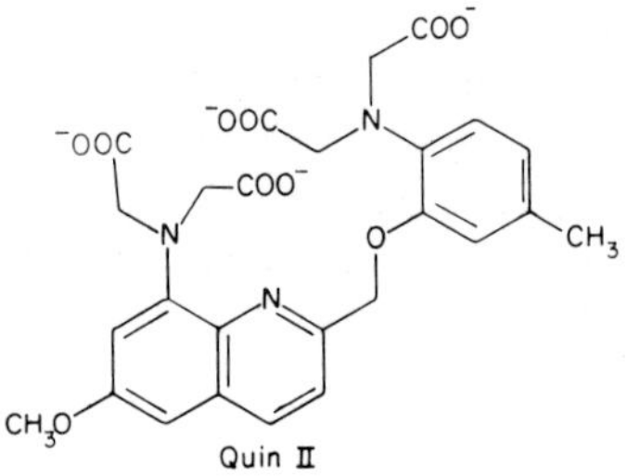

Quin II

Fig. 2.4. Some Ca^{2+} indicator dyes. (A) Absorbing dyes. (B) Fluorescent dyes. See Scarpa (1979) and Tsien (1980)

to anions such as phosphate, the dried tissue is usually dissolved in acid (e.g. Aristar conc. HNO_3). Precipitation of calcium salts and formation of Ca^{2+}–ligand complexes should be avoided. Ca^{2+} contamination of solutions is minimized by acid-washing all vessels and pipettes just before use, by using spectroscopically pure reagents and by using purified water. The latter can be done using a Chelex 100 resin (Campbell and Dormer, 1978). However, salt solutions passed down a Chelex column often displace Ca^{2+} resulting in more Ca^{2+} in the effluent than in the original solution.

Methods employing precipitation of calcium salts

The precipitation of calcium oxalate formed the basis of one of the earliest methods for measuring calcium in serum, being introduced by Kramer and Tisdall in 1921. Modifications of this method (Clark, 1921; Laidlaw and Payne, 1922; Clark and Collip, 1925) are still used in many laboratories. Precipitation of the calcium is induced by the addition of excess ammonium oxalate. After centrifugation the calcium oxalate is dissolved in sulphuric acid and titrated against potassium permanganate. The advantage of oxalate is that it is considerably easier to assay than some of the other insoluble calcium salts such as molybdate (Harrison and Raymond, 1953), picrolonate or carbonate (Weir and Hastings, 1936). However, the method is only really suitable for the measurement of calcium concentrations in the mM range.

Titration using a colourimetric indicator

Several organic compounds exist which change colour when they bind Ca^{2+} (Henry, 1964; Weissmann and Pileggi, 1974). For example, calcium causes eriochrome black T to turn red, whereas calcon or cal-red, the sodium salt of 1-(2-hydroxy-1-naphthylazo)-2-naphthol-4-sulphonic acid, turns from blue to wine-red. These two indicators both produce sharp end-points which can be judged by eye. On the other hand, the indicator dye murexide (ammonium purpurate, Fig. 2.4), which turns from orchid-purple to pink in the presence of calcium, requires the use of a spectrophotometer. A commonly used method is to titrate the calcium with a chelating agent such as EDTA in an alkaline solution in the presence of cal-red as indicator (Weissmann and Pileggi, 1974). Like the precipitation methods, many of these titration methods are only suitable for calcium concentrations in the mM range. Under the appropriate conditions the metallochromic indicators murexide and arsenazo III (Fig. 2.4) can measure calcium concentrations down to 0.1 μM (Ohnishi, 1978; Scarpa, 1972; Gratzer and Beaven, 1977). Care must be taken to standardize the ionic conditions, particularly with respect to pH and Mg^{2+}.

Flame photometry

The most widely used method for the measurement of total calcium in biological samples utilizes the excitation of calcium in a flame, in which the colour of calcium is brick red. Atomic absorption spectrophotometry provides a highly sensitive, specific method for calcium, in which the antagonistic or enhancing effects of other cations, anions and calcium ligands can be minimized.

Flame emission and atomic absorption spectrophotometry both depend on the excitation of electrons in free atoms. The free atoms are obtained by vaporization of compounds in a high-temperature flame. A small proportion of the atoms are excited under these conditions and on returning to ground-state energy photons are emitted. The wavelength of the photon emission is characteristic of the element concerned and the intensity is dependent on the number of atoms excited. This in turn depends on the concentration of the element in the sample. The number of atoms in the excited state (N_{ex}) is given by the equation

$$N_{ex}/N_0 = \exp(-E_j/kT) \tag{2.3}$$

where N_0 = number of atoms at ground-state energy
N_{ex} = number of atoms in the excited state
E_j = energy difference between the states
T = temperature in degrees absolute
k = Boltzmann's constant.

The intensity of the photon emission, which depends on N_{ex}, increases with temperature and is greatest for elements where the energy difference between the excited state and ground-state energy is lowest. The flame temperature of an air–coal gas flame is about 2000 K. At this temperature the ratio of atoms in the excited state to those in the ground state (N_{ex}/N_0) is 1.2×10^{-7} for calcium (λ 422.7 nm), 1.0×10^{-5} for sodium (λ 589 nm) and 7.3×10^{-15} for zinc (λ 213.9 nm). These figures emphasize that at any easily obtainable temperature only a minute fraction of the atoms are excited.

Flame emission photometry provides a highly sensitive method for measuring sodium and potassium, but it is less sensitive for calcium and magnesium where the number of excited atoms may be as much as 100 times less than for sodium (Willis, 1963, 1965). The sensitivity of flame emission spectroscopy is limited by the ability of the detector to distinguish the emission against the background light. Atomic absorption spectrophotometry is limited by the ability of the detector to measure small differences in intensity (signal-to-noise), which is dependent on the noise level of the incident light source.

Atomic absorption is considerably more sensitive for measurement of Ca^{2+} than flame emission. The atoms in the flame are excited by the absorption of light (e.g. for calcium at 422.7 nm) (Willis, 1963, 1965; Raminez-Munoz, 1968; Price, 1972), after the sample has been atomized in an air–acetylene flame (2570 K) or in a nitrous oxide–acetylene flame (3230 K). Anions, like phosphate, reduce atomization and thus decrease the sensitivity for measuring calcium. Interference by these calcium ligands can be eliminated by the addition of La^{3+} or the chelating agent EDTA to the sample. La^{3+} produces the maximum sensitivity for assay of Ca^{2+}. EDTA results in a reduction in the absorbance because of a reduction in the fraction of Ca^{2+} that is atomized. Atomic absorption provides a method of measuring Ca^{2+} concentrations down to 10–100 nM. If the sample volume is 2 ml then the detection limit for Ca^{2+} is about 0.2 ng atom, equivalent to approx. 100 μg of tissue. Micromethods reduce the detection limit to 1 g atom of Ca^{2+}, approaching that required for single-cell analysis.

Atomic absorption spectrophotometry is therefore the most sensitive and specific method for total cell Ca^{2+} estimates.

2.5.2. Free Ca^{2+}

In 1928 Pollack, using the microinjection techniques pioneered by Chambers and developed by Needham and Needham (1925/26) and Reznikoff and Chambers (1926/27), injected the red dye sodium alizarin sulphonate (Fig. 2.4) into a single amoeba in order to measure intracellular Ca^{2+}. Unfortunately this dye is neither specific for Ca^{2+} nor can quantitative data be obtained with it.

In 1957 Hodgkin and Keynes injected radioactive Ca^{2+} into the giant axon of a squid and found that very little diffusion of the Ca^{2+} could be detected when a voltage gradient was applied over a period of 2 hours. The mobility of Ca^{2+} in solution should be approx. 4×10^{-4} cm^2 sec^{-1} V^{-1}, whereas in the axoplasm of the these squid nerves the value was less than 0.9×10^{-5} cm^2 sec^{-1} V^{-1}. Assuming that the discrepancy between these two figures is a reflection of the proportion of Ca^{2+} that is free, then the free Ca^{2+} concentration in the axoplasm must be less than 0.022 of the total. Since the total axoplasmic Ca^{2+} concentration is 0.4 mM then the free axoplasmic Ca^{2+} concentration cannot be more than 8.8 μM. An alternative approach was developed using muscle fibres of the spider crab *Maia squinado* (Portzehl *et al.*, 1964). This null method used Ca/EGTA buffers (see this chapter, section 2.7) to determine the intracellular free Ca^{2+} at which contraction began. The resting free Ca^{2+} was estimated to be 0.3 μM. Mitochondrial O_2 consumption, the activity of Ca^{2+}-regulated enzymes (Table 2.1; Baker, 1972), or K^+ flux (Meech, 1976) can be used similarly as intracellular metabolic indicators of free Ca^{2+}.

'*Raso ligno, parum adeo in tenebris splendet*' (Forskal, 1775). Thus, Forskal described the light emission from the jelly-fish *Aequorea forskalea*, or *Medusa aequorea* as he called it. It was the protein aequorin, responsible for this luminescence, which provided the first method, applicable to a wide range of cell types, for measuring intracellular free Ca^{2+} (Ridgway and Ashley, 1967). Several methods are now available for measuring free Ca^{2+} (Table 2.10), but only three are suitable for intracellular studies (Ashley and Campbell, 1979). These are: (1) Ca^{2+}-activated luminescent proteins, known as photoproteins (Campbell *et al.*, 1979); (2) metallochromic and fluorescent dyes (Scarpa, 1979; Tsien, 1980); (3) Ca^{2+} microelectrodes (Ammann *et al.*, 1979).

Any method for measuring free Ca^{2+} in cells should ideally satisfy seven criteria: (1) It should be specific for Ca^{2+} and sensitive down to at least 10 nM Ca^{2+} in the presence of other physiologically occurring cations such as Mg^{2+} (1–5 mM in cells) and K^+ (100–500 mM in cells). (2) It must be possible to estimate the absolute concentration of Ca^{2+} in cells, even if corrections for interfering ions are necessary. (3) The method must be able to detect, without

Table 2.10. Methods for measuring free Ca^{2+}

Method	Solution or cells*	Useful range	Reference
1. *Precipitation*			
Alizarin sulphonate	Intact cells	?	Pollack (1928)
2. *Bioassay*			
Frog heart	Solution	0.5–1.5 mM	McLean and Hastings (1934)
3. *Polarography*	Solution	0.1–10 mM	Brezina and Zuman (1958); Irving and Watts (1961); Nakagama and Tanaka (1962)
4. *Photoproteins*			
Aequorin	Solution and intact cells	0.1–10 μM	Ridgway and Ashley (1967); Blinks *et al.* (1976); Ashley and Campbell (1979)
Obelin	Solution and intact cells	0.01–10 μM	Campbell (1974); Ashley *et al.* (1976); Hallett and Campbell (1982a, b)
5. *Metallochromic indicators*			
Murexide	Solution and intact cells	20 μM–1 mM	Scarpa (1972); Jöbsis and O'Connor (1966)
Arsenazo III	Solution and intact cells	0.05–20 μM	Kendrick (1976); Scarpa and Brinley (1978)
Antipyrylazo III	Solution and intact cells	0.5–200 μM	Scarpa *et al.* (1979)
6. *Fluorescent indicators (Quin 2)*	Solution and intact cells	0.01–1 μM	Tsien (1980, 1981)
7. *Electrodes*			
Macroelectrodes	Solution	1 μM–10 mM	Ross (1967); Moody *et al.* (1970); Ruzicka *et al.* (1973)
Microelectrodes	Intact cells	0.1 μM–10 mM	Ammann et al. (1979); Owen *et al.* (1977)
8. *Indirect methods* K^+ conductance Ionophore A23187 Ca^{2+} mobility Dialysis of Ca^{2+}- Null-method	Intact cells	0.1–10 μM	Meech (1976); Flatman and Lew (1977); Hodgkin and Keynes (1957) Blaustein and Hodgkin (1969), Portzehl *et al.* (1964)

* Solution or cells refers to whether the method can be used to measure free Ca^{2+} in solution and/or in intact cells.

distorting the signal, changes in free Ca^{2+} in both slow and fast responding cells. For example, in contracting skeletal muscle millisecond responses are required. (4) It must be relatively straightforward to incorporate the Ca^{2+} indicator into the cell without damaging its electrical, morphological or chemical properties. (5) The indicator should diffuse rapidly (Brinley, 1978) and not redistribute itself across intracellular membranes (Blayney et al., 1977). (6) The method should be adaptable to measurement of the distribution of free Ca^{2+} in the cell. (7) The method should be relatively easily available and inexpensive.

No one method satisfies all these criteria in every system. It is therefore necessary to understand the relative merits and problems associated with the three methods in order to select the appropriate one. Results obtained with one technique should always be confirmed with one of the other methods.

Ca^{2+}-activated photoproteins

In 1962 Shimomura and colleagues (1962, 1963) discovered that the luminous jelly-fish *Aequorea forskalea* contains a protein which emits blue light when it binds Ca^{2+}. They showed that the energy for this light emission arises from a chemical reaction within the prosthetic group of the protein (Fig. 2.5) and not from a Ca^{2+}-binding energy. No other cofactors or O_2 are required for this chemiluminescence, in contrast to the many luciferin–luciferase systems first discovered by Dubois at the end of the last century (Harvey, 1952; Campbell and Simpson, 1979). Many other related luminous jelly-fish, hydroids, sea combs and some protozoa (Fig. 2.6) have similar Ca^{2+}-activated photoproteins. So far,

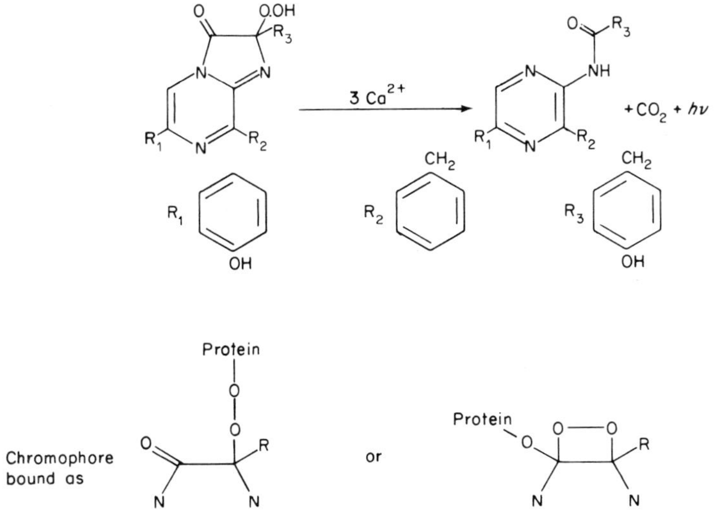

Fig. 2.5. Reaction of Ca^{2+}-activated photoproteins. The chromophore is covalently linked to the protein. Reproduced from Hallett and Campbell (1982) by courtesy of Wascel Dekker, Inc.

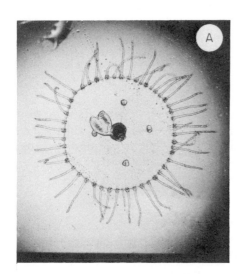
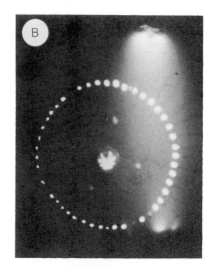

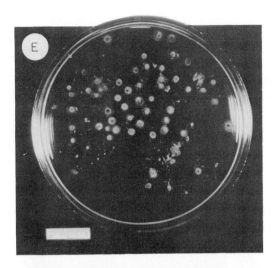

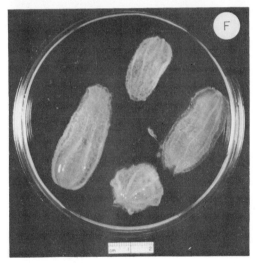

Fig. 2.6. Some luminous organisms with Ca^{2+}-activated photoproteins. (A) *Obelia lucifera* (magnified, bright field)—a cnidarian (Hydrozoa). (B) *Obelia lucifera* (magnified, fluorescence)—a cnidarian (Hydrozoa). (C) *Obelia geniculata* (magnified, fluorescence)—a cnidarian (Hydrozoa). (D) *Aequorea forskalea* (much reduced) (Reproduced by permission of Professor J. R. Blinks)—a cnidarian (Hydrozoa). (E) *Thalassicola* (spp.)—a protozoan (Radiolaria). (F) *Beroe* (spp.) (reduced)—a ctenophore

aequorin, from *Aequorea forskalea*, and obelin, from *Obelia geniculata*, have been used to measure intracellular free Ca^{2+} (Campbell et al., 1979b; Hallett and Campbell, 1982b). In the absence of Ca^{2+} these proteins emit light at a fractional utilization rate of 10^{-6} to 10^{-7} sec^{-1}. Addition of a saturating Ca^{2+} concentration, about 0.1–1 mM, produces a rapid flash of light. The exponential decay constant for aequorin is 1.4 sec^{-1}, and for obelin 4 sec^{-1}, being little affected by temperature over the range 20–37 °C. Although these proteins can be reactivated using a synthetic prosthetic group under experimental conditions, once the protein has luminesced it cannot do so again. Under physiological ionic conditions the properties of these photoproteins are ideally suited for measuring free Ca^{2+} within the range 0.1–10 μM (Table 2.10). Under these conditions the fractional utilization rate is usually between 10^{-6} and 10^{-2} sec^{-1}. Thus 0.1 pmole (approx. 2 ng) of photoprotein (equivalent to approx. 10^{10} total photons) injected into a cell produces 10^4–10^8 photons per sec, easily measurable with the highly sensitive photomultipliers available. It also means that the rate of loss of active protein is small in living cells. The light emission is recorded digitally on a scalar or as an analogue signal on a chart recorder or oscilloscope (Fig. 2.7). The efficiency of such apparatus is rarely better than 1%.

The exponential decay is described by

$$\text{rate of photon emission} = dh\nu/dt = Q\,\text{Ph}P_0\,k\,e^{-kt} \qquad (2.4)$$

where
Q = quantum yield
$\text{Ph}P_0$ = amount of photoprotein at time 0
k = rate constant.
k is dependent on the free Ca^{2+} concentration:

$$k = k_{\text{sat}} K_1 K_2 K_3 [Ca^{2+}]^3 / 1 + K_1 [Ca^{2+}] + K_1 K_2 [Ca^{2+}]^2$$
$$+ K_1 K_2 K_3 [Ca^{2+}]^3 \qquad (2.5)$$

where K_1, K_2, K_3 = the dissociation constants for the Ca^{2+}-binding sites. This is because these proteins apparently need to bind at least three Ca^{2+} in order

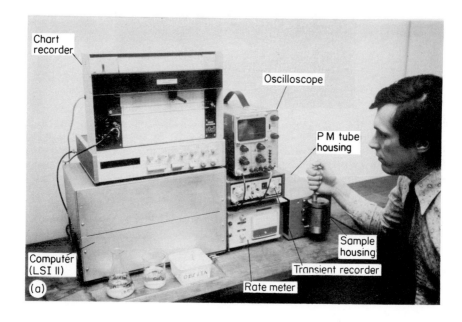

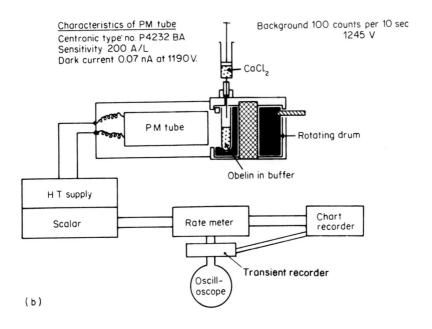

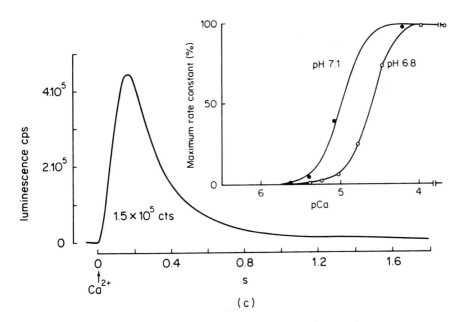

Fig. 2.7. An apparatus for detecting and quantifying chemiluminescence. (a) Apparatus, with the author. (b) Diagram of apparatus. (c) Obelin luminescence and saturating Ca^{2+} (inset shows decreasing rate constant with decreasing Ca^{2+})

to luminesce, though this, and the molecular mechanism involved, have been the source of some controversy (Blinks, 1978; Campbell and Simpson, 1979; Ashley and Campbell, 1979).

Metallochromic indicators

Metallochromic Ca^{2+} indicators are substances which change colour when they bind Ca^{2+} (Scarpa *et al.*, 1978) (Figs. 2.4 and 2.8). Such compounds were

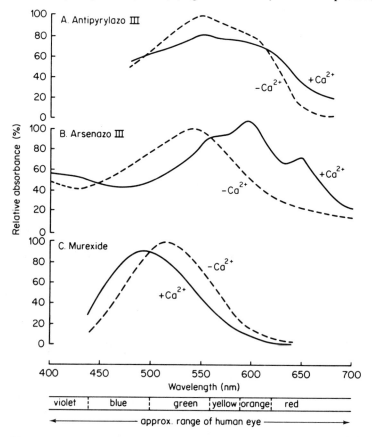

Fig. 2.8. Absorbance spectra of three Ca^{2+} indicator dyes. (A) Antipyrylazo III (50 μM). (B) Arsenazo III (25 μM). (C) Murexide (100 μM). Calculated from the data of Scarpa (1979). Conditions: 50 mM KCl, 10 mM MOPS, ±0.5 mM $CaCl_2$, pH 7.4, 21 °C.

$$\text{antipyrylazo III} \xrightarrow{Ca^{2+}} \text{slightly redder}$$
$$\text{arsenazo III purple} \xrightarrow{Ca^{2+}} \text{blue}$$
$$\text{murexide orchid purple} \xrightarrow{Ca^{2+}} \text{pink}$$

Relative absorbance = % of peak absorbance in the absence of Ca^{2+}

known to the ancient Egyptians and Persians. Alizarin, for example, is an anthroquinone dye which turns purple-red with Ca^{2+}, violet with Mg^{2+}, blue with Ba^{2+}, rose-red with Al^{3+} and dark violet with Fe^{3+}. Three such dyes, arsenazo III, antipyrylazo III and murexide (Figs. 2.4 and 2.8), have the necessary properties (Table 2.11) to provide a method for measuring free Ca^{2+} which has the specificity and sensitivity for studies in intact cells.

The basis of the method is best illustrated by examining the difference spectra between dye and dye–Ca^{2+} (Fig. 2.9). The important features of these spectra are the maxima and minima, at which the greatest sensitivity for detecting Ca^{2+} is obtained. The wavelengths at which no change in absorbance occurs between free and bound dye are known as the isosbestic points. In order to measure a change in free Ca^{2+} the absorbance at one of the maxima or minima can be used. However, in order to minimize non-specific changes in absorbance, particularly with regard to light scattering, and changes in Mg^{2+} and pH, the difference in absorbance at two wavelengths is used (Table 2.11). These wavelengths should be no more than 40 nm apart (Scarpa et al., 1979). This method improves the selectivity for Ca^{2+} but reduces the sensitivity. With arsenazo III, for example, the absorbance difference between 675 nm and 685 nm produces a discrimination between Ca^{2+} and Mg^{2+} of 4000 to 1.

Of the three dyes, arsenazo III is most useful for measurements at low Ca^{2+} concentrations, but antipyrylazo III is a useful middle-range indicator for Ca^{2+} and has a much faster response time. Although murexide was the first of these dyes to be used in living cells (Jobsis and O'Connor, 1966), its main value is in studying Ca^{2+} uptake and release by isolated organelles.

Other dyes such as chlorophosphonazo III (Fig. 2.4) have been used to measure changes in free Ca^{2+} (Budesinsky, 1969; Brown et al., 1976). Like the others they have to be used at concentrations in the range $10\,\mu m - 1$ mM in order to obtain

Table 2.11. Properties of three indicator dyes for Ca^{2+}

Property	Dye		
	Murexide	Antipyrylazo III	Arsenazo III
Molecular weight	284	746	776
Solubility in H_2O (mM)	20	20	50
K_d^{Ca}	1–3 mM	60–500 μM	15–60 μM
Stoichiometry for Ca^{2+} (dye: Ca^{2+})	1	1	2 (? 1)
Approx. $\Delta\varepsilon$ (mM^{-1} cm^{-1})	6	7	25
Relaxation time*	< 2 μsec	approx. 180 μsec	> 2.8 msec
Useful differential (nm)	540, 507	670, 690	675, 685
Useful range for free Ca^{2+}	20 μM–1 mM	0.5 μM–200 μM	50 nM–20 μM

*Relaxation time = time for displacement in Ca^{2+} to decrease to 1/e of the original value, measured by temperature jump.
References: Scarpa (1972, 1979); Kendrick (1976); Ohnishi (1978); Scarpa et al. (1978).

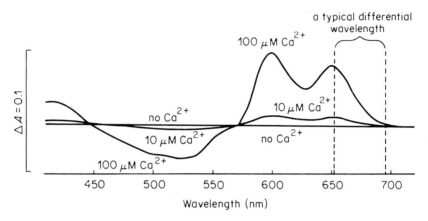

Fig. 2.9. Difference spectrum of arsenazo III. Conditions: 500 mM KCl, 10 mM MOPS, 100 μM arsenazo III, pH 6.8, plus no, 10 μM or 100 μM Ca^{2+}. ΔA = difference in absorbance $\pm Ca^{2+}$. From data of Scarpa (1979)

sufficient sensitivity. These relatively high concentrations often disturb the Ca^{2+} balance of the system being studied.

Fluorescent Ca^{2+} indicators

In view of the considerable increase in the sensitivity of detection of fluorescent or chemiluminescent compounds (Campbell and Simpson, 1979) compared with those that simply absorb, it would be highly desirable to have available a fluorescent Ca^{2+} indicator. By covalently linking a fluorescent chromophore to a Ca^{2+} ligand (quin II, Fig. 2.4) it is possible to synthesise a reagent which on binding Ca^{2+} results either in a large increase in quantum yield of fluorescence (the number of excited molecules that actually release photons) or in a shift in the excitation or emission spectrum (Tsien, 1980). By selecting the appropriate ligand it is possible to produce a fluorescent Ca^{2+} indicator with the necessary selectivity over Mg^{2+} and which is insensitive to changes in pH within the physiological range.

Several chemiluminescent compounds can be detected down to 1 atoml. The coupling of luminol or an acridinium salt to a Ca^{2+} chelator would provide a Ca^{2+} indicator detectable at several orders of magnitude lower than for the fluorescent compounds.

A group of fluorescent antibiotics that are derivatives of tetracycline (Fig. 2.4) have been used to detect changes in Ca^{2+} inside cells (Caswell, 1979; Carbonne, 1979). The binding of divalent diamagnetic cations like Ca^{2+} and Mg^{2+} to chlortetracycline enhances the fluorescence quantum yield and shifts the spectrum. Paramagnetic cations like Mn^{2+} and Co^{2+}, on the other hand, quench the fluorescence. At pH 7.4, in solution, $K_d^{Ca} = 270$ μM and $K_d^{Mg} = 440$ μM, thus the dye is not very selective for Ca^{2+}.

The fluorescence of chlortetracycline has the important property of being dependent on the polarity of the solvent. The dye fluoresces much more strongly in a non-polar, as opposed to a polar (e.g. water), environment. Furthermore, in 70% methanol the dissociation constant for Ca^{2+} is only 9 μM, and for Mg^{2+} is 25 μM (Caswell, 1979). This means that signals from this dye in cells reflect membrane Ca^{2+}, or Ca^{2+} very close to the membrane, rather than being a true indicator of free Ca^{2+}.

A piece of ingenuity by Tsien (1980, 1981) is that by coupling a fluorescent methoxyquinolone to the ester of a Ca^{2+} chelator (Fig. 2.4) a highly sensitive Ca^{2+} indicator can be incorporated into the cytoplasm of small cells. It has been used, for example, to show that the free cytoplasmic Ca^{2+} concentration in resting lymphocytes is 0.12 μM. This offers greater longer-term potential than the null method devised by Murphy et al. (1980). They incubated liver cells in various concentrations of Ca^{2+}–arsenazo III to define the concentration of free Ca^{2+} at which no change in arsenazo signal occurred when the cell membranes were dissolved with the detergent digitonin. Using this method the cytosolic free Ca^{2+} concentration was estimated to be 0.1–0.2 μM; it was unaffected by glucagon and was elevated 2–3-fold by α-adrenergic agonists. The specificity of this method for Ca^{2+}, and its reproducibility with other cell types remain to be established.

Calcium microelectrodes

Glass electrodes for the measurement of membrane potential, H^+ and K^+ have been widely used for many years (Thomas, 1978). In spite of the fact that the first attempt to make a Ca^{2+} electrode was made in 1898 (Luther, 1898) it is only relatively recently that electrodes for measuring intracellular Ca^{2+} have been developed. The idea of using a liquid membrane containing a Ca^{2+} exchanger to develop a potential (Ross, 1967) has lead to the development of a number of Ca^{2+} macroelectrodes (Moore, 1970; Ruzicka et al., 1973; Ammann et al., 1979), some of which have become commercially available. Improvements in the selectivity, detection limit and response time of these electrodes have made it feasible to construct Ca^{2+} microelectrodes for measuring intracellular free Ca^{2+} (Brown et al., 1976; Ammann et al., 1979).

The principle of the technique is that a liquid membrane embedded in polyvinylchloride, containing a Ca^{2+} ion-exchanger or an ionophore, generates a potential when placed in a solution containing Ca^{2+}. The potential generated can be predicted using the Nernst equation providing that all of the constants are known

$$\text{Potential} = E = \text{constant} + \frac{2.303\,RT}{2F}\log_{10} a_{Ca} \qquad (2.6)$$

where a_{Ca} is the *activity* (not concentration) of Ca^{2+}. The presence of an interfering ion (M) alters the form of this equation to

$$E = \text{constant} + \frac{2.303\,RT}{2F}\log_{10}[a_{Ca} + K_{CaM}(a_M)^{2/z}] \qquad (2.7)$$

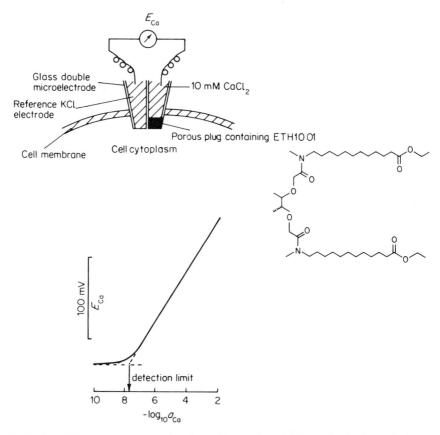

Fig. 2.10. Ca^{2+} microelectrode, from Ammann et al. (1979). The electrode consists of two half cells: a KCl reference electrode to measure the membrane potential, and the Ca^{2+} electrode. The porous plug contains the Ca^{2+} carrier ETH 1001, thereby allowing a potential to be set up across it which is dependent on the Ca^{2+} concentration in the cell. a_{Ca} = activity of Ca^{2+}. The slope of the calibration curve will be mV/$-\log a_{Ca}$

Tetraphenylboron (TPB^-) is present as anion to improve selectivity. Solvent = o-nitro-phenyl octyl ether (o-NPOE) where z is the charge on the cation M, and K_{CaM} is the selectivity coefficient relative to Ca^{2+}. Because of the ratio of Ca^{2+} to Mg^{2+} in the cell, for intracellular measurements the electrode must be 1000 to 10 000 times more selective for Ca^{2+} than Mg^{2+}. A reference electrode, e.g. K^+, is also necessary in intracellular studies to correct for changes in membrane potential which will also be recorded by the Ca^{2+} electrode.

Several different compounds have been employed as cation-exchangers. A popular mixture has been didecyl phosphoric acid + dioctyl phenyl phosphonate. However, these electrodes have poor response times, being 6 sec to several minutes, and have a poor selectivity for Ca^{2+}. A major improvement was made by the employment of neutral charge carriers (Fig. 2.10) (Ammann et al., 1979). These electrodes are sensitive down to 10 nM Ca^{2+} (Table 2.12) with good selectivity

Table 2.12. Properties of some Ca^{2+} electrodes

Electrode	Useful pCa range	Detection limit (pCa)
Orion (macroelectrode)	2–4	5.1
Brown (microelectrode)	2–6	7
Ruzicka (macroelectrode)	2–6	7.3
Simon (microelectrode)	2–7	8.2

References: Ruzicka *et al.* (1973); Brown *et al.* (1976); Ammann *et al.* (1979).

coefficients and reasonably rapid response times for Ca^{2+}, although the latter still need some improvement.

The three methods used in cells

Intracellular free Ca^{2+} has been measured in a wide range of vertebrate and invertebrate cells (Table 2.13, Fig. 2.11). To use the techniques available for measuring free Ca^{2+} in cells it is necessary to gain access to the compartment which is to be studied. In practice this is restricted mainly to the cytoplasm, though some compounds like chlortetracycline may penetrate the mitochondria. Giant cells can be impaled with electrodes. External Ca^{2+} is usually required for the membrane to reseal. A microelectrode filled with aequorin, and containing a light guide, has also been developed (Labeyrie and Koechlin, 1979). Small cells only a few micrometres in diameter are difficult to impale with microelectrodes without damaging the cell. In order to incorporate photoproteins or dyes into small cells three methods could be employed: (1) microinjection; (2) vesicle–cell fusion; (3) lipophilic tag attached to the indicator molecule.

Giant cells can easily be injected with the nl–μl quantities of indicator necessary to produce detectable signals. However, for small cells even if the micropipette does not injure the cell only picolitre amounts could be injected which is insufficient. Liposomes—vesicles prepared from purified phospholipids—have attracted much attention as a possible means of incorporating substances in the cell cytoplasm by liposome–cell fusion. Liposomes containing photoprotein are taken up by cells, but no detectable fusion occurs (Hallett and Campbell, 1980). However, fusion between erythrocyte ghosts containing photoprotein and mammalian cells has been induced by using the fusogenic agent Sendai virus (Hallett and Campbell, 1982a) (Fig. 2.12). Antibodies to specific antigenic sites on the surface of each cell type were used to quantify the amount of photoprotein that had been incorporated into the hybrid cells.

Earlier we saw that any technique for measuring intracellular free Ca^{2+} must satisfy at least seven criteria. How do the three methods available live up to them?

(1) *Specificity and sensitivity*: All three methods can detect Ca^{2+} in the intracellular range $0.1-10\,\mu M$. The sensitivity is determined primarily by the 'signal-to-noise' ratio generated in the measuring device. Chemiluminescence is detected in a light-tight housing against a zero-light background. In absorbance

Table 2.13. Cells in which free Ca^{2+} has been measured

Cell type	Example	Method	Reference
Protozoa	*Spirostomum*	Photoprotein	Ettienne (1970)
	Amoeba	Photoprotein	Taylor et al. (1975)
Slime mould	*Physarum*	Photoprotein	Ridgway and Durham (1976)
Eggs	*Xenopus* (toad)	Photoprotein	Baker and Warner (1972)
	Oryzia (medaka fish)	Photoprotein	Ridgway et al. (1977)
	Marthasterias* (starfish)	Photoprotein	Moreau et al. (1980)
Muscle			
invertebrate	*Balanus** (barnacle)	Photoprotein	Ashley and Ridgway (1970)
		Microelectrode	Ashley et al. (1978)
vertebrate	Heart	Photoprotein	Allen and Blinks (1978)
	Skeletal	Microelectrode	Lee et al. (1980)
		Murexide	Jöbsis and O'Connor (1966)
	Smooth	Photoprotein	Fay et al. (1979); Eubesi et al. (1980)
Nerve			
invertebrate	*Loligo* (squid)	Photoprotein	Baker et al. (1971)
		Arsenazo	Brinley (1978)
	Aplysia (sea hare)	Photoprotein	Stinnakre and Tauc (1973)
		Arsenazo	Gorman and Thomas (1978)
vertebrate	*Rana* (frog)	Microelectrode	Owen et al. (1977)
Photoreceptors	*Limulus* (horseshoe crab)	Photoprotein	Brown and Blinks (1974)
		Arsenazo	Brown et al. (1977b)
		Antipyrylazo	Scarpa et al. (1979)
Secretory cells	*Chironomus*	Aequorin	Rose and Loewenstein (1975, 1976)
	Calliphora (blowfly)	Microelectrode	Berridge (1980)
	β-Cells	Chlortetracycline	Täljedahl (1974)
Kidney	*Necturus*	Microelectrode	Lee et al. (1980)
Red cell ghosts	*Columbia** (pigeon)	Photoprotein	Campbell and Dormer (1978)
Membrane vesicles	Liposomes	Photoprotein	Dormer et al. (1978)
	Liposomes	Arsenazo	Weinstein et al. (1979)
Phagocytes	*Homo* (man)	Chlortetracycline	Naccache et al. (1979a, b)
Lymphocytes	Rabbit	Fluorescent dye (Quin II)	Tsien (1981)
Platelet	*Homo* (man)	Fluorescent dye (Quin II)	Rink and Tsien (1982)

See Ashley and Campbell (1979) and Hallett and Campbell (1981) for further references
Photoprotein is mainly aequorin, except in some cases (*)obelin also.

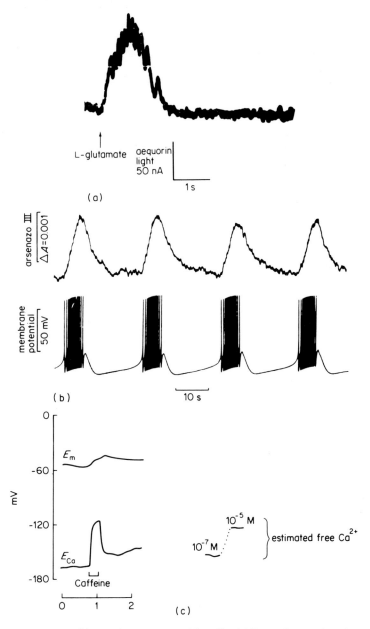

Fig. 2.11. Ca^{2+} transients measured in cells. (a) Barnacle muscle using aequorin. From Ashley and Campbell (1978), reproduced by permission of Elsevier Biomedical Press B. V. (b) *Aplysia* neurones using arsenazo III. From Gorman and Thomas (1978); reproduced by permission of the *Journal of Physiology*. (c) Barnacle muscle using Ca^{2+} microelectrodes. From Ashley *et al.* (1978); reproduced by permission of the *Journal of Physiology*

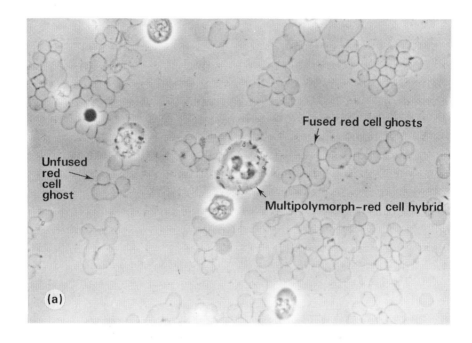

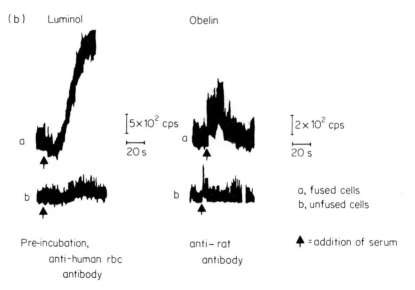

Fig. 2.12. Fusion of erythrocyte ghosts with polymorphs; from Hallett and Campbell (1982). Human erythrocyte ghosts were prepared containing obelin and fused with rat polymorphs using Sendai virus in the presence of UO_2Ac_2 (uranyl acetate), but in the absence of Ca^{2+}. Fusion was demonstrated (A) microscopically and (B) by the ability of anticell antibody and complement to stimulate polymorph chemiluminescence using an antierythrocyte antibody, or to stimulate obelin luminescence using an antipolymorph antibody. Temperautre = 37 °C. (see Hallett and Campbell, 1982)

and fluorescence measurements light scattering and endogenous absorption of fluorescence make these methods inherently less sensitive. However, several cells are known to emit light as a result of the production of peroxide moieties (Allen et al., 1972; Boveris et al., 1980; Foerder et al., 1978). Photoprotein luminescence must therefore be several times greater than this. Ca^{2+} is the only physiologically occurring cation to stimulate photoproteins. Similarly, microelectrodes can be made which are highly specific for Ca^{2+}. The dyes are not always quite so selective.

(2) *Quantification*: Electrodes are best since they measure Ca^{2+} activity directly. Photoproteins are reasonably accurate if the ionic conditions within the cell are known. The problem with the dyes lies in obtaining an absolute value for resting Ca^{2+} rather than measuring just the change in Ca^{2+}.

(3) *Response time*: Photoproteins record distorted Ca^{2+} transients if these are faster than about 10 msec. Some dyes are much better than this (Tsien, 1980), although the response time of substances like arsenazo III may be much slower than is normally realized. Many microelectrodes do not respond quickly enough *in vivo* to detect fast Ca^{2+} transients (Lee et al., 1980), and can be irreversibly affected by exposure to the cell.

(4) *Incorporation and toxicity*: This has been a problem with small cells for all methods. Vesicle–cell fusion may provide a method of incorporating photoproteins and dyes into intact cells. Damage to cells by electrodes can be a problem. Purified photoprotein and dye (Kendrick, 1976) preparations seem to be non-toxic but high concentrations ($> 100\ \mu M$) of arsenazo and chlortetracycline can be toxic. High concentration of dyes may also disturb the Ca^{2+} balance of the cell.

(5) *Diffusion*: The recording of Ca^{2+} transients is not normally limited by the diffusion of the indicator. The diffusion coefficient for aequorin in cells is about $1 \times 10^{-7}\ cm^2\ sec^{-1}$ at 20 °C, which is about eight times slower than it is in free solution and 20 times slower than the diffusion of EGTA ($4 \times 10^{-6}\ cm^2\ sec^{-1}$) in muscle fibres or Ca^{2+} ($6 \times 10^{16}\ cm^2\ sec^{-1}$) in solution (Ashley, 1978). Large cells are ususlly left for 30–60 min to allow the photoprotein to diffuse. The dyes will diffuse faster, but also may leak out of the cell where they will produce a large absorbance signal.

(6) Ca^{2+} *distribution*: Photoproteins are the only indicators used successfully to study Ca^{2+} distribution within the cell. In principle it should also be possible with fluorescent dyes. Electrodes are of no use for this as they only monitor free Ca^{2+} adjacent to the electrode tip. This may be a major problem since in many cells the change in free Ca^{2+} which triggers the cell is restricted to a particular region of the cell.

(7) *Availability*: Many reagent methods are available commercially.

Conclusions

During the years since Ridgway and Ashley (1967) first injected a barnacle muscle fibre with aequorin it has been this technique which has provided most of the new information regarding intracellular free Ca^{2+} (Table 2.13). However, dyes and electrodes are necessary in situations where there is a problem with photoproteins, and to confirm the interpretation of the original experiments. Synthetic fluorescent and chemiluminescent Ca^{2+} indicators offer exciting prospects for the future.

2.5.3. Localization of Ca^{2+} within the cell

If intracellular Ca^{2+} is the trigger for cell activation where does the Ca^{2+} come from? Where and how is it stored in the cell? Where in the cell do changes in Ca^{2+} concentration occur during cell activation? To answer these questions methods are required for estimating both total and free Ca^{2+} at different locations within the cell (Table 2.14).

Subcellular fractionation, followed by total Ca^{2+} analysis or measurement of radioactive Ca^{2+}, has provided some information about Ca^{2+} distribution within the cell (Dormer and Ashcroft, 1974; Dormer et al., 1981) (Table 2.14). To minimize redistribution of Ca^{2+} during homogenization of the cells and isolation of organelles, the homogenization is usually carried out in an EGTA solution containing inhibitors of Ca^{2+} uptake and release. Ruthenium red, for example, blocks mitochondrial Ca^{2+} redistribution.

Precipitation

The appearance of electron-dense deposits observable in the electron microscope after fixation of the tissue in calcium-containing media, but not when EGTA is present, indicates intracellular sites where high concentrations of calcium occur. Such deposits have been observed on the apical surface of the cell membrane of caecal cells but not on other intracellular membranes (Oschman and Wall, 1972), in the axons of the squid *Loligo peallei* (Hillman and Llinas, 1961) and in *Paramecium* (Fisher et al., 1976). A more reliable method is to carry out the

Table 2.14. Methods of Ca^{2+} localization within the cell

Method	Reference
Total Ca^{2+}	
1. Cell fractionation ($^{45}Ca^{2+}$ and total Ca^{2+})	Dormer and Ashcroft (1974)
2. Microscopy	
(a) Precipitation—oxalate	Oschman and Wall (1972)
—pyroantimonate	Hales et al. (1974)
(b) Autoradiography ($^{45}Ca^{2+}$)	Winegrad (1968)
(c) X-Ray microprobe analysis	Chandler (1977); Gupta and Hall (1978)
Free Ca^{2+}	
1. Alizarin sulphonate	Pollack (1928)
2. Aequorin	Rose and Loewenstein (1976)
	Ridgway et al. (1977); Gilkey et al. (1978); Reynolds (1979)

fixation in the presence of a salt which precipitates the calcium. A popular precipitating salt is sodium oxalate which produces observable precipitates in sarcoplasmic reticulum (Pease et al., 1965; Popescu and Diculescu, 1975). The identification of Ca^{2+} in these oxalate precipitates can be done using radioactive Ca^{2+} in combination with autoradiography, as was first demonstrated by Winegrad (1965, 1968) in the sarcoplasmic reticulum of frog muscle.

Potassium pyroantimonate was originally used to localise Na^+ in the cell (Komnick, 1962). However, calcium pyroantimonate is poorly soluble and precipitates have been used in adipocytes, secretory cells, muscle and other cells to locate Ca^{2+} stores within the cell (Hales et al., 1974; Herman et al., 1973). (Fig. 2.13)

There are several problems associated with the use of these precipitation techniques: (1) the sensitivity for Ca^{2+} is unknown; (2) the poor specificity for calcium means that it is essential to know the composition of the precipitate; (3) the precipitating anion may not distribute itself evenly throughout the cell; and (4) changes in the distribution, as well as uptake and release, of Ca^{2+} will certainly occur during fixation.

X-Ray microprobe analysis

The most powerful tool available at present for the identification and estimation of total Ca^{2+} in tissue sections is X-ray microprobe analysis in the electron microscope (Chandler, 1977; Gupta and Hall, 1978). When an electron beam strikes a solid specimen, in addition to the absorption and scattering of some of the electrons a fraction of them transfer energy to the specimen. This leads to the emission of X-rays. The wavelength of the X-ray emission is characteristic for each element and the intensity is related to the quantity of the element in the part of the section analysed. To analyse the amount of an element in a particular part of a cell a monochromator and an X-ray analyser attached to an electron microscope are required (Fig. 2.13). Instruments are available which can detect down to about 10^{-18} g (2.5×10^{-20} g atom) of Ca^{2+} with a special resolution of 50 nm. This apparently high sensitivity for Ca^{2+} sounds considerably less impressive when expressed in terms of tissue concentration since in a thin section it is equivalent to about 4 mmoles/kg wet weight. This means that in order to detect Ca^{2+} at different sites in the cell the local concentration has to be very high. However, X-ray microprobe analysis has proved useful in the analysis of intracellular precipitates (Hales et al., 1974; Herman et al., 1973; Ettienne, 1972; Popescu and Diculescu, 1975) and has confirmed the intracellular sites of some calcium stores, as well as suggesting that the nucleus is high in calcium (Gupta and Hall, 1978). It may also be valuable for studying tissue pathology (Maunder-Sewry and Dubowitz, 1979).

The two major problems with X-ray microprobe analysis are the relatively low sensitivity for Ca^{2+} and the difficulty of specimen preparation. Much of the published work has been carried out on fixed tissue or frozen sections under conditions where large changes in ion distribution have occurred. Redistribution

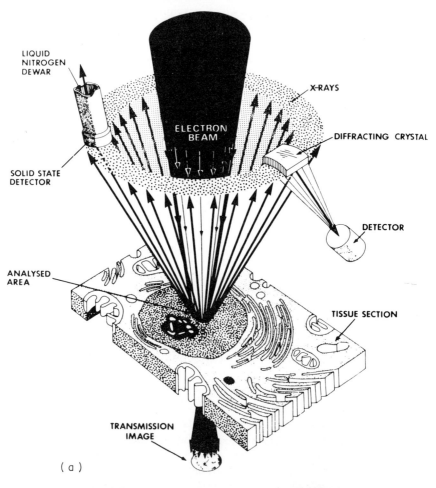

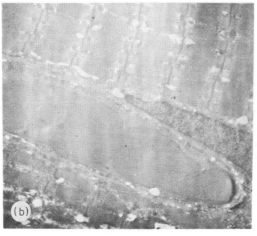

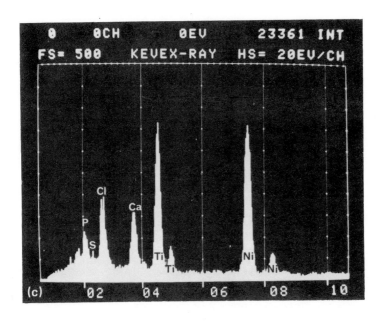

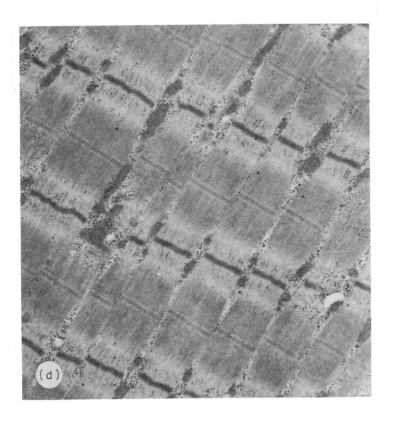

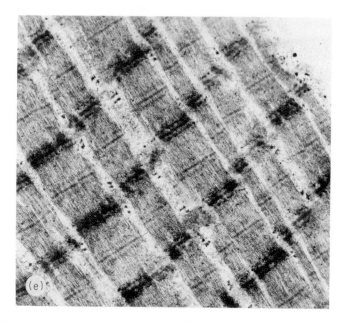

Fig. 2.13. X-ray microprobe analysis and Ca^{2+} distribution. (a) Apparatus (Chandler, 1977). (b) Normal muscle, thick (900 nm) section unstained, × 4830. (c) Energy dispersive spectrum from nucleus of dystrophic muscle. Element peaks are indicated; Ti = grid holder, Ni = grid, height of peak is proportional to concentration of element present. By courtesy of John Wiley and Son and Dr. Caroline Sewry. (d) Normal stained × 19 800 and (e) pyroantimonate stained muscle (× 15 400); by courtesy of Dr. Caroline Sewry. Reproduced by permission of John Wiley & Sons Inc.

of ions can even occur in freeze-dried sections prepared after ultracryotomy as a result of rapid rehydration. Frozen hydrated sections can be used (Gupta and Hall, 1978; Maunder-Sewry, 1980), but the definition of subcellular structure may make identification of Ca^{2+} sites difficult. The physiological significance of data obtained using X-ray microprobe analysis is critically dependent on being able to show that the cells have maintained a normal gradient of K^+, Na^+, Ca^{2+} and anions across the cell membrane.

Image intensification

One of the most exciting developments in the study of intracellular Ca^{2+} has been the application of photoproteins to the study of free Ca^{2+} distribution in cells. Aequorin light emission from inside cells can be visualized using an image intensifier attached to a light microscope (Reynolds, 1979) (Fig. 2.14). Temporal changes in light emission can be recorded on video tape with a spatial resolution of 1–2 μm (Rose and Loewenstein, 1976). The major difficulty lies in injecting sufficient aequorin into a single cell to produce a detectable signal without causing damage to the cell. Using intracellular injection of Ca/EGTA buffers it has been

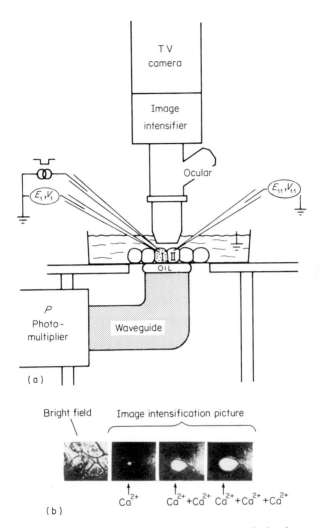

Fig. 2.14. Visualization of intracellular aequorin luminescence. (a) Apparatus. (b) Light image after injection of different quantities of Ca^{2+} in *Chironomus* salivary gland cell. From Rose and Loewenstein (1975); reprinted by permission from *Nature*, Copyright 1975 Macmillan Journals Ltd.

estimated that concentrations as low as 0.5 μM Ca^{2+} can be visualized using aequorin with a spatial resolution of approx. 1 μm (Rose and Loewenstein, 1976). This is, of course, not sensitive enough to detect the free Ca^{2+} in a resting cell.

This method has been used in salivary glands to show that when an increase in free Ca^{2+} occurs close to a gap junction the electrical conductivity between adjacent cells is dramatically reduced (Rose and Loewenstein, 1975, 1976). In fish eggs, on the other hand, fertilization by sperm results in a wave of light, corresponding to an increase in free Ca^{2+}, moving slowly down the cell (Ridgway et al., 1977; Gilkey et al., 1978) and preceding the morphological changes associated with fertilization.

An alternative method for locating free Ca^{2+} changes near the cell surface has been developed by Brinley (1978). A cell is injected with aequorin. An absorbing dye, such as phenol red, is also present, but only near the centre of the cell. Light emission from the cell will then only come from the area near the surface of the cell since the phenol red will absorb most of the light emitted from the centre.

2.6. The measurement of Ca^{2+} fluxes

A change in cytoplasmic Ca^{2+} induced by a cell stimulus is the result of a change in the rate of uptake or release of Ca^{2+} across the cell membrane, or across the membranes of an intracellular organelle. To show how the cytoplasmic Ca^{2+} concentration is regulated it is necessary to study the influx and efflux of Ca^{2+} in cells, and in isolated organelles which act as stores of Ca^{2+} within the cell (Reuter, 1973; Rasmussen and Goodman, 1977; Borle, 1975).

2.6.1. Ca^{2+} pools within the cell

There are four main Ca^{2+} pools in cells: (1) Ca^{2+} bound to the outer surface of the cell—the glycocalyx; (2) free Ca^{2+} in the cytoplasm and in organelles; (3) Ca^{2+} bound to small molecules, proteins, nucleic acids and phospholipids; and (4) crystalline deposits of calcium phosphate and occasionally calcium oxalate, as well as structural complexes of Ca^{2+} with substances like 5-hydroxytryptamine in platelets and adrenaline and ATP in adrenal chromaffin granules.

The glycocalyx consists of glycoproteins in the lipid bilayer on the outer surface of cells. Their high sialic acid content causes these glycoproteins to have low- and high-affinity Ca^{2+}-binding sites ($K_d^{Ca} \mu M-mM$). About 3–5% of cell Ca^{2+} is bound to the glycocalyx; 80% of this Ca^{2+} can be displaced, for example by a^{3+}, within a few minutes (Claret-Berthon et al., 1977). The remainder is more tightly bound, requiring the presence of EGTA for long periods to remove it. The enzyme neuraminidase hydrolyses off neuraminic acid and sialic acid from glycoproteins. This results in loss of the external Ca^{2+} sites. Ca^{2+} release from the inner surface of the cell membrane into the cytoplasm has been suggested as a possible means of triggering some cells. If this is so then the Ca^{2+} sites must either be high-affinity ones, for example a calmodulin (see Chapter 3), or low-affinity in high concentration.

The exchange of external Ca^{2+} with internal pools of Ca^{2+} can be quite fast. For example, Ca^{2+} bound to molecules in the cytoplasm exchanges rapidly with free Ca^{2+}, which in turn exchanges with the external Ca^{2+} in a few minutes. However, Ca^{2+} within organelles or crystalline Ca^{2+} deposits may take many minutes or even hours to exchange with external Ca^{2+}, particularly in resting cells.

2.6.2. Methods of measuring Ca^{2+} flux

Measurement of Ca^{2+} influx or efflux can provide four pieces of evidence in establishing a regulatory role for intracellular Ca^{2+}: (1) a change in Ca^{2+} flux initiated by the primary stimulus provides evidence for an 'active' role for Ca^{2+}; (2) estimation of the size of intracellular Ca^{2+} pools, and their alteration by physiological stimuli or drugs, can help in the identification of their physiological function; (3) measurement of fluxes of Ca^{2+} across individual membranes enable the mechanism of Ca^{2+} uptake or release to be elucidated; and (4) uni- and bidirectional Ca^{2+} pumps or Ca^{2+} channels can be identified.

There are three main methods for measuring Ca^{2+} flux in intact cells, or in isolated organelles: (1) radioactive Ca^{2+} movement (Borle, 1975); (2) absolute Ca^{2+} measurements (Scarpa, 1979; Ashley and Campbell, 1979); and (3) Ca^{2+} currents (Reuter, 1973; Lux and Hofmeier, 1979).

There are two useful radioactive isotopes of Ca^{2+} (Table 1.1)

$$^{45}Ca^{2+} \rightarrow {}^{45}Sc + \beta^- \qquad (t_{1/2} = 165 \text{ days})$$

$$^{47}Ca^{2+} \rightarrow {}^{47}Sc + \beta^- + \gamma \qquad (t_{1/2} = 4.7 \text{ days})$$

$^{47}Ca^{2+}$, being a γ emitter, can be estimated without scintillation fluid, but $^{45}Ca^{2+}$ is most commonly used because of the longer life of stock solutions.

Most studies of radioactive Ca^{2+} uptake and release by cells and organelles require their separation from the incubation medium, either by centrifugation or filtration. This may result in changes in $^{45}Ca^{2+}$ binding. Using cell monolayers attached to a glass scintillator $^{45}Ca^{2+}$ uptake and release can be monitored continuously (Langer et al., 1969). These glass scintillators can also be inserted into giant cells (Caldwell and Lea, 1978) to study continuously $^{45}Ca^{2+}$ uptake inside the intact cell (Ashley and Lea, 1977, 1978). The scintillators are sensitive to β^- and γ radiation, and emit light at the extreme end of the blue region of the spectrum (λ_{max} = approx. 395 nm). A Ca^{2+} solution of 100 μCi/ml produces 2000–3000 cpm against a background of about 100 cpm. More than 90% of the output comes from $^{45}Ca^{2+}$ within 150 μm of the surface of the scintillator probe, which is sufficient to circumvent the problem of the extracellular cleft system in barnacle muscle.

Measurement of radioactive Ca^{2+} uptake or release can be occasionally carried out under steady-state conditions, when the specific activity ($^{45}Ca/Ca_{total}$) of all the calcium pools is the same, and equal to that outside the cell. However, in most studies of transient changes in Ca^{2+} flux this is not the case.

2.6.3. Analysis of data

The analysis of radioactive Ca^{2+} fluxes depends on two main assumptions (Borle, 1975). Firstly, that the measured $^{45}Ca^{2+}$ flux is the net sum of the fluxes in

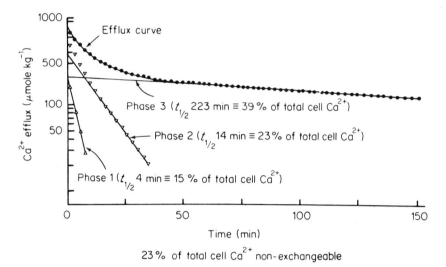

Fig. 2.15. $^{45}Ca^{2+}$ efflux, showing three phases from liver. From Claret-Berthon *et al.*, (1977). Reproduced by permission of the *Journal of Physiology*

two directions, and secondly, that the flux into or out of the cell (or organelle) is the sum of exchanges between different compartments within the cell and can be described by a series of exponentials; for example, for efflux

$$^{45}Ca_t^{2+} = {}^{45}Ca_1^{2+}(1 - e^{-k_1 t}) + \ldots {}^{45}Ca_n^{2+}(1 - e^{-k_n t}) \tag{2.8}$$

where $^{45}Ca_t^{2+}$ = total cell $^{45}Ca^{2+}$ at time t and $^{45}Ca_n^{2+}$ = ^{45}Ca in pool n. Initially, when there is either no intracellular $^{45}Ca^{2+}$ in the case of influx measurements, or no extracellular $^{45}Ca^{2+}$ in the case of efflux, equation 2.8 can be applied to the data analysis. However, once a significant amount of radioactive calcium has accumulated in the pool(s) that were originally non-radioactive, this approximation no longer holds. It is then necessary to describe the turnover of each calcium pool by an equation of the form

$$S_n = A_0 e^{-k_o t} - A_i e^{-k_i t} \tag{2.9}$$

where S_n = specific activity of calcium pool n, A = radioactivity in calcium pool at time 0, o and i represent the outer and inner compartments, respectively, between which the Ca^{2+} flux is being measured.

The number of different compartments revealed by these studies is usually three. If $k_1 > 10 k_2 > 10 k_3$, then to a first approximation the efflux or influx curves will be triphasic, and the rate constants of each compartment can be calculated from the exponential parts of each section of the curve (Fig. 2.15). If the rate constants of the different pools are similar then this simple approach can lead to a gross overestimate of the rate constants of the compartments with slow turnover.

2.6.4. Problems in interpretation

The major problem in interpreting radioactive Ca^{2+} flux experiments is the difficulty of completely equilibrating the various intracellular Ca^{2+} pools to the same specific activity. Furthermore, it is often impossible to rule out changes in the specific activity of Ca^{2+} pools during the experiment, particularly in the cytoplasm. For example, when a large quantity of non-radioactive Ca^{2+} enters the cell during a $^{45}Ca^{2+}$ efflux measurement, there can be a decrease in measured Ca^{2+} efflux under conditions when an actual increase in real efflux has occurred (Baker, 1972). To illustrate this important point let us consider a hypothetical example.

Suppose the rate of Ca^{2+} efflux from the cytoplasm of a cell into the extracellular fluid obeys a Michaelis-Menton relationship with respect to free Ca^{2+} concentration inside the cell. Therefore

$$Ca^{2+} \text{ efflux} = F_{max}[Ca^{2+}]_{free}/(K_m + [Ca^{2+}]_{free}) \quad (2.10)$$

where F_{max} = maximum Ca^{2+} efflux (i.e. when system is saturated with Ca^{2+})
K_m = Michaelis constant for Ca^{2+} (a measure of the affinity of the system for Ca^{2+})
$[Ca^{2+}]_{free}$ = free Ca^{2+} concentration in the cytoplasm.

Consider what happens when there is a trace amount of radioactive Ca^{2+} present:

$$^*Ca^{2+} + E = {}^*CaE \rightarrow {}^*Ca^{2+} \text{ efflux (radioactive)}$$

$$Ca^{2+} + E = CaE \rightarrow Ca^{2+} \text{ efflux (non-radioactive)}$$

$[{}^*Ca^{2+}][E]/[{}^*CaE] = K_m$ and $[Ca^{2+}][E]/[CaE] = K_m[{}^*Ca^{2+}]$ = concentration of radioactive Ca^{2+} (i.e. not specific activity)

Since total free $Ca^{2+} = [Ca^{2+}] + [{}^*Ca^{2+}]$ and if $[E]_{total}$ = total enzyme concentration

then $[{}^*Ca^{2+}]([E]_{total} - [{}^*CaE] - [CaE])/[{}^*CaE] = K_m$

Assuming that $[{}^*CaE] \ll [CaE]$

then $[CaE] = [Ca^{2+}][E]_{total}/(K_m + [Ca^{2+}])$

Therefore solving for $[{}^*CaE]$

$$[{}^*CaE] = [E]_{total}[{}^*Ca^{2+}]\{1 - ([Ca^{2+}]/K_m + [Ca^{2+}])\}/(K_m + {}^*Ca^{2+}) \quad (2.11)$$

Now consider what happens when the concentration of free Ca^{2+} changes from Ca_1^{2+} to Ca_2^{2+} as a result of release into the cell of non-radioactive Ca^{2+}. Since the measured Ca^{2+} efflux is the $^*Ca^{2+}$ efflux and is directly proportional to the concentration of $^*CaE$, then the ratio of measured efflux between state 1 and state 2 is

$$R = [{}^*CaE]_1/[{}^*CaE]_2 = \frac{1 - [Ca^{2+}]_1/(K_m + [Ca^{2+}]_1)}{1 - [Ca^{2+}]_2/(K_m + [Ca^{2+}]_2)} \quad (2.12)$$

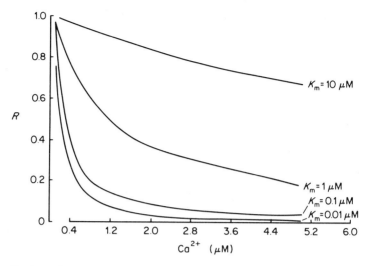

Fig. 2.16. The problem of specific activity. See equation 2.12 for definition of R

Assuming [*Ca^{2+}] does not change (i.e. initial rate when there is no significant loss of radioactive Ca^{2+} from cell), when $[Ca^{2+}]_1 = 0.1$ μM and $[Ca^{2+}]_2$ = 0.1–10 μM, R for four different K_m can be calculated (Fig. 2.16). For a low K_m enzyme a μM increase in free Ca^{2+} from non-radioactive pools produces a large decrease in the measured efflux, and yet the real change in Ca^{2+} efflux is a large increase. For example, if $[Ca^{2+}]_1 = 0.1$ μM, $[Ca^{2+}]_2 = 3$ μM and $K_m = 1$ μM, then the measured radioactive flux of Ca^{2+} *decreases* 3.6-fold, whereas the real Ca^{2+} flux *increases* eight-fold.

These calculations illustrate how difficult it can be to interpret even qualitatively flux experiments when the intracellular free Ca^{2+} concentration is unknown.

2.7. Ca^{2+} buffers

Many of the experiments described in this chapter required a Ca^{2+} chelator at some point, either to reduce free Ca^{2+} inside or outside the cell to < 1 nM or to buffer the free Ca^{2+} in the range 0.1–10 μM. The concept of metal-ion buffering is some 40 years' old (see Bjerrum, 1941; Schwarzenbach, 1954; Schwarzenbach and Anderegg, 1957; Rauflaub, 1960). Calculations are based on the equation

ML $\rightleftharpoons$ M (metal ion) + L (ligand)

pM = $-\log_{10}$ (free metal ion concentration) = pK_d + $\log_{10}$([L]/[ML])

(2.13)

Equation 2.13 is very similar in form to the Henderson-Hasselbach equation for pH buffers.

The aim of a Ca^{2+} buffer is to minimize changes in free Ca^{2+} to < 0.1 pM unit. A buffer is essential when Ca^{2+} contamination is present, or when changes in total Ca^{2+} or Ca^{2+} distribution occur during the experiment. An ideal ligand should satisfy six criteria as a Ca^{2+} buffer: (1) The Ca^{2+} concentration should be adequately buffered over a range likely to be found inside the cell, i.e. 0.1–10 μM. (2) The ligand should be specific for Ca^{2+} and not subject to interference from other ions likely to be present, i.e. Mg^{2+}, Na^+, K^+, Cl^-. (3) The Ca^{2+} buffer should be insensitive to pH changes. (4) The ligand should be non-toxic in biological systems. (5) Unless it is being used as a calcium indicator the ligand should be colourless. (6) The rates of association and dissociation of Ca^{2+} from the ligand should be fast enough to respond quickly to physiological changes.

Several compounds satisfy some, but not all, of these criteria (Fig. 2.17, Table 2.15). As a rule of thumb a Ca^{2+} ligand buffers best when $pK_{app}^{Ca} = pCa$, i.e. the ligand is half-saturated with Ca^{2+}.

Consider a ligand (L) which at pH 7.4 has $K_d^{Ca} = 0.1\ \mu M$. If $[L]_{total} = 10$ mM and pCa = 6, then from equation 2.13 it can be seen that L is 90% saturated with Ca^{2+}.

$$[L]_{total} = [CaL] + [L] = 0.01\ M \quad \text{where L = free ligand}$$
$$[Ca]_{total} = [CaL] + [Ca^{2+}] \quad \text{where } Ca^{2+} = \text{free } Ca^{2+}$$
$$\therefore\ pCa = pK_d^{Ca} + \log_{10}([L]/[CaL]) \quad (2.14)$$
$$\therefore\ 6 = 7 + \log_{10}([L]/[CaL])$$
$$\therefore\ [CaL] = 10[L] = [L]_{total} - [L] = 0.01 - [L]$$
$$\therefore\ [CaL] = 0.00909\ M,\ \text{i.e.}\ [Ca]_{total} \doteq 9.09\ mM$$

Now suppose 100 μM Ca^{2+} contamination is added to the solution.

new $[Ca]_{total} = 0.00919\ M \doteq$ new $[CaL]$
new $[L] \quad\quad = 0.00081\ M$

From equation 2.14 new pCa $= 7 + \log_{10}(0.00081/0.00919) = 5.95$ (1.1 μM). Thus the Ca^{2+} buffer has succeeded in reducing the change in pCa to less than 0.1 pCa unit or to a 10% change in absolute free Ca^{2+}.

Any one of at least five problems can arise when using a Ca^{2+} buffer: (1) how to calculate pCa; (2) the K_{app}^{Ca} may be sensitive to small changes in pH; (3) the buffering capacity may not be adequate for the complete range of Ca^{2+} being studied; (4) calculation of pCa is critically dependent on an accurate value for K_d; (5) calculation of pCa may be affected by other cations or ligands; (6) the ligand may not respond fast enough to a change in Ca^{2+} concentration to buffer it adequately.

For a ligand like citrate, which is more than 95% ionised at physiological pH, the calculation of pCa^{2+} is relatively simple (Fig. 2.18). Citrate buffers in the range 30 μM–3 mM Ca^{2+} and is no good as a buffer over the range 0.1–10 μM Ca^{2+}. EGTA and EDTA are good Ca^{2+} buffers over the range 0.3–1 μM (Fig. 2.19) and can be used up to 3 μM free Ca^{2+}. However, over 90% of the ligand is in the form

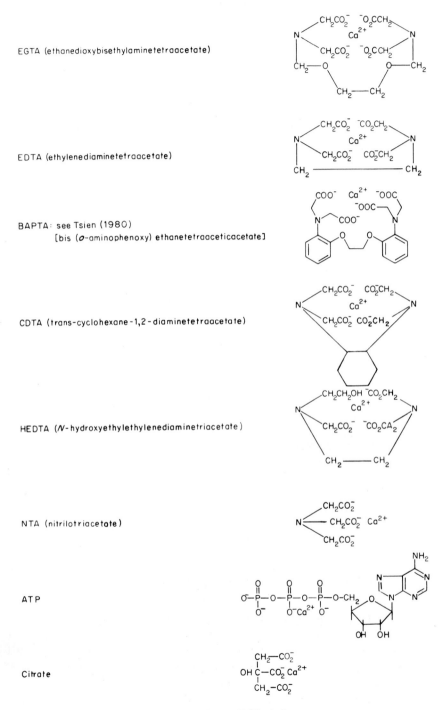

Fig. 2.17. Some Ca^{2+} chelators

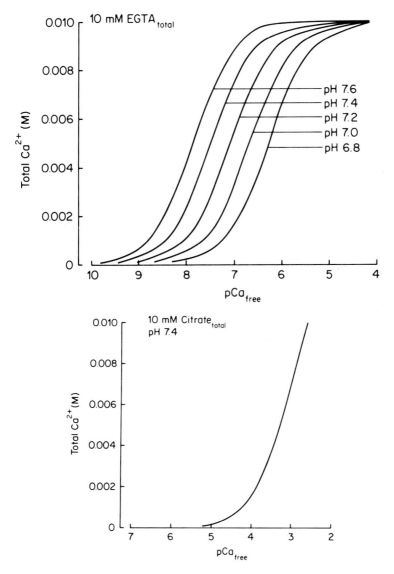

Fig. 2.18. Buffering capacity of (top) EGTA and (bottom) citrate

LH_2^{2-} at pH 7.4. At this pH a change of only 0.2 pH unit can cause a 2–3-fold change in the concentration of free Ca^{2+}. The pH of Ca/EGTA buffers must therefore be maintained within 0.02 pH if accurate data are to be obtained. This problem has led to the design of new high-affinity Ca^{2+} ligands which are much less sensitive to pH near the physiological pH range (BAPTA, Fig. 2.17) (Tsien, 1980). A further problem with EGTA is that addition of Ca^{2+} causes the release of H^+ ions. This can cause massive decreases in pH unless alkali is added (Campbell,

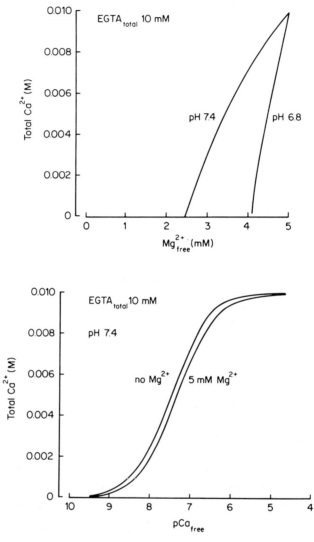

Fig. 2.19. (Top) Effect of Ca^{2+} on free Mg^{2+}. (bottom) Effect of Mg^{2+} on EGTA buffering capacity

1974). Another way of removing the problem of pH sensitivity is to ensure that more than 95% of the ligand is in the form LMg (Fig. 2.20).

To calculate pCa accurately it is necessary to know precisely the concentration ratio of total ligand to total calcium. Calcium salts from standard reagent bottles are rarely defined to more than 95% accuracy. Since an error of only 1% in the total calcium at pH 7.4 can result in an error in the calculation of pCa of 50%, it is necessary to measure the ratio of total calcium to total ligand. This can be done

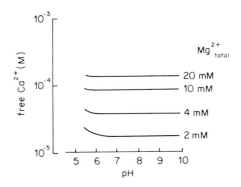

Fig. 2.20. Lack of effect of pH on EDTA in the presence of Mg^{2+}. Data from Wolf (1973), $EDTA_{total} = 1$ mM; $Ca^{2+}_{total} = 0.5$ mM

using an indicator dye such as murexide (Keynes and Lewis, 1956) or by titrating the ligand with calcium and detecting the end-point by measuring the pH (Ashley and Moisescu, 1977).

Several different values for the association constants of various ligands have been reported (Table 2.15), often because of variations in ionic conditions. The association constant can be measured quite easily either by titration using pH as the end-point (Miller and Moisescu, 1976) or by using a Ca^{2+} electrode (Owen, 1976).

Several pH buffers are available which do not bind Ca^{2+} significantly (Good et al., 1966; Ashley and Moisescu, 1977). The binding of Mg^{2+} and K^+ to EGTA can be taken into account by approximation methods (Portzehl et al., 1964; Ashley and Moisescu, 1977). When ATP, P_i and phosphoenol pyruvate are also present, if a value for pCa is assumed, then it is possible to calculate how much of the various cations must be added to achieve this pCa (Vianna, 1975). The best way of coping with several cations and ligands is to use a computer programme (Feldman et al., 1972); these are now available for ten ligands and ten cations. Other ions like Mn^{2+}, Sr^{2+} and Ba^{2+} also bind to ligands like EGTA. This must be taken into account when studying the sensitivity of enzymes to Ca^{2+} (Denton et al., 1978). The problem of activity coefficients can be ignored provided that the ionic conditions are the same as those for the measurement of the association constants.

The forward and backward rate constants for Ca^{2+} and EGTA at pH 7 are $10^{6.3}$ M^{-1} sec^{-1} and $0.4\,sec^{-1}$, respectively (Hellam and Podolsky, 1969). By designing Ca^{2+} ligands that are more fully ionized at physiological pH it should be possible to increase the effective rate at which the ligand binds Ca^{2+} by some 100–1000 times (Tsien, 1980). Although EGTA may respond slowly in the cell to rapid Ca^{2+} transients, experiments with photoproteins in vitro show that free Ca^{2+} can be reduced to < 1 nM within a few milliseconds if sufficient EGTA is added.

Table 2.15. Binding constants of some calcium ligands

L^{4-} –tetrabasic acids

	EGTA			EDTA		CDTA			ATP	
	1	2	3	1	3	3	4	4	5	6
$H^+ + L^{4-}$	9.46	9.46	9.46	10.26	10.26	11.7	11.7	6.5	7.02	6.5
$H^+ + LH^{3-}$	8.85	8.58	8.85	6.16	6.16	6.12	6.12	—	4.02	4.06
$H^+ + LH_2^{2-}$	2.68	2.65	2.67	2.67	2.67	3.52	3.5	—	v.acid	—
$H^+ + LH_3^-$	2.00	2.00	2.00	1.99	1.99	2.43	2.4	—	v.acid	—
$Ca^{2+} + LH^{3-}$	5.33	5.3	—	3.51	—	—	—	1.80	2.13	2.13
$Ca^{2+} + L^{4-}$	11.0	10.97	10.1	10.7	10.59	12.08	12.5	3.6	4.32	4.34
$Mg^{2+} + LH^{3-}$	3.37	3.4	—	2.28	—	—	—	2.00	2.65	—
$Mg^{2+} + L^{4-}$	5.21	5.2	5.4	8.69	8.69	10.32	10.32	4.00	4.65	—
$2Mg^{2+} + L^{4-}$				0.70						

L^{3-} –Tribasic acids

	NTA		HEDTA		citrate
	4	6	3	4	4
$H^+ + L^{3-}$	9.73	9.63	8.10	9.73	5.49
$H^+ + LH^{2-}$	2.5	2.49	5.25	5.33	4.39
$H^+ + LH_2^-$	1.9	1.89	2.04	2.64	3.08
$Ca^{2+} + L^{3-}$	6.41	6.33	8.14	8.00	3.22
$Mg^{2+} + L^{3-}$	5.41	—	5.78	5.20	3.20

These data represent $\log_{10}$ association constants (i.e. $\log_{10} K_a$) = pK_d for various ions used by different workers, where I (Ionic strength) = 0.1 and the temperature = 20°C.
References; 1. Portzehl et al., (1964); 2. Sillen and Martell (1964); 3. Wolf (1973); 4. Raaflaub (1960); 5. Vianna (1975); Reed and Bygrave (1976).
$\log_{10} K_a$ for CaEGTA quoted by Owen (1977) varied from 10.33 to 11.0.—signifies that this was not quoted by the author(s) either because it was not measured or because it was insignificant under the conditions used.

2.8. Other ions

Several other ions besides Ca^{2+}—for example, K^+, Na^+, Mg^{2+}, H^+, Cl^- and amino acids—exist inside cells and may also regulate cell activity (Hodgkin, 1951; Rasmussen et al., 1976; Wacker and Williams, 1968; Wyatt, 1964; Suelter, 1970). The free concentration, or more correctly 'activity', of some of these ions can be measured in living cells by the use of specific microelectrodes (Walker and Brown, 1977; Thomas, 1978). One particularly important ion in cell metabolism is Mg^{2+}. Not only is MgATP the substrate for all known phosphorylating enzymes (kinases), but also this cation may compete with Ca^{2+} for regulatory binding sites on proteins and other macromolecules. Although the affinity constants for Mg^{2+} of these binding sites may be several orders of magnitude less than for Ca^{2+}, since the concentration of free Ca^{2+} in the cytoplasm is probably 0.1 μM, in a stimulated cell these sites need a high degree of selectivity if the free Mg^{2+} concentration is in the mM range (Table 2.16). To assess the significance of the various binding sites for Mg^{2+} it is necessary to estimate the concentration of free Mg^{2+} in the cytoplasm of living cells, and to investigate whether any changes in its

Table 2.16. Free Mg^{2+} concentrations in cells

Method	Cell	Free Mg^{2+} (mM)	Reference
1. Calculation of ligand binding (ATP, ADP, creatine phosphate, proteins, etc.)	Frog muscle	3.4	Nanninga (1961b)
2. Equilibrium reactions Adenylate kinase	Erythrocytes	0.1	Rose (1968)
Aconitase	Rat tissues (liver, brain, etc.)	1.2–2.6	Veloso et al. (1973)
3. Aequorin inhibition	Barnacle muscle	5	Ashley and Ellory (1972)
4. ^{31}P-NMR	Frog muscle	4.4	Cohen and Burt (1977)
5. Indicator dye (eriochrome blue)	Barnacle muscle	6	Brinley and Scarpa (1975); Brinley et al. (1977)

See also Baker and Crawford (1972) for Mg^{2+} mobility in squid giant axons.

concentration occur during cell activation which might mediate the response of the cell to the primary stimulus.

The first attempts to estimate free cytoplasmic Mg^{2+} in cells were based on measurement of the concentration and affinities of Mg^{2+} for the major Mg^{2+}-binding sites (adenine nucleotides, amino acids, phosphate, proteins, etc.), followed by calculation of the free and bound Mg^{2+} from the measured total Mg^{2+} content of the cell (Nanninga, 1961b). This was followed by calculations based on the effect of Mg^{2+} on the equilibrium position of reactions which, because of the high activity of the enzymes catalysing them, are close to equilibrium in the cell. Recently, however, two more direct methods for estimating Mg^{2+} concentrations in cells have been developed. The first involves detecting the ^{31}P-NMR (nuclear magnetic resonanace) signal from free and bound phosphate. From this and a knowledge of the total Mg^{2+}, the free Mg^{2+} can be calculated (Cohen and Burt, 1977). The second method utilizes a metallochromic indicator, eriochrome blue (Scarpa, 1974). A third possibility would be to use a Mg^{2+}-selective microelectrode. Apart from the low value from erythrocytes (Rose, 1968) (Table 2.16) most estimates of the concentration of free Mg^{2+} in cells vary between 1 and 5 mM. The remaining 10–15 mM of cell Mg^{2+} is bound to nucleotides, organic and inorganic phosphate and intracellular proteins; very little (perhaps less than 0.1%) appears to be bound to the cell membrane (Lichtman and Weed, 1973). The free Mg^{2+} concentration in the fluid bathing cells varies from 0.5–1 mM for mammalian serum to 24 mM in sea water. Thus, in mammalian cells there seems to be little or no concentration gradient of free Mg^{2+} across the cell membrane, and even in the cells of marine organisms exposed to sea water there is only a 4–5-fold concentration gradient of Mg^{2+}. This is in contrast to the 10 000–100 000-fold concentration gradient for Ca^{2+} that exists across the cell membrane. One would expect, therefore, that a small increase in the permeability of the cell membrane to Ca^{2+} would cause a very large fractional change in the concentration of cytoplasmic free Ca^{2+}, whereas similar fractional changes in *free* Mg^{2+} would be impossible. However, adrenaline and ACTH have both been reported to affect the Mg^{2+} concentration inside membrane vesicles prepared from adipocytes (Elliot and Rizack, 1974). As with Ca^{2+}, the 'regulatory' role of Mg^{2+} will only be fully understood in terms of 'active' versus 'passive' roles when a reliable method for measuring free Mg^{2+} inside any cell becomes available.

2.9. Conclusions

The characterization of the role of Ca^{2+} as an intracellular regulator requires a combination of experiments using intact cells, cell extracts and isolated proteins. The principle aims of these experiments are: (1) to show that intracellular Ca^{2+} has a role in the observed biological phenomena; (2) to define whether this role for Ca^{2+} is an active or passive one; and (3) to establish the molecular basis of the effects of Ca^{2+} and the control of Ca^{2+} distribution in the cell.

The design and interpretation of experiments is critically dependent on a full understanding of the potential and limitations of the methods used. The central importance of methods for measuring free Ca^{2+} inside the cell highlights the necessity of relating chemical information obtained in the test-tube to conditions in the living cell.

CHAPTER 3

The chemistry of biological calcium

3.1 The value of the chemical approach

There can be few people who are not very familiar with the poor solubility of many calcium salts. Anyone who lives in a hard water area will have observed the precipitation of calcium stearate every time he washes his hands; whilst the white cliffs of Dover are a striking reminder of the importance of calcium carbonate in the skeletal structures of many animals. An understanding of the way in which Ca^{2+} binds to various ligands, particularly those containing oxygen, lies at the heart of understanding its role as a regulator inside cells. In this case one is rarely dealing with the precipitation of calcium salts, but rather the binding of Ca^{2+} to various soluble organic ligands such as proteins, phospholipids and certain small molecules.

Calcium was discovered as an element in 1808 by Humphry Davy. Since that time a considerable amount has been learnt about the chemistry of calcium and the other alkaline earths (Table 3.1).

The binding of Ca^{2+} to organic ligands holds the key to understanding its role as an intracellular regulator. The complexity of these biological ligands, compared with simple inorganic ones such as phosphate and carbonate, has

Table 3.1. Some properties of the group II, and related, elements

Element	Atomic number	Atomic weight	Electron configuration	$E_0(V)$ $(M \rightleftharpoons M^{2+} + 2e)$	Ionic radius (unhydrated) (Å)
Mg	12	24.3	[Ne] $3s^2$	2.37	0.65
Ca	20	40.1	[Ar] $4s^2$	2.87	0.94
Sr	38	87.6	[Kr] $5s^2$	2.89	1.10
Ba	56	137.3	[Xe] $6s^2$	2.90	1.29
Ra	88	226	[Rn] $7s^2$	2.92	1.50
Mn	25	54.9	[Ar] $3d^5 4s^2$	1.18	0.80(Mn^{2+})
La	57	138.9	[Xe] $5d^1 6s^2$	2.52	1.06(La^{3+})

E_0 = ionization potential

provided the chemist with some fascinating structures to analyse. Furthermore, they have highlighted several important properties of the coordination chemistry of calcium. But what is the value to the biologist, interested in the regulatory function of Ca^{2+}, in studying the chemistry of Ca^{2+}? Why, for example, do we want to know how many Ca^{2+}-binding sites there are on troponin C, the protein which mediates the effect of Ca^{2+} to cause muscle contraction? Of what value is a knowledge of the affinity constant of Ca^{2+} for these sites or indeed their three-dimensional structure?

The isolation of Ca^{2+} ligands, together with the characterization of changes in their properties when they bind Ca^{2+}, provides the molecular basis for intracellular Ca^{2+} acting as a regulator. The interaction of these ligands with the substances or structures actually responsible for the cell response forms an essential link in the chain which leads from primary stimulus to the observed phenomenon. A comparison of the equilibrium binding constant for Ca^{2+} with the measured range of intracellular free Ca^{2+} enables the fractional saturation by Ca^{2+} of these sites during cell activation to be calculated. Since reactions in the cell are never at true equilibrium, it is preferable to relate the association and dissociation rate constants for Ca^{2+} to the time course of the physiological response in the intact cell.

These details of the chemistry of Ca^{2+} binding provide further evidence for an 'active' role for Ca^{2+} as defined in Chapter 2 (section 2.2). Yet a study of the chemistry of intracellular Ca^{2+} has far wider implications. Surely the prime attraction of the Natural Sciences is the opportunity it gives the explorer of perceiving the beauty and elegance of Nature from the whole animal down to the level of molecular structures. Like the artist painting a picture or a composer constructing a symphony, the scientist is never fully satisfied until he can step back, absorb and comprehend the whole. A complete understanding of the unique roles of Ca^{2+} in living organisms, and in particular its role as an intracellular regulator, the exploitation during evolution of Ca^{2+} rather than of elements like Mg^{2+}, depend on an understanding of the chemistry of calcium. Whilst we are still some way from a complete picture of the evolutionary significance of Ca^{2+}, sufficient studies on the chemistry of Ca^{2+} binding to biological ligands have been carried out to help us understand the 'active' role of intracellular Ca^{2+} and Ca^{2+}-dependent 'threshold' phenomena (Chapter 2).

3.2 The ligands

3.2.1. What is a ligand?

The word ligand is derived from the Latin *ligare*, meaning to tie or bind. In chemistry, however, the word has a more precise meaning than simply a substance which can bind to another molecule. A chemical ligand is any atom, ion or molecule, capable of acting as a donor in the formation of one or more coordinate bonds. This forms the foundation of the coordination theory developed in the nineteenth century by Werner. The coordinate bond is different from a covalent or ionic bond in that it requires the donation of pairs of electrons from the ligand to the other molecule, usually a cation. Atoms such as N or O are particularly common as ligands since they contain lone pairs of electrons in their outermost occupied orbitals . It is these atoms in biological molecules (Tables 3.2, 3.3, and 3.4) which act as Ca^{2+} ligands. Although some ions can act as ligands, Ca^{2+} is not itself a ligand. The electronic configuration of Ca^{2+} is that of the inert gas argon (Table 3.1) and thus it does not have any spare electrons for coordination. Ca^{2+}, like many transition metals, does, however, contain empty 3d, 4s and 4p orbitals which are able to accept electron pairs from the donor atoms in the ligand.

Early on in the study of coordination chemistry it became apparent that any given metal ion had a favoured number of ligand sites, often leading to a symmetrical three-dimensional complex. The number of ligand atoms which coordinate with a metal ion is called its 'coordination number'. Six coordination is common because of the symmetry of the octahedral arrangement of ligand atoms (Fig. 3.1). Ca^{2+} can have a coordination number of 6 (Fig. 3.1), but because of its relatively large size coordination numbers up to 10 are also possible. Coordination numbers greater than 6 are common for biological Ca^{2+}. This gives the ion greater flexibility over its neighbour in the periodic table, Mg^{2+}, whose smaller size (Table 3.1) usually limits its coordination number to 6. The similarity in size of La^{3+} and Ca^{2+} helps to explain the antagonism between these two ions in biological systems.

Ligands that have more than one ligand atom within the same molecule are called 'polydentate'. These chelating agents, as they are called (from the Greek word meaning claw), form a very important group of strong biological Ca^{2+} ligands (Fig. 3.2) because of a phenomenon known as the 'chelate' effect. Synthetic Ca^{2+} chelators, such as EDTA, used as Ca^{2+} buffers (Chapter 2) and as strong Ca^{2+} binders in biological experiments, exemplify this effect (Fig. 3.1).

Four main parameters characterize the interaction between a cation and a ligand: (1) the non-metallic coordinating element, for example O, N or S; (2) the

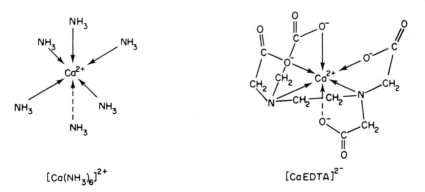

Fig. 3.1. Octahedral binding of Ca^{2+} to ligands. The six ligand atoms form an octahedron.

coordination number; (3) the thermodynamic constants which determine the strength and kinetics of ligand binding; and (4) the stereochemistry of binding. Before considering these parameters in detail with respect to intracellular Ca^{2+}, let us discover what are the substances in cells which can bind Ca^{2+}. These can be conveniently divided into three main groups: inorganic, small organic and macromolecular Ca^{2+} ligands.

3.2.2. Inorganic Ca^{2+} ligands

The calcium salts of a number of biologically occurring anions (Table 3.2) are relatively insoluble. Their precipitates form the main substrate for skeletal structures in many organisms as well as occurring within the cells, for example in the mitochondria, as microprecipitates. Calcium stones occur in various tissues in several diseases. In addition, Ca^{2+} forms complexes with several of these anions in solution. Oxygen-containing anions predominate and protonation of the anion has marked effects on their strength of binding to Ca^{2+}.

These anions occur in cells in mM concentrations. The relative concentrations of protonated species is pH-dependent. For example, more than 80% of phosphate at physiological pH (i.e. 7.4) is HPO_4^{2-}, whereas it is $Ca_3(PO_4)_2$ which is the least soluble calcium salt. This is one reason why calcium phosphate precipitates can often be induced under alkaline conditions which increase the fraction of phosphate in the PO_4^{3-} form. However, the phosphate anion in many naturally occurring Ca^{2+} precipitates is HPO_4^{2-}.

Although the fraction of total cell Ca^{2+} bound to these inorganic anions may be relatively small, this Ca^{2+} can be significant with respect to the cytoplasmic free Ca^{2+} of only 0.1 μM. Furthermore, the passive movement of Ca^{2+} across membranes, as well as transport through certain permeases, for example in mitochondria, may involve complexes of Ca^{2+} and phosphate anions rather than Ca^{2+} alone.

Table 3.2. Some naturally occurring inorganic Ca^{2+} ligands

Ligand	Conc. in mammalian plasma	pK_{Ca}	pK_{Mg}	pK_H
$H_2PO_4^-$	0.11 mM	0.89	—	2.0
HPO_4^{2-}	0.5 mM	2.1	2.0	6.8
PO_4^{3-}	5 nM	6.3	—	12.4
HCO_3^-	25 mM	0.9	0.5	6.2
CO_3^-	0.25 μM	2.2	2.3	10.4
SO_4^{2-}	0.65 mM	1.7	approx. 1.7	1.5

Data from Sillen and Martell (1964) and Walser (1970). $pK = -\log K_d = \log K_a$ (M), no fixed temperature or ionic strength (I). The figures are only intended to indicate the relative affinity of the ligand for the cation; the bigger the pK the higher the affinity.

3.2.3. Naturally occurring small organic Ca^{2+} ligands

This group of ligands includes a wide variety of non-polymeric substances in cells (Fig. 3.2, Table 3.3). Most of them are water soluble (hydrophylic) although a few, such as Ca^{2+} ionophores, are considerably more soluble in organic solvents (hydrophobic). The concentration of these ligands in cells ranges from μM for some phosphorylated metabolites to 5–10 mM for ATP. Many of them also bind Mg^{2+} which markedly reduces the amount of Ca^{2+} bound to them in the cell. Nevertheless, significant quantities of Ca^{2+} can be bound even when the free Ca^{2+} is 0.1 μM. Like the inorganic Ca^{2+} ligands the fraction of the ligand bound to Ca^{2+} is dependent on the protonation state of the ligand, the commonest ligand atom again being oxygen. Under physiological conditions most complexes are soluble. However, in certain special cases this is not so. For

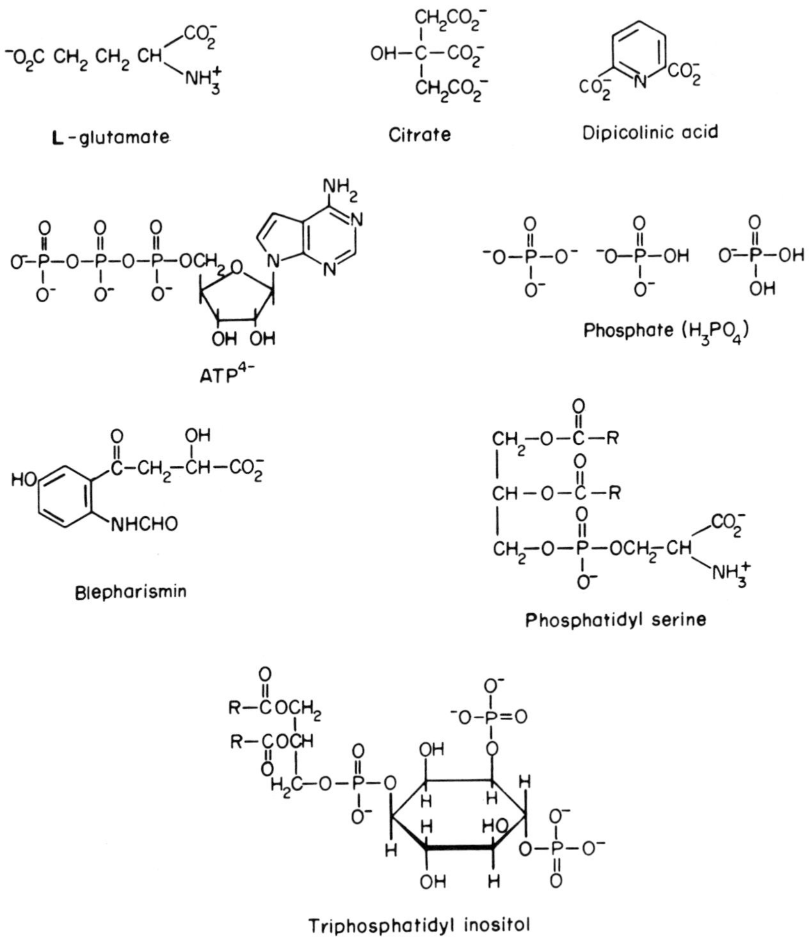

Fig. 3.2. Some naturally occurring Ca^{2+} ligands

Table 3.3. Some naturally occurring small organic Ca^{2+} ligands

Ligand	pK_{Ca}	pK_{Mg}	pK_H
Monobasic acids			
Acetic	1.2	1.3	4.6
Pyruvic	1.9	2.0	2.4
Dibasic acids			
Succinic	1.2	—	5.3; 4.0
Oxalic	3.0	2.6	3.8; 1.4
Dipicolinic	4.6; 2.6	2.3	4.7; 2.2
Tribasic acids			
Citric	3.6	3.4	5.7; 4.4; 2.9
Amino acids			
Gly } neutral	1.4	2.2	9.6; 2.4
Ala	1.2	2.0	9.7; 2.3
Tyr hydrophobic	1.5	—	10.1; 9.0; 2.2
Lys } basic	—	—	10.7; 9.1; 2.0
Arg	2.2	2.2	9.0; 2.1
Asp } acidic	1.6	2.4	9.6; 3.7; 1.9
Glu	1.4	1.9	9.6; 4.2; 2.2
Nucleotides			
AMP^{2-}	1.4	2.1	6.4
ADP^{3-}	2.9	3.2	6.6; 4.2
ATP^{4-}	3.5	4.0	6.7; 4.3
Sugars			
Glucose 1-phosphate	2.5	2.5	6.5; 1.46

Data from Sillen and Martell (1964). $pK = -\log K_d = \log K_a$ (M). Temperature usually, but not always, 25 °C; $I = 0.1$ where available. Figures are only intended to indicate relative affinities for Ca^{2+} and Mg^{2+}. More than one figure represents binding of more than one cation per molecule.

example, the Ca^{2+} salt of dipicolinic acid forms a hydrated precipitate (calcium dipicolinate·$3H_2O$) which causes massive Ca^{2+} accumulation during bacterial spore formation (Ellar *et al.*, 1974), and is released on germination. Calcium oxalate precipitates are found in plants, and Ca^{2+} may play a structural role when it complexes with ATP and amine hormones like 5-hydroxytryptamine in secretory granules. It is an interesting observation that many such granules contain a very active Ca^{2+}-stimulated Mg ATPase, suggesting that they have the capacity to take up Ca^{2+} rapidly (see Chapter 7).

3.2.4. Polymeric and polyvalent electrolytes as Ca^{2+} ligands

This group of macromolecular Ca^{2+} ligands (Tables 3.4 and 3.5) constitutes perhaps the most important one with respect to understanding intracellular Ca^{2+} as a regulator. High-affinity Ca^{2+}-binding proteins mediate many, if not all, of the regulatory effects of intracellular Ca^{2+}. Furthermore, several

Table 3.4. Examples of proteins which bind Ca^{2+}

Protein	Activates (a)/ inhibits (i)	Approx. pK_d^{Ca} or pK_m^{Ca} for each Ca^{2+} site
Intracellular (cytoplasmic, and with Ca^{2+}-binding sites facing cytoplasm)		
Troponin C	a	6–7
Phosphorylase *b* kinase	a	6
Pyruvate kinase	i	3.7
Phosphofructokinase	i	2.7
Adenylate cyclase (cell membrane)	i	2–3
Ca^{2+}-activated MgATPase (cell membrane)	a	6
Ca^{2+}-activated MgATPase (sarcoplasmic reticulum)	a	6.5
Glycerol-3-phosphate dehydrogenase (mitochondria)	a	7
Phosphatidylinositol hydrolase	a	6.8
Intracellular (organelles)		
Calsequestrin (sarcoplasmic reticulum)	—	3.3
Acid protein (sarcoplasmic reticulum)	—	6
Pyruvate dehydrogenase phosphatase (mitochondria)	a	5.4
Pyruvate carboxylase (mitochondria)	i	*ca* 3
Endonuclease (nucleus)	a	3
25-Hydroxycholecalciferol-1-hydroxylase	i	5
	a	4
Extracellular		
α-Amylase	a	6
Trypsin	a	3.2
Prothrombin (factor II)	a	3.3
Transglutaminase (factor XIII)	a	3.6
Acetylcholine receptor (nicotinic)*	a	approx. 3

These figures are only approximate and are intended to indicate the order of magnitude. See Kretsinger (1976a, b) for further examples and references.
* Cohen *et al.* (1974) and Lukas *et al.* (1979).

extracellular degradative enzymes require Ca^{2+} for maximum activity (Table 3.4) and several glycoproteins in the membranes of cells and organelles also bind Ca^{2+}. A few polypeptides have been isolated which have the property of being a Ca^{2+} ionophore; that is, they enable Ca^{2+} to cross hydrophobic regions such as phospholipid bilayers. For example, such a glycoprotein has been isolated from mitochondria (see Chapter 4, section 4.6). A particularly interesting example in this respect is the cyclic peptide alamethicin (Fig. 3.3) isolated from the fungus *Trichoderma viride*. This peptide induces action potentials in black lipid membranes (Mueller and Rudin, 1968) and consists of 18 amino acids, cyclized between 17 of them. The peptide has an unusually high methylalanine content, which also blocks the N-terminal proline. Although alamethicin can transport several different ions across lipid bilayers, it has particularly interesting

Table 3.5. Some naturally occurring non-protein polymeric Ca^{2+} ligands

Phospholipids
    Phosphatidylserine
    Phosphatidic acid
    Phosphatidylinositol
    Triphosphoinositide

Nucleic acids
    Deoxyribonucleic acid (DNA)
    Ribonucleic acid (RNA)
    Poly(A)

Polysaccharides
    Sialic acids on glycoproteins
    Chondroitin sulphate
    Galacturonic acid
    Alginic acid (mururonic acid)
    Fucoidin (poly fucose sulphate)
    Curragenin (poly galactose sulphate)

ionophoretic properties with respect to Ca^{2+} (Martin and Williams, 1975; Williams, 1976). Ca^{2+}-binding proteins that are either sitting in the phospholipid bilayer or which can act as ionophores must contain, in addition to the Ca^{2+}-binding sites, a hydrophobic region which allows them to remain in this non-polar environment.

In addition to proteins several other biological polymers can bind Ca^{2+} (Table 3.5). Included in this list are nucleic acids (Eichhorn et al., 1969; Duane, 1974) and phospholipids which, although not strictly polymeric, in aqueous media form micelles or liposomes each composed of thousands of phospholipid molecules. The polar groups of the phospholipid face out into the hydrophilic medium and the non-polar fatty acid chains face inwards. It is therefore not possible to measure the binding of Ca^{2+} from an aqueous medium to individual phospholipid molecules. Negatively charged phospholipids such as phosphatidic acid, phosphatidyl-serine and -inositol bind Ca^{2+} more strongly than other naturally occurring phospholipids such as lecithin (phosphatidylcholine) (see

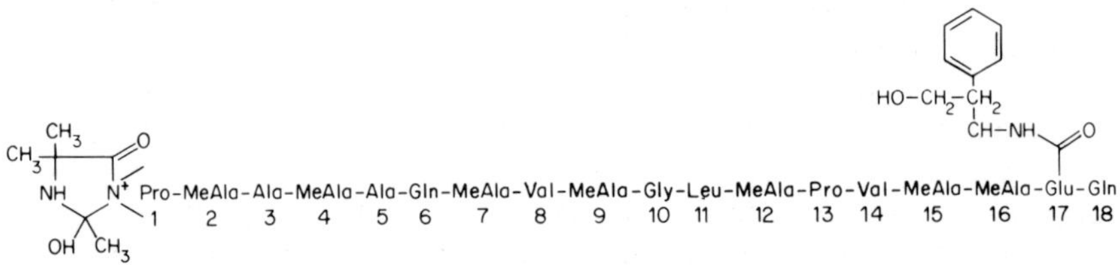

Fig. 3.3. Alamethicin, F30, a peptide of 18 amino acids with high methylalanine (MeAla) content. Proline-1 at the N-terminus is blocked with N-acetylmethylalanine. The γ-carboxyl of glutamate-17 has a β-phenylalaninol attached to it. From Martin and Williams (1975)

Fig. 3.2 for structures) (Joos and Carr, 1967; Hauser and Dawson, 1968; Dawson and Hauser, 1970).

It has been known for some time that the nucleus in many cells contains a substantial fraction, probably at least 40%, of cell Ca^{2+} (Williamson and Gulick, 1944; Naora et al., 1961) (see Table 2.7). How much of this Ca^{2+} is bound to protein and how much to nucleic acid is not known. Ca^{2+} binds to DNA and the various species of RNA (e.g. messenger, ribosomal, transfer) but the affinity for Ca^{2+} is relatively low. At $0.1-10$ μM, the range of cytoplasmic free Ca^{2+} concentration, one would not expect Ca^{2+} to act significantly as a bridge between different parts of a nucleic acid molecule, or between different molecules, through phosphate-Ca^{2+} cross-links. Mg^{2+}, however, may be able to function in this way, though the nuclear content of this cation appears to be much lower than that of Ca^{2+}. It would be useful to know what the free Ca^{2+} concentration is in the nucleus. How much Ca^{2+} is bound to nucleic acid, for example in ribosomes, in the cytoplasm is unknown. Nor is it known whether degradation of nucleic acid by nucleases within the cell can cause significant local changes in free Ca^{2+}.

Without a doubt the most important group of polymeric Ca^{2+} ligands identified so far is a class of high-affinity regulatory Ca^{2+}-binding proteins, including a group of proteins called 'calmodulins'. The first such protein to be discovered was troponin C (Ebashi, 1963, 1974, 1980). The discovery of the interaction of this protein with troponin I in a complex with troponin T provided the molecular basis for understanding the triggering of muscle contraction by Ca^{2+} (see Chapter 5). The discovery of this protein also stimulated a search for similar proteins in non-muscle cells. Several have now been found (Table 3.6), although the *physiological* function of some of them (e.g. phosphodiesterase activator and the Ca^{2+}-binding protein in secretory cells) has yet to be fully established. The proteins in this group exhibit several common properties. They are relatively small molecular weight (10000–20000) and negatively charged at physiological pH, as indicated by the highly acidic pI (4–5) and high proportion of acidic amino acids compared with other proteins (Table 3.6). They contain 2–4 high-affinity Ca^{2+}-binding sites. In some, Mg^{2+} may compete with Ca^{2+} for two of these sites. The interaction of a part of the molecule separate from the cation-binding sites enables Ca^{2+} binding to affect the protein or structure to which the protein is bound (Fig. 3.4). These proteins sometimes also exhibit similarities in sterochemistry of Ca^{2+} binding (Kretsinger, 1976a, b) (see this chapter, section 3.3.5).

What questions can we ask about Ca^{2+} binding to these various inorganic and organic ligands in order to obtain a better understanding of the chemistry and physiology of intracellular Ca^{2+} as a regulator? What must be measured in order to answer these questions? The four most obvious questions are: (1) How much Ca^{2+} is bound to the ligand in the presence of physiological Ca^{2+} concentrations (i.e. 0.1 μM in the resting cell)? What is the rate and extent of the change in this bound Ca^{2+} when cytoplasmic free Ca^{2+} increases in the activated cell to 1–5 μM? (2) How does Ca^{2+} binding to the modulator protein induce a change in

Table 3.6. Amino acid content of some Ca^{2+}-binding proteins compared to some other proteins

Protein	Approx. mol. wt.	pI	Amino acid content (percentage of total)		
			Glu + Asp (acidic)	Lys + Arg (basic)	Acidic − basic
Ca^{2+}-binding proteins					
CBP II (pig brain)	13 000	5.5	36	8	28
Calmodulin (cyclic AMP phosphodies-terase activator, bovine brain)	16 700	3.9	34	9	25
Troponin-like protein (bovine adrenal)	16 000	4.3	33	9	24
Troponin C (rabbit muscle)	18 000	4.2	33	10	23
L-2 (squid brain)	13 000	4.4	32	10	22
S-100 (rabbit brain)	21 000	—	30	10	20
Vitamin D CBP (chick gut)	28 000	4.2	32	12	20
Aequorin	20 000	approx. 4	25	11	14
Parvalbumin (carp muscle)	11 500	4.2	23	13	10
Trypsinogen	24 000	—	17	7	10
Albumin (bovine serum)	68 000	—	23	14	9
CBPI (pig brain)	13 000	5.5	23	14	9
Vitamin D CBP (pig parathyroid)	14 500	—	22	15	7
Vitamin D CBP (pig gut)	9000	4.8	30	23	7
Other proteins					
Rhodopsin	28 600	—	15	8	7
Lactate dehydrogenase (M_4)	144 000	—	19	11	8
Histone IV (calf thymus)	11 300	—	11	25	−14
Mean of 78 proteins	—	—	21	11	10

For further information and references see reviews by Kretsinger (1976a, b). pI = isoelectric point. Acidic = Asp + Glu (usually includes Asp(NH$_2$) and Glu(NH$_2$)); basic = Lys + Arg.

activity of other proteins which actually cause the cell response? (3) What are the ligand atoms binding Ca^{2+} and is it possible to provide a chemical rationalization of the strength and selectivity of the binding with respect to other cations like Mg^{2+}? (4) What is the stereochemistry of Ca^{2+} binding? Does this provide any insights into the evolutionary origin of Ca^{2+}-mediated cell regulation?

The answer to these questions requires a critical examination of the thermodynamics and coordination chemistry of Ca^{2+} binding.

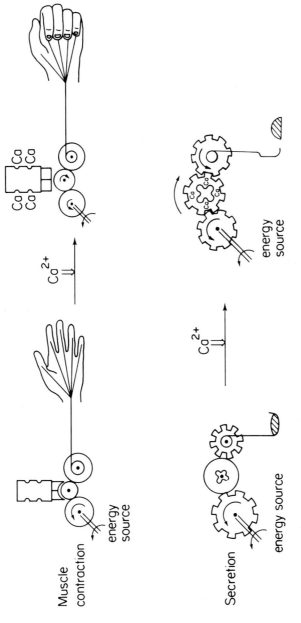

Fig. 3.4. How Ca^{2+}-binding proteins work

3.3. Thermodynamics and coordination chemistry of Ca^{2+} binding

3.3.1. Thermodynamic principles

Since the mid 1950's molecular biology has played an increasingly dominant role in biology. Consideration of the chemistry of individual molecules, which is one of the hall-marks of the molecular biological approach, is entirely reasonable when dealing with a molecule such as DNA which contains at most two sites (or a few more in the case of gene duplication) for a particular process in any one cell. This approach also provides models for the mechanism and modulation of enzymatic catalysis. However, many reactions in the cell involve thousands of millions of molecules. Even in the case of hormone receptors on the surface of the cell, or ion channels, several thousand molecules may be present in each cell. The equations that predict the rate of these reactions in the cell depend on calculating the fractional saturation of ligands using macromolecular thermodynamic principles (see Beck, 1970). When microcompartments, such as the mitochondrion, exist within the cell the number of free atoms of each ion could be small; for example, approx. 20 for H^+ or approx. 50–5000 for Ca^{2+} (assume the radius of the mitochondrion = 1 μm, then the volume = 4 fl; assume pH = 7.4 and free $[Ca^{2+}] = 0.1$–10 μM, then matrix = 20% of total volume). If this is the case then a form of statistical thermodynamics may be required.

The advancement of science is often dependent not only on the discovery of new principles and laws but also on propagating the importance of these discoveries to others. Many physicists and chemists in the nineteenth century became aware of the significance of the laws of thermodynamics, thanks particularly to their propagation by physicists such as William Thomson (later Lord Kelvin). However, it was not until the first quarter of this century that the implications of these laws were brought to the attention of biologists. These laws enable the energetics of Ca^{2+} binding to ligands to be quantified, the energy requirements for Ca^{2+} transport across membranes to be calculated, and the parameters which characterize the strength, selectivity and kinetics of Ca^{2+} binding to be defined.

In solution any metal ion will be surrounded by a sphere of water molecules (Fig. 3.5). Therefore the equation describing Ca^{2+} binding to a ligand (L) is

$$Ca^{2+}(H_2O)_n + L^{x-}(H_2O)_m \rightleftharpoons [(H_2O)_p\text{-CaL-}(H_2O)_q]^{(x-2)-} + (n+m-p-q)H_2O \qquad (3.1)$$

The affinity of the ligand (L) for Ca^{2+}, in other words the equilibrium position of this reaction, is dependent on three main parameters: (1) the strength of the Ca^{2+}–L bond; (2) the change in solvation energies of the components of the reaction; (3) the energetics of any structural change in L as a result of Ca^{2+} binding. An ionic bond between Ca^{2+} and a CO_2^- group in a hydrophobic environment, where Ca^{2+} can retain some of its water of hydration (i.e. $p \geqslant 1$) and where the shape change releases energy, would produce a high-affinity binding site for Ca^{2+}. In the presence of a low Ca^{2+} concentration the equilibrium would lie on the far right-hand side of equation 3.1. The energetics can be quantified as

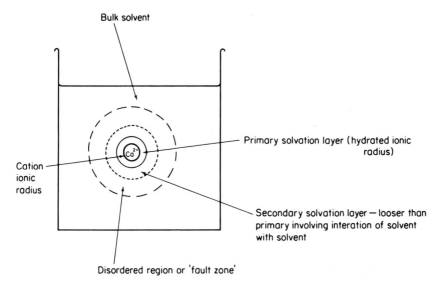

Fig. 3.5. The solvation shells of one Ca^{2+} ion in solution, greatly enlarged. Effective hydrated radii (nm) Mg^{2+}, 0.59; Ca^{2+}, 0.45; Sr^{2+}, 0.37; Ba^{2+}, 0.37.

follows. The energy of any particular state is defined as the Gibbs free energy (G)

$$G = H - TS \tag{3.2}$$

where H = enthalpy, S = entropy.
Hence the change of free energy during a reaction is

$$\Delta G = \Delta H - T\Delta S \tag{3.3}$$

where ΔH is related to the heat of reaction and ΔS, the entropy change, is related to the degree of randomness of the components. The reaction in equation 3.1 can be considered as the combination of three separate ones

(1) Solvation energy of Ca^{2+}

$$Ca^{2+}(H_2O)_n \rightleftharpoons Ca^{2+}(H_2O)_p + (n-p)H_2O \tag{3.4}$$

(2) Solvation energy of ligand

$$L^{x-}(H_2O)_m \rightleftharpoons L^{x-}(H_2O)_q + (m-q)H_2O \tag{3.5}$$

(3) Energy of reaction

$$Ca^{2+}(H_2O)_p + L^{x-}(H_2O)_q \rightleftharpoons [(H_2O)_p\text{-}CaL\text{-}(H_2O)_q]^{(x-2)-} \tag{3.6}$$

Hence ΔG for equation 3.1 = $\sum \Delta G$ for equations 3.4, 3.5 and 3.6

i.e. $\Delta G_{total} = \Delta G^{Ca^{2+}}_{solvation} + \Delta G^{L}_{solvation} + \Delta G^{CaL}_{reaction}$

$\Delta G^{CaL}_{reaction}$ is itself made up of two components, the binding energy between Ca^{2+} and L plus the energy change resulting from any structural change in L or, in the case of a flexible ligand, resulting from restrictions in this flexibility (a negative change in entropy).

With small anions (e.g. OH^-, HPO_4^{2-}, RCO_2^-) one molecule of H_2O in the coordination sphere of a cation can be replaced by one molecule of anion without much steric hindrance. Thus, smaller cations such as Mg^{2+} will bind these anions more strongly than the larger ones (see Table 3.1). The order of affinities (i.e. bond strength predominates) is $Mg^{2+} > Ca^{2+} > Sr^{2+} > Ba^{2+}$. In contrast, for larger anions like SO_4^{2-} one anion molecule may displace several H_2O molecules in the coordination sphere. In this case the strength of the bonds with smaller cations is not able to compensate for the loss of the H_2O molecules. The order of affinities for larger anion binding is therefore $Ba^{2+} > Sr^{2+} > Ca^{2+} > Mg^{2+}$. The ability of Ca^{2+} to retain H_2O of solvation, which often necessitates coordination numbers of 7 or more with protein ligands, is an important factor determining the high affinity of binding sites for Ca^{2+} relative to Mg^{2+}.

The argument changes when comparing the strength of binding of two ligands for the same cation, e.g. Ca^{2+}. Ligands with two or more ligating atoms (i.e. polydentate ligands) in the same molecule form stronger complexes than unidentate ligands. This 'chelate effect' as it is called is due to the more favourable entropy of chelation with polydentate ligands, such as citrate and polymeric biological ligands.

Strong Ca^{2+}–L bonds do not necessarily require an electrostatic interaction. Carbonyl oxygen atoms in ionophores (Trutter, 1975, 1976) and in the peptide chain of proteins can bind Ca^{2+}. Furthermore, there may be steric constraints placed on the ligand, which may dominate the order of cation specificity. This can be described as the so-called 'radius-ratio' effect (Williams, 1970).

$$\text{Hydration energy} = -\left(\frac{A}{r_+ + r_{H_2O}}\right) \quad (3.7)$$

where A = a constant dependent on H_2O dipole and cation charge
r_+ = Pauling ionic radius
r_{H_2O} = H_2O radius.

Whereas energy of ligand–cation interaction $= -\left[\dfrac{B}{r_+ + r_-}\right]$

where B = constant dependent on the charge of the product and the dielectric constant
r_- = ligand radius.

These factors, together with solvation of ligand due to the hydrophilic nature of its outer surface, provide a qualitative explanation for the high selectivity of regulatory Ca^{2+}-binding sites over Mg^{2+} and monovalent cations. The quantitative description of Ca^{2+} binding depends on the change in chemical potential (μ). The chemical potential of Ca^{2+} is

$$\mu_{Ca} = \mu' + RT \log_e f_{Ca} \quad (3.8)$$

where μ' is a function of temperature and pressure only

$$f_{Ca} = \text{fugacity} = \gamma_{Ca} m_{Ca}$$

where γ_{Ca} = activity coefficient = 1 for an ideal solution whereas $\gamma < 1$ for real solutions and m_{Ca} = molality of Ca^{2+} (the molality of a solute = moles/kg solvent). The chemical potential is related to Gibbs free energy by the relation

$$\mu_i = \left(\frac{\delta G}{\delta n_i}\right)_{T,P,j} \quad \text{or} \quad \Delta G = \sum \mu_i n_i \tag{3.9}$$

Ignoring the H_2O in equation 3.1 we can write down the change in chemical potential due to this reaction

$$\Delta \mu = \mu_{products} - \mu_{reactants} = \mu_{CaL} - \mu_{Ca} - \mu_L$$
$$= RT \log_e [f_{CaL}/(f_{Ca} + f_L)]$$

Relating f to a standard state (f_0) of pressure and temperature, a new term activity (a) is defined as $a = f/f_0$. Thus

$$\Delta \mu = \mu^0_{CaL} - \mu^0_{Ca} - \mu^0_L + RT \log_e [a_{CaL}/(a_{Ca} + a_L)] \tag{3.10}$$

At equilibrium $\Delta G = \Sigma \mu = 0$.

Hence $\mu^0_{CaL} - \mu^0_{Ca} - \mu^0_L = \Delta G_0 = -RT \log_e \left(\frac{a_{CaL}}{a_{Ca} a_L}\right) = -RT \log_e K_a$ (3.11)

where K_a = equilibrium association constant = $1/K_d$
and K_d = equilibrium dissociation constant.

Equations such as these enable the equlibrium constant to be measured and how far a reaction is from equilibrium to be calculated. However, some reactions in the cell may be so close to equilibrium that to a first approximation it is possible to use K_d to calculate the concentration of one of the reactants when the others are known. For example, in the case of Ca^{2+} binding to a protein it may be possible to estimate the fractional saturation of the protein ligand by Ca^{2+} from K_d and measurement of free Ca^{2+} and total ligand concentrations. Furthermore, it may be possible to calculate whether sufficient energy is available from a change in Ca^{2+} binding for the protein to do a significant amount of work. In most muscles the energy for contraction is thought to be from ATP hydrolysis and not Ca^{2+} binding. However, in one protozoan, *Vorticella*, an intracellular organelle called a spasmoneme contracts when Ca^{2+} binds to it, the energy of Ca^{2+} binding apparently being sufficient to provide the work done during the movement of the organelle (Amos et al., 1976).

Ca^{2+}–ligand reactions far from equilibrium require a source of energy to maintain this state. The quantity of energy required can be calculated from equation 3.10. The further the reaction is from equilibrium the more energy is necessary to maintain it.

Assuming that, when Ca^{2+} regulates the activity of a cell, the physiological response (α) is directly related to a_{CaL}, then at equilibrium

$$\text{Physiological response} \propto a_{CaL} = K_a a_{Ca} a_L \tag{3.12}$$

In the cell, however, not only is the system not at true equilibrium but also it may not even be in true steady-state. For example, when free Ca^{2+} has been measured in the cell during activation it rarely, if ever, reaches a constant value. Under these conditions the fractional saturation of the ligand by Ca^{2+}, on which the physiological response depends, is dependant on the fugacities of the reactants and products and the rate constants which determine the kinetics of Ca^{2+}–ligand binding.

Rate of Ca-L formation $= k_1 a_{Ca} a_L$
Rate of Ca-L dissociation $= k_{-1} a_{CaL}$
At equilibrium these are equal. Hence

$$\frac{k_1}{k_{-1}} = \frac{a_{CaL}}{a_{Ca} a_L} = K_a \tag{3.13}$$

At non-equilibrium

$$\frac{d(a_{CaL})}{dt} = k_1 a_{Ca} a_L - k_{-1} a_{CaL} \tag{3.14}$$

At any time t the fractional saturation of L by Ca^{2+}, i.e. a_{CaL}, can be defined by solving this equation for a_{CaL} and integrating. Such non-steady-state kinetics of Ca^{2+} binding almost certainly exist in several types of muscle cell (Ashley and Moisescu, 1972) where the rate of change of free Ca^{2+} is fast.

At the present time the free Ca^{2+} concentration has been measured in very few cells. Also the kinetics of Ca^{2+} binding to ligands in the cell have been poorly characterised. Therefore, in order to discuss the characteristics of regulatory Ca^{2+}-binding sites the remaining calculations in this chapter will use equilibrium thermodynamics to illustrate the principles involved. But before doing this it is necessary to reconsider what 'fugacity' is.

3.3.2. The problem of activity coefficients

By the early part of this century it was realized that even the behaviour of strong electrolytes departed from the ideal predicted by the law of Mass Action, which utilizes concentrations of ions in molalities. In 1923 Debye and Hückel produced a theory to explain this non-ideal behaviour, ascribing it entirely to electrical interactions between the ions. The more concentrated the solution and the more charged the ions, the greater were the deviations.

The theory was based on the argument that because of electrostatic interactions between charged ions in solution these ions would not be randomly distributed. Like charges repel and unlike attract each other. Thermal movement of molecules prevents a highly ordered arrangement of ions such as occurs in a crystal. The real situation in solution is therefore a balance between ordering of ions imposed by electrostatic forces and molecular collisions disrupting this ordering. The starting point for the theory was the calculation of the mean distribution of ions in solution based on the Poisson equation [see Robinson and Stokes (1968) and Bates et al., (1970) for details]. This enables an equation for the mean activity coefficient to be derived. Since the chemical potential of an ion depends on its molality and activity coefficient it is then possible to predict the behaviour of salts in solutions of high ionic strength where electrostatic interactions have large effects. In dilute solutions these interactions are less marked as the activity coefficient approaches unity:

In the cell the situation is complex from the physicochemical point of view. The negative zeta potential on the surface of most membranes will tend to attract positive ions like Ca^{2+}. Polyvalent electrolytes such as proteins and nucleic acids, which may total 100 mM or more in some cells, will also have dramatic effects on the activity of ions like Ca^{2+}. Let us therefore consider the relatively simple case of the strong electrolytes NaCl and $CaCl_2$. In solution these salts will be completely dissociated:

$$NaCl = Na^+ + Cl^-$$

$$CaCl_2 = Ca^{2+} + 2Cl^-$$

The activity of $NaCl = a_{NaCl} = a_{Na^+} a_{Cl^+}$

$$= \gamma_{Na^+} \gamma_{Cl^-} m_{Na^+} m_{Cl^-}$$

where γ = activity coefficient, m = molality

and the activity of $CaCl_2 = a_{CaCl_2} = a_{Ca}(a_{Cl^-})^2$

$$= \gamma_{Ca^{2+}} (\gamma_{Cl^-})^2 m_{Ca^{2+}} (m_{Cl^-})^2$$

Since it is not possible to measure the activity coefficient of an individual ion it is useful to define a mean activity coefficient ($\gamma_\pm$).

For $A_x B_y = xA^{y+} + yB^{x-}$, $(\gamma_\pm)^{x+y} = (\gamma_+)^x (\gamma_-)^y$

$\therefore$ $(\gamma_\pm)^2 = \gamma_{Na^+} \gamma_{Cl^-}$ for NaCl

and $(\gamma_\pm)^3 = \gamma_{Ca^{2+}} (\gamma_{Cl^-})^2$ for $CaCl_2$ (3.15)

The Debye-Hückel theory combined with the Poisson-Boltzmann equation produces an equation for the mean activity coefficient ($\gamma_\pm$)

$$\log_e \gamma_\pm = -A |z_+ z_-| I^{1/2} \quad (3.16)$$

where A = constant = $\left(\dfrac{0.002 \pi L^2 q_e^6 \rho_0}{\varepsilon^2 R k^2 T^3} \right)^{1/2}$

where L = Avogadro's number = 6.022×10^{23}
  q_e = charge on the electron = 4.80×10^{-10} e.s.u.
  ρ_0 = density of solvent; $\rho_0^{H_2O}$ = 0.9997 at 10 °C; 0.993 at 37 °C
  ε = dielectric constant = 79 at 10 °C; 80 at 20 °C; 74 at 40 °C (Kaye and Layby, 1959)
  R = universal gas constant = 8.31×10^7 erg K^{-1} mole^{-1}
  k = Boltzmann constant = 1.381×10^{-16} erg K^{-1}
  T = temperature (K)
  z_+ = charge on cation
  z_- = charge on anion

For NaCl $z_+ = 1$; $z_- = 1$; $|z_+ z_-| = 1$
For CaCl$_2$ or MgCl$_2$ $z_+ = 2$; $z_- = 1$; $|z_+ z_-| = 2$
For CaHPO$_4$ $z_+ = 2$; $z_- = 2$; $|z_+ z_-| = 4$

$$I = \text{ionic strength} = 0.5 \sum (m_i z_i^2) \tag{3.17}$$

For example, in a solution mimicking the ionic conditions inside some marine invertebrate cells (KCl = 500 mM; NaCl = 20 mM; MgCl$_2$ = 5 mM)

$$I = 0.5(0.5 + 0.5 + 0.02 + 0.02 + 0.02 + 0.01) = 0.535$$

Whereas if KCl = 150 mM, NaCl = 15 mM; MgCl$_2$ = 2 mM, as perhaps would be found inside a mammalian cell:

$$I = 0.5(0.15 + 0.15 + 0.015 + 0.015 + 0.008 + 0.004) = 0.171.$$

Now suppose for KCl $\gamma_K = \gamma_{Cl}$, therefore $\gamma_\pm = \gamma_K = \gamma_{Cl}$ then for CaCl$_2$ equation 3.15 simplifies to

$$(\gamma_\pm^{CaCl_2})^3 = \gamma_{Ca}(\gamma_{Cl})^2 = \gamma_{Ca}(\gamma_\pm^{KCl})^2$$

Therefore
$$\gamma_{Ca} = (\gamma_\pm^{CaCl_2})^3 / (\gamma_\pm^{KCl})^2 \tag{3.18}$$

Combining equation 3.16 and 3.18, values for the activity coefficients for these two cations can be calculated under various conditions (Table 3.7). These figures act as an indication of the real values for activities which should be used in rate equations for Ca^{2+} binding. In practice the thermodynamic constants are usually measured under defined ionic conditions. Hence the activity coefficients will be automatically incorporated into the binding constants. Molalities, or to a first approximation concentrations, can then be used for any further calculations provided that the conditions of ionic strength and temperature remain the same.

In concentrated electrolyte solutions, in addition to the electrostatic interactions incorporated into the Debye-Hückel equation, actual ion pairs will form.

Table 3.7. Calculated intracellular activity coefficients

	Temperature (°C)	I	A	b KCl	b CaCl$_2$	γ K$^+$	γ Ca^{2+}	m K$^+$	m Ca^{2+}	a K$^+$	a Ca^{2+}
Some marine invertebrate cells	15	0.54	1.2	0.23	0.46	0.68	0.21	0.5 M	0.1 µM	0.34 M	21 nM
Mammalian cells	37	0.17	1.1	0.23	0.46	0.75	0.33	0.15 M	0.1 µM	0.11 M	33 nM

Calculated from equations 3.18 and 3.20. For b see Eqn. 3.20
a = activity = γm; where γ = activity coefficient and m = molality. Assume $\gamma_K = \gamma_{Cl} = \gamma_\pm^{KCl}$ and $\gamma_{Ca} = (\gamma_\pm^{CaCl_2})^3 / (\gamma_\pm^{KCl})^2$. See also Butler (1968) and Bates et al. (1970).

These will be particularly favoured in a medium with a low dielectric constant (poor conductor).

Modification of the Debye-Huckel law produces the equation

$$\log_e \gamma_\pm = \frac{-A|z_+ z_-|I^{0.5}}{1 + BdI^{0.5}} \quad (3.19)$$

where the letters have the same significance as in equation 3.16,

$$B = \left(\frac{8\pi L^2 q_e^2 \rho_0}{100\, \varepsilon RT}\right)^{0.5}$$

and d = average effective diameter of the ions.

When $I > 0.1$ even this equation does not predict completely accurately the mean activity coefficient. Equation 3.19 can be modified according to Guntelberg (1926) and Guggenheim (1935) to provide values for $\gamma_\pm$ which agree well with experimental measurements.

$$\log_e \gamma_\pm = \frac{-A|z_+ z_-|I^{0.5}}{1 + I^{0.5}} + bI \quad (3.20)$$

Assuming that $Bd = 1$ ($d = 3.04$ Å).
b is an adjustable parameter which in aqueous solution equals $0.23|z_+ z_-|$. It is this equation which has been used to calculate values for $\gamma_\pm$ shown in Table 3.7. Whether significant changes in γ_{Ca} occur under different conditions in the cell is not known. However, it has been suggested that large changes in ionic strength might regulate precipitation of calcium salts outside the cell (Walser, 1970) through changes in γ_{Ca}.

Attempts have been made (Whitfield, 1975) to extend these concepts to more complex electrolyte solutions such as sea water. This is still a far cry from biological systems where polyvalent electrolytes, environments of different dielectric constant and zeta potentials on membranes all affect the activity coefficient of an ion such as Ca^{2+}. However, using the simple model and equation 3.16 it is possible to demonstrate the qualitative effect of dielectric constant on the mean activity coefficient of $CaCl_2$ (Fig. 3.6). The possibility that primary regulators might act to change parameters such as dielectric constant without a change in the molality of Ca^{2+} has never been considered. It would be interesting to know whether such changes have any significance in regulating the fractional saturation of ligands by Ca^{2+}.

Although, in theory, Ca^{2+} microelectrodes measure directly the 'activity' of Ca^{2+} (see Chapter 2, section 2.5.2) in biological systems, it is not normally possible to take account of the activity coefficient for Ca^{2+}. In the remainder of this book it will be assumed that, to a first approximation, the activity coefficient is incorporated within the measured thermodynamic constants. Making this assumption it is therefore possible to calculate how much Ca^{2+} is bound to naturally occurring ligands.

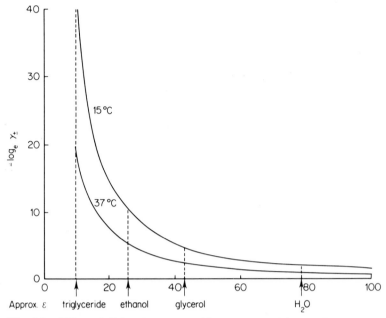

Fig. 3.6. Effect of dielectric constant (ε) on mean activity coefficient ($\gamma_{\pm}$) calculated on the assumption that $\log_e \gamma_{\pm} = -A|z_+ \, z_-|I^{1/2}$ (equation 3.16). These data are only intended to illustrate the order of magnitude of effects of ε on $\gamma_{\pm}$ (see equation 3.20)

3.3.3. How much Ca^{2+} is bound to ligands and how fast does it equilibrate?

Calculations

The physiological significance of Ca^{2+} bound to a ligand depends on the quantity bound and the rate at which this changes when the concentration of free Ca^{2+} changes. In the cell no reactions are ever at true equilibrium. However, in many cases it is possible to estimate the quantity of Ca^{2+} bound to a ligand in the cell by assuming that it is in equilibrium with the free Ca^{2+}. By writing down the equations for Ca–ligand association it is possible to solve the equations

$$\text{total ligand} = \text{free ligand} + \text{bound ligand}$$
$$\text{total } Ca^{2+} = \text{free } Ca^{2+} + \text{bound } Ca^{2+}$$

Providing that the dissociation constant (K_d) for each Ca^{2+} site is known, then the fraction of ligand bound to Ca^{2+} can be calculated. Values for K_d or K_a (the association constant, $= (1/K_d)$ can be found for many ligands in the literature [see, for example, Sillen and Martell (1964) and Kretsinger (1976a, b)]. In the case of enzymes and other proteins it must be remembered that K_d^{Ca} is not the same as K_m^{Ca}. The latter is the concentration of Ca^{2+} necessary for half maximal activation or inhibition. Analogous to the Michaelis constant for the substrate of an enzyme reaction (see Dixon and Webb, 1964) K_m^{Ca} may be very close in value to K_d^{Ca}. However, where two or more Ca^{2+} sites are present with different affinities for Ca^{2+}, and perhaps even exhibiting 'cooperativity' (see below), then K_d and K_m represent two quite different parameters. Various methods have been developed to estimate the number of Ca^{2+} sites and their individual affinities (Table 3.8, Fig. 3.7). Their validity with respect to estimating Ca^{2+} bound to them in the cell depends on carrying out the *in vitro* measurements of K_d at the same ionic strength and temperature as the cell, on making the assumption that the 'activity coefficient' *in vitro* is the same as that in the cell, and on taking into account any interaction of the ligand with other molecules in the cell. The importance of this latter point is well illustrated by troponin C from muscle whose binding sites for Ca^{2+} may increase their affinity for Ca^{2+} by one to two orders of magnitude when the protein is part of its natural complex with troponin I and troponin T.

How much Ca^{2+} can the various ligands (Tables 3.2–3.5, Fig. 3.2) bind in the resting cell or after stimulation? The concentration of free Ca^{2+} necessary to half-saturate any particular Ca^{2+} site is dependent on the K_d^{Ca}. As a rough guide a Ca^{2+} site will be half-saturated when the free Ca^{2+} concentration is equal to K_d^{Ca}, assuming no cooperativity of binding. It is important to note that in the equations which follow (e.g. 3.21–3.23) Ca is the *free* Ca^{2+} concentration. As long as the K_d is known, if the free Ca^{2+} concentration can be measured then the fractional saturation can be estimated. It is not necessary for this calculation to know the absolute amount of Ca^{2+} bound (i.e. released during a cell stimulus). The total amount of Ca^{2+} necessary to activate a given number of binding sites can, however, be estimated if the total ligand concentration is known.

In the resting cell, at a free Ca^{2+} concentration of 0.1 μM, substances with K_d values in the mM range will have a very low fractional saturation with Ca^{2+}. Thus, for example, proteins such as fructose diphosphatase (FDPase) or phosphofructokinase are unlikely to be significantly regulated by direct Ca^{2+} binding, though calmodulin regulation would allow 0.1–5 μM free Ca^{2+} to change the activity of these enzymes. Many of the inorganic (Table 3.2) or small organic (Table 3.3) ligands will have very low saturation by Ca^{2+} since the K_d values are in the mM range and many of them also bind Mg^{2+}. Similar arguments can be made for phospholipids or nucleic acids in the cell. Zwitterionic phospholipids such as phosphatidylcholine, phosphatidylethanolamine and sphingomyelin have little or no Ca^{2+} bound when the free Ca^{2+} is 0.1–1 μM. However, negatively charged phospholipids may bind a significant amount of Ca^{2+} even though the apparent affinity constants are low (Dawson and Hauser, 1970). The affinities for Ca^{2+} follow the sequence phosphatidylserine = phosphatidylinositol < phosphatidic acid < triphosphoinositide. K^+, Na^+, Mg^{2+} and some bases, e.g. local anaesthetics, compete with the Ca^{2+} sites. Biological membranes have potentially a large number of Ca^{2+} sites. With a half saturation at approximately 0.3 mM (Manery, 1969) most of them will be unoccupied on the inner surface (free Ca^{2+} about 0.1 μM) and yet occupied on the outer surface (free Ca^{2+} approx. 1 mM).

Similar arguments hold for nucleic acids (Chang and Carr, (1968). Although the apparent affinity constant for Ca^{2+} binding to DNA is in the mM range, and

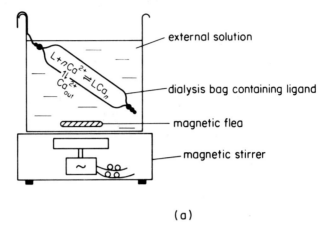

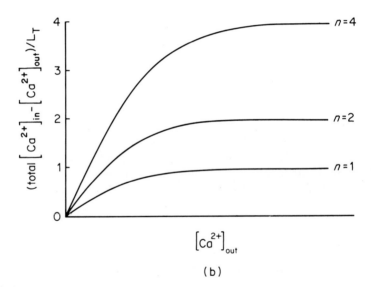

Fig. 3.7. Equilibrium dialysis. (a) Apparatus (thermostat if necessary). Measure total Ca^{2+} inside and outside the dialysis bag by atomic absorption spectrophotometry or radioactive ^{45}Ca per unit volume after equilibrium. When free $[Ca^{2+}]_{in}$ = free $[Ca^{2+}]_{out}$ the ligand is in equilibrium with free Ca^{2+}. Flow systems have been developed for continuous assay (Kretzinger and Nelson, 1977). L = free ligand (eg. protein) impermeable to dialysis tubing; $[L_T]$ = total ligand concentration (known) = $L + LCa_n$; total Ca^{2+} inside dialysis bag (measured) = $nLCa_n$ + free Ca^{2+}; free $[Ca^{2+}]_{in}$ = free $[Ca^{2+}]_{out}$ (fixed or measured). (b) Saturation curves: saturation of ligand by Ca^{2+} enables an estimation of the number (n) of Ca^{2+} sites to be made

Table 3.8. Methods for measuring Ca^{2+} binding to ligands

	Method	Example	Reference
1.	$^{40}Ca^{2+}$ or $^{45}Ca^{2+}$ binding followed by separation equilibrium dialysis	Blood clotting factor XIII	Lewis et al. (1978)
		Troponin C	Potter and Gergely (1975)
		Nucleic acids	Carr and Chang (1971)
	Sephadex G-25	Troponin C	Fuchs (1971)
	Soft β-emission requires no separation	Phospholipid bilayers	Dixon et al. (1949); Kimizuka and Koketsu (1962); Hauser and Dawson (1968)
2.	Measurement of free Ca^{2+} in equilibrium with bound Ca^{2+}		
	Metallochromic indicators (murexide)	ATP	Nanninga (1961a, b)
	Ca^{2+} microelectrodes	EGTA	Owen (1976)
3.	Spectrophotometric changes in ligand induced by Ca^{2+} binding		
	Absorbance	A23187	Pfeiffer et al. (1974)
	Endogenous fluorescence (Trp, Tyr)	A23187	Pfeiffer et al. (1974)
	Fluorescence probe (dansyl) + fluorescence polarization	Troponin C	Potter and Gergely (1975)
	Optical rotatory dispersion and circular dichroism	Transglutaminase A23187	Lewis et al. (1978) Pfeiffer et al. (1974)
4.	Nuclear magnetic resonance		
	Proton	Alamethicin	Urry et al. (1973)
	Proton	Troponin C	Levine et al. (1977, 1978)
	^{43}Ca	Parvalbumin	Parello et al. (1978)
	Paramagnetic probes (e.g. Mn^{2+} and lanthamides) and spin relaxation	Various enzymes, troponin C	Mildvan and Cohn (1970; Martin and Richardson (1979) Cohn, 1963, 1970)
5.	Electron-spin resonance, paramagnetic probes (e.g. Mn^{2+})	Troponin C	Hartshorne and Boucher (1974)
6.	Enzyme activation and inhibition (K_m^{Ca})	Many enzymes	See Kretsinger (1976a, b) for references

cationic proteins compete with Ca^{2+} for the negative phosphate groups, because of the high concentration of nucleic acid in the nucleus one might expect a significant net quantity of Ca^{2+} to be bound to the DNA.

But how can we calculate more precisely the amount of Ca^{2+} bound to a ligand? Let us consider the simplest case of a ligand (L) with one Ca^{2+} binding site.

$$Ca^{2+} + L \underset{}{\overset{K_d}{\rightleftharpoons}} CaL$$

$$K_d^{Ca} = \frac{[Ca^{2+}][L]}{[CaL]}$$

$[L_T]$ = total ligand concentration = $[L] + [CaL]$

Therefore $[CaL]/[L_T]$ = fractional saturation of ligand by Ca^{2+}
$$= [Ca^{2+}]/([Ca^{2+}] + K_d^{Ca}) \quad (3.21)$$

In the case of two independent binding sites

$$Ca^{2+} + L \underset{K_2}{\overset{K_1}{\rightleftharpoons}} CaL + Ca^{2+} \underset{K_1}{\overset{K_2}{\rightleftharpoons}} LCa_2$$
$$LCa + Ca^{2+}$$

$K_1 = ([Ca^{2+}][L])/[CaL] = ([Ca^{2+}][LCa])/[LCa_2]$
$K_2 = ([Ca^{2+}][L])/[LCa] = ([Ca^{2+}][CaL])/[LCa_2]$
$[L_T] = [L] + [CaL] + [LCa] + [LCa_2]$
$\therefore [LCa_2]/[L_T]$ = fractional saturation of active form of ligand (assuming only LCa_2 to be active)
$$= [Ca^{2+}]^2/([Ca^{2+}]^2 + K_1[Ca^{2+}] + K_2[Ca^{2+}] + K_1 K_2) \quad (3.22)$$

If $K_1 = K_2$ then $[LCa_2]/[L_T] = [Ca^{2+}]^2/([Ca^{2+}] + K_d)^2$

Hence for n independent binding sites with equal affinity for Ca^{2+}
$$[LCa_n]/[L_T] = [[Ca^{2+}]/([Ca^{2+}] + K_d)]^n \quad (3.23)$$

As can be seen from Fig. 3.8, this produces curves of different shape for $n = 1, 2$ or 4 when plotting $[LCa_n]/[L_T]$ against pCa ($-\log_{10}[Ca^{2+}]$). This could be of some importance if LCa_n is the only active form since the greater is n the more likely the ligand is to be involved in a 'threshold response' in the cell (see this chapter section 3.5).

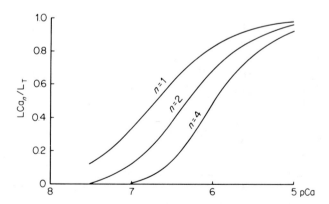

Fig. 3.8. The fractional saturation of a ligand by Ca^{2+}. L = ligand, n = number of Ca^{2+} sites per molecule. Binding is assumed to be non-cooperative. $LCa_n/L_T = [Ca/(Ca + K_d)]^n$, see equation 3.23. Assume $K_d^{Ca} = 0.2\ \mu M$, pCa = $-\log_{10} Ca^{2+}$, L_T = total ligand concentration

Since the concentration of free Ca^{2+} necessary to half-saturate any particular Ca^{2+}-binding site is equal to K_d^{Ca} (Fig. 3.9a), within the normal range of free Ca^{2+} in the cell cytoplasm (i.e. 0.1–5 μM), the proportion of FDPase, or phosphofructokinase bound to Ca^{2+} is negligible.

A complication in estimating Ca^{2+} bound to biological ligands arises from the fact that some of them are not fully ionized at physiological pH, and, even more

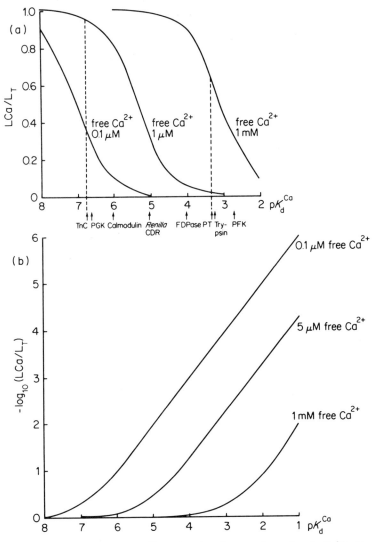

Fig. 3.9. Effect of varying K_d^{Ca} on fractional saturation of each Ca^{2+} site. (a) Linear scale; (b) log scale. $LCa/L_T = Ca/(Ca + K_d)$. The equation deals with each binding site individually. TnC = troponin C, PGK = phosphoglycerate kinase, CDR = calcium dependent regulator, FDPase = fructose diphosphatase, PT = prothrombin, PFK = phosphofructokinase, L_T = total ligand, LCa = Ca bound to ligand; $pK_d^{Ca} = -\log_{10} K_d^{Ca}$

important, many of them bind Mg^{2+}. In order to take into account this binding by Mg^{2+} it is necessary to calculate an apparent K_d^{Ca} (K_{app}^{Ca}) for substitution into equations 3.21–3.23.

$$Ca^{2+} + L \overset{K^{Ca}}{\rightleftharpoons} CaL$$

$$Mg^{2+} + L \overset{K^{Mg}}{\rightleftharpoons} MgL$$

$K^{Ca} = ([Ca^{2+}][L])/[CaL]$; $K^{Mg} = ([Mg^{2+}][L])/[MgL]$
$[L_T] = [L] + [CaL] + [MgL]$
$\therefore [CaL]/[L_T]$ = fractional saturation of ligand by Ca^{2+}
$$= [Ca^{2+}]/[[Ca^{2+}] + K^{Ca}(1 + [Mg^{2+}]/(K^{Mg})] \quad (3.24)$$
Hence $\quad K_{app}^{Ca}/K^{Ca} = 1 + [Mg^{2+}]/K^{Mg}$ (Fig. 3.10) $\quad (3.25)$

Equation (3.24) is sometimes written

$$K^{Ca}[Ca^{2+}]/[1 + ([Ca^{2+}]/K^{Ca}) + ([Mg^{2+}]/K^{Mg})]$$

A similar derivation can be made when pH affects the apparent affinity of the ligand for Ca^{2+}

$$K_{app}^{Ca}/K^{Ca} = 1 + [H^+]/K_H \quad (3.26)$$

From these equations not only can the fractional saturation by Ca^{2+} of a ligand be calculated at any particular concentration of free Ca^{2+}, Mg^{2+} and pH, but also the absolute amount of Ca^{2+} bound can be estimated, provided that $[L_T]$, the total ligand concentration, is known. To a first approximation many intracellular Ca^{2+} ligands are more than 90% ionized in the cell and only this form binds significant Ca^{2+} (e.g. Ca^{2+} binding to $ATP^{4-} \gg ATP H^{3-}$). Assuming a free Mg^{2+} concentration of 2 mM, the K_{app}^{Ca} of various Ca^{2+} ligands can be calculated (Table 3.9). Using this for pK_d in Fig. 3.10b, the fractional saturation by Ca^{2+} of the ligand can be estimated, and thus the quantity of Ca^{2+} bound when free Ca^{2+} is within the physiological range in the cytoplasm, i.e. 0.1–5 μM (Table 3.9). Several conclusions can be made from these calculations. First, in the resting cell (free Ca $\simeq$ 0.1 μM) the amount of Ca^{2+} bound to citrate, ADP and AMP is negligible both compared with free Ca^{2+} and compared with the ligand itself. However, the amount bound to ATP, glutamate or calmodulins may be greater than the free Ca^{2+} concentration. Of course, the glutamate concentration is only as high as 100 mM in certain cells, e.g. some nerves or invertebrate muscles.

Secondly, the most significant portion of rapidly exchangeable bound Ca^{2+} in the cytoplasm is likely to be to high-affinity Ca^{2+}-binding proteins. This probably accounts for the 10 μM rapidly exchangeable Ca^{2+} observed by Baker (1972) in squid axoplasm. This means that the majority of cell Ca^{2+}, which is not free cytoplasmic Ca^{2+} (see Chapter 2), must be in compartments where free Ca^{2+} is > 0.1 mM if it is to bind to low-affinity Ca^{2+} ligands ($pK_d^{Ca} < 4$), or bound to

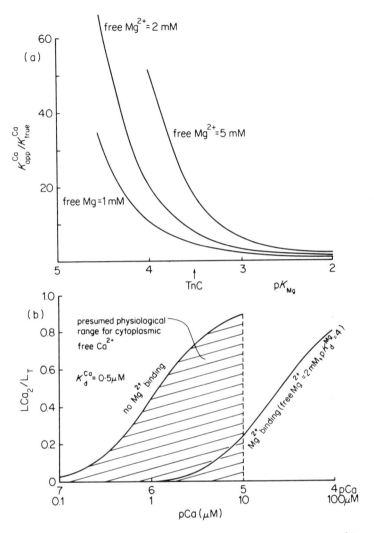

Fig. 3.10. Effect of Mg^{2+} on apparent affinity of ligands for Ca^{2+}. Assume two independent, non-cooperative Ca^{2+} sites which also bind Mg^{2+}. $LCa_2/L_T = [Ca/(Ca + K_{app}^{Ca})]^2$. $K_{app}^{Ca}/K_{true}^{Ca} = (1 + Mg/K_d^{Mg})$ for each binding site (equations 3.24 and 3.25). (a) Effect of Mg^{2+} on K_{app}^{Ca}. TnC = troponin C. (b) Effect of Mg^{2+} on fractional saturation of ligand by Ca^{2+}

high-affinity ligands, or in the form of a precipitate or a kinetically inert form. For a low-affinity Ca^{2+} ligand to bind mM Ca^{2+} in the presence of 0.1–5 μM free Ca^{2+}, the concentration of the ligand would have to be in the molar range (see glutamate, Table 3.9). Thirdly, the fact that at 5 μM free Ca^{2+} the CaATP concentration could be as high as 40 μM raises the possibility of this acting as an inhibitor or activator of intracellular enzymes.

Table 3.9. Estimation of Ca^{2+} bound to some intracellular ligands

Ligand	Approx. intracellular concentration (mM)	pK_d^{Ca}	pK^{Mg}	pK_{app}^{Ca}	$-\log_{10}$ ([LCa]/[L$_T$]) at free Ca^{2+} conc.		Bound Ca^{2+} (μM) at free Ca^{2+} conc.	
					0.1 μM	5 μM	0.1 μM	5 μM
Citrate^{3-}	0.5	3.55	3.41	2.73	4.28	2.75	0.026	0.89
Glutamate$^-$	100	1.43	1.99	1.35	5.66	3.94	0.22	11.5
ATP^{4-}	10	4.10	3.94	2.83	4.16	2.44	0.69	36.3
ADP^{3-}	1	2.80	3.83	1.62	5.36	3.05	4.4 nM	0.89
AMP^{2-}	0.2	1.83	2.09	1.72	5.26	2.54	1.1 nM	0.58
*Calmodulin	20 μM (Ca^{2+} site concentration = 80 μM)	6	?	6	1.0	0.079	8	66.4

Assume free Mg^{2+} = 2 mM

LCa/L$_T$ = [Ca^{2+}]/([Ca^{2+}] + K_{app})
$K_{app}^{Ca}/K_{true}^{Ca}$ = 1 + ([Mg^{2+}]/K^{Mg})
pK values from Sillen and Martell (1964); I = 0.1: teperature = 37°C.

* These figures are only approximate and are intended to indicate the order of magnitude. Calmodulin has four Ca^{2+} sites, no correction has been made for probable cooperativity or Mg^{2+} binding to these sites. K_d^{Ca} for calmodulin is probably nearer 2.4 μM (see Klee et al., 1980 and Means et al., 1982). Mg^{2+} may decrease the apparent affinity for Ca^{2+}

Kinetics

So far we have only considered Ca^{2+} binding under equilibrium conditions. Under some conditions in the cell, particularly where fast transient changes in free Ca^{2+} occur such as in mammalian muscle, this assumption is no longer valid. In order to define the reaction of Ca^{2+} with a ligand under non-equilibrium conditions it is necessary to take into account the 'on' (k_1) and 'off' (k_{-1}) rates of Ca^{2+} binding.

$$L + Ca^{2+} \underset{k_{-1}}{\overset{k_1}{\rightleftharpoons}} CaL$$

$$d[CaL]/dt = k_1[L_t][Ca_t^{2+}] - k_{-1}[CaL_t] \tag{3.27}$$

At equilibrium $d[CaL]/dt = 0$

$$\therefore k_1([L_e][Ca_e^{2+}]) = k_{-1}[CaL_e]$$

k_1 and k_{-1} can be measured using stopped flow or a temperature or pressure jump to cause a small, rapid displacement from equilibrium. Under these conditions a small decrease in [CaL] of Δ[CaL] causes a small increase in [L] and [Ca^{2+}] equal to Δ[CaL].

From equation 3.27

$$-d[\Delta[CaL]]/dt = k_1([L_e] + \Delta[CaL])([Ca_e^{2+}] + \Delta[CaL])$$
$$- k_{-1}([CaL_e] - \Delta[CaL]) \tag{3.28}$$

Since $k_1([L_e][Ca_e^{2+}]) - k_{-1}[CaL_e] = 0$

and $(\Delta[CaL])^2$ is negligible

then $d[\Delta[CaL]]/dt \cong -\Delta[CaL](k_1[L_e] + k_1[Ca_e^{2+}] + k_{-1})$ (3.29)

$1/(k_1[L_e] + k_1[Ca_e^{2+}] + k_{-1})$ is known as the relaxation time (τ) and is the reciprocal of the rate constant for return to equilibrium.

Integration of equation 3.29 produces

$$\frac{\Delta[CaL_t]}{\Delta[CaL_0]} = e^{(-t/\tau)} \tag{3.30}$$

where $t = \tau$ the displacement ($\Delta[CaL_t]$) will have decreased to $1/e$ of its original value ($\Delta[CaL_0]$).

A rapid kinetic analysis has been carried out on very few Ca^{2+} ligands. In fact, the possibility that the rate of dissociation of Ca^{2+} from small ligands might be rate-limiting has often been ignored. However, attempts have been made to study the 'on' and 'off' rates of Ca^{2+} binding to troponin C (Levine et al., 1977, 1978) using proton magnetic resonance. Unfortunately, the assumptions on which equations 3.29 and 3.30 are based are not valid in the cell since Δ[CaL] is large compared with the free Ca^{2+} concentration and the free Ca^{2+} concentration never remains constant. Furthermore, the reaction on which equation 3.27 is based is an oversimplification since Ca^{2+} binding to many biological ligands, such as proteins, involves a rapid binding step followed by a slower conformational change in the ligand.

$$L + Ca^{2+} \underset{k_{-1}}{\overset{k_1}{\rightleftharpoons}} CaL \underset{k_{-2}}{\overset{k_2}{\rightleftharpoons}} CaL^1$$

where CaL to CaL^1 is the conformational change.

An important consequence of considering individual rate constants is that it highlights another chemical difference between Ca^{2+} and Mg^{2+}. In both the binding stage and conformational change the rate of change of solvation of the cation can influence the overall rate constant. The rate of exchange of water with Mg^{2+} is relatively slow compared with Ca^{2+}, Sr^{2+} or Ba^{2+}. Thus, k_1 for Ca^{2+} is mainly diffusion limited, whereas k_1 for Mg^{2+} may be mainly limited by rate of loss of H_2O. In this case k_1^{Mg} can be orders of magnitude less than k_1^{Ca}. It could therefore be important to examine whether in the cell any selectivity of Ca^{2+} over Mg^{2+} could be achieved by a ligand responding to rapid Ca^{2+} transients by this means. A kinetic selectivity might not be obvious from the equilibrium constants of Ca^{2+} and Mg^{2+}.

3.3.4. Solubility of calcium salts

Many calcium salts are only sparingly soluble in water (Table 3.10). Precipitates of calcium phosphate, carbonate and occasionally sulphate form the basis of the skeletal structures of most organisms. Occasionally microprecipitates of calcium salts occur inside cells. This can easily be induced in mitochondria *in vitro* but also is found *in situ* under certain physiological and pathological conditions (Lehninger, 1970; Carafoli, 1974). Since both phosphate and protein concentrations are high in the cell, what prevents large amounts of calcium precipitates occurring within the cell?

The solubility of a pure salt in water can be described by the number of moles which can be dissolved in a defined volume or weight of solvent. However, in a solution containing a mixture of salts this is inadequate to define the conditions for precipitation of a salt where the concentration of total anion may greatly exceed that of the individual cation (Fig. 3.11), in this case calcium. A more useful

Table 3.10. Solubility products of some calcium and magnesium salts*

Salt	$-\log_{10}K_{So}^{Ca}$	$-\log_{10}K_{So}^{Mg}$
$CaCO_3 \rightleftharpoons Ca^{2+} + CO_3^{2-}$	7.3	6.5
$CaHPO_4 \rightleftharpoons Ca^{2+} + HPO_4^{2-}$	6.58	4.5
$Ca_3(PO_4)_2 \rightleftharpoons 3Ca^{2+} + 2(PO_4)^{2-}$	30–35	252
$CaSO_4 \rightleftharpoons Ca^{2+} + SO_4^{2-}$	4.7	—
Calcium oxalate $\rightleftharpoons Ca^{2+}$ + oxalate^{2-}	7.8	—

K_{So} = Solubility product
Data from Sillen and Martell (1964) and Walser (1970) For historical references see Hastings *et al.* (1926); Greenwald (1938, 1941), and Greenwald *et al.* (1940).

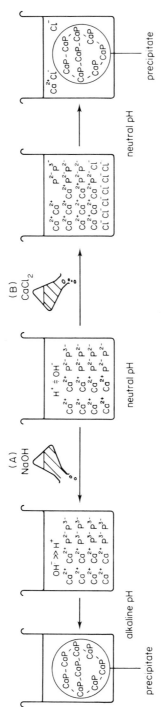

Fig. 3.11. The precipitation of calcium phosphate. $P^{2-} = HPO_4^{2-}$; $P^{3-} = PO_4^{3-}$. This illustrates two ways of precipitating calcium phosphate: (A) by making the solution alkaline thereby increasing PO_4^{3-}, (B) by increasing Ca^{2+}

parameter is the solubility product (K_{So}).

$$Ca^{2+} + X^{2-} \rightleftharpoons CaX \text{ (solution)} \rightleftharpoons CaX \text{ (solid)}$$

$$K_{So} = a_{Ca} a_X$$

$$= \gamma_{Ca} \gamma_X m_{Ca} m_X \quad (3.31)$$

Strictly, the equation for K_{So} is only valid for Ca^{2+} and X in equilibrium with an infinite amount of precipitate. In biological situations, for example bone formation, the rate of precipitation can be determined by kinetic rather than equilibrium considerations (Posner, 1969). In the case of bone the protein substrate is crucial in providing a nucleation site for precipitation of hydroxyapatite. Furthermore, the apparent solubility product for calcium phosphate in equilibrium with microprecipitates in the cell may be at least an order of magnitude greater than the solubility product as normally measured (Williams, 1976). Nevertheless, in spite of these kinetic limitations, relating $a_{Ca}a_X$ to K_{So} provides an indication of the potential for calcium salt precipitation. In particular, precipitation will not occur until the solubility product is exceeded. Let us therefore calculate the minimum concentration of Ca^{2+} necessary to exceed the solubility product for various salts in human plasma and in the cytoplasm of cells (Table 3.11). In order to carry out these calculations a certain number of assumptions have to be made. (1) The activity coefficients for the ions are incorporated in the K_{So}, which is valid since most K_{So} are at $I = 0.1$. (2) Extracellular pH = intracellular pH = 7.4 [may be nearer 7.2–7.3; see Roos and Boron (1981)]. (3) All acid–base reactions are at equilibrium; hence the concentration of cations such as CO_3^{2-} can be calculated from HCO_3^- and H^+ concentrations and pK_a. (4) The intracellular concentration of free anion is calculated on the assumption that it is at electrochemical equilibrium (see

Table 3.11. Potential for precipitation of calcium salts

Anion	Solubility product* (K_{So})	Extracellular		Intracellular	
		Conc. anion in plasma (M)	Conc. Ca^{2+} necessary to exceed K_{So}(M)	Estimated conc. anion in cell (M)	Conc. Ca^{2+} necessary to exceed K_{So}(M)
CO_3^{2-}	$Ca^{2+} \cdot CO_3^{2-} = 5 \times 10^{-8}$	2.5×10^{-7}	2×10^{-1}	6×10^{-9}	8.3
HCO_3^-	—	2.5×10^{-2}	—	3.9×10^{-3}	—
$H_2PO_4^-$	—	1.1×10^{-4}	—	2.6×10^{-5}	—
HPO_4^{2-}	$Ca^{2+} \cdot HPO_4^{2-} = 2.6 \times 10^{-7}$	5×10^{-4}	5.2×10^{-4}	1.2×10^{-5}	2.2×10^{-2}
PO_4^{3-}	$(Ca^{2+})^3 \cdot (PO_4^{3-})^2 = 10^{-30}$	5×10^{-9}	3.4×10^{-5}	1.2×10^{-10}	4.1×10^{-4}
SO_4^{2-}	$Ca^{2+} \cdot SO_4^{2-} = 2 \times 10^{-5}$	6.5×10^{-4}	3.1	1.6×10^{-5}	1.3
$Oxalate^{2-}$	$Ca^{2+} \cdot oxalate = 1.6 \times 10^{-8}$	2.2×10^{-5}	7.2×10^{-4}	5.3×10^{-7}	3.0

Chapter 4) across the membrane of a cell with a membrane potential of $-50\,\text{mV}$ (negative inside).

$$\therefore E_m = -\frac{RT}{F}\log_e([\text{HCO}_3^-]_o/[\text{HCO}_3^-]_i)$$

$$= -\frac{RT}{F}\log_e([\text{HPO}_4^{2-}]_o/[\text{HPO}_4^{2-}]_i) \quad \left(\frac{RT}{F}\log_{10}e = 61.6\right.$$

$$= -\frac{RT}{2F}\log_e([\text{oxalate}^{2-}]_o/[\text{oxalate}^{2-}]_i) \quad \left.\text{at } 37°C \text{ when } E \text{ is in mvolts}\right)$$

$$= -\frac{RT}{2F}\log_e([\text{SO}_4^{2-}]_o/[\text{SO}_4^{2-}]_i) \tag{3.32}$$

This assumption means that CO_3^{2-}, H_2PO_4^- and PO_4^{3-} will not be at electrochemical equilibrium across the cell membrane.

Of course these assumptions may not be wholly valid. However, the estimates (Table 3.11) based on them highlight some interesting points. Since plasma free Ca^{2+} is about 1 mM and intracellular free Ca^{2+} is about 0.1 μM, it can be seen that the solubility products for CaHPO_4 and $\text{Ca}_3(\text{PO}_4)_2$ in plasma seem to be exceeded, whilst that for calcium oxalate is very close to the concentration product in normal plasma. It is on this basis that it is argued that plasma contains a supersaturated solution of these salts and that it is nucleation which controls precipitation. Precipitation of calcium phosphate in bone and mitochondria probably first involves formation of amorphous $\text{Ca}_3(\text{PO}_4)_2$, the most insoluble calcium phosphate salt, followed by crystallization to form hydroxyapatite: $[\text{Ca}_3(\text{PO}_4)_2]_3 \cdot \text{Ca}(\text{OH})_2$. In contrast, in the cytoplasm of cells it is unlikely that, under physiological conditions, the free Ca^{2+} concentration is sufficient to exceed the K_S for the salts illustrated in Table 3.11. After cell injury or inside organelles like mitochondria the K_S for $\text{Ca}_3(\text{PO}_4)_2$ may be exceeded under some conditions.

3.3.5. Coordination chemistry of calcium

'What is so special about calcium?' is a question which biologists have posed for more than a century. We can investigate the effects of Ca^{2+} on living systems, we can try to measure Ca^{2+} inside cells, we can extract Ca^{2+} ligands from cells and measure their thermodynamic properties. All of these investigations can provide evidence for a regulatory role for intracellular Ca^{2+}. However, a real understanding of the 'special' nature of Ca^{2+} as a cell regulator lies in unravelling the coordination chemistry of calcium and identifying how evolutionary forces have driven cells to exploit this chemistry. Five aspects can be listed as being of particular significance: (1) coordination number; (2) ligand atoms coordinating with Ca^{2+}; (3) the number of water molecules contributing –O ligands in the complex with Ca^{2+}; (4) bond distances and angles between Ca^{2+} and ligand atoms; (5) stereochemistry of the Ca^{2+}-binding site and identification of any changes in ligand structure induced by Ca^{2+} binding.

Thanks to the development of high-resolution X-ray crystallography much has been discovered about the three-dimensional structure of proteins and other macromolecules. Some Ca^{2+}-binding proteins have been studied using this technique, notably troponin C and parvalbumin. The binding of Ca^{2+} to small organic ligands such as nucleotides, sugars and ionophores has also been examined. Although X-ray diffraction techniques require crystals of the substance for structural analysis, there is no evidence to suggest that the energetics of crystallization are such as to cause any significant changes in conformation compared with that in solution.

Ca^{2+} binds to many naturally occurring amino acids, sugars, nucleotides and other small organic molecules (Table 3.3). In most of these the coordination number for Ca^{2+} is 8, although in some it is 7 (Table 3.12). This is in contrast to Mg^{2+} which is much more restricted and has a coordination number of only 6. Interestingly, the coordination number for Ca^{2+} is not related to the nature of the oxygen atom or to its electronegativity, this latter parameter being a measure of the tendency of an atom to form an ionic bond. Although several Ca^{2+} ligands contain nitrogen atoms, only in one, glutamate, has the nitrogen atom been shown to be involved in Ca^{2+} ligation. Oxygen is the preferred ligand atom for Ca^{2+}. Introduction of nitrogen into a ligand usually reduces its selectivity for Ca^{2+} over Mg^{2+}. At least two of the oxygen atoms are from water molecules, whilst CO_2^- groups can contribute one or both oxygens in coordinating Ca^{2+}. Ca^{2+} is also much better suited than Mg^{2+} in forming cross-links between two molecules. This could be significant in Ca^{2+} binding between phospholipids or nucleic acids in the cell.

In the primary coordination sphere Ca–O distances vary between 0.229 and 0.265 nm, whereas for Mg–O the bond distances vary over a much narrower range, 0.200–0.216 nm. The stereochemistry of 6 coordination is that of a regular octahedron (see Fig. 3.1). Seven coordination, on the other hand, is usually in the form of a pentagonal bipyramid, whereas 8 coordination is best described as a

Table 3.12. Some aspects of Ca^{2+} coordination to biological ligands

Ca^{2+} complex	Coordination number	Mean Ca–O bond distance (nm)
$CaHPO_4·2H_2O$	8	0.264
Calcium dipicolinate·$3H_2O$	8	0.247
Calcium thymidylate·$2H_2O$	7	0.242
Calcium glutamate·$2H_2O$	8	0.248
Troponin C	8	—
Parvalbumin	7	—

See Williams (1974) and Kretzinger (1976a, b) for references.

distorted square antiprism [see Kretsinger and Nelson (1977) for further details and diagrams]. Ca^{2+} binding to small molecules can cause small changes in the structure of the ligand itself at sites not directly involved in Ca^{2+} binding.

The three-dimensional structures of carp paravalbumin and several low Ca^{2+}-affinity proteins were among the first Ca^{2+}-binding proteins to be described. The coordination number for Ca^{2+} appeared at first sight to be only 6. However, one or two carboxyl groups from glutamate or aspartate coordinate with both oxygen atoms. The true coordination number for Ca^{2+} is therefore 7 or 8 (Table 3.12). Water molecules are not always present, varying between none and three. No ligand atoms besides oxygen have been observed. There is no correlation between pK_d^{Ca} and the number of CO_2^- groups. The acidity of the high-affinity Ca^{2+}-binding proteins reflects the Glu + Asp content rather than signifying strong Ca^{2+} binding. In all protein–Ca^{2+} complexes the carbonyl ($>C=O$) groups in the peptide chain as well as CO_2^- groups from Asp and Glu are involved in Ca^{2+} binding.

On the basis of the three-dimensional structure of parvalbumin a structure for troponin C was proposed (Kretsinger and Nelson, 1977). The two important Ca^{2+}-binding sites both have octahedral symmetry and are sited in two regions signified by Kretsinger as the C–D loop and E–F band (see Fig. 3.12). The letters

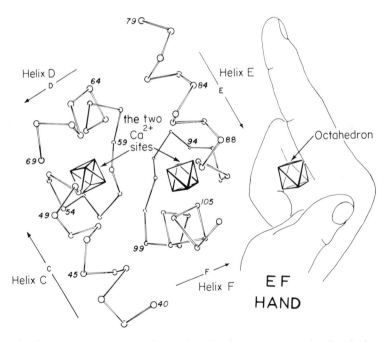

Fig. 3.12. The EF hand of parvalbumin from carp muscle. Octahedra = Ca^{2+} binding sites. Numbers refer to the amino acid sequence (see Enfield et al. (1975) and Table 3.17 for amino acid sequence). From Kretsinger (1976 a,b). Reproduced with permission from the *Annual Review of Biochemistry*. 1976 by Annual Reviews Inc.

signify the six helical (A–F) regions of the protein. It would be interesting to know what happens to the three-dimensional structure when troponin C binds with the other two components in the complete troponin complex since this is known to increase the affinity of troponin C for Ca^{2+}.

3.4. Is Ca^{2+} binding active?

3.4.1. Criteria

In Chapter 2 (section 2.3.3) it was argued that Ca^{2+} binding to biological ligands could be classified as either 'active' or 'passive'. The distinction between these two categories depended on whether changes in Ca^{2+} bound to the ligand occurred during cell activation and were involved in mediating the response of the cell to the primary stimulus. If this is so then Ca^{2+} binding would be considered to be 'active'. 'Passively' bound Ca^{2+} was defined as Ca^{2+} attached to ligands which did not change significantly under conditions when reactions inside or outside the cell were regulated under physiological circumstances. This passive Ca^{2+} need not, however, be non-functional. As has already been pointed out, Ca^{2+} is required for the activity of several extracellular enzymes and for the stability of many biological membranes.

If Ca^{2+} is needed to regulate intracellular proteins, or to induce phenomena such as membrane fusion, then the amount of Ca^{2+} bound to ligands involved in controlling the physiological response must change when the cell is activated. This could, in theory, occur through one of three mechanisms. (1) A change in the number of Ca^{2+}-binding sites as a result of synthesis or degradation of the ligand or the exposure of new sites for Ca^{2+} binding. An example of the latter would be an alteration in protein conformation induced by another intracellular regulator or by protein phosphorylation (Nimmo and Cohen, 1977; Krebs and Beavo 1979). (2) A change in the affinity of Ca^{2+}-binding sites induced by another intracellular regulator. (3) A change in free Ca^{2+}.

There are now many enzymes known which undergo phosphorylation–dephosphorylation cycles, often under the regulation of cyclic nucleotide dependent kinases and Ca^{2+}-activated phosphatases. However, there is no good evidence that this influences directly the affinity or the number of Ca^{2+}-binding sites. The commonest mechanism for increasing the amount of bound Ca^{2+} is through a change in free Ca^{2+} concentration.

The fractional saturation by Ca^{2+} of intracellular ligands such as ATP and citrate is small. However, a ten-fold change in free Ca^{2+} will cause a similar change in Ca^{2+}–ligand concentration. Such a change could be significant if enzymes exist which can be inhibited or activated by μM concentrations of CaATP. Similarly, increases in Ca^{2+} bound to other low-affinity ligands such as phospholipids and nucleic acids will occur when there is an increase in cytoplasmic free Ca^{2+}. At present, there is no evidence to suggest that such changes in Ca^{2+} binding are anything but 'passive'. However, the possibility that there might be an 'active' role for Ca^{2+} bound to these low-affinity sites should not be ignored. They could, for example, affect the rate of protein or nucleic acid breakdown if the Ca^{2+}-bound molecule is degraded at a different rate from the free one. Whilst this might be insignificant for acute changes in cells, occurring over seconds or a few minutes, it might be important in the longer-term regulation of cell structure and function.

A particularly interesting example of a large change in Ca^{2+} bound to a small organic ligand occurs during spore formation in certain bacteria (Ellar et al., 1974; Kornberg et al., 1975). These bacteria include *Bacillus*, *Sporosarcina* and *Clostridium* and are found in the soil. The formation of spores confers great adaptability and survival value, particularly when there are large variations in physical conditions such as temperature and drought. The spores are stable and have a very high Ca^{2+} content, 20 or more times that of the normally growing bacteria (see Table 2.8). The Ca^{2+} seems to be mainly in the form of calcium dipicolinate (see Fig. 3.2) and is released when the spores germinate. Spore formation also occurs in some eukaryotes (e.g. yeasts) but does not seem to involve such an uptake of Ca^{2+} or dipicolinate.

In spite of this unusual example in bacteria the majority of cases of active Ca^{2+} binding in the cell involve proteins. The evidence for this relies almost exclusively on studies of the effect of Ca^{2+} on isolated proteins extracted from cells. Over the last 40 years or so a considerable number of intracellular enzymes have been shown to be modified by Ca^{2+}. This is perhaps not so surprising in view of the number of enzymes which bind Mg^{2+} or MgATP, and the relatively high glutamate and aspartate content of most proteins (Table 3.6). Unfortunately it was not generally realized at the time how low was the free Ca^{2+} concentration in the cytoplasm of most cells. Most of the effects of Ca^{2+} on isolated enzymes only occur at concentrations of Ca^{2+} greater than 0.1 mM (pK_m^{Ca} = approx. 1 mM) (Table 3.13). Such a high Ca^{2+} concentration has never been observed in the cytoplasm of a normal cell.

The binding of Ca^{2+} to several extracellular enzymes can be considered as passive (Table 3.13). However, examples of passive intracellular Ca^{2+} binding, except in the case of the low-affinity ligands already discussed, are more difficult to find. No examples have been found of Ca^{2+} sites with affinities such that they would be saturated with Ca^{2+} at 0.1 μM free Ca^{2+} ($pK_d^{Ca} < 0.1\ \mu$M). This illustrates an important difference from the true metalloenzymes (Lehninger, 1950) which contain transition metal cations at their active centre. In this case the affinity of the ligand for the metal ion is either very high ($pK_d < 0.1\ \mu$M) or kinetically the complex is very stable (i.e. k_{-1} is very slow). A few mitochondrial enzymes with high affinity for Ca^{2+} have been discovered (Table 3.13). Under experimental conditions these can sometimes be regulated in a Ca^{2+}-dependent manner (Denton et al., 1976, 1978). However, the free Ca^{2+} concentration in the mitochondrion is unknown, hence the query as to the physiological significance of this Ca^{2+} binding (Tables 3.13 and 3.14).

Table 3.13. Examples of possible 'passive' Ca^{2+} binding to biological ligands

Example	Activation (a) or inhibition (i)	Reference
1. Extracellular		
α-Amylase (mammalian saliva, plants, bacteria)	a	Stein and Fischer (1960)
Phospholipase A_2	a	Derksen and Cohen (1975)
Prothrombin	a	Suttie and Jackson (1972)
Transglutaminase and other enzymes in blood clotting	a	Lewis et al. (1978)
Formation of Cl complex (complement)	—	Fearon and Austen (1977)
Trypsin(ogen)	a	Abbot et al. (1975)
Collagenase (mammalian and bacterial)	a	Takagi et al. (1977)
Casein in milk	—	Holt et al. (1975)
Albumin	—	McLean et al. (1935); Woodhead (1982)
Haemocyanin (*Helix*) ? passive	—	Burton and London (1972)
Some cell surface receptors	—	Kaplan (1981)
2. Intracellular—'passive' or 'active'?		
Glycerol-3-phosphate dehydrogenase (insect mitochondrial, pK_m^{Ca} = approx. 7)	a	Donnelan and Beechey (1969)
Phosphatidylinositol hydrolase (pK_m^{Ca} = 6.8)	a	Allan and Michell (1974)
Pyruvate dehydrogenase kinase (mitochondrial) pK_m^{Ca} = approx. 6	i	Denton et al. (1980)
NAD-linked isocitrate dehydrogenase (mitochondrial, pK_m^{Ca} = approx. 6)	a	Denton et al. (1978)
Pyruvate dehydrogenase, phosphatase (mitochondrial, pK_m^{Ca} = 5.4)	a	Severson et al. (1974)
Isocitrate dehydrogenase ? (mitochondrial, pK_m^{Ca} = 5)	a	Vaughan and Newsholme (1969)
Phospholipids (e.g. phosphatidylserine)	—	Dawson and Hauser (1970)
Nucleic acids	—	Chang and Carr (1968)

Several intracellular proteins have been discovered which can be activated by Ca^{2+} in the range 0.1–10 μM (Table 3.15). Most of the effects of Ca^{2+} on these proteins are thought to be physiologically relevant and Ca^{2+} binding is therefore considered to be active. The criteria for active Ca^{2+} binding to a protein can be summarised as follows (see also Chapter 2): (1) $pK_d^{Ca} \simeq 0.1$–1 μM under physiological conditions of Mg^{2+} and pH; (2) the rate of Ca^{2+} binding and dissociation is fast enough (or in some cases slow enough) to be compatible with the change in free Ca^{2+} observed in the cell and the time course of the cellular response; (3) Ca^{2+} binding must induce a change in protein conformation which is sufficient to explain the physiological effect of Ca^{2+} observed *in vivo*; (4) the protein must be found in the necessary part of the cell.

Table 3.14. Probable non-physiological effects of Ca^{2+} on intracellular enzymes. Ca^{2+} effect demonstrable on isolated protein but insignificant *in vivo*, Ca^{2+} required in mM range to be effective

Enzyme	Effect* (a or i)	Reference
Hexokinase (muscle)	i	Margreth et al. (1967)
Phosphofructokinase (muscle)	i	Vaughan et al. (1973)
Fructose diphosphatase (muscle)	i	Vaughan et al. (1973)
Pyruvate kinase (muscle)	i	Kachmar and Boyer (1953)
Pyruvate kinase (liver)	i	Gevers and Krebs (1966)
Pyruvate kinase (ascites tumour cells)	i	Bygrave (1966a, b)
Fructokinase (liver)	i	Parks et al. (1957)
Glycolysis in nucleated red cells	i	Ashwell and Dische (1950)
Phosphoenol pyruvate + glucose → glucose 6-phosphate (*E. coli*)	i	Utter and Werkman (1942)
Pyruvate kinase (*E. coli*)	i	Utter and Werkman (1942)
Isocitrate dehydrogenase (liver)	i	DeLuca et al. (1957)
Enolase (yeast)	i	Hanlon and Westhead (1965)
Enolase (red cells)	i	Boszormenyi-Nagy (1955)
Trehalase (muscle)	i	Vaughan et al. (1973)
Microsomal phosphofructokinase (muscle)	i	Margreth et al. (1967)
Pyruvate carboxylase (liver mitochondria)	i	Kimmich and Rasmussen (1969)
Citrate desmolase	i	Dagley and Dawes (1955)
Pyrophosphatase (yeast)	i	Bailey and Webb (1944; Moe and Butler (1972)
Phosphatase (*E. coli*)	i	Pett and Wynne (1933)
Protein kinase	i	Weller and Rodnight (1974)
Adenylate cyclase (liver)	i	Birnbaumer and Rodbell (1969)
$(Na^+ + K^+)$-MgATPase	i	Dunham and Glynn (1961)
Proteolysis of phosphorylase kinase	a	Huston and Krebs (1968); Meyer et al. (1964)

*a = activation; i = inhibition.

3.4.2. Regulatory Ca^{2+}-binding proteins

The discovery in 1965 by Ebashi and Kodama [see Ebashi (1974) and Perry (1974) for reviews] of a Ca^{2+}-binding protein, troponin, which regulates muscle contraction marked what was probably the most important step forward in understanding the molecular mechanism of intracellular Ca^{2+} as a regulator. In fact troponin is composed of three subunits (see Chapter 5) and it is troponin C which binds Ca^{2+}. By the end of the 1970's it became clear that similar proteins exist in most, if not all, eukaryotic cells (Table 3.16). One group apparently present in all eukaryotic cells, and having a highly conserved amino acid sequence, has been named 'calmodulins' (Fig. 3.13). Thus, Ca^{2+}-binding proteins which activate various enzymes and cross-react with troponin C have been discovered in many cells from vertebrates (Dedman et al., 1977; Means et al., 1982) and invertebrates,

Table 3.15. Examples of probable 'active' intracellular Ca^{2+} binding

Example	Activation (a) or inhibition (i)	Reference
Troponin C and calmodulins	a	See Table 3.16
Phosphorylase b kinase (muscle)	a	Ozawa et al. (1967)
Calsequestrin, Ca^{2+} binder inside sarcoplasmic reticulum	—	MacLennan and Wong (1971)
Spasmoneme in *Vorticella*	a	Amos et al. (1976)
Phosphatidylinositol (Ca^{2+} released on hydrolysis)	—	Jones and Michell (1976)
Ca^{2+}-activated MgATPases		
—cell membrane	a	Schatzmann (1966, 1975)
—sarcoplasmic reticulum	a	Ebashi (1960, 1961)
Coelenterate photoproteins	a	Campbell et al. (1979)
Na^+–Ca^{2+} exchanger (cell membrane)	a	Blaustein (1974)
Tryptophan hydrolase	a	Hamon et al. (1977, 1978)
Endonuclease (nucleus)	a	Ishida et al. (1974)
Parvalbumin	—	Benzonana et al. (1972, 1974)
Adenylate cyclase (nucleated erythrocytes only)	i	Campbell and Dormer (1978)

as well as in plant cells, fungi and protozoa. It will be interesting from an evolutionary point of view to find out whether similar proteins exist in prokaryotes. Calmodulin has not been found in bacteria.

The hypothesis has been proposed that these Ca^{2+}-dependent regulator proteins, or calmodulins as most of them are known, mediate the intracellular effects of Ca^{2+} in cell activation. Apart from muscle contraction they have been shown to activate, in the presence of μM Ca^{2+} concentrations, several enzymes as well as membrane phosphorylation, and to affect intracellular structures such as the mitotic apparatus (Table 3.16). Whilst there is no *a priori* reason why Ca^{2+} should not regulate several enzymes directly (Tables 3.4 and 3.13), it is significant that, in the light of the calmodulin hypothesis, re-examination of phosphorylase b kinase, a previously well-studied enzyme (Cohen, 1973, 1974), has resulted in the discovery of another subunit which has the properties of a calmodulin (Cohen et al., 1978). Similar subunits may exist in other Ca^{2+}-activated enzymes in the cell (Table 3.16).

These regulatory Ca^{2+}-binding proteins have several properties in common (Cheung, 1980; Klee et al., 1980): (1) Molecular weight 10 000–20 000 (mol. wt of calmodulin = 16 700 based on amino acid sequence). (2) Composed of about 140–160 amino acids which have a high ratio of Asp + Glu to Lys + Arg (see Table 3.6, calmodulin 148 amino acids). (3) Highly acidic, $pI \simeq 4$. (pI calmodulin = 3.4). (4) Heat-stable. ($t_{1/2}$ calmodulin = 7 min at 100°C). (5) 2–4 Ca^{2+}-binding sites with $K_d^{Ca} \simeq 1\,\mu M$ (calmodulin $K^{Ca} = 2.4\,\mu M$). Some of these sites are highly selective for Ca^{2+} over Mg^{2+}, whilst others probably bind some Mg^{2+} at the

Fig. 3.13. Calmodulin, showing the four Ca^{2+}-binding sites. Modified from Klee et al. (1980). Reproduced, with permission, from the *Annual Review of Biochemistry,* 1980 by Annual Reviews Inc.

Table 3.16. Regulatory Ca^{2+}-binding proteins

Protein	Tissue	Reference
Vertebrates		
Troponin C	Various striated muscles	Ebashi (1974, 1980)
Leitonin	Smooth muscle	Mikawa et al. (1978)
Ca^{2+}-dependent protein kinase (calmodulin)	Smooth muscle	Yagi et al. (1978)
δ Subunit of phosphorylase b kinase	Muscle	Cohen et al. (1978)
Ca^{2+}-dependent activator of cyclic AMP phosphodiesterase (and adenylate cyclase) (calmodulin)	Brain and some other tissues	Watterson et al. (1976); Brostrom et al. (1976, 1978)
Activator of tryptophan mono-oxygenase	Brain	Yamauchi and Fujisawa (1979)
Activator of phospholipase A_2	Platelets	Wong and Cheung (1979)
Modulator of Ca^{2+}-activated Mg + ATPase	Erythrocytes	Jarrett and Penniston (1977a, b)
Inhibitor of microtubule assembly		Marcum et al. (1978)
Calmodulin	Sperm	Jones et al. (1980)
Calmodulin	Islets of Langerhans	Sugden et al. (1979)
Calmodulin	Adrenal medulla	Kuo and Coffee (1976a, b)
Gelsolin	Cytoskeleton of many cells	Weeds (1982)
Invertebrates		
Luciferin-binding protein	*Renilla* (sea pansy)	Cormier (1978)
Calmodulin	*Renilla*	Jones et al. (1974)
Calmodulin	*Lumbricus* (annelid earthworm)	Waisman et al. (1978)
Calmodulin	*Electroplax* (electric eel)	
Plants		
Activator of NAD^+ kinase	Peanut plant	Anderson and Cormier (1978)
Calmodulin	Spinach	Watterson et al. (1980)
Protozoa		
Activator of guanylate cyclase	*Tetrahymena*	Nagao et al. (1979)

See also Means et al. (1982)

concentrations existing within the cell (i.e. 0.1–5 μM free Ca^{2+}, 1–5 mM free Mg^{2+}). (6) Concentration in the cell 1–20 μM (troponin C $\cong$ 100 μM). (7) Show 50% amino-acid sequence homologies with each other (Table 3.17). (8) They cross-react with each other, e.g. some calmodulins can replace troponin C in the Ca^{2+}-dependent activation of myosin ATPase from muscle, and vice versa. However, the efficiency of enzyme activation is usually greatest for the naturally occurring calmodulin from the particular tissue concerned. In a few cases, for example, troponin C cannot mimic the effect of the natural calmodulin. (9) They contain Ca^{2+}-binding sites with very similar stereochemistry. This has been designated as the E−F hand formation by Kretsinger (1976a,b) (see this chapter section 3.3.5 and Fig. 3.12). Calmodulin contains 1 mole of trimethyllysine per mole.

Although the hypothesis that calmodulins mediate the effects most of Ca^{2+}-dependent cell stimuli is an attractive one, the evidence is still sparse and is based almost entirely on the ubiquitous existence of these Ca^{2+}-binding proteins and effects of trifluoperazine in cells. The fractional saturation by Ca^{2+} of three hypothetical Ca^{2+}-binding proteins is illustrated in Fig. 3.8. It is important to realize that, unlike most conditions for enzyme–substrate reactions *in vitro*, the concentration of calmodulin or troponin C in the cell is some 10–100 times that of free Ca^{2+}, and may be even greater if the protein is concentrated at a particular locus in the cell. This means that the absolute quantity of Ca^{2+} bound is considerably greater than the concentration of free Ca^{2+}. If these calmodulins are to be involved in cell activation then not only must it be shown that the affinity of the Ca^{2+} sites is not so high that they are already virtually saturated at the resting free Ca^{2+} concentration (about 0.1 μM), but also they must interact with the enzyme concerned with the response of the cell in question. This latter aspect has been demonstrated in a few cases. However, in many instances the only evidence at present is the existence of a troponin C like protein extractable from the tissue. The active role of a calmodulin can be established by satisfying the following criteria: (1) Extraction of a troponin C like protein from the tissue with the properties outlined above and estimation of intracellular concentration. (2) Specificity and affinity constant for Ca^{2+} binding with the necessary K_d to change its fractional saturation by Ca^{2+} significantly over the physiological range for intracellular free Ca^{2+} (i.e. 0.1–5 μM). (3) Kinetics of Ca^{2+} association and dissociation (i.e. k_1 and k_{-1}) fast enough to satisfy the kinetics of the physiological response. This is a poorly studied area. The kinetic, as opposed to equilibrium, constants for Ca^{2+} binding have been measured in very few calmodulins. (4) Interaction with enzymes, other macromolecules or intracellular structures *in vitro* which can explain the physiological events observed when the intact cell is activated. (5) Intracellular localization of calmodulin using fluorescent or radioactively labelled antibody (Marcum *et al.*, 1978 Means *et al.* 1982) (Fig. 3.14) showing that the protein can be found in the right part of the cell. (6) Modulation of Ca^{2+}-dependent regulation *in vivo* by an antibody to calmodulin (Dedman *et al.*, 1978). (7) Inhibition of cell response by inhibitors of Ca^{2+}–calmodulin, e.g. trifluoperazine, if calmodulin is only loosely bound to site of action.

Whether these criteria can be satisfied or not, the existence of this group of Ca^{2+}-binding proteins certainly needs explaining. If they are not actually involved in mediating the effect of intracellular Ca^{2+} to cause cell activation then either the Ca^{2+} binding is passive or it could be necessary for intracellular Ca^{2+} buffering.

The high content of acidic amino acids together with their amides in most proteins (Table 3.6) means that a mutation in the triplet code converting the amide to the acid would substantially alter the charge and Ca^{2+}-binding properties of

Table 3.17. Amino acid sequences of some Ca^{2+}-binding proteins from muscle

1. *Parvalbumin*

   Carp Ac-Ala-Phe-Ala-Gly-Val-Leu-Asn-Asp-Ala-Asp-Ile-Ala-Ala-
                                              10
   -Ala-Leu-Glu-Ala-Cys-Lys-Ala-Ala-Asp-Ser-Phe-Asp-His-
                        20
   -Lys-Ala-Phe-Phe-Ala-Lys-Val-Gly-Leu-Thr-Ser-Lys-Ser-
                    30
   -Ala-Asp-Asp-Val-Lys-Lys-Ala-Phe-Ala-Ile-[Asp]-Gln-
        40   50
   -[Asp]-Lys-[Ser]-Gly-[Phe]-Ile-[Glu]-Glu-Asp-[Glu]-Leu-
                                    60
   -Lys-Leu-[Phe-Leu]-Gln-Asn-[Phe]-Lys-Ala-Asp-Ala-Arg-
                            70
   -Ala-[Leu]-Thr-Asp-Gly-Glu-Thr-Lys-Thr-[Phe-Leu-Lys]-
                    80
   -Ala-Gly-[Asp]-Ser-[Asp]-Gly-[Asp]-Gly-[Lys]-Ile-Gly-
            90
   -Val-Asp-[Glu]-Phe-Thr-Ala-[Leu-Val]-Lys-Ala-CO_2H
        100   108

2. *Troponin C*

   Bovine cardiac   Ac-Met-Asp-Asp-Ile- Tyr- Lys-Ala - Ala - Val-Glu-Gln- Leu - Thr - Glu-Glu - Gln-Lys-Asn - Glu- Phe-
                                                  10   20
   Rabbit skeletal Ac (Asp,Thr,Gln,Gln)Ala-Glu- Ala - Arg-Ser-Tyr- Leu - Ser- Glu - Glu - Met-Ile-Ala- Glu - Phe -

from Enfield *et al.* (1975)

30 40

Lys-Ala-Ala-Phe-Asp -Ile- Phe -Val-Leu-Gly-Ala-Glu-Asp- Gly -Cys- Ile-Ser -Thr- Lys- Glu -
Lys-Ala-Ala-Phe-Asp -Met- Phe [Asp----- Ala Asp-Gly-Gly - Gly -Asp- Ile-Ser - Val - Lys-Glu-]

50 60

Leu-Gly -Lys- Val-Met-Arg-Met-Leu-Gly-Gln -Asn- Pro-Thr -Pro- Glu-Glu-Leu -Gln-Glu-Met-
Leu-Gly -Thr- Val-Met-Arg-Met-Leu-Gly-Gln -Thr- Pro-Thr -Lys- Glu-Glu-Leu - Asp-Ala- Ile

70 80

Ile -Asp- Glu -Val[Asp-Glu-Asp-Gly-Ser-Gly-Thr-Val- Asp-Phe -Asp- Glu] Phe-Leu-Val-Met-
Ile -Glu- Glu (Val[Asp,Glu,Asp,Gly,Ser, Gly,Thr)Ile- Asp -Phe -Glu- Glu] Phe-Leu-Val Met-

90 100

Met -Val-Arg -Cys- Met -Lys -Asp- Asp- Ser- Lys-Gly-Lys-Ser-Glu-Glu-Glu-Leu- Ser-Asp-Leu-
Met- Val- Arg -Gln- Met- Lys - Glu- Asp- Ala- Lys-Gly-Lys-Ser-Glu-Glu-Glu-Leu- Ala-Glu-Cys-

110 120

Phe-Arg -Met- Phe[Asp -Lys- Asn-Ala-Asp -Gly-Tyr-Ile-Asp -Leu- Glu-Glu-]Leu- Lys-Ile- Met-
Phe -Arg -Ile- Phe[Asp -Arg- Asn-Ala-Asp Gly- Tyr-Ile-Asp Ala Glu-Glu-]Leu- Ala-Glu-Ile-

130 140

Leu-Gln- Ala -Thr- Gly-Glu -Thr-Ile- Thr -Glu-Asp-Asp- Ile-Glu -Glu- Leu-Met-Lys-Asp-Gly-
Phe-Arg- Ala -Ser- Gly-Glu -His-Val- Thr -Asp-Glu-Glu- Ile -Glu -Ser- Leu-Met-Lys-Asp-Gly-

150 160

[Asp-Lys-Asn-Asn-Asp-Gly-Arg-Ile-Asp -Tyr- Asp-Glu] Phe-Leu -Glu-Phe- Met -Lys- Gly-Val-Glu-CO_2H
[Asp-Lys-Asn-Asn-Asp Gly Arg Ile Asp -Phe- Asp-Glu] Phe -Leu -Lys-Met- Met -Glu- Gly-Val-Glu-CO_2H

from van Eerd and Takahashi (1975) and Collins (1976)

☐ = sequence homology

[ ] = Ca^{2+}-binding sites

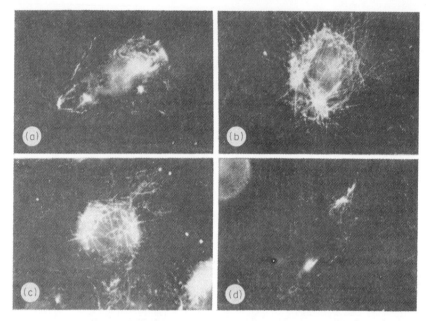

Fig. 3.14. Localization of calmodulin. Fluorescein-labelled antibody to calmodulin during mitosis. From Means and Dedman (1980); reprinted by permission from *Nature*. Copyright 1980 Macmillan Journals Limited

Table 3.18. Genetic code for acidic amino acids

Amino acid	Triplet code
Aspartate (Asp)	GAU;GAC
Asparagine (Asn)	AAU;AAC
Glutamate (Glu)	GAA;GAG
Glutamine (Gln)	CAA;CAG

the protein. As can be seen from Table 3.18, their triplet codes are very similar and a point mutation would convert an amide to its corresponding amino acid. The alternative in the third position of the triplet code is consistent with the 'wobble' hypothesis proposed by Crick in 1967.

3.5. What are the chemical conditions for a 'threshold'?

If a cell, or an 'event' within a cell, is to exhibit a 'threshold' response to a stimulus then the phenomenon requires some form of discontinuity. This discontinuity could involve an 'event' or 'events' not occurring in the unstimulated cell, or it could involve the transformation of the cell from a resting to an activated state without existing at an intermediary level, or for any significant period of time. Examples of the former are the initiation of cell division, a heart beat, and the

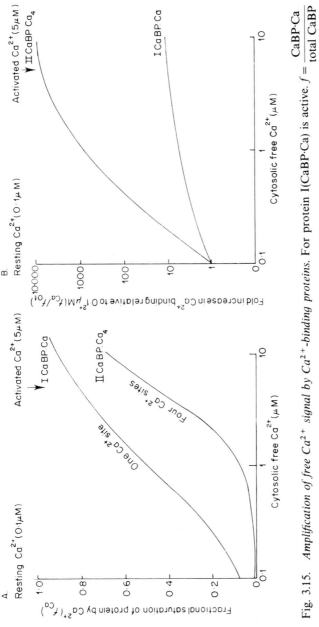

Fig. 3.15. *Amplification of free Ca^{2+} signal by Ca^{2+}-binding proteins.* For protein I(CaBP·Ca) is active. $f = \dfrac{CaBP \cdot Ca}{total\ CaBP} = \dfrac{Ca}{(K_d + Ca)}$. For protein II CaBP·Ca$_4$ is active. $f = \dfrac{CaBP \cdot Ca_4}{total\ CaBP} = \left(\dfrac{Ca}{(K_d + Ca)}\right)^4$. $K_d^{Ca} = 1\ \mu M$ for each Ca^{2+} site

fusion of exocytotic vesicles with the cell membrane. The latter is exemplified by muscle contraction in arthropods, and the activation of intermediary metabolism in cells undergoing a threshold response, for example muscle contraction.

The molecular basis for such thresholds has yet to be characterized. Two possibilities are: (1) a rapid jump in intracellular free Ca^{2+}, for example initiated by the sudden, simultaneous opening of 'Ca^{2+} channels' in the membrane; (2) the coordinated and simultaneous action of several Ca^{2+}-binding units, necessary to produce any response of the structure to which they are attached. Either of these would be aided if the Ca^{2+}-binding protein involved requires the occupation of more than one Ca^{2+} site to be active, e.g. troponin C or calmodulin, and if there is any cooperativity of Ca^{2+} binding, as seems to be the case for calmodulin (Klee et al., 1980). This is illustrated in Fig. 3.15. Protein I is a hypothetical Ca^{2+}-binding protein with only one Ca^{2+} site ($K_d^{Ca} = 1$ μM), whereas protein II has four, non-cooperative sites (K_d^{Ca} for each $= 1$ μM), and all four must be occupied if protein II is to activate the structure to which it is attached. From equations 3.21 and 3.23 the fractional saturation of these two proteins by Ca^{2+} can be calculated (Fig. 3.15A). Although the absolute saturation of protein I by Ca^{2+} is greater than that of protein II at any given cytoplasmic free Ca^{2+} concentration, the increase in saturation of protein II on going from 0.1 μM Ca^{2+} (resting) to 5 μM (activated) is nearly 7000-fold, whereas for protein I it is only 9-fold (Fig. 3.15B). Multiple Ca^{2+}-binding sites therefore provide a means of amplifying the initial signal, i.e. the change in cytosolic free Ca^{2+}. The response of proteins such as protein II to Ca^{2+} will be even more marked if the Ca^{2+} sites exhibit positive cooperativity. Such amplified responses could provide part of the molecular mechanism for generating a threshold response in the cell.

3.6 Diffusion

The movement of a solute down a concentration gradient is known as diffusion. In 1885 Fick published the first Law of Diffusion which quantitatively describes this phenomenon.

$$J = \text{flux (moles cm}^{-2}\text{ sec}^{-1}) = D\frac{\delta c}{\delta x} \qquad (3.33)$$

where D = diffusion coefficient (cm^2 sec^{-1})

$\delta c/\delta x$ = concentration gradient (moles cm^{-4}).

For an ion, D is directly related to the mobility (u) of that ion by an equation described by Nernst in 1888.

$$D = \frac{RT}{F}(u/|z|) \qquad (3.34)$$

$$= 0.013\ u \text{ for } Ca^{2+} \text{ at } 37°C.$$

If $u_{Ca} = 4 \times 10^{-4}$ cm^2 sec^{-1} V^{-1} (Baker, 1972), then $D = 5.2 \times 10^{-6}$ cm^2 sec^{-1}. Whereas D for $CaCl_2$ in solution is quoted by Robinson and Stokes (1968) as

1.335×10^{-5} cm^2 sec^{-1}. D_{Ca} has been measured as 6×10^{-6} cm^2 sec^{-1} (Ashley, 1978) or 7.8×10^{-6} cm^2 sec^{-1}.

Consider a layer of cytoplasm 1 μm (10^{-4} cm) thick close to the surface membrane a cell 20 μm (2×10^{-3} cm) in diameter, with a free Ca^{2+} concentration of 10 μM (10^{-8} moles cm^{-3}), and 0.1 μM (10^{-10} moles cm^{-3}) at the cell centre. To a first approximation let us assume that

$$\delta c/\delta x = \Delta c/\Delta x = 10^{-8}(9 \times 10^{-4}) = 1.11 \times 10^{-5} \text{ moles cm}^{-4}$$

if $D = 10^{-5}$ cm^2 sec^{-1} then $J = 1.11 \times 10^{-10}$ moles cm^{-2} sec^{-1} from this hypothetical sphere of cell cytoplasm.

If net flux from this sphere = $J \times$ area of sphere then net flux = 1.13×10^{-15} moles sec^{-1}

Since the area of sphere = $4\pi r^2 = 1.02 \times 10^{-5}$ cm^2

volume of cell = $\frac{4}{3}\pi r^3 = 4.19 \times 10^{-9}$ cm^3

volume of cell 1 μ form surface = 1.14×10^{-9} cm^3

∴ absolute Ca^{2+} in this surface layer = $10^{-8} \times 1.14 \times 10^{-9} = 1.14 \times 10^{-17}$ moles

Since this is 100 times less than the initial flux of Ca^{2+} out of this surface layer, one would expect Ca^{2+} gradients to exist in a small cell such as this for no more than a few milliseconds to a second, on the basis of free diffusion in the absence of a Ca^{2+} buffering system.

However, the mobility of ^{45}Ca^{2+} in squid axoplasm is 0.9×10^{-5} cm^2 sec^{-1} V^{-1}, and Rose and Loewenstein have shown that Ca^{2+} injected into a cell 'diffuses' only a very small distance. The reason for this apparent discrepancy between Ca^{2+} movement in solution and Ca^{2+} movement in the cell is because of Ca^{2+}-buffering systems in the cytoplasm and organelles such as mitochondria. Hence one may get transient or even long-lasting Ca^{2+} gradients in both large and small cells which are not due to the diffusion of Ca^{2+} in free solution as a limiting factor. It has been argued that the calculations of Hill (1948) concerning the diffusion of substances from the muscle cell surface to the contractile apparatus, slow relative to a contraction, caused some workers to search for an internal store of Ca^{2+} necessary to stimulate a fast twitch.

3.7. Conclusions

The potential number of Ca^{2+} ligands within a cell is very large. Many of them, however, have a relatively low affinity for Ca^{2+} and at physiological concentrations of Mg^{2+} (about 1–5 mM) only a small proportion of the ligand is bound to Ca^{2+} when the free Ca^{2+} concentration is in the range 0.1–10 μM. When cells die they can take up large quantities of Ca^{2+} (Majno, 1964) (see also Chapter 9), presumably bound to these low-affinity Ca^{2+} sites. Nevertheless, more than 99.9% of a cell's Ca^{2+} is to be found elsewhere than free in the cytoplasm. Small molecules, such as ATP and citrate, in the cytoplasm may bind a few μM Ca^{2+}

and high-affinity binding proteins up to 10–20 μM. The remainder of the cell Ca^{2+}, equivalent to several mmoles per litre of cell water, is maintained within organelles, often requiring an energy-dependent Ca^{2+}-uptake mechanism. If the free Ca^{2+} within these organelles is much higher than in the cytoplasm, for example 100 μM, then low-affinity Ca^{2+} sites on nucleic acids, phospholipids and some proteins will bind significant quantities of calcium. In some instances, for example calsequestrin in muscle sarcoplasmic reticulum, special proteins exist to bind this intraorganelle Ca^{2+}. Exactly how much Ca^{2+} is in the form of microprecipitates of calcium phosphate is not known. Release of organelle Ca^{2+} followed by binding to 'calmodulins' (see section 3.4, this chapter) provides an important mechanism for many biological effects mediated by intracellular Ca^{2+}.

The thermodynamic and stereochemical properties of Ca^{2+} and Mg^{2+} have highlighted some important differences between proteins modulated by group II cations and true metalloenzymes, which usually require transition metals such as Mn^{2+}, Zn^{2+} and Mo^{2+}. In the latter case the cation is often strongly attached to the protein as a result of the kinetic or equilibrium constants (Irving and Williams, 1953). Transition metal cations are involved at the active centre of enzymes either in substrate binding or as part of the catalytic mechanism (Vallee, 1955; Vallee and Wacker, 1970; Lehninger, 1950). The affinity of Ca^{2+} and Mg^{2+} for proteins is often much less than for transition metal requiring enzymes. Apart from CaATP and MgATP, the group II cation binding sites are usually at a separate, allosteric site from the active centre of the protein.

The ubiquitous occurrence of low-affinity Ca^{2+}-binding sites in biological molecules means that it is essential to releate data obtained in broken cell preparations, or with isolated proteins, to events in the intact cell.

CHAPTER 4

Intracellular Ca^{2+} and the electrical activity of cells

4.1. Some questions

Let us return to the story of the child who, running down the road to greet its parents, falls over grazing its knees in the process (see Chapter 2, section 2.1). Amongst the many cells that have been activated in the child during this episode several will have undergone changes in their electrical properties. Action potentials will have travelled down nerve axons to stimulate action potentials in muscle cells after release of acetylcholine at the neuromuscular junction, the rate of the heart beat will have increased through changes in the rate of action potentials in the cells in the heart, and noradrenaline acting on certain smooth muscle cells will have caused potential changes across the cell membrane which trigger the cells to contract. In nerve axons and skeletal muscle cells Na^+ is the major current carrier in the generation of such action potentials, whilst K^+ is involved in repolarization of the cell during its recovery. In contrast, Ca^{2+} is an important carrier of the current in excited smooth muscle and heart cells. Can these movements of Ca^{2+} lead directly to increases in cytoplasmic free Ca^{2+}, with consequent chemical changes inside the cell? How does electrical excitability of a skeletal muscle cell provoke Ca^{2+} release from the sarcoplasmic reticulum, a crucial internal Ca^{2+} store in contraction? Do changes in intracellular free Ca^{2+} regulate the electrical properties of the cell membrane through changes in its permeability to other ions and, if so, what role does this play in the response of the cell to a physiological stimulus?

Many articles have been written about Ca^{2+} and the electrical properties of membranes (Brink, 1954; Shanes, 1958a,b; Frankenhauser and Hodgkin, 1957; Baker, 1972; Reuter, 1973; Meech, 1976; Eckert and Brehm, 1979). This book is concerned primarily with the role of intracellular Ca^{2+} as a regulator. It is appropriate, therefore, to consider how changes in membrane potential can regulate the concentration of intracellular free Ca^{2+}, and the converse phenomenon, how changes in intracellular Ca^{2+} concentration can control the permeability of biological membranes to other ions. Intimately concerned with these problems are the mechanisms by which resting cells are able to maintain a very low (about 0.1 μM) intracellular free Ca^{2+} concentration, and how the uptake and release of Ca^{2+} from internal stores can be regulated. But before examining these questions in detail, it is necessary to remind ourselves of the essential electrical properties of biological membranes, the equations which describe them, and how a 'threshold' response can be generated in a cell.

4.2. The electrical properties of membranes

4.2.1. Historical

In 1791 Luigi Galvani unwittingly touched the exposed nerve of a dissected frog's leg with a discharging electric machine, causing the muscle to contract. Some 41 years earlier Vans Gravesande and Adanson discovered independently that some fish can produce large electrical discharges. About 100 years later the experiments of Ringer (1883a,b,c, 1886, 1890) were to provide some of the first clues about the ionic nature of the electrical properties of cells.

Certain cells in the body can be activated electrically and the long cells of nerves and skeletal muscle can conduct these electrical signals over considerable distances, resulting in stimulation of the next cell. In 1902 Overton reported that Na^+ was necessary outside the cell if frog skeletal muscle was to conduct electrical impulses (see also Overton, 1904), and in the same year Bernstein suggested that a high permeability of the cell membrane to K^+ relative to other cations was responsible for the electrical potential across it. Thanks mainly to the work in particular of Hodgkin, Huxley and Katz, as well as Curtis and Cole in the United States, the concept developed that the electrical properties of the cell membrane could be described like any other electrical circuit. The membrane, therefore, has a capacitance, a resistance, and an impedance and, most important in the living cell, a potential across it. The energy for this comes from metabolic reactions in the cell. Unlike metal conductors, where the current is really carried by electrons, across most biological membranes the current is carried by positively charged ions such as Na^+, K^+ and Ca^{2+}, or by negatively charged ions such as Cl^- and HCO_3^-. Working with mitochondria Mitchell (1966, 1968) has shown that H^+ ions can also be an important generator of membrane potential, electrons being carried across the inner mitochondrial membrane by an oxidoreductive cytochrome chain.

We owe much of our fundamental understanding of the ionic nature of electrical membranes, and in particular the role of Ca^{2+}, to studies on giant cells such as the nerve axon of the squid first brought to prominence by J. Z. Young in 1930 (see Young, 1940), as well as to other large nerve and muscle fibres from invertebrates such as the sea hare *Aplysia* and the barnacle *Balanus*. These cells can be several centimetres long and a millimetre or more in diameter, providing an ideal target for the physiologist's microelectrodes and microsyringes.

The 'sodium theory' of the action potentials generated in nerve axons proposed by Hodgkin and Katz (1949a,b), explained the onset of electrical excitability as a rapid increase in the permeability of the cell membrane to Na^+. This was consistent with the observation that nerve axons failed to fire in media without Na^+.

In 1953 Fatt and Katz were working with another giant cell, in this case single muscle fibres from crab. They found that action potentials could be generated in the absence of extracellular Na^+. There seemed also to be a rapid increase in the permeability of the cell membrane to divalent cations, particularly Ca^{2+}. It was later shown that other crustacean muscle cells generate Ca^{2+}-dependent action potentials when the putative natural transmitter, L-glutamate, is added to them (see Ashley and Campbell, 1978). Several other electrically excitable cells generate similar action potentials under physiological or experimental conditions. These potentials are either wholly or partially dependent on Ca^{2+} as the current carrier across the membrane of the cells (Hagiwara, 1973; Gerschenfeld, 1973; Reuter, 1973). Furthermore, the increase in Ca^{2+} permeability which causes this Ca^{2+} current is directly or indirectly responsible for increases in intracellular free Ca^{2+}. It is this free Ca^{2+} which is essential for activation of chemical processes within the cell, as well as for recovery of the cell from the original stimulus.

4.2.2. The cell membrane

If a small glass capillary containing an electrically conducting solution is inserted into a cell, it will usually register a potential of between -20 and -90 mV (negative), and sometimes greater, with respect to a reference electrode in the medium bathing the cell. What precisely is the nature and cause of this resting membrane potential, how can it be modified and what is the role of intracellular Ca^{2+} in these phenomena?

Since Davson and Danielli (Danielli, 1943) first proposed their lipid bilayer theory for the structure of biological membranes, considerable advances have been made in our understanding of the physicochemical properties of biological membranes. Fluorescence, electron spin and nuclear magnetic spin resonance probes have enabled measurements to be made of the movement of lipids and proteins in membranes. As a result it is now thought that, apart from certain specialized areas in the membrane mainly associated with proteins, the phospholipid bilayer is a liquid in which the proteins are suspended, some facing outwards, some inwards, and some traversing the membrane (Fig. 4.1).

Artificial membranes composed only of phospholipids with no proteins in them are highly impermeable to ions, in particular to Ca^{2+}, more so than biological membranes. An illustration of this is the very low luminescence from Ca^{2+}-activated photoproteins entrapped within phospholipid vesicles (liposomes) when incubated for hours, or even days, in the presence of millimolar Ca^{2+} concentrations (Dormer *et al.*, 1978). The insertion of proteins into phospholipid membranes increases their ionic permeability by several orders of magnitude. However, the permeability of biological membranes to Ca^{2+} is still relatively low, particularly when compared with K^+, Na^+ and Cl^- (Table 4.1). The molecular basis of passive ion fluxes across membranes is not known, but may involve either charged pores or leakage around proteins (Fig. 4.1). The

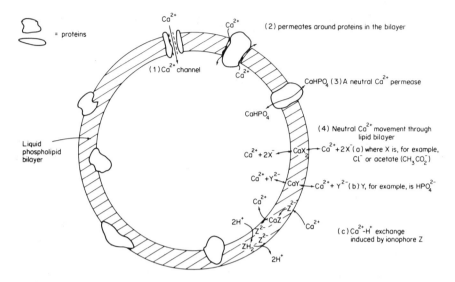

Fig. 4.1. Possible routes for passive movement of Ca^{2+} across biological membranes, i.e. excluding active movement of Ca^{2+} out of the cell via Ca^{2+} activated Mg^{2+} ATPase, Na^+/Ca^{2+} exchange (+ Mg ATP) or H^+/Ca^{2+} exchange (+ Mg ATP)

Table 4.1. The low passive permeability of biological membranes to Ca^{2+} relative to other cations

	Permeability coefficient (cm sec^{-1})				
	P_{Ca}	P_{Na}	P_K	P_{Na}/P_{Ca}	P_K/P_{Ca}
Squid giant axon	2×10^{-9}	3×10^{-8}	1.5×10^{-6}	15	750
Resting cell	Passive Ca^{2+} influx (pmoles. cm^{-2} sec^{-1})				
Rat liver	0.2				
Squid giant axon	0.1				
Guinea pig atrium	0.02				
Frog heart	0.01				
Mast cell	0.01				

The passive influx of Na^+ into squid giant axons is some 400 times that for Ca^{2+}. Stimulation of the cell can increase Ca^{2+} influx by some 10–100 times.

These data are only approximate and are intended to illustrate the order of magnitudes involved. For references see Hodgkin and Keynes (1957), Baker (1972), Foreman et al. (1977), Claret-Berthon et al. (1977).

permeability of membranes to some ions, e.g. K^+, Na^+ and Ca^{2+} can be increased selectively by opening channels or pores, which are presumably either proteins or ionophores. The calculations based on A23187 (see Chapter 2, section 2.3.2) have already highlighted the fact that minute quantities of ionophores dissolved wholly in the phospholipid would produce a relatively high molar concentration in this phase and could transport ions such as Ca^{2+} very rapidly into the cell if available to bind Ca^{2+} in the external medium.

The cell membrane also contains ion pumps for extruding Na^+ and Ca^{2+}. These pumps maintain the cation gradients essential for maintenance of the resting membrane potential, and the very low resting intracellular free Ca^{2+}. Since Na^+ and Ca^{2+} are both pumped out of the cell against an electrochemical gradient, a source of energy is required. For the Na^+ pump this is ATP. In the case of Ca^{2+} this can be ATP, or it may also be the Na^+ gradient under certain circumstances.

4.2.3. Ca^{2+} and the resting membrane potential

Any discontinuity between two solutions, be it a liquid junction or a semipermeable membrane, will have an electrical potential across it, providing that at least one of the ions in the solutions on either side is able to cross the discontinuity, and if the concentrations of permeant anions are different from those of the permeant cations, or if the mobilities of the various anions and cations are different from each other.

The simplest model of the resting cell membrane assumes that, to a first approximation, the two most permeant ions, K^+ and Cl^-, are at electrochemical equilibrium across it. Since the inside of the cell has a higher concentration of impermeable anions, e.g. proteins, nucleic acids and metabolites, than the fluid outside the cell, the concentration of Cl^- is greater outside the cell than inside. Furthermore, there is a large concentration gradient of K^+ across the membrane in the opposite direction, the excess Cl^- outside being electrically balanced by Na^+. At steady state the rate of influx of each ion must equal the rate of efflux, but in the absence of a potential across the membrane, the influx and efflux of each ion will not be equal because of the concentration gradients of K^+ and Cl^-. What happens is that a very small number of charged ions move from one side of the membrane to the other so that the inside of the cell becomes negative with respect to the outside. When the potential is sufficient to counterbalance the tendency of the ions to move down their concentration gradients, then the system is at equilibrium. This simple model is known as the Donnan equilibrium and is described by the Nernst equation:

$$E_m = \frac{RT}{F} \log_e ([K_o^+]/[K_i^+]) = \frac{RT}{F} \log_e ([Cl_i^-]/[Cl_o^-]) \qquad (4.1)$$

where E_m = membrane potential will respect to the inside (i.e. will be negative when $[K_i^+] > [K_o^+]$)
 R = universal gas constant = $8.314 \, J\,K^{-1}\,mole^{-1}$
 T = temperature in K
 F = Faraday constant = $96495 \, C\,mole^{-1}$
 i = inside the cell
 o = outside the cell.

All ions will eventually equilibrate across the membrane to obey the Donnan equilibrium, unless an ion pump exists to prevent it.

It is important to remember that, for K^+ or Cl^- at least, the number of ions that have to move across the membrane to generate a sufficient potential for the system to reach Donnan equilibrium is only a very small fraction of the ions present on either side. In fact, it is possible to calculate this if the membrane potential and capacitance are known. Consider a membrane with a capacitance (C) of $1\ \mu F\ cm^{-2}$ [most biological membranes have a capacitance in the range 0.5–$1.5\ \mu F\ cm^{-2}$; see Mueller and Rudin (1969)].

The charge (Q) necessary to produce a potential of $-50\ mV$ (negative inside) is given by

$$Q = CV \qquad (4.2)$$

$$\therefore Q = 10^{-6} \times 50 \times 10^{-3} = 5 \times 10^{-8}\ C\ cm^{-2}$$

For a monovalent ion (e.g. K^+) 1 mole is equivalent to F coulombs

$$\therefore Q \equiv 5 \times 10^{-8}/96495 \equiv 5.2 \times 10^{-13}\ moles\ cm^{-2} \qquad (4.3)$$

In a spherical cell of diameter $20\ \mu m$ (2×10^{-3} cm) the volume $= \frac{4}{3}\pi r^3 = 4.2 \times 10^{-9}\ cm^3$ or $4.2\ pl$ and the surface area $= 4\pi r^2 = 1.3 \times 10^{-5}\ cm^2$. From equation 4.3 the number of moles of K^+ moving out of such a cell to produce $E_m = -50\ mV$ is therefore 6.8×10^{-18}. Assuming that the intracellular K^+ concentration $= 150\ mM$, then the total K^+ in the cell $= 6.3 \times 10^{-13}$ moles. Hence the amount of K^+ moved is only a tiny fraction of the total.

The significance of this is that when the same calculation is done for Ca^{2+} the concentration change inside the cell *is* significant compared with the resting free Ca^{2+} concentration (see this chapter, section 4.3.3).

In many living cells the measured membrane potential is very close to that estimated from equation 4.1. However, measurement of the Na^+ and Ca^{2+} concentrations on either side of the cell membrane has shown that these ions are a long way from Donnan equilibrium. This is because Na^+ and Ca^{2+} pumps maintain gradients of these ions against an electrochemical force, at the expense of energy supplied by the cell's metabolism. In fact, the apparent Donnan equilibrium state of K^+ and Cl^- is misleading since the membrane of the resting cell really exists in a steady-state rather than a true thermodynamic equilibrium. This is illustrated by the fact that poisoning of the cell with CN^- or by inhibition of the sodium pump with the cardiac glycoside ouabain, results in leakage of K^+ from the cell and movement of Na^+ and Ca^{2+} into it.

For a true Donnan equilibrium (Fig. 4.2) the membrane potential is not dependent on the 'mobilities' of the various anions and cations across the membrane. In the living cell, however, the membrane potential exists because of differences between the permeability of the membrane to the various ions, in particular K^+ and Na^+. The equation that describes the membrane potential by considering the three major permeant ions is the Goldman constant field

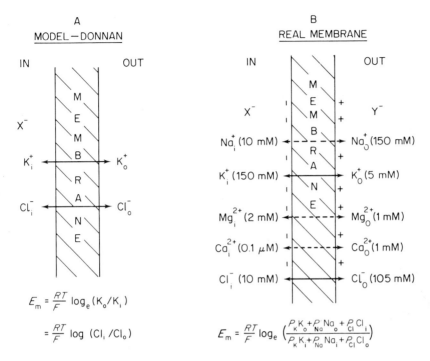

Fig. 4.2. The electrical potential across semipermeable membranes: A = model membrane (Donnan potential); B = real biological membrane (diffusion potential). Ionic concentrations are only approximate and are intended to illustrate the order of magnitude involved for a mammalian cell. E_m = membrane potential inside relative to outside

equation (Goldman, 1943; Hodgkin and Katz, 1949)

$$E_m = \frac{RT}{F} \log_e \left[\frac{P_K[K_o^+] + P_{Na}[Na_o^+] + P_{Cl}[Cl_i^-]}{P_K[K_i^+] + P_{Na}[Na_i^+] + P_{Cl}[Cl_o^-]} \right] \quad (4.4)$$

where P_x = permeability coefficient of the ion concerned.

Where then does Ca^{2+} fit into these equations? The permeability of biological membranes to Ca^{2+}, relative to other ions (Table 4.1), is sufficiently low that Ca^{2+} makes a negligible contribution to the resting membrane potential.

In the absence of an active transport mechanism pumping Ca^{2+} out of the cells, one would expect Ca^{2+} eventually to equilibrate across the membrane in accordance with the Donnan equilibrium. Calculation of the predicted intracellular Ca^{2+} concentrations under such conditions for various known extracellular Ca^{2+} concentrations reveals some startling figures.

If the Ca^{2+} ion is at equilibrium with the membrane potential then

$$E_{Ca} = E_m = \frac{RT}{2F} \log_e ([Ca_o^{2+}]/[Ca_i^{2+}]) \quad (4.5)$$

or
$$E_{Ca} = (RT/2F \log_{10} e)\log_{10}([Ca_o^{2+}]/[Ca_i^{2+}])$$

The relationship between $\log_{10}([Ca_o^{2+}]/[Ca_i^{2+}])$ and E_{Ca} is shown in Fig. 4.3. In mammals body temperature is approximately 36–40 °C (309–313 K) and the extracellular free Ca^{2+} concentration is about 1 mM. The intracellular free Ca^{2+} concentration in the resting cell is usually in the range 10–100 nM. This means that the actual $[Ca_o^{2+}]/[Ca_i^{2+}]$ ratio is between 10^5 and 10^4, and the predicted membrane potential with which this would be in equilibrium is between $+120$ and $+160$ mV (*positive* inside). This is the reverse of the actual membrane potential which is -20 to -90 mV (*negative* inside). Similarly, in cells from marine invertebrates, e.g. squid giant nerves, the extracellular Ca^{2+} concentration is about 10 mM whereas the intracellular Ca^{2+} is 100 nM (Baker, 1972); i.e. $[Ca_o^{2+}]/[Ca_i^{2+}] = 10^5$, predicting a Ca^{2+} equilibrium potential (E_{Ca}) of about $+143$ mV at 15°C. The actual membrane potential is about -90 mV (*negative* inside).

These calculations emphasize two important characteristics about the Ca^{2+} gradient across the cell membrane: (1) The enormous electrochemical force on Ca^{2+} places this ion in a unique position to act as an intracellular regulator. Small changes in the permeability of the cell membrane to Ca^{2+} caused by physiological stimuli would cause large fractional changes in cytosolic free Ca^{2+}. The situation is analogous to the effect of a water leak in a ship floating on the surface of the sea compared with a submarine several hundred metres under water. On the surface the pressure forcing sea water into the ship is relatively low and the pumps on the ship can easily cope with it. However, at the very high pressures under the sea a small leak would result in a considerably larger quantity of water being taken in by

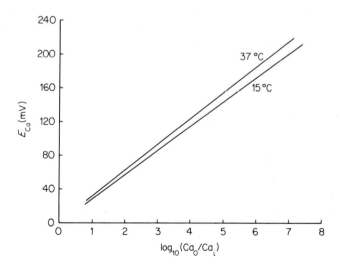

Fig. 4.3. Relationship between the Ca^{2+} equilibrium potential and the predicted Ca^{2+} gradient across the cell membrane. $E_{Ca} = [RT/(2F \log_{10} e)] \log_{10}(Ca_o/Ca_i)$. o = outside cell; i = inside cell

the ship in the same time interval. (2) A Ca^{2+} gradient of this sort can only be maintained if there is a Ca^{2+} pump, or if the Ca^{2+} gradient is directly coupled to another ion gradient, which is supplied with the energy necessary to maintain it far from the Ca^{2+} equilibrium potential.

Another way of looking at this is to calculate the intracellular free Ca^{2+} concentration that would be in electrochemical equilibrium with the known membrane potential and extracellular Ca^{2+} in the absence of a Ca^{2+} pump (Fig. 4.4). For a membrane potential of -50 mV (*negative* inside) the calculated intracellular free Ca^{2+} concentration in a mammalian cell, at electrochemical equilibrium under these conditions, would be about 50 mM. For cells from marine invertebrates the cytoplasmic free Ca^{2+} concentration would be as high as 500 mM. These values are some 0.5–5 million times that of the actual free Ca^{2+} concentration in the cytoplasm. Although Mg^{2+} does not seem to be at electrochemical equilibrium either (see Chapter 2, section 2.8), the electrochemical force is considerably less than that for Ca^{2+}.

The membrane potential of resting cells may be influenced by extracellular Ca^{2+} (Brink, 1954; Shanes, 1958a, b; Brecht and Gebert, 1966). Removal of

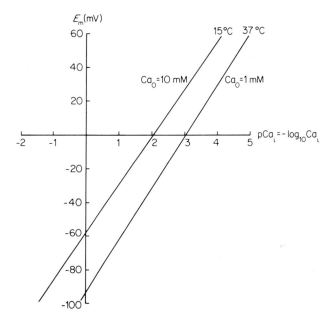

Fig. 4.4. The predicted intracellular free Ca^{2+} concentration at equilibrium with the membrane potential (E_m).

$E_m = \dfrac{RT}{(2F \log_{10} e)} \log_{10} (Ca_o/Ca_i)$.

$E_m = 28.6 \log_{10} (Ca_o/Ca_i)$ at 15°C (e.g. marine invertebrate cells).

$E_m = 30.8 \log_{10} (Ca_o/Ca_i)$ at 37°C (e.g. mammalian cells) extracellular Ca^{2+} experimentally can induce spontaneous action potentials in electrically excitable cells (Frankenhauser and Hodgkin, 1957) and affect motor end-plates (Del Castillo and Stark, 1952). These conditions expose Ca^{2+}-binding sites, probably on the outside of the cell, which under normal circumstances are 'passive' as defined in Chapter 2 (section 2.2.2). Nevertheless, changes in intracellular free Ca^{2+}, induced by physiological stimuli, can influence the permeability of some cell membranes to K^+ and Na^+, and thus influence the electrical properties of the membrane. It is to these 'active' roles for intracellular Ca^{2+} that we will now turn our attention. First, we must examine the criteria laid down in Chapter 2 (section 2.3.3) which are necessary in order to establish an active role for Ca^{2+} in modifying the electrical properties of cells.

4.2.4. How to establish an 'active' relationship between intracellular Ca^{2+} and the electrical properties of cells

There can be few people who have not at some time or other had the opportunity to watch the plethora of different animals to be found in rock pools when the tide has gone out. How fascinating it is to watch a sea anemone entrap something within its tentacles, to see the darting movements of the shrimps, the attempts of the crabs to pinch as they are picked up, and the rapid retraction of gastropod molluscs as they are pulled off the rocks. Many of these movements in invertebrates involve an 'active' role for Ca^{2+} in the nerves and muscles excited during the animal's response. The criteria established in Chapter 2 (section 2.2.2) for demonstrating an 'active', as opposed to a 'passive', role for Ca^{2+} in cell activation were established primarily for phenomena which involve changes in the concentration of Ca^{2+} within the cell, followed by chemical changes which elicit the observed response of the cell. However, when examining excitable cells, such as nerves, it is necessary to define another type of active role for Ca^{2+}—an electrical role. Why some cells which transmit signals over long distances should use Na^+ as the current carrier for the action potential (e.g. squid and mammalian nerve axons) and others (e.g. *Aplysia* nerve axons) should use Ca^{2+}, is still not clear. Could it be that when Ca^{2+} is used it not only results in electrical excitability but also a change in free Ca^{2+} concentration inside the cell occurs which plays a crucial role in either maintaining excitability or in recovery before the next excitation of the nerve can occur? In order to solve this problem two questions must be answered: (1) Can changes in membrane potential cause changes in intracellular free Ca^{2+} concentration, either by direct passage of Ca^{2+} through the membrane from outside or by stimulating release from intracellular stores? (2) Do changes in intracellular free Ca^{2+} regulate the permeability of the cell membrane, or intracellular membranes, to other ions and thereby alter the electrical properties of the cell in a way which is essential for it to respond to a physiological stimulus?

In conjunction with these questions, it is also appropriate to ask, what are the properties and molecular mechanisms of the Ca^{2+} pumps in the cell membrane and intracellular organelles, which restore the intracellular free Ca^{2+} after the primary stimulus of the cell has ended. Characterization of these pumps involves measurement of Ca^{2+} flux in intact cells and isolated organelles, the identification of the energy source of the pump, e.g. ATP or another ion gradient (e.g. Na^+), and the isolation of the proteins and ionophoretic substances followed by reconstitution of the pump in artificial membranes (e.g. black lipid membrane or liposomes). A particularly useful tool in the investigation of the Na^+ pump has been ouabain which specifically inhibits it. Unfortunately, no such specific compound has yet been discovered for Ca^{2+} pumps, although several pharmacological substances do seem to inhibit them. The use of monoclonal antibodies specific for a particular Ca^{2+}-activated MgATPase may have some important applications in this regard in the future.

4.3. Ca^{2+} and excitable cells

4.3.1. What is an excitable membrane?

Consider what happens to a cell when an electrode is inserted into it, a second electrode is placed in the external fluid and a current is passed between them. The result will be a movement of charge across the membrane of the cell. If the cell is depolarized, a net positive charge (e.g. K^+) will move into the cell or negative charge (e.g. Cl^-) will move outwards. In the case of hyperpolarization the reverse will occur. When the battery is disconnected from the electrodes the membrane will behave like a charged capacitance (C), discharging through a resistance (R) until the original membrane potential has been established, when the cell will again be in electrochemical steady-state (Fig. 4.5). The relationship between voltage (V) and time (t) for this simple electrical circuit is

$$(V_t - V_1) = (V_0 - V_1)\exp(-t/RC) \qquad (4.6)$$

when $t = 0$, $V_t = V_0$

and when $t = \infty$, $V_t = V_1 = E_m$ resting

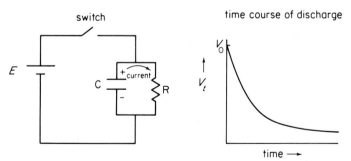

Fig. 4.5. The cell membrane as an electrical circuit. An electrotonic change is illustrated (see equation 4.6); the charge to produce the initial change across a biological membrane is carried by ions such as Na^+ or Ca^{2+}

Physiologically a change in membrane potential occurs either when there is a change in the permeability coefficient of one or more of the ions, or when there is a change in concentration of an ion on either side of the membrane. In non-excitable membranes a quantity of charge moves across the membrane until a new potential has been established, as predicted by the Goldman constant field equation (equation 4.4). The return of the membrane potential to its original value after the stimulus results in a flow of charge in the reverse direction as described by equation 4.6. This type of 'non-excitable' change in membrane potential is known as an electrotonic change. Such a change is induced in liver cells by adrenaline and other hormones (Friedmann and Park, 1968), and by acetylcholine in some secretory cells (Petersen, 1976).

Excitable membranes, on the other hand, have the fascinating property of becoming transiently unstable when depolarized beyond a particular potential, the 'threshold voltage'. When an excitable membrane reaches this point it continues to depolarize. It may reach a potential of $+40$ mV (*positive* inside) at which point the membrane begins to repolarize, eventually returning to the original resting potential ready for the next stimulus. This type of membrane response is known as an action potential. The comparison between action potentials and electrotonic depolarizations is illustrated in Fig. 4.6. Action potentials enable long cells, such as those of nerve and muscle, to transmit signals rapidly from one end of the cell to another without any loss in signal. In some cells, e.g. nerve cells in the sea hare *Aplysia*, rhythmic bursting occurs (Fig. 4.6d). But

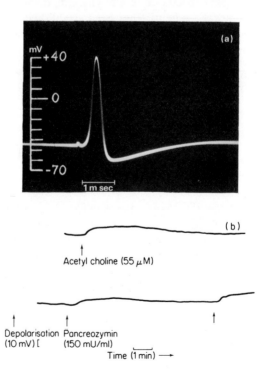

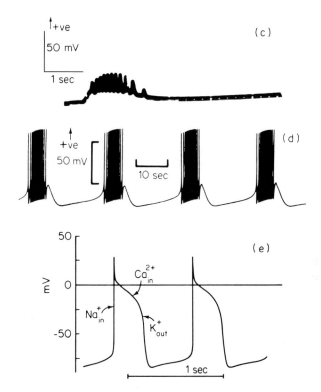

Fig. 4.6. Action potentials and electrotonic changes induced by cell stimuli. (a), squid giant axon; (b), exocrine pancreas; (c), barnacle muscle; (d), *Aplysia* neurone; (e), mammalian heart. (a) Electrical stimulation. From Hodgkin and Huxley (1945); reproduced by permission of the *Journal of Physiology*. (b) Acetylcholine. From Petersen and Matthews (1972); reproduced by permission of Birkhauser Verlag. (c) L-Glutamate. From Ashley and Campbell (1978); reproduced by permission of Elsevier Biomedical Press B. V. (d) Electrical stimulation. From Gorman and Thomas (1978); reproduced by permission of the *Journal of Physiology*. (e) Heart beat. From Draper and Weidmann (1951) and Noble (1962) reproduced by permission of the *Journal of Physiology*

electrically excitable membranes, capable of generating action potentials under physiological circumstances, are not restricted to cells where the function of the action potential is to transmit a signal over long distances. Several small cells, for example in the endocrine pancreas and heart myocardial cells (Fig. 4.6e), also exhibit 'spiking' membrane potential changes, the characteristic of an action potential. In small cells action potentials act as a switch which takes the cell through a 'threshold' (see Chapter 2, section 2.2).

Two questions now arise: (1) Is there a role for intracellular Ca^{2+} in the generation of action potentials? Is Ca^{2+} involved in provoking the instability in the membrane which produces the rapid depolarization phase of the action potential? (2) Do changes in intracellular Ca^{2+} play any part in the repolarization phase?

The first action potentials to be fully investigated were those transmitted down the giant axon of the squid, *Loligo forbesi* (see Hodgkin and Katz, 1949a,b; Hodgkin, 1951; Hodgkin and Huxley, 1945; Aidley, 1971). Hodgkin, Huxley and Katz proposed that these and other nerve membranes contain voltage-dependent ion channels. Once the membrane potential passes the threshold voltage, a rapid increase in Na^+ permeability occurs because of the opening of voltage-dependent Na^+ channels. This produces a further depolarization, which in turn, but more slowly, opens voltage-dependent K^+ channels. This opening of K^+ channels, together with the shutting of the Na^+ channels, enables the membrane to repolarize. Eventually the K^+ channels also shut and the membrane returns to its resting state. This 'sodium theory', so-called, explains adequately the electrical excitability of several other cells besides squid giant axons; for example, myelinated nerves and the skeletal muscle of vertebrates. In all of these cells conduction of impulses can be blocked by removal of extracellular Na^+ (Overton, 1902, 1904) and by tetrodotoxin (Fig. 4.7), a toxin extracted from snake venom which blocks Na^+ channels in mammalian and some invertebrate cells.

In 1953 Fatt and Katz were working with another excitable invertebrate cell—crab muscle. They found that replacement of external Na^+ by tetraethylammonium ions actually potentiated the excitable membrane response, in contrast to the inhibition which would be expected in cells with action potentials dependent on Na^+. Action potentials independent of, or only partially dependent on, external Na^+ (Fig. 4.8) have since been found in several other excitable cells from invertebrates and vertebrates (Hagiwara, 1973; Reuter, 1973) (Table 4.2). In these cells the action potential can be inhibited or blocked, not by replacing external Na^+, but by removing external Ca^{2+}. The explanation for this is that these excitable cells contain voltage-sensitive Ca^{2+} channels which open once a threshold potential across the membrane is reached. Ca^{2+}, not Na^+, is the current carrier through these channels. Heart cells have Na^+ and Ca^{2+} channels which open during an action potential. Both channels are required for a normal heart beat. Even in excitable cells, like squid giant axons, where Na^+ is the major current carrier through Na^+ channels, Ca^{2+} channels also exist (Baker, 1972).

Is there really a clear distinction between cells which generate action potentials *in vivo* and those which do not, particularly since experimentally action potentials can be generated in some cells not normally regarded as being 'excitable'? Could

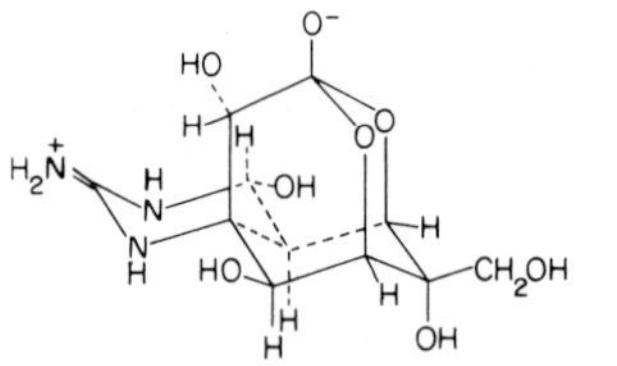

Tetrodotoxin a blocker of the voltage-dependent Na^+ channel

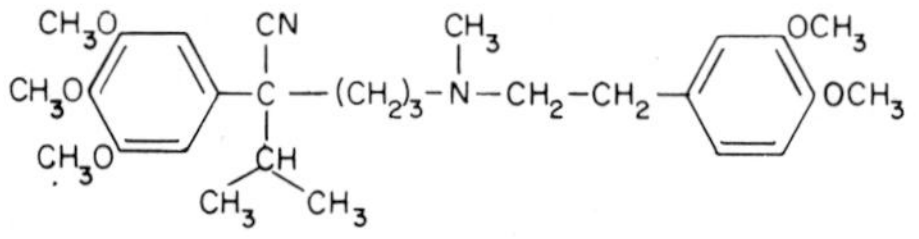

D600 a blocker of the voltage-dependant Ca^{2+} channel

Fig. 4.7. Na^+- and Ca^{2+}-channel blockers

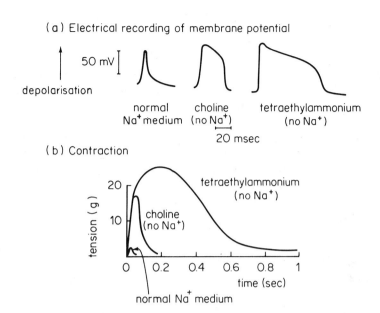

Fig. 4.8. Na^+-independent action potential in crab muscle. Experiment shows (a) potentiation, i.e. not inhibition, of action potential, and (b) contraction when external Na^+ is removed. From Fatt and Katz (1953)

intracellular Ca^{2+} play a role in determining whether a cell is 'excitable'? Two pieces of evidence suggest that this might be so. Firstly, Frankenhauser and Hodgkin (1957) found that changes in intra- *or* extracellular Ca^{2+} concentration affected the electrical properties of squid giant axons, and yet Ca^{2+} ions did not carry any significant current during the action potential. Secondly, injection of EGTA into barnacle muscle reduces intracellular free Ca^{2+}, and sensitizes the muscle to stimulation, making it more likely to the generate action potentials. These observations have lead to the idea that displacement of Ca^{2+} from the inner surface of the membrane is the mechanism by which opening of voltage-dependent ion channels occurs. Unfortunately, the influence of extracellular Ca^{2+} on the resting membrane potential is not entirely consistent with this hypothesis (Frankenhauser and Hodgkin, 1957). This problem highlights a most exciting area for the biochemist, to discover the molecular basis of ion channels in biological membranes, and in particular to show whether lipid-soluble ionophores or physical pores in the membrane are responsible for them.

4.3.2. Ca^{2+} as a current carrier across membranes

In excitable cells where Na^+ is the major carrier of current during the onset of the action potential, one would expect a process activated within the cell to be dependent on Na^+ in the external medium. In contrast, in cells where Ca^{2+} carries a significant amount of the current during an action potential, removal of external

Table 4.2. Cells with action potentials initiated by Na^+ or Ca^{2+} currents

Example	Reference
1. All Na^+	
Vertebrate myelinated nerve axon	
Squid giant axon	Hodgkin (1951)
Crab (*Maia*) nerve axon	
Insect nerve axon	Aidley (1971)
Vertebrate skeletal muscle and some vertebrate smooth muscles	Reuter (1973)
2. All Ca^{2+}	
Snail (*Helix*) nerve ganglion	Meeves (1968)
Sea Hare (*Aplysia*) nerve ganglion	Junge (1967)
Vertebrate presynaptic nerve terminal	Kusano (1970)
Squid giant synapse (presynaptic)	Katz and Miledi (1969)
Vertebrate β-cell	Dean and Matthews (1970a, b)
Crustacean muscle (e.g. barnacle and crab)	Fatt and Katz (1953); Hagiwara (1973)
Some vertebrate smooth muscles	Bülbring and Tomita (1970); Reuter (1973)
Post-synaptic	Kusano et al. (1975)
3. Na^+ and Ca^{2+}	
Vertebrate heart muscle (e.g. Purkinje fibres)	Reuter (1973)
Sea hare (*Aplysia*) giant nerve	Gorman and Thomas (1978)
Crayfish X organ	Iwasaki and Sato (1971)
Paramecium	Eckert and Brehm (1979)
Some smooth muscle	Anderson et al. (1971)

Ca^{2+} will inhibit or even block the spike potentials (Fatt and Katz, 1953; Hagiwara, 1973; Reuter, 1973) (Table 4.2).

On the basis of such experiments, excitable cells can be grouped into three categories: (1) cells where a Na^+ current is responsible for the initiation of the action potential, but where some Ca^{2+} may enter through Na^+ channels; (2) cells where a Ca^{2+} current is mainly responsible for initiation of an action potential; (3) cells where both Na^+ and Ca^+ currents play a role in the generation of an action potential.

Two questions arise from these observations: (1) What is the physiological significance of Ca^{2+} currents? (2) By what mechanism are these Ca^{2+} currents regulated? But before relating the answers to these questions to the role of intracellular Ca^{2+} as a regulator, we must first discover what the evidence is for Ca^{2+} currents in cells.

Evidence for Ca^{2+} currents

(1) Action potentials initiated by Ca^{2+} currents (Table 4.2) will be dependent on the presence of extracellular Ca^{2+}, but should be independent of extracellular Na^+. Even in cells which have both Ca^{2+} and Na^+ currents the action potential should still be sensitive to changes in extracellular Ca^{2+} concentration. This

provides at least a partial explanation for the original observations of Ringer (1883–1886) on frog heart.

(2) Ca^{2+} currents through specific Ca^{2+} channels are not inhibited by tetrodotoxin (Baker, 1972). But such an experiment is not meaningful when the Na^+ channels are not inhibited by tetrodotoxin, for example in some coelenterate excitable cells.

(3) Ca^{2+} currents through tetrodotoxin-sensitive Na^+ channels will be blocked by tetrodotoxin, but the onset of such Ca^{2+} currents will be much faster than through the slower Ca^{2+}-specific channels (Baker et al., 1971; Baker, 1972).

(4) Measurement of $^{45}Ca^{2+}$ taken up during an action potential makes it possible to calculate whether a sufficient quantity of Ca^{2+} has moved to provide the charge necessary for depolarization of the membrane.

Let us consider, for example, the mammalian ventricular myocardium.

The estimated Ca^{2+} influx per impulse ≈ 0.2 pmoles cm^{-2} (Reuter, 1973). 1 mole of $Ca^{2+} \equiv 2 \times 96494$ C (Faraday constant). Therefore the charge carried by Ca^{2+} influx $= 0.04 \mu C\, cm^{-2}$. Since membrane capacitance $= C \approx 1\, \mu F\, cm^{-2}$ and $V = Q/C$ then in the absence of any significant Ca^{2+} outward current, this amount of Ca^{2+} influx would cause a depolarization of about 40 mV, sufficient to account for the major part of membrane depolarization. If an intracellular electrode is coupled to a specially designed electrical circuit it is possible to fix the potential across a membrane at a specific value, even if it would normally generate a 'spiking' response to depolarization. Using this so-called 'voltage clamp' [see Aidley (1971) for reference], the current passing across the membrane can be measured at various times after the clamp has been applied. To discover how much of this current is due to Ca^{2+} one simply removes Ca^{2+} from the external medium and re-measures the current. It is thus possible to demonstrate the dependence of membrane currents, at various applied voltages, on extracellular Na^+ or Ca^{2+} and also, actually to calculate whether each ion current would move sufficient charge to depolarize the membrane under physiological conditions. The voltage-clamp technique has played a vital part in establishing the existence of a role for Ca^{2+} currents in excitable cells.

(5) Pharmacological substances, with defined specificities, have always formed an important part of the experimental biologist's armoury. The snake venom toxin, tetrodotoxin (Fig. 4.7), a potent blocker of Na^+-generated action potentials (Watanabe et al., 1967; Hille, 1972), blocks Ca^{2+} entry through the voltage-dependent Na^+ channel (Baker, 1972; Hagiwara, 1973). Action potentials arising from the opening of specific voltage-dependent Ca^{2+} channels are not inhibited by tetrodotoxin. On the other hand, verapamil and its methoxy derivative D600 (1–100 μM), and also probably nifidepine (Fig. 9.20), do inhibit the Ca^{2+} channels and Ca^{2+}-dependent action potentials. These compounds do not, however, significantly affect Na^+ currents. The effect of these substances on Ca^{2+} channels explains at least some of their therapeutic actions on the human heart (see Chapter 9, section 4.4) (Fleckenstein, 1974).

Several cations inhibit, relatively specifically, voltage-sensitive Ca^{2+} channels. For example, Mg^{2+}, and La^{3+}, Mn^{2+}, Co^{2+} and Ni^{2+} at mM concentrations inhibit Ca^{2+} channels in squid giant axons and several other excitable cells (Baker, 1972; Reuter, 1973). However, in the ciliated protozoan *Paramecium*, the voltage-sensitive Ca^{2+} channels are not blocked by these cations (Eckert and Brehm, 1979). Tetraethylammonium ions, on the other hand, inhibit only the K^+-activated channel associated with repolarization of the membrane.

Tissues where these experimental procedures have been used include squid giant axons (Baker, 1972), barnacle muscle (Hagiwara, 1973), heart muscle in the presence or absence of adrenaline (Davis, 1931; Pappano, 1970; Reuter, 1973), squid giant synapse (Katz and Miledi, 1969), and *Paramecium* (Eckert and Brehm, 1979).

In presynaptic nerve terminals, under conditions where the Na^+ channel is blocked with tetrodotoxin and the outward K^+ current is blocked with tetraethylammonium ions, regenerative responses in membrane potential which are Ca^{2+}-dependent can also be demonstrated.

Distinction between the Ca^{2+} channel and other ion channels

The membranes of excitable cells contain at least three cation channels for Na^+, K^+ and Ca^{2+}, respectively. The molecular basis of these channels is not known, nor is it known how a change in voltage across the membrane is able to induce a selective, transient increase in permeability to a particular cation. However, the measurement of intracellular free Ca^{2+}, in conjunction with Ca^{2+} currents, has enabled some of the properties of the Ca^{2+} channel to be defined (Baker, 1972; Baker *et al.*, 1973a, b). In particular, the time course, the effect of pharmacological agents and the significance of Ca^{2+} channels in the regulation of intracellular free Ca^{2+} have been investigated. Whilst the properties of Ca^{2+} channels have been examined in many different cells (Table 4.2), the giant axon of the squid particularly has provided some important insights into the properties of Ca^{2+} channels in excitable membranes, in spite of the fact that its physiological significance in these cells has not yet been fully established.

Baker and his co-workers have distinguished two phases of the Ca^{2+} current during an action potential in the giant axon of the squid. The fast phase is inhibited by tetrodotoxin and has all the characteristics expected if a small quantity of Ca^{2+} can enter through the Na^+ channels. In axons stimulated by one electrical pulse these Na^+ channels remain open for some time after repolarization of the membrane. Thus, the fraction of Ca^{2+} entering through the Na^+ channel could in theory be considerable. However, even when the external medium contains 112 mM Ca^{2+}, the Na^+ conductance is about 100 times that of Ca^{2+}. So under physiological conditions Ca^{2+} is unlikely to contribute significantly to the action potential itself. The slow phase, however, is not blocked by tetrodotoxin, but is by D600, La^{3+}, Mn^{2+} and Co^{2+}. As well as being distinguishable from the Na^+ channel by its time course and pharmacological properties, this slow Ca^{2+} channel, although still occurring over only a few milliseconds, can also be distinguished from the K^+ channel. For example, although the time course of activation of the Ca^{2+} channel is similar to that for K^+, tetraethylammonium ions, which inhibit the K^+ channel which opens during

Table 4.3. Comparisons between Na^+, K^+ and Ca^{2+} channels

Experimental procedure	Ca^{2+} channel	Na^+ channel	K^+ channel
Speed of activation	Slow (peak current after several msec)	Very fast (peak 1 msec)	Slow (peak 4 msec)
Speed of inactivation	Very slow (several min recovery)	Fast (4 msec recovery)	Slow (50 msec recovery)
Tetrodotoxin (10 μg/ml)	No effect	Blocks	No effect
Tetraethylammonium (1 μg/ml)	No effect	No effect	Blocks
Mg^{2+} (50–100 mM)	Inhibits	No effect	Little effect
La^{3+} (4 mM)	Inhibits	No effect	Blocks
Co^{2+}, Mn^{2+}, Ni^{2+} (5–50 mM)	Inhibits	No effect	Little effect
D600 and iproveratril (0.2 mM)	Inhibits ((−)-isomer)	No effect of (−)-isomer, (+)-isomer may inhibit	No effect
Maintained depolarization (voltage clamp) + extracellular ions	Requires Ca^{2+}	Requires Na^+	No requirement, except for intracellular K^+

For references see Baker (1972, Aidley (1971), Reuter (1973), Baker et al. (1973a, b).

an action potential and hence delay repolarization, have no effect on Ca^{2+} entry (Baker et al., 1973a, b).

The conclusions from experiments with squid axons, and other excitable cells (Table 4.2), is that the major fraction of the Ca^{2+} current across the cell membrane, under physiological conditions, is through channels highly selective for Ca^{2+}. These Ca^{2+} channels are slower to be activated and inactivated than Na^+ channels, and have pharmacological properties which distinguish them from both the Na^+ and the K^+ channels (Table 4.3).

In summary, the following parameters provide evidence for the existence of a Ca^{2+} channel: (1) Ca^{2+} current detectable under voltage-clamp conditions. (2) The time course of the Ca^{2+} current (activation and inactivation) is distinct from that of Na^+ and K^+. (3) Increase in intracellular free Ca^{2+} detectable by Ca^{2+} indicators, e.g. aequorin or arsenazo III. This may require tetanic stimulation (Baker et al., 1971). (4) Inhibited by D600 and iproveratril (verapamil), but not tetrodotoxin or tetraethylammonium ions (though not always). (5) Inhibited by La^{3+}, Mn^{2+}, Co^{2+} and high Mg^{2+} (though not always).

Problems

Although the criteria for establishing the existence of a Ca^{2+} channel appear to be relatively straightforward there are, unfortunately, several problems in interpreting experiments where the parameters described above are examined.

(1) Removal of ions from the bulk of the external medium can be carried out rapidly. However, slow diffusion from unstirred layers near the surface of the cell can obscure a requirement for a particular ion, for example Na^+ or Ca^{2+}.

(2) Alterations in external ion concentrations may cause secondary changes inside the cell. For example, the replacement of extracellular Na^+ by Li^+ or choline results in an increase in intracellular free Ca^{2+} in cells with a Na^+-dependent Ca^{2+} efflux mechanism in the cell membrane (see this chapter, section 4.6.3). Furthermore, some cations also affect cell functions besides electrical ones. Li^+, for example, inhibits adenylate cyclase in some cells and accentuates the requirement for Na^+ of hormone receptors (Campbell and Siddle, 1978).

(3) Free Ca^{2+} may not be uniformly distributed in the cell cytoplasm in an excited cell.

(4) Ions, and Ca^{2+} in particular, maintain membrane stability. Experimental manipulations may disturb this normally passive Ca^{2+}.

(5) Verapamil (iproveratril) and its methoxy derivative D600, inhibit the activation of many cells (Table 4.4). Yet the Ca^{2+} dependency of these effects is

Table 4.4. Cells affected by verapamil (iproveratril) and D600

Cell	Effect of verapamil or D600	Reference
(A) *Effects thought to be due to blockage of Ca^{2+} channels*		
Luminescent cells in *Obelia*	Blocks K^+-induced luminescence	Campbell and Hallett (1982a)
Squid giant axon	Blocks slow Ca^{2+} current	Baker et al. (1973a)
Barnacle muscle	Blocks L-Glu-induced contraction	Ashley and Campbell (1978)
Vertebrate heart	Blocks Ca^{2+} channel	Cranefield et al. (1974)
Beating heart cells in culture	Inhibits	McCall (1976)
Guinea pig atrium	Inhibits tension	Robinson and Sleator (1977)
Portal vein	Inhibits electrical and mechanical activity	Jetley and Weston (1976)
Frog sartorius muscle	Effects excitation–contraction coupling	Bondi (1978)
Pituitary	Inhibits $^{45}Ca^{2+}$ efflux and release of growth hormone	Schofield and Bicknell (1978)
Rabbit lacrimal gland	Inhibits secretion	Botelho and Dartt (1980)
Pancreatic β-cells	Inhibits glucose-induced insulin secretion	Malaisse et al. (1977)
(B) *Effects probably not due to blockage of Ca^{2+} channel*		
Pigeon erythrocytes	Inhibits adrenaline-induced cyclic AMP	Campbell and Siddle (1978)
Adrenal cortex (Y-1 cells)	Inhibits steroidogenesis	Warner and Carchman (1978)
Hepatocytes	α-Blocker	Blackmore et al. (1978)
Kidney	Stimulates gluconeogenesis	Gordon and Ferris (1971)
Embryonic bone	Inhibits parathyroid hormone *in vitro*	Herrman–Erlee et al. (1977)
Lymphocytes	Inhibits transformation by phytohaemagglutinin	Jensen et al. (1977)

not always well documented. Several effects of D600 have been shown to be competitive with Ca^{2+} (Katz and Miledi, 1969; Kohlhardt et al., 1971), but D600 can also inhibit both Na^+ and Ca^{2+} passage through the fast Na^+ channel (Baker et al., 1973a). D600 may also have effects independent of its effects on Ca^{2+} channels; for example, it sometimes acts like an α- or β-adrenergic blocker (Campbell and Siddle, 1978; Blackmore et al., 1978). Concentrations used have varied from 0.5 mg/l (approx. 1 μM) to 0.5 g/l (approx. 1 mM) (Kohlhardt et al., 1971; Baker et al., 1973a). A further problem arises because of the apparently different effects of the two optically active isomers of D600 (Batzri et al., 1973; Bayer et al., 1975; Jensen et al., 1977). Most workers use a racemic mixture of D600. Yet it seems that it is only the $(-)$-isomer that is specific for the Ca^{2+} channel. The $(+)$-isomer may block the fast Na^+ channel. Thus, although the inhibition of cell activation by a physiological stimulus is consistent with the existence of Ca^{2+} channels being opened, it is essential to demonstrate the Ca^{2+} dependency of this inhibition. Furthermore, in interpreting an inhibition by D600 one should be very wary in extrapolating from electrically excitable cells to non-excitable ones.

4.3.3. Can opening of Ca^{2+} channels lead directly to an increase in cytosolic free Ca^{2+}?

The calculations based on equation 4.2 (see section 4.2.3) showed that the amount of K^+ that must move across a membrane to establish a potential of -50 mV is negligible compared with the K^+ concentrations on either side of the membrane. A similar argument holds for Na^+. Intracellular Na^+ concentration is in the range 5–50 mM and the external Na^+ is some ten times higher. Hence the quantity of Na^+ moving into a cell during an action potential causes a negligible net change in the concentration of intracellular Na^+. However, because the concentration of free Ca^{2+} in the cytoplasm of a resting cell is only about 0.1 μM, the same may not be true for Ca^{2+}. Let us consider a hypothetical situation:

Suppose the membrane potential is depolarized by 50 mV.

Since $Q = \dfrac{CV}{ZF}$ if all of this is due to Ca^{2+} movement $(Z = 2)$ $Q_{Ca} = 2.8 \times 10^{-13}$ moles cm^{-2} (assuming $C = 1$ μF cm^{-2}). For a spherical cell of diameter 20 μm (surface area $= 1.3 \times 10^{-5}$ cm^2) this means that the total Ca^{2+} moving into the cell $= 3.64 \times 10^{-18}$ moles. Since the volume of this hypothetical cell $= 4.2$ pl this is equivalent to a concentration change in Ca^{2+} of about 1 μM; i.e. it would increase the free Ca^{2+} about ten times if all the Ca^{2+} entering were uniformly distributed throughout the cell's cytoplasm, and if none were removed by internal Ca^{2+} buffers. If, as is likely, this Ca^{2+} influx is restricted to a narrow region close to the cell membrane, then it could cause the free Ca^{2+} concentration to increase by as much as 10 μM. In squid axons an increase in intracellular free Ca^{2+} has been observed using aequorin luminescence (Fig. 4.9). Although this required somewhat extreme experimental conditions (repetitive stimulation of 200 sec^{-1} and an extracellular Ca^{2+} concentration of 112 mM), the increase in free Ca^{2+} appeared to be directly related to the quantity of Ca^{2+} entering the cell during

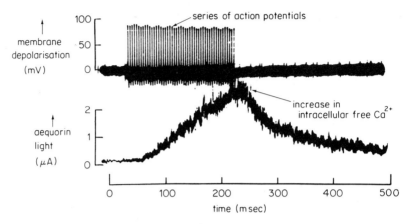

Fig. 4.9. Increase in cytosolic free Ca^{2+} induced by tetanic stimulation. Squid giant axon in artificial sea water containing 112 mM Ca^{2+} at 20°C; free Ca^{2+} change detected by aequorin luminescence. From Baker and Reuter (1975)

each individual stimulus. What is more, the increase in intracellular free Ca^{2+} could be prevented by Ca^{2+}-channel blockers (Baker et al., 1973a, b). Similarly, in *Aplysia* bursting neurones an increase in intracellular Ca^{2+}, measured using arsenazo III, has been observed associated with each burst of action potentials (Gorman and Thomas, 1978). However, these experiments do not prove that it is only Ca^{2+} moving 'electrogenically' which is responsible for the rise in intracellular free Ca^{2+}. Once the Ca^{2+} channels have opened, the electrochemical gradient may pull Ca^{2+} into the cell which could be compensated for electrically by K^+ moving out or Cl^- moving into the cell. The net result would be a small movement of charge but a relatively large movement of Ca^{2+}.

Estimation of the Ca^{2+} current (I_{Ca}) occurring during an action potential allows its effect on net intracellular Ca^{2+} concentration to be calculated.

Net change in intracellular Ca^{2+} concentration = $\Delta[Ca_i^{2+}]$

$$= \frac{1}{2Fv} \int_0^T I_{Ca}\, t\, dt \qquad (4.7)$$

where v = volume of cell in which the Ca^{2+} is distributed
I_{Ca} = Ca^{2+} current (a function of time)
t = time
T = time Ca^{2+} channel stays open.

If $\Delta[Ca_i^{2+}]$ can be estimated by an independent method (assuming it to be uniform) it is then possible to calculate v, the volume to which the Ca^{2+} increase is restricted (Beeler and Reuter, 1970; Bassingthwaite and Reuter, 1972). This can be done using voltage-clamp techniques to measure the reversal potential (E_R). E_R = the potential at which the net current across the membrane is zero, i.e. when inward current = outward current. If we assume that at this point Ca^{2+} across the membrane is transiently at electrochemical equilibrium then

$$E_R = \frac{RT}{2F} \log_e ([Ca_o^{2+}]/[Ca_i^{2+}]) \quad (4.8)$$

$$= 30.8 \log_{10}([Ca_o^{2+}]/[Ca_i^{2+}]) \text{ at } 37\,°C.$$

If $E_R = +50\,mV$, then $\log_{10}([Ca_o^{2+}]/[Ca_i^{2+}]) = 1.62$.
Hence $[Ca_o^{2+}]/[Ca_i^{2+}] = 42$.

If $[Ca_o^{2+}] = 1\,mM$ = extracellular free Ca^{2+} concentration, then the calculated $[Ca_i^{2+}]$ = Ca^{2+} concentration in electrochemical equilibrium at the top of the spike, in the absence of any inactivation of Ca^{2+} channel or movement of other ions (e.g. K^+ to cause repolarization), $\simeq 24\,\mu M$.

By applying a voltage clamp for defined times and measuring E_R for each condition, $[Ca_i^{2+}]$ can be estimated. It is then possible to plot $\Delta[Ca_i^{2+}]$ versus I_{Ca} and thus calculate v (see equation 4.7). For cardiac muscle v turns out to be only about 1% of the cell's volume.

Unfortunately, these calculations are a gross oversimplification as far as many cells are concerned. This is because no account is taken of the effect of increases in intracellular free Ca^{2+} to inactivate the Ca^{2+} channel directly or to activate K^+ conductance (Eckert and Brehm, 1979; Meech, 1979). In spite of this limitation some conclusions can be drawn from these theoretical and experimental observations: (1) Sufficient Ca^{2+} can enter the cell electrogenically, as a result of the opening Ca^{2+} channels, to contribute significantly to depolarization during an action potential, and to change the concentration of intracellular free Ca^{2+} significantly. (2) The increase in intracellular free Ca^{2+} which results from the transient increase in the permeability of the cell membrane to Ca^{2+} may be, but is not necessarily, greater than can be accounted for by simple electrogenic movement. (3) The transient increase in intracellular free Ca^{2+} may be restricted to a narrow zone near to the cell membrane. (4) The Ca^{2+} entering during a series of action potentials may summate, as has been shown in squid axons (Baker et al., 1973a) and *Aplysia* neurones (Gorman and Thomas, 1978). This has been given the name 'staircase phenomenon' (Niedergerke, 1963; Wood et al., 1969; Beeler and Reuter, 1970).

These conclusions lead one to ask whether transient increases in free Ca^{2+} close to the inner surface of the cell membrane have any effects on the electrical properties of this membrane. In particular is this Ca^{2+} causing activation of K^+ channels responsible for repolarization in single action potentials or resting phases in bursting neurones?

A cell in which Ca^{2+}-dependent action potentials play an important part in the physiological response, and which has provided some important quantitative information about voltage-dependent Ca^{2+} channels is the unicellular ciliated protozoan *Paramecium* (Eckert et al., 1976; Brehm and Eckert, 1978). This organism responds to touch, moving away from an object which it has hit by coordinated reversal of ciliary movement. The primary signal produces a membrane depolarization which moves from one end to the other, analogous to that in metazoan electrically excitable cells, the bulk of the current being carried by Ca^{2+}. Measurement of intracellular free Ca^{2+} with aequorin has shown that an increase in free Ca^{2+} in the cell is induced by this electrical excitation of the cell membrane. This increase in cytoplasmic Ca^{2+} is required for the ciliary response (Fig. 4.10) and the ultimate recovery of the electrical properties of the membrane. The early inward current is fully activated within 3 msec and decreases over the following 5 msec, being some three times slower than Na^+ currents in excitable cells such as squid axons. The membrane current obeys Ohm's Law and is dependent on the concentration of extracellular and intracellular free Ca^{2+}.

$$I_{Ca} = g_{Ca}(E_m - E_{Ca}) \tag{4.9}$$

where $I_{Ca} = Ca^{2+}$ current
g_{Ca} = membrane conductance to Ca^{2+} (really change in conductance)
E_m = membrane potential
E_{Ca} = Ca^{2+} equilibrium potential.

E_{Ca} starts at about $+120$ mV (initially $[Ca_o^{2+}] = 1$ mM; $[Ca_i^{2+}] = 0.1$ μM) but becomes less +ve as $[Ca_i^{2+}]$ increases. I_{Ca} is sensitive not only to the concentration of extra- and intracellular Ca^{2+} but also to the concentration of other cations outside the cell. The resting membrane potential is sensitive to surface charge, which itself is dependent on extracellular cation binding. The protozoan is able to compensate for alterations in surface potential, a mechanism often found in such a fresh water organism. This allows the cell membrane to be excited in different ionic environments by the same physiological stimulus.

Mutants have been isolated which have no ciliary reversal on depolarization. In these mutants no voltage-dependent Ca^{2+} current can be detected. However, ciliary activity can be stimulated by addition of a detergent, such as Triton, which

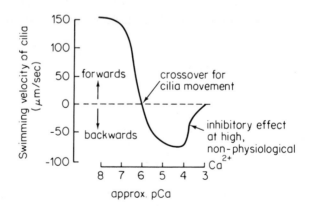

Fig. 4.10. Increase in free Ca^{2+} in *Paramecium* alters direction of ciliate movement. Triton-extracted cilia + MgATP + Ca^{2+} (i.e. cell membrane made permeable to Ca^{2+} but contractile apparatus intact). From Naitoh and Kaneko (1972) copyright © 1972 by the American Association for the Advancement of Science

increases non-specifically the concentration of intracellular Ca^{2+}. Furthermore, removal of normal cilia by dissection causes loss of the Ca^{2+} current, which returns on regeneration of the cilia. These observations confirm the physiological significance and location of Ca^{2+} channels within the cilia. They raise an interesting question, relevant to other localized electrical activity in cells such as the endplates in nerve and muscle: How do these cells maintain polarity in a fluid membrane? Why, for example, do the Ca^{2+} channels in *Paramecium* not diffuse throughout the cell membrane?

The maximum Ca^{2+} current, I_{Ca}, is approximately 20 nA. Since there are about 15 000 cilia per one *Paramecium* the Ca^{2+} current per cilium is 1.3 pA. The electromotive force at peak current, taking into account the change in intracellular free Ca^{2+} concentration, is about 75 mV. Therefore, the calculated activated Ca^{2+} conductance (g_{Ca}) per cilium is about 17 pS (S = a Siemen, the SI equivalent of a mho), calculated from equation 4.9. This is within an order of magnitude of activated channels in molluscan neurones, the Na^+–K^+ channel in frog muscle and Na^+ in squid axon. If a single Ca^{2+} channel has a conductance of 0.1 pS when activated, like that of molluscan neurones, then there must be about 170 channels per cilium or 30 channels per μm^2. This figure is similar to the 10–100 Ca^{2+} channels per μm^2 estimated for molluscan neurones, and the 2.5–500 Na^+ channels per μm^2 estimated for other nerves.

Our present understanding of the role of Ca^{2+} channels in *Paramecium* is illustrated in Fig. 4.11. Although these Ca^2 channels have many similar characteristics to those investigated in metazoans (Table 4.2), great difficulty has been found in producing significant inhibition of voltage-dependent Ca^{2+} currents in *Paramecium* using Mn^{2+}, Co^{2+}, La^{3+} or D600. They are not blocked by tetrodotoxin. A similar calculation to that for cardiac muscle (equations 4.7 and 4.8), where the volume of a cilium is approximately 3×10^{-16} l and integration of I_{Ca} gives a value for net charge of movement of Ca^{2+} into the cell, shows that whilst the Ca^{2+} channels are open the Ca^{2+} current would produce a net change of about 12 μM Ca^{2+} in each cilium.

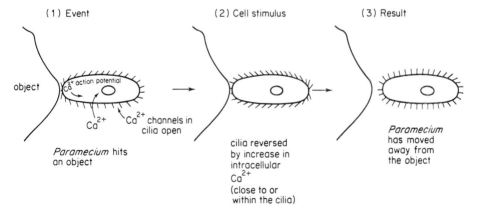

Fig. 4.11. The role of intracellular Ca^{2+} in reversing protozoan ciliate movement

It can be concluded that Ca^{2+} currents play an important part in the physiological response of many cells and that they frequently lead to significant elevations in the concentration of cytoplasmic free Ca^{2+}.

4.3.4. How does the Ca^{2+} movement through Ca^{2+} channels stop?

All electrical excitability phenomena involve two phases—activation and recovery. This usually occurs quite rapidly (msec to sec) and is necessary if the electrical properties of the cell are to be restored so that it can respond to another stimulus. The inactivation, or arrest, of Ca^{2+} entry into the cell through the Ca^{2+} channel could occur by three separate mechanisms: (1) The increase in intracellular free Ca^{2+}, together with the depolarization of the membrane, could allow Ca^{2+} to reach its equilibrium potential. (2) An increase in the conductance of another ion, e.g. K^+, might produce an outward current. This would counteract the inward current of Ca^{2+}, repolarize the membrane, and, since the Ca^{2+} channels are voltage sensitive, shut these channels. (3) The Ca^{2+} channel could be shut or inactivated by some means independent of membrane potential.

Of course, a different combination of these mechanisms may be responsible for recovery in each individual cell type. Injection of EGTA into *Paramecium*, molluscan neurones, and other cells which have Ca^{2+}-dependent action potentials usually prolongs the inward current and hence delays the recovery of the membrane potential. This suggests that, whatever the mechanism of shutting off inward Ca^{2+} movement, inactivation of the Ca^{2+} channels is dependent on an increase in intracellular free Ca^{2+}. The Ca^{2+} conductance in many cells, including barnacle muscle, molluscan neurones and tunicate eggs, is sensitive to changes in intracellular free Ca^{2+} (Eckert and Brehm, 1979), being activated in the range 0.01–0.1 μM and inhibited when the intracellular Ca^{2+} concentration is greater than 1 μM. In *Paramecium* the most important rapid effect of the increase in intracellular free Ca^{2+} seems to be inactivation of the Ca^{2+} channels. The mechanism causing this is as yet unknown, and is not related directly to the voltage sensitivity of this channel. The relative importance of this type of inactivation in molluscan neurones and other cells, as against the role of Ca^{2+}-activated K^+ conductance (see below), has been a controversial issue for some years (Lux and Hofmeier, 1979; Meech, 1979). It is therefore appropriate now to consider what are the effects of the increase in intracellular free Ca^{2+} on the electrical properties of membranes, and how this might explain recovery of electrical excitability and other responses of the cell.

4.4. Effects of intracellular Ca^{2+} on the electrical properties of membranes

4.4.1. Possible physiological significance

Electrophysiologists have identified physiological responses of cells which involve either an increase or a decrease in the permeability of the cell membranes to ions. These phenomena include the action of excitatory and inhibitory transmitters released from the nerve terminal and affecting the next nerve, the secretion of ions and fluids by specialist cells in skin, kidney and salivary glands, and the response of photosensitive plant and animal cells to light. Hopefully, after examining the tables in various chapters of this book, the reader will be able to compile a longer list of such phenomena. The question arises, are these physiological responses the result of a direct effect of the primary stimulus on the permeability of the cell membrane to individual cations (e.g. Na^+, K^+ or Cl^-), or is the effect transmitted via a secondary signal, i.e. a change in intracellular free Ca^{2+} concentration?

Fast (msec) excitatory or inhibitory post-synaptic potentials in nerves are thought to be of direct ionic origin, i.e. Na^+ or Cl^-, respectively. However, at the dendrites of some nerves slower (many msec to sec) excitatory and inhibitory potentials also occur. These slower potentials could be caused by changes in intracellular cyclic AMP, cyclic GMP (Greengard, 1976) or intracellular Ca^{2+}. Many electrical changes in the membranes of other cells are slow enough to be mediated by chemical changes within the cell, rather than by very fast permeability changes induced in the membrane directly by the cell stimulus. Similar changes in membrane potential can also be found in non-excitable cells. For example, insulin induces K^+ efflux from liver and adipose tissue. Insulin also stimulates K^+ uptake by muscle and causes hyperpolarization (Zierler, 1972). Glucagon and exogenous cyclic AMP hyperpolarize liver cells (Dambach and Friedmann, 1974). In contrast, noradrenaline depolarizes brown fat cells (Horwitz et al., 1969). Acetylcholine depolarizes cells in the exocrine pancreas, an effect which can be mimicked by intracellular Ca^{2+} injection or ionophore A23187 (Iwatsuki and Petersen, 1977). Acetylcholine also affects the permeability of membranes to water. It has been known for a long time that changes in extracellular Ca^{2+} concentration cause similar changes in the water permeability of cells (Baptiste, 1935; Fukuda, 1935; Adelman and Moore, 1961; Lipicky and Bryant, 1963). Whether these effects are mediated by direct effects of Ca^{2+} on the outside surface of the cell or through intracellular Ca^{2+} changes is not known.

Changes in intracellular Ca^{2+} affect the electrical activity of several cells

Table 4.5. Possible sites of action of intracellular Ca^{2+} in electrically excitable cells

	Cell type	Proposed effect of intracellular Ca^{2+}
1.	Molluscan bursting neurones	Delays onset of next burst of action potentials
2.	Muscle	Required for Ca^{2+} release from sarcoplasmic reticulum
3.	Electronically connected neurones and heart cells	Disconnects electrical conductivity between cells
4.	Nerve terminals	Stimulates neurotransmitter release
5.	Ciliated protozoa	Stimulates cilia reversal
6.	Photoreceptors	Mediates light-induced electrical changes
7.	Light-emitting cells in some luminous organisms	Activates intracellular luminescent reaction

The Ca^{2+} for all but one of these effects, i.e. photoreceptor activation, is thought to arise directly from activation of a Ca^{2+} channel in the cell membrane and not from intracellular Ca^{2+} release.

(Table 4.5). Let us examine three of these, which may be brought about by an increase in cytoplasmic free Ca^{2+}.

4.4.2. Intracellular Ca^{2+} and K^+ conductance

K^+ conductance is the reciprocal of the resistance of the membrane to K^+. Therefore, from Ohm's Law

$$\Delta g_K = \text{change in potassium conductance} = I_K/(E_m - E_K)$$

where I_K = net potassium current
E_m = membrane potential
E_K = potassium equilibrium potential

A voltage-induced increase in K^+ conductance is responsible for repolarization of the cell membrane during the downward phase of the action potential in squid axons. Squid axons seem to be unusual in that they contain only this type of K^+ channel. Other electrically excitable cells usually also contain a K^+ channel that is activated when there is an increase in intracellular free Ca^{2+} (Table 4.6) (Meech and Standen, 1975; Meech, 1976, 1979). This Ca^{2+}-induced increase in K^+ conductance is much slower to be activated than the voltage-dependent K^+ conductance (channel). It is also slower to desensitize. This time course is partly a reflection of the duration of the intracellular free Ca^{2+} 'transient'. Similar Ca^{2+}-activated increases in K^+ permeability may exist in non-excitable cells such as erythrocytes (Flatman and Lew, 1977; Meech, 1976).

Table 4.6. Cells in which membrane conductance may be regulated by intracellular Ca^{2+}

Cell type	Ion conductance change	Reference
Neurones–snail (*Helix*)	K^+	Lux and Hofmeier (1979)
sea hare (*Aplysia*)	K^+	Meech (1976)
sea hare (photoresponse)	K^+	Brown and Brown (1972); Brown et al. (1975)
cat spinal motor neurone	K^+	Krnjevic and Lisiewicz (1972)
autonomic nerves	K^+?	Gabella and North (1974)
Muscle–cardiac Purkinje fibres	K^+	Isenberg (1975)
vertebrate smooth	K^+	Bülbring and Tomita (1977)
vertebrate skeletal	K^+	Meech (1976)
crustacean	K^+	Krnjevic (1974)
Electroreceptors—skate	K^+	Clusin et al. (1975)
Photoreceptors—invertebrate	Na^+	Hagins (1972)
vertebrate	Na^+	Hagins (1972)
Salivary gland—insect	Na^+, Cl^-	Berridge (1976b)
Ciliated protozoa—*Paramecium*	K^+	Brehm et al. (1978); Brehm and Eckert (1978)
Mammilian erythrocyte (?physiological)	K^+	Simons (1975)

For further references see Meech (1979), Eckert and Brehm (1979) and Lux and Hofmeier (1979).

The evidence for Ca^{2+}-induced increases in K^+ permeability is based on: (1) dependence of K^+ conductance on extracellular Ca^{2+} and Ca^{2+} current, together with effects of Ca^{2+}-channel blockers; (2) activation of K^+ conductance by injection of Ca^{2+} into the cell; (3) inhibition of K^+-conductance activation by injecting EGTA into the cell; (4) the ineffectiveness of tetraethylammonium ions, which inhibit the voltage-dependent K^+ conductance.

In molluscan neurones, for example, a component of the increased K^+ conductance associated with a complete action potential is dependent on Extracellular Ca^{2+}, and can be considerably reduced by D600 and Co^{2+}. Even in non-excitable cells, where it is better to use the term Ca^{2+}-activated increase in K^+ permeability, metabolic poisons like F^- and iodoacetate, which increase K^+ permeability in erythrocytes, require the presence of extracellular Ca^{2+} (Gardos, 1958; Romero and Whittam, 1971; Blum and Hoffman, 1972). Microinjection of Ca^{2+} into *Aplysia* and *Helix* neurones increases K^+ permeability. Because of the rapid removal of the injected Ca^{2+} by internal buffers, the injection of Ca/EGTA is more effective than $CaCl_2$ alone. Outward movements of K^+ induced by intracellular Ca/EGTA (free Ca^{2+} concentration about $1 \mu M$) have been observed in molluscan neurones (Meech, 1976) and cardiac Purkinje fibres (Isenberg, 1975), as well as in erythrocyte ghosts resealed in the presence of Ca/EGTA buffers (Simons, 1975).

Attempts have been made to estimate the intracellular Ca^{2+} concentration required to activate K^+ permeability, by using Ca/EGTA buffers injected into the cell or by measuring the mean free Ca^{2+} concentration with aequorin or arsenazo III (Fig. 4.12). The minimum free Ca^{2+} concentration necessary can also be estimated by a null method which requires measurement of the membrane

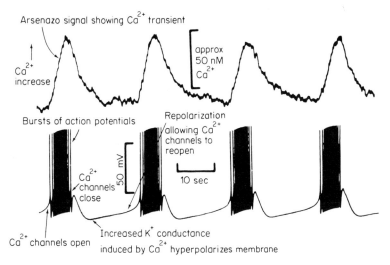

Fig. 4.12. The role of intracellular Ca^{2+} in *Aplysia* bursting neurones. From Gorman and Thomas (1978); reproduced by permission of the *Journal of Physiology*

potential, fixed by voltage clamp, at which there is no increase in Ca^{2+}-dependent K^+ conductance. Assuming that this potential is close to the Ca^{2+} equilibrium potential, otherwise Ca^{2+} will move into the cell, then in *Aplysia* neurones the calculated resting free Ca^{2+} is about 50 nM (Meech, 1976). Whilst it is extremely difficult to take into account any heterogeneity in the distribution of free Ca^{2+} in the cytoplasm after one or a burst of action potentials, it is likely that about 1 μM free Ca^{2+} is necessary, under physiological conditions, to activate the K^+ conductance sufficiently.

The conclusions from these experiments are that many excitable cells contain two K^+ conductance pathways. The first is activated rapidly by membrane depolarization, and is inhibited by tetraethylammonium ions. The second is activated more slowly by Ca^{2+} and is insensitive to tetraethylammonium ions. However, it has been pointed out by Heyer and Lux (1976) that an alternative explanation for the apparent Ca^{2+}-activated K^+ conductance exists. They observed an inverse relationship between Ca^{2+} and K^+ conductance in some cells. This suggested to them that there might be a direct coupling between Ca^{2+} and K^+. Baker (1972) showed that in squid axons the voltage-dependent Ca^{2+} channels are distinct from the fast K^+ channels. But is it possible that the Ca^{2+} channel is the same as the *slow* K^+ channel? Depolarization could increase the permeability of this channel to Ca^{2+}, increasing the concentration of intracellular free Ca^{2+}. This, in turn, would transform the channel from a Ca^{2+}- to a K^+- conducting channel. The resolution of this question is now a chemical problem. These hypothetical channels must be isolated, characterized chemically, and reconstituted in artificial lipid membranes. *Paramecium* could prove very useful since it is possible to strip the cilia, which contain most of the Ca^{2+} channels, off the protozoan.

The hyperpolarization induced by increased K^+ conductance plays a vital role in regulating the frequency of bursts in action potentials in pacemaker cells. For example, in *Aplysia* a new burst cannot occur until the hyperpolarization has ended (Fig. 4.12). Whilst Ca^{2+}-mediated changes in the permeability of ions other than K^+ may occur, the relationship between intracellular Ca^{2+} and K^+ permeability seems to be a property common to many cells. The following criteria need to be satisfied if a physiologically significant effect of intracellular Ca^{2+} on K^+ permeability is to be established: (1) The physiological stimulus, be it electrical or chemical, must cause an increase in K^+ permeability. (2) This effect of the stimulus should be mimicked by experimental conditions which increase intracellular free Ca^{2+} (see Chapter 2, section 2.3). (3) An increase in free Ca^{2+} detectable by aequorin, arsenazo III or another Ca^{2+} indicator should correlate with the K^+ conductance increase induced by injection of Ca/EGTA buffers into the cell. (4) In excitable cells there should be a correlation between Ca^{2+}-channel activation and K^+-conductance activation, the latter occurring after the former. (5) The molecular basis of the effects of intracellular Ca^{2+} should be established.

The first four criteria have been satisfied in several cell types (Tables 4.5 and 4.6). However, virtually nothing is known in molecular terms about how intracellular

Ca^{2+} might regulate K^+ permeability. For example, is there a role for regulatory Ca^{2+}-binding proteins (see Chapter 3, section 3.4.2) in this phenomenon?

4.4.3. Intracellular Ca^{2+} and communication between adjacent cells

Ever since it was first discovered that electrical impulses could be transmitted from one cell to another, arguments raged as to whether this occurred directly through an electrical conducting pathway between the cells, or whether there were chemical agents which transmitted these signals (Dale, 1934). The evidence for chemicals communicating signals between cells is now very strong. Neurotransmitters released by a nerve terminal activate the next nerve or, in the case of the neuromuscular junction, a muscle cell. Neurosecretory and endocrine cells, on the other hand, secrete hormones into the blood which act on other cells some distance from the releasing cell. However, during the last 40 years many examples of direct electrical contact between cells in the animal and plant kingdoms have been discovered (Loewenstein, 1966, 1975; Loewenstein and Penn, 1967; Gilula and Epstein, 1976; Rose and Loewenstein, 1971, 1976). As with the cell membrane, electric currents between such cells are carried by ions and not by electrons *per se*. But the similarity between the intercellular electrical conducting pathway and that across the membrane of one cell seems to end here. The elegant work of Loewenstein, Rose and their colleagues has shown that much larger molecules with molecular weights in the range 300–1500, including intracellular regulators like cyclic AMP, can also pass from one cell to another. In the salivary gland of the midge *Chironomus* the upper limit of permeability is 1400–1600 daltons, each channel having a diameter of about 1.4 nm. Macromolecules such as proteins and nucleic acids do not readily pass between cells.

The region responsible for this intercellular communication pathway has been identified using the electron microscope (Fig. 4.13) and is called the gap junction. In the region of this gap junction the two cell membranes are very close together, being separated by a small gap of a few nanometers. After enzyme digestion of the

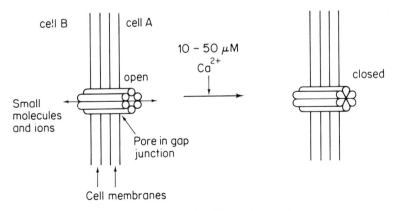

Fig. 4.13. The effect of Ca^{2+} on the gap junction

cells followed by treatment with detergent (Goodenough, 1975), a subcellular fraction enriched in 'gap junctions' can be isolated. Using this method two major proteins have been identified electrophoretically from mammalian liver gap junctions—a hydrophobic protein of molecular weight about 25 000 and a hydrophilic protein of molecular weight about 18 000 (Gilula and Epstein, 1976).

Cells which communicate through gap junctions include several that are electrically excitable, such as invertebrate nerves and mammalian heart muscle, as well as non-electrically excitable cells of epithelia, liver and adipose tissue. The transmission of ions and metabolites has also been demonstrated between cells in culture. This latter transfer is remarkably independent of the cell type or species of origin. Thus, cells prepared from the lens of a rabbit eye can form a communicating junction between mouse liver cells or human fibroblasts. The only major restriction seems to be that it is impossible to establish such communicating pathways between cells from different phyla (Gilula and Epstein, 1976). The coupling of adjacent cells is usually bidirectional, although crayfish giant nerves seem to be at least one exception to this rule.

This form of cell–cell communication plays an important function in transmitting electrical impulses rapidly from one arthropod nerve to the next, and along the epithelia of coelenterates. For example, in sea anemones electrical messages sent to the foot of the animal tell it to detach itself so that it can move to another support. In endocrine and exocrine gland cells (Petersen, 1976) (Chapter 7, section 7.2.4) electrical communication between cells seems to play a role in enabling the whole organ to secrete in response to a physiological stimulus. It is also essential for synchronous beating in heart muscle cells. However, the physiological function of metabolite transfer and of molecules up to 1500 molecular weight from cell to cell is less clear. Furthermore, gap junctions exist in many non-electrically excitable cells like the liver where their physiological function is not understood. It is possible that they play a role in the growth and differentiation of tissues, and perhaps help to determine the ultimate size and shape of organs in metazoans. Many experimental conditions, particularly those which change free Ca^{2+} and pH in the cell, have been found which regulate the permeability of gap junctions to ions and other substances. Changes in cell–cell communication may be important in tissue development (Loewenstein, 1979). However, what is lacking is any firm evidence that physiological regulators such as hormones and neurotransmitters, or drugs, can affect gap junctions in this way and thus affect tissue function. Nevertheless, the existence of this communicating pathway between cells and the ability to regulate it experimentally by changes in intracellular Ca^{2+} warrant a closer examination of its properties, in the hope that it might throw more light on its physiological and evolutionary significance.

At the end of the last century Ringer and Sainsbury (1894) found that removal of Ca^{2+} from the fluid bathing many tissues caused them to disaggregate into single cells. One preparation in which this phenomenon was first investigated was the group of cells formed after fertilization of a sea urchin egg (Herbst, 1900; Dan, 1947; Steinberg, 1958). The requirement for external Ca^{2+} if cells are to remain adhered to each other has been exploited in the preparation of many isolated cell suspensions for tissue culture experiments (Willmer, 1965). Removal of extracellular Ca^{2+} also abolishes the communication pathway between adjacent cells (Nakos et al., 1966; Loewenstein, 1966, 1975; Gilula and Epstein, 1976). The role of extracellular Ca^{2+} in the formation and maintenance of the structure of gap junctions is probably passive (see Chapter 2, section 2.3.3) in that it is unlikely that changes in extracellular Ca^{2+} have a physiological function in determining whether gap junctions form or not. Yet strong evidence exists that changes in *intra*cellular Ca^{2+} can regulate the permeability properties of gap junctions, a potentially active role for Ca^{2+}.

The salivary gland of the midge *Chironomus* has proved very useful in investigating the effect of intracellular Ca^{2+} on gap junctions. The cells are easily identifiable, are accessible and are relatively large (approx. 100 μm diameter). Microelectrodes and micropipettes can be inserted into adjacent cells and substances injected into them without causing significant damage to the cell.

The first clue that the permeability of gap junctions might be regulated by changes in intracellular Ca^{2+} came from experiments where one cell was deliberately damaged mechanically. This resulted in blockage of the communication pathway between this cell and its immediate neighbours, but only if Ca^{2+} was present in the extracellular medium (Fig. 4.15). Many other experiments have confirmed that an increase in intracellular Ca^{2+} can seal gap junctions. Intercellular communication has been followed in three ways: (1) By measuring the potential between an electrode inside one cell and one in the external medium, when a change in membrane potential is induced in the cell next to it. (2) By observing the rate of movement of dyes from cell to cell. Dyes like fluorescein (molecular weight 332) and Chicago sky blue (molecular weight 993) are fluorescent and so their movement can easily be observed using a fluorescence microscope. Fluorescent compounds of higher molecular weight can be obtained by tagging fluorescein to a substance not normally fluorescent. (3) By measuring the passage of radioactively labelled compounds from cell to cell. This can be done by autoradiography or, after isolation of individual cells, by scintillation counting.

The evidence that an increase in intracellular free Ca^{2+} seals gap junctions in *Chironomus* salivary gland cells, making the junctions impermeable to both ions and organic molecules, can be summarized as follows:

(1) The transfer of fluorescein between cells is prevented when a hole is made in the cells on either side of the cell originally injected only if Ca^{2+} is present in the medium bathing the cells.

(2) Conditions which increase intracellular free Ca^{2+}, e.g. ionophore A23187, substances which release Ca^{2+} from mitochondria (CN^-, ruthenium red, dinitrophenol), and microinjection of Ca^{2+}, inhibit the movement of fluorescein and electrical conductivity between cells. Microinjection of Ca/EGTA buffers has shown that the permeability of the channel between cells decreases with increasing free Ca^{2+} concentration over the range 0.1–10 μM. Between 10 and 50 μM Ca^{2+} the channel is virtually completely blocked. In view of the very poor Ca^{2+}-buffering capacity of EGTA under these conditions (see Chapter 2, section 2.7) i.e.

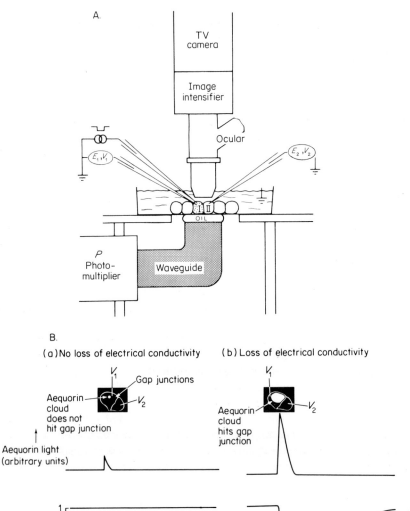

Fig. 4.14. Demonstration of the effect of Ca^{2+} on the gap junction. Electrical conductivity is measured by the ratio (V_2/V_1) of the potential between two electrodes in adjacent cells. (A) Diagram of apparatus. (B) Ca^{2+} microinjected into cell (V_1); Aequorin, visualized by image intensification, monitors cytoplasmic free Ca^{2+}. (a) No change in V_2/V_1 ($=1$) since Ca^{2+} does not reach the gap junction. (b) $V_2 = 0$ since Ca^{2+} reaches gap junction. From Rose and Loewenstein (1976); reproduced by permission from *Nature*, Copyright © 1976 Macmillan Journals Limited

when pCa > 5 μM and pH ≈ 7.2–7.4, the upper value estimated for free Ca^{2+} by this method can only be approximate.

(3) The most convincing evidence comes from a direct correlation between measured free Ca^{2+} in the cell and junctional permeability. Using cells injected with aequorin, changes in free Ca^{2+} in different parts of the cell can be detected and quantified by amplifying the light image from the cell with an image intensifier (see Chapter 2, section 2.5.3). The spacial resolution of this technique is about 1–2 μm and the detection limit for Ca^{2+} is about 0.5 μM. It has not therefore been possible so far to detect the glow of aequorin from a resting cell using image intensification. In spite of this limitation, conditions which increase intracellular free Ca^{2+} e.g. microinjection of Ca/EGTA, addition of A23187 or mitochondrial inhibitors, cause a visible cloud of light in the cell. Contact between adjacent cells through junctions is inhibited, or blocked, only when the light reaches the surface next to the adjoining cell (Fig. 4.14) (Rose and Loewenstein, 1976, 1979). Small injections of Ca^{2+} into the cell produce only a transient, localized light cloud and no inhibition of coupling between the cells. In this case the Ca^{2+} is presumed to be rapidly taken up by intracellular buffers such as the mitochondria.

How fast is the action of Ca^{2+} on the junction and how quickly does the junction return to its original state after the free Ca^{2+} in the cytoplasm has returned to normal?

Fluorescein takes about 1 sec to cross the junction, placing an upper limit on the time taken to close the gap junctions. By having aequorin in both the Ca^{2+}-injected cell and its neighbouring cell it is possible to show that no detectable increase in free Ca^{2+} occurs in the non-injected cell. Since the Ca^{2+} gradient across the junction is about 10–100-fold this leads to the interesting conclusion that not only does the effect of Ca^{2+} on the junction occur within a few milliseconds but also that this could be another example of an increase in free Ca^{2+} blocking its own permeability. The recovery of the junction is much slower, taking a minute or two after the free Ca^{2+} has apparently returned to normal.

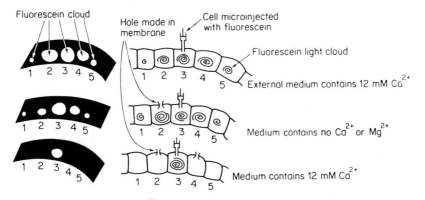

Fig. 4.15. Effect of intracellular Ca^{2+} on fluorescein transfer between cells. Cell Number 3 microinjected with fluorescein. From Oliveira-Castro and Loewenstein (1971), reproduced by permission

A problem in the interpretation of these experiments is that the increase in intracellular free Ca^{2+} causes the cell to depolarize. However, this depolarization is neither sufficient nor necessary for sealing of the junctions. For example, depolarization induced by high K^+ or voltage clamp produces no immediate change in free Ca^{2+} near the junction nor does it affect its permeability properties. Furthermore, by avoiding effects of injected Ca^{2+} on non-junctional membranes, sealing of gap junctions can be demonstrated without depolarization of the cell occurring.

Do the junctions seal to ions at a different free Ca^{2+} concentration from that required to seal them against larger, organic molecules? The answer is 'yes'. A concentration of about 50 μM free Ca^{2+} in the cell reduces the permeability of the junctions for both ions and fluorescein to 10% or less of normal. However, at lower free Ca^{2+} concentrations (0.1–10 μM) a more selective change in permeability can be detected. For example, 1 μM free Ca^{2+}, estimated using aequorin luminescence, reduces the passage of fluorescein between cells, but has little or no effect on electrical conductivity between the cells. This could be explained by there being one channel for ions and another for organic molecules, or by Ca^{2+} regulating the size of the pores in the junction (Fig. 4.13). However, the most likely explanation is that the number of pores which are open depends on the concentration of free Ca^{2+}. The closing of 50% of the pores would substantially reduce the rate of movement of fluorescein between two cells without much affecting the ability of the junctions to transmit a change in membrane potential from one cell to another. If it were possible to measure directly ion currents and conductances across the gap junction itself, then an effect of Ca^{2+} on pore number could be detected.

Ca^{2+} can uncouple electrical synapses (Baux et al., 1978) but the evidence for intracellular Ca^{2+} affecting the permeability properties of gap junctions in mammalian cells is less well documented (Penn, 1966; Loewenstein, 1975, 1979). An increase in Ca^{2+} inside heart Purkinje cells (DeMello, 1974) and human lymphocytes (Oliveira-Castro and Barcinski, 1974) uncouples the communicating link between cells. This effect of Ca^{2+} can be prolonged by addition of caffeine and can be mimicked by microinjection of Sr^{2+} but not Mg^{2+}. Injection of Na^+ into Purkinje cells also induces uncoupling between cells (DeMello, 1974, 1975). It would be interesting to know whether this effect is caused by Na^+ increasing intracellular free Ca^{2+} either through the Na^+-dependent efflux mechanism in the cell membrane (Baker, 1972; Blaustein, 1974; Blaustein et al., 1974), or through Na^+-induced release of Ca^{2+} from mitochondria (Crompton et al., 1976).

If Ca^{2+} can regulate gap junctions in any cell, why do increases in intracellular free Ca^{2+} induced by the primary stimulus acting on these cells, for example action potentials in the heart or 5-hydroxytryptamine on fly salivary gland, not close these junctions. The answer probably lies in the location of the increase in free Ca^{2+}. So long as intracellular buffers prevent Ca^{2+} increasing next to the junctions, they will remain open. Only when it becomes possible to study local changes in free Ca^{2+} in small cells, with indicators like aequorin, will this problem be fully resolved.

The mechanism by which Ca^{2+} exerts its effect on cell–cell junctions is unknown. The rapidity of its effect, the reasonably fast recovery on returning free Ca^{2+} to approximately normal, together with the fact that the free Ca^{2+} must be increased very close to the junction, argue for a direct effect of Ca^{2+} binding to the junction. It is, therefore, unnecessary to invoke other intracellular regulators such as cyclic AMP or cyclic GMP causing protein phosphorylation, or Ca^{2+}-activated proteolysis, though these cannot be ruled out. It is tempting to suggest a role for a specialized Ca^{2+}-binding regulatory protein (see Chapter 3, section 3.4.2).

What is the physiological significance of the effect of Ca^{2+} on cell–cell junctions? To the author's knowledge no physiological regulators have been shown to regulate the permeability of gap junctions by a Ca^{2+}-dependent mechanism. Loewenstein has made the interesting suggestion that it plays an important role in minimizing tissue injury. Injury to one cell, in the absence of a switch-off mechanism, could lead to massive loss of ions and metabolites from the adjoining cells through the gap junctions. Because of the large Ca^{2+} gradient across the cell membrane, one of the earliest changes in the injured cell will be a very large increase in free Ca^{2+} switching off the junctions (see Chapter 9). Under these circumstances Ca^{2+} can be considered to be mediating a threshold phenomenon (see Chapter 2, section 2.2). It would be interesting to know whether physiological substances exist which regulate the permeability of these junctions in either a 'threshold' or a 'graded' fashion, and thereby regulate tissue responsiveness to primary stimuli or affect tissue development.

Before leaving this subject it would be well to point out that in spite of the convincing nature of arguments for the role of Ca^{2+} in this phenomenon other hypotheses have been proposed. In particular, it has been suggested that regulation of gap junctions by intracellular pH is the major mechanism involved. The chemical basis of sealing of the junctions should resolve this controversy.

4.4.4. Intracellular Ca^{2+} and the response of photoreceptors to light

In addition to the supply of energy through photosynthesis, visible light (about 390–760 nm) has effects on bacteria, eukaryotic unicellular organisms as well as on plants and animals. Light absorbed by specialized pigments in photoreceptive cells causes chemical or electrical changes in these cells. These are then communicated to other cells, e.g. nerves, thereby eliciting the response of the organ to light (Erlanger, 1976). Microorganisms exhibit at least four such phenomena: (1) phototropism—light-orientated growth; (2) photophobaxis—accumulation of organisms in an area of dark; (3) photokinesis—light-induced changes in speed of movement; (4) phototaxis—light-induced changes in positioning of cells.

Light causes migration of insects and marine animals. It also affects behaviour in man through photopigments. In plants, too, light induces behavioural responses. Think, for example, of the effect of light and dark on blooming flowers, on seed and spore germination, and on the growth and development of leaves, stems and roots.

Is there a role for intracellular Ca^{2+} in any of these responses of photoreceptive cells to light? The most convincing evidence for such a role comes from studies on the eyes of vertebrates and a few invertebrates (Hagins, 1972; Hagins and Yoshikami, 1974). Vision in animals begins with the absorption of photons by specialized molecules in the photoreceptive cells. These cells have the remarkable ability to respond both to very dim and to bright light, sometimes over six orders of light intensity. In the commonest type of photoreceptor the membrane containing the photosensitive pigment is confluent with the cell membrane. These include vertebrate cones responsible for colour vision and respond to bright light, molluscan rods, photoreceptors in the horseshoe crab *Limulus* and insect ommatadia. Less common, but including vertebrate rods responsible for detecting very low light intensities, are photoreceptors where the photosensitive pigment is within a structure (disc) which is separate from the cell membrane.

The 11-*cis* optical isomer of an aldehyde of vitamin A, retinal, (Fig. 4.16) is the substance which absorbs the incident photons. It is bound to a protein called opsin, the complex being known as rhodopsin. The light reaction converts *cis*-retinal eventually to *trans*-retinal. This activates the photoreceptor so that it can pass on the information that it has received light to the central nervous system. Three major questions arise: (1) How does absorption of photons by retinal in one part of the cell activate the synapse with the next cell, which may be several micrometres away from the site of light absorption? (2) How are photoreceptors able to adapt to such a wide range of light intensities? (3) What is the role of intracellular Ca^{2+} in these two aspects of photoreceptor response.

The transmission of the message of light activation from the membranes containing retinal to the synapse between the photoreceptor and the next cell could occur in three ways: (1) A transducer or intracellular second messenger could be produced which diffuses to the synaptic end of the cell, and there stimulate transmitter release. (2) A change in membrane potential could occur which, by electrotonic spread, could produce sufficient change in potential at the synaptic end of the photoreceptor to stimulate release of transmitter. In cells too long for sufficient change by electrotonic spread, a propagated action potential would be required to travel down the cell to its terminal. (3) After absorption of light by the first disc, a signal could be passed from disc to disc, being amplified at each stage in an analogous way to the stages of a photomultiplier, so that at the end a large release of Ca^{2+} from an intracellular store occurs in the synaptic terminal.

Passive diffusion of a second messenger, for example cyclic AMP or Ca^{2+}, from one end of the cell to the other would be too slow to account for the rapid activation of the optic nerve after absorption of light by the photoreceptor. However, electrotonic spread of a change in membrane potential can occur in cells less than 20 μm long. Larger photoreceptors, which exist in some invertebrates, generate action potentials (Hagins, 1972).

Electrical activity in the retina has been known for more than 50 years (Granit, 1933). Voltage-clamp experiments, together with experimental manipulation of extracellular Na^+, have established that changes in the membrane potential of

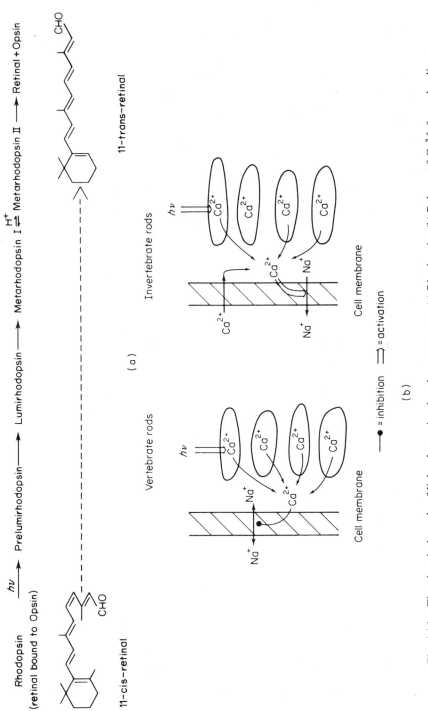

Fig. 4.16. The chemical result of light absorption by photoreceptors. (a) **Rhodopsin**. (b) Release of Ca^{2+} from the discs

photoreceptors are caused primarily by changes in the permeability of the cell membrane to Na^+ (Tomita, 1970; Hagins, 1972). As predicted by the Goldman field equation (equation 4.4), an increase in Na^+ permeability P_{Na} depolarizes the membrane. This is what happens when light hits an invertebrate photoreceptor. If the depolarization is sufficient to activate voltage-dependent Na^+ channels, then an action potential occurs. In vertebrate photoreceptors the situation is different. These cells are already highly permeable to Na^+ in the dark. Illumination of the cells *decreases* Na^+ permeability and thus hyperpolarizes the cell. This creates a real puzzle. How does a hyperpolarization in one part of the cell activate the synapse, some tens of micrometers away from the light-absorbing area? Could the photoreceptors be continuously secreting, a process which would be blocked by hyperpolarization? Stimulation of neurosecretion in other cells requires a depolarization, followed by an increase in free Ca^{2+} inside the nerve terminal (see Chapter 7, section 7.2).

What is the evidence that the chemical change in vertebrate rhodopsin (Fig. 4.16) causes a change in intracellular free Ca^{2+} which in turn causes a change in Na^+ permeability in the cell membrane? It is based mainly on effects of removal of extracellular Ca^{2+}, increasing intracellular free Ca^{2+} by ionophoretic injection of Ca^{2+} or ionophore A23187, and also on studies with subcellular fractions of photoreceptive cells (Hagins, 1972; Yoshikami and Hagins, 1977). For example, increases in intracellular Ca^{2+} induced by ionophoretic injection of Ca^{2+}, microinjection of Ca/EGTA, or by activation of Na^+ –Ca^{2+} exchange across the cell membrane (see this chapter, section 4.6.3), hyperpolarize vertebrate rods by inhibiting Na^+ conductance, thus mimicking the normal response to light.

The evidence is a little more convincing for invertebrate photoreceptors. In invertebrate rods the Na^+ conductance is high when the extracellular Ca^{2+} concentration is experimentally made low ($< 10\ \mu M$) and decreases as the Ca^{2+} concentration outside the cell increases. At 20 mM Ca^{2+} the Na^+ conductance in both resting cells and in illuminated cells is virtually abolished (Hagins, 1972). Prolonged exposure of the cells to low Ca^{2+}, e.g. in the presence of EGTA, depletes intracellular Ca^{2+} stores and increases the amount of light energy required to produce a given increase in Na^+ conductance.

The photoreceptors of barnacles or the horseshoe crab, *Limulus*, are large enough to be injected with Ca^{2+} indicators, such as aequorin (Brown and Blinks, 1974) or metallochromic indicator dyes (Brown *et al.*, 1977a, b). These direct studies with invertebrate photoreceptors have shown that illumination of these cells causes an increase in intracellular free Ca^{2+}. Since detection of the Ca^{2+} indicator involves measuring light, ingenious methods involving flash illumination have had to be developed to study the change in intracellular Ca^{2+}: The aequorin response is inhibited by EGTA inside the cell. Changes in extracellular Ca^{2+} concentration have rapid effects on the electrical responsiveness of photoreceptor membranes and on the free Ca^{2+} transient. These effects of extracellular Ca^{2+} vary between different invertebrates. For example, in barnacle photoreceptors removal of extracellular Ca^{2+} virtually abolishes the aequorin response and depolarization, whereas in horseshoe crab photoreceptors the effect is much less marked. It has been concluded that at least part of the Ca^{2+} comes from outside the cell, and the rest comes from internal stores. One problem is that the peak in intensity of aequorin light emission occurs after the membrane response.

The results of these experiments pose a number of questions: (1) How strong is the evidence for light-induced changes in free Ca^{2+} in other photoreceptors? (2) What is the mechanism by which light induces an increase in intracellular Ca^{2+}? (3) How is the Ca^{2+} removed from the cytoplasm so that the photoreceptor can recover and be ready to respond to more light? (4) How does intracellular Ca^{2+} affect the Na^+ permeability of the cell membrane, and how does this explain the difference between vertebrate and invertebrate photoreceptors? (5) What is the role of Ca^{2+} and cyclic nucleotides in light–dark adaptation?

Little has been done on intracellular Ca^{2+} in photoreceptors other than those already discussed. Nor is the mechanism understood by which an increase in cytoplasmic free Ca^{2+} could cause a change in the permeability of the cell membrane to Na^+. However, some experiments have been carried out to investigate the mechanism of release of Ca^{2+} from its store in the photoreceptor cell.

Release of intracellular Ca^{2+}

The discs from vertebrate photoreceptors can be isolated by high-speed centrifugation after breaking the cells. They contain a large amount of Ca^{2+}, about 11 moles of Ca^{2+} per mole of rhodopsin. Illumination of the cells before isolation diminishes the amount of Ca^{2+} in the isolated discs (Hendricks et al., 1974; Szuts and Cone, 1974). More direct evidence that the discs form the intracellular Ca^{2+} store which releases Ca^{2+} into the cytoplasm of the cell on illumination is obtained by loading the discs with $^{45}Ca^{2+}$. Not only does light induce a release of radioactive Ca^{2+} from the isolated discs (Poo and Cone, 1973; Kaupp and Junge, 1977; Mason et al., 1974), but also the actual amount of Ca^{2+} released correlates with the fraction of rhodopsin that is bleached (Smith et al., 1977). Isolated discs, or sonicated vesicles prepared from them, can reaccumulate the Ca^{2+} if ATP is present (Neufeld et al., 1972; Ostwald and Heller, 1972; Mason et al., 1974; Hemminki, 1975). This occurs at μM Ca^{2+} concentrations and in bleached vesicles at three times the rate of discs isolated from dark-adapted cells. Data from isolated discs should however be treated with some caution (Yoshikami and Hagins, 1977) since there have been difficulties in reproducing Ca^{2+} release from these discs (Szuts and Cone, 1974).

Is there enough Ca^{2+} in the discs to cause a sufficient change in free Ca^{2+} to block the Na^+-conducting channels? Using flash spectrophotometry, and arsenazo III as a Ca^{2+} indicator, Kaupp et al., (1979) found that for $1-10\%$ rhodopsin bleaching the free Ca^{2+} concentration rose by 2 μM. On average one

Ca^{2+} was released for every two rhodopsins bleached. Therefore, the discs do contain sufficient Ca^{2+} to change cytoplasmic free Ca^{2+} in the μM range when rhodopsin is bleached. How the Ca^{2+} is released is not known. Since ionophore A23187 abolishes the light-induced Ca^{2+} release the Ca^{2+} cannot be simply bound to the outside of the disc membranes. Illumination appears to lower the affinity of the binding sites for Ca^{2+} within the discs (Kaupp et al., 1979). It is unlikely that light simply induces a Ca^{2+} pore in the disc membrane, unless rhodopsin itself is able to transport the Ca^{2+} through the hydrophobic phase of the disc membranes, as has been suggested by Cone (1972).

Dark and light adaptation

One of the striking features of vertebrate photoreceptors is their ability to adapt to the general level of light intensity, enabling the eye to provide the brain with an image in both bright and dim light. Without this ability the eye would either be hopelessly dazzled in bright light, or would not be able to see at night. Light-induced changes in cyclic nucleotides (see Chapter 6 for structures and metabolism) have been known to occur for some time [(see Berridge (1976b) and Rasmussen and Goodman (1977) for references)]. However, a proper kinetic analysis has yet to be carried out to show whether the time course of the changes in cyclic AMP or cyclic GMP is fast enough for them to be involved in the primary response of photoreceptors. Furthermore, cyclic AMP added externally to cells produces no effect on the membrane potential (Miller, 1978). The most likely function of cyclic GMP, and possibly cyclic AMP, is mediation of the adaptation of photoreceptors to varying ranges of light intensities. The question then arises: Do the cyclic nucleotides interact directly with the Ca^{2+} release−accumulation process, or do they modify rhodopsin itself and thereby alter the number of photons required to produce a given degree of bleaching?

The concentrations of cyclic AMP and cyclic GMP in photoreceptors are decreased by illumination. This is caused by an activation of the enzyme phosphodiesterase (Bitensky et al., 1975). The effect of light is more pronounced on cyclic GMP because of the greater specificity of photoreceptor phosphodiesterase for this nucleotide, and the inhibition of retinal guanylate cyclase by light (Pannbacker, 1973). The mechanism by which light induces these changes is not known. It is tempting to invoke a role for Ca^{2+}, in view of the prolonged elevation of intracellular free Ca^{2+} after illumination has ceased (Brown and Blinks, 1974) and the discovery of Ca^{2+}-dependent proteins which activate cyclic nucleotide phosphodiesterase (see Chapter 6). However, the first experiments with these activator proteins from brain produced little effect on photoreceptor phosphodiesterase.

How do cyclic nucleotides, and cyclic GMP in particular, mediate the dark−

light adaptation of photoreceptors? Since nearly all effects of cyclic nucleotides inside cells occur through activation of a protein kinase followed by protein phosphorylation (see Chapter 6, sections 6.1 and 6.2), the possible effects of cyclic GMP in photoreceptors are: (1) phosphorylation of rhodopsin, increasing either its responsiveness to photons or its ability to cause Ca^{2+} release from discs: or (2) phosphorylation of Na^+ channels making them more sensitive to Ca^{2+}. An increase in cyclic GMP in dark-adapted photoreceptors would therefore increase the sensitivity to light, reducing the threshold necessary to stimulate the next cell in the neuronal pathway.

Light-induced phosphorylation of rhodopsin has been reported (McDowell and Kühn, 1977), but no effects of cyclic nucleotides regulating this phosphorylation have yet been found (Bitensky et al., 1975). Using single perfused retinal rods Lipton et al. (1977) found that ionophore A23187, like illumination, produces a hyperpolarization of the cell. Increasing the cyclic GMP in the cell, e.g. by addition of exogenous dibutyryl cyclic GMP or the phosphodiesterase inhibitor IBMX (see Fig. 9.23 for structure), causes depolarization. The effect of dibutyryl cyclic GMP is blocked by a high concentration of extracellular Ca^{2+}, dibutyryl cyclic AMP or ionophore A23187. These results support the contention that cyclic GMP decreases the uptake of Ca^{2+} by the discs after its release induced by illumination. However, the precise role of cyclic GMP in dark–light adaptation is unlikely to be completely understood until: (1) cytoplasmic free Ca^{2+} is measured in small photoreceptor cells; (2) the molecular mechanism of Ca^{2+} release is defined; (3) the molecular basis of Ca^{2+} on Na^+ conductance in vertebrate and invertebrate photoreceptors is defined; (4) the way in which a change in membrane potential enables the message 'light has stimulated the photoreceptive cell' to be passed on to the next cell and thence to the brain.

Photoreceptors, therefore, seem to provide a nice example of a 'threshold phenomenon' as outlined in Chapter 2, section 2.2. Light will either activate a single photoreceptor cell or not. This threshold response requires a sufficient increase in intracellular free Ca^{2+}. In vertebrate rods this Ca^{2+} comes mainly from the discs inside the cells. In some invertebrate photoreceptors external Ca^{2+} can provide much of this Ca^{2+}. The amount of light necessary to induce the threshold response is dependent on the concentration of cyclic GMP, and possibly of cyclic AMP, inside the cell.

Animals and plants, as well as unicellular organisms, are able to respond to many other sensory stimuli besides light (Table 4.7). The sensory cells are often activated by an electrical change caused by the primary stimulus. How do these cells communicate the fact that they have been activated to the rest of the organism? In animals this usually involves stimulation of neurotransmitter release so that the message can be transmitted to the central nervous system. In view of the key role for intracellular Ca^{2+} in neurosecretion (Chapter 7) it is likely that a role for intracellular Ca^{2+} in mediating the responses of many sensory cells will eventually be discovered.

Table 4.7. Examples of sensory receptors

Receptor	Stimulus	Example
Photoreceptors	Light	Eyes of vertebrates and some invertebrates, ocelli in coelenterates
Chemoreceptors	Pheromones	Insect sex sensor
	C3a, C5a, by-products of activation of complement pathway	Polymorphs in the inflammatory response
	Gaseous or soluble substances	Taste and smell
	Amino acids	Bacterial tumbling
Thermoreceptors	Temperature	Temperature regulation in mammals
Phonoreceptors	Sound	Ear
Tango receptors	Pressure	Receptors on insect antennae
	Touch	Skin and hairs of mammals, coelenterate nematocysts
	Pain	Vertebrate nerves
Statoreceptors	Gravity orientation	Statocysts in invertebrates (e.g. in crabs and jelly-fish)
Stretch receptors	Stretch	Muscles of some fish (e.g. rays)

4.5. Intracellular Ca^{2+} and bioluminescence

The ability of organisms from many phyla to emit light has fascinated man since classical times [see Harvey (1952) and Herring (1978) for references]. Aristotle and Pliny were two of the earliest writers to describe luminescence in Nature, and the writings of many famous travellers and naturalists abound with accounts of the phenomenon. Most luminous animals live in the sea. Some of them (Fig. 4.17), for example hydroids, jelly-fish, dinoflagellates and worms, can be found on the beach at night, with the aid of a 'KCl gun' to activate the nervous system of the animals. Luminous bacteria and fungi emit light continuously. Other luminous organisms flash spontaneously or when touched. Although the electrophysiology of many of the light-emitting cells has not been studied, the neuronal stimulation together with the threshold aspect of flashing would suggest that it would be well worthwhile examining whether intracellular Ca^{2+} plays a role in its regulation.

What is bioluminescence? Many physicochemical phenomena exist which involve the emission of photons and therefore come under the heading of luminescence (Table 4.8). These various types are distinguished from each other by a prefix, depending on the energy source. Chemiluminescence is the form of luminescence in which the energy required to excite the light-emitting molecule comes from a chemical reaction. Bioluminescence is used to describe such chemiluminescence when it occurs in living organisms.

The light organs of luminous animals can emit many tens of millions of photons per second when stimulated. Each individual luminous cell may itself emit millions of photons during a light flash from the organ. However, a much

Table 4.8. Some examples of Luminescence

Type of luminescence	Energy source
Photoluminescence (fluorescence, phosphorescence)	absorption of light
Pyroluminescence	Heat
Thermoluminescence	Heat after previous exposure to radiation
Electroluminescence	Electric discharge
Triboluminescence	Crystal structure change
Radioluminescence	Radioactive decay
Sonoluminescence	Sound
Chemiluminescence (including bioluminescence)	Chemical reaction

lower intensity chemiluminescence is found in some other cells. In this case each cell only emits at most a few hundred photons per second. An example of this low-intensity bioluminescence is found in phagocytic cells, and some fertilized eggs, after the generation of oxygen free radicals (Allen et al., 1972). The chemiluminescence may be from singlet oxygen ($^1O_2$) or from the reaction of one or more of the species with organic constituents of the cell. Lipid peroxidation also gives rise to low-intensity chemiluminescence from cells. Mitochondria chemiluminesce, and several mammalian tissues have been found to emit low levels of light (Boveris et al., 1980) because of internal oxygen radical production. Intracellular Ca^{2+} is thought to be involved in provoking polymorph chemiluminescence (Hallett et al., 1981; Hallett and Campbell, 1982a) (see Chapter 7, section 7.3.3), but is it involved in cells that are visibly luminous and where the luminescence is thought to have a function in the animal?

Two classical experiments have dominated the investigation of the chemistry of bioluminescence. The first was carried out by Robert Boyle in 1667 who showed, using his newly developed 'bell jar', that air (oxygen) was necessary for bacteria to luminesce. The second was carried out by Dubois in the 1880's who showed, using a luminous beetle, *Pyrophorus*, and a luminous rock-boring bivalve, *Pholas* (Fig. 4.17), that two organic components were necessary for luminescence. One was relatively heat-stable, the luciferin, and the other was heat-labile, the luciferase. The chemistry of the luminescence of several other luminous animals has been elucidated (Fig. 4.18) (see Herring, 1978), although the reactions remain unknown in many others. Ca^{2+} is required to activate photoproteins in coelenterates and radiolarians, and external Ca^{2+} is required for chemical stimulation of the luminous cells from these organisms (Fig. 4.19) (Campbell et al., 1979).

Over the last century there have been many reports that changes in the ionic composition of the medium affect the intensity of the light or the rate of flashing from luminous animals [for references see Harvey (1952)]. For example, immersing the animals in 'high-K^+' sea water often stimulates a bright

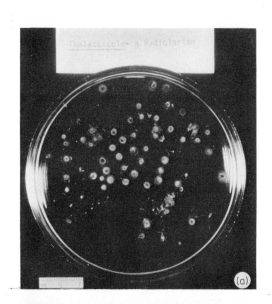

(a)

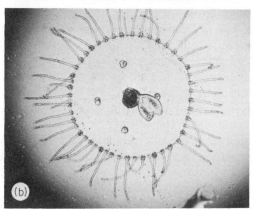

(b)

(c)

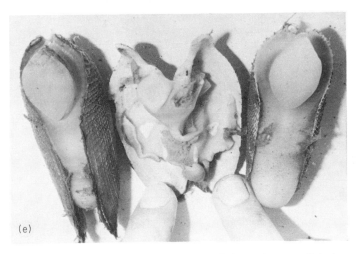

Fig. 4.17. Some luminous animals. (a) *Thalassicola*—a radiolarian (Protozoa). (b) *Obelia lucifera* (magnified)—a hydrozoan (Cnidaria). (c) *Pionosyllis*—a syllid worm (Polychaeta). (d) *Chaetopterus*—a polychaete worm (Polychaeta). (e) *Pholas dactylus*—a boring bivalve (Mollusca)—note white luminous organs in the opened animal

luminescence. This occurs in scale worms, coelenterates, protozoans, annelids and several luminescent organisms from other phyla. Unfortunately, many of the experiments described in Harvey's classic book on bioluminescence lack any conceptual framework. However, over the last 20 years the increasing evidence for Ca^{2+} currents in excitable cells (see this chapter, section 4.3) and for Ca^{2+} as an intracellular regulator, has focussed attention on the possible role of this cation in regulating bioluminescence (Table 4.9).

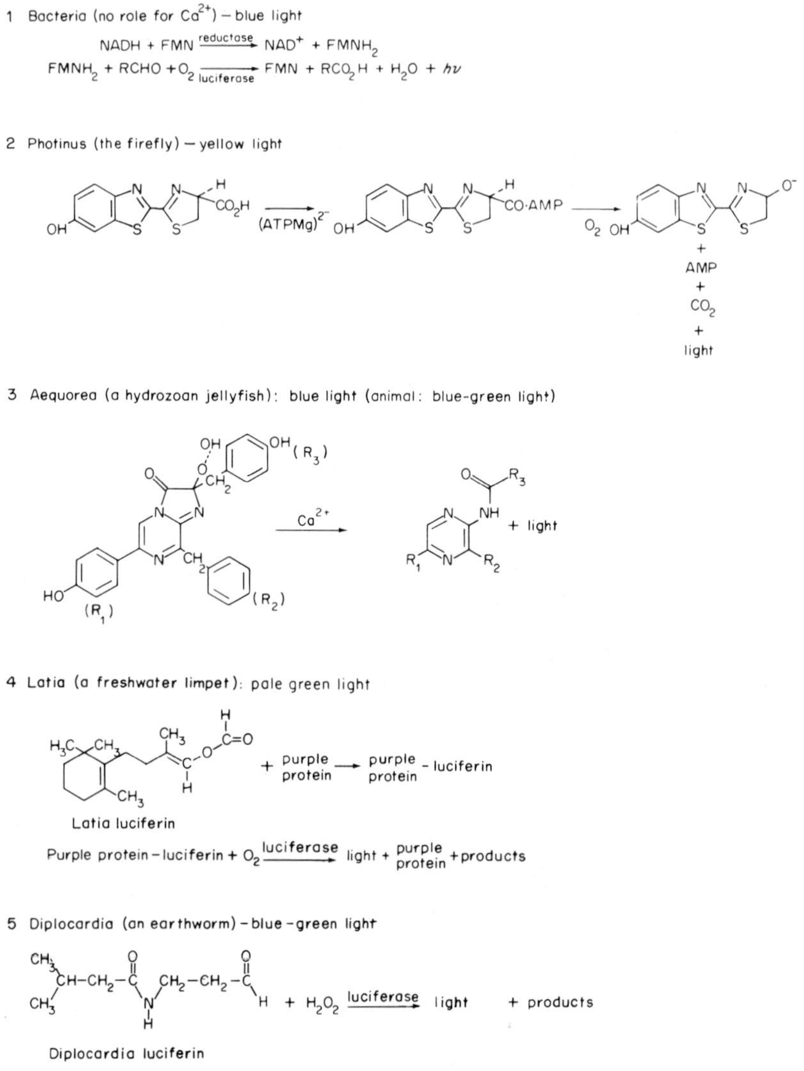

1 Bacteria (no role for Ca2+) – blue light
NADH + FMN reductase NAD+ + FMNH2
FMNH2 + RCHO + O2 luciferase FMN + RCO2H + H2O + hν
2 Photinus (the firefly) – yellow light
(ATPMg)2−
O2
AMP
CO2
light
3 Aequorea (a hydrozoan jellyfish): blue light (animal: blue-green light)
(R1)
(R2)
(R3)
Ca2+
+ light
4 Latia (a freshwater limpet): pale green light
+ purple protein → purple protein - luciferin
Latia luciferin
Purple protein – luciferin + O2 luciferase light + purple protein + products
5 Diplocardia (an earthworm) – blue-green light
+ H2O2 luciferase light + products
Diplocardia luciferin

Fig. 4.18. Some bioluminescent reactions

In luminous annelids, for example, the photocytes exhibit a Ca^{2+}-dependent action potential, which is inhibited by Mn^{2+} but not by tetrodotoxin. High-K^+ sea water stimulates luminescence only in the presence of extracellular Ca^{2+}. Similar observations have been made with echinoderms and the sea pansy *Renilla*, where ionophore A23187 also stimulates luminescence. Thanks to the work of Cormier and his colleagues the molecular basis of Ca^{2+} action in *Renilla* is one of the best understood. A Ca^{2+}-binding protein with high affinity Ca^{2+} sites (K_d of approximately 0.14 μM) has been identified which

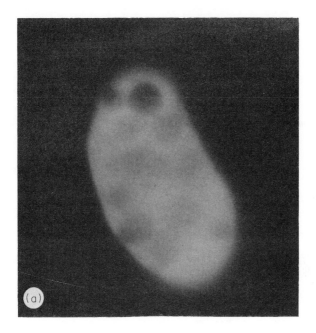

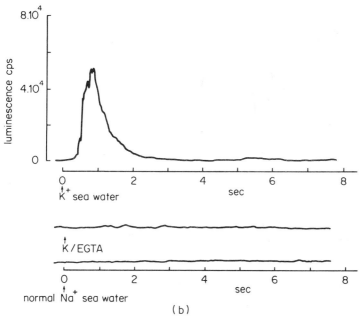

Fig. 4.19. Luminescence of cells from *Obelia* requires external Ca^{2+}. Cells isolated from coensarc of *Obelia geniculata* approx. 5 μm across, and may be fragments of the original cells. (a) Fluorescence. (b) Effect of K^+ ± external Ca^{2+}. From Campbell *et al.* (1979); reproduced by permission of Elsevier Biomedical Press B. V.

Table 4.9. Some luminous organisms in which intracellular Ca^{2+} may regulate light emission

Phylum	Organism	Reference
Protozoa	*Noctiluca*—a dinoflagellate	Hamman and Seliger (1972)
	Thalassicola—a radiolarian	Campbell *et al.* (1981)
Coelenterata	*Aequorea*—a hydrozoan	Shimomura *et al.* (1962, 1963)
	Obelia—a hydrozoan	Campbell *et al.* (1979)
	Renilla—an anthozoan	Cormier (1978); Henry and Ninio (1979)
	Mnemiopsis—a ctenophore	Ward and Seliger (1974)
Annelida (Polychaeta)	*Harmothoë*—a Polynoid scale worm	Herrera (1977)
	Acholoe—a scale worm	Bassot and Bilbaut (1977); Herrera (1977)
Echinodermata	*Ophiopsila*—a brittle star	Brehm *et al.* (1973); Case and Strause (1978)
	Amphipholis—a brittle star	
Arthropoda	*Photinus*—a firefly	Case and Strause (1978)
Chordata	*Porichthys*—a toad fish	Baguet (1975)

For other references see Herring (1978) and Harvey (1952).

contains the luciferin and allows it to bind to the luciferase in the presence of Ca^{2+}. Light emission requires the presence of oxygen. In other luminescent coelenterates like *Aequorea*, *Obelia* and *Mnemiopsis*, it is possible to isolate a so-called photoprotein, which, in the absence of molecular oxygen, emits light when it binds Ca^{2+}. Under ionic conditions thought to exist within the cell, i.e. about 150–500 mM K^+, 5 mM Mg^{2+}, pH 7.4, the apparent dissociation constant for these photoproteins binding Ca^{2+} may be as high as 0.01–0.1 mM. In the photocyte if μM concentrations of Ca^{2+} in the cell cytoplasm are sufficient to produce light, then only a small proportion of the photoprotein will be consumed during each flash. An interesting feature in some coelenterates is the presence of a green fluorescent protein which changes the quantum yield and luminescence spectrum (Campbell *et al.*, 1979). One possibility is that this protein is packaged with the photoprotein inside vesicles which, on cell activation, move to the cell membrane, partially fusing with it, thus exposing the photoprotein to the very high Ca^{2+} concentration (about 11 mM) outside the cell. Although so-called 'lumisomes' have been observed in cell fragments (Campbell *et al.*, 1979), there is as yet no evidence from electron microscopy that they exist in the real cell.

No mechanisms have yet been proposed for any luminescent organism which explain completely the response of the intact animal. In particular, the very rapid flashes, sometimes with rate constants of 5–100 sec^{-1}, would require an incredibly rapid transient burst of Ca^{2+} in the cell if this were to provide the complete explanation of the physiological response. Nevertheless, the photocytes of luminescent animals could provide useful models for studying particular aspects of the role of intracellular Ca^{2+} as a regulator, particularly since these cells have the unique property of allowing the site of stimulation within the cell to be observed using image intensification (Reynolds, 1978, 1979) (see Chpater 2).

4.6. The regulation of intracellular Ca^{2+} through internal stores and pumps

4.6.1. The function of intracellular Ca^{2+} stores and pumps

Suppose you are on a supertanker carrying oil from oil-fields in the Middle East to Europe. On arrival the oil will be pumped out of the tanks, being replaced with air from the atmosphere around the ship. Similarly, after the tanks are washed with water this water will also be replaced with air from outside the ship. Now consider a submarine several hundred metres under the surface. The captain, wishing to surface, must replace the water in the ballast tanks with air from a store of compressed air within the ship. Once on the surface these stores of compressed air can be replenished, which requires a supply of energy, so that these manoeuvres can be repeated.

So far, in this chapter, we have considered how changes in the electrical properties of the cell membrane can cause changes in intracellular free Ca^{2+}, and the converse process of how changes in intracellular free Ca^{2+} can regulate the electrical properties of the cell. Yet in many cells, for example muscle, there is an internal Ca^{2+} store analogous to the submarine's compressed air supply. This is the main source of the increase in cytoplasmic Ca^{2+} during contraction and not Ca^{2+} from outside the cell. As with the cell membrane, the release and uptake of Ca^{2+} from these stores within the cell frequently involves changes in the electrical properties of the membranes in which they are enclosed. Furthermore, if the cytoplasmic Ca^{2+} is to return to its original value after the stimulus has ended, the Ca^{2+} must be pumped back into the internal store. If the Ca^{2+} originally came from outside the cell it must be pumped out again. This process must be finely controlled if the cell is to be able to respond to a series of stimuli without impairment of function. What are these intracellular Ca^{2+} stores and Ca^{2+} pumps? How are they regulated? What are their respective roles in regulating cytoplasmic free Ca^{2+} and free Ca^{2+} within the organelles of the cell?

It has already been pointed out (Chapter 2, section 2.4) that more than 99.99% of the cell Ca^{2+} is bound. In theory, a change in any of the pools which make up this bound Ca^{2+} (Table 4.10) could significantly alter cytoplasmic free Ca^{2+}. For example, a decrease in pH would decrease the apparent affinity for Ca^{2+} of substances like ATP, citrate and glutamate (see Chapter 3, section 3.3.3) and thereby cause them to release Ca^{2+}. Although these soluble Ca^{2+} buffers do bind significant quantities of Ca^{2+}, perhaps as much as 10 μM (Baker, 1972), it is the membrane-bound pools of Ca^{2+} which seem to be regulated by physiological activators of cells. The function of these pools of Ca^{2+}, together with the Ca^{2+} pumps associated with them and on the cell membrane, can be summarized as follows: (1) A particular internal store may be activated by a primary stimulus to release its Ca^{2+} into the cytoplasm. (2) Ca^{2+} pumps limit the area of cytoplasm which has an increase in free Ca^{2+} when the cell is activated or damaged. (3) Ca^{2+} pumps remove Ca^{2+} from the cytoplasm after the stimulus to the cell is over, allowing it to recover. (4) In the resting cell a Ca^{2+} pump on the cell membrane is required to maintain the steady-state cytoplasmic free Ca^{2+}

concentration of about 0.1 μM in the presence of a small, but significant, passive net flux of Ca^{2+} into the cell down its electrochemical gradient. The same argument applies to the maintenance of the internal stores in the resting cell.

4.6.2. Intracellular Ca^{2+} stores and the regulation of cytoplasmic free Ca^{2+}

There are several potential sources of Ca^{2+} inside the cell which could be released into the cytoplasm (Table 4.10). Furthermore, regulation of free Ca^{2+} within the organelles themselves, for example in the nucleus and mitochondria, may play a key role in controlling some metabolic reactions. What is the free Ca^{2+} concentration within these organelles? How do they maintain a large net Ca^{2+} gradient between the inside of the organelle and the cytoplasm? What are the Ca^{2+} ligands responsible for binding Ca^{2+} and allowing accumulation against the electrochemical gradient existing across the membrane of the organelle?

The nucleus, mitochondria and endoplasmic reticulum together contain more than 90% of the cell Ca^{2+} (Table 2.7), equivalent to about 2 mM free Ca^{2+}, if all of it were released into the cytoplasm. In order to raise the free Ca^{2+} to 5 μM, an increase of about 20 times the resting free Ca^{2+}, only 0.25% of the Ca^{2+} within these orgenelles would have to be released. For a Ca^{2+} store containing only 10 μmoles/l of cell water more than 50% of this would be required. In fact, although

Table 4.10. Sites of significant bound Ca^{2+} within the cell

Ca^{2+} store	Example	General reference
Organelles		
Sarcoplasmic reticulum	Muscle	Endo (1977); Martonosi (1972); Martonosi et al. (1978)
Endoplasmic reticulum	Probably most non-muscle cells	Hales et al. (1974)
Mitochondria	All cells	Carafoli and Crompton (1976)
Secretory granules	Nerve terminals	Israel et al. (1980)
	Chromaffin granules	Douglas (1974)
Ca^{2+} granules	Platelets	Holmsen (1972)
	Invertebrates with shells	Simkiss (1974)
Discs	Photoreceptors	Yoshikami and Hagins (1977)
Nucleus	All cells	Maunder et al. (1977)
Molecules		
Ca^{2+}-binding proteins	Parvalbumin in some calmodulin muscles S-100 in brain structural proteins	Kretzinger (1976a, b); Cheung (1980)
Nucleic acids	Any cell	Chang and Carr (1968)
Phospholipids	Any cell	Dawson and Hauser (1970)
ATP, amino acid, citrate	All cells	See Chapter 3, section 3.3.3

the free Ca^{2+} in the cytoplasm may never exceed 5–10 μM during activation of the cell, the total Ca^{2+} which must be released for the Ca^{2+}-binding proteins is likely to be at least ten times this (i.e. 50 μM Ca^{2+}, equivalent to about 2.5% of the stored Ca^{2+}).

Endoplasmic reticulum

This is the name given to the system of tubes and vesicles observed inside the cell under the light or electron microscope. It has no direct contact with the cell membrane, though it often comes very close. This membranous system is very well organized in muscle, where it is called the sarcoplasmic reticulum. It is identifiable from the Golgi appartus, and when it has ribosomes attached it is known as the rough endoplasmic reticulum. It was in muscle that the well-organized endoplasmic reticulum (Bennet and Porter, 1953) was first identified as a major Ca^{2+} store crucial to the contraction of the muscle (see Chapter 5, section 5.3.3). The key observation was that a vesicle preparation could be isolated from muscle homogenates which accumulated Ca^{2+} in the presence of ATP (Ebashi and Lipmann, 1962; Weber et al., 1966). These vesicles were the 'relaxing factor' in muscle homogenates which when added to contracted myofibrils caused them to relax (March, 1951; Ebashi, 1961). Similar systems have been isolated from virtually all types of muscle. Up to 90% of the protein of the vesicles is a Ca^{2+}-activated Mg^{2+}-ATPase responsible for pumping Ca^{2+} into the vesicles. In the intact reticulum a binding protein, calsequestrin, is responsible for chelating the Ca^{2+} inside the vesicles, which may rise as high as 5 mmoles/l of intravesicular water. Under experimental conditions the broken reticulum contains less calsequestrin and oxalate is necessary to precipitate Ca^{2+} within the vesicles in order to obtain a large net uptake of Ca^{2+}. Less is known about the mechanism of release *in vivo* (Endo, 1977). The sarcoplasmic reticulum is so close to the cell membrane that the action potential may itself induce a transient increase in Ca^{2+} permeability. Alternatively, the small amount of Ca^{2+} entering the cell during the action potential may stimulate release of more Ca^{2+} from the sarcoplasmic reticulum (Fabiato and Fabiato, 1977). This controversy highlights the need for a better *in vitro* preparation to find out precisely how an electrical change in the cell membrane induces a rapid, transient change in the permeability of an intracellular membrane to Ca^{2+}. No other tissues besides muscle contain such an extensive Ca^{2+} store in their endoplasmic reticulum. However, sufficient data exist in other cells to suggest that the endoplasmic reticulum acts as a Ca^{2+} store in many tissues (Table 4.11). The principal pieces of evidence for this are: (1) the identification of a Ca^{2+}-activated Mg^{2+}-ATPase in the microsomal fraction of a cell homogenate, which is not affected by inhibitors of Ca^{2+} transport in mitochondria; (2) the presence of a MgATP-dependent Ca^{2+} uptake into microsomal vesicles, and its inhibition or stimulation of Ca^{2+} release by caffeine; (3) accumulation of Ca^{2+} precipitates in cells by oxalate or pyroantimonate together with identification of Ca^{2+} using X-ray microprobe analysis.

Table 4.11. Possible endoplasmic reticulum Ca^{2+} stores in non-muscle cells

Tissue	Evidence	Reference
Vertebrates		
Adipose tissue	X-ray microprobe analysis	Hales et al. (1974)
	Microsomal Ca^{2+}-MgATPase	Bruns et al. (1976, 1977)
Liver	Ca^{2+} binding, ATP-dependent	Moore et al. (1975, 1976)
Kidney	Ca^{2+} pumping	Moore et al. (1974); Moore and Landon (1979)
Islets of Langerhans	X-ray microprobe analysis	Herman et al. (1973)
	Microsomal (Ca^{2+})-ATPase	Montague et al. (1976)
Salivary gland	Microsomal Ca^{2+} uptake	Selinger et al. (1970)
Platelets	X-ray microprobe analysis	Daimon et al. (1978)
	Microsomal (Ca^{2+})-ATPase and accumulation	Statland et al. (1969)
	Ca^{2+} uptake and ATPase in membranes	Robblee et al. (1973)
Brain	X-ray microprobe analysis (glial cells)	Gambetti et al. (1975)
	Microsomal Ca^{2+} binding, ATP-dependent	Blitz et al. (1977); Nakamuru and Schwartz (1977)
Fish eggs	Aequorin light transient	Gilkey et al. (1978)
Invertebrates		
Sea urchin egg	Isolation of a Ca^{2+} accumulation system	Kinoshita and Yazaki (1967)
Starfish egg	Isolation of a Ca^{2+} accumulation system	Moreau et al. (1980); Moreau and Guerrier (1979)
Squid giant axon	Oxalate precipitation in the cell	Henkart et al. (1978)
Unicellular		
Amoeba	Vesicular store	Reinhold and Stockem (1972)

Further data are required to show how the Ca^{2+} can be released when the cell is activated by a physiological stimulus, such as a hormone or neurotransmitter. A major problem with studies on subcellular fractions is to be sure that any Ca^{2+}-activated Mg-ATPase and Ca^{2+}-accumulating system in the microsomal fraction is from the endoplasmic reticulum and not the cell membrane.

Mitochondria

Mitochondria were first identified by light microscopy in the nineteenth century. Thanks to the work of Keilin, Krebs and others, by the 1930s mitochondria had been recognized as the main supply of energy (ATP) in aerobic eukaryotic cells. These organelles measure up to 1 μm in length and 0.2–0.5 μm in diameter in most cells. It has been known for some time that isolated mitochondria and those inside cells, under the appropriate conditions, take up or release substantial amounts of Ca^{2+} (Chance, 1965; Lehninger, 1962, 1970; Carafoli and Crompton, 1976; Bygrave, 1976, 1977, 1978). It has been suggested that as much as 50 % of the Ca^{2+} in a normal cell may be inside the mitochondria, though a more realistic upper estimate is 20–30 %. Several enzymes isolated from the matrix of the mitochondria can be regulated by Ca^{2+} (Bygrave, 1966a, b, 1967; Severson et al., 1976; Denton et al., 1978, 1980), though the physiological significance of these in vitro observations is still unclear. Let us therefore examine the evidence for Ca^{2+} uptake and release by mitochondria, the possible mechanisms involved, and the physiological significance they might have.

The first clue that mitochondria might take up Ca^{2+} was discovered by Axelrod and co-workers in 1941, who showed that addition of Ca^{2+} to a tissue homogenate increased the rate of respiration, i.e. O_2 uptake. Although they explained their data through a Ca^{2+}-activated breakdown of NAD, several other workers (see Siekovitz and Potter, 1953; Lehninger, 1970) demonstrated that Ca^{2+} increased the respiration of isolated mitochondria. Lehninger and co-workers showed that this effect of Ca^{2+} was not the same as that of uncouplers such as dinitrophenol, but did redirect the energy of substrate oxidation from the respiratory chain away from ATP synthesis. Although net Ca^{2+} uptake by subcellular fractions containing mitochondria was reported in the 1950s, the first proper demonstration in vitro was carried out by Vasington and Murphy in 1961, supporting the in vivo findings of Thiers et al. (1960). This uptake of Ca^{2+} exhibited several properties which distinguished it from uptake by the endoplasmic or sarcoplasmic reticulum (Fanburg and Gergely, 1965). For example oligomycin, dinitrophenol, sodium azide and ruthenium red, which have no effect on the sarcoplasmic reticulum, inhibited or blocked mitochondrial Ca^{2+} uptake. Furthermore Lehninger (1962) showed that oxalate, required for maximal Ca^{2+} uptake by sarcoplasmic reticulum in vitro, had no effect on mitochondrial Ca^{2+} uptake. This Ca^{2+} uptake could be stimulated by ATP, and the mitochondria had a different affinity for Ca^{2+} than the reticulum.

Quite early on in these studies (see Peachy, 1964; Chance, 1965; Drahota et al., 1965) it was found that millimolar Ca^{2+} concentrations resulted in such a large uptake of Ca^{2+} by mitochondria that swelling, calcium phosphate precipitation, and irreversible uncoupling of the respiratory chain from ATP synthesis occurred. Since Ca^{2+} concentrations in the range 50–200 μM did not seem to impair mitochondrial structure, enzyme activities or respiration, but produced an easily measurable Ca^{2+} uptake, this was the concentration used for many years in experiments on Ca^{2+} uptake by isolated mitochondria. Unfortunately, it is now clear that these Ca^{2+} concentrations are between 500 and 2000 times higher than that of free Ca^{2+} in the cytoplasm of a resting cell. At lower, physiological Ca^{2+} concentrations, uptake is much less. For example, at 0.1 μM free Ca^{2+} the initial rate of Ca^{2+} uptake is nearly a fiftieth of that at 2 μM.

In the resting cell (cytoplasmic free $Ca^{2+} \approx 0.1\ \mu M$) there is a steady-state between the uptake and release of Ca^{2+} by the mitochondria. When the free Ca^{2+} surrounding the mitochondria is increased, for example to 5 μM, there is a rapid uptake of Ca^{2+} which continues until a new steady-state is reached. Studies with isolated mitochondria have shown that the free Ca^{2+} concentration at this new steady-state is virtually the same as that before the mitochondria were exposed to the small burst of Ca^{2+} (Fig. 4.21b). Mitochondria can therefore be considered to be intracellular Ca^{2+} buffers. A balance point is reached when uptake of Ca^{2+} equals release. If the free Ca^{2+} is lowered below the balance point, net Ca^{2+} release occurs, whereas above the balance point net Ca^{2+} uptake occurs (Fig. 4.21a). In isolated mitochondria the balance point is usually at about 0.3–0.5 μM free Ca^{2+}, a little higher than that expected for the resisting cell.

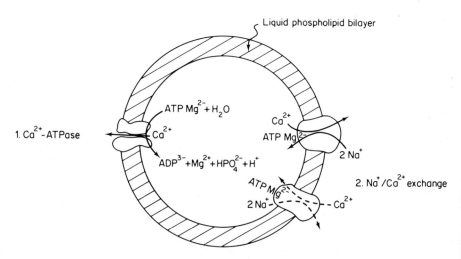

Fig. 4.20. The two major Ca^{2+} efflux pathways in eukaryotic cells

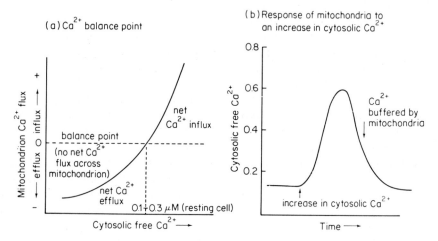

Fig. 4.21. Mitochondria as an intracellular Ca^{2+} buffer. (a) Ca^{2+} balance point. (b) Response of mitochondria to an increase in cytosolic Ca^{2+}

How can this uptake and release of Ca^{2+} by mitochondria be measured? What is the mechanism of uptake and release of Ca^{2+}? What is the concentration of free Ca^{2+} inside the mitochondria? Can isolated mitochondria be induced to release significant quantities of Ca^{2+} under conditions mimicking those in an activated cell? What is the biological significance of mitochondrial Ca^{2+}?

Incubation with Ca^{2+} of mitochondria isolated from most animal and plant cells, though not apparently from insect flight muscle or yeast, and a source of energy results in an increase in O_2 utilization which is directly related to Ca^{2+} uptake. Ca^{2+} uptake can be measured more directly by atomic absorption spectrophotometry, radioactive Ca^{2+} fluxes, or by measurement of free Ca^{2+} in the suspension with aequorin, indicator dyes such as murexide, or a Ca^{2+} microelectrode (see Chapter 2, section 2.5, for details). In $^{45}Ca^{2+}$ experiments the free Ca^{2+} concentration can be kept constant throughout if required. Results are usually expressed as nmoles Ca^{2+} taken up per mg of mitochondrial protein. Whilst this is, of course, a perfectly valid way of expressing results, more information might be extracted from published data if estimates of intramitochondrial H_2O and Ca^{2+} uptake per mitochondrion were also made. When examining such data it is important to check that the mitochondria were fully coupled, i.e. a P:O ratio before Ca^{2+} of about 3. The recovery of mitochondrial enzymes, e.g. glutamate dehydrogenase, at the end of the experiment preferably should be at least 90%, and the free Ca^{2+} concentrations used should be within the expected physiological range for the cell, i.e. 0.1–5 μM. The maximal initial rate of Ca^{2+} uptake is usually between 1 and 50 nmoles Ca^{2+} per mg mitochondrial protein per sec, the total uptake being about 100 nmoles Ca^{2+} per mg mitochondrial protein. The concentration of Ca^{2+} causing half maximal rate of uptake is usually 1–10 μM, though values of 30 μM (Baker, 1976) and 100 μM (Bygrave, 1978) have been reported. One hesitates to call this a K_m for Ca^{2+} since the shape of the Ca^{2+} dose–response curve does not always obey simple Michaelis-Menton kinetics.

Many experiments with mitochondria are carried out in the absence of a permeant anion. Although permeant anions such as NO_3^- and CNS^- have little effect on Ca^{2+} uptake, phosphate and acetate increase both the initial rate and total Ca^{2+} uptake by mitochondria between two- and ten-fold. Interestingly, in the presence of Mg^{2+} and ADP or ATP, calcium phosphate precipitates can form inside the mitochondria without causing swelling. The total Ca^{2+} taken up under these conditions can be as high as 3 μmoles Ca^{2+} per mg mitochondrial protein.

How is the energy from the respiratory chain directed towards Ca^{2+} uptake? In the presence of ADP and phosphate and above the balance point, Ca^{2+} uptake occurs in preference to ATP synthesis. Is ATP itself the energy source or is there a high-energy intermediate which can fuel Ca^{2+} transport? The most important step forward in understanding the mechanism of ATP synthesis and ion transport in mitochondria was made by Mitchell (1966, 1968) when he proposed his chemiosmotic hypothesis. The respiratory chain oxidizes NADH and as a result pumps H^+ ions out of the mitochondria, generating both a pH gradient and a membrane potential across the inner mitochondrial membrane. This is described by Mitchell's equation

$$\text{Proton motive force} = \text{p.m.f.} = \Delta(\text{pH}) - F\psi \qquad (4.10)$$

where ψ = membrane potential and F = Faraday constant.
This is now more correctly expressed as an electrochemical potential, $\Delta\tilde{\mu}_{H^+} = \Delta\psi - (2.3RT/F \times \Delta\text{pH})$.

It is this electrochemical potential which provides the driving force for Ca^{2+} uptake. Loss of the pH gradient and membrane potential by inhibiting the respiratory chain with antimycin A or CN^-, or uncouplers such as dinitrophenol, inhibits Ca^{2+} uptake and hence causes Ca^{2+} release from mitochondria. When ATP is used experimentally as the energy source, Ca^{2+} uptake is blocked by oligomycin but not by respiratory-chain inhibitors (Fig. 4.22). Ca^{2+} uptake is also inhibited by La^{3+} (Mela, 1968). It has been reported that a maximum of 2 Ca^{2+} are taken up per 'energy-rich' site in the respiratory chain. In other words, oxidation of NADH provides sufficient energy for 6 Ca^{2+} to be transported into the mitochondrion (Fig. 4.22). Here we are faced with a dilemma. Since 1 NADH produces 6 H^+ the stoicheiometry for $H^+:Ca^{2+}$ must be 1:1, which is not electrically neutral. The problem can be resolved if Ca^{2+} goes into the mitochondrion as $(Ca_2)^{4+}HPO_4^{2-}$ on a Ca^{2+}-phosphate symport (Moyle and Mitchell, 1977), restoring the stoicheiometry to $2H^+/Ca^{2+}$. The discovery of a 3 Na^+ for 1 Ca^{2+} exchange (Crompton et al. 1976) may provide a necessary link

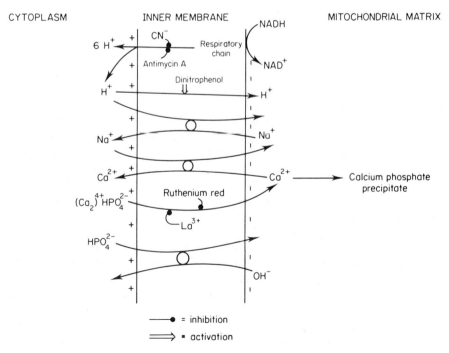

Fig. 4.22. Some ion-transport mechanisms in mitochondria

between Ca^{2+} and H^+ (Fig. 4.22) and explain why Ca^{2+} does not have to be at its equilibrium potential across the mitochondrial membrane. Although this Na^+–Ca^{2+} system exists in heart, brain and brown-fat mitochondria, it has not been found in liver.

An acidic glycoprotein, molecular weight 35 000 and rich in Glu and Asp residues, has been isolated from liver mitochondria (Carafoli, 1974). It has two classes of Ca^{2+}-binding site—one low affinity the other high affinity. The latter has a K_d for Ca^{2+} of about 1 μM. This Ca^{2+} binding is inhibited by La^{3+} and ruthenium red. Antibodies produced to this protein inhibit mitochondrial Ca^{2+} uptake (Parfili et al., 1976). This protein could therefore provide part of the molecular basis for Ca^{2+} transport in mitochondria. It is but one of a class of Ca^{2+}-binding proteins with ionophore-like properties associated with membrane Ca^{2+} transport and Ca^{2+}-activated MgATPases (Blondin et al., 1977; Shamoo and Goldstein, 1977).

The uptake of Ca^{2+}, however, presents us with a problem. The chemiosmotic theory predicts that a simple uniport for a divalent cation such as Ca^{2+}, across the mitochondrial membrane maintaining a maximum potential of -180 mV, would produce a concentration gradient of about a million fold at electrochemical equilibrium. This would mean that the mitochondrial-matrix free Ca^{2+} concentration would be about 0.1 M, some one to ten thousand times greater than expected from measurements of enzyme activity and the known inorganic phosphate content. This dilemma cannot be overcome by the more complex symport proposed by Moyle and Mitchell (1977) transporting $(Ca_2)^{4+} HPO_4^{2-}$, nor by the fact that the true 'equilibrium free Ca^{2+}' in the mitochondria is nearer 2 mM based on a measured membrane potential of -130 mV (equation 4.11).

There has therefore to be an efflux pathway for mitochondrial Ca^{2+}. Two have, so far, been discovered. One, found in heart, brain and brown-fat mitochondria exchanges Ca^{2+} for Na^+ and the other, found in liver, exchanges Ca^{2+} ultimately for H^+. This means that in resting cells there is a slow, continuous cycle of Ca^{2+} across the membrane utilizing energy at the expense of the proton electrochemical potential. The Ca^{2+} cycle enables a steady state gradient of $Ca^{2+} \ll 10^6$ to be maintained across the mitochondrial membrane. Furthermore, regulation of Ca^{2+} distribution can be achieved by altering kinetic parameters. Like many cycling systems (Newsholme and Start, 1973), only a small fractional change in Ca^{2+} movement in one direction can result in a large change in net flux.

The steady-state cycling of Ca^{2+} across the inner membrane of mitochondria results in no net movement of ions. When net Ca^{2+} accumulation occurs entry is balanced by protons from the respiratory chain, the stoichiometry being $2H^+$ out per Ca^{2+} in. In the absence of a weak acid, pH gradients of up to -2 units can build up experimentally with a concommitant decrease in $\Delta\psi$, $\Delta\tilde{\mu}_H$. being approximately constant. In the cell, however, acetate or phosphate will be present which can move into the mitochondrial matrix, thereby dissipating the pH gradient and allowing $\Delta\psi$ to increase, and more Ca^{2+} to move into the mitochondria.

Some years ago it was proposed that the inhibition by Ca^{2+} of pyruvate kinase outside the mitochondria, and pyruvate carboxylase inside the mitochondria, could be physiologically relevant if a physiological stimulus provoked net Ca^{2+} release from the mitochondria (Krebs et al., 1963). This could result in activation of pyruvate carboxylase and inhibition of pyruvate kinase, the Ca^{2+} acting as a switch converting glucose metabolism from glucose breakdown (i.e. glycolysis) to glucose synthesis (i.e. gluconeogenesis). Unfortunately the Ca^{2+} concentrations required for these enzymes to be significantly inhibited in vitro seem to be too high for it to be physiological important (Table 4.12). However other enzymes, such as pyruvate dehydrogenase phosphatase, are affected by μM Ca^{2+} (Table 4.12). Under experimental conditions, these enzymes can be regulated by Ca^{2+} and mitochondrial metabolism.

Plant mitochondria in addition to oxidation of intramitochondrial NADH can also oxidize exogenous NADH. This latter pathway is not proton translocating but is activated by Ca^{2+}.

The question arises whether this Ca^{2+} regulation of enzymes, demonstrable under experimental conditions and in mitochondrial isolated from cells exposed to agents such as ionophore A23187, is physiologically relevant. Pyruvate metabolism, for example, in adipose tissue is regulated by adrenaline and insulin (see Chapter 6). Are the effects of the hormones mediated by changes in mitochondrial Ca^{2+}? The major problem in answering such questions is lack of knowledge of the free Ca^{2+} concentration inside mitochondria.

If all the Ca^{2+} in a mitochondrion were free it would produce a concentration of up to 1–20 mM within the mitochondrial matrix. More than 97% of mitochondrial Ca^{2+} would have to be bound if the mitochondrial matrix free Ca^{2+} were to be less than 10 μM. Free Ca^{2+} inside mitochondria is probably in the range 10–100 μM (Carafoli and Crompton, 1976; Bygrave, 1978). If this is correct it poses two problems. Firstly, all the sites inside mitochondria potentially regulated by Ca^{2+} (Table 4.12) would be saturated, and this Ca^{2+} would be regarded as playing a passive role in the chemistry of the cell (Chapter 2, section 2.3.3). Only when mitochondrial Ca^{2+} was markedly reduced physiologically or

Table 4.12. Some mitochondrial enzymes affected by Ca^{2+}

Enzyme	Effect of Ca^{2+}	Reference
Pyruvate carboxylase	Inhibition	Krebs et al. (1963); Gevers and Krebs (1966)
Pyruvate dehydrogenase phosphatase	Activation	Seversen et al. (1976)
Pyruvate dehydrogenase kinase	Inhibition	Denton (1977)
α-Oxoglutarate dehydrogenase phosphatase	Activation	Denton et al. (1978)
Succinate oxidase	Activation	Ogata et al. (1972)
α-Glycerophosphate dehydrogenase	Activation	Hansford and Chappell (1967)

experimentally, e.g. with ionophore A23187 + EGTA, would any effects on Ca^{2+}-dependent enzymes be demonstrated. Secondly, a thermodynamic problem arises, for under these circumstances of a low free Ca^{2+} in the matrix, the Ca^{2+} cannot be in electrochemical equilibrium (see equation 4.5 and Fig. 4.3). An energy-dependent Ca^{2+}-efflux mechanism is necessary as well as one for influx (Heaton and Nicholls, 1976; Nicholls, 1978) as discussed previously.

Let us examine the argument a little more closely. The electrical potential across the mitochondrial membrane, based on $^{81}Rb^+$ or $^{42}K^+$ distribution in the presence of valinomycin, is between -110 and -150 mV, negative inside (Scarpa and Graziotti, 1973; Bygrave, 1978). This is less than the -180 mV originally predicted by Mitchell (1966, 1968).

If $E_{mit} = -130$ mV and $[Ca^{2+}_{cyt}]$ = cytoplasmic free Ca^{2+} concentration = 0.1 μM, then, in the absence of any energy-dependent efflux of Ca^{2+}, the Ca^{2+} will eventually reach electrochemical equilibrium as predicted by the Nernst equation

$$E_{mit} = \frac{RT}{2F \log_{10} e} \log_{10}([Ca^{2+}_{cyt}]/[Ca^{2+}_{mit}]) \quad (4.11)$$

At 37 °C, $E_{mit} = 30.8 \log_{10}([Ca^{2+}_{cyt}]/[Ca^{2+}_{mit}]) = -130$ mV

and $[Ca^{2+}_{mit}]/[Ca^{2+}_{cyt}] = 16626$

∴ $[Ca^{2+}_{mit}] = 1.66$ mM.

This value is high, but not impossible, in view of the large mitochondrial Ca^{2+} content and small water space inside. If the free Ca^{2+} inside is less than 1.66 mM, as is likely from enzyme studies, then an energy-dependent efflux mechanism for Ca^{2+} must exist.

Now suppose that the membrane potential across the mitochondrial membrane vanishes. By how much would the cytoplasmic Ca^{2+} be reduced if all of this were due to electrogenic movement of Ca^{2+} into the mitochondria? Some calculations can be made on the basis of the following assumptions: (1) a mitochondrion is a hollow sphere radius of 0.25 μm and volume $= \frac{4}{3}\pi r^3 = 6.5 \times 10^{-17}$ l; (2) if the membrane capacitance of the mitochondrial membrane $= 1$ μF cm^{-2} and the surface area $= 4\pi r^2 = 7.9 \times 10^{-9}$ cm^2, then the capacitance of 1 mitochondrion $= 7.9 \times 10^{-15}$ F.

Therefore the number of g atoms of Ca^{2+} per mitochondrion equivalent to

$$-130 \text{ mV} = \frac{CV}{2F} = \frac{7.9 \times 10^{-15} \times 0.13}{2 \times 96495} = 5.3 \times 10^{-21}.$$

Since Avogadro's number $\approx 6 \times 10^{23}$ atoms mole^{-1}, this is equivalent to only 3200 Ca^{2+} ions and a change in matrix free Ca^{2+} of 36 μM.

We now have the data to calculate the decrease in cytoplasmic Ca^{2+} if the depolarization of all the mitochondria in the cell was caused by an inward Ca^{2+} current into the matrix of the mitochondria. Consider a liver cell with 1000 mitochondria. If the radius of the cell is 10 μm, then the volume of the cell is approximately 4 pl. If 5.3 tipomoles (1 tipomole = 10^{-21} moles, from the Welsh *tipyn* meaning small) of Ca^{2+} are required per mitochondrion to depolarize -130 mV, then the total change in cytoplasmic Ca^{2+} will be

$$\frac{5.3 \times 10^{-21} \times 1000}{4 \times 10^{-12}} M \doteq 1 \, \mu M.$$

Furthermore, if free $[Ca^{2+}_{mit}] = 1.7$ mM, then the number of ions of free Ca^{2+} per mitochondrion $= 1.7 \times 10^{-3} \times 6.5 \times 10^{-17}$

$$= 1.1 \times 10^{-19} \text{ moles (110 tipomoles)}$$
$$= 66\,000 \text{ ions}$$

If free $[Ca^{2+}_{mit}]$ $= 100 \, \mu M$ then it would be 3900
$= 10 \, \mu M$ then it would be 390.

Under the latter conditions normal thermodynamic equations start to become invalid and a form of statistical thermodynamics is required. These order of magnitude calculations show that if the free Ca^{2+} concentration within a mitochondrion is maintained at a high level because of the membrane potential, then a loss in potential might result in a detectable local change in cytoplasmic free Ca^{2+}.

The increased rate of Ca^{2+} uptake into isolated mitochondria occurring over the first few minutes is comparable with the rates of Ca^{2+} buffering observed in cells using aequorin or arsenazo III as Ca^{2+} indicators (Rose and Loewenstein, 1976; see also Ashley and Campbell, 1979). Isolated mitochondria can reduce the free Ca^{2+} concentration to less than $0.1 \, \mu M$, which is below the threshold for myofibril contraction in muscle. Poisoning mitochondria or adding ionophore A23187 causes Ca^{2+} release *in vivo* and *in vitro*.

The possible physiological roles for mitochondrial Ca^{2+} uptake and release can be summarized as follows: (1) acts as a Ca^{2+} buffer under physiological conditions, and after cell injury; (2) regulates cytoplasmic free Ca^{2+}, and in particular localizes increases in Ca^{2+} during cell activation; (3) required for regulation of Ca^{2+}-sensitive mitochondrial enzymes; (4) production of heat by Ca^{2+} cycling, analogous to substrate cycling and turnover of the Na^+ pump; (5) required for transport of anions and NADH into the matrix of the mitochondrion (Harris and Berent, 1969; Vinogradov et al., 1972); (6) required for calcification in cells involved in formation of skeletons and regulation of whole body Ca^{2+} (DeLuca et al., 1962).

Several criteria must be satisfied in order to establish a physiological role for mitochondrial Ca^{2+} in cell regulation: (1) cell activation must be dependent on changes in intracellular Ca^{2+}; (2) mitochondria should be observed in the electron microscope near to the site of activation within the cell; (3) substances known to release Ca^{2+} from mitochondria should interfere with the action of the normal cell stimulus, in a way not explained by effects on cell ATP, mitochondrial ATP or intracellular pH; (4) a change in free Ca^{2+} near to mitochondria should be demonstrated using aequorin and image intensification (Chapter 2, section 2.5.3); (5) demonstration in the intact cell and in the broken cell of the mechanism (i.e.

Table 4.13. Physiological substances acting on cells affecting Ca^{2+} content or transport in isolated mitochondria

Tissue	Substance	Reference
Liver	Thyroxine	Herd (1978)
	Steroid hormones	Kimberg and Goldstein (1966)
	17 β-Oestradiol and dehydrocorticosterone	Bygrave (1978)
	Dexamethasone	Kimura and Rasmussen (1977)
	Insulin	Dorman et al. (1981)
	Glucagon	Yamazaki (1975)
	Glucagon, adrenaline, dibutyryl cyclic AMP	Hughes and Barritt (1971, 1978)
	Prostaglandins	Malmström and Carafoli (1975)
Heart	Electrical activity	Carafoli (1974)
Adipose tissue	Adrenaline (but not insulin)	Denton (1977)
Salivary gland	Adrenaline	Dormer and Ashcroft (1974)
Exocrine pancreas	Acetylcholine	Dormer et al. (1980)
Kidney	Vitamin D	Scarpelli (1965)
	Parathyroid hormone	Caulfield and Schrag (1964)

second messenger) by which the primary stimulus causes the change in mitochondrial Ca^{2+}.

Several naturally occurring substances effect mitochondrial Ca^{2+} (Table 4.13). How do these substances affect mitochondrial Ca^{2+}? In the case of hormones and neurotransmitters acting on the cell surface it is necessary to invoke a 'second messenger'. Borle (1972) reported that cyclic AMP stimulated Ca^{2+} uptake in kidney mitochondria. This has not been reproduced by others (Scarpa et al., 1976) and has since been refuted. Another possible candidate is Na^+ (Carafoli, 1974; Carafoli et al., 1977; Baker, 1976; Crompton and Carafoli, 1979) which stimulates release of Ca^{2+} from rat heart, squid axon and other cell mitochondria, but not apparently those from liver. The relationship is sigmoidal with respect to Na^+ concentration, with half maximal release between 5 and 10 mM Na^+. K^+ does not stimulate release nor is the effect of Na^+ inhibited by ruthenium red. The problem is that there is no convincing evidence in intact cells that physiological agents change the free Na^+ concentration in the cytoplasm and thereby elicit Ca^{2+} release. Experimental conditions (e.g. cardiac glycosides) which elevate intracellular Na^+ may act in this way.

Developmental studies provide an interesting alternative approach to investigating the possible significance of mitochondrial Ca^{2+} transport (Bygrave, 1978). For example, in blowfly flight muscle development of mitochondrial Ca^{2+} transport coincides with the requirement for insect flight. In foetal liver mitochondrial Ca^{2+} transport is small but develops soon after birth and before endoplasmic reticulum Ca^{2+} transport. Hormones like insulin may, through effects on protein synthesis, also have long-term effects on mitochondrial Ca^{2+}.

The circumstantial evidence for a physiological role for mitochondrial Ca^{2+} in cell regulation is therefore considerable. A membrane-permeant ester of a fluorescent Ca^{2+} indicator, analogous to quin II (Rink and Tsien, 1982) and perhaps using carnitine, would be a very valuable tool for resolving many of the problems of mitochondrial Ca^{2+}.

4.6.3. Ca^{2+} efflux from cells

The arguments concerning mitochondrial Ca^{2+} and the Ca^{2+} equilibrium potential made in the previous section apply equally to Ca^{2+} efflux from the cell membrane. Because of the enormous electrochemical gradient of Ca^{2+} across the cell membrane an energy-dependent Ca^{2+}-efflux mechanism is required to maintain the resting free Ca^{2+} at about 0.1 μM, and to remove Ca^{2+} (up to 100 μM in total) after a cell stimulus.

The mammalian erythrocyte and the giant axon of the squid have enabled the two known mechanisms for Ca^{2+} efflux in cells to be characterized Fig 4.20 (see page 190). A Ca^{2+}-activated MgATPase responsible for Ca^{2+} efflux in red cells, and possibly requiring no counterion, was first characterized by Schatzmann (1966, 1973, 1975). In contrast, a Na^+-dependent Ca^{2+} efflux was first fully characterized by Baker, Blaustein and colleagues in arthropod and molluscan nerves (Baker and Blaustein, 1968; Baker et al., 1969). Some evidence for similar systems in other cells has since been obtained (Tables 4.14 and 4.15). In some microorganisms (pro- and eukaryote) a $Ca^{2+}-H^+$ antiport system has been discovered (Stroobant et al., 1980).

Before examining the properties of these two known Ca^{2+} efflux mechanisms let us examine the energy requirements of such pumps. If W is the work required

Table 4.14. Examples of Cells with Ca^{2+}-activated MgATPase in the cell membrane

Cell	Reference
Mammalian erythrocytes	Schatzmann (1975)
Nucleated erythrocytes	Campbell and Dormer (1978)
HeLa cells	Borle (1968)
L-cells	Lamb and Linsay (1971)
Squid giant axon	DiPolo (1978)
Bacteria (*Bacilli*, *Azotobacter*, *Escherichia*)	Rosen and McClees (1974)
Neurospora	Stroobant et al. (1980)

See also Godfraind-De Becker and Godfraind (1980)

Table 4.15. Cells with possible Na^+–Ca^{2+} exchange system in the cell membrane

Cell	Evidence*	Reference
Nerve		
Squid giant axon	1, 2, 3, 4, 5, 6, 7	Baker et al. (1969); Baker (1972); Baker (1976)
Crab peripheral nerve	1, 2, 3	Blaustein (1974)
Desheathed vagus nerve	1 (?), 6	Kalix (1977)
Brain	1, 2, 6	Stahl and Swanson (1972)
Muscle		
Vertebrate skeletal (frog)	2, 4	Cosmos and Harris (1961)
Vertebrate heart	1, 2, 3, 8	Niedergerke (1963); Glitsch et al. (1969)
Smooth	1, 2, 3, 8	Goodford (1967); Reuter et al. (1973)
Invertebrate (crab and barnacle)	1, 2, 3, 4, 5, 8	DiPolo (1973); Ashley et al. (1976)
Secretory cells		
Nerve endings (synaptosomes and giant nerves)	1, 2, 3, 6, 8	Blaustein and Weissman (1976)
Anterior and posterior pituitary	2, 8	Dicker (1966); Fleischer et al. (1972)
Pancreatic β-cells	8	Hales and Milner (1968)
Adrenal medulla	2, 8	Douglas (1974)
Other cells		
Obelia luminescence	8	Campbell et al. (1979)
Hepatocyte	1, 2	Judah and Ahmed (1963, 1964)
Gut	1	Schachter et al. (1970)
Photoreceptor (horseshoe crab)	1, 2	Llinas and Blinks (1972); Brown and Blinks (1974)
Paramecium	1, 2	Eckert et al. (1976)
Ca^{2+}-*transporting cells*		
Intestinal serosal membrane	1	Schachter et al. (1970)
Kidney tubule	1, 2	Blaustein (1974)
Chick chorioallantoic membrane	1	Terepka et al. (1976)
Bone	1	Krieger and Tashjian (1980)

*Evidence for Na^+–Ca^{2+} exchange
(1) $^{45}Ca^{2+}$ efflux: decreased by removal of extracellular Na^+; increased by increasing intracellular Na^+.
(2) Ca^{2+} influx: increased by increasing intracellular Na^+ (^{45}Ca or aequorin measurement).
(3) Stoicheiometry of Ca^{2+}:Na^+ is 2 or 3.
(4) Concomitant Na^+ flux opposite to Ca^{2+} flux.
(5) Decrease extracellular Ca^{2+} produces decreased Na^+ efflux.
(6) Na^+-dependent Ca^{2+} efflux relatively insensitive to intracellular ATP concentration.
(7) Ca^{2+} flux sensitive to E_m.
(8) Ouabain (increases intracellular Na^+) or a change in extracellular Na^+ produces an effect on a physiological stimulus dependent on extracellular Ca^{2+}. N.B. Need to rule out Na^+-dependent release of Ca^{2+} from mitochondria.

to transport 1 mole of Ca^{2+} from inside the cell to the outside,

$$W = \frac{1}{J}(RT\log_e([Ca_o^{2+}]/[Ca_i^{2+}]) - 2E_m F) \text{ cal mole}^{-1} \quad (4.12)$$

where 1 cal = J joules and $J = 4.184$ J cal^{-1}
E_m = membrane potential.
If $E_m = -50$ mV
and $[Ca_o^{2+}] = 1$ mM
and $[Ca_i^{2+}] = 0.1$ μM
Then $W = 8$ kcal mole^{-1}.
If $[Ca_o^{2+}] = 10$ mM
then $W = 9.4$ kcal mole^{-1}. (4.13)

For the reaction $ATP \rightleftharpoons ADP + P_i$ (really MgATP)

$$\Delta G' = \Delta G_o - RT\log_e \frac{[ATP]}{[ADP][P_i]} \quad (4.14)$$

$\Delta G_o = -7.2$ kcal mole^{-1} (Benzinger et al., 1959).
Thus, if, in the cell, ATP = 5 mM, ADP = 1 mM, and P_i = 1 mM, (i.e. reaction not at equilibrium) then $\Delta G' = -12.5$ kcal mole^{-1}. (4.15)

Hence the hydrolysis of 1 ATP provides sufficient energy for only 1 Ca^{2+} to move out across the cell membrane. If, however, the energy were to come from an ion gradient (e.g. Na^+) and not ATP, then the work (W) done by n moles of Na^+ is

$$W = n[RT\log_e([Na_o^+]/[Na_i^+]) - E_m F]/J \text{ cal} \quad (4.16)$$

If this supplies the energy for 1 mole of Ca^{2+} efflux, then also

$$W = (RT\log_e([Ca_o^{2+}]/[Ca_i^{2+}]) - 2E_m F)/J \text{ cal} \quad (4.17)$$

$$\therefore \frac{RT}{F}\log_e([Na_o^+]/[Na_i^+])^n - \frac{RT}{F}\log_e([Ca_o^{2+}]/[Ca_i^{2+}]) = (n-2)E_m$$

$$\left(\frac{[Na_o^+]}{[Na_i^+]}\right)^n \left(\frac{[Ca_i^{2+}]}{[Ca_o^{2+}]}\right) = \exp[(n-2)E_m F/RT] \quad (4.18)$$

In a squid giant axon, if $[Ca_i^{2+}] = 0.1$ μM, $[Ca_o^{2+}] = 10$ mM and $[Na_i^+] = 10$ mM, $[Na_o^+] \div 500$ mM; $E_m = -90$ mV (negative inside) then $n = 2.2$ at 15 °C.

In a mammalian cell, on the other hand,
if $[Ca_i^{2+}] = 0.1$ μM, $[Ca_o^{2+}] = 1$ mM
and $[Na_i^+] = 10$ mM, $[Na_o^+] = 150$ mM; $E_m = -50$ mV (negative inside) then $n = 2.3$

Thus, in order to maintain the intracellular free Ca^{2+} concentration at about 0.1 μM using the Na^+ gradient alone, 3 Na^+ must enter the cell for every Ca^{2+}

extruded in vertebrate or invertebrate cells. Alternatively, the energy could come from 2 Na$^+$ and 1 ATP.

These conclusions tell us something about the energetics necessary for the mechanism of Ca^{2+} pumping, but not how much of the cell's energy metabolism is required to maintain the Ca^{2+} gradient. To estimate this the actual flux of Ca^{2+} must be known. At steady-state, influx must equal efflux. For Na$^+$ ions, relating membrane flux (Ling, 1965) to measured ATP turnover or O$_2$ utilization suggests that in red cells, muscle and probably other cells, between 20 and 50% of the cell's metabolic energy is required by the (Na$^+$ + K$^+$)-MgATPase (Whittam, 1961; Caldwell, 1968). Similar calculations first made for Ca^{2+} by Ling (1965) are invalid because he assumed a mM Ca^{2+} concentration within the cells. Recalculation using data from Schatzmann (1973) and Baker (1972) shows that less than 1% of the cell's energy is required in a resting cell to maintain the Ca^{2+} gradient across the membrane. However, if the Ca^{2+} efflux system is activated ten- to one hundred fold when intracellular free Ca^{2+} increases ten- to one hundred fold after cell stimulation then the rate of ATP consumption will increase similarly and could then require a significant proportion of the cell's energy supply. This may account for the decrease in cell ATP induced experimentally by substances such as A23187.

Assuming Ca^{2+} efflux (Ca$^{2+}_{out}$) is first-order with respect to free Ca$^{2+}_i$,

then
$$d[Ca^{2+}_{out}]/dt = -k_{out}[Ca^{2+}_i] \tag{4.19}$$

If 1 Ca^{2+} requires 2 ATP then the fraction of the energy supply used by the Ca^{2+} pump is

$$\frac{2k_{out}[Ca^{2+}_i]}{\text{total rate of ATP turnover in the cell}} = \frac{k_{out}[Ca^{2+}_i]}{3 \text{ moles } O_2 \text{ utilized per min}} \tag{4.20}$$

since 1O$_2 \equiv$ 6 ATP in the mitochondria.

Cell membrane Ca^{2+}-activated MgATPase

This enzyme has been extensively studied in the mammalian erythrocyte (Schatzmann, 1975), where it was first discovered (Dunham and Glynn, 1961). Schatzmann and his coworkers were the first to show that red cells, or sealed red cell ghosts prepared by reversible osmotic lysis, loaded with ^{45}Ca^{2+} would only pump Ca^{2+} out if MgATP was present inside. The kinetics of Ca^{2+} efflux with respect to MgATP and Ca^{2+} correspond to those of the (Ca^{2+})-MgATPase in broken cells. An approximate K_m^{Ca} of 4 μM for Ca^{2+} pumping was estimated by sealing Ca/EGTA buffers (see Chapter 2, section 2.7) inside the ghosts. This means that at 0.1 μM free Ca^{2+} the enzyme is less than 3% saturated with Ca^{2+} (Chapter 3, section 3.3.3).

Measurement of the rate of appearance of inorganic phosphate orginally suggested that 2 Ca^{2+} were transported per ATP split. However, under more carefully controlled conditions, a value of 1 Ca^{2+} per ATP was found, which is consistent with the energetic restrictions (equations 4.13 and 4.14). This distinguishes it from the (Ca^{2+})-MgATPase in muscle sarcoplasmic reticulum which pumps 2 Ca^{2+} per ATP (see Chapter 5, section 5.3.3). The Ca^{2+}-activated MgATPase in red cells is insensitive to extracellular Ca^{2+} or Na^+. It is not inhibited by ouabain, but can be inhibited by SH-group antagonists. A compound analogous to ouabain for the Na^+ pump would be an extremely useful experimental tool.

If the Ca^{2+} pump were electrogenic, in other words had no counterion, then it would contribute < 0.1 mV to the membrane potential of the cell. To illustrate this let us assume a flux of 0.1 pmoles Ca^{2+} sec^{-1} cm^{-2} in the resting cell.

Converting this to charge, efflux = 0.2 F pCs^{-1} cm^{-2}, or A cm^{-2}
$$= 1.9 \times 10^{-8} \text{ A cm}^{-2}$$

If the membrane resistance = 1000 Ωcm^2 = R

since $\quad\quad V = IR$ \hfill (4.21)

then $\quad\quad V = 0.019$ mV = contribution to membrane potential

Even if Ca^{2+} efflux increased tenfold the contribution of Ca^{2+} to the membrane potential would be negligible.

Convincing evidence for a Ca^{2+}-activated MgATPase in cells other than the erythrocyte in sparse (Table 4.14). This is because of the difficulty of purifying the cell membrane and of detecting a small activity of (Ca^{2+})-MgATPase against a high background of Ca^{2+}-independent MgATPase, even when ouabain is present to inhibit the $(Na^+ + K^+)$-activated MgATPase. Contamination of subcellular fractions with (Ca^{2+})-MgATPases from endoplasmic reticulum and mitchondria also cause problems in interpretation. Furthermore, sealed cell 'ghosts' can only be prepared from erythrocytes.

Is the Ca^{2+} pump reversible?

Under experimental conditions the Na^+ pump can be reversed and ATP is formed. A similar experiment has yet to be demonstrated for the Ca^{2+} pump. Presumably there is a very small net influx of Ca^{2+} through the Ca^{2+} pump. A large change in the [ATP]/[ADP][P_i] ratio would increase this influx without necessarily having much effect on efflux (e.g. measured by $^{45}Ca^{2+}$). Yet under these conditions the net flow of Ca^{2+} out of the cell through the pump would be decreased. This can be described quantitatively by the Haldane equation (Haldane, 1930).

There is, as yet, no convincing evidence for physiological effectors regulating intracellular free Ca^{2+} through effects on the Ca^{2+} pump, although the action of substances like adenosine at nerve terminals may involve such a mechanism. Calmodulin (see Chapter 3, section 3.4.2) has been isolated from human erythrocytes. This protein activates the (Ca^{2+})-MgATPase in red cell membranes (Bond and Clough, 1973; Jarrett and Penniston, 1977a, b; Gopinath and Vincenzi, 1977). Since the reported experiments have so far been carried out using relatively high Ca^{2+} concentrations, the physiological significance of this activator has yet to be established. In other cells calmodulin could provide a means of activating the Ca^{2+} pump in order to remove cytoplasmic Ca^{2+} rapidly after a stimulus.

Na^+-dependent Ca^{2+} efflux

Interactions between Na^+ and Ca^{2+} have been known to exist in muscle for nearly a century (Ringer, 1883a, b, c). Na^+ inhibits competitively Ca^{2+} uptake in heart (Zutgar and Niedergerke, 1958), and replacement of extracellular Na^+ by Li^+ causes an increase in intracellular free Ca^{2+} and the muscle contracts (Brinley, 1968; Ashley et al., 1974). This is associated with an increased Na^+ efflux. It was the slightly unexpected observation in axons of squid giant nerve and crab nerve that replacement of extracellular Na^+ by Li^+, choline or dextrose stimulated Na^+ efflux which lead to the discovery of a Na^+-dependent Ca^{2+}-efflux mechanism (Baker and Blaustein, 1968; Baker et al., 1969), and which can lead to an elevation of intracellular free Ca^{2+} under experimental conditions.

The Na^+ efflux has several surprising properties. (1) It is not inhibited by ouabain, even at concentrations 10 000 times those necessary to inhibit the Na^+ pump which is responsible for the bulk of the Na^+ efflux. (2) It is dependent on the presence of extracellular Ca^{2+}, and inhibited by La^{3+}. These conditions have no effect on Na^+ efflux through the Na^+ pump. (3) There is a correlation between Na^+ flux in one direction and Ca^{2+} flux in the other, such that 2–3 Na^+ are equivalent to 1 Ca^{2+} (Baker et al., 1969; Baker, 1972; Blaustein, 1974; Baker and McNaughton, 1976a, b). (4) Manipulation of intra- and extracellular Na^+ leads to changes in intracellular Ca^{2+} detectable by aequorin (Baker, 1972; Ashley et al., 1974) or glass scintillation probe (Ashley and Lea, 1977, 1978).

These experiments confirm the energetic restrictions placed on Na^+–Ca^{2+} exchange (see equation 4.18). Most of the energy for Ca^{2+} efflux by this route comes from the Na^+ gradient, though there may be a direct requirement for ATP (Baker and McNaughton, 1976a). If 3 Na^+ move per 1 Ca^{2+}, then there is a thermodynamic necessity for ATP (equation 4.18). Such a mechanism would involve the movement of net positive charge into the cells. Such an electrogenic exchange should be sensitive to changes in membrane potential (E_m). This is indeed the case. Depolarization (less negative inside the cell) inhibits Na^+ dependent Ca^{2+} efflux, whereas hyperpolarization enhances it (Blaustein, 1974). Since Ca^{2+} efflux is proportional to $\exp[(n-2)E_m F/RT]$ (see equation 4.18), for a 25 mV depolarization there should be an 'e'-fold change in Ca^{2+} efflux. This prediction has been borne out by observation, but only in CN^--poisoned axons. In unpoisoned axons changes in membrane potential do affect Ca^{2+} efflux but not in the manner predicted by this equation.

The situation in squid giant axon is complicated by the existence of three components of Ca^{2+} flux, as measured by $^{45}Ca^{2+}$ (Baker and McNaughton,

1976a, b): (1) activation by extracellular Ca^{2+} and not simply a Ca^{2+}–Ca^{2+} exchange; (2) activation by extracellular Na^+, a Na^+–Ca^{2+} exchange; (3) Ca^{2+} efflux occurring in the absence of external Ca^{2+} and Na^+.

Changes in internal ATP concentration affect the 'K_m' for extracellular Na^+. However, maximal Ca^{2+} efflux is unchanged at saturating concentrations of external Na^+, even when the internal ATP concentration is reduced to a very low level. On the other hand, when EGTA/Ca^{2+} is injected into the axon, the addition of CN^- which blocks mitochondrial ATP synthesis, markedly reduces both the external Ca^{2+}- and Na^+-dependent components of Ca^{2+} efflux. These components can be restored by injection of ATP into the cell. The reason for ATP requirement is therefore still unclear. The ATP could provide the additional energy for Ca^{2+} efflux (see equation 4.18) so that the energy requirement is satisfied by $2 Na^+ + 1$ ATP. Alternatively, ATP need not necessarily be hydrolysed. If could, for example, determine the affinity of the system for Ca^{2+} and Na^+, or it could be required to maintain the integrity of the efflux mechanism. A problem in investigating the role of ATP in this system arises from the great difficulty in reducing cell ATP to less than $1 \mu M$, which is necessary to be well below the K_m for MgATP of most ATPases. Studies with non-hydrolysable analogues of ATP, such as AMP-PNP (adenylyl imidodiphosphate) and AMP-PCP (adenylyl (β,γ-methylene)-diphosphonate), are equally inconclusive since these preparations are often contaminated with ATP. Perhaps the best evidence for ATP hydroloysis in this system would be to demonstrate its ability to synthesize ATP after reversing the direction of the Na^+ and Ca^{2+} fluxes.

The molecular mechanism of Na^+–Ca^{2+} exchange is unknown. An ionophoretic model has some attractions, particularly since the bromine derivative of the ionophore X-537A can be used to generate a Na^+-dependent Ca^{2+} gradient across an organic phase (Malaisse and Couturier, 1978).

The criteria for demonstrating a Na^+-dependent Ca^{2+} efflux in cells are outlined in Table 4.15. Erythrocytes do not have such a Ca^{2+} efflux mechanism. Three major differences from the (Ca^{2+})-MgATPase of red cells stand out: (1) Na^+–Ca^{2+} movement is bidirectional, and is therefore potentially better suited to causing an increase in intracellular Ca^{2+} by regulating its activity; (2) extracellular Na^+ has no effect on (Ca^{2+})-MgATPase; (3) ATP is the sole energy source for (Ca^{2+})-MgATPase and is an absolute requirement for Ca^{2+} efflux by this mechanism.

The physiological significance of Na^+–Ca^{2+} exchange is not fully understood nor is it known whether physiological effectors can alter intracellular Ca^{2+} by interacting directly with this process. The reason for this is that fairly drastic experimental procedures (e.g. zero Na^+ or 112 mM Ca^{2+}) are required to measure Na^+–Ca^{2+} exchange. Even in squid nerve it seems that, under physiological conditions, the majority of the Ca^{2+} efflux occurs by a (Ca^{2+})-MgATPase (DiPolo, 1978) and not by Na^+–Ca^{2+} exchange. The fact that effects of cyclic AMP on Na^+ and Ca^{2+} flux have been found suggests that it might be worthwhile searching for physiological effectors (Bittar et al., 1976; Cheng and Chen, 1975).

4.7. Transcellular Ca^{2+} transport

Transport of Ca^{2+} across cells is necessary for absorption of Ca^{2+} by the small intestine, for Ca^{2+} resorption by the kidney tubule and by the chorioallantoic membrane during egg formation in birds (Terepka et al., 1976).

It was originally thought that Ca^{2+} uptake from the intestine was passive. This would inevitably lead to an increase in intracellular free Ca^{2+} and an increase in mitochondrial Ca^{2+}. This has not been found. The inhibition by cytochalasin B and evidence from X-microprobe analysis in the electron microscope suggest that a vesicular uptake mechanism may be involved (see Chapter 7). Uptake could be by pinocytosis, followed by transport of the vesicles across the cell and release of the Ca^{2+} by exocytosis at the serosal surface of the cell.

Hormones such as vitamin D and parathyroid hormone enhance Ca^{2+} transport across the gut and kidney. In the gut vitamin D induces the formation of a Ca^{2+}-binding protein, which belongs to the troponin C–calmodulin class of proteins (see Chapter 3, section 3.4.2). Although it is likely that this protein is involved in transcellular Ca^{2+} transport its precise role has not been established.

4.8. Conclusions

Much has still to be learnt about the interactions of Ca^{2+} with the electrical properties of both organelles and the cell itself during activation. Several mechanisms have been identified which can, under experimental conditions at least, lead to changes in intracellular free Ca^{2+}. The evidence for these mechanisms has relied heavily on systems which are relatively easy to study, such as invertebrate giant nerves and mammalian red cells. The experimental conditions often involve gross disturbances in the chemical and electrical properties of the cells. Future progress requires the demonstration that physiological regulators, like action potentials, neurotransmitters and hormones, can control cell function through these mechanisms, and that the molecular basis of ion channels and pumps be established. In excitable cells the existence of Ca^{2+}-dependent 'threshold' activation under physiological conditions provides support for the concept of 'threshold' phenomena which was developed in Chapter 2.

CHAPTER 5

Calcium and cell movement

5.1. Types of cell movement

In several chapters we have considered a little boy running along the pavement to greet his parents as one example of phenomena in which many different cell types are activated. Amongst those activated will be various types of muscle cell. For instance, the contraction of skeletal muscle provides the means of movement of the arms and legs, whilst the increased rate of contraction of the heart cells, together with effects on smooth muscle, enables more oxygen to be supplied to skeletal muscle by increasing blood flow. The study of the mechanisms involved in these contractions has an important place in the development of electrophysiology and biochemistry, but in particular provides one of the best-characterized examples of the role of intracellular Ca^{2+} as a regulator.

Important as muscle contraction is in many animals, it is but one of Nature's examples of cell movement (Table 5.1). Some 2500 to 3000 million years ago small appendages appeared on certain prokaryotic cells enabling them to move so that they could find food more easily, and respond to beneficial or detrimental environmental conditions. In eukaryotes movement enabled the organism to escape from predators, as well as being necessary for the male gamete to find its female partner for fertilization to take place.

There are two major forms of cell movement in eukaryotes—flagellate and amoeboid (Willmer, 1960). Movement can be provoked by a variety of electrical, chemical and mechanical stimuli; (Table 5.2). Let us examine the evidence for intracellular Ca^{2+} playing a role in mediating the effect of a primary stimulus on eukaryotic cell movement, and the movement of organelles within the cell (Table 5.3).

5.1.1. Prokaryote motility

Blue-green algae and myxobacteria are able to move along a solid surface by a gliding motion. However, these two groups, unlike the other two major groups of prokaryotes, have no visible locomotor organelles on the surface. In contrast, the spirochetes have an axial filament wrapped around the cell and attached at each end of the bacterium, whereas motile eubacteria have one or more flagellae (Fig. 5.1). The energy for movement comes ultimately from ATP hydrolysis, and the single-stranded protein responsible is known as flagellin.

Table 5.1. Examples of cell movement

Cell type	Example	Reference
1. *Prokaryotes*		
Flagellate bacteria	*Sphaerotilus*	Stanier et al. (1964)
Spirochetes	*Leptospira*	Stanier et al. (1964)
2. *Flagellate or ciliate eukaryotes*		
Protozoan phytoflagellates	*Euglena*	Barnes (1974)
Protozoan zooflagellates	*Trypanosoma*	Barnes (1974)
Protozoan ciliates	*Paramecium*	Brehm and Eckert (1978)
Spermatozoa	*Homo sapiens*	Ham (1965)
Ciliate epithelial cells from the alimentary canal	*Homo sapiens*	Ham (1965)
3. *Amoeboid*		
Many protozoa	*Amoeba*	Barnes (1974)
Polymorphs and macrophages	*Homo sapiens*	Klebanoff and Clark (1978)
Slime mould	*Dictyostelium*	Chi and Francis (1971)
Cells during differentiation—embryonic blastula (morphogenetic)	Coelenterate cnidoblasts	Lenhoff and Loomis (1961)
	Fibroblasts	Gail et al. (1973)
Wound healing	Epithelial cells	Gabbiani and Montaudon (1977)
4. *Contractile cells*		
Invertebrate muscle	*Balanus*	Ashley (1978)
Vertebrate muscle	Mammals	Needham (1971)
Spasmoneme	*Vorticella*	Amos et al. (1976)
'Contractile' organs in plants	Venus fly trap	

Table 5.2. Some stimuli of movement

Primary Stimulus	Example	Reference
Action potential	Electrically coupled muscle cells in Cnidarians	Lenhoff and Loomis (1961)
Spontaneous action potentials	Heart muscle	Winegrad (1961); Reuter (1973)
Neurotransmitters and action potentials	Vertebrate and invertebrate muscle	Krnjevic (1974); Gerschenfeld (1973)
Touch	The protozoan *Paramecium*	Brehm and Eckert (1978)
Chemotactic	Cyclic AMP and slime mould *Dictyostelium*	Bonner (1971)
	Complement products e.g. C5a on polymorphs	Gallin and Rosenthal (1973, 1974); Lynne and Mukherjee (1978)
	Motile algae and protozoa	Levandowsky and Hauser (1978)
	Nutrients on chemotactic bacteria	Koshland (1979)
Spontaneous	Flagellate and ciliate cells	Brokaw (1972)

Table 5.3. Examples of movement inside cells

Phenomenon	Example	Reference
Cytoplasmic streaming including protoplasm and organelle movement	*Nitella* and several other algae and protozoa	Piddington (1976); Kamiya (1959); Kamiya and Kuroda (1956); Jahn and Bovee (1969); Allen (1974)
	The slime mould *Physarum*	Durham and Ridgway (1976)
	The foraminiferan *Allogromia*	Jahn and Rinaldi (1959)
	Axoplasmic flow in nerve axon	Ochs *et al.* (1977)
Contractile vacuoles	Various protozoa	Barnes (1974); Piddington (1976)
Secretion	Movement of secretory vesicles	Shaw and Newby (1972)
Endocytosis	Movement of endocytotic vesicles towards lysosomes	Allison *et al.* (1971); Allison and Davies (1974)
Cell division	Movement of chromosomes along spindle during mitosis and meiosis	Kimball (1974)
Cell polarity	Location of vesicles at one pole of exocrine cells	Case (1978)

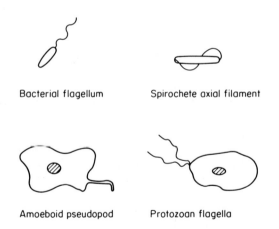

Fig. 5.1. Some structures responsible for cell movement

Mostly the movement of bacteria appears to be random with no postulated role for intracellular Ca^{2+}. However, in certain special cases bacteria can move in an apparently directed manner. For instance, motile purple bacteria can move along a gradient of light intensity, a phenomenon known as phototaxis. Others exhibit aerotaxis, which causes them to move along an oxygen gradient. Perhaps the most interesting example as far as the theme of this book is concerned is that of chemotactic bacteria (Koshland, 1979). *Bacillus subtilis*, for example, is attracted by certain amino acid nutrients and repelled by others. The concentration gradient across an individual bacterium is too small for any direct chemical receptor mechanism to respond to. The bacteria have evolved a way of enhancing the 'apparent' concentration gradient across themselves. They do this by their ability to 'tumble'. The motion of these bacteria involves a smooth movement in one direction, followed by a 'tumble' and then a movement off in another direction. Chemotaxis occurs because if the bacteria head off in an unfavourable direction the time between each tumble is reduced, whereas in the favourable direction it is increased. Response to either an attractant or a repellant is thought to involve changes in membrane potential across the membrane of the bacteria.

In the presence of A23187 and Ca^{2+} concentrations greater than 0.1 μM, the bacteria become incessant tumblers (Ordal, 1977), whereas when the external Ca^{2+} < 10 nM tumbling is virtually absent. It is therefore possible that tumbling is brought about by an increase in intracellular free Ca^{2+} in the range 0.1–1 μM caused by a depolarization of the membrane potential when repellant molecules bind to sufficient numbers of receptor molecules.

5.1.2. Eukaryote motility

There are many fundamental differences in morphology and chemistry between pro- and eukaryotic organisms, one of which is the mechanism of movement (Table 5.1). The flagella or cilia of eukaryotes may look superficially like bacterial flagella; however, the major protein involved is quite different as is the fine structure, exhibiting the universal '9 + 2' arrangement (Fig. 5.7) to be found in protozoan cilia and flagella as well as in gut epithelial cells or the tails of animal sperm. Some eukaryotes, on the other hand, exhibit a form of movement known as amoeboid, which is not found in prokaryotes. It is easily observable in protozoa attached to a substrate, and is caused by directed cytoplasmic streaming. It can also be observed in some mammalian cells. For example, phagocytic cells such as polymorphonuclear leukocytes (polymorphs) respond to chemotactic stimuli. The result is the infiltration of sites of inflammation by these white cells. Two well-known examples are wounds infected by bacteria, and the synovial fluid of patients with rheumatoid arthritis. Although amoeboid and flagellate movement appear to be fundamentally different, some interesting protozoa, e.g. *Naegleria gruberi*, can transform from one to the other under certain conditions (see Chapter 8).

The major protein in eukaryote flagella is tubulin, and is the same as that found in microtubules in other parts of the cell. On the other hand, in microfilaments—the structure responsible for the cell's cytoskeleton and amoeboid movement—the major protein is actin. Both microtubules and microfilaments may be regulated by changes in intracellular Ca^{2+} (Tables 5.4 and 5.6). In some of these examples movement occurs spontaneously as a result of internal programming. In others, movement is triggered by an external stimulus (Table 5.2), which may be physical or chemical in origin. In several of these latter cases the cell changes rapidly from one state to another and thus can be considered to have passed

Table 5.4. Some possible functions of microtubules

Ciliate and flagellate movement in eukaryotes*
Centriole and basal body structure
Formation of the mitotic spindle*
Some examples of cytoplasmic transport*
Movement of secretory granules towards the cell membrane*
Connective tissue in arthropods
Sensory transduction
Cell shape and polarity—mobility and positioning of membrane components*, eg.
	capping in lymphocytes and the end-plate in muscle
	—Orientation of cell structures, e.g. development of cell walls in plants or collagen fibres in animal tissues

---

* Where evidence exists for intracellular Ca^{2+} being a regulator.
For references see Tilney and Gibbins (1969), Durham (1974), Pollard and Weihing (1974), Soifer (1975), Stephens and Edds (1976); Albertini and Anderson (1977).

through a 'threshold' (see Chapter 2). Changes in intracellular Ca^{2+} may play a role in regulating both spontaneous movement and movement activated by an external stimulus. Sometimes oscillations of movement occur. In these cases one might expect the intracellular Ca^{2+} concentration to oscillate in response to the cyclical electrical or chemical changes (Ridgway and Durham, 1976). Examples of this are amoeboid movement in protozoa, the beating of heart muscle cells, and cytoplasmic streaming in slime moulds.

To understand how changes in intracellular Ca^{2+} might bring about these movements it is necessary first to examine the structure and mechanism of action of the two major proteins involved—tubulin and actin.

5.2. The molecular basis of eukaryotic cell movement

Four distinct intracellular fibrous structures have been recognised in the electron microscope and which are thought to play a role in eukaryotic cell movement: (1) microtubules 18–25 nm in diameter (Rebhun, 1972; Stephens and Edds, 1976) (Fig. 5.2); (2) tonofilaments 8–10 nm in diameter; (3) microfilaments 4–7 nm in diameter (Pollard and Weihing, 1974; Clarke and Spudich, 1977) (Fig. 5.3); (4) actomyosin (Mannherz and Goody, 1976). These structures have been identified in most eukaryotic cells by normal fixation and staining techniques, as well as by specific histochemical and fluorescent protein or antibody methods. They are responsible for most types of movement in eukaryotic cells but are not found in prokaryotes.

Microfilaments seem to be the non-muscle equivalent of actomyosin, being composed of actin and myosin, but they are not so regularly organized as they are in muscle. They may also contain the proteins necessary for the regulation of contraction by Ca^{2+} and in particular calmodulin, originally designated as a troponin C-like protein, and gelsolin: Microtubules, on the other hand, are more symmetrically organized structures, consisting of a helical arrangement of globular proteins forming a cylindrical tube several micrometres long. In cilia and

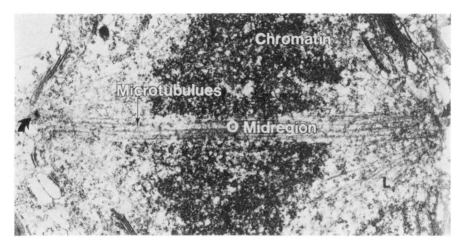

Fig. 5.2. Microtubules. From McDonald *et al.* (1977); reproduced from the *Journal of Cell Biology* by copyright permission of the Rockfeller University Press

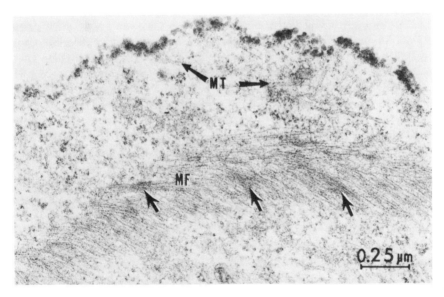

Fig. 5.3. Microfilaments (MF) and microtubules (MT). From McDonald *et al.* (1977); reproduced from the *Journal of Cell Biology* by copyright permission from the Rockfeller University Press

flagella the tubules are arranged in a 9 + 2 conformation (Fig. 5.7), whereas inside the cell microtubules exist individually in less well organized groups. Tubulin, the major protein of microtubules, is quite distinct from actin.

Connections between microtubules and microfilaments have been observed in the electron microscope. This has led to the suggestion that the combined action of both types of fibrous element may be required for cell motility and intracellular movements in non-muscle cells (Durham, 1974), though this is not entirely consistent with the changes in protein composition which occur in *Naegleria* during amoeboid-flagellate transformation (see Chapter 8). Let us therefore examine the chemistry of these structures and in what way Ca^{2+} may regulate their activity.

5.2.1 Microtubules and tubulin

Microtubules are cylinders which can form in the cell within a few minutes, or over a period of several hours. They may be anything from one to several micrometres long, are 18–25 nm in diameter, and have a wall thickness of 5 nm (Stephens and Edds, 1976). This uniformity in structure is caused by the wall of the tubule being composed of a helical arrangement of 4–5 nm globular subunits. The structure can also be thought of as thirteen longitudinal protofilaments slightly skewed to each other. Tubulin (6 S), the major protein, is a dimer and was first isolated in 1967 (see Taylor and Condeelis, 1979). It is dumbell-shaped, consisting of two subunits α- and β-tubulin, each with a molecular weight of about 58 000. Tubulin is an acidic protein but is also abundant in hydrophobic amino acids. It is therefore not surprizing that tubulin binds both Ca^{2+} and Mg^{2+} (Rosenfeld 1976; Solomon, 1977). Two classes of binding site have been identified (Solomon, 1977): (1) a single high-affinity (K_d^{Ca} approx. 3 μM, K_d^{Mg} approx. 50 μM) site where the binding is inhibited by KCl; (2) 16 low-affinity sites (K_d^{Ca} approx. 0.28 mM).

The low-affinity sites are unlikely to be of any physiological significance, though the Mg^{2+} stimulation of Ca^{2+} binding to these sites has been reported and may be the cause of some of the Ca^{2+}-induced microtubule depolymerization observed *in vitro* (Rosenfeld, 1976). But what about the high-affinity site? As we shall see, direct effects of Ca^{2+} on microtubules have been demonstrated *in vitro* but whether they can occur under physiological conditions is open to question. From equation 3.25 we can calculate that $K_{app}^{Ca} = K_d^{Ca}\{1 +([Mg^{2+}]/K_d^{Mg})\} = 0.1$ mM, at a free Mg^{2+} concentration of 2 mM, approximately that in the cell. This means that at 1 μM free Ca^{2+} less than 5% of the protein will have Ca^{2+} bound to it (see Fig. 3.9A and Fig. 3.10). 1 mM Ca^{2+} would be required to saturate more than 90% of the protein. This calculation makes it unlikely that direct binding of Ca^{2+} to tubulin has any great significance in the cell, unless a substance can be found which markedly enhances its affinity for Ca^{2+} (e.g. calmodulin or calmodulin-activated phosphorylation).

Each tubulin dimer binds two molecules of GTP, one tightly the other loosely. Tubulin also has GTPase activity. Dephosphorylation and phosphorylation of the GTP on each binding site may regulate depolymerization and polymerization of the tubules (Gaskin et al., 1974). The β-subunit can also be phosphorylated by a protein kinase associated with isolated microtubules and which does not require cyclic AMP, though addition of cyclic AMP dependent protein kinase also phosphorylates tubulin. It would be interesting to know what this phosphorylation does to the affinity of the tubules for Ca^{2+} and calmodulin.

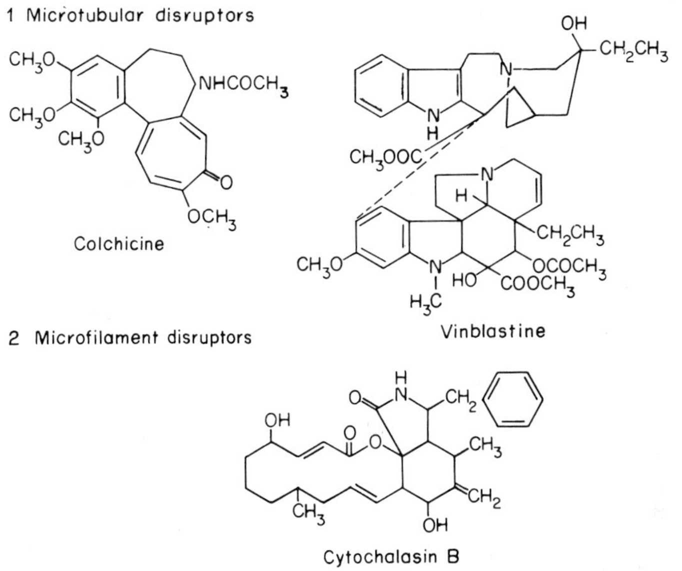

1 Microtubular disruptors

Colchicine

Vinblastine

2 Microfilament disruptors

Cytochalasin B

Fig. 5.4. Disruptors of microtubules and microfilaments

An important pharmacological property of tubulin is its ability to bind certain alkaloids such as colchicine, vinblastine and vincristine (Fig. 5.4). They cause breakdown of microtubules *in vitro* and *in vivo* (Margulis, 1973). Each tubulin dimer binds one molecule of colchicine or two of vinblastine. These alkaloids have been used to provide evidence for involvement of microtubules in several phenomena, though rarely help in investigating whether Ca^{2+} regulates microtubule assembly and disassembly.

Other proteins

Several other proteins have been found associated with isolated microtubules (Table 5.5). The two most important of these are dynein—a 500 000 molecular weight MgATPase, which plays an important role in linking microtubules to other structures in the cell—and calmodulin. Using immunofluorescence this Ca^{2+}-binding protein (Chapter 3, section 3.4) has been observed with microtubules in mitotic cells (Means and Dedman, 1980) and also binds to them *in vitro*.

Assembly and disassembly

The formation (polymerization) and depolymerization of microtubules has been observed using conventional fixation techniques, and has been demonstrated *in vitro* under various conditions. Several problems exist concerning the interpretation of such experiments: (1) The fixation or isolation procedures may only reveal very stable structures; labile microtubules may not be revealed. (2) The

Table 5.5. Some proteins associated with microtubules

Tubulin, the major protein
Dynein, a MgATPase, Ca^{2+}-activated
Calmodulin
Tau factor
Microtubule-associated proteins
Nucleotide phosphatases (ATP and GTP)
Protein kinase
  — cyclic AMP dependent
  — independent of cyclic AMP
Adenylate kinase
Nucleotide transphosphorylase

Evidence based on cytochemical and immunolocalization, but mainly on data from microtubules isolated from cells.
For references see Voglmayr et al. (1969), Langford (1977), Gaskin et al. (1974), Taylor and Condeelss (1979), Means and Dedman (1980).

method of isolation of tubulin affects the conditions required for polymerization. (3) Specific proteins required to stimulate polymerization or depolymerization may be absent, or present in insufficient quantities, in isolated tubules to induce physiological responses. (4) Tubulin may require a 'nucleus' in order to form microtubules.

Tubulin can partially polymerize *in vitro* to form 'ring intermediates' (Fig. 5.5), which can be detected by changes in turbidity or viscosity, as well as by sedimentation analysis and electron microscopy. The rings have a sedimentation coefficient of 30 S, compared with 6 S for tubulin, and can polymerize to form full tubules. Formation of the rings occurs within 5 min with GTP present, but in its absence may take several hours, though no evidence exists that changes in GTP concentration in the cell can regulate microtubule formation. The rings can be stabilized in D_2O and dissociate within 1–2 min when colchicine is added. Ca^{2+} at mM concentrations also stimulates disassembly but more slowly than with colchicine, the half-time of dissociation with Ca^{2+} being about 15 min.

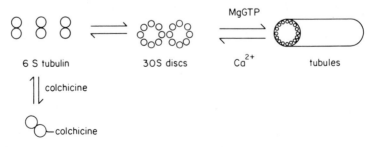

Fig. 5.5. *In vitro* action of Ca^{2+} and MgGTP on microtubules

Role of Ca^{2+} in microtubule assembly and function

Two main sites for Ca^{2+} regulation of microtubules exist, whether direct or mediated by a Ca^{2+}-binding protein such as calmodulin: (1) Tubulin–tubulin interactions, and thus assembly and disassembly of microtubules. (2) The interaction of microtubules with other intracellular structures such as microfilaments, chromosomes, secretory vesicles or the inside surface of the cell membrane. Here the dynein ATPase 'arms' attached to the tubulin dimers may act as a bridge between the tubules and these other structures (Durham, 1974).

The effect of changes in extra- and intra-cellular Ca^{2+} on microtubules observed, after fixation, in the electron microscope together with isolation of polymerized tubules from such cells provides perhaps the most compelling evidence that changes in intracellular Ca^{2+} can affect microtubules in the cell.

Six pieces of evidence support the hypothesis that changes in intracellular Ca^{2+} regulate microtubule structure *in situ*. (1) Extracellular Ca^{2+} competitively alleviates the inhibition by vinblastine and vincristine of catecholamine release from the adrenal medulla (Poisner and Cooke, 1975). (2) Extracellular Ca^{2+} affects microtubules in *Arbacia* eggs (Tilney and Gibbins, 1969), the pituitary, platelets (Steiner, 1978) and in certain invertebrates (Tilney and Gibbins, 1969). (3) Raising the extracellular Ca^{2+} concentration increases the proportion of non-tubular aggregates isolated from cells (Lagnado *et al.*, 1975). (4) The heliozoan *Actinosphaericum eichorni* contains long, thin axopodia with highly organized microtubules. Ionophore A23187, in the presence of Ca^{2+}, causes shortening of the axopodia and disruption of axopodia microtubules (Fig. 5.6) (Schliwa, 1976). Concentrations as low as 10 μM extracellular Ca^{2+} are effective, whereas Mg^{2+} + A23187 has no effect. Readdition of EGTA to chelate the Ca^{2+} produces re-extension of the axopodia. (5) Addition of rat liver mitochondria to brain tubulin reverses the inhibition of microtubule assembly caused by Ca^{2+} (Fuller *et al.*, 1975). Similarly, in brain (Nakamaru and Schwartz 1971) and sea urchin eggs (Petzelt and Ledebur-Villiger, 1973) uptake of Ca^{2+} by an endoplasmic reticulum ATPase *in vitro* can also aid microtubule assembly. (6) Microtubules in amputated nerve axons are only preserved if EGTA is present (free Ca^{2+} < 0.3 μM). These microtubules are disrupted by 25–50 μM Ca^{2+}, a condition which also causes degeneration of the axon (Schlaepfer and Bunge, 1973).

Unfortunately, *in vivo* observations are both small in number, are restricted to only a few cell types, and have rarely been made in the presence of a physiological primary stimulus. Furthermore, in at least one system, glucose, which stimulates a Ca^{2+}-dependent secretion of insulin from pancreatic β-cells, causes a decrease in polymerized tubulin isolated from these cells, an effect that is independent of extracellular Ca^{2+} (Pipeleers *et al.*, 1976).

What is the molecular basis of effects of Ca^{2+} on microtubule structure?

Soon after polymerization of tubulin had been demonstrated *in vitro*, concentrations of Ca^{2+} in the range 0.1–10 mM were shown to inhibit this polymerization and to stimulate disassembly of tubules that had already formed.

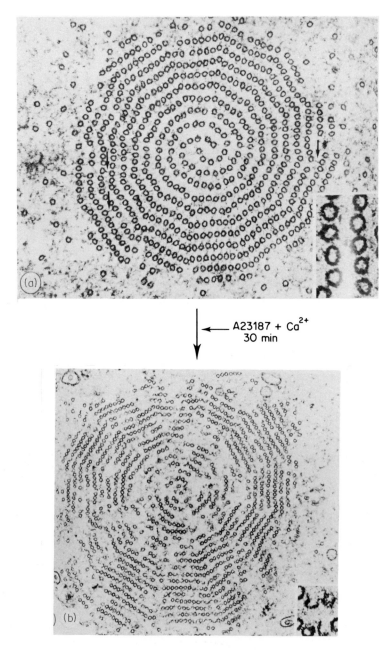

Axoneme of *Actinosphaericum*

Fig. 5.6. Disruption of microtubules *in vivo* by Ca^{2+}. Effect of A23187 + Ca^{2+} on the microtubules of the axopodium of *Actinosphaericum*. × 59,500. From Schliwa (1976), reproduced from the *Journal of Cell Biology* by copyright permission of the Rockefeller University Press

That this was not likely to be of any physiological significance was confirmed by several experiments.

(1) Ca^{2+} concentrations some 10–100 times those in the cell cytoplasm are required to stimulate disassembly (Langford, 1977). Half-maximum inhibition of polymerization can be observed in a medium containing 4 mM Mg^{2+}, 40 mM Na^+ or K^+, and 0.2 mM Ca^{2+}. Lower concentrations of Ca^{2+} (approx. 30 μM) cause significant microtubule disassembly only when the Mg^{2+} concentration is increased to about 20 mM (Olmsted and Borisy, 1975), some 4–20 times that in the cell (see Chapter 2, Table 2.16).

(2) Disruption of the mitotic apparatus isolated from marine eggs requires 1 mM Ca^{2+} (Haga et al., 1974).

(3) Several workers have been unable to observe any effect of EGTA (0.1 μM–1 mM) on the rate or extent of microtubule polymerization in vitro (Olmsted and Borisy, 1975). Addition of EDTA to tubulin can inhibit polymerization, though this seems to be caused by chelation of Mg^{2+}.

(4) The affinity of the Ca^{2+}-binding sites on tubulin itself is too low to be physiologically significant (see above).

The first report that concentrations of Ca^{2+} within the expected intracellular range could affect microtubules was made by Weisenberg (1972), who showed that, at about 10 μM Ca^{2+}, microtubule reassembly from tubulin was inhibited. Although similar results have been obtained by Haga et al. (1974) and Rosenfeld (1976), others have failed to reproduce effects at low Ca^{2+} concentrations under physiological ionic conditions (Olmsted and Borisy, 1975).

It has been proposed that several proteins, so-called microtubule-associated proteins (Langford, 1977), and in particular the 'tau' factor, are required for initiation and growth of microtubules. However, the protein most likely to be present in variable amounts in different preparations of tubulin is calmodulin. Not only has immunofluorescence using antibodies to calmodulin shown calmodulin to be associated with microtubules in the cell, but also calmodulin can inhibit assembly and promote disassembly of microtubules in vitro (Marcum et al., 1978; Welsh et al., 1979; Means and Dedman, 1980). This effect requires about 10 μM Ca^{2+}, does not occur when $Ca^{2+} < 1$ μM, and requires more than catalytic amounts of calmodulin. Troponin C was also effective, but was less potent than calmodulin. These results explain the observations of Nishida and Sakai (1977) who found that with crude brain extracts, which are known to contain reasonably large quantities of calmodulin, inhibition of microtubule assembly could be produced by adding 10 μM Ca^{2+} whereas purified tubulin required 1 mM Ca^{2+} to prevent assembly.

Microtubules therefore play an important role in many physiological processes (Table 5.4). Dissociation induced by Ca^{2+} + calmodulin is the best-established effect of Ca^{2+} on microtubules and is likely to be important in the function of the mitotic spindle and the formation of cilia and flagella. This latter role may explain its presence in the flagellum of sperm (Jones et al., 1980). But what controls the movement of flagellate or ciliate cells? We have already seen in Chapter 4 that movement of the ciliated protozoan *Paramecium* can be controlled through increases in intracellular free Ca^{2+} brought about by action potentials in the cell membrane. The ends of the microtubules of cilia and flagella interact with structures inside the cell which are responsible for controlling their movement. Since microtubules can exist in more than one state (Gillespie, 1975), the interaction of these different states with intracellular structures and its regulation by Ca^{2+} and cyclic nucleotides are likely to hold the key to understanding fully the mechanism of ciliate and flagellate cell movement in eukaryotes.

$$\text{Tubulin subunits} + \text{calmodulin} \underset{+Ca^{2+}}{\overset{-Ca^{2+}}{\rightleftharpoons}} \text{microtubule A} \underset{\text{cyclic AMP} + \text{protein kinase}}{\overset{+Ca^{2+}}{\rightleftharpoons}} \text{microtubule B} \quad (5.1)$$

Ca^{2+} and the beating of flagella and cilia

Flagella and cilia provide the apparatus for propulsion of some cells and the movement of fluid or mucus along the surface of others. At one time it was thought that flagella and cilia were passive appendages being waved to and fro from a base within the cell. This idea was discarded some time ago when it was found that isolated flagella and cilia can still beat. Addition of Triton to sea urchin sperm tails dissolves the membrane leaving the $9+2$ microtubule axoneme exposed. Addition of ATP initiates bending of the flagellum. On the other hand if the axoneme is treated briefly with a protease, like trypsin, and ATP then added it disintegrates into 'microtubule doublets'. Such observations have lead to the 'sliding filament' theory of flagellate and ciliate movement (Brokaw, 1972), for which there is also electron-microscopic evidence. The flagellum of a sperm, for example, is thus able to propogate an undulating, non-sinusoidal wave, which projects the sperm approximately along the line of the flagellum (Fig. 5.7). The sliding filament model is based on an analogy with actomyosin in muscle (see below), where flagellum tubulin plays the part of actin and dynein plays the part of

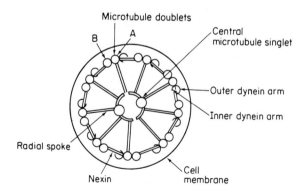

Fig. 5.7. Cross-section of a eukaryotic flagellum, showing the $9+2$ arrangement of microtubules

myosin. Sliding between doublets is generated by the MgATPase in the dynein molecules which form the cross-bridges between the microtubule doublets (Gibbons and Gibbons, 1974) (Fig. 5.7). In the absence of ATP the flagellum or cilium is in rigor. But how is the motility of these moving appendages controlled?

The spontaneous beating of flagella and cilia requires Mg^{2+} (Nichols and Rikmenspoel, 1978), but does not seem to need Ca^{2+} (Walter and Satir, 1979). However, changes in intracellular Ca^{2+} do seem to play an important role in causing alterations in rhythm or direction of movement (Holwill and McGregor, 1975; Brehm and Eckert, 1978; Walter and Satir, 1979), presumably through calmodulins known to be found in flagella (Jamieson *et al.*, 1979), or through connections to contractile proteins at the base of the organelles (Sleigh, 1979). For example, ciliary arrest can be produced by electrical or chemical means in many metazoan epithelia. In the fresh water mussel *Elliptio* and the marine mussel *Mytilus* this arrest requires Ca^{2+} and can be mimicked by ionophore A23187 and Ca^{2+} (Satir, 1975). We have already seen in Chapter 4 that the reversal of the protozoan *Paramecium* when this organism hits an obstruction is caused by an influx of Ca^{2+} which reverses the direction of the beating cilia (Brehm and Eckert, 1978). A similar phenomenon occurs in the flagellate trypanosome *Crithidia oncopelti* (Holwill and McGregor, 1975), which normally swims by a tip-base propagated wave in the flagellum. On hitting an object, or in the presence of A23187 + Ca^{2+}, the wave is reversed and the organism is able to retreat from the obstruction.

It can be concluded that changes in intracellular Ca^{2+} regulate two examples of threshold phenomena (see Chapter 2) associated with microtubules. The first is the breakdown by Ca^{2+} of a recognisable intracellular structure composed of microtubules and the second is a change in the direction of spontaneous movement. Both of these seem to be mediated by calmodulin.

5.2.2. Microfilaments and actin

Actin-myosin microfilaments observed under the electron microscope (Fig. 5.3) are branched, filamentous structures which form the basis of the cytoskeleton of eukaryotic cells. Contraction of this cytoskeleton is the mechanism by which amoeboid, and other types of movement not caused by microtubules, occurs (Table 5.6). Assembly and disassembly of microfilaments can in some circumstances occur within a few seconds, whilst under other conditions the microfilaments may remain stable for many hours or even days. The major protein in these filaments is actin, which has a molecular weight of about 45 000 and is quite distinct from tubulin. The amount of actin in cells can be measured relatively easily by quantitative densitometry in association with fluorescence-labelled heavy meromyosin which binds to actin, or by using a labelled antibody to actin. These 'actin indicators' are added to glycerinated cells, allowing the relatively large and normally impermeant molecules to enter the cell. Actin can also be measured in cell extracts. It contains the relatively unusual amino acid, N^{τ}-methylhistidine (Fig. 5.8), also found in myosin, which enables the actin

Table 5.6. Some possible functions of microfilaments

Amoeboid cell movement, including chemotaxis and morphogenetic movement*
Cytoskeleton and cell shape
Acrosome in sperm*
Movement of intracellular granules, e.g. nerve axons or exocytosis*
Phagocytosis and pinocytosis*
Cytoplasmic streaming*
Cleavage furrow in dividing eukaryotic cells*
Platelets and clot retraction*
Cell fusion*

* Where evidence that intracellular Ca^{2+} may be a regulator.
For references see Hatano, (1970), Durham (1974), Pollard and Weihing (1974), and Clarke and Spudich (1977).

1-Methylhistidine   3-Methylhistidine

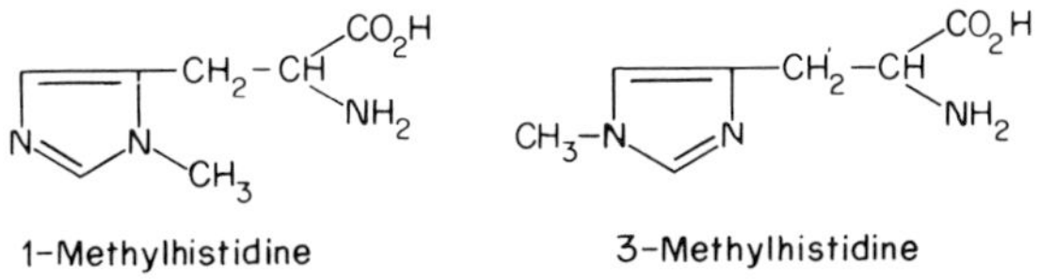

Fig. 5.8 N^{τ}-Methylhistidine in actin = 3-Methylhistidine, N-Methylhistidine

to be quantified. Significant quantities of actin have been found in many eukaryotic cells including mammalian blood cells, vertebrate tissues, cultured cells, vertebrate and invertebrate eggs, sperm, algae and protozoa and slime moulds (Pollard and Weihing, 1974; Clarke and Spudich, 1977). The quantities of actin measured in cell extracts vary from 0.2 to 2% of the total cell protein, whereas *in vivo* estimates have produced the surprisingly high value of around 10%. Action is now thought to be a constituent of all eukaryotes, and an actin-like protein has even been found in some bacteria.

There appear to be at least three distinct genes for actin. Muscle produces α-actin for actomyosin (see below), whereas in non-muscle cells this protein is designated as either β- or γ-actin. All actins bind 1 mole of ATP per monomer, and also bind Ca^{2+} and Mg^{2+}. However, the K_a^{Ca} is approximately four times that for Mg^{2+} and hence Ca^{2+} binding at free Ca^{2+} in the range 0.1–1 μM is insignificant when free Mg^{2+} is in the mM range. The most important properties of actin are (1) its ability to polymerize to form long filaments, (2) its ability to bind myosin and activate myosin ATPase, and (3) its ability to bind tropomyosin and Ca^{2+}-regulatory proteins like troponin.

In the cell, myosin, a protein of molecular weight between 140 000 and 240 000, is usually bound with actin, though its physicochemical properties are more variable than those of actin from different sources. At least five classes of myosin have been characterized from different cells (Pollard and Weihing, 1974; Clarke and Spudich, 1977). In non-muscle cells myosin constitutes about 0.3–1.5% of the total cell protein. Myosin has a subunit structure that can be broken up by proteolytic enzymes, and consists of two identical heavy chains and four light

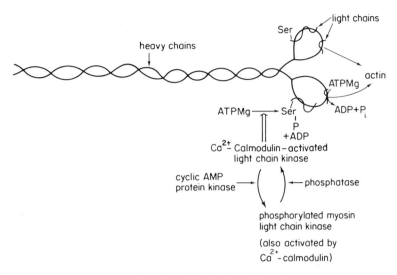

Fig. 5.9. Phosphorylation of myosin

chains of different size. The most important properties of myosin are (1) its ability to bind to actin (2) its MgATPase which is activated by actin and (3) its ability to be phosphorylated by a calmodulin-activated protein kinase. The actin-binding site and MgATPase are found on the heavy chains, whereas phosphorylations occur on the light chain (Fig. 5.9).

Other proteins

Several other proteins associated with actin and myosin have been discovered either by using labelled antibody techniques or by studies on these proteins after isolation from muscle or non-muscle cells (Tables 5.7 and 5.9). Of particular importance has been the discovery of tropomyosin which binds to actin and troponin in muscle, as well as troponin C like proteins, the so-called 'calmodulins' (Chapter 3). The presence of this latter group of proteins provides the molecular basis for possible effects of intracellular Ca^{2+} on the formation and contraction of non-muscle microfilaments. As we shall see there are two distinct

Table 5.7. Some proteins associated with microfilaments

Actin, the main protein
Myosin, contains MgATPase
Tropomyosin
Calmodulin
Myosin light chain kinase
Ca^{2+} sensitive actin-binding proteins (α-actinin, gelsolin, actinogelin, vinculin, villin, fragmin)
β-Actinin (Ca^{2+} insensitive)

See Pollard and Weihing (1974), Mannherz and Goody (1976), Means and Dedman (1980), and Weeds (1982) for references.

mechanisms by which Ca^{2+} can regulate the contraction of actin-myosin. The kinase responsible for phosphorylation of myosin has been isolated from several muscle and non-muscle cells and is activated by calmodulin (Hathaway and Adelstein, 1979; Yerna *et al.*, 1979).

Regulation of microfilaments by Ca^{2+}

Much evidence exists that changes in intracellular free Ca^{2+} can regulate cytoplasmic streaming and amoeboid movement through effects on microfilaments (Table 5.8). Many of the experiments have been carried out using cells large enough to be microinjected, and where movement of organelles in the cytoplasm is easy to observe in the light microscope. The organisms include algae such as *Nitella*, protozoa such as *Amoeba* and *Chaos*, and slime moulds such as *Physarum* and *Dictyostelium*. These latter two myxomycetes contain a vegetative structure known as a plasmodium which is enveloped by an outer membrane and contains many hundreds of nuclei. The fan-shaped plasmodium flows out in an amoeboid fashion engulfing microorganisms and bits of decaying plant material. As with many examples of amoeboid movement it occurs spontaneously but may be increased or inhibited by chemotactic stimuli (Durham and Ridgway, 1976). The first hint that contractile mechanisms in non-muscle cells might be regulated in a way similar to that in muscle came from one of these slime moulds, *Physarum polycephalum*, when Loewy (1952) extracted an actomyosin-like complex from it, which contained ATPase activity and whose viscosity was markedly affected by addition of ATP.

Amoeboid movement occurs through the development of pseudopods from the cytoplasm which allow the cell to creep along the surface to which it is loosely

Table 5.8. Intracellular Ca^{2+} and contraction of microfilaments

Cell	Effect caused by Ca^{2+}	Reference
The amoeba *Amoeba proteus*	Increase in cytoplasmic streaming	Heilbrunn (1923, 1943, 1958); Allen (1974)
The slime mould *Physarum polycephalum*	Increase in cytoplasmic streaming and chemotaxis	Durham and Ridgway (1976)
Macrophages	Movement and endocytosis	Allison *et al.* (1971)
Polymorphs	Chemotaxis	Naccache *et al.* (1979a, b)
Protozoa	Contractility	Taylor and Condeelis (1979); Ettienne (1970)
Mammalian nerve	Axoplasmic transport	Ochs *et al.* (1977)
The foraminiferan *Allogromia laticellaris*	Protoplasmic movement	Jahn and Rinaldi (1959)
The amoeba *Chaos carolinensis*	Cytoplasmic contractions and amoeboid movement	Taylor and Condeelis (1979)
Fertilized eggs	Contraction of cleavage furrow	(see Chapter 8)
The giant plant cell *Cara*	Inhibition of cytoplasmic streaming	Ashley *et al.* (1981); Williamson and Ashley (1982)

attached. The idea that changes in intracellular Ca^{2+} might regulate protoplasmic flow and amoeboid movement is more than 50 years' old (Heilbrunn, 1923, 1958; Pantin, 1926; Chambers and Reznikoff, 1926; von der Wense, 1933). Pollack's observation in 1928 of a 'shower of red crystals' in an amoeba injected with alizarin sulphonate was interpreted as an increase in cytoplasmic Ca^{2+} which was a necessary preliminary for pseudopod formation.

Injection of Ca^{2+} into the plasmodium of *Physarum* stimulates 'fountain' cytoplasmic streaming, whereas EGTA inhibits this phenomenon. Similarly, injection of Ca/EGTA buffers, or ionophore A23187 added externally, induces contractions in many cells, including amoebae (Hitchcock, 1977). The active range for stimulation of contractions by intracellular free Ca^{2+} is 0.1–10 μM. In *Chaos*, for example, the threshold for cytoplasmic streaming is 0.7 μM. Aequorin has been injected into several of these cells (Reynolds, 1979; Cobbold, 1980), and increases in intracellular free Ca^{2+} have been demonstrated under some, but not all, experimental conditions. In *Physarum* cyclical changes in free Ca^{2+} have been detected which correlate with the polarity and extent of cytoplasmic streaming (Durham and Ridgway, 1976). However, the evidence in amoebae is less convincing (Reynolds, 1979; Cobbold, 1980).

Manipulation of intracellular Ca^{2+} does affect cytoplasmic streaming,—sometimes increasing it, sometimes inhibiting it—but is it the controlling factor in spontaneous movements? Changes in intracellular Ca^{2+} are probably most important in mediating the effects of regulators of amoeboid movements such as chemotactic agents (see section 5.4). An increase in intracellular Ca^{2+} appears to be crucial to aggregation of platelets, which is provoked by several stimuli, including thrombin, and prevents profuse bleeding from damaged blood vessels.

There are two main mechanisms for regulating microfilament structure: (1) the extent of polymerization and cross linking (2) the sliding of myosin over actin.

In non-muscle cells a large proportion of the actin is in monomeric or G (globular) form. G actin polymerizes to form helical, filamentous, F actin (5.2). Further interaction between F actin filaments produces a multifilamentous complex which forms a gel. This provides the molecular basis for Heilbrunn's ideas concerning the control of gelation in the cytoplasm by Ca^{2+}. Addition of Ca^{2+} to axoplasm of squid nerve solubilizes the 'gel' (Hodgkin and Katz, 1949a; Gilbert 1975, 1977). Since the original experiments with axoplasm required mM concentrations of Ca^{2+} they were thought not to be physiologically relevant. However, more recently, it has been shown that optimal conditions for gel formation in extracts from *Dictyostelium* (Condeelis and Taylor, 1977) or from Ehrlich ascites tumour cells (Mimura and Asano, 1978) are mM Mg^{2+}, EGTA and physiological pH, and that gel formation can be inhibited by more than 50 % by about 1 μM Ca^{2+}. As with studies on isolated microtubules, the presence of calmodulin, polymerization-stimulating proteins such as α-actinin, and gelsolins may be necessary to show effects of Ca^{2+} on gel–sol conversion in broken cells. This could explain why Ca^{2+} does not apparently inhibit gel formation of actin and myosin purified from macrophages (Stossel and Hartwig, 1976).

This physical phenomenon of gel–sol conversion, and its regulation by Ca^{2+}, may be important not only in amoeboid movement but in allowing membrane–membrane fusion (Mimura and Asano, 1978). However, on its own it cannot explain amoeboid movement. It is interesting that the force generated by the Ca^{2+}-activated actin-myosin ATPase is greatly reduced, and can be abolished, in the gel state (Taylor et al., 1973; Condeelis and Taylor, 1977).

Myosin and tropomyosin bind to the F actin filaments, but not significantly to the monomeric G actin. The myosin MgATPase is activated when it binds to actin and it is the hydrolysis of the ATP which provides the energy for cross-bridge formation, which is thought to be the molecular basis of the sliding of actin against myosin to produce movement. In non-muscle cells a kinase catalyses the phosphorylation of a serine residue on the light chain of the myosin. This kinase requires calmodulin + Ca^{2+} for activity (Hathaway and Adelstein, 1979; Yerna et al., 1979). It is this phosphorylation which may provide the stimulus for activating the contractile apparatus, though direct effects of Ca^{2+}-binding proteins have yet to be ruled out. Contraction then causes movement of the cell membrane since actomyosin is probably attached to the inner surface. This contraction would also cause movement of any granules that may be attached to the filaments (Ostlund, 1977). Further modulation of this mechanism may occur through phosphorylation of the myosin heavy chain, as well as by cyclic nucleotide dependent kinases and phosphatases which can regulate phosphorylation and dephosphorylation of the myosin. A key question is whether this phosphorylation–dephosphorylation can be shown to be regulated in the intact cell by Ca^{2+}, rather than simply demonstrating it in cell extracts.

Conclusions

Microfilaments composed of actin, myosin and a few other proteins may play three major roles in cell activation. (1) They form the basic framework of the cytoskeleton, having the physical characteristics of a gel. An increase in free Ca^{2+} in the μM range may depolymerize the multifilamentous F actin, thereby converting the gel to a sol and allowing freer random movement of particles. A directed breakdown, followed by repolymerisation, would be necessary to explain amoeboid movement by this mechanism. (2) The actin–myosin complex is responsible for contraction in non-muscle cells. Contraction may be regulated by a Ca^{2+}-binding protein interacting with tropomyosin, but more likely by a calmodulin activated myosin light chain kinase. (3) The microfilaments may be connected through the myosin light chain to microtubules. Particles attached to such microtubules would move when the microfilaments contract.

There are five potential ways in which Ca^{2+} could regulate microfilaments and thereby activate these phenomena in the cell: (1) Actin polymerization, (2) Myosin filament formation; (3) Actin cross linking and solation; (4) Interaction of actin with other proteins, e.g. myosin and capping proteins; (5) Actin-myosin contraction.

In vitro effects of Ca^{2+} in the μM range have been clearly demonstrated on all but the first of these. The effects of Ca^{2+} on protein cross linking and gelation appear to be modulated by a series of Ca^{2+} binding proteins which do not belong to the calmodulin group, though a subunit calmodulin has yet to be ruled out. These actin binding proteins have molecular weights in the range 40 000–150 000, in contrast to the 16 700 of calmodulin. An increase in free Ca^{2+} concentration from $0.1\,\mu M$ to $1\,\mu M$ can break actin cross links either by regulating the formation of actin bundles or by altering the gel network. Actin cross links formed by α-actinins are broken when the free Ca^{2+} is greater than $1\,\mu M$ and are rapidly reformed when Ca^{2+} returns to $0.1\,\mu M$. Vinculin, originally isolated from smooth muscle (see Weeds, 1982), may act similarly. In contrast when Ca^{2+} causes solation of the gel network by binding to gelsolin the restoration of the network is relatively slow after the free Ca^{2+} concentration has returned to the original level.

Ca^{2+}, in the μM range, regulates myosin filament formation and actin-myosin (actomyosin) contraction in non-muscle cells through a different protein from those concerned with actin cross links or gelation, namely a calmodulin-activated myosin light-chain kinase.

These *in vitro* observations provide the potential molecular basis for effects of intracellular Ca^{2+} on the cytoskeleton. Experiments are now required to establish the role of these proteins in the intact cell.

5.2.3. Actomyosin and Troponin

Muscles are recognizable by their characteristic morphology, consisting of a regular array of thin and thick filaments (Fig. 5.10). The basic unit is known as

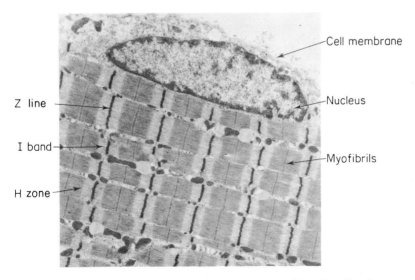

Fig. 5.10. Myofibrils in skeletal muscle; by courtesy of Dr. Caroline Sewry

the sarcomere, composed of two interdigitating filaments which, in cross-section, give the appearance of an hexagonal array and which are repeated many times in parallel and in series in each muscle fibre (Fig. 5.11). Unlike most cells, the skeletal muscle fibre is multinucleate, formed by the fusion of large numbers of 'lined-up' myoblasts. The contractile fibres, the myofibrils, can therefore be many centimetres long and can contain many thousands of sarcomeres.

All muscles contain four main protein complexes which make up the sarcomere (Table 5.9). These are actin, myosin, tropomyosin, and the Ca^{2+}-dependent regulatory system, troponin. It is the differences between these which confer the individual properties of a particular type of muscle (Lehman, 1976).

Actin

As in non-muscle cells, the thin filaments of muscle are made up of polymers of F actin. F actin itself is a polymer, made from G actin plus ATP:

$$n \text{ (G actin–ATP)} \xrightarrow{\text{high ionic strength}} \text{F actin–ADP} + n P_i \text{ polymer} \quad (5.2)$$

In skeletal muscle the single chain G actin is a protein of molecular weight approximately 42 000, consisting of 376 amino acids with an N^τ-methylhistidine

Table 5.9. Proteins important in muscle contraction

Protein	Approximate molecular weight
1. *Contractile apparatus*	
Actin	42 000 (monomer)
Myosin	470 000
Tropomyosin	70 000
Troponin T	37 000–40 000
Troponin I	22 000–24 000
Troponin C	18 500
2. *Sarcoplasmic reticulum proteins responsible for regulation of Ca^{2+}*	
Ca^{2+}-activated MgATPase	102 000
Calsequestrin	45 000
Acidic protein	55 000
3. *Other important muscle proteins*	
Parvalbumin (fish)	11 000
Phosphorylase	184 000 (dimer); approx. 400 000 (tetramer)
Phosphorylase kinase	1 300 000
Myosin light chain kinase (smooth muscle)	120 000
+	+
Calmodulin	16 700

Data are for fast white skeletal muscle except where stated, usually rabbit.
See Mannherz and Goody (1976) and Ebashi (1960) for further references.

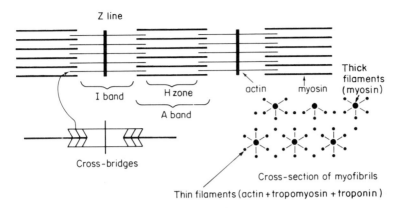

Fig. 5.11. Periodicity of actomyosin

(Fig. 5.8) at position 73. There has been some controversy concerning the high affinity Mg^{2+}/Ca^{2+} site on G actin. In the cell this site is most probably occupied by Mg^{2+} as in non-muscle cells (see above). Exchange and dissociation of nucleotide occur only when this divalent cation site is unoccupied. The N-terminus is acetylated.

Although the actins from different sources may have slightly different primary sequences, their three-dimensional structure is very similar. Circular dichroism has shown that only about 30% of the protein is α-helical. By using antibodies which bind to G actin subunits of the actin polymer it has been shown that the thin filaments of muscle actin have a periodicity of about 40 nm (Fig. 5.12). This is produced by two strands of actin polymers twisting in a helical manner, each strand containing 7 G actins per unit period.

Tropomyosin

Each of the two grooves formed by the actin thin filament contain another protein, tropomyosin. This protein was first discovered by Bailey and his coworkers (1946, 1948). However, its significance in the regulation of skeletal muscle contraction was discovered by Ebashi (1963) who called his preparation 'viable tropomyosin', and which, he showed, conferred Ca^{2+} sensitivity on the actomyosin MgATPase. Tropomyosin is a rod-shaped protein, about 41 nm in length, with a molecular weight of about 70 000. It consists of two nearly identical polypeptide chains which are almost completely α-helical, producing a coiled core. Tropomyosin is found with all muscle actins and with many non-muscle ones. In vertebrate skeletal and heart muscle it is this protein which binds the troponin complex (see below) to which the Ca^{2+} to stimulate contraction actually binds. In some muscles the chains of tropomyosin are not identical, the α- and β-chains being found in the ratio 4:1.

Myosin

The thick filaments consist entirely of a large protein, myosin, which varies considerably between different cells. It is this protein which contains the MgATPase of actomyosin, and which forms the cross-bridges which provide the molecular mechanism for the sliding-filament theory proposed originally by Huxley and Niedergerke in 1954.

In rabbit fast skeletal muscle myosin is a protein of molecular weight about 470 000, composed of two heavy (200 000 molecular weight each) and four light (molecular weight 20 000 each) chains. In other cells the light chain pattern can be quite different from this. Proteolytic digestion of myosin with papain and trypsin has shown that the protein consists of a highly α-helical rod which contains the heavy chains and two head groups in which the light chains are found. The globular head groups contain the MgATPase activity, the actin interaction sites, and, in those myosins that can be phosphorylated, the serine residue which in some non-striated muscle cells may be phosphorylated by a calmodulin-regulated myosin light chain kinase (Fig. 5.12).

X-Ray diffraction and electron microscopy have shown that the thick filaments in the muscle sarcomere are composed of myosin molecules packed such that the head groups of each molecule stick out from the filament and form a helical arrangement with an axial spacing of 14.5 nm. These form the cross-bridges with the actin thin filaments.

Myosin isolated from muscle and non-muscle cells often has other proteins associated with it besides actin (Mannherz and Goody, 1976). These include C protein, paramyosin, and myosin light chain kinase. Their function is still not completely known, nor is that of α- and β-actinin which is sometimes found with actin.

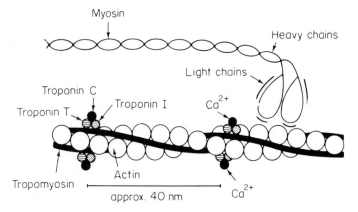

Fig. 5.12. Actomyosin. From Adelstein and Eisenberg (1980); reproduced with permission from the *Annual Review of Biochemistry*. 1980 by Annual Reviews Inc.

Regulatory Ca^{2+}-binding proteins

Ca^{2+} is the trigger for contraction in all types of muscle, though, as we shall see, the precise mechanism by which the Ca^{2+} acts can be different. In muscle the energy for contraction is derived from the hydrolysis of ATP. Ca^{2+} acts by binding to one of the components of the contractile apparatus, thereby regulating the rate of this hydrolysis and stimulating contraction. Four types of Ca^{2+} regulation have been identified in different muscles (Table 5.10) (Ebashi, 1980).

During the 1950's many workers found it difficult to accept the idea that intracellular Ca^{2+} was the universal trigger for muscle contraction. The sceptics pointed to the fact that, whilst Ca^{2+} was effective on crude preparations like glycerinated muscle fibres or natural actomyosin (also known as myosin B), Ca^{2+} did not stimulate contraction or the ATPase in actomyosin reconstituted simply from purified actin and myosin. The first clue as to how Ca^{2+} causes muscle to contract was uncovered by Ebashi in 1963. He found that an 'active tropomyosin' complex when added to a purified actomyosin preparation could restore its sensitivity to Ca^{2+}. He called the Ca^{2+}-binding component troponin (Greek $\tau \rho o \pi o \sigma$ = a turn). It was soon shown by Hartshorne and Mueller (1968) that troponin was made up of more than one protein. Three subunits were isolated by Greaser and Gergely in 1971 and called troponin T, troponin I and troponin C, though they have been known by other names (Table 5.11).

Table 5.10. Four types of Ca^{2+} regulation in muscle

Ca^{2+}-binding site	Example	Reference
Troponin C	Vertebrate skeletal and cardiac muscle	Ebashi (1980)
Leiotonin	Vertebrate smooth muscle	Mikawa et al. (1977; 1978)
Calmodulin on myosin light chain kinase	Vertebrate smooth muscle	Chacko et al. (1977); Nairn and Perry (1979)
Myosin	Molluscan muscle	Kendrick-Jones et al. (1970)
	Decapod muscle	Lehman (1977); Lehman et al. (1972)

Table 5.11. Other names for troponin subunits

Present name	Functional name	Other names
Troponin T	Tropomyosin-binding protein	Troponin I (fraction I) Troponin B (binding protein) Troponin P (promoting protein)
Troponin I	Inhibitory protein	Troponin II (fraction II) Troponin B (fraction B) Inhibitory factor
Troponin C	Ca^{2+}-binding protein	Troponin III (fraction III) Troponin A (fraction A) Ca^{2+}-sensitizing factor

For references see Ebashi (1974) and Hartshorne and Pyun (1971).

Troponin C is the Ca^{2+}-binding component. Antibodies to troponin can be produced by injecting a purified preparation into rabbits and these can be used to localise the troponin complex on the actin filament. Each sarcomere, the basic contractile unit of the muscle myofibrils, contains one troponin complex of molecular weight approx. 80 000 containing one molecule of each component (Fig. 5.12).

Many interactions between the three subunits of troponin have been demonstrated *in vitro*. However, only some of them are involved in the cell to regulate contraction. Troponin I is the subunit which interacts with troponin C, troponin T, actin and tropomyosin, all of which are essential for normal activation of contraction by Ca^{2+}. Ca^{2+} binds to troponin C which interacts with troponin I and this affects the actomyosin. An interaction between tropomyosin and troponin T is also essential.

Muscle cells are found in all animals, from coelenterates to chordates. But how many of them contain troponin? This question has been mainly investigated in Bilateria, animals with bilateral symmetry, which are divided into Deuterostomia and Protostomia (Kendrick-Jones *et al.*, 1970; Lehman *et al.*, 1972; Lehman, 1976). Deuterostomia are animals with bilateral symmetry where the embryonic blastomere becomes the anus; they include Chaetognatha, Echinodermata, Hemichordata and Chordata. These animals appear to have a troponin-mediated contractile system. However, this is not necessarily the case in the muscle of the Protostomia, which are animals with bilateral symmetry but where the embryonic blastomere becomes the mouth; for example, in Arthropoda and Mollusca. Until 1970 it was thought that troponin–tropomyosin would be found in all types of muscle. However, Kendrick-Jones and coworkers showed that scallop striated muscle contains no troponin and the Ca^{2+}-regulatory sites are on the myosin itself. This Ca^{2+}–myosin system is widely distributed in the Protostomia, and in molluscs it is the predominant mechanism by which Ca^{2+} regulates contraction. However, in insects and crustaceans (Arthropoda) both systems exist, but the troponin-linked one appears to be more important. Why the myosin-linked mechanism should have evolved in molluscs is not known. It is interesting that this, together with the special case of *Vorticella* and its relatives, are the only contractile systems known where Ca^{2+} directly binds to the contractile apparatus, apparently without the need for a specialized regulatory Ca^{2+}-binding protein.

The other type of muscle cell where the Ca^{2+}-dependent mechanism is different from the troponin–tropomyosin-linked one is in vertebrate smooth muscle. Here an actomyosin–tropomyosin complex exists but is regulated either by phosphorylation, through a calmodulin-activated myosin light chain kinase, or by leiotonin, a protein similar to, but not the same as, troponin (Tables 5.10 and 5.12).

Leiotonin consists of only two subunits: subunit A, a neutral protein of molecular weight about 80 000; and subunit C, an acidic protein of molecular weight about 20 000 which binds Ca^{2+} (Mikawa *et al.*, 1978; Hirata *et al.*, 1980). Originally it was thought that smooth muscle myosin was activated by the

Table 5.12. Muscle regulatory Ca^{2+}-binding proteins

Protein complex	Location	Approximate molecular weight	Subunits	Binds to	Ca^{2+} sites
Troponin	Skeletal (and cardiac) muscle	80 000	T, I, C	Actin and tropomyosin	4(3)
Leiotonin	Smooth muscle	100 000	A, C	Actin	2
Calmodulin	Smooth muscle	16 700	1	Myosin light chain kinase	4

Modified from Ebashi (1980).

phosphorylation mechanism and that relaxation of the muscle occurred as a result of dephosphorylation by a specific phosphatase (Chacko *et al.*, 1977; Nairn and Perry, 1979). However, the isolation of a troponin-like complex, named leiotonin by Ebashi's group, raises the question as to which mechanism is the physiological one.

Chicken gizzard and guinea pig taeniae coli have provided much of the information regarding the role of Ca^{2+} in smooth muscle contraction. Using chicken gizzard Ebashi has provided evidence for actin regulation of myosin MgATPase and reconstitution of a Ca^{2+}-activatable complex from actin, Myocin, tropomyocin and leiotonin. With guinea pig taeniae coli, as with skeletal muscle, no evidence for significant phosphorylation of myosin, observed by radioautography, could be obtained when contraction was initiated by acetylcholine or K^+. This suggests that leiotonin may be the physiologically more important mechanism for initiating contraction and perhaps the calmodulin-regulated kinase may provide a secondary regulation of contraction.

In any event, leiotonin can be clearly distinguished from both troponin C and calmodulin. Leiotonin does not activate cyclic AMP phosphodiesterase whereas calmodulin does, and troponin C does not enable the Ca^{2+}- dependent smooth muscle contractile apparatus to be reconstituted. A particularly intriguing feature of leiotonin is the fact that the molar ratio of leiotonin to tropomyosin necessary for maximal activation by Ca^{2+} is about one-tenth that of troponin to tropomyosin.

5.3. Muscle

5.3.1. The problems of muscle contraction

Let us return to the example described at the start of Chapter 2, that of the little boy running down the road. Before he decides to run many of his muscles will be relaxed, although his heart is continuously undergoing a contraction–relaxation cycle as it beats. As a result of the activation of a large number of nerves many of his muscles contract and his heart beat and blood flow increase, perhaps some ten fold.

Muscle is the tissue responsible for the movement of all metazoan animals. It provides one of the clearest examples of a change in intracellular Ca^{2+} mediating a threshold response in a cell (see Chapter 2). Having crossed the threshold the strength and frequency of contraction may be regulated. In some muscles this can only be done by controlling the number of individual fibres which contract, whilst in others the extent of contraction can be varied over a wide range in each cell under the influence of external factors.

Muscle fibres may consist of a large number of small cells each only a few micrometres across, or they may be formed by fusion of many cells to produce a single, multinuclear cell which can be anything from several hundred micrometres to many centimetres long (Table 5.13). The contraction of muscle is the result of the sliding of actin and myosin filaments stimulated by an increase in the concentration of Ca^{2+} in the cytoplasm of the cell (Huxley and Niedergerke, 1954). This is true of all muscles. However, the precise nature of the contractile mechanism and, more particularly, how Ca^{2+} causes contraction vary in different muscles.

Muscle can be classified in five different ways on the basis of its: (1) phylogeny; (2) morphology, particularly whether the organization of the contractile apparatus results in a 'striated' or 'smooth' appearance (Hanson and Lowy, 1960); (3) electrical properties, particularly with respect to the role of Na^+ and Ca^{2+} in action potential generation; (4) function and physiology; and (5) biochemistry, particularly with regard to the properties of the actin and myosin, and the mechanism by which Ca^{2+} causes contraction.

Experimentally it is possible to induce in many muscles either a 'twitch' or a prolonged contraction, the latter usually being called a tetanus. The experimenter often finds it easier to study the excitation of muscle under isometric conditions, i.e. where no change in length occurs but where the degree of activation is measured by the force generated. Normally, however, in the animal the most important characteristics of a particular muscle are the speed with which it contracts and relaxes, together with the strength and change in length brought about by the contraction. These parameters can vary considerably between different muscles (Close, 1972). For example, a single twitch of skeletal muscle may last only 0.1 sec, an animal's heart beat may be anything from 0.2–2

Table 5.13. Size of some muscle 'cells' in which free Ca^{2+} has been measured

Muscle cell	Dimension (length × diameter)	Approx. volume
Giant muscle in barnacle	20 mm × 1 mm	> 12 μl
Giant muscle in crab	1.2 mm × 0.5–1 mm	0.5 μl
Amphibian skeletal	7 mm × 100 μm	70 nl
Rat and human skeletal	150 μm × 10–30 μm	45 pl
Frog cardiac	250 μm × 10 μm	10 pl
Frog smooth	250 μm × 4.5 μm	4 pl

Data from Hallett and Campbell (1982b).

Table 5.14. Primary stimuli and secondary regulators of some muscles

Organisms	Primary stimulus	Secondary regulator*
Coelenterates	Electrical via conducting pathway	?
Nematodes—(e.g. ascaris)	Acetylcholine	GABA (i)
Annelids (e.g. leech or earthworm)	Acetylcholine	GABA (i)
Arthropods (e.g. barnacle)	L-Glutamate	GABA (i)
Insects (e.g. locust)	L-Glutamate	GABA (i)
Molluscs (e.g. catch muscle in *Mytilus*)	Acetylcholine	5-hydroxytryptamine, dopamine (relaxes)
Vertebrates—skeletal	Acetylcholine	?
smooth (several)	Prostaglandins	Adenosine
vas deferens	Catecholamine (α)	Catecholamine (β) (relaxes)
uterine	Oxytocin, prostaglandins	Prostaglandins
ileum	Histamine	Prostaglandins
cardiac	Electrical	Acetylcholine (i) β-Adrenaline (a)

*i = inhibitory; a = activates; GABA = γ-aminobutyric acid.

sec, whilst insect flight muscle contracts many times a second. In contrast, several muscles, including those responsible for the posture of an animal, can undergo a tonic or prolonged contraction. Tonic contraction of smooth muscle may last many seconds or even minutes, whilst the muscle responsible for keeping the shells of molluscs such as mussels, oysters and scallops, closed can remain in tonic contracture for several hours.

The duration and strength of a contraction are determined both by the nature and duration of the primary stimulus, as well as by the properties of muscle itself. The primary stimulus may be chemical, such as a neurotransmitter at a neuromuscular junction, or it may be electrical conducted via action potentials generated in other cells, for example in coelenterates (Table 5.14).

Myogenic hearts, found in vertebrates, have the unique distinction that excitation occurs spontaneously in cells within the heart itself. In mammalian heart excitation is conducted through the ventricle by a system known as the Purkinje fibres.

Innervation of muscles may be uni- or multiterminal, or it may be polyneuronal. For example, each mammalian muscle cell has only one neuromuscular junction, whilst arthropod muscle cells have many junctions as a result of multiterminal or polyneuronal innervation. In the latter case some of the nerves may be inhibitory. For example, in some arthropods γ-aminobutyric acid (GABA) antagonizes the natural agonist, L-glutamate. Other muscles also have secondary regulators. For example, in mammalian heart the strength and rate of beating can be increased by adrenaline. How does Ca^{2+} fit into this apparently complex array of different muscles? Several questions arise: (1) Does the energy for muscular contraction come from Ca^{2+} binding? (2) Does Ca^{2+} play a role in the generation of action potentials? (3) Can a change in cytoplasmic Ca^{2+} account for the duration and strength of contraction in all muscles? (4) What is the source of the Ca^{2+} for contraction and how is it regulated? (5) What is the molecular mechanism by which Ca^{2+} triggers contraction? (6) Does Ca^{2+} play a role in secondary regulation of muscle contraction by natural or pharmacological substances?

The first of these questions is easily answered. The hydrolysis of ATP, catalysed by the myosin ATPase and not by Ca^{2+} binding itself, provides the energy for muscle contraction. Although there is a buffer system for maintaining ATP concentrations in muscle—namely creatine or arginine phosphate—metabolism of glucose or fatty acid is the ultimate source of energy (Campbell and Hales, 1976). Heilbrunn's failure to recognize the thermodynamic significance of ATP may well have accounted for his difficulty in getting his Ca^{2+} hypothesis accepted (Heilbrunn, 1943, 1956). Before considering the other questions let us first examine the evidence that an increase in cytoplasmic Ca^{2+} is the event which triggers the contraction of all muscles.

5.3.2. Evidence for the role of intracellular Ca^{2+}

Although a role for Ca^{2+} in muscle contraction was assumed from the results of experiments by Ringer at the end of the last century, and by others at the beginning of this century (Mines, 1910; Davis, 1931; Gillespie and Thornton, 1932; Weise, 1934), it was not until the 1940's that direct evidence was obtained for Ca^{2+} acting within the muscle cell itself (Table 5.15). Following the observations by Kamata and Kinoshita (1943), Heilbrunn and Wiercinski (1947) were able to show that, of the major physiologically occurring cations, only the injection of Ca^{2+} into muscle caused contraction (Fig. 5.13). This experimental approach has since been adopted by many others who have shown that microinjection of Ca^{2+} into vertebrate and invertebrate muscles stimulates contraction. For example, arthropod muscles, such as those from crab or barnacle, have provided the physiologist with large single muscle fibres easily microinjected without damage to the cell. These fibres can be several centimetres long and 1–2 mm wide and can be cannulated (Ashley, 1978). Both Sr^{2+} and Ba^{2+} can mimic the action of Ca^{2+}, albeit at higher concentrations. However, these cations do not occur naturally in sufficient concentrations for them to be physiologically significant.

An important step forward was made in 1964 by Portzehl, Caldwell and Ruegg, who used the newly available Ca^{2+} chelator EGTA to fix the Ca^{2+} concentration inside muscle fibres of the crab, *Maia squinado*. Since contractions occurred whenever the buffered free Ca^{2+} concentration was greater than 0.3 µM, they concluded that the concentration of free Ca^{2+} in resting muscle must be less than this.

Although Jöbsis and O'Connor (1966) reported a direct demonstration of a change in intracellular free Ca^{2+} in toad muscle using murexide, the first comprehensive quantitative studies on intracellular free Ca^{2+} were carried out

Table 5.15. Some landmarks in defining the role of intracellular Ca^{2+} in muscle contraction

Year	Observation	Reported by
1940	Ca^{2+} can act on isolated muscle protoplasm	Heilbrunn
1942	Ca^{2+} is a potent activator of isolated myosin ATPase	Bailey; Needham
1943	Injection of Ca^{2+} into muscle causes local contraction	Kamata and Kinoshita
1949	Increased efflux of radioactive Ca^{2+} caused by electrical stimulation	Woodward
1961	Release of internal Ca^{2+} causes contraction	Bianchi
1962	Muscle homogenate contains an ATP-dependent Ca^{2+}-uptake system	Ebashi and Lipmann
1963	'Active tropomyosin' leads to discovery of troponin C	Ebashi
1964	Injection of Ca^{2+} buffers into muscle defines range of cytoplasmic free Ca^{2+}	Portzell, Caldwell and Ruegg
1967	Direct measurement of intracellular free Ca^{2+} using aequorin	Ridgway and Ashley
1970	Discovery of direct action of Ca^{2+} on some invertebrate myosins	Kendrick-Jones, Lehman and Szent-Gyorgyi
1971	Discovery of calsequestrin as Ca^{2+} binder in sarcoplasmic reticulum	MacLennan and Wong
1977	Discovery of leiotonin, a Ca^{2+} regulatory protein in smooth muscle	Mikawa et al.
1977–1978	Discovery of calmodulin regulation of myosin phosphorylation in smooth muscle	Dabrowska and Hartshorne

by Ashley and Ridgway (Ridgway and Ashley 1967; Ashley and Ridgway 1968, 1970) using aequorin-injected barnacle muscle fibres. Since these pioneering experiments, by using photoproteins, metallochromic indicators and microelectrodes it has been shown that the free Ca^{2+} concentration in resting muscle is in the range 0.1–0.3 μM (Table 5.16). Stimulation electrically (Fig. 5.14) or by neurotransmitter (Fig. 5.15) causes an increase in cytoplasmic free Ca^{2+} up to about 1–5 μM. An increase can be detected before the onset of contraction. The duration of the Ca^{2+} transient correlates with the duration of both the impulse and the contraction, though, because Ca^{2+} dissociation from the myofibrils may be rate-limiting, the muscle often remains contracted when the free Ca^{2+} concentration has returned almost to its baseline (Fig. 5.15). Prevention of the Ca^{2+} transient with EGTA also prevents contraction.

The final pieces of evidence for Ca^{2+} being the trigger for muscle contraction came when the sites of action and storage of Ca^{2+} within the cell were identified. Ca^{2+} activation of a myosin ATPase was first demonstrated by Bailey and Needham in the 1940's. More convincing evidence for the physiological significance of this observation was obtained in 1966 by Hasselbach, who showed that the MgATPase in barnacle muscle was maximally activated by 2–3 μM Ca^{2+}, whereas at concentrations of Ca^{2+} in the range for resting muscle, i.e.

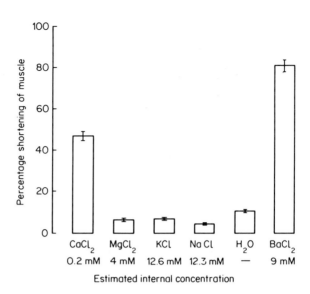

Fig. 5.13. Injection of Ca^{2+} into muscle triggers contraction. Muscle fibres from the frog, *Rana pipens*, were immersed in isotonic Ringer solution, pH 6, with no Ca^{2+} (to prevent contraction when the muscle was impaled with the micropipette). The fibre was injected with approx. 10% of its volume of various salt solutions ($CaCl_2$, $MgCl_2$, KCl, NaCl, $BaCl_2$ and H_2O). $MgCl_2$, KCl, NaCl and H_2O all caused an initial distension of about 10–15%. Measurement of muscle length was made a few minutes after this. Shortening of the muscle induced by $CaCl_2$ occurred within 5–15 sec of injection. Data calculated from Heilbrunn and Wiercinski (1947)

Table 5.16. Intracellular free Ca^{2+} in muscle

Tissue	Method	Resting (μM)	Contracting (μM)	Reference
Crab muscle	EGTA	0.1	1–5	Portzehl et al. (1964)
Barnacle muscle	Aequorin	0.1–0.3	1–5	Ashley and Ridgway (1970)
Frog skeletal	Aequorin	< 0.3	5–10	Rudel (1979)
Frog skeletal	Arsenazo III	?	5	Parker (1979)
Amphibian skeletal	Aequorin	< 0.2	5	Blinks et al. (1978)
Toad smooth	Aequorin	?	approx. 5	Fay et al. (1979)
Mammalian smooth	Aequorin	—	—	Nering and Morgan (1981)
Frog heart (atrial trabecular)	Aequorin	< 0.3	1–3	Allen and Blinks (1978)
Rabbit heart (ventricular)	Electrode	0.1–0.3	1–5	Lee et al. (1980)

Data are expressed as molarities; see Chapter 3 for conversion to activities.

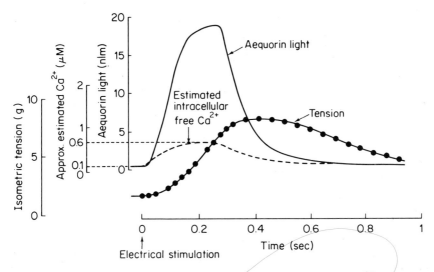

Fig. 5.14. Correlation of muscle contraction with intracellular free Ca^{2+}. Isolated barnacle (*Balanus nubilus*) muscle fibre injected with aequorin. Fibre stimulated electrically with an intracellular electrode. Note tension rises after initiation of Ca^{2+} rise. Temperature approx. 10°C. From Ashley and Moisescu (1972); reprinted by permission from *Nature*. Copyright © 1972 Macmillan Journals Limited. See also Ashley *et al.* (1976) and Ashley (1978)

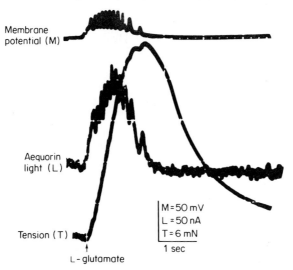

Fig. 5.15. Contraction, membrane response, and free Ca^{2+} change induced by a neurotransmitter. Isolated barnacle muscle fibre injected with aequorin. Contraction stimulated by 1 mM L-glutamate applied externally. Note tension rises after initiation of Ca^{2+} rise. From Ashley and Campbell (1978); reproduced by permission of Elsevier Biomedical Press B. V.

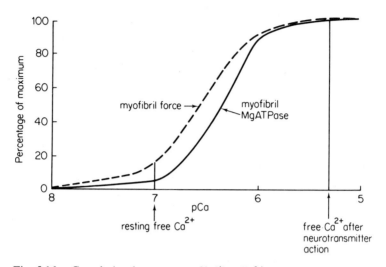

Fig. 5.16. Correlation between myofibrillar Ca^{2+}-activated MgATPase and force. Data are extrapolated from Endo (1972) and Hellam and Podolsky (1969) for frog muscle and are only interacted to illustrate the concomitant rise in force and MgATPase in isolated myofibrils as the free Ca^{2+} (Ca/EGTA) is increased

< 0.1 μM, the MgATPase had < 5% of its maximal activity (Fig. 5.16). Three years later Hellam and Podolsky (1969), using frog muscle, demonstrated a similar quantitative relationship between free Ca^{2+} concentration and the force developed by isolated myofibrils (Fig. 5.16). Similar relationships between Ca^{2+}-activated MgATPase, force, and free Ca^{2+} concentrations within the range 0.1–10 μM have since been demonstrated in many types of muscle from organisms belonging to a wide range of phyla.

A change in cytoplasmic Ca^{2+} concentration is therefore the universal trigger for contraction in all types of muscle (Table 5.14). Furthermore, under physiological conditions, the duration of the contraction, be it 'twitch' as in skeletal muscle or 'tonic' as in the anterior byssus rectractor muscle of molluscs, is directly related to the time which the cytoplasmic Ca^{2+} remains above the threshold level.

5.3.3. Regulation of intracellular free Ca^{2+} in muscle

How much Ca^{2+} is required to stimulate contraction? Where does it come from and how is it regulated?

The concentration in the cell of the Ca^{2+}-binding proteins responsible for muscle concentration (Table 5.12) is about 50–100 μM. Assuming 2–4 Ca^{2+}-binding sites per mole each with a K_d of 1 μM, then if the free Ca^{2+} is 1 μM and in equilibrium with the protein, then approx. 200 μM Ca^{2+} will be bound (see equation 3.21 and Fig. 3.8). Could this Ca^{2+} come from outside the cell? If so, then one would expect removal of extracellular Ca^{2+} rapidly to prevent concentration. This is the case with heart and smooth muscle in vertebrates, and in many invertebrate muscles. However, mammalian skeletal muscle can contract for some time in the absence of extracellular Ca^{2+}, whereas amphibian muscle requires a few minutes before contraction stops (Lüttgau, 1963; Brecht and Gebert, 1966; Constantin, 1968; Shanes, 1958; Frank, 1960). $^{45}Ca^{2+}$ uptake can in fact be demonstrated in many types of muscle including mammalian skeletal muscle (Bianchi and Shanes, 1959). In frog approximately 0.2 pmoles Ca^{2+} enter per cm^2 per sec at each impulse. Assuming a diameter of 100 μm (Table 5.13), this is equivalent to approx. 400 nM per twitch, i.e. 1/500 of that required. Similarly, in heart, Ca^{2+} entry during each action potential is only about 1 μM (see Chapter 4). It is concluded that external Ca^{2+} cannot supply the Ca^{2+} required for contraction. What then is the reason for the requirement of extracellular Ca^{2+} and the Ca^{2+} influx? There are five possible explanations: (1) Neurotransmitter release at the nerve terminal requires Ca^{2+}; (2) end-plates require Ca^{2+} for stability; (3) excitable cells often require Ca^{2+} for stability; (4) the small Ca^{2+} entry may trigger a much larger release of Ca^{2+} from inside; (5) Ca^{2+} entry is necessary to balance the Ca^{2+} efflux (Woodward, 1949) which occurs because of the elevated cytoplasmic Ca^{2+}, otherwise the cell would gradually be depleted of Ca^{2+} and muscle fatigue would occur, as can happen in prolonged tonic contractions.

A further argument was based on calculations made by A. V. Hill in 1948. Diffusion of Ca^{2+} from the surface of the cell into the interior would be too slow to account for the very fast twitches that occur in vertebrates.

The sarcoplasmic reticulum

Some 50 years ago Lawaczek (1928) and Hermann (1932) proposed that Ca^{2+} release from inside the cell was involved in the contraction of the heart. In 1951 March discovered a factor in muscle homogenates which modified contraction. Four years later, the relaxing factor in muscle homogenates was shown to be vesicular, required Mg^{2+} and ATP, and removed Ca^{2+} from the surrounding fluid (Kumagai *et al.*, 1955; see also Ebashi and Lipmann, 1962; Hasselbach and Makinose, 1963; Weber *et al.*, 1964, 1966). These vesicles are fragments of the tubules first observed by Verati in 1902 using a light microscope and in more detail by Bennett and Porter in 1953 using an electron microscope; the Ca^{2+} was localized by Constantin *et al.* (1965). It is the release of Ca^{2+} from this sarcoplasmic reticulum that provides more than 90% of the Ca^{2+} required for contraction.

Several proteins have been isolated from the sarcoplasmic reticulum and provide a molecular explanation for its function (MacLennan and Wong, 1971; MacLennan *et al.*, 1972; MacLennan and Campbell, 1979) (Table 5.9, Fig. 5.17). The (Ca^{2+})-MgATPase and proteolipid are firmly embedded in the phospholipid membrane and can be detached only by using detergent. The other proteins are more loosely attached and can be solubilized without using a detergent.

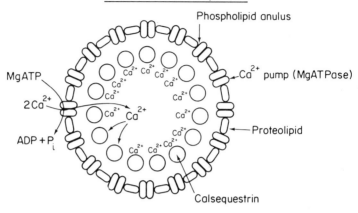

Fig. 5.17. Structure and mechanism of Ca^{2+} pumping into the sarcoplasmic reticulum (SR)

(1) (Ca^{2+})-MgATPase

About 70% of isolated whole sarcoplasmic reticulum and more than 90% of the membrane protein consists of this enzyme, although *in vivo* these values may be slightly less. Each protein molecule is surrounded by a ring of 30 phospholipid molecules (Bennett et al., 1978). By estimating the recovery of enzyme activity after alteration of this lipid by detergent, followed by dialysis to reconstitute the lipid anulus, the lipid specificity has been defined. ATP hydrolysis is measured by following the rate of P_i or ADP formation. Ca^{2+} accumulation in vesicles is usually carried out in the presence of oxalate. This traps the Ca^{2+} in the form of a calcium oxalate precipitate inside the vesicles, which contain fewer Ca^{2+}-binding sites and are more permeable than the original sarcoplasmic reticulum. Optimal activity is obtained with a zwitterionic phospholipid such as dioleoylphosphatidylethanolamine or phosphatidyllecithin. Although the enzyme complex contains 30 molecules of lipid per molecule of MgATPase, the overall molar ratio of phospholipid to MgATPase is 90:1 in the whole membrane.

The isolated enzyme is a globular protein, diameter 6.3 nm, molecular weight 102 000. In the membrane a particle measuring 4×6 nm is on the surface connected to another section of the protein 8–9 nm long which traverses the membrane. The substrate is $ATPMg^{2-}$, with ATP hydrolysis to ADP and inorganic phosphate providing the energy for Ca^{2+} accumulation against the electrochemical gradient which exists across the membrane of the sarcoplasmic reticulum. The maximum rate of hydrolysis is about 4000 moles of ATP per mole enzyme per min. There are two high-affinity Ca^{2+} sites per molecule (K_d^{Ca} approx. 0.3 μM) exposed to the cytoplasm. The enzyme also has 1–3 low-affinity sites on the inner surface. The molecular mechanism of Ca^{2+} pumping has been worked out from studies on phosphate–water–ATP exchange reactions and the isolation of a phosphorylated enzyme intermediate (Fig. 5.17). The first stage is the binding of 2 Ca^{2+} and 1 MgATP to the enzyme. Hydrolysis of ATP occurs, producing a phosphorylated aspartate residue on the enzyme. Ca^{2+} is moved across the bilayer either through a mobile or a fixed pore, which could be a 34 amino acid β-helix, or by rotation of the molecule (Tada et al., 1978). Mg^{2+} is required for dephosphorylation of the enzyme. The enzyme can also be reversed (Hasselbach, 1978) experimentally. The release of $2Ca^{2+}$, through the ATPase, produces 1 ATP.

Mitochondrial inhibitors of Ca^{2+} transport, such as oligomycin and dinitrophenol, do not affect the sarcoplasmic reticulum, ruling out H^+ as a counterion for Ca^{2+} transport. Ouabain also is without effect. However, modification of the CO_2^- group at the active centre by dicyclohexylcarbodiamide does inhibit the (Ca^{2+})-MgATPase in the sarcoplasmic reticulum. It is not yet clear whether the enzyme generates a membrane potential, or what the counterion for Ca^{2+} is in the cell.

(2) *Calsequestrin*

Under the electron microscope the reticulum appears to contain a fibrous matrix within the tubules, particularly near the terminal cisternae and the A–I junction (Fig. 5.11). This is caused by the protein calsequestrin which has been identified at these sites using immunofluorescence. In contrast, the (Ca^{2+})-MgATPase is found all over the reticulum. Calsequestrin binds most of the Ca^{2+} inside the tubules. About 20% of the protein inside reticulum vesicles is calsequestrin. It is a glycoprotein with a molecular weight of 44–63 000 and contains approx 400 amino acids, two glucosamines and three mannose per molecule. The Asp + Glu content is 52% of the total. This very high percentage (see Table 3.6 for comparison with other proteins) explains why calsequestrin can bind up to 45 Ca^{2+} per molecule. The mean affinity constant for each site is 500 μM at physiological ionic strength. These properties clearly distinguish calsequestrin from calmodulin and other regulatory Ca^{2+}-binding proteins (Table 5.12).

(3) *Proteolipid, acidic proteins and high-affinity Ca^{2+}-binding proteins*

The role of these proteins has not been so well characterized compared with the Ca^{2+} pump or calsequestrin. Proteolipid is an intrinsic protein in the bilayer, with a molecular weight of 12 000. It constitutes 1 % of the vesicle protein. A group of about three acidic proteins of molecular weight 20 000–32 000 have been isolated and have high-affinity Ca^{2+}-binding sites ($K_d^{Ca} = 1\,\mu M$). Another high-affinity Ca^{2+}-binding protein has been isolated; this is located inside the tubules and has a molecular weight of about 55 000. It has 25 Ca^{2+} sites with an overall K_d^{Ca} of 0.1 mM, and one with a higher affinity and a K_d^{Ca} of about $5\,\mu M$. It contains no detectable carbohydrate. Since it makes up about 10 % of the vesicle protein it may bind some Ca^{2+} inside the tubules.

Development of the sarcoplasmic reticulum begins in the embryo and continues after birth. However, (Ca^{2+})-MgATPase and calsequestrin synthesis do not parallel each other. The MgATPase is synthesized on membrane-bound polysomes, a phospholipid anulus forms round it which develops into the complete reticulum membrane. Calsequestrin is synthesized on the rough endoplasmic reticulum, secreted into the lumen, and transported to the Golgi. Golgi vesicles containing calsequestrin fuse with the membranes containing the ATPase to form the complete reticulum. Glycosylation of calsequestrin occurs in the lumen of either the Golgi or rough endoplasmic reticulum.

Ca^{2+} release into cytoplasm

Ca^{2+} can be released through the MgATPase in the sarcoplasmic reticulum, thereby resynthesizing ATP (Hasselbach, 1978). However, this is not the mechanism which is activated physiologically. Release of Ca^{2+} from the reticulum occurs through the proteolipid bilayer. If a special Ca^{2+} channel is necessary then, on the basis of estimates of the number of ion channels in nerves, 0.1–1 % of the membrane protein would be sufficient to provide sufficient Ca^{2+} for contraction. Based on measurements of the release of ^{45}Ca and free Ca^{2+} in skinned muscle fibres and fragmented sarcoplasmic reticulum vesicles, two possible stimuli for Ca^{2+} release have been identified (Endo, 1977) (Table 5.17): (1) depolarization of the reticulum membrane as a result of a direct interaction with electrical changes in the cell membrane; (2) Ca^{2+}-induced release caused by a local increase of Ca^{2+} arising from the influx of Ca^{2+} across the cell membrane.

The latter mechanism was first proposed by Bianchi (1968) and demonstrated by both Ford and Podolsky (1970) and Endo (1977). The reticulum from heart and smooth muscle can be stimulated to release Ca^{2+} by Ca^{2+} concentrations as low as $10\,\mu M$. In the latter tissue, contraction can be stimulated by neurotransmitters such as acetylcholine without causing an action potential, but rather by opening a receptor-dependent Ca^{2+} channel. However, in vertebrate skeletal muscle as much as 0.1–1 mM Ca^{2+} is necessary to stimulate Ca^{2+} release from the reticulum under experimental conditions. Birefringence studies on intact skeletal muscle support the hypothesis that in this case reticulum membrane depolarization of about 100 mV is the most likely trigger for Ca^{2+} release. A randomly

Table 5.17. Mechanisms for Ca^{2+} release from sarcoplasmic reticulum

Intracellular stimulus	Properties in skinned fibres or vesicles					Probable muscle cells
	Preloading by Ca^{2+} requirement	Increase in Mg^{2+}	ATP	Local anaesthetics	Sugars (40 mM)	
Electrical depolarization	Not required	No effect	Little or no effect	Small stimulation	Inhibition	Muscles with fast action potentials e.g. vertebrate skeletal
Local increase in free Ca^{2+}	Required	Inhibition	Stimulation	Inhibition	Little or no effect	Heart, smooth, invertebrate

From Endo (1977).

dispersed K^+/Na^+ channel (McKinley and Meissner, 1978) and a voltage-dependent K^+ channel (Miller, 1978) have been discovered in sarcoplasmic reticulum.

Relaxation

Relaxation of muscle occurs as a result of a reduction in cytoplasmic free Ca^{2+} to less than about 0.5 μM. Once the stimulus is over Ca^{2+} is pumped back into the sarcoplasmic reticulum. This lowers the cytosolic free Ca^{2+}. Tension in the muscle decreases as the Ca^{2+} dissociates from its binding protein attached to the filaments (Fig. 5.15). In fast skeletal muscle this means that approx. 200 μM Ca^{2+} is pumped back into the reticulum within 100–200 msec. The fact that, in this time, *in vitro* preparations of reticulum can only take up the equivalent of about 40 μM Ca^{2+} has lead to the suggestion that binding of Ca^{2+} to the outside of the reticulum is the cause of the initial rapid decrease in cytoplasmic free Ca^{2+}. This bound Ca^{2+} is then pumped into the lumen of the reticulum, requiring about 5–20 turnovers of the pump to remove all of the Ca^{2+}.

Some problems

Huxley and Taylor (1958) and Winegrad (1965) have provided evidence obtained with intact muscle that Ca^{2+} is released from the terminal cisternae of the sarcoplasmic reticulum, taken up in the longitudinal tubule and transferred back to the original site (Fig. 5.18c). However, skinned muscle fibres and fragmented vesicles have provided little evidence to support this. A further problem is the role of mitochondria in muscle Ca^{2+} balance, and why the cytoplasm of some muscles contains another group of high-affinity Ca^{2+}-binding proteins called the parvalbumins (Benzonana *et al.*, 1972, 1974; Kretsinger, 1976a,b).

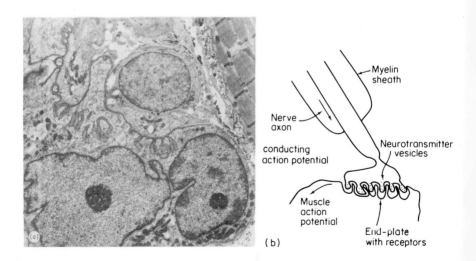

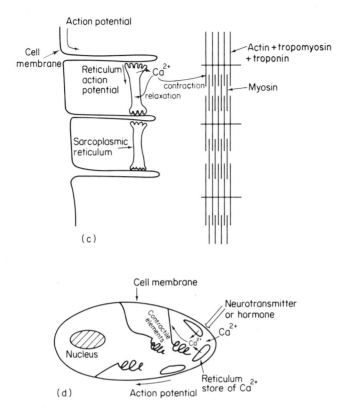

Fig. 5.18. Stimulation of Ca^{2+} release from muscle sarcoplasmic reticulum. (a) Electron micrograph of a neuromuscular junction (by courtesy of Dr. Caroline Sewry). × 4200. (b) Diagram of a neuromuscular junction. (c) T tubules and the release of Ca^{2+} from sarcoplasmic reticulum in skeletal muscle. (d) The contraction of smooth muscle

Mitochondrial Ca^{2+} uptake and release may be important in injured muscle cells and in pharmacological circumstances (see Chapter 9). However, it does not seem to play a major part in the contraction–relaxation cycle initiated by a primary stimulus. The parvalbumins are a group of Ca^{2+}-binding proteins found in chordates, and particularly in fish. They may be necessary for rapid relaxation of certain muscles where the sarcoplasmic reticulum is not close enough or active enough to decrease the free Ca^{2+} concentration close to the fibres quickly enough.

5.3.4. The mechanism by which Ca^{2+} stimulates contraction

There are three ways in which Ca^{2+} can provoke actomyosin to contract depending on the type of muscle concerned (Table 5.18). Two of these involve

Table 5.18. Ca^{2+} regulation in different muscles

Ca^{2+} site	Muscle example	Actomyosin state before Ca^{2+}	Tropomyosin	Role of Ca^{2+}
Troponin C	Vertebrate striated	Primed	Required	Derepressor
Leitonin	Vertebrate smooth	Relaxed	Required	Activator
Myosin	Molluscan	Primed	Not required	Derepressor

From Ebashi (1980).

special Ca^{2+}-binding proteins—troponin C in many striated muscles and leiotonin in smooth muscle. The third, found in some invertebrates, requires the Ca^{2+} to bind directly to the myosin. Two further differences in mechanism between different muscles exist. The first concerns the role of tropomyosin which is required for regulation of all muscles, except where Ca^{2+} interacts directly with myosin. The second concerns the nature of the action of Ca^{2+}. In striated muscle, whether it has troponin C or myosin linked regulation, Ca^{2+} acts by derepressing a contractile mechanism which is ready to contract but is prevented from doing so because of tropomyosin being in the wrong position. In smooth muscle, on the other hand, leiotonin acts as an activator of the contractile mechanism. The situation is analogous to the difference between two cars that are ready to move, one with a conventional clutch and gear box, the second an automatic with a fluid fly wheel. In the conventional car the engine is revving hard but the car cannot move until the foot is taken off the clutch, whereas in the automatic a positive activation of the accelerator causes the car to move. Studies on the amino acid sequence (Table 3.17) and three-dimensional structure of troponin C using X-ray crystallography, tyrosine fluorescence, electron spin resonance, proton nuclear magnetic resonance and fluorescent probes have lead to an understanding of how this Ca^{2+}-binding protein works (Kretsinger, 1976a, b; Levine *et al.*, 1977, 1978; Johnson *et al.*, 1978, 1980).

Cardiac and skeletal muscle troponin C have much sequence homology between them. The proteins consist of eight α-helical regions which can be thought of as four pairs (Fig. 5.19A), designated the E–F hands by Kretsinger (see Fig. 3.12). Skeletal muscle troponin C has four Ca^{2+} sites corresponding to these four regions. However, in cardiac troponin C the region nearest the N-terminus has fewer Asp and Glu residues and hence this troponin C only has three sites for Ca^{2+} (Leavis and Kraft, 1978). Several values for the affinity constants of the Ca^{2+}-binding sites have been reported. Although there is some variation between the absolute values there is general agreement that in skeletal muscle sites III and IV are of high affinity and can also bind Mg^{2+}, whilst sites I and II are of lower affinity and do not bind Mg^{2+} (Fig 5.19A). Proton NMR and fluoresence studies have shown that it is the high-affinity sites which require Ca^{2+} in order to maintain the backbone and ordering of the helices in a stable form. An increase in free Ca^{2+} from 0.1 to 5 μM causes at least three of the Ca^{2+} sites to be more than 90% saturated, the mean increase being equivalent to the

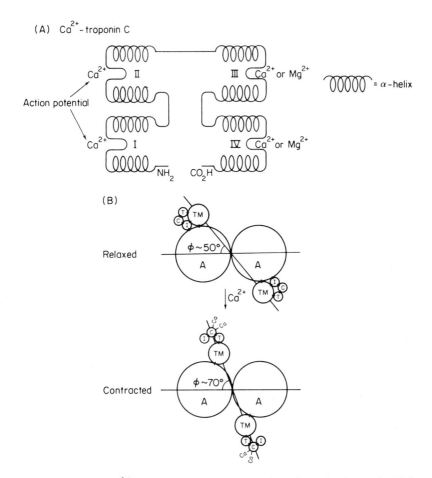

Fig. 5.19. How Ca^{2+}-troponin C allows contraction of muscle. A = actin; TM = tropomyosin; T = troponin T; I = troponin I; C = troponin C (proteins not drawn to scale). From Potter and Gergeley (1974), reproduced by permission of Elsevier Biomedical Press B. V.

occupation of two Ca^{2+} sites (Table 5.19). The binding of Ca^{2+} to sites I and II causes a change in the interactions between hydrophobic residues rather than altering the α-helical arrangement. These latter alterations cause a change in the configuration of the troponin complex, mainly through an interaction with troponin I. This change in the complex shifts tropomyosin in the actin grooves (see Figs. 5.11 and 5.19 B), thereby allowing the actin and myosin filaments to slide together.

So far we have assumed that the cytosolic free Ca^{2+} is in equilibrium with troponin C. Strictly this is never the case, but often holds to a first approximation. However, this is not true for a twitch contraction. In this case Ca^{2+} binding will be limited mainly by diffusion, whereas contraction is dependent on

Table 5.19. Fractional saturation of Ca^{2+} sites on troponin C

Parameter	\multicolumn{4}{c}{Ca^{2+} site (numbered from N-terminus; see Fig. 5.19)}			
	IV	III	II	I
K_d^{Ca} (μM)	0.002	0.05	5	0.2
K_d^{Mg} (mM)	0.2	0.2	—	—
K_{app}^{Ca} (μM)	0.022	0.55	5	0.2
Percentage saturation at 0.1 μM free Ca^{2+}	82	15	2	50
Percentage saturation at 5 μM free Ca^{2+}	99.6	90	50	96

Assume no cooperativity of Ca^{2+} binding. Data calculated, assuming free $[Mg^{2+}]$ = 2 mM, from equations 3.24 and 3.25; K_d values from Levine et al. (1978). Fractional saturation of each Ca^{2+} site = $[Ca^{2+}]/\{[Ca^{2+}] + K_d^{Ca}(1 + [Mg^{2+}]/K_d^{Mg})\}$. The mean increase in fractional saturation of all sites on going from 0.1 to 5 μM is 46% (i.e. 2 Ca^{2+} sites). This agrees well with the figure based on the model of Ashley and Moisescu (1972) and measurement of free Ca^{2+} using aequorin. The affinity of troponin C for Ca^{2+} may be increased by as much as tenfold when bound to troponin I in the cell.

the rate of conformational change, first in troponin C, and then in the other proteins of the contractile apparatus. The half-time for the conformational change in troponin C, as a result of Ca^{2+} binding to the 'low'-affinity sites, is about 70 msec (Levine et al., 1977). The high-affinity sites already have Ca^{2+} on them in resting muscle. This is fast enough to account for the onset of a twitch. Relaxation is dependent on a slower release of Ca^{2+}. These observations have been confirmed in vivo by Ashley and Moisescu (1972, 1977; Ashley 1978). By correlating intracellular free Ca^{2+}, measured with aequorin, with the rate of force development they showed that, in barnacle muscle at least, tension development is dependent on 2 Ca^{2+} binding to each force-developing unit, probably consecutively.

Although the stimulation of contraction of smooth muscle is most probably due to Ca^{2+} binding to leiotonin, another mechanism of regulation has been discovered (Nairn and Perry, 1979; Barany and Barany, 1980). Smooth muscle contains an 80 000 molecular weight kinase which catalyses the phosphorylation of a serine on the light chain of myosin, the phosphate coming from ATP. This kinase forms a complex with calmodulin, the molecular weight of the total complex being about 100 000. This kinase only has significant activity when Ca^{2+} is bound to the calmodulin. In vitro maximum activation occurs at about 10 μM Ca^{2+} and is inhibited by phenothiazines (see Chapter 9) (Cassidy et al., 1980). Phosphorylation of myosin activates the myosin MgATPase. Furthermore, the myosin cannot be phosphorylated by cyclic AMP dependent protein kinase. Several workers have argued that a Ca^{2+}-dependent phosphorylation is the primary mechanism for the stimulation of contraction in smooth muscle. Further evidence is required to show that this phosphorylation occurs in the cell and whether the time course relates to contraction. An alternative function of phosphorylation is in secondary regulation, or in enabling smooth muscle to be activated independently by two different types of primary stimulus.

Calmodulin-activated phosphorylation of myosin seems to play a crucial role in allowing myosin filaments to form in non-muscle cells (Scholey et al., 1980).

5.3.5. Secondary regulators

We have just examined the mechanism by which a primary stimulus, be it a neurotransmitter or direct electrical impulse, causes muscle to contract. In several muscles, however, this primary response can be modified by the action of secondary regulators which include neurotransmitters, hormones and drugs. For example, four types of secondary regulation exist in mammalian heart: inotropy—influence on the strength of contraction; chronotropy—influence on the heart beat rate; dromotropy—influence on the conduction velocity; and bathmotropy—influence on the excitability and stimulus threshold. Secondary effectors can have a positive influence on all or some of these parameters; for example, β-adrenergic agonists such as circulating adrenaline or noradrenaline from nerve endings, or drugs such as digoxin. In contrast, acetylcholine release from parasympathetic nerves is inhibitory on these parameters in the heart.

In smooth muscle the situation is complicated by the fact that different muscles may react differently to the same substances. For example, prostaglandins such as PGE_1 or $PGE_{2\alpha}$, endoperoxides and thromboxanes, can stimulate contraction in some smooth muscles and yet inhibit others (Bolton, 1979). Adrenaline, on the other hand, usually relaxes the mammalian smooth muscles on which it acts, mainly through a β-rather than an α-receptor mediated mechanism.

In invertebrate muscles activated by L-glutamate, γ-aminobutyric acid (GABA)-releasing nerves are inhibitory (Table 5.14).

These secondary regulators of muscle contraction can act by one of two mechanisms: (1) interference with electrical excitability, or influx of ions such as Ca^{2+}, across the cell membrane; (2) modification of intracellular processes directly involved in Ca^{2+}-dependent regulation. Examples of the first type include GABA in invertebrates which competes with L-Glu for end-plate receptors, acetylcholine influencing Na^+ and K^+ permeability in the heart and α-effects of adrenaline, which may involve effects on Ca^{2+}, Na^+ and K^+ permeability of the cell membrane. Adrenaline also increases the Ca^{2+} current during a cardiac action potential via a β-receptor mediated mechanism (Reuter, 1974). It is not necessary to postulate the existence of a 'second messenger' acting inside the cell to mediate the effects of these types of secondary regulator. Although a 'second-messenger' mechanism may be involved, resulting in phosphorylation of proteins on the inner surface of the membrane, most of the effects of the first type of secondary regulator can be explained by a direct interaction between the effector, bound to its receptor, and the cell membrane.

In the second type of secondary regulation—for example, the action of adrenaline on the heart or smooth muscle—it is necessary to invoke an intracellular 'second messenger' such as cyclic AMP or cyclic GMP (see Chapter 6). As long ago as 1913 Mines proposed a major involvement for Ca^{2+} in the action of adrenaline on the heart. Measurement of intracellular free Ca^{2+} using aequorin or microelectrodes (Chapter 2) has shown that adrenaline influences the Ca^{2+} transient during a contraction (Allen and Blinks, 1978; Marban et al., 1980). In frog heart, positive inotropic agents such as isoprenaline and acetylstrophanthidin increase the maximum cytosolic free Ca^{2+} concentration, as well as increasing the rate of Ca^{2+} removal (Fig. 5.20). In ferret heart, the increase in intracellular free Ca^{2+} induced by removal of external Na^+ is reduced by adrenaline. These observations are consistent with the action of adrenaline *in situ* to strengthen contraction and to accelerate relaxation. The latter is important if time is to be given for diastolic filling when the heart rate is also increased, for example during exercise.

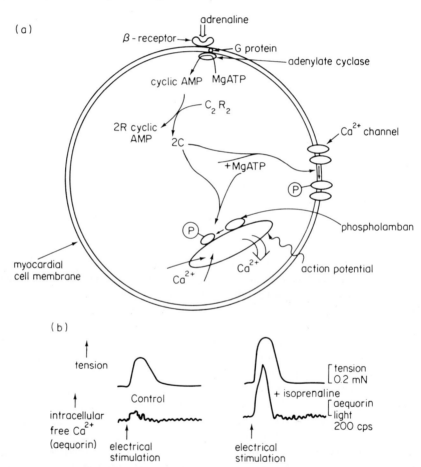

Fig. 5.20. How adrenaline could have a positive inotropic and chronotropic effect on the heart. (a) Diagram. C = catalytic subunit of protein kinase; R = regulatory subunit of protein kinase. (b) Increased intracellular Ca^{2+} transient in frog atrial trabeculae + isoprenaline (2.5 nM). From Allen and Blinks (1978). Reprinted by permission from *Nature*, **273**, 509–513. Copyright © 1978 Macmillan Journals Limited

The β-effect of adrenaline is mediated by cyclic AMP, which activates a protein kinase in the cytoplasm of the cell (for mechanism see Chapter 6). The protein kinase (subunit C of cyclic AMP dependent protein kinase) stimulates phosphorylation of a 22 000 molecular weight protein consisting of two 11 000 subunits and called phospholamban (Wray et al., 1973). This protein is attached to the membrane of the sarcoplasmic reticulum in heart muscle cells (Katz et al., 1975) and when phosphorylated activates the (Ca^{2+})-MgATPase. Under these conditions the sarcoplasmic reticulum will have more Ca^{2+} available for release during the contractile phase of the heart beat, and will remove the Ca^{2+} faster during the relaxation phase (Fedelesova and Ziegelhöffer, 1975). A calmodulin-activated protein kinase has also been shown, in vitro, to phosphorylate phospholamban (Le Peuch et al., 1979).

Another mechanism by which adrenaline influences cardiac muscle is through phosphorylation of troponin I (Barany and Barany, 1980). Phosphorylation of myosin, tropomyosin and troponin T have also been detected in vitro and in vivo using labelled phosphate or $[^{32}P]ATP$ in skeletal and cardiac muscle. Vertebrate skeletal muscle is not usually subject to secondary regulation, and this phosphorylation probably plays a role in the primary contractile mechanism. However troponin I in the heart contains an additional 26 amino acids at its N-terminus, compared with its counterpart in skeletal muscle. Adrenaline stimulates phosphorylation of the serine at position 20 of cardiac troponin I. This phosphorylation can be detected within a few seconds in vivo, and correlates with an increase in both cyclic AMP and the strength of contraction. Phosphorylation is catalysed by the catalytic subunit of cyclic AMP dependent kinase (see Chapter 6) in vivo, though other kinases also work in vitro. Conflicting reports on the function of this phosphorylation have been published. Whether it helps troponin $C-Ca^{2+}$ to strengthen contracture, or whether it affects the overall sensitivity of the actomyosin ATPase to Ca^{2+}, remains to be established. A key question is what effect it has on the kinetics of Ca^{2+} binding and dissociation from troponin, as well as the kinetics of the interaction between the troponin complex and tropomyosin.

Cyclic AMP may also play an important role in the action of thyroxine on the heart, since this hormone increases the apparent sensitivity of adenylate cyclase to other hormones (Elkeles et al., 1975).

The action of adrenaline in smooth muscle is less well understood. The relaxing effect of adrenaline is probably similar to the cyclic AMP dependent increase in speed of relaxation in the heart beat. In contrast to heart muscle, phosphorylation of proteins may decrease the cytoplasmic free Ca^{2+} transient by increasing Ca^{2+} efflux from the cell, as well as by activating the endoplasmic reticulum (Ca^{2+})-MgATPase (Bülbring and Tomita, 1970, 1977). Cyclic AMP dependent protein kinase does not seem to phosphorylate myosin, in contrast to the calmodulin-activated enzyme.

It can be concluded that secondary regulators of muscle contraction which interact with the Ca^{2+}-dependent mechanism (i.e. mechanism 2 above) can do so in two ways: (1) modification of the Ca^{2+} transient, the increase in cytoplasmic

Ca^{2+}, induced by the primary stimulus; (2) modification of the sensitivity of the contractile apparatus to Ca^{2+}.

Prolonged exercise, or sustained contraction of muscle, may involve a form of secondary regulation through elevation of tissue CO_2 concentration and a decrease in intracellular pH. In barnacle muscle CO_2 increases intracellular free Ca^{2+}, possibly by releasing Ca^{2+} from the sarcoplasmic reticulum (Lea and Ashley, 1978).

5.4. Chemotaxis in eukaryotes

Chemotaxis is the movement of cells in one direction along a chemical gradient. The term was first used by Pfeffer to describe the migration of plant cells, but it has since been described in a wide variety of phyla (Table 5.20). The type of cell movement which is activated may be flagellate or ciliate, as in the case of sperm or the marine dinoflagellate *Crypthecodinium*, or it may be amoeboid, as occurs when polymorphs, monocytes and macrophages invade sites of infection in vertebrates. Evidence for Ca^{2+} playing a role in the chemotactic response has been obtained in several cells (Table 5.20). Unfortunately, in many instances it is difficult to decide which biological role of Ca^{2+} is involved (see Chapter 1), i.e. whether it is structural, electrical or regulatory. We have seen in Chapter 4 that the reversal of cilia movement in *Paramecium*, occurring as a result of the protozoan touching an obstacle, is triggered by a Ca^{2+}-dependent electrical activation in the membrane. This causes a local increase in intracellular free Ca^{2+} which changes the direction of cilia movement. Several chemotactic stimuli on

Table 5.20. Examples of chemotaxis in eukaryotes

Cell type	Chemotactic stimulus	Reference
Physarum	Nutrients	Durham and Ridgway (1976)
Dictyostelium	Cyclic AMP	Bonner (1971)
Hydra	Glutathione	Lenhoff and Bovaird (1959)
Sperm (*Tubularia*)	?	Miller and Brokaw (1970)
Polymorphs and macrophages	C3a, C5a, kallikrein	Gallin and Rosenthal (1973); Sorkin *et al.* (1970); Lynne and Mukherjee (1978)
Crypthecodinium, a massive dinoflagellate	O_2	Hauser *et al.* (1978)
Various algae and protozoa	Nutrients	Levandowsky and Hauser (1978)
Fibroblasts	Chemotactic factors	Gail *et al.* (1973)

References selected where there is evidence for intracellular Ca^{2+} mediating effect of primary stimulus.
Note: C3a may not act directly on polymorphs but via mast cells

flagellates or ciliates probably act in a similar manner. Similarly, a random movement of 'amoeboid'-like cells can be activated and converted into a directed movement by a chemotactic stimulus. Two good examples are the movement of polymorphs in the inflammatory response, and the movement of slime mould cells as an essential preliminary to differentiation.

In the inflammatory response in vertebrates, phagocytic cells such as polymorphs attach to capillary walls and move into the site of infection or inflammation in order to remove the infective agent. Several factors have been discovered experimentally which provoke polymorph chemotaxis (Table 5.21). Those derived from the complement pathway are likely to be particularly important *in situ*. Synthetic N-acylated peptides also have potent chemotactic activity in the range 0.1–10 nM. Extracellular Ca^{2+} and Mg^{2+} are required for the maximum chemotactic response (Becker and Showell, 1972).

Experiments using A23187 have been equivocal, probably because of problems with cell viability, though it has been reported to stimulate chemotaxis (Lynne and Mukherjee, 1978). Both $^{45}Ca^{2+}$ influx and efflux are stimulated by the chemotactic factors C5a and kallikrein (Gallin and Rosenthal, 1973, 1974). There appears to be a correlation between Ca^{2+} influx and chemotaxis. For example, La^{3+} and nordihydroguainetic acid, an antioxidant, inhibit both Ca^{2+} flux and chemotaxis in polymorphs (Boucek and Snyderman, 1976; Naccache et al., 1979a, b). On this evidence one might expect that chemotactic stimuli activate polymorphs by increasing intracellular free Ca^{2+} through an increased influx of Ca^{2+} from outside, and possibly through release from an intracellular Ca^{2+} store. However, a slightly puzzling piece of data has been produced by Naccache and coworkers (1979a, b). They preincubated polymorphs with the fluorescent compound chlortetracycline (see Chapter 2). Addition of chemotactic concentrations of C5a or N-formyl-Met-Leu-Phe caused a large decrease in fluorescence indicating a release of membrane Ca^{2+}

Table 5.21. Some naturally occurring chemotactic stimuli of polymorphs

Source	Chemotactic factors
Bacteria	Low molecular weight peptides
Complement	C3a, C5a, $\overline{C567}$, C3 convertase formed by C3bB with presence of D
Hageman factor and plasma coagulation	Kallikrein, plasminogen activator, C3a, fibrinogen and fibrin products
Leukocytes	Peptides, lymphokines, transfer factor
IgG	Leucogressin
Macrophages	5000 molecular weight peptide
Lipid	Arachidonic and eicosapentaenoic acids
Miscellaneous	Cyclic AMP, prostaglandin E_1, mellitin from bee venom

From Klebanoff and Clark (1978). Secondly regulation by adrenaline, histamine and prostaglandins is similarly dependent on either cyclic AMP or cyclic GMP in a manner similar to exocytosis in these cells (see Chapter 7, section 7.2.7).

This effect did not require the presence of extracellular Ca^{2+}. N-formyl-Met-Leu-Phe is a synthetic chemotactic peptide which mimics bacterial chemotactic factors. The Ca^{2+}-activated photoprotein obelin has been incorporated into erythrocyte ghost–polymorph hybrids which respond to N-formyl-Met-Leu-Phe (Hallett and Campbell, 1982a, b). Cytosolic free Ca^{2+}, estimated from the obelin luminescence, is about 0.3 μM and rises to about 0.7 μM within 30 sec of addition of the chemotactic peptide, where it apparently remains for several minutes.

Secondary regulation of chemotaxis may occur through changes in intracellular cyclic AMP or cyclic GMP. Cyclic AMP plays an extracellular role in the differentiation of the slime mould *Dictyostelium*. It is a chemotactic stimulus. In this organism, chemotaxis is not dependent on extracellular Ca^{2+}, though aggregation and development of the fruiting body are. However, ionophore A23187 accelerates aggregation and cyclic AMP increases $^{45}Ca^{2+}$ efflux (Chi and Francis, 1971), suggesting that the chemotactic stimulus may act by activating the release of Ca^{2+} from an internal store into the cytoplasm.

Thus chemotaxis may be activated by intracellular Ca^{2+} coming from outside the cell or from an internal store. The molecular mechanism by which Ca^{2+} acts could involve a calmodulin interaction between microtubules and microfilaments, depending on whether the cell movement is flagellate or amoeboid.

5.5. The special case of the spasmoneme

Certain ciliated protozoa, usually attached to a substrate, contain a contractile organelle called a spasmoneme (Fig. 5.21) which causes the organism to retreat when touched or stimulated electrically. Some of these protozoa are only a few micrometres in diameter; however, giant species such as *Vorticella*, *Carchesium* and *Zoothamnium* contain spasmonemes which can be 140 μm in diameter and 1 mm long. It is these species which have provided the experimental evidence for a unique role for Ca^{2+} in contraction and a new class of Ca^{2+}-binding proteins named spasmins (Amos *et al.*, 1976; Routledge, 1978).

The spasmoneme has a rubbery texture and can contract to a third of its original length at a rate of $100-200$ mm sec^{-1}, generating a force of 5×10^4 N m^{-2}. This compares well with muscle, and is considerably faster than, for example, the movement of sperm, which is about 1 mm sec^{-1}.

Addition of $1-10$ μM Ca^{2+} to cells made permeable in glycerine stimulates contraction. Other alkaline earths also stimulate but are less potent than Ca^{2+}, i.e. $Ca^{2+} > Sr^{2+} > Ba^{2+}$. The threshold is about 0.4 μM Ca^{2+}. Addition of EGTA causes relaxation. The unique feature of this contractile system is that MgATP is not the energy source, rather it is Ca^{2+} binding itself. About 60% of the spasmoneme consists of an acidic protein with a molecular weight of about 20 000. It has a pI of 4.7 and a Glu + Asp content of 20%, a little lower than regulatory Ca^{2+}-binding proteins but greater than other proteins (Table 3.6). The spasmoneme contains no detectable actin or tubulin. Each spasmin molecule can bind 2 Ca^{2+}, with a K_d of approx. 1 μM (Routledge, 1978). X-ray microprobe

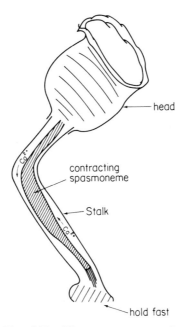

Fig. 5.21. The spasmoneme of *Vorticella*

analysis (Chapter 2) has shown that about 8.5 mmoles of Ca^{2+} bind per kg of wet tissue.

After electrical stimulation of the cell tension begins to develop within a few milliseconds. On the basis of Hill's (1948) argument, diffusion of Ca^{2+} from the cell membrane to the organelle would be some 10–100 times too slow to stimulate contraction at the required rate. The spasmoneme consists of filaments of spasmin arranged longitudinally, with a series of membranous tubules which run in close proximity to the filaments. It is probably these tubules which, like the sarcoplasmic reticulum in muscle, release and take up Ca^{2+} during a contraction–relaxation cycle.

The special feature of the spasnomene is that Ca^{2+} binding, rather than ATP hydrolysis, appears to be the immediate source of energy, as well as being the trigger, for contraction in protozoa which contain this organelle.

5.6. Conclusions

In this chapter we have examined many examples of Ca^{2+}-dependent cell movement based on either an actin or tubulin contractile system. Many fit into the category of 'threshold phenomena' (Chapter 2) where an electrical, mechanical or chemical stimulus initiates a change in the cell from one state to another. An increase in intracellular free Ca^{2+} is the major, if not the only, trigger of this event. The Ca^{2+} comes mostly from reticulum stores in the cell, though occasionally extracellular Ca^{2+} contributes significantly to raised cytoplasmic Ca^{2+}. Ca^{2+} currents in excitable cells provide some of this Ca^{2+} and the resulting electrical change may be sufficient to release Ca^{2+} from internal stores.

Secondary regulation of the strength, speed and threshold level may be mediated by modifying the electrical behaviour of the cell, or by modifying the response of intracellular Ca^{2+} to a primary signal. In both cases protein phosphorylation, activated by cyclic nucleotide dependent kinases, plays an important role in determining the ultimate response of the cell.

CHAPTER 6

Intracellular Ca^{2+} and intermediary metabolism

6.1. How and why is intermediary metabolism regulated?

All cells require a source of energy in order to remain alive and to carry out any special functions they may have. Plants and photosynthetic bacteria derive their energy from sunlight. Animals, on the other hand, derive their energy from the oxidation of the three main components of their food—protein, carbohydrate and lipid. Nutrients in excess of the animal's requirement are stored in the form of protein, glycogen or triglyceride in various cells of the body. In addition to their energy needs, cells also require substances such as phospholipids, amino acids and nucleotides for membrane, protein and nucleic acid synthesis necessary for maintenance of cell structure and replication. Furthermore, the function of certain cells requires the synthesis of special substances, for example actin and myosin in muscle, or adrenaline in the adrenal medulla. The chemical pathways involved in providing the carbon skeletons for such compounds, as well as providing energy for the cell, are known collectively as intermediary metabolism.

6.1.1. Intercellular pathways

In unicellular organisms the provision of amino acids, sugars, fatty acids, purines and pyrimidine bases, which form the building blocks for proteins, glycogen, cell walls, cell membranes and nucleic acids, must all be synthesized within the one cell. In animals this is no longer necessary. The changing needs of cells in the various tissues of an animal are coordinated through a series of intercellular pathways (Campbell and Hales, 1976) (Fig. 6.1). Body fluids such as the blood, or in invertebrates the haemolymph, transport nutrients from one tissue to another. Hence, there is a kind of symbiotic relationship between the tissues, each depending on the other for the ultimate survival and reproduction of the whole.

The direction and magnitude of these intercellular pathways vary under different physiological conditions. Let us take, for example, the case of the little boy running down the road, which was first described in Chapter 2. He has just had tea, so glucose oxidation is providing most of the energy for the cells in his

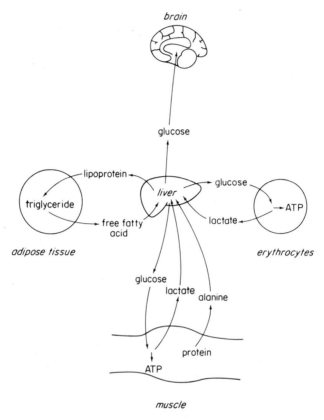

Fig. 6.1. Some intercellular pathways

body. Before his tea, oxidation of free fatty acid by tissues such as liver, kidney cortex, adipose tissue and muscle was contributing significantly to his energy supply. Also glucose will be stored as glycogen and triglyceride in several cell types after the meal. The boy begins to run. His blood flow may increase ten times and the energy requirements of his muscle increase similarly. Glycogen breakdown occurs in muscle, stimulated by adrenaline released by the adrenal medulla. Lactate is produced, which is metabolized by the heart and reconverted to glucose in the liver.

How are these intercellular pathways coordinated so as to provide individual tissues with the substances they require? What is the role of intracellular Ca^{2+} in this regulation?

6.1.2. Mechanisms of regulation

The net flow of substances between tissues through intercellular pathways is controlled by regulating the metabolism of individual tissues. In theory the flux of a metabolic pathway is affected by an alteration in the activity of any enzyme within it. However, in practice, many of the reactions within a cell are close to equilibrium so that a change in enzyme activity has only a small effect on the flux of the overall pathway. Changes in direction or flux of a pathway are usually the results of changes in the rates of a small number of reactions far from equilibrium. These reactions are catalysed by so-called rate-limiting step enzymes. In situations where the cell needs to alter the direction of the pathway, for example in liver from glucose oxidation to glucose synthesis, pairs of enzymes such as phosphofructokinase and fructose diphosphatase form substrate cycles at the rate-limiting steps. These cycles enable relatively small changes in intracellular regulators to cause a change in direction of the pathway or a large change in flux (Newsholme and Gevers, 1967; Newsholme and Start, 1973; Newsholme and Crabtree, 1976). Rate-limiting step enzymes can be regulated in three ways: (1) control of substrate supply; (2) activation or inhibition of the enzyme; (3) control of enzyme concentration.

There are three ways in which substrate supply to a pathway can be altered. Firstly, the concentration of the initial substrate in the fluid bathing the cells may be changed as a result of an intake of food or the release or uptake of the substance by other tissues. Hydrophobic substances like fatty acids simply diffuse across the lipid bilayer. Hydrophilic substances like sugars, amino acids and nucleotide bases require a carrier protein known as a permease. Providing this mechanism is not saturated it will respond to a change in the concentration of substrate in the external fluid. Secondly, the permease may be regulated, for example by a change in intracellular Ca^{2+} induced by a hormone. Glucose uptake by cells (Clausen, 1975, 1980; Clausen et al., 1975) and metabolite uptake by mitochondria (Mason, 1974; Bygrave, 1978) may represent Ca^{2+} dependent mechanisms of this sort. Thirdly, one of the substrates in the pathway may be involved in another pathway, and a change in this second pathway will therefore affect the first. The glucose-fatty acid cycle proposed by Randle and coworkers in the 1960's is dependent on this mechanism. For example, an increase in glycolysis will increase the glycerol phosphate concentration in the cell and thereby increase triglyceride synthesis from fatty acyl CoA (FA CoA) and glycerol phosphate. Equilibrium reactions can also play a part where cofactors are common to two pathways; for example, NAD/NADH and the lactate dehydrogenase–alcohol dehydrogenase reactions (Krebs, 1971).

Rate-limiting enzymes may be regulated by ions and metabolites, or by covalent modification of 'active' and 'inactive' forms. There have been many reports of Ca^{2+} inhibiting, or activating, enzymes and pathways in cell-free extracts (Table 6.1). These effects form an important part of several hypotheses concerning the physiological role of intracellular Ca^{2+} as a regulator (Gevers and Krebs, 1966; Bygrave, 1967; Rasmussen and Goodman, 1977). Unfortunately, many of the experiments on which these ideas were based used Ca^{2+} concentrations in the range 0.1–10 mM, several thousand times that in the cytoplasm of a resting cell (Chapter 2). These non-physiological effects of Ca^{2+} (Tables 3.14 and 6.1B) can be explained by Ca^{2+} competing with another cation,

Table 6.1. Some effects of Ca^{2+} on enzymes of intermediary metabolism

Enzyme	Tissue example	Effect* of Ca^{2+}	Reference
(A) *Physiological*			
Phosphorylase kinase	Muscle	a	Ozawa et al. (1967); Brostrom et al. (1971)
Pyruvate dehydrogenase phosphatase	Adipose tissue mitochondria	a	Severson et al. (1974; 1976)
Pyruvate dehydrogenase kinase	Adipose tissue mitochondria	i	Cooper et al. (1974)
NAD-linked isocitrate dehydrogenase	Heart mitochondria	a	Denton et al. (1980)
FAD-linked glycero phosphate dehydrogenase	Muscle	a	Denton et al. (1980)
Kynurenine aminotransferase	Kidney mitochondria	a	Mason (1974)
Succinate dehydrogenase	Liver mitochondria	a	Axelrod et al. (1941)
(B) *Non-physiological*			
Glycolysis in cell extract	Yeast and ascites	i	Ashwell and Dische (1950); Bygrave (1966a, b)
Hexokinase	Muscle	i	Wimhurst and Manchester (1970, 1972)
Phosphofructokinase	Muscle	i	Margreth et al. (1967)
Enolase	Erythrocytes	i	Boszormenyi-Nagy (1955)
	Muscle	i	Boyer et al. (1942, 1943)
Pyruvate kinase	Liver, muscle	i	Wimhurst and Manchester (1970, 1972); Kachmar and Boyer (1953)
Citrate desmolase	Liver	i	Dagley and Dawes (1955)
NAD-linked isocitrate dehydrogenase	Heart mitochondria	i	Vaughan and Newsholme (1969)
	Kidney	i	DeLuca et al. (1957)
Glutamate dehydrogenase	*Blastocladiella*	i	Le John et al. (1969); Dugal and Gopel (1975)
Pyruvate carboxylase	Liver mitochondria	i	Gevers and Krebs (1966)
Proteolysis of phosphorylase *b* kinase	Skeletal muscle, cardiac muscle	a	Krebs and Fischer (1956)
Phosphoenol pyruvate + glucose → glucose 6-phosphate	*E. coli*	i	Utter and Werkman (1942)
Hexose diphosphatase	*Clostridium*	i	Pett and Wynne (1933)

* a = activates; i = inhibits. Non-physiological includes all enzymes where Ca^{2+} > 10 μM, usually 0.1–1 mM, necessary for effects. Physiological is good effects at < 10 μM Ca^{2+}. Mitochondrial free Ca^{2+} not known, but assumed to be 1–10 μM.

such as Mg^{2+} or Mn^{2+}, for a site on the enzyme, by $ATPCa^{2-}$ being inhibitory, or by a low-affinity Ca^{2+} site of no significance at free Ca^{2+} concentrations in the range 0.1–10 μM. Many of the physiological affects of intracellular Ca^{2+} (Table 6.1A), for example activation of phosphorylase b kinase, require a special Ca^{2+}-binding protein, such as calmodulin, attached to the enzyme. Regulation of enzymes by metabolites such as AMP, glucose 6-phosphate or citrate rarely involves Ca^{2+} directly. However, covalent modification of the enzyme by a Ca^{2+}-dependent mechanism may alter the sensitivity of the enzyme to one or more of the metabolite effectors. Such is the case with phosphorylase, glycogen synthetase and pyruvate dehydrogenase. These enzymes exist in phosphorylated and dephosphorylated forms, which have different affinities for AMP, glucose 6-phosphate and acetyl-CoA, respectively. Other types of covalent modification of enzymes are acetylation of ε-Lys residues, methylation of the γCO_2H of Glu or Asp, nucleotide attachment to tyrosine or ADP-ribosylation of arginine (Chock et al., 1980). The best established Ca^{2+}-dependent covalent modification of enzymes is the phosphorylation and dephosphorylation of serine residues in enzymes like phosphorylase b kinase, controlled by changes in kinase and phosphatase activity (Nimmo and Cohen, 1977).

A change in enzyme concentration can be brought about by altering either its rate of synthesis from mRNA, or its rate of breakdown by proteases. This provides a relatively long-term mechanism occurring over many minutes or hours, in contrast to the acute regulation by ions and metabolites which can occur within a few seconds. In principle, Ca^{2+} could regulate enzyme synthesis by inducing transcription of the gene to form mRNA or by altering the factors controlling mRNA translation, for example by phosphorylation of ribosomal proteins. Neutral proteases that are activated by Ca^{2+} have been found in the cytoplasm of several cells, including muscle (Duncan, 1978). However, like the proteolytic activation of phosphorylase b kinase (Krebs and Fischer, 1956) these neutral proteases, in the absence of any calmodulin regulation, require mM Ca^{2+} for activation.

6.1.3. The need for 'second messengers'

In 1902 Bayliss and Starling injected an extract of jejunal mucosal membrane into the pancreatic vein of a dog. They succeeded in stimulating the secretion of digestive juices from the pancreas. Such chemical stimuli were called hormones, from the Greek ὁρμάω, meaning I excite or arouse, (Starling, 1905). Since that time many other naturally occurring chemical substances have been discovered which stimulate cells, including neurotransmitters released from nerve endings (Dale, 1934). Hormones and neurotransmitters can be broadly defined as substances released by one cell type which affect the behaviour of another (Huxley, 1935). They can be divided into three main structural groups, depending on whether they are derived from amino acids, cholesterol or arachidonic acid (Table 6.2, Fig. 6.2), though some plant hormones such as ethylene do not conform to this simple classification.

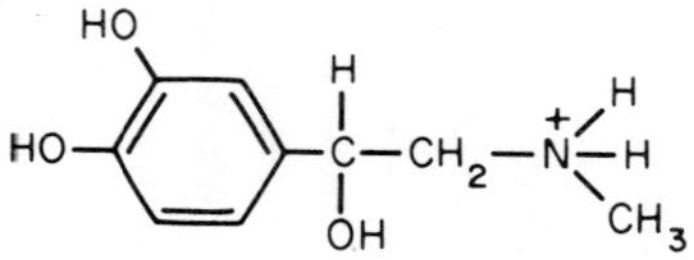

Thyroxine

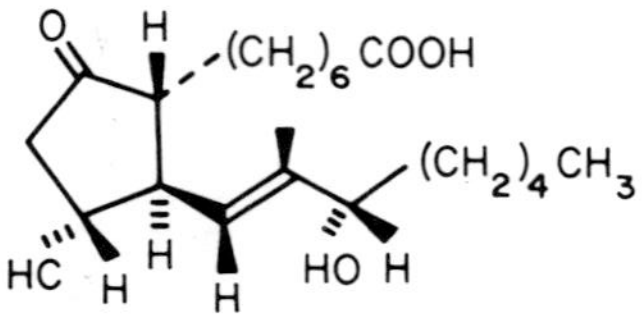

prostaglandin E_1

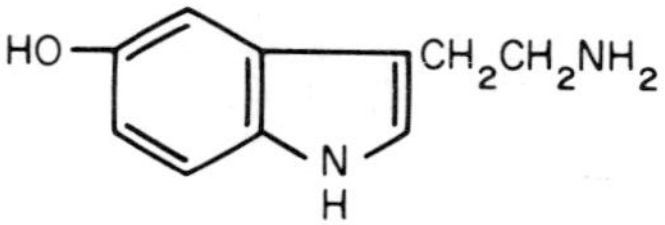

Serotonin (5-hydroxytryptamine)

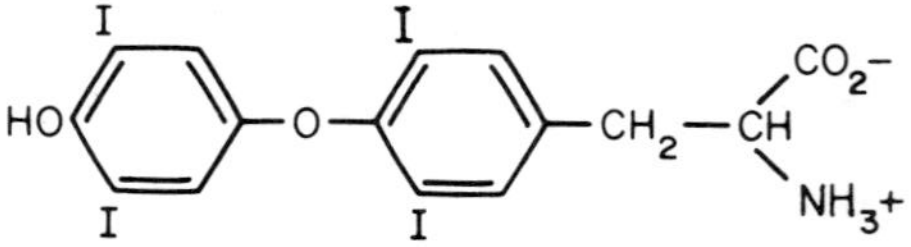

adrenaline (epinephrine)

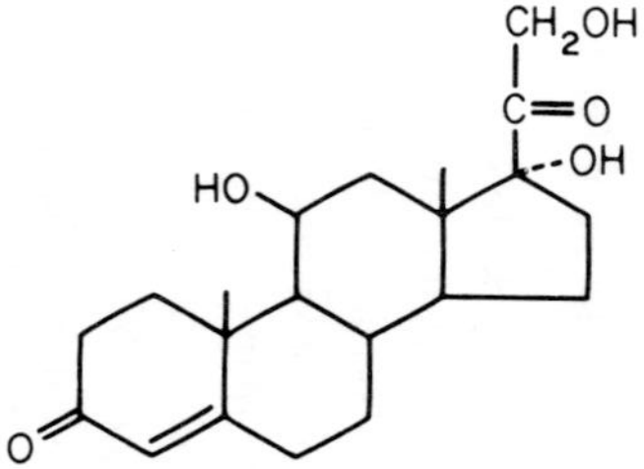

cortisol

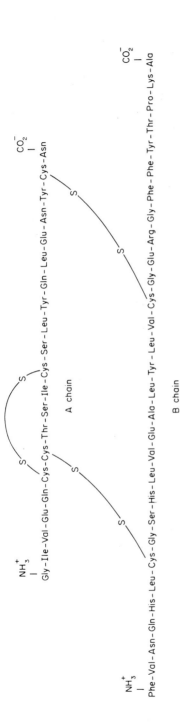

Fig. 6.2. Hormones which regulate intermediary metabolism. Some evidence exists for these hormones working by, or interacting with, intracellular Ca^{2+} and/or cyclic nucleotides

Table 6.2. Occurrence of hormones and neurotransmitters in various phyla

Substance	Phylum
1. *Amino acid derivatives*	
(A) Small molecules	
Adrenaline/noradrenaline/dopamine	All phyla with a nervous system
Acetylcholine	Protozoa
5-Hydroxytryptamine	Plants (seaweeds)
Iodoacetic acid	Porifera (scleroproteins–iodine)
Thyroxine and Triiodothyronine	Cnideria (scleroproteins, diiodothyronine)
	Protochordata
	Chordata
Histamine	Prokaryotics
	Plants
	Porifera
	Cnideria, Annelida, Arthropoda, Mollusca, Echinodermata, Chordata
Adenosine	Chordata
ADP	Chordata
(B) Polypeptides	
Insulin	Chordata
Glucagon	Chordata
Gut hormones	Chordata
Endorphins	Chordata
Various metabolic hormones	Arthropoda, Mollusca, Echinodermata
Gonadotrophins	Mollusca, Chordata
Prothoracotrophic hormone	Arthropoda
Eyestalk hormones	Arthropoda
Pituitary hormones	Chordata
2. *Cholesterol derivatives*	
(A) Steroids	Plants (oestrone, testosterone, progesterone)
	Prokaryotics
	Platyhelminthes (sex), Mollusca, Echinodermata, Chordata
(B) Ecdysones	Plants
	Cnideria, Annelida, Nemertini, Arthropoda, Echinodermata
(C) Juvenile hormone	Arthropoda
3. *Arachidonic acid and other fatty acid derivatives*	
Prostaglandins thromboxanes, endoperoxides	Cnideria (coral), Chordata

Plants have some hormones, e.g. ethylene, which do not fit into this classification. See Hanström (1939), Carlisle and Knowles (1959), Highnam and Hill (1977), Janakidevi *et al.* (1966), and Blum (1967) for references.

Primary chemical messengers are found in all animals and plants, and have even been isolated from some unicellular protozoa (Janakidevi et al., 1966; Blum, 1967). Several of them regulate the intermediary metabolism of tissues, including hormones such as adrenaline, insulin and glucagon released from vertebrate endocrine glands, neurotransmitters like noradrenaline released from nerves terminating in tissues such as the liver, and neurohormones released from neurosecretory cells in invertebrates (Table 6.3).

For the other phenomena dealt with in this book it has usually been possible to identify primary regulators responsible for initiating the response of a cell, and secondary regulators responsible for modifying this response, strengthening it, weakening it or altering the threshold level at which the cell is activated. This simple classification appears at first sight to be invalid when considering intermediary metabolism. Tissues metabolizing glucose, fatty acid or amino acids

Table 6.3. Some effectors of intermediary metabolism

Effect	Hormone or neurotransmitter	Phylum (tissue source of hormone or transmitter)
1. *Stimulation*		
Glycogen breakdown and hyperglycaemia	Adrenaline	Chordata (adrenal medulla)
	Noradrenaline	Chordata (peripheral nerve endings)
	Glucagon	Chordata (pancreas)
	Gut hormones	Chordata (intestine)
	Eyestalk hormone	Arthropoda: Crustacea
	Corpora cardiaca hormones	Arthropoda: Insecta
	Brain hormone	Polychaeta
Glucose utilization	Insulin	Chordata (pancreas)
Triglyceride breakdown (lipolysis)	Adrenaline	Chordata (adrenal medulla)
	Noradrenaline	Chordata (peripheral nerve endings)
	Adipokinetic hormone	Arthropoda (corpora cardiaca)
Protein degradation	Cortisol	Chordata (adrenal cortex)
Glycogen synthesis	Insulin	Chordata (pancreas)
Lipid synthesis	Insulin	Chordata (pancreas)
2. *Inhibition*		
Glycogen breakdown	Insulin	Chordata (pancreas)
	Adipokinetic hormone	Arthropoda (corpora cardiaca)
Lipolysis	Insulin	Chordata (pancreas)
Lipogenesis	Adrenaline	Chordata
	Eyestalk hormone	Arthropoda: Crustacea
3. *Regulation of tissue sensitivity to stimulatory or inhibitory effectors*		
Reduces insulin sensitivity	Cortisol	Chordata (adrenal cortex)
Increases glucagon	Cortisol	Chordata (adrenal cortex)
Potentiates glycogenolytic hormones	Thyroid hormones (T_3/T_4)	Chordata (thyroid)

all have a physiologically significant basal activity in the metabolic pathways concerned. Primary-acting hormones and neurotransmitters can either stimulate or inhibit this basal activity. Secondary regulation involves modifications in the sensitivity of the responding tissue to these primary activators or inhibitors, for example by steroid or thyroid hormones. Does this mean that the 'threshold concept' (Chapter 2) does not apply to the regulation of intermediary metabolism?

Most of the biochemical data concerning intermediary metabolism have been obtained from measurements of many thousands, if not millions, of cells, and are not able to take account of any heterogeneity in the cell population. It is the belief of this author that in order to understand fully the role of intracellular Ca^{2+} in the regulation of intermediary metabolism it is essential that this heterogeneity in the responses of individual cells be quantified (see Chapter 10). Two particular questions can be asked: (1) Is the physiologically significant basal activity of a tissue only to be found in a few cells? (2) Does the graded response of a tissue like liver or adipose tissue to different plasma concentrations of hormone involve a graded response in each of the individual cells or the 'switching on' of different numbers of cells? Although these questions remain to be answered much has been learnt about the molecular mechanisms which potentially might enable a hormone to regulate the metabolism of a cell through a 'graded' or 'threshold' response.

To the early workers the puzzle of how hormones and neurotransmitters regulate intermediary metabolism was that they seemed effective on intact cells, but not on cell extracts (Hechter, 1958). The reason for this is that they must first bind to a receptor protein either on the surface of the cell or, in the case of steroid and thyroid hormones, in the cell cytoplasm. These latter two groups of receptors act on the nucleus, ultimately affecting the concentration of other proteins in the cell. Hormones like adrenaline and insulin, on the other hand, bind to receptors on the cell surface, there being on average 1000 to 10 000 receptors per cell. For example, the β-adrenergic receptor is a protein of molecular weight about 75 000, and cells such as nucleated erythrocytes contain about 2000 per cell. Adrenaline and insulin both affect the membrane potential of cells, as well as stimulating glucose transport in cells such as those of adipose tissue and muscle. However, these effects are insufficient to explain completely the effect of these hormones on metabolism within the cell. A 'second messenger', acting within the cell, is therefore required. Three second messengers regulate intermediary metabolism— cyclic AMP, cyclic GMP and Ca^{2+}. The idea that changes in intracellular Ca^{2+} might be necessary for the action of adrenaline is more than 50 years' old (Lawaczek, 1928; Hermann, 1932; von der Wense, 1933). However, it was not until Sutherland's work on glycogen breakdown in the 1950's (Sutherland, 1961–1962) that the other major group of intracellular messengers, the cyclic nucleotides cyclic AMP and cyclic GMP (Fig. 6.3), was discovered. Cyclic CMP has also been found but as yet has no established function. There are many phenomena, apart from intermediary metabolism, where changes in cyclic nucleotides and intracellular Ca^{2+} are thought to play a role in their regulation (Rasmussen et al., 1972;

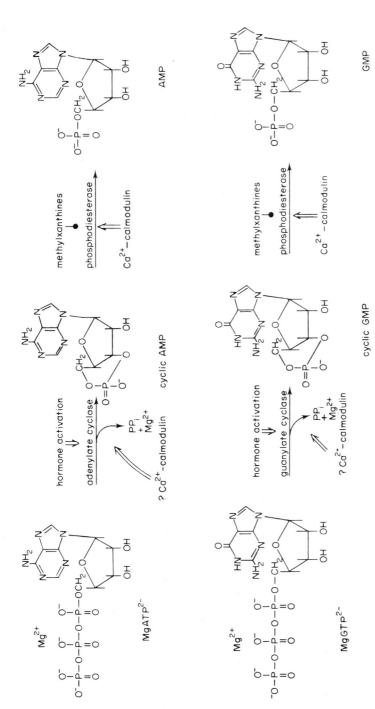

Fig. 6.3. Cyclic nucleotide synthesis and breakdown. ⇒ = activation, —● = inhibition

Berridge, 1976a, b; Rasmussen and Goodman, 1977). What then is the relationship between cyclic AMP, cyclic GMP and intracellular Ca^{2+}?

6.2. Cyclic nucleotides and intracellular Ca^{2+}

Many hormones and neurotransmitters affect the concentration of cyclic AMP and cyclic GMP in animal and plant tissues (Robison et al., 1971; Cook et al., 1975; Goldberg and Haddox, 1977) (Table 6.4). Many of these cellular regulators also affect intracellular free Ca^{2+}. In the other chapters of this book it is argued that intracellular Ca^{2+} is usually the trigger for cell activation. Cyclic nucleotides are required either to enable the primary stimulus to act over an appropriate time interval, or to mediate the action of a secondary regulator. In contrast, in

Table 6.4. Effects of hormones and neurotransmitters on cyclic AMP

Effector	Tissue or cell type	Effect on tissue cyclic AMP
Adrenaline/noradrenaline (β)	Liver, adipose tissue, various muscles, skin	Increase
Glucagon	Liver	Increase
Adrenocorticotrophin (ACTH)	Adrenal cortex, adipose tissue	Increase
Luteinizing hormone	Ovary	Increase
Thyroid-stimulating hormone	Thyroid	Increase
Vasopressin	Renal medulla, bladder	Increase
Parathyroid hormone (PTH)	Renal cortex, bone	Increase
Melanocyte-stimulating hormone (MSH)	Melanophores	Increase
PGE_1	Platelets, thyroid, lung	Increase
Histamine (H_2)	Parietal cells, brain	Increase
5-Hydroxytryptamine	Salivary gland (insect), brain	Increase
Adenosine	Adipose tissue, brain, renal cortex, liver, tumour cells (Ehrich)	Decrease
$PGE_1/PGF_{2\alpha}$	Adipose tissue, bladder, renal tubule	Decrease Inhibits adrenaline
Adrenaline (α)	Smooth muscles, several tissues	Inhibits cyclic AMP increase
Melatonin	Melanophores	Inhibits MSH increase
Insulin	Adipose tissue, liver	Decrease and inhibits cyclic AMP increase
ADP	Platelets	Inhibits PGE_1 increase

Decrease = lowers basal concentration. Inhibition = reduction in level induced by a stimulus. See Robison et al. (1971) for references.

intermediary metabolism, both Ca^{2+} and cyclic AMP play a direct role in the activation of metabolic pathways by a primary cell stimulus.

Hence there are three ways in which intracellular Ca^{2+} and cyclic nucleotides can interact: (1) regulation of intracellular cyclic nucleotide concentration by Ca^{2+}; (2) regulation of intracellular Ca^{2+} by cyclic nucleotides; (3) coregulation of enzymes by Ca^{2+} and cyclic nucleotides.

6.2.1. Regulation of intracellular cyclic nucleotide concentration by Ca^{2+}

Cyclic nucleotides are found in all healthy cells. The cytosolic concentration of cyclic AMP in most unstimulated cells is 0.1–1 μM and rises to a maximum of 10–50 μM in response to a physiological stimulus. Cyclic GMP is usually some ten-fold lower in concentration. Both cyclic nucleotides are formed from their respective nucleotide triphosphate and are then broken down to the 5′-nucleotide monophosphate (Fig. 6.3). Tissue concentrations are regulated by hormones and transmitters (Tables 6.4 and 6.5), and by electrical and physical stimuli.

Adenylate cyclase is the enzyme, on the inner surface of the cell membrane, responsible for catalysing the formation of cyclic AMP from $ATPMg^{2-}$ (Ross and Gilman, 1980). The enzyme has a molecular weight of 160 000–250 000, depending on the cell, and an associated MgGTP-binding protein, which also appears to have GTPase activity. The hormone receptors on the outside surface of the cell are 1000 to 10 000 in number per cell and lock on to the cyclase only when bound to an agonist. The active form of the enzyme is when GTP is bound to its binding protein (Fig. 6.4). This active form is maintained when a hormone is bound to its receptor. When the receptors become unoccupied the GTP is hydrolysed to GDP, which dissociates from the protein, thereby inactivating the cyclase. The mechanism of activation of guanylate cyclase is less well understood (Goldberg and Haddox, 1977). The molecular weight of this enzyme can vary from 150 000 to 900 000 and, in contrast to adenylate cyclase, it is found in both

Table 6.5. Effects of hormones and neurotransmitters on cyclic GMP

Effector	Tissue or cell type	Effect on cyclic GMP
Acetylcholine (muscarinic)	Many cells	Increase
Histamine (H_1)	Gut, brain	Increase
Adrenaline (α)	Many cells including liver cells	Increase
Prostaglandins	Polymorphs	Increase
$PGE_1/PGF_{2\alpha}$	Ovary, vein	Decrease
Light	Retina	Decrease
Adrenaline (β)	Many cells	Decrease
Glucose	Heart	Decrease
Histamine (H_2)	Heart	Decrease
Luteinizing hormone	Ovary	Decrease

See Goldberg and Haddox (1977) for references.

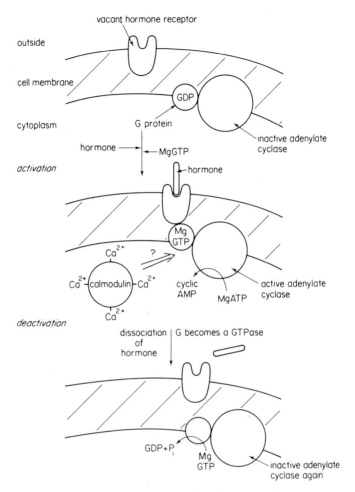

Fig. 6.4. Mechanism of activation of adenylate cyclase

the soluble and particulate fractions of a tissue homogenate. Thus, whilst the activation of adenylate cyclase can be explained by a direct interaction of the hormone–receptor complex through the phospholipid bilayer, activation of a cytoplasmic guanylate cyclase would require a second messenger.

The breakdown of cyclic AMP and cyclic GMP is catalysed by selective phosphodiesterases (Fig. 6.3) which are mainly cytoplasmic. However, hormone-sensitive, membrane-bound phosphodiesterases have been reported.

Does a change in intracellular Ca^{2+} alter the concentration of cyclic AMP or cyclic GMP by regulating the cyclase or phosphodiesterase?

There have been many reports that disturbances in extra- or intracellular Ca^{2+} influence cyclic nucleotide concentrations in the cell (Tables 6.6 and 6.7) (Berridge, 1976b; Rasmussen and Goodman, 1977). Several problems arise in the interpret-

Table 6.6. Ca^{2+} and some reported effects on tissue cyclic AMP

Tissue	Stimulus of tissue cyclic AMP	Effect of external EGTA	Effect of A23187	Reference
Adipose tissue	ACTH	Decrease	—	Bär and Hechter (1969a, b)
	Adrenaline (β)	Decrease	—	Schimmel (1973, 1976)
		No effect	No effect	Siddle and Hales (1980)
Liver	Glucagon	No effect	No effect	Hales et al. (1977)
	Glucagon	Increase	—	Rosselin et al. (1974)
	Glucagon	Decrease	—	Pointer et al. (1976)
	Vasopressin Phenylephrine	No effect	—	Whitton et al. (1977)
	Insulin inhibition of glucagon	Abolised	—	Claus and Pilkis (1976)
	Adrenaline (α)	Increase	—	Chan and Exton (1977)
Heart	Adrenaline (β)	No effect	—	Namm et al. (1968)
	Noradrenaline	Increase	—	Harary et al. (1976)
Adrenal	ACTH	Decrease	—	Pointer et al. (1976)
Vesicular secretory cells	Hormonal or electrical	No effect	—	See Chapter 7
Brain	K^+ or electrical	Decrease	—	Zanella and Rall (1973)
Nucleated erythrocytes	Adrenaline (β)	No effect	Decrease	Campbell and Siddle (1976)
Kidney	Parathyroid hormone	No effect	—	Nagata and Rasmussen (1968; 1976)
Salivary gland (insect)	—	Decrease	—	Berridge (1976b)
Barnacle muscle	Li^+ ($noNa^+$), tetanic stimulation	—	—	Baker and Carruthers (1980)

—, indicates not reported in the reference quoted.

ation of these experiments. Firstly, the experimental conditions used affect cell viability and this has seldom been taken into account (Siddle and Hales, 1980). Secondly, hormones like ACTH require extracellular Ca^{2+} for interaction of the receptor with adenylate cyclase. Thirdly, incubation with EGTA for many minutes or hours is often required to show effects. Fourthly, there have been very few attempts to correlate measurement of intracellular Ca^{2+} directly with cyclic nucleotide concentrations (Beam et al., 1977; Campbell and Dormer, 1978).

The situation becomes even more confused when one examines the reports of Ca^{2+} affecting the enzymes responsible for controlling cyclic nucleotide synthesis

Table 6.7. Ca^{2+} and some reported effects on tissue cyclic GMP

Tissue	Stimulus of tissue cyclic GMP	Effect of removing extracellular Ca^{2+}	Reference
Adipose tissue	Insulin, adrenaline, A23187	Inhibits	Fain et al. (1975)
Liver	Insulin, adrenaline, A23187	Inhibits	Fain et al. (1975); Pointer et al. (1976)
Kidney	Acetylcholine, histamine, bradykinin	Inhibits	De Rubertis and Craven (1976)
Platelets	Acetylcholine, A23187	Inhibits	Haslam and Say (1975)
Polymorphs	Acetylcholine, A23187	Inhibits	Ignarro and Colombo (1973)
Parotid	A23187	Inhibits	Wen et al. (1978)
Artery	Acetylcholine	Inhibits	Goldberg and Haddox (1977)
Brain	Noradrenaline	Inhibits	Goldberg and Haddox (1977)
Thyroid	Acetylcholine, A23187	Inhibits	Goldberg and Haddox (1977)
Ductus deferens	Acetylcholine, adrenergic agonists	Inhibits	Schultz et al. (1973, 1975); Schultz and Hardman (1975)
Barnacle muscle	Tetanic stimulation, insulin	—	Baker, and Carruthers (1980)
Pineal gland	Noradrenaline	Inhibits	O'Dea and Zatz (1976)
Pancreatic acini	Acetylcholine	Inhibits	Lopatin and Gardner (1978)

See Goldberg and Haddox (1977) for further examples and references.

and breakdown (Tables 6.8 and 6.9). Much of the early work showing that Ca^{2+} inhibited adenylate cyclase has no physiological significance. Concentrations of Ca^{2+} between 0.2 and 1 mM were usually required for half maximal inhibition, there being little detectable inhibition in the physiological range for intracellular free Ca^{2+}, i.e. $< 10\,\mu M$. Even in nucleated erythrocytes, where μM Ca^{2+} does appear to inhibit adenylate cyclase, no physiological inhibitor of cyclic AMP production in the cells has been discovered (Campbell and Siddle, 1978). These non-physiological effects of Ca^{2+} must be caused by CaATP or CaGTP inhibition, Ca^{2+}-phospholipid or Ca^{2+} binding to a low-affinity site on the enzyme, or GTP-binding protein. A similar problem exists with guanylate cyclase which can be activated by Mn^{2+} or Ca^{2+}, but not at concentrations found in the cell.

The discovery in 1967 by Cheung of a troponin C like protein, calmodulin, which activated phosphodiesterase in the presence of Ca^{2+}, provides a potential mechanism by which μM Ca^{2+} may reduce the concentration of cyclic AMP or cyclic GMP in the cell. Like all calmodulins the phosphodiesterase activator binds $4\,Ca^{2+}$ per molecule (see Chapter 3), each site with a K_d^{Ca} of $1-2\,\mu M$. Ca^{2+}

Table 6.8. Effects of Ca^{2+} on adenylate cyclase in broken cells

Tissue	Effect on enzyme	Reference
Non-physiological ($Ca^{2+} > 0.1$ mM required for significant effect)		
Brain	Inhibition	Bradham et al. (1970); Bradham (1972); Brostrom et al. (1978)
Pituitary	Inhibition	Deery (1974)
Liver	Basal inhibition	Rosselin et al. (1974)
	Glucagon-inhibits	Ray et al. (1970)
	Adrenaline activates	Ramwell and Shaw (1970)
Adipose tissue	Inhibition	Birnbaumer and Rodbell (1969)
Exocrine pancreas	Inhibition	Williams (1972, 1980)
Muscle: skeletal and heart	Inhibition	Severson et al. (1972)
Bladder	Inhibition	Hynie and Sharp (1971); Bockaert et al. (1972)
Adrenal	Basal inhibition	Taunton et al. (1969)
Glial tumour	Inhibition	Brostrom et al. (1976)
Trypanosomes	Inhibition	Voorheis and Martin (1981)
Possible physiological ($Ca^{2+} < 10$ μM effective)		
Adipose tissue	EGTA inhibits activation	Lefkowitz et al. (1970)
Liver	EGTA inhibits activation	Leoni et al. (1978)
Brain	EGTA inhibits, Ca^{2+} activates	Bradham et al. (1970); Bradham, (1972); Brostrom et al. (1978); Cheung et al. (1975)
Renal medulla	EGTA inhibits, Ca^{2+} activates basal and arginine-vasopressin	Campbell et al. (1972)
Chick embryo fibroblasts	Ca^{2+} activates	Watterson et al. (1976)
Nucleated erythrocytes (e.g. pigeon)	Inhibition	Hanski et al. (1977); Campbell and Dormer (1978)
Trypanosomes	Activation	Voorheis and Martin (1981)

Non-physiological Ca^{2+} range usually 1–10 mM. Animals include rat, pig, frog, dogfish, chicken.

increases the helical content of the calmodulin thereby enhancing the interaction with the phosphodiesterase. Calmodulin can also activate adenylate cyclase and guanylate cyclase (Kakiuchi et al., 1981) at μM Ca^{2+}, which explains the activating effects of Ca^{2+} observed in some experiments (Tables 6.6 and 6.8).

It can be concluded that increases in intracellular Ca^{2+} are often associated with an increase in cyclic GMP, and that disturbances in cell Ca^{2+} often alter the response of tissue cyclic AMP to hormones. However, the evidence that *physiological* cell stimuli modify cyclic nucleotide concentration in cells by increasing, or decreasing, intracellular free Ca^{2+} is poor.

In order to provide the evidence for this: (1) the physiological stimulus must change the cytosolic free Ca^{2+}; (2) the change in free Ca^{2+} must occur before the change in cyclic nucleotide concentration; (3) inhibition of the change in cytosolic

Table 6.9. Effects of Ca^{2+} on guanylate cyclase

Tissue	Effect of Ca^{2+} on enzyme	Reference
High Ca^{2+} effects (approx. mM)		
Heart	Soluble: stimulated Membrane: inhibited	Kimura and Murad (1975)
Lung	Soluble: stimulated Membrane: inhibited	Chrisman et al. (1975)
Muscle, brain, liver	Soluble and membrane: inhibition of Mn^{2+} stimulation	Sulakhe et al. (1976a, b)
Ductus deferens	Soluble: stimulated	Schultz et al. (1973)
Low Ca^{2+} effects (approx. 10 µM)		
Fibroblasts	Membrane: stimulated	Wallach and Pastan (1976)
Muscle, brain, liver	Soluble and membrane: stimulated	Sulakhe et al. (1976a, b)

free Ca^{2+}, e.g. with EGTA, should have the predicted effect on the cyclic nucleotide concentration; (4) the Ca^{2+}-sensitive enzyme or enzyme complex must be identified, i.e. cyclase or phosphodiesterase ($\pm$ calmodulin); (5) the thermodynamic constants for Ca^{2+} interacting with the enzyme in the broken cell preparation must correlate with the time scale and concentration range for free Ca^{2+} found in the living cell.

6.2.2. Regulation of intracellular Ca^{2+} by cyclic nucleotides

Regulation of intracellular Ca^{2+} by cyclic nucleotides provides a mechanism by which secondary regulators (see Chapter 2 for definition) can determine the threshold response of a cell, or modify the Ca^{2+}-dependent activation of the cell by prolonging or curtailing the response through an alteration in the Ca^{2+} transient. Cyclic nucleotides could alter intracellular Ca^{2+} by one of three mechanisms: (1) an effect on the passive permeability of the cell membrane or a specific ion channel to Ca^{2+}; (2) an effect on an intracellular store; (3) an effect on a Ca^{2+} pump which extrudes Ca^{2+} from the cell.

Apart from bacteria, which contain a cyclic AMP acceptor protein, and the chemotactic effect of cyclic AMP on slime moulds, all effects of cyclic nucleotides are mediated by protein phosphorylation (Krebs and Beavo, 1979; Nimmo and Cohen, 1977). Cyclic AMP protein kinase is a tetramer composed of two catalytic and two regulatory subunits (Fig. 6.5). An increase in cyclic AMP pulls the equilibrium to the right, thereby increasing the concentration of dissociated catalytic unit, the active form. Cyclic GMP protein kinase is one protein and is activated by binding cyclic GMP.

Many soluble and membrane-bound proteins have been shown *in vitro* and *in vivo* to be phosphorylated when there is an increase in cyclic nucleotide. For example, phosphorylation of phospholamban on the sarcoplasmic reticulum in

$$2 \text{ cyclic AMP} + C_2R_2 \rightleftharpoons 2C + 2R \text{ cyclic AMP}$$
$$\downarrow$$
$$\text{catalyses phosphorylation}$$

$$\text{cyclic GMP} + C-R \rightleftharpoons C-R-\text{cyclic GMP}$$
$$\downarrow$$
$$\text{catalyses phosphorylation}$$

Fig. 6.5. Mechanism of activation of cyclic nucleotide kinase. N.B.: C_4R_4 has also been reported. C–R for cyclic GMP seems to be one protein unit. Early reports of Ca^{2+} inhibition of protein kinase have been refuted because of the high, non-physiological Ca^{2+} concentrations used. The question now arises whether Ca^{2+}-calmodulin or Ca^{2+}-activated proteolysis could regulate these protein kinases *in vivo*

certain muscles activates the (Ca^{2+})-MgATPase (see Chapter 5). This increases the rate of removal of cytoplasmic Ca^{2+} during recovery of the cell from a stimulus, e.g. an action potential, and probably also increases Ca^{2+} in the store for a following stimulation. Reports that cyclic AMP may regulate directly Ca^{2+} release from mitochondria have since been refuted (Scarpa et al., 1976). Cyclic AMP may also regulate the Na^+-dependent efflux system which is found in many invertebrate muscles and nerves, and in some vertebrate cells (see Chapter 4).

To show that a cyclic nucleotide can modify intracellular Ca^{2+} physiologically several criteria must be satisfied. (1) The physiological stimulus must increase cell cyclic AMP or cyclic GMP. (2) The stimulus must change the concentration of cytosolic free Ca^{2+}, or modify the Ca^{2+} transient. (3) There must be a correlation between cyclic nucleotide, intracellular free Ca^{2+} and the physiological event. (4) The site of action, and phosphorylation by the cyclic nucleotide kinase, must be identified. (5) The effect of cyclic AMP dependent phosphorylation on a (Ca^{2+})-MgATPase, or uptake and release from an intracellular store, should be demonstrable in broken cell preparations.

6.2.3. Coregulation of enzymes by Ca^{2+} and cyclic nucleotides

Several enzymes can be regulated by both Ca^{2+} and cyclic nucleotides. For example, during muscle contraction activation of phosphorylase kinase by Ca^{2+} plays an important part in providing the additional energy needs of the muscle, whereas adrenaline activates the same enzyme during prolonged exercise through a cyclic AMP dependent phosphorylation. Coregulation of enzymes by Ca^{2+} and cyclic nucleotides therefore enables the enzyme to be activated or inhibited by two

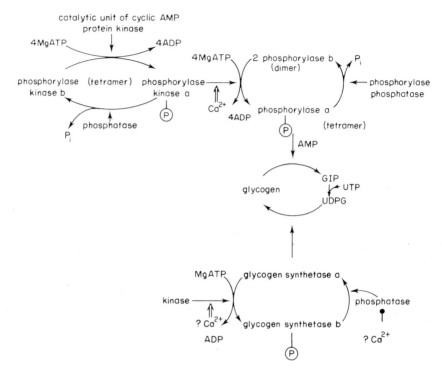

Fig. 6.6. Activation and inhibition of phosphorylase and glycogen synthetase. ⇒ = activation, —• = inhibition

or more independent cell regulators, or alternatively provides a mechanism by which a secondary regulator can modify the response of the enzyme to a primary cell regulator.

There are many types of covalent modification of proteins now known (Chock et al., 1980) including ADP ribosylation, guanylation and phosphorylation. It is in this latter mechanism that coregulation of enzymes by Ca^{2+} and another intracellular regulator was first established. The two best known examples are phosphorylase and glycogen synthetase. There are at least four possible ways in which Ca^{2+} and enzyme phosphorylation could interact to coregulate the same enzyme: (1) phosphorylation could affect directly the affinity of the enzyme, or a regulatory protein, for Ca^{2+}; (2) phosphorylation could affect indirectly the affinity of the enzyme, or a regulatory protein, for a Ca^{2+}-binding protein, thereby modifying protein–protein interactions; (3) Ca^{2+} bound to the enzyme, Ca^{2+} activation or Ca^{2+} inhibition of a kinase or phosphatase, could alter the rate of phosphorylation or dephosphorylation of the enzyme; (4) Ca^{2+} and phosphorylation could modify the activity of the enzyme independently.

Let us now examine some examples of intermediary metabolism where intracellular Ca^{2+} and cyclic nucleotides may mediate the effects of primary and secondary cell stimuli.

6.3. Glucose metabolism

Activated cells require more energy than those at rest (Table 6.10). For example, phagocytosing cells, or cells moving via chemotaxis, require energy and use glucose as the major fuel. Chemotaxis of polymorphs activates glycolysis, which can be inhibited by 2-deoxyglucose, an inhibitor of glucose utilization (Klebanoff and Clark, 1978). Similarly, glycogen degradation is increased after phagocytosis, and this seems to be independent of cyclic AMP. Likewise, transformation of lymphocytes (see Chapter 8) activates glycogen breakdown and glycolysis.

During the normal daily fasting–feeding cycle in man, glucose provides energy not only for blood cells, but also for the renal medulla and the brain (Campbell and Hales, 1976) at all times. However, in other tissues it is only in the fed state that glucose is the major source of energy. Glucose comes from the diet in the form of starch, sucrose, and cellulose in animals such as ruminants with bacteria able to degrade plant cell walls. Glucose can also be generated intracellularly by the breakdown of glycogen, or in the liver and kidney from lactate and amino acids by the gluconeogenic pathway (Fig. 6.7). The intertissue glucose pathways (Figs. 6.1 and 6.7) are under hormonal and neurotransmitter control. For example, glucagon, released by the pancreas during fasting, stimulates glycogen breakdown and gluconeogenesis in the liver in order to maintain the plasma glucose concentration. Adrenaline released from the adrenal medulla and noradrenaline released from peripheral nerve endings do likewise, as well as increasing glucose oxidation in peripheral tissues. Insulin, on the other hand, is released after a meal and increases the formation of glycogen and triglyceride from glucose.

The increased energy needs of an activated tissue, under the influence of hormones and neurotransmitters, during the fasting–feeding cycle can be met by a change in glucose metabolism mediated by cyclic nucleotides or intracellular Ca^{2+} in any one of four ways: (1) modification of glucose transport in and out of the cell; (2) a change from net glycogen synthesis to glycogen breakdown or vice versa; (3) modification of the rate of glucose to pyruvate (glycolysis) and the pentose phosphate pathway; (4) a change from glucose utilization to glucose synthesis (gluconeogenesis) or vice versa, in the liver and kidney.

Cations have long been known to influence the metabolism of glucose and the action of hormones such as insulin (Battacharya, 1961). The question now arises

Table 6.10. Examples of cell activation that require energy

Phenomenon	Example	Physiological stimulus
Exercise	Muscle contraction	Action potential
	Heart muscle modulation	Adrenaline
Chemotaxis	Polymorph movement	C3a, C5a
Phagocytosis	Macrophages, polymorphs	Bacteria
Secretion	Insulin from β cell	Glucose
	Neurotransmitter from nerve terminal	Action potential
Cell division	Lymphocyte transformation	Antigen

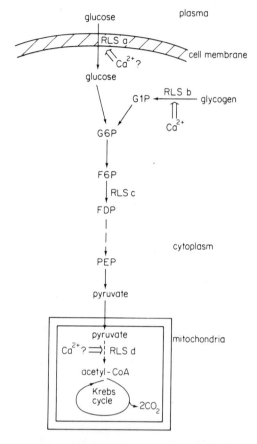

Fig. 6.7. Possible sites for regulation of glucose oxidation by Ca^{2+}. This diagram is illustrative of muscle and adipose tissue. RLS = rate-limiting step (N.B. a is not RLS in liver), ⇒ = activation. a = glucose transport; b = phosphorylase; c = phosphofructokinase; d = pyruvate dehydrogenase. G1P = glucose-1-phosphate; F6P = fructose-6-phosphate; FDP = fructose diphosphate; PEP = phosphoenolpyruvate; Pyr = pyruvate; AcCoA = acetyl coenzyme A

whether these manipulations of external ions can be explained through changes in intracellular Ca^{2+}.

6.3.1. Glucose transport across the cell membrane

Glucose and other sugars are transported in and out of cells by a special group of proteins found in the cell membrane. Two main groups have been identified

(Silverman, 1976). The first, exemplified by red cells, is a facilitated diffusion mechanism with relatively low selectivity for the sugar and is independent of Na^+. The second, found in the gut and intestine, is an active transport system working against a glucose concentration gradient, exhibiting high selectivity for certain sugars, and utilizing the Na^+ gradient to provide the energy for transport. The first mechanism is found in liver, muscle and adipose tissue. In liver, glucose transport is not a rate-limiting step in glucose metabolism, the intracellular concentration being close to that outside the cell. However, in muscle and adipose tissue the concentration of glucose inside the cell is some 10–100 times lower than that in the plasma. Insulin increases the rate of glucose uptake in these two tissues and thus increases glucose metabolism. The K_m of the uptake system for glucose in approx. 5–10 mM, plasma glucose being 4–6 mM. Insulin increases glucose transport by increasing the V_{max} within 2–3 min. The effect of insulin is inhibited by free fatty acids, and growth hormone plus cortisol, in the plasma.

Several pieces of evidence have been cited in support of the idea that an increase in cytosolic Ca^{2+} can increase glucose uptake into cells (Bihler, 1972; Clausen, 1980) (Table 6.11). Several of these experiments involved the transport of the non-metabolizable sugar 3-O-methylglucose rather than glucose itself. (1) Removal of extracellular Ca^{2+} inhibits insulin-stimulated glucose uptake in diaphragm, several other muscles and adipose tissue, but has little effect on basal glucose uptake. (2) There is a correlation between increased $^{45}Ca^{2+}$ efflux and the increase in sugar transport in contracting muscle stimulated by electrical impulses, caffeine and metabolic poisons. Similar observations have been made for insulin on sugar transport in adipocytes. (3) A23187 causes a Ca^{2+}-dependent increase in sugar transport in muscle and several other cells. (4) Several of the actions of insulin can be explained by an increase in cytoplasmic Ca^{2+} (Czech, 1977). Insulin stimulates muscle contraction, under hypertonic conditions. Insulin also stimulates the release of Ca^{2+} from biological and artificial membranes (Hauser and Dawson, 1968; Clausen, 1980).

Table 6.11. Cells where intracellular Ca^{2+} may regulate glucose transport

Cell type	Reference
Skeletal muscle (frog)	Holloszy and Narahara (1965, 1967)
Smooth muscle	Clausen et al. (1975)
Heart muscle	Schudt et al. (1976)
Invertebrate muscle (barnacle)	Baker and Carruthers (1980)
Adipocyte	Krahl, (1966); Bonne et al. (1978); Aktar and Perry (1979)
Avian erythrocyte	Carruthers and Simons (1978)
Thymocyte	Reeves (1977)
Lymphocyte	Reeves (1975)
Exocrine pancreas	Clausen (1980)

Evidence based on effect of insulin, A23187$\pm Ca^{2+}$ on glucose or 3-O-methylglucose transport (see Clausen, 1980), or correlation between $^{45}Ca^{2+}$ flux and sugar transport.

In spite of these experiments there is no direct evidence that a physiological stimulus, such as insulin or an action potential, increases glucose uptake into cells by a mechanism dependent on an increase in intracellular Ca^{2+}.

In barnacle muscle Baker and Carruthers (1980) found that a high insulin concentration, some million times that in human plasma, increased sugar transport some 2.5 fold and yet cytoplasmic free Ca^{2+} decreased. However, an external medium in which the Na^+ was replaced by Li^+ had no effect on sugar transport, in spite of the large increase in cytoplasmic free Ca^{2+} which occurred under these conditions. Furthermore, many of the experimental conditions used for other cells (Table 6.11) cause changes in intracellular ATP, H^+, Na^+, K^+, Mg^{2+} and metabolites as well as in Ca^{2+}.

To show that the effect of a physiological stimulus on sugar transport in cells is caused by an increase in intracellular Ca^{2+} several criteria must be satisfied: (1) an increase in intracellular free Ca^{2+} must be demonstrated before the increase in sugar transport occurs; (2) inhibition of the increase in cytoplasmic free Ca^{2+}, e.g. with intracellular EGTA, should inhibit the increase in sugar transport; (3) agents which increase intracellular free Ca^{2+}, e.g. A23187, should also increase sugar transport with the same time course and effect on K_m and V_{max} as the physiological stimulus; (4) the intracellular Ca^{2+}-binding site must be identified and characterized (e.g. calmodulin), and a molecular explanation of the effect of Ca^{2+} provided.

It is interesting to note that insulin also increases the Na^+-dependent active transport of certain amino acids into cells.

6.3.2. Glycogen breakdown and synthesis

Glycogen is a branched polymer of glucose and is found in most animal cells. Its breakdown to glucose 1-phosphate is catalysed by two enzymes. One, phosphorylase, acts on the $\alpha(1-4)$ links and the other, the debranching enzyme, acts on $\alpha(1-6)$ links (Fig. 6.8). Glycogen is synthesized by a reaction which is not

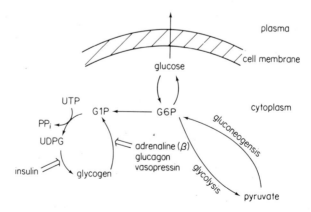

Fig. 6.8. Hormonal activation of hepatic glycogen synthesis and breakdown. ⇒ = activation

simply the reverse of degradation. Glucose in the form of UDPG is required, the reaction being catalysed by the enzyme glycogen synthetase (Fig. 6.8). Conditions which active glycogen breakdown often inhibit glycogen synthesis, and vice versa. There are two main mechanisms for the regulation of phosphorylase or glycogen synthetase. (1) Cofactor regulation: AMP for phosphorylase and glucose 6-phosphate for glycogen synthetase. (2) Regulation by phosphorylation of the enzyme (Fig. 6.6): the phosphorylated form of phosphorylase is about 100 times more active than the dephosphorylated form and has a higher affinity for AMP. In contrast the dephosphorylated form of the synthetase is the active form with a higher affinity for glucose 6-phosphate than the phosphorylated form.

The phosphorylation of a serine, 14 residues from the N-terminus of phosphorylase b (865 residues), converts the enzyme from a dimer to a tetramer. Mg ATP is the substrate and the reaction is catalysed by phosphorylase kinase (Fig. 6.6). This enzyme has four subunits $(\alpha, \beta, \gamma, \delta)$ with molecular weights of approx. 145 000, 138 000, 45 000 and 17 000, respectively. The complete kinase is $(\alpha\beta\gamma\delta)_4$ with a molecular weight of approx. 1.2×10^6. Phosphorylation of the β-subunit of the kinase activates the enzyme, and is catalysed by the catalytic subunit of cyclic AMP dependent protein kinase. This protein kinase also phosphorylates the α-subunit but more slowly. A similar situation of phosphorylation at more than one site occurs with glycogen synthetase. In this case, site I requires cyclic AMP protein kinase and inactivates the synthetase, whereas site II requires a separate, cyclic AMP independent, kinase. Interestingly, the phosphatase for this second site on glycogen synthetase is also the phosphatase for the β-subunit of phosphorylase kinase, which is phosphorylase phosphatase (Nimmo and Cohen, 1977; Krebs and Beavo, 1979).

In 1955 Fischer and Krebs found that mM concentrations of Ca^{2+} activated phosphorylase in muscle homogenates. However, this was subsequently shown to be the result of irreversible cleavage of a peptide from phosphorylase b kinase,

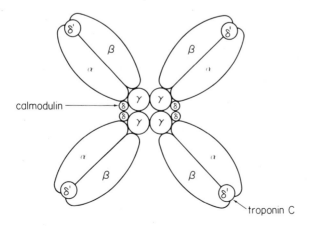

Fig. 6.9. Subunit structure of muscle phosphorylase kinase

caused by activation of a protease by Ca^{2+} (Meyer et al., 1964; Drummond and Duncan, 1966; Huston and Krebs, 1968). Glycogen synthetase is altered similarly by high Ca^{2+} (Belocapitow et al., 1965) but in this case the enzyme is inactivated. Fischer and Krebs also found that conversion of phosphorylase b to a was inhibited by EDTA. The first real clue that physiological concentrations of Ca^{2+} might activate phosphorylase was discovered when it was found that EGTA inhibited b to a conversion, and that this inhibition was alleviated specifically by Ca^{2+}.

It was subsequently shown that about 10–20 μM Ca^{2+} was required for maximum activation of phosphorylase kinase (Ozawa et al., 1967; Brostrom et al., 1971). In fact, this enzyme is virtually inactive without Ca^{2+}, as has been demonstrated by addition of EGTA. The explanation of the Ca^{2+} activation of phosphorylase kinase has been elegantly elucidated by Cohen and his colleagues (Cohen et al., 1978; Picton et al., 1981). The δ-subunit is a calmodulin, with only two amino acid differences from the brain Ca^{2+}-binding protein, calcineurin. The δ-subunit is very tightly bound to the enzyme complex, there being little inhibition of the Ca^{2+} activation of the enzyme by trifluoperazine, or by other phenothiazines. The enzyme is, however, inhibited by antibodies to calmodulin. A second Ca^{2+}-binding protein site, δ', has also been identified on the enzyme complex, between the α- and β-subunits (Fig. 6.9), as opposed to the γ-subunit for δ binding. In this case, addition of calmodulin to the isolated enzyme increases the activation by Ca^{2+}, and this effect of Ca^{2+} *can* be inhibited by trifluoperazine. Skeletal muscle troponin C, but not cardiac troponin C or parvalbumin, can also activate this δ' site on phosphorylase kinase. Although the affinity of the δ' site for troponin C is much less than for calmodulin ($K_{app}^{calmodulin} = 0.01$ μM, $K_{app}^{troponin\ C} = 1$ μM) it is the Ca^{2+}–troponin C activation of this site which may be physiologically important during muscle contraction.

The concentration of phosphorylase kinase in muscle is about 0.75 μM containing the equivalent of 3 μM calmodulin. This about one-third of the total calmodulin concentration in skeletal muscle, in contrast to the concentration of troponin C which is about 100 μM. This means that full activation of phosphorylase kinase would require 12 μM Ca^{2+} if all four sites on the calmodulin are occupied, plus a further 12 μM Ca^{2+} for the δ'-subunit site. These figures agree fairly well with early estimates of a Ca^{2+}-binding capacity of about 18 μM for the complete kinase without δ' (Cohen, 1974). Since up to 0.2 mM total Ca^{2+} may be released by the sarcoplasmic reticulum during a contraction (see Chapter 5), there would be sufficient Ca^{2+} to activate the kinase. It is difficult to be precise about the concentration of free Ca^{2+} necessary for activation *in vivo*, since extrapolation from data obtained with the purified enzyme will depend on the phosphorylation state, troponin C, free Mg^{2+} and pH *in vivo*.

Ca^{2+}–Calmodulin may also stimulate phosphorylation of glycogen synthetase kinase, thereby helping to inactivate the synthetase itself (Srivastava et al., 1979). How insulin stimulates the synthetase is still a puzzle.

What then is the evidence that glycogen breakdown *in the cell* is activated by intracellular Ca^{2+}? Five experimental approaches have been used (Table 6.12).

Table 6.12. Activation by Ca^{2+} of glycogen breakdown *in vivo*

Tissue	Stimulus	Reference
Skeletal muscle	Electrical	Namm et al. (1968); Friesen et al. (1969)
Heart muscle	Adrenaline, inotropic drugs	Namm et al. (1968)
Smooth muscle	Electrical and catecholamines	Namm (1971); Diamond (1973)
Liver	α-Adrenergic agonists	Assimacopoulos-Jeannet et al. (1977)
	Vasopressin	Stubbs et al. (1976); Whitton et al. (1977)
	Oxytocin	Keppens et al. (1977)
	Angiotensin	Blackmore et al. (1978)
Adipose tissue	Insulin stimulates synthesis Adrenaline stimulates breakdown	Hope-Gill et al. (1976)
Nervous tissue	Electrical	Lansdowne and Ritchie (1971)
Phagocytes	Particles	Klebanoff and Clark (1978)

(1) The increase in glycogen degradation cannot be explained by AMP or cyclic AMP. For example, electrical activation of skeletal and smooth muscle, and the action of α-adrenergic stimuli, vasopressin, oxytocin and angiotensin on the liver, cause no detectable increase in cyclic AMP. (2) Removal of extracellular Ca^{2+} in some cells inhibits the activation of glycogen degradation. Although this has no immediate effect in most skeletal muscles, in smooth muscle, nervous tissue and in the action of α-adrenergic agents, vasopressin and angiotensin on the liver, inhibition of phophorylase activation has been demonstrated. In the liver some 30–60 min are required in order to deplete intracellular Ca^{2+} stores and observe this inhibition. The effect of glucagon on the liver seems to be only partially dependent on Ca^{2+}. (3) A23187 activates glycogen breakdown, in a cyclic AMP independent manner. (4) Activation of glycogen breakdown occurs concomitantly with an increased release of Ca^{2+} from the cell, e.g. in the liver. (5) The site of the action of Ca^{2+} is known, i.e. the calmodulin subunit of phosphorylase *b* kinase.

Whilst this array of evidence sounds convincing, it is by no means definitive. Inhibitory effects of experimental procedures can often be accounted for either by ATP depletion or removal of Ca^{2+} already bound to the kinase in the resting cell. Effects of AMP are often not ruled out. Furthermore, in spite of the evidence for an increase in cytoplasmic Ca^{2+} in the electrical activation of skeletal muscle, its role in regulating glycogen metabolism *in vivo* in non-muscle cells has not been convincingly demonstrated.

Two key experiments are required: (1) direct evidence using a Ca^{2+} indicator that cytosolic Ca^{2+} increases before activation of phosphorylase; (2) inhibition of the increase in cytosolic Ca^{2+}, e.g. with intracellular EGTA, should prevent activation of phosphorylase.

A further problem is how insulin, which increases glycogen synthesis through a cyclic AMP independent activation of glycogen synthetase, works. It was reported some 15 years ago (Belocapitow et al., 1965) that Ca^{2+} affected glycogen synthetase. But the proposal that it increases intracellular Ca^{2+} (Czech, 1977; Clausen, 1980) is not consistent with its action to oppose the action of adrenaline or glucagon on glycogen breakdown.

Nevertheless, the evidence is good for cytosolic Ca^{2+} activating glycogen breakdown in contracting muscle. In resting muscle most of the phosphorylase is in the b form. Action potentials, generated at regular intervals by pulses of acetylcholine at the muscle end-plates, cause a prolonged rise in cytosolic Ca^{2+} in the range $1-5\mu M$. This Ca^{2+} stimulates both contraction through troponin C, and activates phosphorylase kinase mainly through the δ' site, probably troponin C. Activation of phosphorylase kinase leads to conversion of up to 70% of the phosphorylase to the a form within a few seconds. The consequential increase in glycogen breakdown provides some of the energy required for contraction, particularly if this is prolonged.

Initially most of the phosphorylase kinase is in the dephosphorylated form. During exercise adrenaline is released from the adrenal cortex into the blood stream. Adrenaline (β) increases cyclic AMP in muscle, leading to a phosphorylation of phosphorylase kinase and a protein inhibitor of phosphorylase phosphatase. The phosphorylated form of the kinase is much more active than the dephosphorylated form, but is hardly activated at all by troponin C. Under these conditions it is the δ-subunit, calmodulin, which is the dominant site for Ca^{2+} activation of phosphorylase kinase. Ca^{2+}–calmodulin also stimulates phosphorylation of glycogen synthetase (Srivastava et al., 1979).

6.3.3. Glycolysis and glucose oxidation

Glycolysis is the pathway which converts glucose to lactate:

$$\text{glucose} + 2\text{ ADP} + 2\text{P}_i \rightarrow 2\text{ lactic acid} + 2\text{ ATP}$$

This pathway provides the energy for tissues such as red and white blood cells and the renal medulla, as well as contributing significantly in exercising muscle. The lactate is released and converted back to glucose in the liver and kidney cortex. In aerobic cells, however, most of the pyruvate, the substance before lactate in glycolysis, is oxidized to CO_2 via the citric acid (Krebs) cycle. In these cells glycolysis and pyruvate oxidation provide most of the energy in the fed state. In the fasting state fatty acid oxidation becomes the major body fuel (Campbell and Hales, 1976), though the brain still oxidizes glucose. The opposition between glucose oxidation and fatty acid oxidation has been termed the glucose-fatty acid cycle (Randle et al., 1966; Hales, 1968). How is this cycle regulated?

The four main rate-limiting enzymes of glucose oxidation are hexokinase (glucokinase in the liver), phosphofructokinase, pyruvate kinase, and pyruvate dehydrogenase. An increase in flux through one reaction caused by an increase in substrate concentration can modify other reactions within the cell. For example, an increase in fatty acid oxidation leads to an increase in acetyl-CoA and citrate. Acetyl-CoA inhibits pyruvate dehydrogenase, whilst citrate inhibits phosphofructokinase. There is a further mode of regulation, namely the activation and inhibition of enzymes by hormones. Insulin increases glucose oxidation, not only by increasing the uptake of glucose by certain cells, but also by relieving the inhibition of phosphofructokinase and activating pyruvate dehydrogenase. In contrast, adrenaline can inhibit glucose oxidation when fatty acid is the major substrate, for example in liver in the fasting state. Some of these hormonal effects can be explained by changes in substrate supply and changes in the intracellular concentration of regulatory metabolites. However, this is not the complete answer. After activation in cells such as phagocytes, moving cells or secreting cells, a rapid activation of glycolysis is often observed which cannot be explained simply through changes in substrate or metabolite concentrations. Is intracellular Ca^{2+} the regulator in these cases?

It has been known for nearly 50 years that high concentrations of Ca^{2+} inhibit glycolysis in tissue slices and that mM Ca^{2+} inhibits glycolysis in tissue homogenates (Dickens and Greville, 1935; Geiger, 1940; McIlwain, 1952; Ashwell and Dische, 1950). This, together with the observation that mM Ca^{2+} inhibits several glycolytic enzymes (see Tables 3.14 and 6.1) such as phosphofructokinase and pyruvate kinase, lead to the suggestion that an increase in intracellular Ca^{2+} inhibited glycolysis *in situ* (Bygrave, 1966 a, b; 1967). Furthermore, it was suggested that in the liver the release of mitochondrial Ca^{2+}, caused for example by glucagon or noradrenaline, would activate pyruvate carboxylase in the mitochondria and inhibit pyruvate kinase in the cytoplasm (Gevers and Krebs, 1966). Unfortunately these proposals were made before it was generally realized that the cytoplasmic free Ca^{2+} never exceeded about 10 μM.

More recently it has been found that phosphofructokinase, pyruvate kinase and pyruvate dehydrogenase can be regulated by a phosphorylation–dephosphorylation mechanism (Krebs and Beavo, 1979). The phosphatase of pyruvate dehydrogenase phosphate is activated by Ca^{2+} in the range 0.01–10 μM, whereas the kinase is inhibited (Severson *et al.*, 1974, 1976). This is true for heart, kidney, adipose tissue and liver at saturating Mg^{2+}. At low Mg^{2+} the heart phosphatase is inhibited by Ca^{2+}.

Three enzymes make up the pyruvate dehydrogenase catalytic complex, found in the mitochondria:

(1) pyruvate decarboxylase

pyruvate + thiamine pyrophosphate (TPP) $\rightarrow CO_2$ + OH-ethyl-TPP

(2) dihydrolipoate acetyl transferase

OH-ethyl-TPP + lipoate $\rightarrow$ acetyl-hydrolipoate + TPP

(3) dihydrolipoate dehydrogenase

acetyl-hydro-lipoate + CoA $\rightarrow$ dihydrolipoate + acetyl-CoA

dihydrolipoate + NAD^+ $\rightarrow$ lipoate + NADH + H^+

The enzyme complex is regulated by several metabolites. For example, an increase in acetyl-CoA/CoA or NADH/NAD ratio inhibits the enzyme complex. In bacteria simple feedback inhibition is the major regulatory mechanism, whereas in mammalian cells control of phosphorylation plays a major role. Three serine residues on the α-subunit of the dehydrogenase can be phosphorylated by a Mg^{2+}-dependent kinase, utilizing $ATPMg^{2-}$ as the substrate. Like glycogen synthetase it is the phosphorylated from of pyruvate dehydrogenase which is inactive. Pyruvate dehydrogenase kinase is inhibited by several metabolites, including ADP (competitively with ATP), pyruvate (non-competitively with ATP), phosphate, pyrophosphate and thiamine pyrophosphate. A Mg^{2+}-dependent phosphatase activated by Ca^{2+} (apparent $K_m^{Ca} = 1$ μM; $K_m^{Mg} = 0.5$ mM) converts the enzyme to the active form. Insulin activates this dephosphorylation in adipose tissue, within 5–10 min. The effect of insulin is inhibited by ruthenium red, Ni^{2+} and Mn^{2+}. Activation of pyruvate dehydrogenase persists in mitochondria isolated from insulin-treated cells, and can be diminished in these isolated mitochondria by depleting them of Ca^{2+} with A23187 and EGTA. Furthermore, physiological concentrations of insulin (approx. nM) do seem to increase $^{45}Ca^{2+}$ uptake in adipocytes. Unfortunately attempts to demonstrate an effect of insulin on mitochondrial Ca^{2+} *in vivo* have failed (Denton, 1977; Denton *et al.*, 1980), although the inhibition by adrenaline of the insulin effect on the dehydrogenase may be due to a change in mitochondrial Ca^{2+}. A major stumbling block is the lack of knowledge of the free Ca^{2+} concentration in normal mitochondria *in situ*. It is interesting that α-oxoglutarate dehydrogenase in the citric acid cycle may also be regulated by Ca^{2+}, but not via a dephosphorylation mechanism of the sort found in pyruvate dehydrogenase.

The widespread occurrence of phosphorylation of regulatory enzymes, together with regulation of enzymes by calmodulin, provides a strong argument for re-examining the possible role for intracellular Ca^{2+} in the regulation of glycolysis (i.e. glucose → pyruvate) and pyruvate oxidation (i.e. pyruvate → CO_2). The following questions arise: (1) What is the primary physiological activator or inhibitor? Is there a threshold phenomenon either in the whole cell population or at the level of individual cells being switched on at different ratios of hormone agonists and antagonists? (2) Does a disturbance in cell Ca^{2+} mimic or modify the hormonal effect on the overall pathway and on the intermediates? Cross-over points have been used for many years to identify sites of regulation in cells. The theory is that activation of an enzyme causes pathway intermediates to decrease before the regulated step and to increase after it. A plot of percentage change in intermediate concentrations enables the regulated step to be identified. (3) Can an effect of μM Ca^{2+}, e.g. via calmodulin on a kinase, be demonstrated with the isolated enzyme? (4) Can a change in cytoplasmic Ca^{2+} be observed in cells exposed to the stimulus?

6.3.4. Gluconeogenesis

Gluconeogenesis is the synthesis of glucose from smaller precursor molecules such as lactate, amino acids and glycerol (Fig. 6.10). Only the liver and the kidney, in chordates, have the necessary enzymes to circumvent the non-equilibrium steps in glycolysis. The rate-limiting steps in gluconeogenesis are transport of

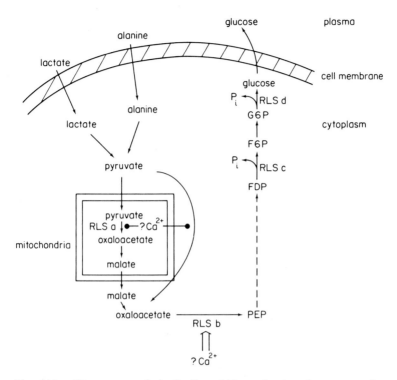

Fig. 6.10. Gluconeogenesis in the liver; kidney also has the enzymes for gluconeogenesis. RLS = rate-limiting step, ⇒ = activation, —● = inhibition. (N.B.: reactions are not complete, cofactors have been left out). a = pyruvate carboxylase; b = phosphoenolpyruvate carboxykinase; c = fructose diphosphatase; d = glucose-6-phosphatase. The pathway is activated by adrenaline (α), glucagon and vasopressin, and inhibited by insulin

metabolites across the mitochondrial membrane, pyruvate carboxylase, phosphoenolpyruvate carboxykinase, fructose diphosphatase and glucose-6-phosphatase. Tissues other than the liver and kidney contain some of these enzymes, but *not* glucose-6-phosphatase. For example, muscle contains fructose diphosphatase. The substrate cycle formed between this enzyme and phosphofructokinase provides greater flexibility of metabolic control (Newsholme and Start, 1973).

In the fasting state a rise in plasma free fatty acid and glucagon, together with a fall in insulin, causes the metabolism of the liver and kidney cortex to switch from glycolysis to gluconeogenesis. In liver the main substrates are lactate, pyruvate and alanine, whereas in the kidney they are lactate, pyruvate, glutamine and various citric acid cycle intermediates. An elevation of plasma adrenaline, for example in exercise and stress, together with noradrenaline released from nerve terminals also activates gluconeogenesis. Other hormones which activate hepatic gluconeogenesis *in vitro* at concentrations found in plasma include vasopressin, oxytocin and angiotensin II, acting over the range nM–µM. Similarly, parathyroid hormone is a potent activator of renal gluconeogenesis *in vitro*. The use of substrates which feed into the gluconeogenic pathway at different points, together with the measurement of the intermediates of the pathway to identify cross-over points, have shown that the steps between pyruvate and phosphoenolpyruvate (Fig. 6.10) contain the site of hormonal regulation. Some evidence also exists for control at the fructose diphosphatase (FDPase)–phosphofructokinase step (Kneer *et al.*, 1974).

What is the intracellular mediator of these hormonal effects?

Glucagon and adrenaline elevate liver cyclic AMP, and parathyroid hormone elevates kidney cyclic AMP. Furthermore, cyclic AMP added to the external medium can activate gluconeogenesis from lactate (Friedmann and Rasmussen, 1970). However, cyclic AMP is not thought to be the intracellular regulator which directly alters the activity of the rate-limiting gluconeogenic enzymes. The evidence for this is as follows: (1) There is a poor correlation between tissue cyclic AMP, activation of cyclic AMP dependent protein kinase, and gluconeogenesis, apart from animals such as dog where the effect of adrenaline on gluconeogenesis seems to be a β-receptor activation of adenylate cyclase. (2) Isoproterenol (isoprenaline), a potent β-adrenergic agonist, increases cyclic AMP in liver but not gluconeogenesis in isolated hepatocytes. (3) Propranolol, a β-adrenergic receptor blocker, blocks the effect of adrenaline on cyclic AMP but not its effect on gluconeogenesis. (4) Phentolamine, an α-adrenergic receptor blocker, does block the effect of adrenaline on gluconeogenesis but does not block its effect on cyclic AMP. (5) Vasopressin, angiotensin II and oxytocin do not raise liver cyclic AMP yet they do activate gluconeogenesis. (6) The effect of insulin to inhibit basal, glucagon- and adrenaline-stimulated gluconeogenesis in liver, cannot be explained through cyclic AMP. Insulin does not inhibit vasopressin- or angiotensin-induced gluconeogenesis. (7) Experimental manipulation of intracellular Ca^{2+} can affect the rate of gluconeogenesis without changing cellular cyclic AMP. (8) Tetracaine (see Chapter 9) inhibits glucagon-induced gluconeogenesis but not the rise in cyclic AMP.

What then is the intracellular second messenger for hormonal activation of hepatic or renal gluconeogenesis? If it is not cyclic AMP is it cyclic GMP or is it Ca^{2+}? Whilst a role for cyclic GMP cannot be ruled out (Goldberg and Haddox, 1977), since α-adrenergic responses often lead to increases in cyclic GMP, most of the evidence points towards intracellular Ca^{2+}: (1) Removal of extracellular Ca^{2+} inhibits both basal and parathyroid hormone elevated renal gluconeogenesis (Krebs *et al.*, 1963; Nagata and Rasmussen, 1970; Kurokawa and Rasmussen, 1973; Kurokawa *et al.*, 1973). Renal gluconeogenesis *in vitro* is also sensitive to changes in pH and HPO_4^{2-}. Hepatic glucose synthesis, on the other hand, is much less sensitive to removal of extracellular Ca^{2+}. Liver cells or perfused liver require an incubation without Ca^{2+} or with EGTA of 30–60 min in order to inhibit hormonal stimulation of gluconeogenesis (Friedmann and Park, 1968; Friedmann and Rasmussen, 1970; Tolbert and Fain, 1974; Fain *et al.*, 1975). In contrast, an increase in extracellular Ca^{2+} increases gluconeogenesis. (2) An increase in $^{45}Ca^{2+}$ efflux is provoked by glucagon, and also exogenous cyclic AMP, before the rise in glucose output is detectable (Friedmann, 1974). The other hepatic gluconeogenic hormones also increase $^{45}Ca^{2+}$ efflux. Parathyroid hormone increases $^{45}Ca^{2+}$ influx and efflux in the renal cortex. (3) A23187 activates renal gluconeogenesis (Mennes et al., 1978) but appears to inhibit it in the liver (Fain et al., 1975). The problems of interpreting data from A23187 experiments have been discussed in Chapter 2. (4) Tetracaine, a local anaesthetic, and the Ca^{2+}-channel blocker D600 inhibit glucagon- and cyclic AMP induced gluconeogenesis, and elevated $^{45}Ca^{2+}$ efflux in the liver.

These experimental data have lead to the hypothesis that hormonal activation of gluconeogenesis is mediated by an increase in intracellular free Ca^{2+}. In liver this Ca^{2+} comes from an intracellular store, whereas in kidney cortex it comes from both outside the cell and an internal store. What could this store be? An endoplasmic reticulum and mitochondrial store of Ca^{2+} have been identified in liver (see Chapter 4). The mitochondrial store seems to be sensitive to hormones (Babcock et al., 1979). It has been suggested that glucagon and parathyroid hormone might induce release of mitochondrial Ca^{2+} through a cyclic AMP mediated mechanism. The early reports by Borle (1968, 1972, 1973, 1974) of such an effect of cyclic AMP on isolated mitochondria have been refuted. However, more recently, this has been re-examined and still remains a possibility (Juzu and Holdsworth, 1980).

We have already seen that although exogenous cyclic AMP can stimulate ^{45}Ca efflux, it is not necessary for physiological activation of gluconeogenesis. Some other factor must be required. Could it be cyclic GMP, or could it be a derivative of arachidonic acid (Barritt, 1981) (see this charter, sections 6.4.3 and 6.4.4)?

The site of action of the intracellular Ca^{2+} is somewhere between pyruvate and phosphoenolpyruvate. Ca^{2+}, albeit at the relatively high concentration of 50–100 μM, activates the transport of substances such as oxaloacetate and α-oxoglutarate across the membrane of isolated mitochondria (Robinson and Chappell, 1967; Haslam and Griffiths, 1968; Harris and Berent, 1969; Mason, 1974). Ca^{2+}, in the physiological range (0.1–10 μM), does not affect isolated phosphoenolpyruvate carboxykinase (Roobol and Alleyne, 1973). Suggestions that the increase in cytoplasmic Ca^{2+} might switch off adenylate cyclase and pyruvate kinase seem unlikely in view of the mM Ca^{2+} required in vitro. However, because of the discovery of calmodulin and enzyme phosphorylation it would seem worthwhile to re-examine the apparently non-physiological effects of Ca^{2+} on isolated enzymes of gluconeogenesis (Bygrave, 1967; Wimhurst and Manchester, 1970). α-Adrenergic and glucagon stimulation of liver metabolism also causes phosphorylation of at least 12 cytoplasmic proteins by a mechanism independent of cyclic AMP. A23187 increases phosphate incorporation into them (Garrison, 1978).

In summary, there is much circumstantial evidence that the hormonal regulation of glycogenolysis, gluconeogenesis and pyruvate oxidation requires changes in intracellular free Ca^{2+} (Figs. 6.6, 6.7, 6.10). Yet many problems and questions still remain.

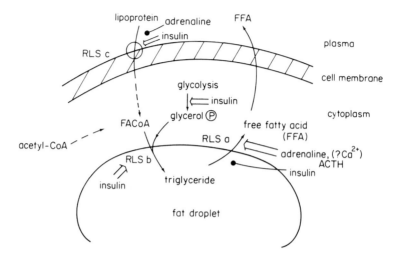

Fig. 6.11. Lipolysis and lipogenesis in adipose tissue. ⇒ = activation, RLS = rate-limiting step; a = triglyceride lipase; b = triglyceride synthetase; c = lipoprotein lipase (on capillary wall and cell surface)

(1) No direct measurement of intracellular free Ca^{2+} has been carried out in parallel with hormonal effects on intermediary metabolism.

(2) Experimental conditions may expose Ca^{2+} sites which belong to one of class I–III Ca^{2+} functions not directly related to intracellular Ca^{2+} as a regulator (Chapter 1); for example, parathyroid hormone activated and basal gluconeogenesis in kidney require extracellular Ca^{2+}.

(3) The molecular basis of many of the proposed effects of intracellular Ca^{2+} is poorly understood, in particular what is the role of protein phosphorylation?

(4) What is the intracellular source of Ca^{2+} for certain hormones?

(5) What is the role of the change in membrane potential induced by hormones such as glucagon, adrenaline and insulin (Horwitz et al., 1969; Zierler, 1972; Dambach and Friedmann, 1974)?

(6) What is the role of other ions such as H^+, K^+, Na^+, Mg^{2+} whose fluxes across the cell membrane are often altered by hormones?

(7) What is the mechanism of action of insulin (Czech, 1977)? Insulin does lower cyclic AMP but neither this nor cyclic GMP can explain its actions, nor does insulin usually require extracellular Ca^{2+} for its acute effects (Claus and Pilkis, 1976). Yet insulin inhibits gluconeogenesis and inhibits ^{45}Ca release (Foden and Randle, 1978).

(8) Several of the proposed mechanisms of cell activation require a hitherto unidentified 'second messenger' which releases Ca^{2+} into the cytoplasm. What is this 'second messenger'?

(9) Is it possible to rationalize the roles of cyclic AMP, cyclic GMP and intracellular Ca^{2+}, perhaps through the different time scales over which they act?

(10) What is the heterogeneity of cell response in a tissue like the liver? Do different cells exhibit a threshold to different concentrations of hormone?

(11) Hormones and transmitters have been found in plants and in invertebrates which regulate glucose and intermediary metabolism. Are these effects mediated by intracellular Ca^{2+}?

6.4. Lipid metabolism

The extraction of pro- and eukaryotic organisms in organic solvents such as ether and chloroform solubilizes a fraction of the cells known as 'lipid'. Two of its major components are triglyceride and phospholipid, both esters of glycerol (Figs. 6.12 and 6.13). The lipid fraction also contains free fatty acid, steroids, prostaglandins (Fig. 6.17) and fat-soluble vitamins. The rates of synthesis and breakdown of many of these lipids change following cell activation. For example, triglyceride breakdown can provide the necessary additional energy for activated aerobic cells, whilst additional phospholipids are required to form membranes of dividing cells. The metabolism of lipids is controlled by substrate supply, the concentration of metabolites such as ATP and CTP, and the primary and secondary regulators of cells (Tables 2.2, 6.2 and 6.3). These regulators include chemical, mechanical and electrical stimuli.

Do changes in intracellular Ca^{2+} mediate any of the modifications in lipid metabolism caused by primary or secondary regulators? Do changes in lipid metabolism alter intracellular Ca^{2+} or its action?

6.4.1. Lipolysis

The major fuel stored in most animals is triglyceride. Although it is found in all aerobic cells its is the specialized cells in adipose tissue which contain most of the body fat stores. Two morphologically distinct types of adipose tissue exist— white and brown. The cells of white adipose tissue contain one main fat droplet (Fig 6.14) and produce free fatty acid as the energy source for many aerobic tissues in the fasting phase of the daily cycle. The cells of brown adipose tissue, on the other hand, contain several large fat droplets and many mitochondria. These cells provide heat, particularly important in naked newborn animals. In both types of adipose tissue lipolysis the hydrolysis of triglyceride, eventually to free fatty acid and glycerol, is regulated by hormones and neurotransmitters. Brown adipose tissue lipolysis is activated by noradrenaline. White adipose tissue, *in vitro*, can be stimulated by a bewildering range of hormones, few of which act at concentrations normally found in the blood (Hales *et al.*, 1979). The major physiological activators of white adipose tissue lipolysis, in mammals, are adrenaline, released from the adrenal medulla into the blood, and noradrenaline from nerve terminals. Their effects on lipolysis are inhibited by insulin, a hormone which also reduces basal lipolysis in isolated adipocytes. In birds glucagon is the main lipolytic stimulus to adipose tissue. Avain adipose tissue, however, is insensitive to insulin. Other possible physiological regulators of

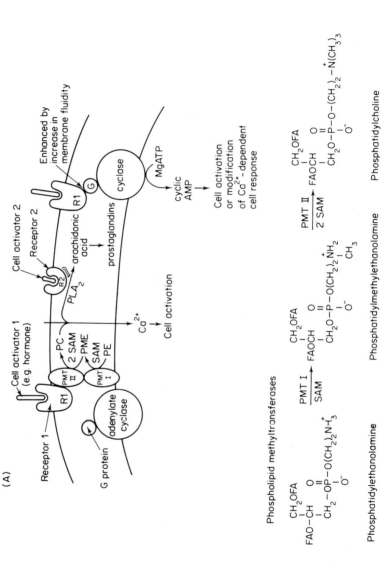

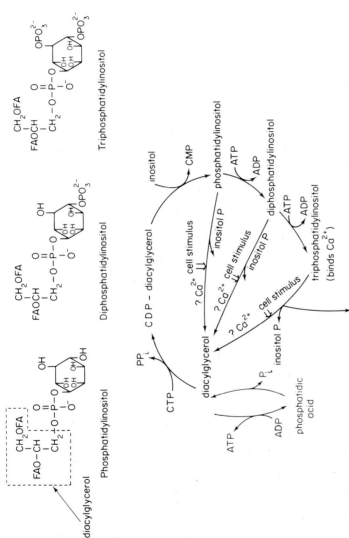

Fig. 6.12. Phospholipid metabolism. (A) Possible consequences of phospholipid methylation. Two types of receptor are postulated: R1 activates phospholipid methyltransferases 1 and II (PMT I and II); R2 activates phospholipase A_2 (PLA$_2$). SAM = S-adenosylmethionine. From Hirata and Axelrod (1980). (B) Phosphatidyl inositol metabolism. $\Rightarrow$ = activation, FA = fatty acid

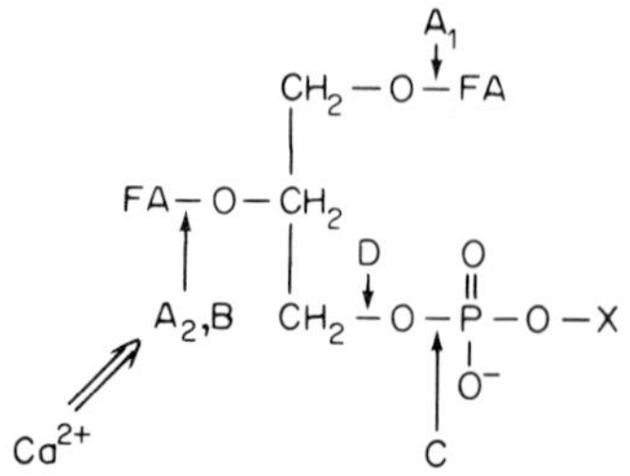

$$\begin{array}{c}
\overset{A_1}{\downarrow}\\
CH_2-O-FA\\
FA-O-CH_2\\
| \quad \overset{D}{\downarrow}\\
\overset{\nearrow}{A_2,B}CH_2-O-\overset{O}{\overset{\|}{P}}-O-X\\
\phantom{Ca^{2+}aaaaaaaa}\overset{\uparrow}{|}\\
Ca^{2+}O^-\\
C
\end{array}$$

Fig. 6.13. Phospholipases A_1, A_2, B, C, and D are illustrated. A_2 is active in eukaryotic cell membranes and is activated by Ca^{2+}. The product of the reaction catalysed by phospholipase A_2, lysolecithin, can be hydrolysed to release its second fatty acid by the enzyme lysophosphatidase

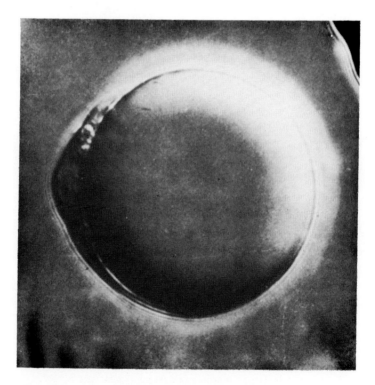

Fig. 6.14. A mammalian adipocyte. Note small ring of cytoplasm surrounding the large fat droplet, which can be 100–200 μm in diameter. From Hales *et al.* (1979); reproduced by permission of The Biochemical Society

adipose tissue lipolysis *in vivo* are growth hormone, glucocorticoids as stimulators, and prostaglandins E_1 and E_2 as antilipolytic agents.

The physiological activation of white adipose tissue lipolysis is mediated through a cyclic AMP dependent activation of protein kinase (Fig. 6.5). This enzyme catalyses the phosphorylation of the triglyceride lipases, thereby activating the group of enzymes responsible for triglyceride hydrolysis. Although the activation of lipolysis, for example by adrenaline acting on rat epididymal fat, satisfies Sutherland's postulates for cyclic AMP, this is not the complete story (Siddle and Hales, 1974a). Firstly, the time course of lipolysis *in vitro* does not correlate completely with that of cyclic AMP (Fig. 6.15). Secondly, activation of triglyceride lipase, in broken cells, by cyclic AMP and protein kinase is about an order of magnitude lower than it has to be in the cell. Thirdly, the inhibition of lipolysis by insulin does not correlate with the ability of this hormone to inhibit cyclic AMP accumulation (Fig. 6.15).

As a result of these discrepancies many workers have attempted to implicate an additional intracellular regulator in the activation or inhibition of lipolysis—either Ca^{2+} or cyclic GMP. No reproducible effects of EGTA or A23187 on lipolysis have been obtained (Schimmel, 1973, 1976; Siddle and Hales, 1980). Effects of homones on Ca^{2+} flux have been reported (Alm *et al.*, 1970) but the

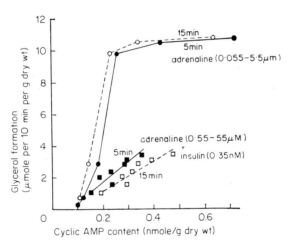

Fig. 6.15. Non-correlation of adipocyte cyclic AMP with insulin inhibition of lipolysis. Rat adipocytes were incubated with adrenaline (0.055–5.5 µM), (●, o) or with adrenaline (0.55–55 µM) + insulin (0.35 nM) (■, □), to produce a range of lipolytic rates. Lipolysis was measured by the rate of glycerol release. The cyclic AMP content of the adpocytes was measured at 5 min (●, ■) and 15 min (o, □). Isolated rat adipocytes 23 mg dry wt/ml, 37°C. Cyclic AMP assayed by radioimmunoassay, glycerol by change in NADH absorbance. From Siddle and Hales (1974a); reproduced by permission of the Biochemical Society

significance of these observations remains to be established. The possible role of intracellular Ca^{2+} has been confused by the reports of inhibition of adenylate cyclase and protein kinase (Birnbaumer and Rodbell, 1969; Weller and Rodnight, 1974) (see Tables 3.14 and 6.8) and activation of triglyceride lipase (Rizack, 1964) which were carried out at non-physiological Ca^{2+} concentrations, in the mM range. However, local anaesthetics which displace membrane Ca^{2+} can inhibit adrenaline-stimulated lipolysis (Kissebah et al., 1974a, b, 1975; Siddle and Hales, 1974b), and a vesicular store of intracellular Ca^{2+} has been found in adipocytes (Fig. 6.16). Ca^{2+} may modulate insulin and oxytocin action in adipocytes (Bonne et al., 1977). Furthermore, preincubation of these cells with physiological concentrations of insulin (pM–nM) increases Ca^{2+} bound to membranes isolated from these cells (McDonald et al., 1976, 1978) and can decrease the activity of the Ca^{2+}-activated MgATPase when added to isolated plasma membranes (Pershadsingh and McDonald, 1979). In order to establish that intracellular Ca^{2+} is involved in the physiological regulation of lipolysis, the answers to two key questions are required: (1) Can a change in cytosolic free Ca^{2+} be demonstrated, in the intact cells, which correlates with a change in the rate of lipolysis? (2) Can an effect of Ca^{2+} in the range $0.1–10\mu M$ be demonstrated on the lipase in broken cells, for example through calmodulin, a kinase or a phosphatase. These questions must be answered not only for adipose tissue but in other cells activated via a Ca^{2+}-dependent mechanism under conditions where fatty acid oxidation is a major contributor to the cell's energy supply.

Fatty acids can be converted to ketone bodies, acetoacetate and β-hydroxybutyrate, in the liver. Ketone bodies are an important fuel for the brain during prolonged starvation. Their production is primarily regulated by the supply of free fatty acid to the liver. As yet no convincing evidence for a role for intracellular Ca^{2+} in the control of ketone body formation has been found, though effects of external Ca^{2+} on liver slices have been reported (Mellanby and Williamson, 1963).

6.4.2. Lipogenesis

The synthesis of triglyceride requires fatty acyl-CoA, formed from acetyl-CoA, and glycerol phosphate (Fig. 6.11), formed via glycolysis or by a glycerokinase, active in liver but not very active in adipose tissue. Insulin stimulates lipogenesis acutely, mainly by providing more glycerol phosphate through glycolysis, and more acetyl-CoA through its activation of pyruvate dehydrogenase. The problems in assessing the role of Ca^{2+} in these two processes have already been discussed. The possible role of Ca^{2+} in regulating triglyceride synthetase or lipoprotein lipase, the enzyme which allows tissues to take up triglyceride, might well repay examination.

6.4.3. Phospholipid metabolism

Phospholipids are the major lipid component of all biological membranes and some bind Ca^{2+} (Table 3.5). They are grouped in classes, depending not only on

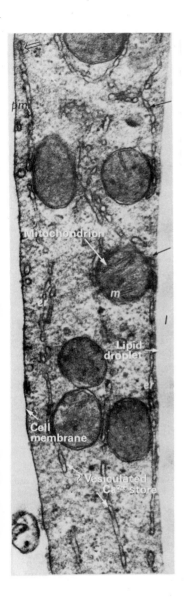

Fig. 6.16. A possible vesiculated store of Ca^{2+} in the adipocyte. (left) Electron micrograph of rat adipocyte cytoplasm, approx. × 32 000. (below) Electron micrograph of rat adipocyte cytoplasm after pyroantimonate staining, approx. × 102 000. From Hales *et al.* (1974), reproduced by permission of the *Journal of Cell Science*. pm = plasma membrane, er = endoplasmic reticulum, m = mitochondrion l = lipid

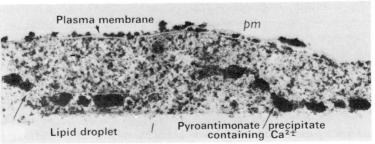

the base attached to the phosphate moeity (Fig. 6.12) but also on the fatty acid moeity. The major phospholipid in cells is phosphatidylcholine, lecithin, and has a low affinity for Ca^{2+}. In many cells it is not uniformly distributed across the bilayer, the outer layer having a higher concentration than that facing the cytoplasm. In contrast, phosphatidylserine, which does bind Ca^{2+}, may be higher on the inner surface, and phosphatidylethanolamine is found in highest concentration in the surface facing the external medium.

Regulation of phospholipid metabolism is required physiologically to control and modify membrane turnover, to alter the ionic permeability of the membrane and to regulate proteins within the membrane. These alterations are provoked by receptors on the cell surface and electrical events in the membrane. Three particular reactions may be important for intracellular Ca^{2+}:(1) methylation of phosphatidylethanolamine may alter the permeability of the plasma membrane to Ca^{2+} (Hirata and Axelrod, 1980); (2) breakdown of phosphatidylinositol or triphosphoinositide may release Ca^{2+} into the cytoplasm or open Ca^{2+} channels in the membrane (Jones and Michell, 1978); (3) arachidonic acid production may be stimulated by a Ca^{2+}-dependent activation of phospholipase A_2 (Barritt, 1981).

Methylation of phosphatidylethanolamine to phosphatidylcholine is catalysed by two methyltransferases, S-adenosylmethionine being the methyl donor (Fig. 6.12). The enzymes are asymmetrically distributed in the membrane. Methyltransferase I is on the inner, cytoplasmic surface, whereas as methyltransferase II is in contact with the outer surface of the cell. This means that during methylation there is a translocation of phospholipid from the inner to the outer surface. Methylation of phosphatidylethanolamine is stimulated by many primary cell stimuli (Table 6.13), for example β-adrenergic agonists, occupation of IgE receptors on mast cells, the action of concanavalin A on lymphocytes, and chemotactic peptides on phagocytic cells. By using S-[^{3}H]adenosylmethionine the transient effect of these stimuli on phospholipid methylation can be detected within 30 sec.

There are four possible consequences of methylation of phospholipids. Their relative importance has yet to be fully established and will depend on the cell concerned and type of response: (1) An increase in mobility of hormone and neurotransmitter receptors, leading to interaction with proteins such as GTP-protein binding adenylate cyclase on the inner surface of the plasma membrane. (2) Alteration of receptor number on the outer surface of the cell. (3) An increase in Ca^{2+} permeability of the plasma membrane. This may lead to activation of phospholipase A_2, a membrane-associated lipase. The juxtaposition of phosphatidylcholine and this enzyme may lead to the production of arachidonic acid. (4) An increase in prostaglandin synthesis as a result of production of arachidonic acid.

The compound 5'-deoxyisobutylthio-3-deaza adenosine (3-deaza-SIBA) inhibits phospholipid methylation (Hirata and Axelrod 1980). It also inhibits $^{45}Ca^{2+}$ influx and histamine secretion in mast cells. Phospholipid methylation may also regulate membrane-bound enzymes such as the Ca^{2+}-activated MgATPase in the cell membrane and endoplasmic reticulum. Attractive as this

Table 6.13. Activation of phospholipid metabolism in cells

Cell type	Stimulus	Cellular effect
(A) *Stimulation of phospholipid methylation*		
Mast cells	IgE receptor occupancy	Histamine release
Phagocytes	Chemotactic stimulus	Chemotaxis
Fibroblasts	Bradykinin	Adenylate cyclase activation
Reticulocytes	β-Adrenaline	Adenylate cyclase activation
Lymphocytes	Concanavalin A	Cell division
(B) *Stimulation of phosphatidylinositol turnover*		
Smooth muscle	α-Adrenaline, 5-hydroxytraptamine, (muscarinic) acetylcholine, histamine	Contraction
Liver	Angiotensin, vasopressin, α-adrenaline	Glycogenolysis
Pineal gland	Noradrenaline	Secretion
β-Cell of pancreatic islets	Glucose	Insulin secretion
Submaxillary gland	Electrical	Saliva
Insect flight muscle	Glutamate	Contraction
Insect salivary gland	5-Hydroxytryptamine	Saliva

For references see (A) Hirata and Axelrod (1980); (B) Michell *et al.* (1977), Michell (1975), Jones and Michell (1978), Billah and Michell (1979), Fain and Berridge (1979), Garcia-Sainz and Fain (1980).

hypothesis may seem, the methylation of phospholipids is not always associated with cell activation, even when this is mediated by a change in intracellular Ca^{2+}. For example, thrombin and prostaglandins both stimulate platelet aggregation without increasing phospholipid methylation. The general importance of this phenomenon in cell activation therefore remains to be established.

Whilst occupation of the β-adrenergic receptor can activate phospholipid methylation, the α-adrenergic receptor leads to an increase in the turnover of another phospholipid—phosphatidylinositol (Fig. 6.12) (Michell, 1975; Michell *et al.*, 1977; Jones and Michell, 1978). There are, in fact, three phospholipids involved—phosphatidylinositol, diphosphatidylinositol and phosphatidylinositol-4, 5-biphosphate (triphosphatidylinositol). They are all derivatives of phosphatidic acid (Fig. 6.12). A cycle exists in the membrane of cells involving the degradation and resynthesis of these phospholipids. Activation of the enzymes catalysing synthesis or degradation could increase or decrease the incorporation of radioactive $^{32}P_i$ or [^{3}H]inositol into the phosphatidylinositol fraction, depending on the experimental conditions. This is what happens, for example, in the α-adrenergic activation of smooth muscle and liver. The increase in breakdown of inositides in stimulated cells is usually studied after preloading them with $^{32}P_i$ or [^{3}H]inositol. The latter can be a problem in cells that are poorly permeable to inositol.

Turnover of phosphatidylinositol is increased in many cells by electrical, neurotransmitter or hormonal stimulation (Table 6.13). Most of these phenomena associated with this turnover are thought to be activated ultimately through an increase in intracellular free Ca^{2+}. On the other hand, β-adrenergic activation of the parotid, the actions of glucagon on the liver, PGE_1 on many cells including platelets, ACTH on the adrenal cortex, secretin on the pancreas and follicle-stimulating hormone on the ovary, are primarily dependent on cyclic AMP and do not increase phosphatidylinositol turnover. The close correlation between Ca^{2+} and the increase in turnover of this group of phospholipids, which when it occurs can be detected within 30–60 sec of cell activation, has lead to the idea that it might be the event in the membrane which leads to the change in cytosolic Ca^{2+}.

Four possibilities exist which require investigation. (1) Phosphatidylinositol and phosphatidylinositol-4, 5-biphosphate are negatively charged and bind Ca^{2+} more avidly than most other phospholipids (Dawson and Hauser, 1970). Their degradation would be expected to release any bound Ca^{2+} into the cytosol. (2) The release of Ca^{2+} bound to phosphatidylinositol could be the signal required to open a membrane Ca^{2+} channel, thereby allowing Ca^{2+} into the cell. This is supported by the observation that reduction of cytosolic Ca^{2+} to < 50 nM using EGTA or citrate inside giant cells leads to hyperexcitability and the opening of voltage-dependent Ca^{2+} channels (see Chapter 4). (3) The fatty acid moiety of phosphatidylinositol in several of the activated tissues (Table 6.13) is rich in arachidonic acid. The production of this fatty acid can lead to increased production of prostaglandin (Fig. 6.17) which in turn might release Ca^{2+} from stores in the cell or increase Ca^{2+} permeability across the cell membrane. (4) Several of the products of phospholipid degradation act as a Ca^{2+} ionophore, thereby bringing about an increase in cytosolic Ca^{2+} (Michell, 1975; Kirtland and Baum, 1972; Murphy et al., 1979; Barritt, 1981). These include *myo*-inositol 1,2-cyclic phosphate, phosphatidic acid, 1,2-diacylglycerol, several prostaglandins, endoperoxides, thromboxanes and leukotriene formed from released fatty acids like arachidonic acid.

Several experiments have shown that increased phosphatidylinositol turnover precedes Ca^{2+}-dependent activation of cells, for example smooth muscle, liver and salivary gland (Billah and Michell, 1979; Fain and Berridge, 1979), and that the phospholipid turnover is not itself Ca^{2+}-dependent. (1) No acute effect of removal of extracellular Ca^{2+} on phosphatidylinositol is observed, even in cells where removal of external Ca^{2+} blocks the cell response. (2) No effect of A23187 on phosphatidylinositol is observed. (3) No inhibition by the Ca^{2+}-channel blockers, D600, La^{3+} and nifedipine or local anaesthetics has been found on phosphatidylinositol turnover, even in cells like smooth muscle where these agents block contraction. (4) Prolonged exposure of cells to a stimulus can lead to a diminution of the cell response and Ca^{2+} influx. In insect salivary gland the Ca^{2+} response can be restored by adding *myo*-inositol, but not choline or ethanolamine, to the incubation medium. (5) Phosphatidylinositol turnover does not require an increase in cyclic AMP or cyclic GMP in the cell.

However, at least one group of workers (Cockcroft et al., 1980) have reported that in phagocytes and platelets, for example, the increased turnover of phosphatidylinositol is a consequence and *not* a cause of the increase in intracellular Ca^{2+}. They found, using the chemotactic peptide N-formyl-met-leu-phe and the Ca^{2+} ionophore ionomycin, that the effect on phosphatidylinositol occurred a few seconds after the presumed increase in cytosolic Ca^{2+} and *was* dependent on external Ca^{2+}. The explanation of the apparent discrepancy between these observations and those of Michell and colleagues may lie in the different experimental conditions used, in particular the preincubation of the tissue with radioactive precursor.

No significant changes in absolute phosphatidylinositol concentration have been reported in these experiments. In order to establish the *physiological* significance of this phenomenon of inositide turnover it will be necessary to: (1) demonstrate a net change in phospholipid concentration in the membrane using, for example, NMR or ESR; (2) establish whether this change is a cause or a consequence of any detectable change in free Ca^{2+}; (3) demonstrate the release of Ca^{2+} from the membrane or a 'Ca^{2+} ionophore' in broken cells; (4) establish the role of phospholipids in the structure of Ca^{2+} channels; (5) identify the structure and mechanism of any natural Ca^{2+} ionophore.

6.4.4. Prostaglandins, endoperoxides and thromboxanes

Prostaglandins are cyclopentane derivatives of the polyunsaturated fatty acids λ-linolenic acid, arachidonic acid and eicosapentaenoic acid (Fig. 6.17). They can cause either contraction or relaxation of a particular smooth muscle, they can stimulate platelet aggregation, and they have many other metabolic effects in other tissues (Horton, 1969; Ramwell and Shaw, 1970; Samuelsson, 1978). Investigations into the synthesis of this group of compounds led to the discovery of two other groups of related compounds—the thromboxanes and endoperoxides (Fig. 6.17). These are much less stable than prostaglandins but may have important physiological activity. The most recent new compound to be discovered is PGI_2, prostacycline. The synthesis of several of these compounds can produce an endogenous chemiluminescence from cells because of the involvement of peroxidation, enabling certain aspects to be studied in intact cells.

Receptors for prostaglandins and related compounds are found in the cell membrane, but may also exist in the cell. The dissociation constants are of the order 1–10 nM. The distinctive feature of this group of cell stimuli is that their synthesis is stimulated by another hormone or neurotransmitter and the prostaglandin produced then acts locally on a few surrounding cells. For example, action of thyroid-stimulating hormone on the thyroid produces PGE_1 which in turn can stimulate cyclic AMP formation, increase glucose oxidation, and increase iodine incorporation into thyroglobulin and thyroxine. Many of their effects can be explained, at least partially, by production of cyclic AMP through activation of adenylate cyclase. However, this does not explain all of the actions of prostaglandins, nor does it explain how their synthesis is activated.

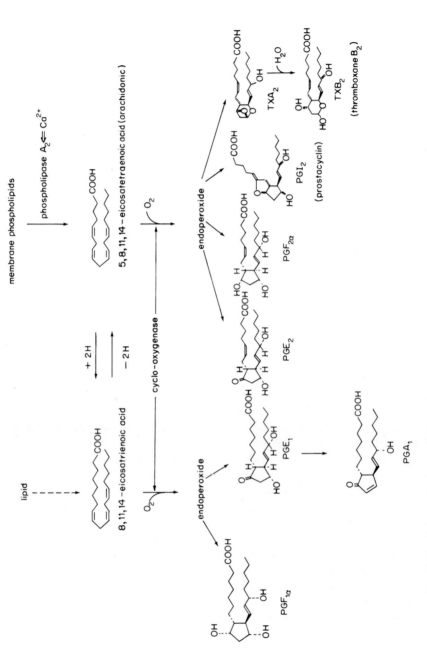

Fig. 6.17. Prostaglandin, endoperoxide and thromboxane synthesis. See Samuelsson (1978) for details

Table 6.14. Some cells where intracellular Ca^{2+} may regulate prostaglandin synthesis

Cell	Reference
Adrenal cortex (cat)	Laycock et al. (1977)
Intestine (rabbit)	Bikhazi et al. (1977)
Macrophage	Gemsa (1979)
Myocardium	Limas and Cohn (1973)
Medulla (rat)	Zenser and Davis (1978)
Smooth muscle (guinea pig)	Diegel et al. (1980)
Anterior pituitary (rat)	Betteridge (1980)
?Bone + complement (see Chapter 9, section 9.1.6)	Raisz et al. (1974)

Evidence based on effects of EGTA and A23187. The prostaglandin released is E_1, E_2, or $F_{2\alpha}$ in most of these examples.

There are two possible roles for intracellular Ca^{2+} in prostaglandin metabolism (Barritt, 1981): (1) Intracellular Ca^{2+} could stimulate prostaglandin synthesis. Production of prostaglandins is dependent on a supply of unsaturated fatty acid as substrate. Phospholipase A_2 catalyses the release of arachidonic acid from phospholipid, in contrast to phospholipase C which hydrolyses phosphatidylinositol to 1,2-diacylglycerol and inositol phosphate. Although original experiments with Ca^{2+} used mM concentrations (Newich and Waite, 1973; Derksen and Cohen, 1975), more recent evidence suggests that physiological Ca^{2+} concentrations (i.e. 10 μM) may significantly activate phospholipase A_2 (Pickett et al., 1976; Frei and Zahler, 1979; Cheung, 1980). Furthermore, release of prostaglandins in several tissues has been found to be Ca^{2+}-dependent (Table 6.14). (2) Prostaglandins could increase cytosolic free Ca^{2+}. Prostaglandins have been shown to increase the permeability of pure phospholipid vesicles (Weissmann et al., 1979) and may act as Ca^{2+} ionophores in biological membranes (Allen and Rasmussen, 1971; Horton, 1969; Kirtland and Baum, 1972; Carsten and Miller, 1978). The active metabolite of arachidonic acid could, for example, be PGE_2, $PGF_{2\alpha}$, a thromboxane or a leukotriene.

Prostaglandins are certainly produced by many cells after activation by the primary stimulus. Much needs to be done on the mechanism of stimulation of synthesis and action of this class of compounds before the role of intracellular Ca^{2+} can be established. In particular, their true physiological function has often not been defined. Do they activate 'threshold phenomenon'? Do they prolong Ca^{2+}-dependent cell activation after the Ca^{2+} transient is over? Do they act as feedback regulators? Do they stimulate cells not activated by the original stimulus, but close to the activated cell? The answers to these questions require techniques for studying changes in intracellular Ca^{2+} and other chemical changes both in intact cells and isolated single cells.

6.5. Conclusions

Much still needs to be learnt about the physiology of hormones and other primary stimuli of cells which regulate intermediary metabolism, particularly in invertebrates and plants. Many such examples of metabolic regulation may appear to show a continuously graded response to increasing concentrations of agonist or antagonist, and not to exhibit a 'threshold' phenomenon (Chapter 2). However, the precise behaviour of individual cells in tissues such as the liver or adipose tissue has yet to be investigated. Furthermore, in those tissues which contain opposing reactions such as glycolysis–gluconeogenesis, lipolysis–lipogenesis, glycogen degradation–synthesis, threshold points must exist where the direction of metabolism is changed from one to the other. It is under conditions where this occurs physiologically that the best chance of establishing a role for intracellular Ca^{2+} lies. A crucial experiment will be the correlation between a change in free Ca^{2+}, measured directly in the intact cell, and a change in the direction of intermediary metabolism.

This chapter has been concerned mainly with the acute effects of hormones on metabolism. The longer-term actions of hormones such as insulin, steroids and thyroid hormones on protein concentration may also involve Ca^{2+}-dependent mechanisms (Krahl, 1966; Jacobs and Krahl, 1973; Bresciani et al., 1973; Goswami and Rosenberg, 1978), though relatively little attention has been paid to this important aspect of intracellular Ca^{2+}.

CHAPTER 7

Endocytosis and exocytosis: The uptake and release of substances from cells

7.1. What are endocytosis and exocytosis?

All cells, prokaryotic or eukaryotic, release and take up substances from the fluid surrounding them. Many lipid-soluble compounds cross the outer membrane of cells by diffusion, whereas many water-soluble compounds require a special permease. This is the sole mechanism in prokaryotes. Eukaryotes have evolved another way of getting certain materials across the cell membrane, involving the incorporation of the substance within a membrane-enveloped vesicle during the process. It is the formation and release of these vesicles which characterize endo- and exocytosis.

Endocytosis, which includes pinocytosis and phagocytosis, is the process by which materials are taken up into cells by invagination of the cell membrane culminating in the formation of a boundary membrane in the form of a vesicle enclosing the engulfed material. Exocytosis, also known as emiocytosis, is the process by which materials within granules or droplets are released from the cell by the fusion of the granule membrane with the cell membrane, followed by the discharge of its contents into the external fluid. Endocytosis includes phenomena such as the ingestion of food by amoebae, and the phagocytosis of bacteria by macrophages and polymorphs as part of an animal's defence mechanism. In animals, exocytosis includes the release of hormones, neurotransmitters and enzymes from specialized secretory cells. For example, muscle movements are caused by neurotransmitters released from nerve terminals, whilst the digestion of food is carried out by enzymes released from the pancreas into the gastrointestinal tract. A particularly fascinating example of secretion, yet to be established as true exocytosis, is found in the sea firefly, *Cypridina hilgendorfi* (Harvey, 1952). This crustacean, when fearful of attack by a predator, releases a small organic molecule, a luciferin, from one group of cells and an enzyme, a luciferase, from another group. The combination of the luciferin and the luciferase with O_2 produces a cloud of light in the water which distracts the predator, enabling the *Cypridina* to escape. Secretion of substances from eukaryotic cells is not restricted to animals. Plant cells also release substances including, for example, the components of their cell walls and hormones.

There are three reasons for considering these apparently different phenomena of endo- and exocytosis in the same chapter: (1) Both involve vesicles and their fusion with another membrane as part of the complete process. (2) Both endo- and exocytosis can be found in the same cell type as part of one phenomenon. For example, the ingestion of food by amoebae ends with the excretion of undigested material by exocytosis (Chapman-Andresen, 1977). Macrophages and polymorphs, on the other hand, secrete several enzymes and other proteins upon addition of a phagocytic or other stimulus (Page et al., 1978). Conversely, the release of hormones and neurotransmitters by exocytosis is followed by uptake of the vesicle membrane back into the cell by a mechanism mimicking endocytosis. This has been demonstrated, for example, in pancreatic β-cells and the adrenal medulla. It enables the cell to maintain its size and the composition of its outer membrane. (3) Sometimes the response of a cell to a given stimulus involves first endocytosis, and then exocytosis from a different part of the cell. For example, the release of thyroxine from thyroid follicular cells first requires endocytosis of thyroglobin from the centre of the follicle.

Both endo- and exocytosis require ATP as a source of energy. A stimulus which activates either process usually also activates intermediary metabolism to provide the necessary substrates for this additional energy.

Can they be regarded as 'threshold phenomena' and are the effects of the stimuli which induce them mediated through an increase in cytosolic Ca^{2+}?

About 80 years ago Locke and Overton independently showed that removal of external Ca^{2+} prevented the transmission of impulses both from one nerve to another and from nerve to muscle. In 1928 Houssay and Molinelli demonstrated a similar requirement for Ca^{2+} in the splanchnic nerve. Harvey and McIntosh (1940) confirmed that external Ca^{2+} was required for the stimulation of release of acetylcholine from cholinergic nerves (Fig. 7.1). More recently many examples of exocytosis have been shown to be inhibited or abolished by removal of external Ca^{2+} (Table 7.1), though in several cases prolonged incubation in a 'Ca^{2+}-free' medium is necessary to demonstrate such inhibition. A requirement for Ca^{2+} in endocytosis has been less well established (Table 7.2). External Ca^{2+} is required for endocytosis in amoebae and, like many secretions, is inhibited by high external Ca^{2+}. In mammalian phagocytic cells removal of Mg^{2+} and Ca^{2+} inhibits, or can prevent, phagocytosis (Allison and Davies, 1974; Stossel, 1973, 1974). Readdition of Ca^{2+} restores the response. However, under many of the experimental conditions used, readdition of Mg^{2+} without Ca^{2+} also increased particle uptake. The production of oxygen radicals and the secretion of lysosomal enzymes, both of which can be stimulated by particles, show a much clearer requirement for external Ca^{2+}. The chemiluminescence of polymorphs responding to a phagocytic stimulus, for example, and which is a measure of oxygen radical formation within the cells, is prevented in a Ca^{2+}-free medium (Hallett et al., 1981) (see Chapter 4 and Fig. 7.13).

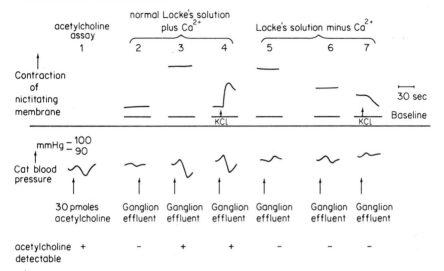

Fig. 7.1. Removal of external Ca^{2+} stops acetylcholine secretion from nerve terminals. From Harvey and MacIntosh (1940); reproduced by permission of the *Journal of Physiology*. System: perfused superior cervical ganglion of the cat; medium = Locke's solution $\pm Ca^{2+}$. (A) Measure of ganglionic excitation = contraction of nictitating membrane in the eye (upper traces). (B) Measure of acetylcholine in effluent = blood pressure in the cat. (1) Acetyl choline assay, producing decreased blood pressure. (2) No stimulation, $+Ca^{2+}$, no ganglion excitation, therefore no acetylcholine. (3) Maximum preganglionic stimulation, $+Ca^{2+}$, therefore large acetylcholine release detected. (4) KCl-provoked preganglionic stimulation, $+Ca^{2+}$, therefore large acetylcholine release. (5) No stimulation, $-Ca^{2+}$ showing spontaneous firing of the ganglion but no detectable acetylcholine release (a slight increase in blood pressure, instead of a decrease necessary to show the presence of acetylcholine). (6) Maximum preganglionic stimulation, $-Ca^{2+}$, showing no detectable acetylcholine release. (7) KCl, $-Ca^{2+}$, showing no acetylcholine release.
Conclusion: removal of external Ca^{2+} provokes excitation in nerves but prevents secretion of the neurotransmitter from the nerve terminal

Are these requirements for external Ca^{2+} in exocytosis and endocytosis because an increase in cytosolic Ca^{2+} is the trigger for these phenomena, or is it because Ca^{2+} has a structural or electrical role independent of internal Ca^{2+}? What exactly is exocytosis, and what is the evidence for intracellular Ca^{2+} provoking it?

7.2. Exocytosis—vesicular secretion

7.2.1 What is secretion?

Substances are released from cells for two principle reasons. They may be waste-products or substances harmful to the cell, or they may play a more

Table 7.1. Inhibition of exocytosis by removal of external Ca^{2+}

Substance released	Stimulus	Reference
1. *Neurotransmitter*		
Acetylcholine from parasympathetic nerves	Action potential from axon	Harvey and MacIntosh (1940); Del Castillo and Katz (1956); Dodge and Rahamimoff (1967)
Noradrenaline from sympathetic nerves	Acion potential	Hukovic and Muscholl (1962)
Amino acids from central nervous system	Action potential, K^+	Bradford (1970)
Inhibitory amino acids (GABA, Gly, taurine)	K^+ and electrical	Krnjevic (1974)
Invertebrate transmitters (L-Glu, dopamine) (stimulatory and inhibitory)	K^+ and electrical	Miledi and Slater (1966) Gerschenfeld (1973)
2. *Neurohormones*		
Adrenaline from adrenal medulla	Acetylcholine	Poisner and Douglas (1966); Douglas and Poisner (1963); Douglas (1974); Rubin (1974)
Posterior pituitary hormones—oxytocin vasopressin	K^+ and electrical	Dreifuss et al. (1971) Douglas and Poisner (1964)
Invertebrate neurosecretions	Electrical	Berlind and Cook (1968); Berlind (1977)
Somatostatin from brain	Elecrical	Iversen et al. (1978)
3. *Endocrine secretions*		
Insulin from β-cells	Glucose, amino acids, K^+, glucagon + glucose, ouabain Adrenaline inhibition	Hales and Milner (1967); Milner and Hales (1968) Grodsky and Bennet (1966)
Histamine from mast cells	Antigen to IgE receptor	Mongar and Schild (1958); Foreman et al. (1976)
Anterior pituitary hormones (GH, LH, TSH, FSH)*	K^+ or hypothalamic releasing hormones	Vale et al. (1967); Samli and Geshwind (1968); Jutisz and de la Llosa (1970); Gershwind (1970)
Renin from kidney	Adrenaline	Davis and Freeman (1976); Park and Malvin (1978)

Thyroid hormones from thyroid (partial inhibition when no external Ca^{2+})	TSH Acetylcholine inhibition	Williams (1972) Dumont et al. (1977); Dumont and Mockel (1977); Dumont and Lamy (1980)
Protein release from platelets	Thrombin, collagen, antigen	Sneddon (1972)
4. *Degradative enzymes*		
Amylase and proteases from exocrine pancreas	Acetylcholine Pancreozymin Cholecystokinin	Hokin (1966) Kanno (1972); Case and Clausen (1973) Williams (1980)
Amylase from parotid	K^+	Harfield and Tenenhouse (1973); Dormer and Ashcroft (1974)
Lysosomal enzyme release from polymorphs and monocytes	Particulate and non-particulate (e.g. leucocidin)	Woodin (1968); Smith and Ignarro (1975)
Cortical granule release in fertilized eggs	Sperm	Gilkey et al. (1978)
5. *Structural components*		
Polysaccharides from plant cell walls	Basal and auxin(?)	Morris and Northcote (1977)
6. *Substrates*		
Lipoprotein (VLDL)** from liver	Basal	Le Marchand and Jeanrenaud (1977)
7. *Extrusive organelles*		
Trichocyst discharge in *Paramecium*	Sensory	Fulton (1963); Hausmann (1978)
Nematocyst discharge in coelenterates	Sensory	Lenhoff and Bovaird (1959)

* GH = growth hormone; LH = luteinizing hormone; TSH = Thyroid-stimulating hormone; FSH = follicle-stimulating hormone.
** VLDL = very low density lipoprotein.
Many of the experiments quoted in this table involved preincubation of tissue or cells for more than 30–60 min with no Ca^{2+} to show inhibition.

Table 7.2. Effect of removal of external Ca^{2+} on endocytosis

Effect	Reference
1. *Inhibition*	
Inhibition of particle uptake by macrophages	Van Oss and Stinson (1970); Stossel (1973)
Inhibition of emulsion uptake by polymorphs	Stossell et al. (1973, 1974)
Inhibition of particle uptake by polymorphs	Quie et al. (1977)
Inhibition of pinocytosis induced by polyanions	Gordon and Cohn (1973)
Inhibition of phagocytosis by amoebae	Rabinovitch and De Stefano (1971)
2. *No inhibition*	
No effect on endocytosis of colloid in thyroid	Rubin (1970); Williams (1972)
Low external Ca^{2+} enhances and raised Ca^{2+} inhibits phagocytosis in amoebae	Brandt and Freeman (1967)

positive role, for example by interacting with other cells so as to benefit the whole organism. The substances released may be liquids, such as the fluid from salivary glands or the release of water from the contractile vacuoles of freshwater protozoa (Calkins, 1933; Smith, 1974), the latter enabling these unicellular organisms to maintain their osmotic balance. The substances released may be solids, as in excretory food vesicles of protozoa (Calkins, 1933) or the calcium granules from the hepatopancreas of many invertebrates (Simkiss, 1964, 1974); they may be protein-bound or crystalline substances which dissolve rapidly or become free of the binding protein once secreted into the external fluid, for example hormones and neurotransmitters released from neurosecretory cells, endocrine cells or nerve terminals (Rubin, 1974; Rahamimoff et al., 1975; Highnam and Hill, 1977); or they may even be gaseous, for example the gas bubbles found in the shells of some protozoa which enable them to rise to the surface (Heilbrunn, 1943), or the swim bladders of certain fishes.

The function of these various secretions can be classified into six main categories (Table 7.3); many occur by exocytosis (Table 7.4). Most fluid, ion and lipid secretions probably do not involve a vesicular-exocytotic mechanism, though in many cases this has yet to be ruled out. Fluid and ions can be secreted from the same tissue which releases substances by exocytosis. Both occur, for example, in the exocrine pancreas and salivary glands. This can complicate the interpretation of experiments designed to investigate the role of intracellular Ca^{2+} in exocytosis, particularly since some non-vesicular secretions may also be regulated by intracellular Ca^{2+}. A further problem is that these different secretions may come from different cell types within the same tissue, for example the exocrine pancreas, or from different surfaces of the same cell type.

Although the presence of granules in secretory cells has been known for fifty years or more, it was not until 1956 that Del Castillo and Katz, measuring minature potentials at muscle end-plates, showed that the stimulus for muscle contraction was released from the nerve terminals in discrete packets or 'quanta'. In the same year Bennett had proposed that secretion might occur by a reversal of

Table 7.3. Functions of secretions from cells

1. *Regulation of intra- and extracellular environment*
   Removal of waste-products and toxic substances (e.g. bile)
   Ionic and osmotic regulation (vacuoles in protozoa, salt glands in birds, kidney)
   Temperature regulation (sweat glands)
   Antibiotics in bacteria
   Formation of pearls in bivalve molluscs

2. *Secretion of substances affecting other organisms*
   Toxins (bitter substances in plants protect against herbivores, snake venoms, bee and wasp stings)
   Attractants (insect pheromones, pericarp of fruit, plant perfumes, animal sex glands, chemiluminescence)
   Repellents (foul-smelling substances, chemiluminescence)
   Protectants (waxy coat of some plants, mucus)

3. *Structural components*
   Skeleton calcium (bone, shells, perisarc of coelenterates)
   Cell walls (plant, bacteria, protozoa)
   Fibrous tissue (collagen, mucopolysaccharides)

4. *Substances for metabolism of other cells*
   Liver (glucose, lipoproteins, vitamins)
   Kidney (glucose)
   Mammary glands (milk)
   Muscle (amino acids, lactate)

5. *Regulators of cell function*
   Hormones (amines, polypeptides, steroids, plant hormones)
   Neurotransmitters and neurosecretions
   Cyclic AMP (serum, urine, slime mould)

6. *Digestive juices and degradative secretions*
   Gastric mucosa (H^+)
   Exocrine pancreas (proteases and nucleases, etc.)
   Salivary gland (amylase)
   Insectivorous plants (proteases, etc.)
   Polymorphonuclear leucocytes (lysosomal enzymes)

pinocytosis, the term 'exocytosis' being devised by De Duve in 1963. During recent years there has been much controversy about the validity of the 'vesicle hypothesis' for secretory cells (Marchbanks, 1975; Israel et al., 1979). The main reason for this is that when using rapid labelling techniques the acetylcholine released from a nerve terminal by an action potential does not seem to come from the main vesicle fraction. This apparent paradox highlights two contrasting features of different secretory cells—the amount of substance secreted and the time course of release. At nerve terminals less than 1% of the total transmitter store may be released by one action potential, the whole response being complete within a few milliseconds. Whereas the binding of antigen to IgE receptors on mast cells can release 30% of the total cell histamine over a period of seconds to minutes. Substances used *in vitro* like A23187 can stimulate release of more than 90% of the histamine. In contrast, the release of enzymes from the exocrine

Table 7.4. Examples of vesicular and non-vesicular secretions

1. *Secretion of substances from vesicles—exocytosis*
   Neuro-secretion (e.g. from nerve terminals or neurosecretory cells)
   Polypeptide hormone release (e.g. from endocrine cells)
   Lipoprotein and albumin release from the liver
   Enzyme release from the exocrine pancreas
   Enzyme release from salivary glands
   Enzyme and complement component release from macrophages
   5-Hydroxytryptamine, ADP and protein release from platelets
   Milk proteins from mammary gland
   H^+ secretion from gastric mucosa
   Phagocytic vacuole release
   Mucus secretion from goblet cells
   Polysaccharide secretion from plant cell walls

2. *Secretion of substances through the cell membrane—non-vesicular*
   Steroid hormones by endocrine glands
   Substrate and metabolite release
   Bile secretion by the liver
   Cation and anion secretion
   Fluid secretion from exocrine glands (e.g. pancreas and salivary glands)
   Release of substances from bacteria

Note: a vesicular mechanism for some of the phenomena in group 2 has not always been ruled out.

pancreas can continue for many minutes, or an hour or more *in vitro*, whilst the release of insulin from the endocrine pancreas, stimulated by glucose, is biphasic, at least *in vitro*. The first phase of insulin release has a latency of about a minute and returns to the baseline within 5–10 min. This is followed by a second, slower phase which can continue for 30–60 min or longer. These differences between cells mean that in fast-responding cells like nerve terminals the substance released may originate from a population of vesicles close to the surface of the cell. Cells with a longer time course for secretion may mobilize vesicles initially in the interior of the cell. The 'vesicle hypothesis' is supported by several different experimental approaches: (1) light and electron microscopy showing particle movement and vesicle–cell membrane fusion (Fig. 7.2); (2) radioactive tracer studies and autoradiography; (3) quantal release, for example a cholinergic granule may contain about 10 000 molecules of acetyl choline; (4) other substances released concomitantly with the major component correspond to substances found inside isolated granules; (5) NMR, providing evidence for membrane fusion; (6) incorporation of the vesicle membrane into the cell membrane, followed by uptake of ferritin into emptied granules after secretion has occurred.

Secretory vesicles from vertebrates are usually 50 nm – 1 μm in diameter. In some protozoa larger vacuoles can be found.

In 1953 Blaschko and Welch, and Lund, showed that adrenaline was stored in the adrenal medulla in vesicles, named chromaffin granules. These granules, like those from many secretory cells, have been isolated and their composition

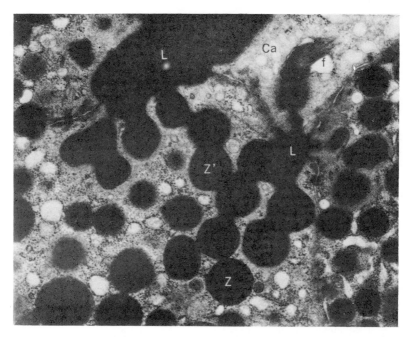

Fig. 7.2. Fusion of a secretory granule with the cell membrane. Exocrine pancreas. L = lumen into which enzymes are released; Z = zymogen granule. From Ichikawa (1965); reproduced from the *Journal of Cell Biology* by copyright permission of the Rockefeller University Press

defined (Table 7.5). The membrane of most secretory vesicles, exemplified by that of chromaffin granules, contains cholesterol and an array of phospholipids. The protein composition of the membrane is usually much simpler than that of the cell membrane, containing relatively few proteins. Two important components are MgATPase and ion pumps. Chromaffin granules contain an electron transport chain and a H^+ pump which is able to maintain the pH inside the granule at 5–6, as elegantly demonstrated by ^{31}P-NMR of internal nucleotide phosphate (Njus and Radda, 1978). With MgATP outside there is also a membrane potential of about 30 mV (positive inside) across the granule membrane. Adrenaline (RNH_3^+) is taken up into the granule by H^+ exchange. Other types of secretory granule may contain different ion pumps. For example cholinergic vesicles from the electric organ of *Torpedo* contain an ATP-dependent Ca^{2+}-uptake system (Israel *et al.*, 1980). This could provide a mechanism for rapidly regulating the cytoplasmic Ca^{2+} transient in nerve terminals. Ca^{2+} uptake also occurs in adrenal chromaffin granules (Serck-Hanssen and Christiansen, 1973).

The matrix of secretory vesicles usually has four main components: (1) amine; (2) polypeptide; (3) divalent cation; (4) nucleotide. Depending on the cell, one or more of these components is the active substance released. In the β-cell of the

Table 7.5. Composition of chromaffin granules in the adrenal medulla

Substance	Percentage dry weight of granule
Granule membrane	
Lipid—cholesterol	4.5
phospholipid	15
Protein—dopamine β-hydroxylase	9.7
chromomembrin B	
MgATPase	
phosphatidylinositol kinase	
cytochrome b-561	
NADH: acceptor oxidoreductase	
Matrix	
Catecholamine—mainly adrenaline (72%), some noradrenaline and dopamine	19.1
Nucleotides—mainly ATP (70%), some ADP, AMP, GTP, GDP and UDP	16.8
Ca^{2+}	0.12
Ascorbate	0.063
Protein—chromogranin A	32.3
dopamine β-hydroxylase	
Mucopolysaccharide	0.6–2.2

Data from Winkler (1976) and Njus and Radda (1978).

islets of Langerhans it is the polypeptide insulin, in platelets it is the nucleotide ADP. The exocrine pancreas secretes many enzymes simultaneously. Immunofluorescence has shown that each cell contains all of the enzymes. Whether this is true of each vesicle has yet to be fully established. Platelets contain at least two types of vesicle, releasing different substances.

The divalent cation in many secretory granules is Mg^{2+} or Ca^{2+}. In several species, islet β-cells contain Zn^{2+} which forms a crystal hexamer with insulin.

The nucleotide can play an important structural role in the granule. For example, ATP and adrenaline form a semicrystalline complex in the chromaffin granule (Berneis et al., 1969, 1971). If an increase in intracellular Ca^{2+} is the trigger for exocytosis, could it act by causing fusion of the granule membrane with the outer membrane of the cell? This might explain a rapid phase of secretion, but slower secretions may be regulated at sites further back in the pathway. We owe much of our knowledge of this pathway (Fig. 7.3) to the work of Palade and his coworkers (Jamieson and Palade, 1967). A stimulus of secretion could in theory act at any one of six points on the pathway: (1) Synthesis of hormone, or prohormone in the case of most polypeptides. (2) Degradation of hormone, as has been proposed, for example, for parathyroid hormone (Habener, 1976; Habener et al., 1977). (3) Formation of the granule; this occurs in the Golgi apparatus which may also fix the polarity of the cell. (4) Movement of the granule to the cell surface. In small cells this is only a few

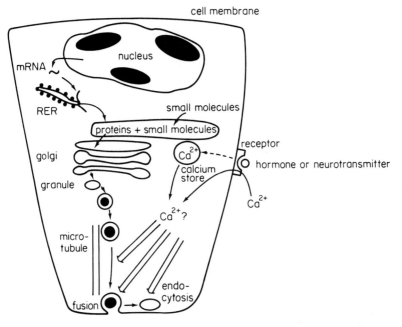

Fig. 7.3. The pathway of exocytosis

micrometres, but in nerves the granules may have to be transported from the cell body down the axon to the nerve terminal (Dahlstrom, 1971). (5) Fusion of the granule membrane with the cell membrane. (6) Endocytosis of spent granule membrane enabling fusion of more granules with the cell membrane.

Do changes in intracellular Ca^{2+} regulate any of these stages in secretion? Several other questions also arise. When a secretory cell is activated how are the correct materials released in the appropriate quantities at the required rate? What determines from which surface of a cell that has a polarity the secretion occurs? Can all exocytotic phenomena be classified as 'threshold' (see Chapter 2)? In particular, is intracellular Ca^{2+} involved in controlling any of these physiological aspects of secretory cells? In order to answer these questions it is necessary to determine first what are the stimuli of secretion.

7.2.2. The stimuli of exocytosis

The physiological stimuli of secretion by exocytosis can be classified into three main groups, based on the nature of the stimulus: electrical, chemical and mechanical (Table 7.6). In many cells the effects of the primary stimuli may be enhanced or inhibited by one or more secondary regulators. For example, lysosomal enzyme secretion from polymorphs, activated by occupancy of certain surface receptors, is enhanced by the α-effect of adrenaline and prostaglandin $F_{2\alpha}$, and inhibited by the β-effect of adrenaline and prostaglandin E_1. In contrast, in

Table 7.6. Some stimuli of exocytosis

Cell	Substance secreted	Primary stimulus	Secondary regulator
1. Electrically activated			
Nerve terminal	Neurotransmitter	Action potential	Adenosine (i)
Neurosecretory cells	Neurohormone	Action potential	?
Retinal rods	Neurotransmitter	Light → electrical change	Dark adaptation (a)
2. Chemically activated			
Anterior pituitary	Prolactin	Hypothalamic releasing hormone	Dopamine (i), oestrogen (i)
	ACTH	Hypothalamic releasing hormone	Cortisol (i)
	TSH	Hypothalamic releasing hormone	T_3/T_4 (i)
Thyroid	Thyroxine	TSH	Acetylcholine (i)
Adrenal medulla	Adrenaline	Acetylcholine (muscarinic)	
Mast cell	Histamine	Antigen	β-Adrenaline (i)
Exocrine pancreas	Enzymes	Acetylcholine, cholecystokinin	
β-Cell	Insulin	Glucose, amino acids	Glucagon (a), adreneline (i)
Platelet	ADP, (I) enzymes (II)	Damaged blood vessel, exposed collagen fibres, thrombin	β-Adrenaline (i), PGE_1 (i), adenosine (i), α-adrenaline (a)
Polymorphs	Lysosomal enzymes	Complement opsonized particles (C3b), immunoglobulin (Fc), chemotactic peptides (C5a, C$\overline{567}$, kallikrein)	β-adrenaline (i), PGE_1 (i), PGE_2 (i), histamine H_2 (i)
Mammalian parotid	Amylase	Noradrenaline (β)	α-Adrenaline (a), $PGF_{2\alpha}$ (a), acetylcholine (a)
3. Mechanical			
Protozoa (*Paramecium*)	Trichocyst	Sensory	—
Coelenterates	Nematocyst	Sensory	—

i = inhibits primary stimulus; a = enhances primary stimulus. β-Adrenaline = adrenaline bound to β-adrenergic receptor; α-adrenaline = adrenaline bound to α-adrenergic receptor.

mast cells the only possible physiological secondary regulator of antigen-induced histamine secretion is the inhibitory β-effect of adrenaline, mediated through cyclic AMP. This seems to be supported by evidence that some asthmatics may have circulating antibodies to this β_2-receptor in the lung.

There are some tissues, such as the adrenal medulla and exocrine pancreas, where only primary stimuli have so far been identified. This has lead to the idea that there may be two types of cell activation—uni- or monodirectional and bidirectional (Goldberg et al., 1974; Berridge, 1976b) (see Chapter 10). Unidirectional secretory cells were proposed to be stimulated by only a few substances which usually act through intracellular Ca^{2+}, which in some cases could be potentiated by increasing cyclic AMP. In bidirectional secretory cells one stimulant was proposed to activate the cell through an increase in intracellular Ca^{2+}; another agent could cause recovery from, or antagonize, the main stimulant by a cyclic AMP dependent mechanism. It was proposed that unidirectional secretions included nerves, adrenal medulla, anterior pituitary, exocrine and endocrine pancreas and salivary gland, whereas platelets and mast cells were examples of bidirectional systems. Yet neurotransmitter release may be modified by adenosine (Kuroda, 1978), and other secondary regulators of these systems are known (Table 7.6). Whether regulators of other unidirectional systems are discovered or not, it is clear that this hypothesis, as originally proposed, does not apply to all secretory cells.

As in the other chapters of this book, the actions of physiological agents on secretory cells have been divided into primary and secondary regulators (Table 7.6). It is proposed that the primary stimulus of exocytosis acts by increasing cytoplasmic free Ca^{2+}. Secondary regulators act by modifying this change in intracellular Ca^{2+} or by altering its action, for example by cyclic nucleotide induced protein phosphorylation. In addition, there may be longer-term regulation through alteration of receptor number on the cell surface or protein concentration inside the cell.

Is exocytosis a 'threshold phenomenon'? Certainly the fusion of each granule with the cell membrane is an all-or-nothing event. However, a basal secretion can be detected from most secretory cells *in vitro*, and the detection of minature end-plate potentials at the post-synaptic membrane suggests that small numbers of vesicles are continuously fusing with the cell membrane *in situ*. Nevertheless, the activation of exocytosis by an action potential or a mechanical stimulus can be considered as a threshold event since it involves the rapid transformation of the cell from one state (resting) to another (activated). Whether a similar state of affairs exists for individual cells activated by a chemical stimulus remains to be established. It is the belief of this author that this is so, at least in some small cells, and that the transition is initiated by a dramatic increase in cytoplasmic free Ca^{2+}.

Several problems must be tackled before the role of intracellular Ca^{2+} in a secretory cell can be investigated:

(1) Is the secretion truly exocytotic? Tyramine-induced secretion at certain nerve terminals may not be exocytotic. In the thyroid T_4/T_3 release is presumed

to be by exocytosis from a secondary lysosome containing the cleaved thyroglobin, but this has still to be definitely demonstrated.

(2) Where is the site of regulation in the exocytotic pathway? In several secretory cells regulation occurs at the level of prohormone synthesis, prohormone to hormone conversion or hormone degradation. The elevation of parathyroid hormone secretion by a lowered plasma free Ca^{2+} (Sherwood *et al.*, 1970) is thought to be mediated through a reduction in hormone degradation in the cell. Lipoprotein secretion from the liver, enhanced for example by insulin, may not be mediated by a Ca^{2+}-dependent threshold but through substrate supply. Cells, such as the β-cell in the islets of Langerhans, which exhibit more than one phase of secretion may be regulated at more than one site in the exocytotic pathway. In this case, intracellular Ca^{2+} is proposed to activate both granule movement and fusion with the cell membrane.

(3) Have the primary and secondary regulators been correctly identified? Many cells respond to agents *in vitro* which have no physiological role *in situ*. Prostaglandins, in particular, are difficult to study *in vivo* because of their apparent action as 'local' hormones.

(4) A primary stimulus may activate more than one process in the same cell. Examples are the thyroid, platelets, polymorphs, fertilized eggs and the parotid. In the thyroid TSH stimulates iodine into thyroglobulin and thyroxine release, whereas acetylcholine inhibits secretion but also activates iodine incorporation into protein and prostaglandin synthesis. At one time it was thought that the action of TSH, and in Graves disease antibodies to the TSH receptor, acted solely through an increase in intracellular cyclic AMP, whereas the effects of acetylcholine were thought to be mediated by an increase in intracellular Ca^{2+} and cyclic GMP (Dumont and Lamy, 1980). But to elucidate fully the role of intracellular Ca^{2+} in the thyroid follicular cell it is necessary to examine separately each step in the pathway from amino acid to thyroid hormone release. Platelets are complicated by the fact that shape changes and pseudopod formation accompany the first phase of secretion and a second phase, stimulated by another set of stimuli, consolidates the clot. In polymorphs not only are endocytosis and exocytosis occurring simultaneously in the same cell, but also peroxide formation and chemotaxis may be taking place. Since intracellular Ca^{2+} has been implicated as the mediator of all these phenomena, it is necessary to discover how they can be regulated independently.

The other process may not necessarily occur at the same time as secretion. For example, exocytosis of cortical granules in fertilized eggs occurs many minutes before cell division. Furthermore, two different types of secretion can occur in the same cell. For example, acetylcholine and the α-affect of adrenaline stimulate K^+ and H_2O release, but little or no protein, from parotid cells. Whereas the β-effect of adrenaline stimulates amylase secretion but little or no H_2O and K^+. Extracellular Ca^{2+} is required for the acetylcholine and α-adrenaline effects, but exocytotic release of amylase involves cyclic AMP and release of Ca^{2+} from an internal store.

(5) Are there species differences? For example, acetylcholine inhibits secretion of the hormones from dog and human thyroid, whereas in other species α-adrenaline receptors may be inhibitory.

7.2.3. Is intracellular Ca^{2+} the trigger for exocytosis?

After muscle, secretory cells are probably the most extensively studied cell type so far as intracellular Ca^{2+} is concerned. The term stimulus–secretion coupling, analogous to excitation–contraction coupling in muscle, has been used to describe secretory exocytosis (Douglas, 1974).

The evidence for intracellular Ca^{2+} being the trigger for exocytosis has been based on four main approaches: (1) manipulation of extracellular Ca^{2+}, and replacement of Ca^{2+} by Sr^{2+} or Ba^{2+}; (2) manipulation of intracellular Ca^{2+}; (3) inhibition of exocytosis by Ca^{2+} antagonists; (4) measurement of radioactive Ca^{2+} flux.

Direct measurement of free Ca^{2+} has been made in only a few secretory cells; for example, fertilized eggs (Gilkey *et al.*, 1978) and the presynaptic terminal of giant nerves (Llinas and Nicholson, 1975).

External Ca^{2+}

The earliest experiments concerned with the role of Ca^{2+} in a secretory cell were carried out by Locke in 1894. He showed that if frog sartorius muscle, still attached to its nerve, was soaked in a Ca^{2+}-free solution for 15–20 min it failed to contract when the nerve was stimulated, but did contract if the electrical stimulus was applied to the muscle itself. This experiment was carried out before a vesicular mechanism for secretion had been discovered, and even before neurotransmitters were known. The discovery of acetylcholine as the transmitter between certain nerves, and between nerve and vertebrate muscle, (Loewi, 1921; Dale, 1934) enabled Harvey and MacIntosh in 1940 to show that Ca^{2+} was required for acetylcholine release in the cervical sympathetic ganglion of the cat. Using the effect of acetylcholine on blood pressure as an assay, Harvey and MacIntosh showed that when transmission of an impulse, initiated electrically or with K^+, between two nerves was prevented by removal of external Ca^{2+}, no transmitter release occurred, whilst conduction of action potentials down the nerve axon was not blocked under these conditions (Fig. 7.1). Using a more sophisticated assay—the measurement of miniature end-plate potentials at the muscle end-plate—Katz and his coworkers showed that removal of external Ca^{2+} abolished the large, quantal release of acetylcholine from stimulated nerve terminals but not from unstimulated terminals. A similar requirement for external Ca^{2+} has been shown for invertebrate nerve terminals and other types of vertebrate nerve (Table 7.1). Purinergic nerves are found in the gastrointestinal tract and release ATP. The requirement for Ca^{2+} in these nerves is not so well documented (Burnstock, 1972, 1977; Burnstock and Wong, 1978). Several problems have arisen in the interpretation of experiments involving removal of external Ca^{2+} (Tables 7.1 and 7.7).

(1) Some stimuli have been found not to require external Ca^{2+} (Table 7.7). In secretory cells dependent on external Ca^{2+} the concentration of Ca^{2+} in the fluid surrounding the cell must be reduced to below 10 μM in order to block the stimulation of secretion. Before the advent of Ca^{2+} chelators there was often sufficient Ca^{2+} contamination in media to allow some response to stimuli; however, this does not appear to explain the examples shown in Table 7.7. In one case—tyramine-induced secretion of noradrenaline from splenic nerve—the release does not seem to be by exocytosis. Compounds such as cyclic AMP, theophylline and mitochondrial inhibitors are sometimes able to circumvent the normal mechanism by which the stimulus induces an increase in intracellular Ca^{2+} by releasing Ca^{2+} from internal stores such as mitochondria. Ba^{2+} seems to act on islet β-cells, at least partially, as a Ca^{2+} analogue. Of the remaining examples, platelets, anterior pituitary ACTH, thyroxine from the thyroid, and amylase from the parotid, extensive treatment with Ca^{2+} chelators may be necessary to deplete the internal stores of Ca^{2+}, which are likely to be the source of the Ca^{2+} for stimulation of secretion in these cells.

(2) A requirement for external Ca^{2+} (Table 7.1) could reflect one of the non-regulatory roles of Ca^{2+} (see Chapter 1). Ca^{2+} is necessary for maintenance of the viability and structure of secretory cells. Prolonged exposure to a Ca^{2+}-free medium can irreversibly damage cells. This possibility can be ruled out if readdition of Ca^{2+} immediately leads to activation of exocytosis as it does in neurosecretion, mast cells and the adrenal medulla. However, Ca^{2+} currents are also required in the electrical activity of some secretory cells, for example in nerve terminals and in β-cells of the islets of Langerhans. Removal of external Ca^{2+} would prevent any Ca^{2+}-dependent action potentials being generated (see Chapter 4).

(3) The time required for Ca^{2+}-free media, usually containing EGTA, to be effective has lead to some confusion. Removal of external Ca^{2+} immediately prevents the stimulation of secretion from nerve terminals, β-cells, mast cells and the adrenal medulla (Fig. 7.4). This suggests that if intracellular Ca^{2+} is the trigger for secretion it is coming from outside the cell. Some of the early experiments on tissue slices required several hours in Ca^{2+}-free medium to prevent the stimulation of secretion. This long time interval was not accountable for simply by the need for Ca^{2+} to diffuse out of the lump of tissue. Even cells or acini isolated from exocrine glands, such as the pancreas and salivary gland, require 30–60 min in EGTA to inhibit significantly the stimulation of secretion. In these cells, and also possibly in the thyroid and some pituitary cells, the Ca^{2+} for initiating secretion may come from an internal store.

(4) The significance of effects of Ca^{2+}-free media on basal secretion, i.e. secretion in the absence of the stimulus, is not always clear. In many secretory cells little or no effect on basal secretion of removing external Ca^{2+} is seen acutely (Table 7.7). However, prolonged exposure to EGTA does inhibit; for example, quantal release of acetylcholine at unstimulated nerve terminals (Katz and Kopin,

Table 7.7. Lack of inhibition of exocytosis by removal of external Ca^{2+}

Substance released	Stimulus	Reference
1. *Transmitters and neurohormones*		
Noradrenaline from splenic nerve	Tyramine (? not exocytosis)	Chubb *et al.* (1972)
	Supramaximal stimuli	Ziance *et al.* (1972); Schneider (1972)
Acetylcholine from nerve terminals	Basal	Del Castillo and Katz (1956)
Adrenaline from adrenal medulla	Cyclic AMP, theophylline	Peach (1972)
	Metabolic inhibitors, Ba^{2+}	Douglas (1974)
5-Hydroxytryptamine from brain	Electrical	Katz and Kopin (1969)
Acetylcholine from frog neuromuscular junction	Mitochondrial inhibitors	Rahamimoff *et al.* (1975)
2. *Hormones*		
Parathyroid hormone from parathyroid	Decrease in external Ca^{2+}	Habener (1976)
5-Hydroxytryptamine and ADP from platelets	High thrombin	Lüscher *et al.* (1972)
Histamine from mast cells	Slow reacting substance, polyamine 48/80	Douglas (1974)
ACTH from anterior pituitary	Hypothalamic releasing hormone	Katsumi *et al.* (1969); Milligan and Kraicer (1974)
Insulin from β-cells	Ba^{2+}	Hales and Milner (1967)
Thyroxine from thyroid	TSH	Rubin (1974)
3. *Enzyme*		
Amylase from parotid	Adrenaline (β-effect)	Batzri and Selinger (1973); Kanagasuntheram and Randle (1976)
Cortical granule release in eggs	Sperm	Gilkey *et al.* (1978)

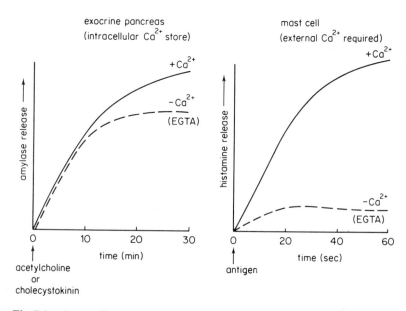

Fig. 7.4. Acute effect of removing external Ca^{2+} on exocytosis. Ca^{2+} removed at time 0

1969; Rahamimoff et al., 1975). The question arises whether basal secretion occurs by precisely the same mechanism as in the stimulated cell, and simply reflects the low fractional saturation of the internal Ca^{2+} sites at the resting cytoplasmic free Ca^{2+} concentration of about 0.1 μM. It seems a reasonable hypothesis that a low rate of secretion can occur by a mechanism independent of the Ca^{2+}-activated one, not requiring the involvement of a special Ca^{2+}-binding protein. However, there could be a structural requirement for Ca^{2+} in this basal secretion.

The relationship between extracellular Ca^{2+} concentration and stimulation of secretion has been investigated in several cells (Fig. 7.5). Concentrations of Ca^{2+} above 2–5 mM often inhibit the stimulation of secretion. Rapidly responding cells exhibit a sigmoid relationship between secretion and the Ca^{2+} concentration, below these inhibitory concentrations of Ca^{2+}. Histamine secretion from mast cells shows an approximate square-law relationship (i.e. secretion versus $[Ca^{2+}]^2$ is linear) (Foreman et al., 1976), whereas transmitter release from the squid giant synapse exhibits a fourth-power relationship with respect to external Ca^{2+} concentration (Baker, 1975; Rahamimoff et al., 1975). A high power law relationship is also observed at mammalian neuromuscular junctions and the vas deferens.

Cells which secrete over longer periods when stimulated tend to show lower power law relationships with respect to external Ca^{2+}. In the β-cell it is approximately linear (Curry et al., 1968), whereas in the case of pancreozymin secretion from the pancreas there is a reciprocal relationship between the rate of secretion(s) and the external Ca^{2+} concentration (i.e. $1/s$ versus $1/[Ca^{2+}]$ is linear) (Kanno, 1972).

What is the explanation of these relationships and do they provide any insights into the mechanism by which Ca^{2+} activates secretion?

Consider a simple model:

Assume n Ca^{2+} sites, each of which can compete with Mg^{2+}, binding to a receptor R.

For one site

$$Ca^{2+} + R_1 \overset{K^{Ca}}{\rightleftharpoons} CaR_1$$

and

$$Mg^{2+} + R_1 \overset{K^{Mg}}{\rightleftharpoons} MgR_1$$

$\therefore$ $[CaR]$ = concentration of occupied sites

$$= A[Ca^{2+}]/(1 + [Ca^{2+}]/K^{Ca} + [Mg^{2+}]/K^{Mg}) \quad (7.1)$$

N.B. Equation (7.1) can be written $[CaR] = B[Ca^{2+}]/([Ca^{2+}] + K^{Ca}_{app})$ where $K^{Ca}_{app} = K^{Ca}(1 + ([Mg^{2+}]/K^{Mg}))$.

A or B = constant; $[Ca^{2+}]$ = free Ca^{2+} concentration outside the cell.

If there are n identical sites then the cell response $\propto [CaR]^n$

$$\text{Response} = k[CaR]^n = kA[Ca^{2+}]^n/(1 + [Ca^{2+}]/K^{Ca} + [Mg^{2+}]/K^{Mg})^n \quad (7.2)$$

$\therefore$ a plot of log response versus $\log[Ca^{2+}]$ will give a slope of n at low saturation of the Ca^{2+} sites.

If $n = 4$, then this model predicts a 10 000-fold increase in rate of transmitter release for a 10-fold increase in intracellular free Ca^{2+} concentration, assuming that intracellular free Ca^{2+} is directly related to the extracellular Ca^{2+} concentration when the cell is stimulated.

An alternative simple model might be

Response = $k_1 [Ca_i^{2+}]^{n_1}$

where k_1 = constant

when n_1 = number of internal Ca^{2+} sites occupied

and $[Ca_i^{2+}]$ = internal free Ca^{2+} concentration

$\quad = k_2 [CaR]^{n_2}$ where $[CaR] = Ca^{2+}$ bound outside the cell

$\quad = k_2 A[Ca_o^{2+}]^{n_2}/(1 + ([Ca_o^{2+}]/K^{Ca}) + [Mg_o^{2+}]/K^{Mg})^{n_2} \quad (7.3)$

where n_2 = number of external Ca^{2+} sites occupied

$[Ca_o^{2+}]$ = external free Ca^{2+}; $[Mg_o^{2+}]$ = external free Mg^{2+}.

$\therefore$ response $\propto [Ca_o^{2+}]^{n_1 + n_2}$

(for example measurement of end-plate potentials at the neuromuscular junction).

The clarification of such models requires the measurement of intracellular free Ca^{2+} during secretion, and its correlation with extracellular Ca^{2+} and the rate of secretion.

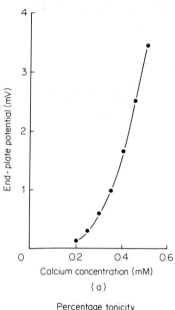

(a)

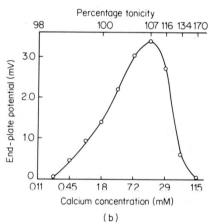

(b)

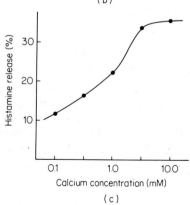

(c)

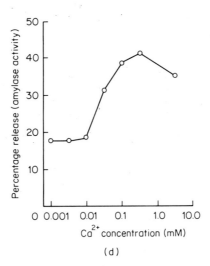

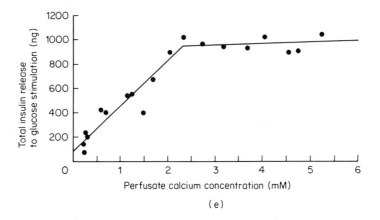

Fig. 7.5. Relationship between external Ca^{2+} concentration and the rate of exocytosis; All systems illustrated require external Ca^{2+} for stimulation of secretion. (a) Transmitter release from nerve terminal assessed by end-plate potential. From Dodge and Rahamimoff (1967); reproduced by permission of the *Journal of Physiology*. (b) Transmitter release assessed by motor end-plate potential. From Del Castillo and Stark (1952); reproduced by permission of the *Journal of Physiology*. (c) Histamine secretion from mast cells. From Mongar and Schild (1958); reproduced by permission of the *Journal of Physiology*. (d) Amylase secretion from exocrine pancreas. From Heisler *et al.* (1972); reproduced by permission of Elsevier Biomedical Press B. V. (e) Insulin secretion from pancreatic β-cells. From Curry *et al.* (1968); reproduced by permission of The American Physiological Society

Replacement of Ca^{2+} by other cations

As long ago as 1911 Mines showed that Sr^{2+} or Ba^{2+}, but not Mg^{2+}, could maintain neuromuscular transmission in the absence of external Ca^{2+}. The conclusion that the chemistry of Sr^{2+} and Ba^{2+} is similar enough to Ca^{2+} to replace it in exocytosis seems to be borne out in several other cells which require external Ca^{2+} (Rubin, 1974); for example, mast cells (Foreman et al., 1976), adrenal medulla (Douglas, 1974), and pancreatic β-cells (Hales and Milner, 1967). The affinity of the cells for Sr^{2+} is usually lower than for Ca^{2+}, concentrations of Sr^{2+} some ten-fold higher than Ca^{2+} being required. Sr^{2+} does not appear to act by releasing Ca^{2+} from internal stores. However, Ba^{2+} actually stimulates secretion of adrenaline from the adrenal medulla, or insulin from β-cells, in the absence of the physiological primary stimulus, suggesting that Ba^{2+} might cause release of internal Ca^{2+}. In contrast to Sr^{2+} and Ba^{2+}, Mg^{2+} usually antagonizes the effect of Ca^{2+} as it does in many cells (Engbaek, 1952).

Manipulation of intracellular Ca^{2+}

An increase in intracellular Ca^{2+} stimulates exocytosis (Table 7.8). This has been done in three ways: (1) microinjection of Ca^{2+} into the cell; (2) Ca^{2+} ionophores A23187, X537A or ionomycin; (3) substances releasing Ca^{2+} from intracellular stores. A particularly ingenious method of manipulating intracellular free Ca^{2+} in secretory cells has been developed by Baker and Knight (1978) (see Chapter 2). Using a high-voltage discharge (about 40 kV) lasting a few μsec across a suspension of cells from the adrenal medulla, 'leaky' cells are produced. Incubation of the cells in potassium glutamate + $ATPMg^{2-}$ enables the secretion of adrenaline and dopamine β-hydroxylase from the chromaffin granules to be stimulated by Ca^{2+} in the range 1–10 μM.

Using these various methods it is possible to stimulate exocytosis not only in cells which normally require external Ca^{2+} for the physiological response (e.g. adrenal medulla, mast cells and β-cells), but also in cells which do not require external Ca^{2+} for a normal, acute response, (e.g. stimulation of amylase from parotid, enzyme release from exocrine pancreas, ADP and 5-hydroxytryptamine release from platelets, and cortical granule release in fertilized eggs) (see Tables 7.7 and 7.8).

There are some problems in the interpretation of these experiments: (1) Does the microinjection of Ca^{2+}, or use of ionophore, mimic the physiological response in terms of time course and amount of material secreted? (2) Do these experimental procedures lead to an increase in cytosolic free Ca^{2+} in the same range and in the same part of the cell as the physiological response? (3) Are the effects reversible and are they specific for Ca^{2+}? Microinjection of Mg^{2+} or K^+ has usually been found to be ineffective, and ionophores such as nigericin or valinomycin, which make membranes permeable to Na^+ or K^+ but not to Ca^{2+}, do not usually stimulate exocytosis (Pressman, 1976). To demonstrate specific activation of exocytosis it is essential to show that cytosolic enzymes like lactate dehydrogenase are not released concomitantly with the granule components.

Table 7.8. Increasing intracellular Ca^{2+} stimulates exocytosis

Cell	Effect	Reference
Microinjection of Ca^{2+} (pressure, ionophoresis or electrical discharge*)		
Presynaptic terminal of squid giant synapse	Increase in miniature end-plate potentials	Katz and Miledi (1965, 1967); Miledi (1973); Llinas and Nicholson (1975)
Mast cell	Stimulates granule secretion	Kanno et al. (1973)
Adrenal medulla*	Stimulates adrenaline release	Baker and Knight (1978)
Pancreatic acinar cells	Acetylcholine-like effect on membrane potential	Iwatsuki and Petersen (1977)
Paramecium	Stimulates trichocyst discharge	Plattner (1974)
Sea urchin eggs	Cortical granule release	Baker et al. (1980)
Ionophores (mainly A23187)		
Neurohypophysis	Stimulates oxytocin release	Nordmann and Currell (1975); Nordmann and Dyball (1975)
Mast cell	Stimulates histamine release	Cochrane and Douglas (1974)
Adrenal medulla	Stimulates adrenaline release	Garcia et al. (1975)
β-Cell (pancreas)	Stimulates insulin release	Kahl et al. (1975); Wollheim et al. (1975); Ashby and Speake (1975)
β-Cell (pancreas)	Stimulates contractile activity at cell boundary	Somers et al. (1976)
Platelets	Stimulates ADP, 5-hydroxytryptamine, ATP and protein release	Feinman and Detwiler (1974); Feinstein and Fraser (1974)
Polymorphs and monocytes	Stimulates lysosomal enzyme release	Schell-Frederick (1974); Smith and Ignarro (1975)
Salivary gland (parotid and submaxillary)	Stimulates amylase release	Butcher (1975); Thomson and Williamson 1976
Exocrine pancreas	Stimulates enzyme release	Iwatsuki and Petersen (1977); Selinger and Naim (1970); Selinger et al. (1974)
Intracellular Ca^{2+} release (induced by dinitrophenol, CN^-, ruthenium red, methylxanthines)		
Nerve terminal	Increase in miniature end-plate potentials at muscle end-plate	Rahamimoff et al. (1975)
Adrenal medulla	Stimulates adrenaline release	Douglas (1974)

See Grenier et al. (1974) for problems with thyroid hormone secretion.

Ca^{2+} antagonists

A plethora of ions and organic compounds thought to act as Ca^{2+} antagonists have been shown to inhibit many examples of exocytosis. Local anaesthetics, Ca^{2+}-channel blockers (see Chapter 4) like verapamil, D600 and nifedipine, Mn^{2+}, Ni^{2+} and La^{3+} usually block secretion in cells dependent on external Ca^{2+}, such as histamine from mast cells, adrenaline from the adrenal medulla or insulin from pancreatic β-cells (Miledi, 1971; Foreman et al., 1976; Devis et al., 1975; Douglas, 1974; Baker, 1975; Malaisse et al., 1976, 1977a,b; Flatt et al., 1980). Whilst these experiments are consistent with Ca^{2+} being the trigger for exocytosis, the effect of the experimental procedures on intracellular free Ca^{2+} has rarely been studied, nor has the reversibility or effect on cell viability always been determined. The assumption that these agents act at the cell membrane may not always be valid in view of the observation by Dormer and Ashcroft (1974) that Ni^{2+} inhibits adrenaline-stimulated amylase secretion from the parotid, a process not acutely dependent on external Ca^{2+}. Tetrodotoxin, which blocks voltage-dependent Na^+ channels in vertebrates and some invertebrates, does not inhibit release of transmitter at nerve terminals.

The discovery of a high-affinity Ca^{2+}-binding protein, calmodulin, in secretory cells (Sugden et al., 1979; Hutton et al., 1981) has lead to experiments using trifluoperazine and other phenothiazines which block the action of Ca^{2+}-calmodulin on myosin light chain kinase (Levin and Weiss, 1976, 1978). These phenothiazines, in the range 10–100 μM, inhibit insulin secretion and several other examples of exocytosis (Gagliardino et al., 1980), including exocytosis in sea urchin eggs (Baker and Whitaker, 1979).

Ca^{2+} fluxes

A 5–10-fold increase in $^{45}Ca^{2+}$ influx usually occurs when secretory cells, dependent on external Ca^{2+}, are acted on by the physiological stimulus (Table 7.1). This occurs, for example, in intact and broken off nerve terminals (synaptosomes), in K^+-and acetylcholine-stimulated adrenal medulla, after K^+ stimulation of the neurohypophysis, in glucose-stimulated pancreatic β-cells and in antigen-activated mast cells (Dalquist, 1974; Douglas, 1974; Rubin, 1974; Eto et al., 1974; Malaisse, 1977; Foreman et al., 1977). Inhibition of this Ca^{2+} influx correlates with inhibition of secretion by Ca^{2+} antagonists such as La^{3+}, Ca^{2+}, D600 and local anaesthetics.

In cells such as those of the exocrine pancreas, not acutely dependent on external Ca^{2+} but where an internal Ca^{2+} store appears to be required for stimulation of exocytosis, a rapid *increase* in $^{45}Ca^{2+}$ efflux is observed (Williams, 1980; Dormer et al., 1981), induced by the physiological stimulus. It has been proposed that this increased efflux is the result of an elevated cytosolic free Ca^{2+} concentration, rather than simply an activation of the enzyme responsible for transporting Ca^{2+} out of the cell. In many cells like the exocrine pancreas an increase in $^{45}Ca^{2+}$ *influx* is also observed when external Ca^{2+} is present, even though this is not necessary for activation of exocytosis. What is the reason for this influx, which is analogous to that seen in skeletal muscle (see Chapter 5)?

Continuous stimulation of exocrine cells *in vitro* results in a depletion of total cell Ca^{2+} of about 30–50% within 30 min. *In situ* a Ca^{2+} influx would be necessary to balance the Ca^{2+} efflux, and thereby prevent the cells becoming depleted of Ca^{2+}.

Stimulation of $^{45}Ca^{2+}$ flux provides supporting evidence for intracellular Ca^{2+} being the trigger for exocytosis. It also provides evidence for the source of this Ca^{2+}. Yet the conclusions from these experiments may be limited by any one of three problems:

(1) More than one Ca^{2+} pool inside the cell. Incubation of cells with $^{45}Ca^{2+}$ can only load 30–60% of the cell Ca^{2+} within 2 hours. At least three pools of Ca^{2+} have been found in several secretory cells (pool A, rapid turnover, $t_{1/2} \approx 1-5$ min; pool B, medium turnover, $t_{1/2} \approx 30$ min; pool C, slow turnover, $t_{1/2} \approx 1-2$ hours). For a cell with total $Ca^{2+} = 2$ mmoles/kg cell water, a pool containing as little as 2.5% of the total would elevate total cytoplasmic Ca^{2+} by 50 μM, and could be easily missed in $^{45}Ca^{2+}$ experiments.

(2) Specific activity. The specific activity of cytosolic free Ca^{2+} is not known in most experiments.

(3) Activation of Ca^{2+} efflux mechanism. Ca^{2+} efflux may occur by $Ca^{2+}-Ca^{2+}$ or $Ca^{2+}-Na^{+}$ exchange, (Ca^{2+})-MgATPase, or possibly through vesicle–cell membrane fusion since granules usually contain Ca^{2+}. An increase in $^{45}Ca^{2+}$ efflux is consistent with, but not sufficient to prove, an increase in cytosolic free Ca^{2+}.

Cytosolic free Ca^{2+}

Cytosolic free Ca^{2+} has been measured using aequorin in giant synapse of the squid and in fertilized eggs (Llinas and Nicholson, 1975; Gilkey et al., 1978). Most of the attempts to detect changes in cytosolic free Ca^{2+} in small cells have used the lipophilic fluorescent indicator chlortetracycline (Naccache et al., 1979a, b; Taljedahl, 1974; Williams, 1980). The decrease in fluorescence observed on stimulation of polymorphs or exocrine pancreas illustrates how difficult it is to interpret experiments using this indicator. Chlortetracycline remains mainly in the membranes of cells and its ability to detect changes in cytosolic free Ca^{2+} has not been established. The compounds of Tsien (see Chapter 2) have been used in platelets and offer much greater potential.

In summary, these various experimental approaches strongly support the hypothesis that an increase in cytosolic free Ca^{2+}, induced by the primary stimulus, is the trigger for exocytosis. The question now arises as to how this increase in free Ca^{2+} occurs, how Ca^{2+} acts to stimulate exocytosis, and how the secondary regulators modify the response of the cell to the primary stimulus.

7.2.4. Ca^{2+} and the electrophysiology of secretory cells

Until the late 1950's relatively little had been done on the electrical properties of secretory cells (Lundberg, 1958). The development of microelectrodes and

Table 7.9. Some electrical properties of secretory cells

1. *Secretory cells which can exhibit spike potentials*
   Nerve endings
   Neurosecretory cell terminals
   β-Cell
   Adrenal cortex
   Adrenal medulla
   Pituitary
   Neoplasia (endocrine cells in culture)

2. *Secretory cells apparently non-excitable*
   Liver
   Parotid
   Exocrine pancreas
   Anterior and posterior pituitary
   Mast cells
   Polymorphs

3. K^+-, *external* Ca^{2+}-*dependent, secretion*
   All cells exhibiting spike potentials
   Parotid
   Endocrine pancreas
   Platelets
   Pituitary
   Mast cell

For references see Rubin (1970, 1974), Petersen (1976), Douglas (1974), Iwatsuki and Petersen (1977), Biales et al. (1977), Tischler et al. (1976).

potential sensitive dyes has enabled the resting potential, and changes in membrane potential, to be measured in secretory cells (Table 7.9, Fig. 7.6). The resting potential in most of these cells is in the range -40 to -70 mV (negative inside) (Petersen, 1976), though a value as low as -20 mV has been reported for pancreatic β-cells (Dean and Matthews, 1970a, b, 1972; Matthews, 1975). The action of a primary stimulus on a secretory cell can cause an increase in

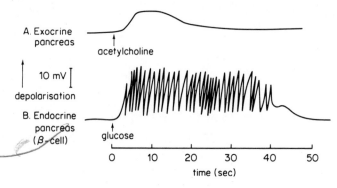

Fig. 7.6. An electrotonic and excitable membrane response in two secretory cells. (A) Exocrine pancreas. (B) Endocrine pancreas (β-cell)

membrane potential (hyperpolarization), as is the case for β-adrenaline on the parotid, or acetylcholine acting on gastric mucosa, lacrimal gland or exocrine pancreas. Alternatively, the cell may respond by a decrease in membrane potential (depolarization); such is the case for α-adrenaline acting on the parotid or acetylcholine acting on the adrenal medulla (Douglas et al., 1961). In some cells this depolarization leads to a spiking membrane response, analogous to an action potential in nerve axons. Examples of such secretory cells are nerve endings and pancreatic β-cells (Fig. 7.6). Several of these cells which generate the 'spiking membrane response' show properties characteristic of a slow voltage-dependent Ca^{2+} channel (see Chapter 4), in that secretion and the electrical response require extracellular Ca^{2+}, and are inhibited by Mg^{2+}, Mn^{2+}, Co^{2+}, Ni^{2+}, La^{3+}, D600 and nifedipine (Miledi, 1971; Baker, 1975). Usually secretory cells are not inhibited by tetrodotoxin (Katz and Miledi, 1969), which blocks the voltage-dependent Na^+ channel, although neurosecretory cells in invertebrates do contain such Na^+ channels (Berlind, 1977).

One of the most interesting examples of a secretory cell producing 'spiking' when activated was discovered by Dean and Matthews (1970a, b). D-Glucose, which has to be metabolized by the β-cell in order to stimulate insulin secretion, induces a rapid spiking in β-cells within a second of addition. The threshold for spiking is about -16 mV and each spike is 1–4 mV in amplitude (Fig. 7.6). More than 80 % of the cells showed this response to a glucose concentration producing maximum insulin secretion, as well as in response to other stimuli including mannose, leucine and tolbutamide, though the shape of the spikes was not always the same for each stimulus. This phenomenon is consistent with the 'threshold' hypothesis outlined in Chapter 2. Since the threshold glucose concentration varies from cell to cell, as the glucose concentration is increased so the number of spiking cells increases (Fig. 7.7). Adrenaline, which inhibits glucose-induced insulin secretion, also blocks the glucose-induced 'action potentials'. On the other hand, removal of external Ca^{2+}, which inhibits glucose-induced insulin secretion, produces a slow, continuous firing in the presence of glucose, rather than rhythmic bursts of action potentials. Electrical activity, induced by D-glucose, tolbutamide or 'current injection' is blocked by D600 and Mn^{2+} (Matthews, 1975, 1977).

The electrical activity of secretory cells therefore parallels secretion. The question arises, do the electrical changes in secretory cells play a key role in the activation of exocytosis or are they secondary to the increase in intracellular free Ca^{2+}, or vesicle–cell membrane fusion (Dean and Matthews, 1970a, b; Meissner and Schmelz, 1974; Atwater and Beigelman, 1976; Beigelman et al., 1977; Atwater et al., 1978).

Depolarization of most vertebrate cells can be produced experimentally by increasing the external K^+ concentration. High K^+ stimulates many secretory cells including pancreatic β-cells and nerve terminals. However, histamine from mast cells, protein secretion from exocrine glands, and prolactin from adenohypophyseal cells are not stimulated by K^+ even though the cells are depolarized. Furthermore, although K^+ stimulation of secretion requires

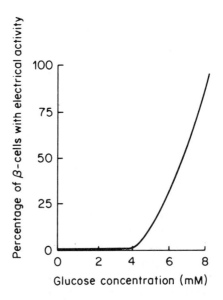

Fig. 7.7. Relationship between glucose concentration and the number of 'spiking' β-cells. From Beigelman et al. (1977); reproduced by permission of Masson S. A., Paris

external Ca^{2+}, it does not always generate spike potentials in the same way as the physiological stimulus (Dean and Matthews, 1970a, b; Meissner and Atwater, 1976). Also, in the case of insulin secretion, high external Mg^{2+} inhibits secretion but does not abolish the action potentials. Changes in membrane potential could be caused by an increase in intracellular free Ca^{2+} (see Chapter 4). For example, injection of Ca^{2+} into exocrine pancreatic cells increases K^+ permeability and one might expect large permeability changes to occur at the site of fusion of granules with the cell membrane, which would inevitably lead to changes in membrane potential.

The mechanism by which glucose stimulates insulin secretion is confused by the fact that a glycolytic metabolite of glucose produced within the cell activates exocytosis rather than an external glucoreceptor. This idea is supported by the observation that inhibitors of glucose metabolism inhibit both insulin secretion and the generation of spike potentials.

The significance of these electrical changes in secretory cells will not be fully established until it is possible to correlate membrane potential with secretion in single cells.

One possible function of the electrical changes is that they could enable one cell to stimulate another since the cells in many multicellular secretory tissues are in electrical contact with each other through gap junctions between adjoining cells (Petersen, 1976). Whether this leads to an increase in cytosolic free Ca^{2+} remains to be established.

7.2.5. The maintenance of cytosolic free Ca^{2+} and the source of Ca^{2+} for exocytosis

As in all cells the cytosolic free Ca^{2+} concentration in the unstimulated secretory cell is thought to be about 0.1 μM. The remainder of Ca^{2+}, 2–5 mmoles/kg cell water, is bound within the cell. Some secretory cells, for example platelets and invertebrate hepatopancreas (Simkiss, 1961, 1964), contain much higher total cell Ca^{2+} because of Ca^{2+}-containing granules within the cell. The electrochemical gradient of Ca^{2+} across the cell membrane is maintained by an energy-dependent efflux pathway balancing the passive flux into the cell. What is this efflux mechanism? Is it a Ca^{2+}-activated MgATPase or a Na^+-dependent Ca^{2+} efflux?

Replacement of external Na^+ by Li^+ or choline stimulates many secretory cells, including adrenaline release from the adrenal medulla (Douglas, 1974), insulin from pancreatic β-cells (Hales and Milner, 1967; Malaisse, 1977), and transmitter release from intact nerve terminals and synaptosomes (Blaustein, 1974). Furthermore, removal of external Na^+ can enhance the response of the cell to an agonist, whereas depletion of intracellular Na^+ leads to an inhibition of secretion. An increase in Na^+ within the cell, induced for example by ouabain inhibition of the Na^+ pump, stimulates exocytosis from cells in the pituitary, adrenal medulla and endocrine pancreas. These effects, on exocytosis, of external and internal Na^+ require extracellular Ca^{2+}. The simplest interpretation of these observations is that these secretory cells contain a Na^+-dependent Ca^{2+}-efflux pathway analogous to some invertebrate nerves and muscles. Removal of external Na^+, or an increase in internal Na^+, would therefore lead to an increase in cytosolic free Ca^{2+}, which in turn stimulates exocytosis.

However, the criteria for such a mechanism in secretory cells have not yet been satisfied (see Chapter 4). Furthermore, veratridine, a compound which inhibits Na^+ permeability in nerves, stimulates insulin secretion in the *absence* of external Ca^{2+} (Lowe et al., 1976). A similar lack of requirement for external Ca^{2+} has been demonstrated for the stimulation by ouabain of acetylcholine release at the neuromuscular junction (Baker and Crawford, 1975). Thus these experimental conditions may lead to release of Ca^{2+} from internal stores rather than Na^+-Ca^{2+} exchange across the cell membrane.

The possible importance of intracellular Ca^{2+} stores in secretory cells was pointed out by Bianchi in 1968. Attempts have been made to localize such stores using oxalate and pyroantimonate precipitation, followed by identification of Ca^{2+} using X-ray microprobe analysis (see Chapter 2) (Herman et al., 1973; Howell, 1977) (Fig. 7.8). Precipitates were observed in granules and mitochondria, on the inner surface of the cell membrane, in nuclei and in the cytoplasmic matrix. No evidence for alteration in these deposits after activation of exocytosis has been obtained.

In theory an increase in cytosolic free Ca^{2+}, leading to activation of exocytosis, could occur in four ways: (1) An increase in the flux of Ca^{2+} into the cell, initiated by the opening of a Ca^{2+} channel in the cell membrane or through a

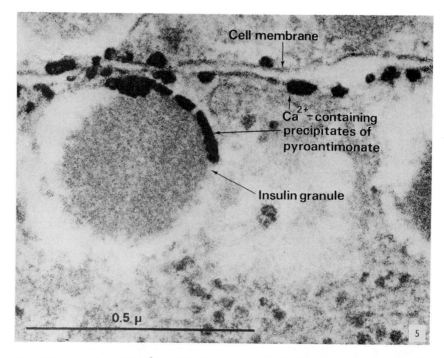

Fig. 7.8. Location of Ca^{2+} inside the pancreatic β-cell. Tissue incubated with pyroantimonate. From Herman et al. (1973); reproduced by permission of Academic Press

Ca^{2+} ionophore; (2) a release of Ca^{2+} from an internal store; (3) an inhibition of Ca^{2+} flux out of the cell; (4) an inhibition of Ca^{2+} flux into organelles within the cell. Exocytosis which depends on external Ca^{2+} is likely to involve the first of these (Table 7.1). In cells which can generate action potentials a voltage-dependent Ca^{2+} channel may be the means by which Ca^{2+} permeability across the cell membrane is increased. It has been suggested that a mechanism by which Ca^{2+} channels could be opened is through release of Ca^{2+} from phosphatidylinositides, particularly triphosphoinositide (Michell, 1975). The binding of acetylcholine to muscarinic receptors, for example in adrenal medulla, results in a rapid breakdown of phosphatidylinositol in the membrane. In view of the small amount of Ca^{2+} likely to be bound on the inner surface of the cell membrane, at a free Ca^{2+} concentration of 0.1 μM, it is unlikely that such Ca^{2+} release will contribute significantly to the free Ca^{2+} necessary to activate exocytosis itself. Other proposed mechanisms for increasing Ca^{2+} influx include phospholipid methylation (Hirata and Axelrod, 1980) and compounds related to prostaglandins or arachidonic acid (Barritt, 1981) (see Chapter 6 section 6.4.3 and 6.4.4).

It is unlikely that a rapid increase in intracellular free Ca^{2+} could be obtained by inhibition of Ca^{2+} efflux from the cell or by inhibition of Ca^{2+} influx into organelles. However, release from an internal store must occur in secretory cells which do not require external Ca^{2+} acutely. The increase in cytosolic free Ca^{2+} must occur at the appropriate surface of the cell, which may be different from the hormone or transmitter receptor sites. In exocrine pancreas and parotid the primary stimuli of secretion cause a redistribution of internal Ca^{2+}, measured using $^{45}Ca^{2+}$ (Dormer and Ashcroft, 1974; Dormer et al., 1981). The second messenger necessary to release this internal store of Ca^{2+} has yet to be identified.

After the stimulus is over, the cytosolic Ca^{2+} is removed by pumping it back out of the cell or into the original internal store. Alternatively, it has been suggested that in nerve terminals, mitochondria or even the secretory granule itself, known to contain a Ca^{2+} pump, may act as a temporary store of Ca^{2+}, thereby limiting the period of stimulation of exocytosis to only a very short time.

7.2.6. How does cytoplasmic Ca^{2+} initiate secretion?

To answer this question it is first necessary to identify the point in the pathway for secretion (Fig. 7.3) that is regulated by the primary stimulus, and then to provide a molecular explanation for this regulation. There are four possible sites: (1) the amount of substance stored in the cell, regulated through synthesis or degradation; (2) granule formation; (3) granule movement to the cell membrane; (4) fusion of secretory granules with the cell membrane. The first of these may be an important long-term regulatory mechanism in many secretory cells. In the parathyroid gland, where an increase in external Ca^{2+} inhibits parathyroid hormone secretion, intracellular Ca^{2+} and cyclic AMP may regulate hormone degradation within the cell (Habener et al., 1977; Brown et al., 1980).

In most exocytotic cells, however, an increase in intracellular Ca^{2+} activates either movement of granules or granule–membrane fusion, or both.

Secretory granule movement

Anyone who has observed live secretory cells in the light microscope will have been struck by the random dancing movements of the granules. This is Brownian motion, observed with all intracellular organelles, and caused by their bombardment by atoms and molecules from the surrounding fluids. The phenomenon was first observed by the botanist Robert Brown in 1827 whilst watching pollen grains under the microscope. The energy of any particle, or granule, is governed by the Boltzmann distribution

$$N_i = N_o \exp(-E_i/kT) \tag{7.4}$$

where N_i = number of particles with energy state ε_i
N_o = total number of particles
k = Boltzmann constant
T = temperature (K)
$E_i = \frac{1}{2}mv^2$ (i.e. kinetic energy)
where m = mass of particle
v = velocity of particle

$$m = \tfrac{4}{3}\pi r^3 \rho \tag{7.5}$$

where r = radius of particle
ρ = density of particle

But Brownian motion can be treated as a three-dimensional walk (Matthews, 1970); therefore, by Einstein's (1906) equation (see also Green and Casley-Smith, 1972):

$$\bar{\Delta} = \left(\frac{kTt}{3\pi r \eta}\right)^{1/2} \tag{7.6}$$

where $\bar{\Delta}$ = mean displacement of a particle from its original position along a particular axis
t = duration of the random Brownian motion
η = viscosity coefficient

From equation 7.6 it can be seen that granule displacement is inversely proportional to both the square root of the radius of the granule and the viscosity of the cytoplasm. Matthews (1970), for example, has calculated that an insulin granule, radius 100 nm and 5 μm from the centre of the β-cell, itself of radius 10 μm, would take about 33 sec to reach the cell membrane by random Brownian motion, assuming the viscosity of the cytoplasm (η) = 0.06 P. A similar granule only 10 nm from the cell surface would take only 0.13 msec.

To the naked eye, activation of cells with easily visible granules appears to provoke granule movement, and in melanocytes at least this movement appears to be caused by Ca^{2+} (Dikstein et al., 1963). Attempts have been made to demonstrate granule movement in secretory cells using a laser combined with interference spectrometry (Piddington, 1976). The idea that Ca^{2+} might regulate the fluidity of the cytoplasm is an old one (Heilbrunn, 1928, 1937, 1956, 1958). Addition of Ca^{2+} to axoplasm from giant nerves causes liquefaction (Hodgkin and Katz, 1949a; Hodgkin and Keynes 1957; Gilbert, 1975, 1977). This effect can be reversed by EGTA, and is not caused by Mg^{2+} or K^+. Unfortunately, the concentrations of Ca^{2+} used in these experiments were in the mM range, several thousand times the expected free concentration in the cytoplasm of the intact cell. A lowering in viscosity would cause an increase in random movement of granules (equation 7.6), and thus increase the rate at which granules reach the cell membrane.

An alternative mechanism is a directed, as opposed to random, movement of granules. Experimentally Ca^{2+} can induce rapid transport of vesicular material down the highly organized structure of nerve axons (Hammerschlag et al., 1977). Removal of external Ca^{2+} inhibits axoplasmic transport of vesicles (Ochs et al., 1977; Iqbal and Ochs, 1978), but there is little evidence for a physiological stimulus activating this process. So how could a natural stimulus in secretory cells cause a directed movement of vesicles?

Microtubules and microfilaments have been observed in many secretory cells in the electron microscope. Furthermore, the major proteins, tubulin, actin and myosin, responsible for these structures have been extracted from them (Pollard and Weihing, 1974; Clarke and Spudich, 1976) (see Chapter 5). Secretory granules have also been observed apparently attached to these intracellular structures, for example at the nerve terminal and in the pancreatic β-cell (Smith et al., 1970; Lacy and Malaisse, 1973; Gray, 1975). Substances such as colchicine and vinblastine (Fig. 5.4) inhibit many secretions (Margoulis, 1973), including insulin from the pancreatic β-cell (Lacy and Malaisse, 1973; Malaisse, 1977), enzymes from polymorphs (Hoffstein et al., 1977), lipoprotein from the liver (Le Marchand and Jeanrenaud, 1977) and collagen from fibroblasts.

The interpretation of experiments using cytochalasin B (Fig. 5.4), which disrupts microfilaments, is more difficult since stimulatory effects, inhibitory effects or no effect have been reported in different secretory cells.

Perhaps the function of microfilaments or microtubules in the regulation of secretion is different in different cell types. For example, in contrast to the pancreatic β-cell, histamine secretion from mast cells, stimulated by antigen binding to IgE on the cell surface or by a pharmacological substance known as 48/80 or by A23187, is inhibited by cytochalasin B but not by colchicine.

The evidence that an increase in intracellular Ca^{2+}, induced by the primary stimulus, causes a directed movement of vesicles or granules towards the cell surface is still poor. Muscle actin filaments bind to secretory granules isolated from the pituitary but Ca^{2+} has no apparent effect when added to this experimental system (Ostlund, 1977). Glucose, a primary stimulus for insulin secretion from pancreatic β-cells, increases tubulin polymerization, but this effect, unlike exocytosis, does not require external Ca^{2+} (Pipeleers et al., 1976). Yet in these same cells time-lapse photography has shown that glucose and A23187 cause a contraction at the edge of the cells (Orci et al., 1974; Somers et al., 1976), a phenomenon observed in several secretory cells when stimulated. This would be consistent with an activation of a myosin ATPase, perhaps by calmodulin. In contrast, a role for Ca^{2+} in microtubule function is supported by the fact that increasing external Ca^{2+} relieves the inhibition of secretion from adrenal medulla (Poisner and Cooke, 1975) and pituitary tumour cells (Gautvik and Tashjian, 1973) caused by colchicine.

The resolution of this problem of 'random' versus 'directed' movement of granules and the involvement of intracellular Ca^{2+} requires two things: (1) a direct demonstration and characterization of the effect of the primary stimulus on granule or vesicle movement in the intact cell; (2) a cell-free system mimicking granule or vesicle movement in the cell and which responds to μM concentrations of Ca^{2+}. Even if granule movement is enhanced by a cell stimulus the granules still have to fuse with the cell membrane.

Membrane fusion

It has been calculated that the average energy of particles through Brownian motion alone (equation 7.4) would not be enough to allow sufficient membrane fusion during the activation of exocytosis (Poste and Papahadjopoulos, 1976). Furthermore, most biological membranes have a net negative charge on their surface caused by negative phospholipids like phosphatidylserine, protein carboxyl groups, and sialic acid in membrane glycoproteins. Using electrophoresis it has been shown that chromaffin granules from the adrenal medulla contain about 11 000 negative sites per granule, about 4200 of which could bind Ca^{2+}. The Gouy-Chapman theory (see Njus and Radda, 1978) predicts a 'diffuse layer' of counterions near the surface of membranes, which in the case of a negatively charged surface cuases an accumulation of cations and a depletion of anions near the surface. For example,

$$[H^+]_{surface} = [H^+]_{bulk} \exp(-e\psi_0/kT) \tag{7.7}$$

where e = charge on the electron
ψ_0 = negative surface potential
k = Boltzmann constant
T = temperature (K)

For a solution of univalent ions

$$\psi_0 = \frac{2kT}{e} \sinh^{-1}(\sigma/c) \tag{7.8}$$

where $c = \sqrt{2\varepsilon k Tn/\pi}$
and σ = surface charge density
ε = dielectric constant of the solvent
n = concentration of ions

Equation 7.8 predicts that ψ_0, the electrical potential at the surface, will increase as σ increases and decrease as n increases, n being directly related to the ionic strength.

If exocytosis is to occur then it may be necessary to reduce the repulsion between the two negatively charged surfaces, namely the secretory granule and the inner surface of the cell membrane, to allow membrane fusion. Ca^{2+} could neutralize this charge repulsion (Dean, 1975).

The physiological examples of membrane fusion can be divided into three main groups: (1) vesicle–cell fusion, necessary for exocytosis; (2) vesicle–vesicle fusion, for example formation of secondary lysosomes; (3) cell–cell fusion, for example skeletal muscle formation, polykaryocytes in bone, and fertilization of eggs by sperm. The first of these is the type of membrane fusion involved in exocytosis. However, the similarity between this and the other types of membrane fusion, together with evidence that Ca^{2+} may be required (Table 7.10), has led to many attempts to show that intracellular Ca^{2+} is the trigger for membrane fusion. Several experimental models have been used to investigate this possibility, including fusion between secretory vesicles and other vesicles, fusion between pure phospholipid vesicles (liposomes) and cell–cell fusion induced by various fusogens (Table 7.11).

In 1966 Okada and Murayama reported that external Ca^{2+} was required for cell fusion induced by a group of membrane-enveloped viruses, the paramyxoviruses (see Chapter 9), for example Sendai virus, but not for viral infection. Since this report there have been many conflicting arguments regarding the possibility that Ca^{2+} can initiate cell–cell fusion (Poste, 1970, 1972; Poste and

Table 7.10. Requirement for Ca^{2+} in membrane fusion

System	Response	Reference
Cell–cell fusion		
Erythrocytes	Fusion induced by Sendai virus	Hart et al. (1976); Volsky and Loyter (1978)
	Fusion induced by fusogens	Ahkong et al. (1973)
	Fusion induced by A23187	Ahkong et al. (1975)
	Fusion induced by polyethylene glycol	Blow et al. (1979)
Ehrlich ascites cells	Fusion induced by Sendai virus	Okada and Murayama (1966); Yanovsky and Loyter (1972)
Myogenic cells	Fusion induced by A23187	Schadt and Pelks (1975)
Muscle myoblasts	Spontaneous fusion	Shainberg et al. (1969)
Erythrocyte ghosts	Fusion induced by Sendai virus	Apostolov and Poste (1972)
Intracellular vesicle fusion		
β-Cell (pancreas)	Insulin granules fusing with the cell membrane	Berger et al. (1975)
Medaka fish eggs	Cortical granules fusing with each other and the cell membrane after sperm fertilization	Gilkey et al. (1978)
Mast cells	Fusion of histamine granules with cell membrane	Lagunoff (1973)
Vesicle fusion in broken cell preparations		
β-Cell insulin granules	Fusion with each other and membranes	Dahl and Gratzl (1976)
β-Cell granules + cell membranes	Release of insulin	Lazarus and Davis (1977)
Liver Golgi vesicles	Fusion with each other	Gratzl and Dahl (1976)
Pancreatic zymogen granules	Interaction with membranes	Milutinovic et al. (1977)
Liposomes	Fusion with each other	Papahadjopoulos (1977); Sun et al. (1978)
Phospholipid films	Fusion with each other	Blioch et al. (1968)

Criteria for fusion mainly light or electron microscopy, but sometimes release of substance. Ca^{2+} (mM) required extracellularly for cells, in other cases μM Ca^{2+} required. Percentage fusion in some cases (e.g. liposomes) is very small.

Table 7.11. Some stimuli of cell–cell fusion

Paramyxoviruses, e.g. Sendai virus
Polyethylene glycol
Diacylglycerol
Lysolecithin
Glycerol monooleate
Retinol
Phospholipase C
Ionophore A23187
Dimethylsulphoxide

For references and other fusogens see Poste and Allison (1973), Hart *et al.* (1976), Maggio *et al.* (1976).

Allison, 1973; Hart *et al.*, 1976; Volsky and Loyter, 1978; Blow, 1978; Blow *et al.*, 1979; Pasternak, 1980).

A rapid influx of $^{45}Ca^{2+}$ has been observed with many fusogens (Table 7.10) (Impraim *et al.*, 1979; Blow *et al.*, 1979; Fuchs *et al.*, 1980). In addition, Sendai virus stimulates the release of histamine from mast cells, γ-aminobutyrate from brain slices, and corticotropin release from pituitary cells, presumably as a consequence of the Ca^{2+} influx. However, both cell–cell and virus–cell fusion have been observed in the absence of external Ca^{2+}. Furthermore, 1-day-old virus, rather than the 3-day-old virus normally used, which fuses with the membrane of cells apparently does not induce a change in Ca^{2+} influx or efflux (Impraim *et al.*, 1979). Impraim and coworkers have proposed that the Ca^{2+} influx induced by mature virus is a consequence, rather than the cause, of membrane fusion. Sendai virus does increase the concentration of free Ca^{2+} in erythrocyte ghosts before fusion occurs (Fig. 7.9). However, the ghosts can fuse, with more than 90% efficiency, with polymorphs (Fig. 7.10) in the absence of external Ca^{2+} and under conditions which do not apparently induce a large increase in intracellular free Ca^{2+}.

Six problems arise in the interpretation of many of the experiments where the possible role of intracellular Ca^{2+} in cell–cell fusion has been investigated. (1) Cell fusion can be difficult to distinguish from cell aggregation and also difficult to quantify, particularly the rate at which cells fuse, and the total number of cells fused. (2) Two separate fusion events, with different possible Ca^{2+} requirements, may be induced by a virus, namely virus–cell fusion and cell–cell fusion. (3) The lack of measurement of intracellular free Ca^{2+} (but see Fig. 7.9). (4) Differences between different cell types. (5) Whether fusion occurs by a virus or another molecule acting as a bridge between two cells, or whether fusion occurs at a site separate from the viral attachment site. (6) Leakage of cell constituents and cell lysis. Cell leakage is a major source of confusion, particularly since it may be exacerbated by experimental conditions such as removal of external Ca^{2+}. If intracellular Ca^{2+} is necessary for cell–cell or virus–cell fusion, it may be required for the maximal rate of fusion, rather than being an absolute requirement under natural conditions. Other viruses, for example the arbor virus Semliki Forest

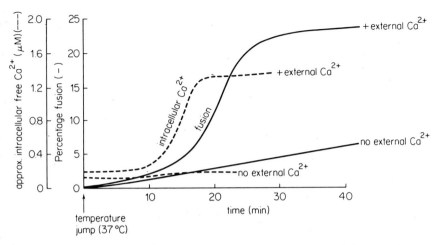

Fig. 7.9. Intracellular free Ca^{2+} during cell fusion induced by Sendai virus. Erythrocyte ghosts were incubated with Sendai virus (1500 haemagglutinating units) at 0 °C for 10 min. The temperature then jumped (↑) to 37 °C. Medium contained balanced salts + 0.25 mM uranyl acetate (UO_2Ac_2). Intracellular Ca^{2+} was assessed by obelin luminescence, and fusion by counting under a light microscope. Note that with no external Ca^{2+} there is no rise in intracellular free Ca^{2+} yet fusion is observable, albeit at a reduced rate compared to that when Ca^{2+} is present. Unpublished data from Hallett, Fuchs and Campbell

virus, also induce cell fusion. It is not known whether intracellular Ca^{2+} is required in these other types of virus-induced fusion.

Isolated secretory vesicles are relatively stable (Hillarp, 1958) and attempts have been made to induce their fusion *in vitro* (Table 7.10). Unfortunately, considerable variation in the reproducibility of such experiments has been reported between different laboratories.

In a study of lysosomal enzyme release from polymorphs induced by leucocidin, vitamin A or streptolysin O, Woodin and Wieneke (1963) showed that, as with other secretory stystems, Ca^{2+} and a source of energy are required inside the cell for secretion to occur. They proposed that, because Ca^{2+} and ATP can stabilize many isolated membrane preparations, removal of Ca^{2+} and ATP from internal membranes, initiated by activation of a (Ca^{2+})-MgATPase in the granules, enabled fusion of the granule with the cell membrane to occur. This proposal is difficult to reconcile with the hypothesis that an *increase* in cytosolic free Ca^{2+} is the trigger for exocytosis.

Three mechanisms have been proposed to explain how the increase in cytosolic Ca^{2+} could induce fusion of secretory vesicles with the inner surface of the cell membrane.

(1) Binding of Ca^{2+} to the negative surface of the membrane, (equation 7.8) could reduce the repulsion between the two membranes (Njus and Radda, 1978). Ca^{2+} reduces, for example, the electrophoretic mobility of chromaffin granules (Siegel, 1978).

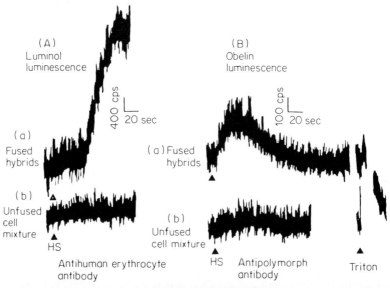
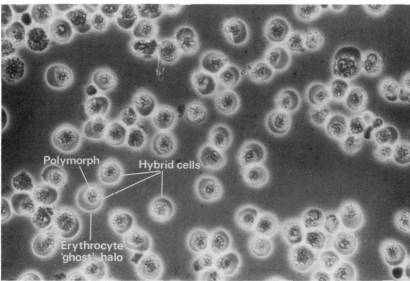

Fig. 7.10. Immunological demonstration of cell fusion. From Hallett and Campbell (1982b). Reprinted by permission from *Nature*. Copyright © 1982 Macmillan Journals Limited. Human erythrocyte 'ghosts' containing obelin were fused with rat polymorphs, in the absence of external Ca^{2+}, using Sendai virus. Complement (HS = human serum) was activated by antibody bound specifically to (A) erythrocyte antigens or (B) to a polymorph antigen, the ectoenzyme 5'-nucleotidase. (A) Luminol (10 μM) amplified chemiluminescence activated by a complement + antibody to human erythrocytes (B) Obelin luminescence activated by complement + anti-5'-nucleotidase antibody. (a) Fused hybrids (below) (b) non-fused cell mixture of 'ghosts' and polymorphs. Light measured as luminescence counts per sec (cps)

(2) Ca^{2+} could induce a change in phospholipid structure leading to inverted micelle formation or breakdown of the two bilayers at the point of contact between the vesicle and the cell membrane (Fig. 7.11). One fusogenic agent, produced perhaps in the cell by a Ca^{2+}-activated phospholipase, which could do this is lysolecithin (Poste and Allison, 1973); another is diacylglycerol (Allan *et al.*, 1976a, b; Allan and Michell, 1977) which could be produced by a Ca^{2+}-activated phosphatidase.

(3) Calmodulin. Precisely how Ca^{2+}–calmodulin could induce membrane fusion is not known. What is known is that phenothiozines like trifluoperazine, which block Ca^{2+}–calmodulin activation of several isolated enzymes, also block the stimulation of several secretory cells. Furthermore, not only has calmodulin been found in synaptic vesicles (De Lorenzo *et al.*, 1979) and in secretory cells

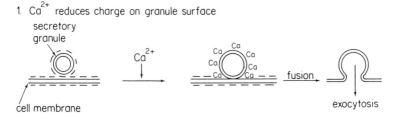

Fig. 7.11. Possible mechanism of Ca^{2+}-induced phospholipid change during membrane fusion. These mechanisms are still to be established

(Kuo and Coffee, 1976; Sugden et al., 1979; Hutton et al., 1981), but also exocytosis in nerve terminals and mast cells is accompanied by a rapid, Ca^{2+}-dependent phosphorylation of specific proteins including membrane proteins (Krueger et al., 1977; Schulman and Greengard, 1978; Sieghart et al., 1978). The elegant system developed by Baker and Knight (1978) may enable the factors required for vesicle–membrane fusion to be defined.

Release of substances from vesicles

The final stage of secretion is the release of substances from the vesicles which have fused with the cell membrane. Many of these substances are bound, or are even crystalline, within the original vesicle; for example, adrenaline–ATP in the chromaffin granule, histamine–ATP in the mast cell, 5-hydroxytryptamine–ATP in the platelet, insulin crystals in the pancreatic β-cell, or acetylcholine bound to intravesicular proteins. It is difficult to dissolve thyroid hormones in aqueous media at concentrations greater than 0.1–1 mM. This is perhaps why the thyroid cell has had to evolve a means of storing thyroxine at high concentration in a soluble form, coupled to the protein thyroglobulin. However, most products or complexes from secretory granules seem to dissolve very rapidly. Whilst it was reported that Ca^{2+} might influence the binding of oxytocin and vasopressin to neurophysin (Smith and Thorn, 1965; Ginsburg et al., 1966), there is no evidence that this is normally a rate-limiting step in secretion nor that it can be regulated by intracellular Ca^{2+}.

7.2.7. Secondary regulators of exocytosis

In many secretory cells, exocytosis, activated by a primary stimulus, is subject to secondary regulation (Table 7.6). Secondary regulators can either enhance or inhibit exocytosis, though in several cell types only one class of secondary regulator has so far been discovered (Table 7.6). An important feature of a physiological secondary regulator is that it requires the presence of the primary stimulus. Little or no effect will be observed on basal secretion. For example, enteroglucagon and gut peptides activate insulin secretion but only in the presence of glucose, the primary stimulus. Similarly, enzyme secretion from polymorphs initiated by the immune system can be enhanced by α-adrenergic agonists, $PGF_{2\alpha}$ or cholinergic receptors, and inhibited by β-adrenergic receptor occupation, PGE_1 or histamine H_2 receptor occupation.

Whilst the precise mechanism by which these secondary regulators act is not yet known, in most cases changes in intracellular cyclic AMP and cyclic GMP seem to be necessary (Table 7.12). Some primary stimuli also alter cyclic nucleotide concentrations, perhaps as a means of prolonging or curtailing the cell response. A role for cyclic AMP, presumably through phosphorylation of intracellular proteins modifying the response of secretory cells to primary stimuli, is supported by effects of cyclic AMP or theophylline on isolated cells (Table 7.13).

Table 7.12. Effects of primary and secondary regulators on cyclic nucleotides in secretory cells

Cell	Effector	Effect on cyclic AMP	Effect on cyclic GMP
Nerve terminals (brain)	Adenosine (si)	Increase	—
Retinal rods	Dark adaptation (sa)	Increase	Increase
	Light (p)	Decrease	Decrease
Anterior pituitary	Releasing factors (p)	Increase	—
	Dopamine (si)	—	—
	PGE_1, PGE_2 (si)	Increase	—
	Hypothalamic releasing factor (p)	—	Increase
Thyroid	TSH (p)	Increase	—
	Acetylcholine (si)	—	Increase
Adrenal medulla	Acetylcholine (p)	Increase	—
Mast cell	Adrenaline (β) (si)	Increase	—
Exocrine pancreas	Acetylcholine (p)	—	Increase
	Secretin (p)	Increase	—
β-Cell (pancreas)	Glucose 3–10 mM (p)	No effect	—
	Glucose > 10 mM (p)	Increase	—
	Glucagon (sa)	Increase	—
	Adrenaline (α) (si)	Decrease	—
Platelet	ADP (p), collagen (p)	Decrease	Increase
	Adrenaline (α) (sa)	Decrease	Increase
	Adrenaline (β) (si), PGE_1 (si)	Increase	—
Polymorph	Opsonized particles (p)	—	Increase
	Adrenaline (α), $PGF_{2\alpha}$, acetyl choline (sa)	Decrease	Increase
	Adrenaline (β), PGE_1, histamine H_2 (si)	Increase	—
Mammalian parotid	Noradrenaline (β) (p)	Increase	—
	Adrenaline (α), acetylcholine	—	Increase

p = primary stimulus; si = secondary inhibitor; sa = secondary activator. —, indicates no effect or not investigated.
References: Goldberg et al. (1974), Berridge (1976b), Goldberg and Haddox (1977), Rasmussen and Goodman (1977), Kuroda (1978), Weissmann et al, (1980).

Table 7.13. Effects of exogenous cyclic AMP or theophylline on secretory cells

Cell	Agent	Effect
β-Cell (pancreas)	Dibutyryl cyclic AMP + glucose or theophylline + glucose	Stimulate insulin release
Anterior pituitary	Dibutyryl cyclic AMP or theophylline	Stimulate growth hormone, TSH, ACTH, prolactin and luteinizing hormone secretion
Mast cell	Dibutyryl cyclic AMP	Inhibits antigen-stimulated histamine release
Exocrine pancreas	Dibutyryl cyclic AMP	Stimulates enzyme secretion
Platelet	Cyclic AMP or dibutyryl cyclic AMP	Inhibits platelet activation
Polymorph	Dibutyryl cyclic AMP	Inhibits enzyme release induced by immune system
Mammalian parotid	Dibutyryl cyclic AMP	Stimulates amylase secretion
Thyroid	Dibutyryl cyclic AMP	Stimulates T_4 secretion

In principle, secondary regulators could act in any one of four ways: (1) They could alter the level at which a threshold is evoked in the cell by the primary stimulus. (2) They could modify the cytosolic Ca^{2+} transient by activating or inhibiting Ca^{2+} pumps, thereby altering the magnitude or time course of the cell response to the primary stimulus. (3) They could interact with the mechanism by which intracellular Ca^{2+} activates exocytosis. (4) They could act at a rate-limiting step independent of the action of intracellular Ca^{2+}.

Most secondary regulators are presumed to interact with the Ca^{2+}-dependent exocytotic mechanism. Cyclic AMP stimulation of the mammalian parotid, for example, can be reduced or abolished by incubation of the cells with EGTA. However, in most cases direct evidence is lacking. Measurement of Ca^{2+} transients and precise characterization of the molecular basis of Ca^{2+} and cyclic nucleotide are required.

Nevertheless, the effects of secondary regulators mostly fall into two broad categories: (1) Cyclic AMP enhances the primary response, for example, glucagon on pancreatic β-cells and dark adaptation on retinal rods. (2) Cyclic AMP inhibits and cyclic GMP enhances the primary response, for example in polymorphs, mast cells, platelets and nerve terminals.

It is not yet clear precisely which category cells from tissues like the exocrine pancreas and the pituitary fit into. The mechanism selected by evolutionary forces will presumably have depended on four main factors: (1) Where the increase in cytosolic Ca^{2+} has to occur. (2) By what molecular mechanism the secondary regulator works. (3) What other phenomena are exhibited by the cell concerned. (4) Whether the source of Ca^{2+} for exocytotic activation is external or internal.

The evidence that intracellular Ca^{2+} is the universal trigger for exocytosis is strong. Direct measurement of cytosolic free Ca^{2+} in intact cells is likely to resolve many of the problems. These problems are well illustrated by the apparent complexity of the response of the polymorph in inflammation. This cell contains two morphologically identifiable granules: (A) azurophilic granules containing microbicidal enzymes like neutral proteases, myeloperoxidase, lysozyme and acid hydrolases; (B) 'specific' granules containing lysozyme, lactoferrin and vitamin B_{12} binding proteins. Opsonized particles, immune complexes, the chemotactic peptides C5a and N-formyl-Met-Leu-Phe, and A23187 + 10 μM Ca^{2+} experimentally stimulate secretion from both types of granule. On the other hand, phorbal esters, plant lectins like concanavalin A, or A23187 + 1 μM Ca^{2+} stimulate secretion only from the 'specific' granules. Furthermore, other processes that can be activated in the polymorph by these and other stimuli, and which are thought to be Ca^{2+}-dependent, include phagocytosis, oxygen radical formation, chemotaxis, and the production of endoperoxides, prostaglandins, thromboxanes and leukotrienes.

How interdependent are these different processes and, if an increase in intracellular Ca^{2+} is the trigger for them, how is the cell able to regulate them separately? One question yet to be answered is whether all the cells exhibit all these phenomena at the same time.

7.3. Endocytosis

7.3.1. The phenomena of phagocytosis and pinocytosis

Endocytosis is the uptake of substances into vesicles within the cell. These vesicles are formed by invagination of the cell membrane (Fig. 7.12). Phagocytosis and pinocytosis are special examples of endocytosis. Phagocytosis is the uptake of large particles like bacteria, which may be several micrometres in diameter, and was first observed by Metchnikoff (1905). Pinocytosis, on the other hand, is the uptake of non-particulate material into vesicles inside the cell and was first reported by Lewis in the early part of this century. Pinocytosis can be classified into two categories, depending on the size of the vesicle formed. Macropinocytosis causes the formation of vesicles 0.3–2 μm in diameter. It can be observed in the light microscope; for example, in thyroid follicular cells when they endocytose the thyroglobulin in the follicle. Micropinocytosis can only be observed in the electron microscope since the vesicle may be as small as 50–70 nm in diameter. This phenomenon is assoicated with a special protein, clathrin, and the formation of a so-called 'coated pit' on the cell surface which precedes vesicle formation. All eukaryotic cells exhibit some form of endocytosis. However, some cells are particularly active. For example, in protozoa endocytosis is an essential means of obtaining food (Chapman-Adresen, 1977). Endocytosis by polymorphonuclear leukocytes (polymorphs), macrophages and other cells of the reticuloendothelial system plays a key role in the inflammatory response of vertebrates, enabling the organism to destroy foreign organisms and to remove dead tissue (Page et al., 1978; Stossel, 1974). Platelets can also exhibit phagocytosis (Mustard and Packham, 1968). In diseases such as pneumoconiosis phagocytosis of dust particles by fibroblasts may play an important part in pathogenesis.

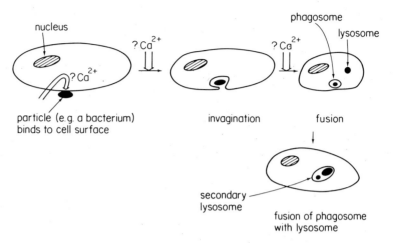

Fig. 7.12. Endocytosis

In contrast to phagocytosis, most eukaryotic cells are pinocytosing all the time, and in some cells this process has been shown to be stimulated by hormones. For example, insulin stimulates pinocytosis in fat cells (Barnett and Ball, 1960), and endocytosis of thyroglobulin in the thyroid is stimulated by thyroid stimulating hormone. Endocytosis is also important in secretory cells (Fig. 7.3). It is the mechanism by which the secretory granule membrane is removed from the cell membrane after its fusion as part of the exocytosis.

Is endocytosis a 'threshold phenomenon'?

Phagocytosis is provoked by the binding of chemical stimuli to receptors on the outside surface of the cell. Polymorphs, for example, can phagocytose latex beads but the uptake is enhanced if the particles are first coated with immunoglobulin complexes or complement component C3b, after activation of the complement pathway. This process is called opsonization and activates the phagocytic response by binding to Fc and C3b receptors on the cell surface. A similar process happens naturally when infecting bacteria are coated with specific IgG *in vivo*. In the absence of particles no phagoctosis can occur. So this phenomenon can be considered to exhibit a 'threshold' for each individual cell. A cell can of course take up more than one particle. It remains to be established whether the activation of pinocytosis by chemical stimuli can be considered as a threshold phenomenon, since individual cells have not been studied to see if they can be transformed abruptly from one state to another.

7.3.2. Is intracellular Ca^{2+} the trigger for endocytosis?

Removal of external Ca^{2+} has been shown to inhibit some, but not all, examples of endocytosis (Table 7.2). The physiological significance of these effects is not clear since removal of external Mg^{2+} also inhibits phagocytosis (Allison *et al.*, 1963; Stossel, 1973). Ca^{2+} could, in some cases, be required for binding of the particles to the receptor sites on the surface of the cell. Furthermore, phagocytic stimuli initially depolarize then hyperpolarize the membrane. If this depolarization is necessary for phagocytosis to take place, then this too could be deleteriosuly affected by the removal of external Ca^{2+}. A further problem for the Ca^{2+} hypothesis is that no change in intracellular free Ca^{2+} has been seen in cells exposed to a phagocytic stimulus (Hallett and Campbell, 1982b), and yet a prolonged rise in cytosolic Ca^{2+} has been observed in polymorphs incubated with the chemotactic peptide N-formyl-Met-Leu-Phe. It is still possible, however, that particles induce a local, transient increase in cytosolic Ca^{2+}. Ca^{2+}, in the mM range, actually inhibits endocytosis in freshwater amoebae. These organisms normally live in water which contains 10–200 μM Ca^{2+}.

The complete phagocytotic process involves six phases (Fig. 7.12). (1) Binding of the particle to glycoproteins on the cell surface, which may be specific receptors like Fc or C3b. (2) Invagination of the particle, a process involving pseudopod formation, contractile activity at the inner surface of the cell membrane, and is inhibited by the microfilament disrupter cytochalasin B. (3) Vesiculation of the cell membrane to form a 'phagosome'. (4) Fusion of the phagosome with intracellular granules (lysosomes). (5) Digestion of the particle, (6) Excretion of undigested material.

Intracellular Ca^{2+} could in theory be involved in the activation of any one of these processes, although Ca^{2+} has been reported not to be required for fusion of the phagosome with lysosomes in *Acanthamoeba*. If intracellular Ca^{2+} is involved then its most likely roles are in stimulating pseudopod formation in the invagination of the particle, or membrane fusion to form the phagosome. However, few of the criteria for intracellular Ca^{2+}, and laid down in Chapter 2, have yet to be satisfied.

Secondary regulators of particle uptake have yet to be clearly defined, although it has been reported that cyclic AMP inhibits phagocytosis. Whether cyclic nucleotides can regulate intracellular Ca^{2+} in phagocytes remains to be established.

7.3.3. Intracellular Ca^{2+} and other phenomena associated with endocytosis

Several other phenomena are associated with phagocytosis in, for example, polymorphs and macrophages. The evidence for intracellular Ca^{2+} being the mediator of these other effects is better than it is for phagocytosis itself. They include secretion of enzymes (this chapter, section 7.2), chemotaxis induced by C5a after opsonisation of particles with C3b or C567 (Chapter 5), granule movement, increased glycogen breakdown and glycolysis to satisfy the increased energy requirement of the cell (Chapter 6), increased hexose monophosphate pathway, and oxygen radical formation. The last of these is necessary for efficient digestion and killing of bacteria inside the phagolysosome. The release of oxygen radicals may also play an important part in the pathogenesis of diseases such as rheumatoid arthritis where large numbers of phagocytes are found surrounding areas of tissue damage (Chance et al., 1979).

What are oxygen radicals?

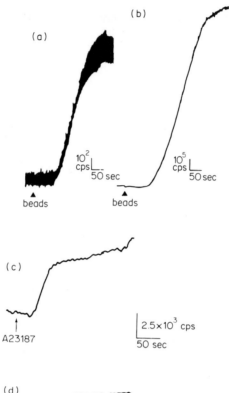
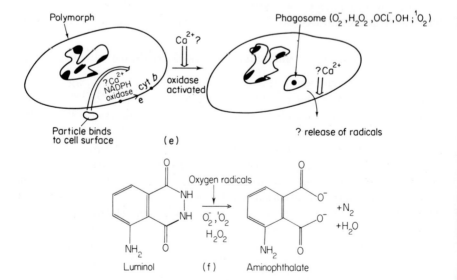

Oxygen radicals, peroxides and chemiluminescence

Polymorphs and macrophages when exposed to a phagocytic stimulus exhibit a rapid uptake of oxygen, the so-called 'respiratory burst' (Klebanoff and Clark, 1978). This is not inhibited by cyanide and therefore cannot be explained by increased activity of the respiratory chain in mitochondria. The respiratory burst is caused by the activation of an NADPH oxidase plus possibly a b cytochrome in the cell membrane, resulting in formation of superoxide anion (O_2^-). This oxidation of NADPH, together with increased activity of the glutathione–NADPH cycle, is the cause of the increase in the hexose monophosphate pathway during phagocytosis. The chemistry of oxygen radical and peroxide production is complicated. The species include O_2^-, ·OH, OCl^-, H_2O_2 and $^1O_2$ and can be produced by the following reactions:

$$NADPH + O_2 \xrightarrow{oxidase} NADP^+ + O_2^- \text{ (superoxide anion)}$$

$$2O_2^- + 2H^+ \xrightarrow{spontaneous} H_2O_2 + {}^1O_2 \text{ (singlet oxygen)}$$

$$O_2^- + H_2O_2 \xrightarrow{spontaneous} OH^- + O_2 + \cdot OH \text{ (hydroxyl radical)}$$

$$H_2O_2 + Cl^- \xrightarrow{myeloperoxidase} H_2O + OCl^- \text{ (hypochlorite anion)}$$

$$H_2O_2 + OCl^- \xrightarrow{spontaneous} H_2O + Cl^- + {}^1O_2$$

Several of these species, for example O_2^- and ·OH, have unpaired electrons and are therefore highly reactive. Singlet oxygen is an excited state of O_2 where the outer electron pair is in the singlet spin state, instead of the normal triplet state. These various peroxide moieties be scavenged by proteins, for example albumin and caeruloplasmin, or by enzymes, for example

$$2O_2^- + 2H^+ \xrightarrow{superoxide\ dismutase} O_2 + H_2O_2$$

$$2H_2O_2 \xrightarrow{catalase} 2H_2O + O_2$$

A large production of these moieties is provoked within 20–30 sec in polymorphs not only by a phagocytic stimulus, but also by immune complexes, chemotactic peptides and formation of the terminal complement attack complex C56789 (Hallett *et al.*, 1981). The response can be detected in the intact cell because $^1O_2$ (singlet oxygen), and the reaction of some of the oxygen species with unsaturated hydrocarbon chains, produces light (Fig. 7.13a) (chemilumines-

Fig. 7.13. Polymorph chemiluminescence. From Hallett *et al.* (1981); reproduced by permission of Blackwell Scientific Publications Limited. stimulus (↑) = latex beads on rat polymorphs or A23187. (a) Endogeneous chemiluminescence (i.e. no luminol). (b) Chemiluminescence in the presence of luminol (1.1 µM). (c) Chemiluminescence in the presence of luminol and A23187 (1.9 µM) in DMSO (0.1 %). (d) Chemiluminescence in the presence of luminol and DMSO (0.1 % v/v). External Ca^{2+} (1 mM) present in all cases. Light measured in luminescence counts per sec (cps). (e) Possible mechanisms for oxygen radical production. (f) The 'luminol reaction'

cence). The light is not visible to the naked eye, comes from the cells and not from the surrounding fluid, and can be enhanced several thousand-fold by addition of an organic compound such as luminol which is chemiluminescent when it is oxidized, for example, by O_2^- (Fig. 7.13b). Each cell emits only a few hundred photons. Chemiluminescence, induced by particles, is prevented by removal of external Ca^{2+} (Fig. 7.14a, b), and initiated by A23187 without apparently needing particles (Fig. 7.13c). The response can be detected within 30 sec of adding the stimulus and is inhibited by trifluoperazine (Fig. 7.14), which can act as a local anaesthetic at high concentrations, but does inhibit the action of calmodulin in the range 1–50 μM in broken cell preparations (see Chapter 3). An increase in free Ca^{2+} has also been detected after complement activation, but not phagocytosis, in cell hybrids produced by fusion using Sendai virus, and containing the photoprotein obelin (Hallett and Campbell, 1982b). Chemotactic peptides also increase intracellular free Ca^{2+} in these hybrids.

It is not yet known whether the production of peroxide moieties is a threshold phenomenon at the level of individual cells. Chemiluminescence is barely detectable in unstimulated cells, and the light emission in individual stimulated cells is too faint to be detected even by image intensification. Nor is it yet known whether histamine, adrenaline and prostaglandins, which act as secondary regulators of chemotaxis and exocytosis in polymorphs, also regulate peroxide formation in these cells.

Although Ca^{2+}–calmodulin activation of NADPH oxidase could explain the apparent requirement for intracellular Ca^{2+} in oxygen radical formation, Ca^{2+} apparently has no effect on this enzyme in isolated granules (Rossi *et al.*, 1976).

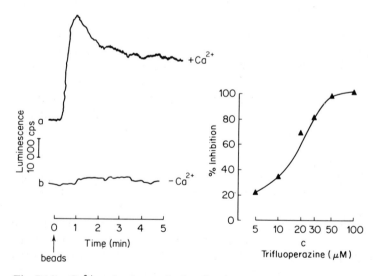

Fig. 7.14. Ca^{2+} and polymorph chemiluminescence. From Hallett *et al.* (1981); reproduced by permission of Blackwell Scientific Publications Limited. Stimulus (↑) = latex beads. (a) No external Ca^{2+}. (b) +external Ca^{2+} (1 mM). (c) Inhibition by trifluoperazine

One problem is that formation of fused phagolysosomes may result in a change in pH, which may not remain constant. Many of the reactions involving peroxide moieties, including chemiluminescence, are particularly sensitive to H^+ concentration.

The chemiluminescence response is usually dependent on external Ca^{2+}, but the respiratory burst has been detected in its absence. Furthermore, Sendai virus can induce some chemiluminescence from polymorphs without external Ca^{2+} being present. The possibility therefore exists that some internal store of Ca^{2+} may be necessary. The identity of this store and the molecular mechanism by which intracellular Ca^{2+} initiates the 'peroxide' (oxygen radical) response remain to be established. A further problem in experimental interpretation is the possible structural requirement (see Chapter 1) for Ca^{2+} in binding stimuli (e.g. particles) to the cell membrane.

The evidence that intracellular Ca^{2+} is the initiator of many of the phenomena associated with phagocytosis is quite convincing. Less convincing is the evidence of a role for intracellular Ca^{2+} in particle uptake itself, or in other examples of endocytosis such as pinocytosis.

It is now necessary to apply the criteria laid down in Chapter 2 to each part of the endocytotic process in a system where this can be quantified in each cell.

7.4. Non-vesicular secretions

The release of substrates and metabolites from cells occurs through specific permeases in the cell membrane or by diffusion. In addition, some cells release specialized fluids and lipid-soluble substances, apparently through a non-vesicular, non-exocytotic mechanism (Table 7.4). The stimuli for these non-vesicular secretions in both vertebrates and invertebrates are usually hormones or neurotransmitters (Table 7.14). The secretions may occur in cells which also exhibit exocytosis, for example the secretion of saliva fluid (non-vesicular) and amylase (vesicular) from the mammalian parotid. In other tissues the non-vesicular secretion is from different cells from those exhibiting exocytosis, for example the secretion of alkaline fluid (non-vesicular) from the centroacinar cells of the exocrine pancreas in parallel with the release of degradative enzymes (vesicular) from the acinar cells.

A further interesting feature in several of these non-vesicular secretory cells is that the primary stimulus may act on one surface of the cell whilst the secretion occurs at another surface. This is the case for the stimulation by secretion of $NaHCO_3$ from the centroacinar cells in the exocrine pancreas, and the secretion of fluid in mammalian salivary glands stimulated by noradrenaline or acetylcholine, though in this latter tissue there is also a release of K^+ at the serosal surface (Fig. 7.15). A similar polarity of the cell is seen in the bladder, where antidiuretic hormone acting at the serosal surface stimulates uptake of ions and water on the other side of the cell at the mucosal, or apical, surface. In such cells a mechanism is required which enables the hormone or neurotransmitter to

Table 7.14. Stimuli of some non-vesicular secretions

Cell	Primary stimulus	Secretion
Mammalian salivary gland	Adrenaline (α), acetylcholine	Fluid containing NaCl
Insect (*Calliphora*) salivary gland	5-Hydroxytryptamine	Fluid containing KCl
Exocrine pancreas (centroacinar cells)	Secretin	Fluid containing $NaHCO_3$
Gastric mucosa	Gastrin	Acid fluid
Sweat gland	Acetylcholine	Fluid containing NaCl
Salt gland in birds	Acetylcholine	Fluid containing NaCl
Bladder	Antidiuretic hormone	Absorption of Na^+ and H_2O
Adrenal cortex		
zona fasciculata	ACTH	Glucocorticoids
zona glomerulosa	ACTH, antiogensin II, 5-hydroxytryptamine	Aldosterone
Insect prothoracic gland	Prothoracotropic brain hormone	α-Ecdysone

All secretions are assumed to be non-vesicular though some, e.g. H^+ from gastric mucosa (Jacobson *et al.*, 1975), may be vesicular. Also included is ion absorption in the bladder.

increase the intracellular messenger concentration, be it cyclic nucleotide or Ca^{2+}, at the appropriate surface of the cell.

Of the non-vesicular secretions listed in Table 7.14 little or nothing is known of any secondary regulators. As a result they have been classified as monodirectional systems (Berridge, 1976b). If secondary regulators do exist then prostaglandins or adenosine, produced locally, are the most likely candidates. Another unknown characteristic of these cells is whether they exhibit a threshold response under physiological conditions. The presumption at present is that they do not, though the tissue response has yet to be studied at the level of single cells or a single acinus.

The stimulation of cyclic AMP formation by 5-hydroxytryptamine acting on insect salivary gland, ACTH acting on the adrenal cortex, and antidiuretic hormone acting on the bladder, has lead to the hypothesis that a cyclic AMP protein kinase induces phosphorylation of proteins inside the cells that are responsible for activation of the response. However, the poor correlation between tissue cyclic AMP, under certain experimental conditions, and the rate of secretion has led to a search for an alternative, or additional, intracellular messenger.

The removal of external Ca^{2+} inhibits, and can abolish, the stimulation of many non-vesicular secretions (Table 7.15). Unfortunately, there are several problems in the interpretation of these experiments. For example, in the fly salivary gland it is necessary to incubate the tissue for several minutes in a Ca^{2+}-free medium before an inhibition of the rate of secretion, initiated by 5-hydroxytryptamine, is observed. In contrast, the bladder and adrenal cortex require external Ca^{2+} for maintenance of the normal structure of the tissue. Toad bladder disaggregates rapidly into individual cells in a Ca^{2+}-free medium, whereas in the adrenal cortex

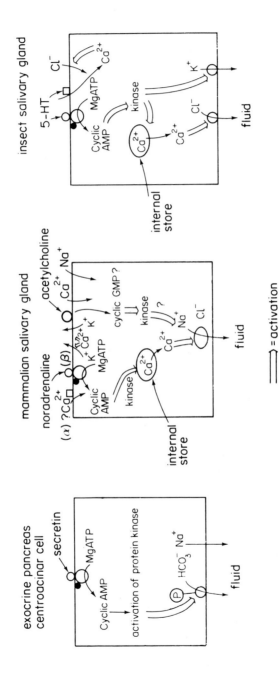

Fig. 7.15. Fluid secretion—some possible mechanisms

Table 7.15. Revomal of external Ca^{2+} inhibits non-vesicular secretions

Secretion	Stimulus	Reference
KCl from insect salivary gland	K^+ or 5-hydroxytryptamine	Prince et al. (1973); Berridge et al. (1975)
NaCl from mammalian salivary gland	Adrenaline (α)	Batzri and Selinger (1973)
	Acetylcholine	Douglas and Poisner (1963)
H^+ from gastric mucosa	Gastrin	Jacobson et al. (1975)
Cortisol from adrenal cortex	ACTH	Birmingham et al. (1953, 1960); Farese (1971a, b)

Prolonged incubation in EGTA may be necessary to demonstrate maximum inhibition.

Ca^{2+} is required for transduction of the effect of ACTH-receptor complex, through a presumed GTP-binding protein, to adenylate cyclase.

The question therefore arises, does intracellular Ca^{2+} play an active role in mediating the response of cells to a physiological stimulus that provoke the secretion of specialized fluids or lipophilic hormones by a non-exocytotic mechanism?

7.4.1. Fluid secretions

Fluids released by cells are electrolyte solutions which regulate the external ionic environment, and flush out substances secreted through exocytosis from cells within the same gland. The salivary glands of mammals release saliva rich in Na^+ and amylase, whereas the fluid from insect salivary gland is rich in KCl. The stimulation of fluid secretion usually involves the active extrusion of one ion followed by passive movement of a counterion (Fig. 7.15). The resulting osmotic imbalance between the inside and outside of the cell causes a net movement of H_2O across the cell. In mammalian parotid, Na^+ efflux at the duct surface is stimulated by acetylcholine, whereas K^+ efflux is stimulated at the serosal surface, the counterion for either Na^+ or K^+ being Cl^-. In the centroacinar cells of the pancreas it is HCO_3^- efflux which is stimulated, in this case by secretin, the counterion being Na^+. In contrast, in the fly salivary gland the efflux of both K^+ and Cl^- is stimulated by the primary stimulus, 5-hydroxytryptamine.

Insect salivary gland

Fluid secretion from the luminal surface of the salivary gland of the blowfly *Calliphora erythrocephala* is stimulated by 5-hydroxytryptamine, acting at the serosal surface membrane. The fluid secretion is caused by activation of both K^+ and Cl^- efflux at the luminal surface. Passive influx of K^+ and Cl^- at the serosal surface maintains the ionic content of the cells.

The interaction of 5-hydroxytryptamine with its receptor causes both a rapid increase in intracellular cyclic AMP, through activation of adenylate cyclase, and an increase in Ca^{2+} influx. Changes in membrane and transepithelial potential also occur. Cyclic AMP seems to be mainly responsible for activating K^+ efflux at the luminal surface (Berridge, 1976b). However, cyclic AMP does not stimulate Ca^{2+} influx, nor is cyclic AMP directly responsible for the activation of Cl^- efflux.

Depletion of intracellular Ca^{2+} stores inhibits the effect of 5-hydroxytryptamine. Furthermore, A23187 can stimulate a rapid burst of secretion in the absence of external Ca^{2+}, and cyclic AMP added externally, like 5-hydroxytryptamine, increases Ca^{2+} efflux. It has therefore been proposed that 5-hydroxytryptamine stimulates Cl^- efflux by an increase in cytosolic free Ca^{2+} released from the mitochondria by cyclic AMP, presumably via protein kinase (Berridge, 1976b). The resting cytosolic free Ca^{2+} in these cells is about $0.1\ \mu M$ and is increased to about $1\ \mu M$ by a maximal concentration of 5-hydroxytryptamine, as shown using a Ca^{2+}-sensitive microelectrode (Berridge, 1980). Although the mitochondria contain a reasonable proportion of the total cell Ca^{2+} it remains to be established whether they are the source of the Ca^{2+} for Cl^- secretion. An increase in Ca^{2+} influx also occurs and is associated with an increase in the breakdown of phosphatidylinositol in the salivary gland (Fain and Berridge, 1979). Prolonged exposure to 5-hydroxytryptamine results in a fall off in Ca^{2+} entry and a reduction in the rate of secretion. Incubation in a medium containing inositol enables the response to 5-hydroxytryptamine to be restored. It has therefore been proposed that in addition to activation of adenylate cyclase another 5-hydroxytryptamine receptor initiates the opening of a Ca^{2+} channel, through the breakdown of a small fraction of membrane phosphatidylinositol. This Ca^{2+} entry is necessary to maintain the stimulation of Cl^- secretion after the intracellular Ca^{2+} store has released its Ca^{2+} in the first phase of secretion. Whether the Ca^{2+} acts via calmodulin remains to be established.

Mammalian parotid

Activation of the acetylcholine or α-adrenergic receptors in parotid cells leads to increased fluid secretion at the luminal surface and an increased K^+ efflux at the serosal surface. These receptors stimulate little or no amylase release (exocytosis) in contrast to the β-adrenergic receptor which stimulates amylase release, but little or no fluid secretion.

The effects of acetylcholine and α-adrenergic receptors are rapidly abolished by removal of external Ca^{2+}. In addition, there is a hyperpolarization of the cell membrane which cannot be wholly explained by the increase in K^+ efflux at the serosal surface, and therefore may also involve an increased Na^+ influx. Intracellular cyclic GMP concentrations are also increased. It is proposed that there is an increase in cytosolic free Ca^{2+}, caused by the increased influx of Ca^{2+} at the serosal surface. This activates K^+ efflux and guanylate cyclase. There follows an activation of Na^+ efflux at the luminal membrane into the ducts of the salivary gland. This hypothesis is able to explain how fluid and enzyme secretion can be controlled independently whilst both requiring an increase in cytosolic free Ca^{2+}. If this is to be so, then the increase in free Ca^{2+} must occur at different sites in the cell, depending on whether the stimulus is acetylcholine or noradrenaline. The molecular mechanisms by which these various ion fluxes are stimulated remain to be established.

Centroacinar cells of the exocrine pancreas

The stimulation by secretin of HCO_3^- secretion from the centroacinar cells of the pancreas is thought to be mediated by cyclic AMP. There is, as yet, no evidence of a role for intracellular Ca^{2+}.

Bladder

Earlier in this chapter a parallel was drawn between endo- and exocytosis. Similarly, a parallel exists between the release and uptake of fluids. The stimulation of Na^+ and H_2O uptake at the mucosal surface of the bladder by antidiuretic hormone acting on the serosal surface was originally thought to be mediated entirely by cyclic AMP. Antidiuretic hormone increases toad bladder cyclic AMP, and the effect on Na^+ and H_2O movement is mimicked by exogenous cyclic AMP and potentiated by theophylline (Argy et al., 1967). Yet antidiuretic hormone and cyclic AMP stimulate $^{45}Ca^{2+}$ efflux from the bladder cells. It is tempting to propose that cyclic AMP acts to release Ca^{2+} from an internal store close to the mucosal membrane and thereby increase Na^+ and H_2O uptake.

7.4.2. Steroid hormone secretion

Steroid hormones have a variety of roles in the function and development of both vertebrates and invertebrates. In vertebrates glucocorticoids released from the adrenal cortex regulate glucose and amino acid metabolism, mineralocorticoids like aldosterone regulate Na^+ and K^+ balance, and sex steroids control the secondary sexual characteristics. On the other hand, in insects ecdysones released from the thoracic or prothoracic glands regulate moulting.

Lipid globules have been observed in several steroid-secreting cells in the electron microscope, but these globules do not seem to be enclosed within a phospholipid membrane. The presumption is that steroid hormone secretion does not occur by exocytosis (Fawcett et al., 1969), and that the rate-limiting step(s) is in the synthetic pathway from cholesterol (Fig. 7.16). The most well-studied of these steroid secretions is the stimulation of glucocorticoid release from the mammalian adrenal cortex caused by the pituitary hormone ACTH. The relative importance and mechanism of action of intracellular Ca^{2+} and cyclic AMP in this stimulation is still not fully established (Tait et al., 1980).

The mammalian adrenal cortex consists of three main groups of cells. Under the adrenal capsule is the zona glomerulosa, then the zona fasciculata and then, next to the medulla, the zona reticularis. Aldosterone is released from the zona glomerulosa. In man the major glucocorticoid is cortisol, released from the zona fasciculata and the zona reticularis. In contrast, in rat, rabbit and some other mammals, corticosterone is the major glucocorticoid, being released from all three zones.

ACTH stimulates glucocorticoid release by increasing the flux of two rate-limiting steps, 20α-hydroxycholesterol to pregnenolone which occurs in the mitochondria, and 11β-hydroxylation of progesterone to corticosterone (Fig. 7.16). In addition, ACTH may enhance release of steroid from lipid droplets. The original idea that ACTH stimulates steroid synthesis by activating phosphorylase, thereby increasing the glucose 6-phosphate concentration and thus NADPH for steroidogenesis through the hexose monophosphate pathway, proved to be incorrect.

The increase in pregnenolone synthesis caused by ACTH seems to be due mainly to an increased association of mitochondrial cholesterol with cytochrome

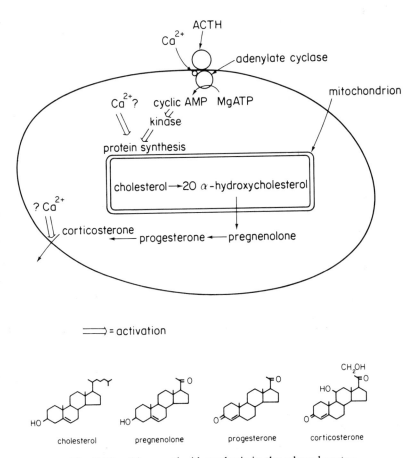

Fig. 7.16. Glucocorticoid synthesis in the adrenal cortex

P-450 rather than an increase in enzyme activity (Tait *et al.*, 1980). Protein synthesis is required for maximal stimulation by ACTH and this is thought to be mediated through a cyclic AMP dependent protein kinase phosphorylating ribosomes, and increasing the rate of translation of mRNA.

ACTH activates adenylate cyclase in zona fasciculata cells, and this requires external Ca^{2+} (Carchman *et al.*, 1971). It was once thought that a cyclic AMP induced phosphorylation of proteins inside these cells was entirely responsible for the ACTH activation of steroidogenesis. However, several experiments have since shown that there are problems with this simple idea. Firstly, the intracellular cyclic AMP concentration does not completely correlate with the increased rate of steroid release. This is best demonstrated experimentally using a specially prepared analogue of ACTH, *o*-nitrophenylsulphenyl (NPS)-ACTH, or using parts of the ACTH molecule such as $ACTH_{5-24}$ or $ACTH_{6-24}$ and comparing them with the activity of native $ACTH_{1-59}$ (Bristow *et al.*, 1980). The effect of NPS-ACTH is approximately 70 times more potent on steroidogenesis than ACTH, whereas maximal concentrations of NPS-ACTH cause only a three-fold increase in cyclic AMP compared with a 100–200-fold increase for the native hormone. Secondly, addition of external cyclic AMP is not sufficient to produce maximal stimulation of steroidogenesis, nor to activate the flux between 20α-hydroxycholesterol and pregnenolone (Fig. 7.16). Another intracellular messenger is required to explain the complete action of ACTH on glucocorticoid secretion.

It has been proposed that there are two types of ACTH receptor on the cells, one activates adenylate cyclase, the other increases intracellular Ca^{2+}. Unfortunately, the evidence for intracellular Ca^{2+} is confused by the passive role that external Ca^{2+} plays in the coupling of the ACTH–receptor complex to adenylate cyclase. Furthermore, data on Ca^{2+} fluxes are equivocal, and A23187 or X-537A do not seem to stimulate steroid secretion (Neher and Milani, 1978a, b). However, ACTH-induced steroid release from the zona fasciculata is inhibited by Ca^{2+} antagonists such as verapamil, tetracaine, La^{3+} and ruthenium red.

ACTH causes membrane depolarization and in a K^+-depleted medium can generate action potentials in cells (Matthews and Saffran, 1973) that are not blocked by tetrodotoxin. Though this membrane depolarization is not essential for ACTH stimulation of secretion, the possibility that ACTH releases Ca^{2+} from internal stores and increases Ca^{2+} influx through a normally voltage-independent Ca^{2+} channel needs to be investigated.

The evidence for intracellular Ca^{2+} mediating aldosterone secretion from the zona glomerulosa cells is also somewhat circumstantial. Aldosterone secretion, like glucocorticoid secretion, is stimulated by ACTH, and also by 5-hydroxytryptamine, angiotensin II and an elevation in serum K^+. *In vitro* maximal stimulation by K^+ occurs at about 8 mM external K^+. Cyclic AMP plays a major role in the action of ACTH and 5-hydroxytryptamine, but not in the action of angiotensin II and K^+. The effects of these latter two stimuli may be mediated through intracellular Ca^{2+}.

In invertebrates the stimulation of secretion of ecdysones also seems to involve cyclic AMP (Vedeckis et al., 1976). A role for intracellular Ca^{2+} remains to be established.

7.5. Overall conclusions

Intracellular Ca^{2+} seems to play a key role in mediating the effects of primary stimuli on exocytosis, endocytosis, fluid secretion and possibly steroidogenesis.

Cyclic nucleotides are required to modify endo- and exocytosis when a cell responds to a secondary regulator, and may either enhance or inhibit the response depending on the cell concerned. This remains a working hypothesis and several problems remain. (1) Has the step in the secretory pathway, acted on by a primary or secondary regulator, been correctly identified? (2) Is it possible to demonstrate directly a change in cytosolic free Ca^{2+} at the site of regulation of secretion? (3) In cells which exhibit more than one type of response, or which respond independently to different stimuli, how is the particular cell response selected if they all involve an increase in intracellular Ca^{2+}? Is the cell able to do this by localising changes in cytosolic free Ca^{2+} to specific areas? (4) How is the increase in intracellular Ca^{2+} caused? (5) In cells which require external Ca^{2+} how is Ca^{2+} influx increased? In cells which release Ca^{2+} from an internal store can this store be definitely identified and the mechanism of release defined? (6) How does an increase in cytosolic Ca^{2+} evoke secretion? Is calmodulin or 'gelsolin' required?

The answers to these questions will not only enable detailed descriptions of individual secretory cells to be made, but also it should then be possible to rationalize these phenomena as a whole, together with their evolutionary significance.

CHAPTER 8

The reproduction and development of cells

8.1. Ca^{2+} and cell growth

It has long been known that changes in the concentration of Ca^{2+} in the fluid bathing cells can influence their rate of growth or their ability to divide. There have been reports of Ca^{2+} being required for the growth and development of both pro- and eukaryotic organisms (Table 8.1), as well as acting as a stimulus for cell division in some instances. Nearly 50 years ago Heilbrunn and Wilbur (1937) showed that addition of sodium citrate to the medium surrounding eggs of the marine annelid worm *Nereis limbata* inhibited breakdown of the germinal vesicles induced by agents such as ultraviolet light or excess potassium chloride. These so-called parthenogenetic stimuli, from the Greek word *parthenos* meaning virgin, can stimulate the cells to divide in the absence of sperm. Immersion of the eggs for 5–6 min in the medium containing citrate was necessary for maximum inhibition. The normal response of the cells could be restored by readdition of Ca^{2+}. In the same year, Mazia reported that a release of Ca^{2+} could be detected in homogenates of recently fertilized eggs of the sea urchin *Arbacia*. From these experiments Heilbrunn concluded that the trigger for cell division was the release of Ca^{2+} from the calcium proteinate gel in the cell cortex.

In spite of these observations made so many years ago [see Dalcq (1925) and Heilbrunn (1937, 1943) for further references], it was the discovery of cyclic AMP in the late 1950s which focussed the attention of many cell biologists on the role of intracellular second messengers in provoking cell division. Cyclic AMP and cyclic GMP have been measured at many stages in the life cycle of cells (Rebhun, 1977). Cyclic AMP can be depressed or elevated, as can cyclic GMP. Cyclic AMP can activate cells to divide or inhibit the action of other stimuli, whilst some mutant cells appear to divide quite happily when they contain no cyclic AMP dependent protein kinase. Thus, no generalizations concerning the role of cyclic nucleotides in cell growth and development are possible and in many cases cyclic AMP, at least, is not essential for the normal cell cycle (Coffino *et al.*, 1975). Attention has therefore returned to intracellular Ca^{2+}, in the hope that this might be a universal trigger for cell division and that changes in cellular cyclic nucleotides might be rationalized through an interaction with intracellular Ca^{2+}.

Table 8.1. Extracellular Ca^{2+} and cell growth

Cell	Effect	Reference
Prokaryotes		
Various spore-forming bacteria	Removal of Ca^{2+} required for germination	Ellar *et al.* (1974)
Marine bacteria	Ca^{2+} required for growth	Hutner (1972)
Blue-green algae	Ca^{2+} required for growth	Hölm-Hansen (1968); Fogg *et al.* (1973)
Plants		
Many plants	Ca^{2+} required for root or whole plant growth	Burström (1968); Wyn-Jones and Lunt (1967); Hewitt and Smith (1974)
Leafy plants	Ca^{2+} inhibits abscission (loss of leaves)	Poovaiah and Leopold (1973)
Animals		
Hydroids	Ca^{2+} required for development	Pasteels (1935)
Various animal cells	Ca^{2+} potentiates growth	Rubin and Koide (1976)
Rat cells in tissue culture	Ca^{2+}, Mg^{2+}, K^+ required for growth	Shooter and Grey (1952)
Rat development	Plasma Ca^{2+} required	Perris *et al.* (1968)
Tissue culture cells	Ca^{2+} required for cell proliferation	Balk (1971); Balk *et al.* (1973)
Thymus	Parathyroidectomy induces involution of the tissue	Whitfield *et al.* (1973)
Thymic lymphoblasts	Ca^{2+} can be mitogenic	Whitfield *et al.* (1973)
Fibroblasts	Ca^{2+} can induce growth	Dulbecco and Elkington (1975)
Mouse L cells	Cell growth dependent on Ca^{2+} concentration	Birch and Pirt (1971)
Lymphocytes	Mitogenic activation requires Ca^{2+}	Whitney and Sutherland (1972)

The triggering of cell division by an external stimulus, or as a result of internal programming, is a good example of the type of threshold phenomenon defined in Chapter 2 (section 2.2). Many cells undergo a natural cycle which ends in division of the cell into two daughter cells, when the cycle begins again. Some cells, however, will only divide if provoked by a stimulus, for example the fertilization of an ovum by sperm or the stimulation of lymphocytes in the immune response. In these cases a chemical or electrical change in the cell must pass through a threshold if the cell is to be set on course for division. The growth and development of organisms, as well as individual tissues within the organism, involve many such threshold phenomena (Table 8.2). Of particular interest from a chemical standpoint are those cases where, once activated, the cell is set on a path from which it does not deviate, even when the original stimulus has long gone. In view of the evidence presented in the earlier chapters of this book it is perhaps not surprising that there has been much speculation over the role of

Table 8.2. Threshold phenomena associated with tissue growth and development

Phenomenon	Example	Reference
Spore germination	Various bacteria	Ellar et al. (1974)
Mitosis	Sperm-fertilized eggs	Gilkey et al. (1978)
	Parthenogenetic activation of eggs	Steinhardt et al. (1974)
	Lymphocyte activation	Greene et al. (1976)
	Thymocyte activation	Burgoyne et al. (1970)
	Many eukaryotic cells	Rebhun (1977)
Meiosis	Gamete formation	Kimball (1974); Highnam and Hill (1977)
Cell transformation (may sometimes result in cell division)	Sperm capacitation	Singh et al. (1978)
	Oocyte maturation	Moreau et al. (1980)
	Amoeboid to flagellate conversion in *Naegleria*	Willmer (1977); Fulton (1977)
	Virus-induced release of inhibition in tissue culture	Dulbecco (1975); Dulbecco and Elkington (1975)
	Pleiotypic response of animal cells	Hershko et al. (1971)

intracellular Ca^{2+} as the possible primary trigger of this type of cell activation (Berridge, 1976a, b; Rebhun, 1977).

Before examining the evidence for such a role for intracellular Ca^{2+} in stimulating the reproduction and development of cells and tissues, it might perhaps be propitious to review the biology of these phenomena so that we may first identify potential sites for the involvement of Ca^{2+}.

8.2. The biology of cellular reproduction and development

8.2.1. The life cycle of cells

No cell can live for ever. It must either grow and divide to form daughter cells, or it will eventually die (Sheldrake, 1974). This may take only a few hours, whilst in other cases the cell may survive for several hundreds of days without dividing. Cells can therefore be classified into three main groups: (1) continuously dividing cells, for example epithelial cells; (2) quiescent cells which can be activated to divide by a physiological or pharmacological stimulus, for example lymphocytes in the immune response; (3) quiescent cells which can no longer divide, the most extreme example being the mammalian erythrocyte which no longer contains a nucleus.

In 1951 Howard and Pelc, studying the uptake of radioactive phosphorus (^{32}P) into the DNA of continuously dividing cells, showed that DNA synthesis only took place during a discrete period of the cell cycle, some time before division occurred. Since DNA consists of a chain of nucleotides linked by phosphate–sugar bonds, where the major bases are adenine, guanine, cytosine and *thymine*, in contrast to those of RNA which are adenine, guanine, cytosine and *uracil*, radioactive thymine or the nucleoside thymidine can be used to measure DNA synthesis in cells, and radioactive uracil can be used to measure RNA synthesis. The use of [^{3}H] thymidine in a follow-up of Howard and Pelc's experiment has lead to the identification of four phases in the life cycle of continuously dividing cells (Fig. 8.1) [see Pelc and Howard, 1956 Baserga (1976) for details].

(1) G_1 phase. This is a discrete period immediately after a cell is formed in which synthesis of specific RNA and protein molecules prepare the cell for DNA synthesis. This may last from 1 to 40 hours, or even longer in some cases.

(2) S phase. This phase follows G_1. It usually lasts between 4 and 24 hours and is the only phase of the cell cycle in which significant DNA synthesis occurs. The chromosomal proteins, including the histones, are also synthesized during this phase. It is interesting to note that this phase is much longer in eukaryotes than the equivalent phase in bacteria; for example, for *Escherichia coli* it is only 20–30 min.

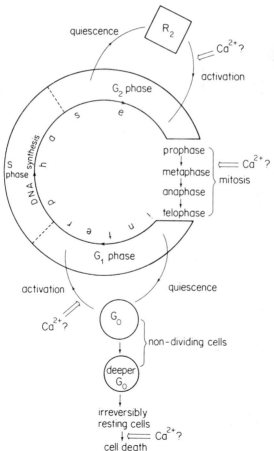

Fig. 8.1. The cell cycle in eukaryotes. ⇒ = activation
Note: evidence for G_0 much stronger than R_2

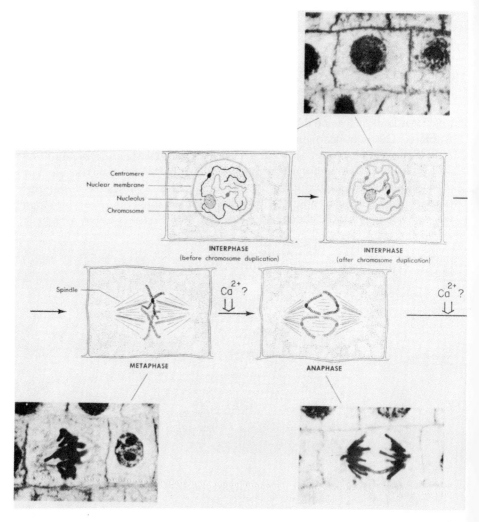

Fig. 8.2. The stages of mitosis. From Kimball, *Biology*, © 1974, Addison-Wesley,

(3) G$_2$ phase. This is a discrete period after DNA synthesis has occurred when certain RNA molecules and proteins, for example microtubular proteins for the mitotic spindle, are synthesized in preparation for cell division. This phase often lasts only 1–5 hours, though it can last for up to 2 days in organisms such as *Hydra*.

(4) Mitosis. Cell division itself may take only up to an hour and can be separated into four distinct phases (Figs. 8.1 and 8.2; see below).

The time of the complete cell cycle from one mitosis to the next varies considerably with different cells. For example, it may take only 10 hours in antibody-producing cells but up to 48 hours in the liver. The cycle may be

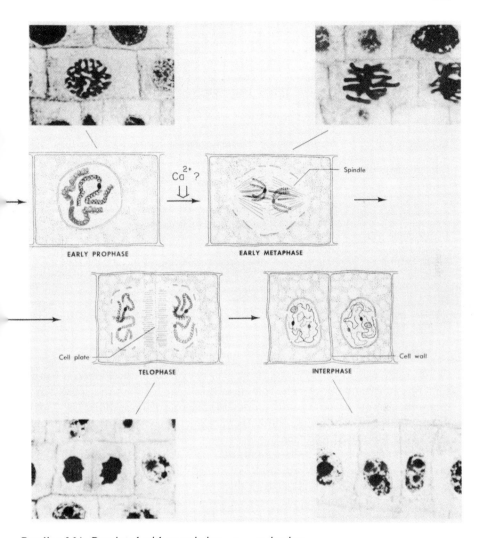

Reading MA. Reprinted with permission. ⇒ = activation

arrested in either the G_1 or G_2 phase to produce quiescent cells, but not in the S phase or during mitosis once this has begun. Many quiescent cells undergo morphological and chemical changes not normally associated with either the G_1 or G_2 phases of continuously dividing cells. This has lead to the classification of two other phases, namely G_0 and R_2 (Fig. 8.1). Quiescent cells can be awakened to rejoin the cell cycle by a variety of stimuli (Table 8.3) which include viruses, hormones and several chemical agents. The question we are particularly concerned with is, what is the role of intracellular Ca^{2+} in the cell cycle? The following possibilities need to be examined: (1) The role of intracellular Ca^{2+} in the passage of the cell from one phase of the cycle to the next; that is, G_1 to S, S to

Table 8.3. Some primary stimuli of cell growth, transformation and division

System	Stimulus
Invertebrates	Various neurosecretory growth-promoting hormones
Mammals	Various trophic hormones, e.g. TSH, ACTH, GH, FSH, Prolactin
Plants	Various growth hormones, e.g. auxins and cytokinins
Tissue injury	Chalones
Egg activation	Sperm, various parthenogenetic agents, e.g. ultraviolet light, chemicals
Gamete maturation	1-Methyladenine in starfish
	Progesterone in amphibians
	Luteinizing hormone on mammalian follicles
	Spermatogenic hormones
	Prostaglandins and sperm capacitation
Lymphocyte transformation	Antigen
	Plant lectins
Tissue culture	Reactivation, after contact inhibition, by foetal calf serum, insulin, trypsin
	Virus transformation, e.g. Rous sarcoma

TSH = thyroid stimulating hormone; ACTH = adrenocorticotropic hormone; GH = growth hormone; FSH = follicle-stimulating hormone.

G_2, G_2 to mitosis, G_1 to G_0 and G_2 to R_2. (2) The role of intracellular Ca^{2+} in the four phases of mitosis. (3) The role of intracellular Ca^{2+} in the activation of quiescent cells out of the G_0 or R_2 phases. (4) The role of intracellular Ca^{2+} in the differentiation of cells from G_0 into other phases, from which the cell may not be able to be stimulated to divide.

The primary stimulus for these events may be internal, arising naturally as part of the programme for the cell cycle, or it may be external, as in the case of activation of many quiescent cells (Tables 8.2 and 8.3). The present chapter is concerned with examining the evidence for intracellular Ca^{2+} mediating the effects of these stimuli.

8.2.2. Sexual reproduction

One of the prime characteristics of living organisms is their ability to reproduce. Two distinct methods have evolved for this, both requiring division and thus multiplication of cells. One, sexual reproduction, requires two parent cells, whilst the other, asexual reproduction, requires only one parent.

Unicellular organisms such as bacteria and protozoa can reproduce asexually by fission, producing two apparently identical cells. Yeasts, and several multicellular organisms including hydroids and many plants, can reproduce by budding. A further type of reproduction, found for example in fungi and mosses, is the formation of millions of spores, which remain dormant until germinated. Many bacteria and yeasts can also form spores, but this is more a protective

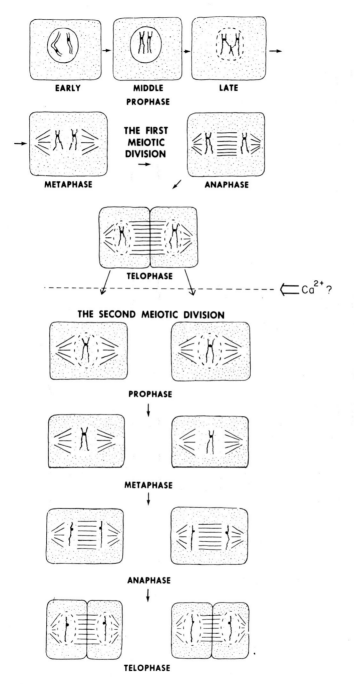

Fig. 8.3. The stages of meiosis. From Kimball *Biology*, © 1974, Addison-Wesley, Reading MA. Reprinted with permission. ⇒ = activation

mechanism against severe environmental conditions such as drought, rather than a genuine method of reproduction.

Virtually all prokaryote cells contain only one chromosome. They are haploid. Eukaryotic cells not only contain several chromosomes but also each cell contains two copies of each. They are diploid. The fruit fly *Drosophila* contains four pairs of chromosomes, humans 23 pairs, and the crayfish 100 pairs. Sexual reproduction in eukaryotes requires a form of cell division, distinct from mitosis, namely meiosis. The result of meiosis is the formation of gametes which contain only half the total number of chromosomes found in the original cells. The fertilization of the female gamete, the ovum, by the male gamete, the sperm, produces a zygote with the normal complement of chromosomes. In the case of multicellular organisms the zygote undergoes a series of cell divisions to produce an embryo which will eventually develop into a mature animal or plant. Intracellular Ca^{2+} could, in theory, play a role in provoking or regulating any one of four phases of this process: (1) meiosis (Fig. 8.3); (2) maturation of the gametes, necessary if fertilization and zygote formation are to occur; (3) fertilization and zygote formation, including stimulation of the first cell division; (4) differentiation of cells, necessary for embryonic development.

8.2.3. Cell transformation

We have already seen that the cell cycle of continuously dividing cells involves the transformation of the cell from one state to another, and that quiescent cells can be transformed to rejoin the cycle by various stimuli (Tables 8.2 and 8.3). Some types of cell can transform from one state to another without it necessarily leading to cell division. During the 1960's Tomkins and his group (Hershko *et al.*, 1971) noticed that the transformation of cells, in the presence or absence of cell division, caused a group of chemical changes which were apparently independent of cell type or stimulus. The term 'pleiotypic response' was invented to describe this phenomenon. The group of chemical changes encompassed by the pleiotypic response includes enhancement of protein and RNA synthesis, stimulation of intermediary metabolism, for example glycogen breakdown, as well as ionic changes in the cell. The primary stimulus for such a response is normally either a change in nutrient(s) or a hormone. Could cyclic nucleotides or intracellular Ca^{2+} be the second messenger mediating the effect of a primary stimulus which causes a pleiotypic response?

Another fascinating aspect of cell transformation not associated with cell division occurs in the protozoan *Naegleria gruberii* (see Willmer, 1970, 1977). This unicellular organism exists in an amoeboid form until activated by a change in the medium when, after a lag of a few minutes, an appendage appears transforming the organism into a flagellate. Under the microscope this cell transformation is quite dramatic as the relatively passive amoeboid protozoa suddenly become flagellate and start spinning and then quickly swim away. Is a change in intracellular Ca^{2+} the internal trigger mediating this type of 'threshold' transformation (Fulton, 1977)?

8.2.4. Embryonic growth and tissue development

After a series of cell divisions the zygote of most animals develops into a blastula, formed by morphogenesis and movement of some of the cells. It consists of three layers. The outer layer or ectoderm and the inner layer or endoderm sandwich a middle layer, the mesoderm. Some animals, notably the coelenterates, do not have a true mesoderm but are diploblastic with a jelly-like, almost acellular, mesogloea between the ecto- and endodermal cell layers. The cells from these three layers divide and differentiate into the various tissues to form a recognizable embryo and eventually an adult (Table 8.4). In plants, too, cell differentiation is necessary for formation of the xylem and phloem cell layers, as well as organized structures such as leaves and flowers associated with the mature plant.

The chemical basis of morphogenesis and tissue differentiation is still poorly understood. The primary regulators appear to be a combination of internal substances produced as a result of preprogramming in the DNA, and external substances including nutrient gradients, hormones and chalones, the latter being associated with inhibition of cell growth. Tissue differentiation is also necessary in normal tissue repair after injury.

The paucity of knowledge concerning the primary stimuli which cause cells to progress towards, for example, a nerve cell or an hepatocyte, means that it is difficult to investigate the intracellular regulators involved. Nevertheless, the apparent importance of intracellular Ca^{2+} in other types of cell transformation suggests that an investigation of its role in cell differentiation and tissue development might be fruitful.

Table 8.4. Fate of germinal layers in normal development

Ectoderm (outer layer) gives rise to:
  1. Epithelium and derivatives (skin, hair, mammary glands)
  2. Central and peripheral nervous systems
  3. Pituitary
  4. Amine Precursor Uptake and Decarboxylation (APUD) endocrine cells (gut, islets, thyroid C cells)

Mesoderm (middle layer) gives rise to:
  1. Adrenal cortex
  2. Muscle
  3. Lymphatic system and spleen
  4. Adipose tissue
  5. Kidney
  6. Connective tissue (cartilage, bone, dentine)
  7. Blood cells
  8. Testis and ovaries

Endoderm (inner layer) gives rise to:
  1. Internal epithelium (alimentary canal)
  2. Derivatives of internal epithelium (liver, thyroid, thymus)
  3. Prostate
  4. Exocrine pancreas

8.2.5. Dormancy

The spores and seeds of many species of plants lie dormant until germinated by absorption of water. This type of dormancy is rare in the animal kingdom, though interesting examples are to be found in some freshwater shrimps and in the freshwater sponge *Spongilla lacustris* (Ostrom and Simpson, 1978). This sponge forms gemmules, which consist of binucleate cells surrounded by an acellular coat. A rise in the water temperature in the spring results in release of Ca^{2+} from the cells, mitosis to form four mononucleate cells, followed by the hatching of a young sponge.

Another interesting example of dormancy is found in some eubacteria (Stanier *et al.*, 1964). Among these bacteria are those of the genera *Bacillus*, *Clostridium* and *Sporosarcina*. Under certain extreme environmental conditions such as

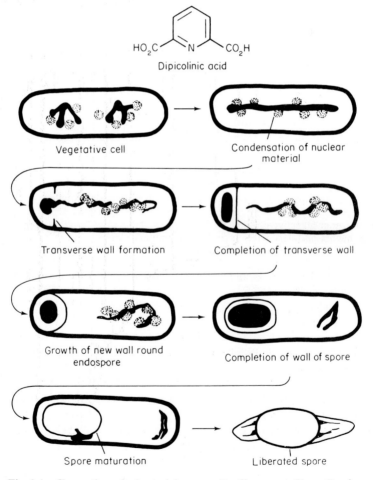

Fig. 8.4. Formation of a bacterial spore—*Bacillus cereus*. From Stanier *et al.* (1968)

drought, or sometimes spontaneously, these bacteria form endospores. These endospores have a very low water content, and are highly refractile. They are remarkably heat-resistant, being able to withstand temperatures of up to 100 °C in some cases, as well as being less sensitive to ultraviolet light and X-rays than the normal vegetative cell.

The onset of sporulation is heralded by the appearance of two nuclear bodies which then fuse together to form a rod-shaped 'nucleus' (Fig. 8.4). This then breaks into several pieces, one of which becomes enclosed by a transverse wall at one end of the cell, and which will eventually form the bacterial spore. This end of the cell develops a thick coat and an external exosporium. The maturing spore becomes highly refractile, a process which is associated with accumulation of dipicolinic acid, Ca^{2+} and dehydration. Dipicolinic acid (Fig. 8.4) is virtually absent from the normal vegetative cell, yet 5–10 % of the dry weight of the spore is made up of this substance. Its presence is essential for formation of stable spores which are resistant to high temperature. In low-Ca^{2+} media spores can form, but they have a low content of dipicolinic acid and are relatively unstable. The mature spore is eventually released after lysis of the cell.

Germination of bacterial spores consists of rehydration, followed rapidly by loss of Ca^{2+} and dipicolinic acid (Ellar *et al.*, 1974). Within 30–60 min the spore has developed into a normal, vegetative cell. Intracellular Ca^{2+} and dipicolinic acid obviously play an important part in spore formation and the reverse process, germination. However, the precise nature of their respective roles is still not clear.

8.3. Ca^{2+} and mitosis

8.3.1. The cell biology of mitosis

The process by which a eukaryotic cell divides to produce two daughter cells is called mitosis. Mitosis is heralded by the disappearance of the nuclear membrane and the appearance of the chromosomes. It consists of four morphologically recognizable phases (Figs. 8.1 and 8.2).

(1) Prophase. This is the onset of mitosis where the nucleoli disappear, the nuclear membrane begins to disappear and the chromosomes, after appearing, begin to shorten and condense. The number of chromosomes is of course twice that of the normal cell since DNA synthesis has already occurred during the S phase of the cell cycle. The duplicate copies are attached at a constricted region called the centromere.

(2) Metaphase. During this phase the mitotic spindle composed of microtubules appears, along which the chromosomes move to a point midway between the two poles of the spindle.

(3) Anaphase. The duplicated chromosomes separate at the centromere and each chromosome moves back along the microtubules to its respective pole.

(4) Telophase. This is almost the reverse of the prophase in that the chromosomes disappear, and the nuclear membrane and nucleoli reappear. However, an important difference from a simple reversal of the prophase is the

formation of a dividing structure along the equatorial plane of the mitotic spindle. In plants this consists of a rigid plate, whereas in mammalian cells and other eukaryotic cells without a cell wall it consists of a cleavage furrow being pulled inwards by a belt of microfilaments.

The time taken to go through these phases may be only a few minutes for some cells or several hours for others. Remembering that the decision to proceed with mitosis has been taken by the cell during the G_1 or G_2 phase, as a result of preprogramming or an external stimulus, could intracellular Ca^{2+} be involved in regulating this process?

8.3.2. The potential role for intracellular Ca^{2+} in mitosis

In theory, intracellular Ca^{2+} could play a part in regulating any one of three stages involved in cell division. (1) The decision of the cell to leave the G_0 or R_2 phase and rejoin the preprogrammed cell cycle. (2) The formation and disassembly of the mitotic apparatus including the movement of the chromosomes along the mitotic spindle during the metaphase and anaphase. (3) Formation and completion of the cleavage furrow.

The first of these may occur as a result of preprogramming of the DNA, although several examples exist where it is triggered by an external primary stimulus. Two examples of the latter are lymphocyte activation and egg fertilization. In both of these processes there is quite convincing evidence that the internal trigger initiating cell transformation is an increase in intracellular Ca^{2+}.

The other two stages may be provoked or regulated by physiological, or pharmacological substances, though normally the rate of progress is predetermined by the programme within the DNA. Where changes in intracellular Ca^{2+} are necessary for the mitotic process to proceed, it may be possible to block, or change, the rate of mitosis by interfering with intracellular Ca^{2+} experimentally.

Four models for the movement of the chromosomes during mitosis have been proposed (see Rebhun, 1977): (1) Bridges, composed of the MgATPase dynein, between the microtubules of the spindle exert sliding forces to move the chromosomes. (2) Shortening of the microtubules caused by disassembly causes movement of the chromosomes towards the poles. (3) A 'zippering' between adjoining microtubules provides the force for microtubule movement. (4) A cooperation between microtubules as the substrate and microfilaments as the provider of the force to move the chromosomes.

Experimentally, Ca^{2+} can promote disassembly of microtubules (see Chapter 5). Although nothing is known of the free Ca^{2+} concentration in the nucleus it has been proposed (Rebhun, 1977) that a low free Ca^{2+} concentration in the cell allows microtubule association and that a rise promotes disassociation. A more convincing argument can now be formulated in view of the demonstration, using fluorescence-labelled antibodies to calmodulin, that this Ca^{2+}-dependent regulatory protein is associated with the mitotic apparatus during all phases of mitosis (Fig. 8.5) (see Means and Dedman, 1980). Fluorescence intensity is

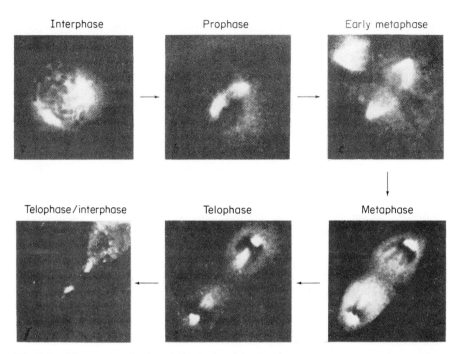

Fig. 8.5. Movement of calmodulin during mitosis. From Means and Dedman (1980). Reprinted from *Nature* by permission. Copyright © 1980 Macmillan Journals Limited. Fluorescence = fluorescein-labelled antibody to calmodulin

greatest at the poles and projects in the direction of the chromosomes. The fluorescence, and hence calmodulin, does not seem to cross the equatorial plane to which the chromosomes move during metaphase. Calmodulin seems to associate preferentially with microtubules which join the chromosomes to the pole of the mitotic spindle.

A role for calmodulin in chromosome movement during mitosis is supported by the observation that, *in vitro*, calmodulin + about 10 μM Ca^{2+} prevents microtubule assembly and promotes disassembly. The site of action of the calmodulin may be 6 S tubulin. What is needed now is a direct demonstration that changes in Ca^{2+} bound to the microtubule calmodulin occur during mitosis, and that the Ca^{2+} –calmodulin is playing an active as opposed to a passive role in this process. This will require the measurement of the free Ca^{2+} concentration in the area of the nucleus during the development of the mitotic apparatus and the movement of the chromosomes.

More direct evidence exists for Ca^{2+} regulating cleavage furrow formation. Not only has calmodulin been observed in association with the actomyosin microfilaments, but also microinjection of Ca^{2+} into several cells, for example eggs, causes the furrow to form.

Ca^{2+} can cause structural changes in chromatin in solution (Li *et al.*, 1977). However, no evidence exists at present that this direct action of Ca^{2+} mediates any of the structural changes in chromosomes which cause them to become visible during prophase and to disappear during telophase.

In summary, Ca^{2+} could play a role in at least six places during the provocation of cell division. (1) Ca^{2+} could trigger the cell from G_0 to G_1, or R_2 to G_2 phases and may activate RNA and DNA synthesis (Burgoyne et al., 1970; Ishide et al., 1974). (2) Changes in Ca^{2+} binding to DNA could cause some of the chromatin changes associated with mitosis. (3) Ca^{2+} might play a part in breakdown of the nuclear membrane, perhaps by causing vesiculation or activating a phospholipase. (4) Ca^{2+}, through activation of calmodulin, may control assembly and disassembly of microtubules. (5) Ca^{2+}, through ATPases known to increase in activity during mitosis and through calmodulin–microfilament–microtubule interactions, may stimulate movement of the chromosomes along the mitotic spindle. (6) Ca^{2+}, through calmodulin–microfilament interactions, may stimulate formation of the cleavage furrow.

It may not have escaped the reader's notice that a crucial piece of evidence for these proposals would be provided by measurement of changes in intracellular free Ca^{2+}, particularly in the region of the mitotic apparatus and cleavage furrow. As we shall see, apart from certain eggs, this has yet to be carried out.

8.3.3. Lymphocyte activation

During the last century it became clear that man and other higher animals possessed a defence mechanism which was able to attack and destroy infective microorganisms. Within this mechanism are three crucial lines of attack: (1) the production of antibodies which specifically bind to one or more antigens on the foreign body; (2) the lysis of foreign cells mediated by killer T lymphocytes or by complement, the latter being activated either by antibody–antigen binding or by a pathway not dependent on antibody; and (3) phagocytosis and destruction of the foreign cell. Of prime importance in these defence mechanisms is the ability of the immune system to be activated to produce highly specific antibodies to foreign substances. What provokes the production of these antibodies, in response to a foreign substance in the body? What goes wrong in autoimmune diseases where antibodies are apparently produced to the host's own proteins?

The commonest antibody molecule belongs to the class IgG, and is composed of two heavy and two light chains with a total molecular weight of about 150 000. It is produced by lymphocytes. These cells can be found not only in the peripheral circulation but also in the lymph nodes, thymus and spleen. Cells within the thymus are called thymocytes, but on leaving become T or thymus-influenced lymphocytes. It is these cells which make up most of the circulating lymphocytes and which are initially stimulated to divide by antigen, and are responsible for cell-mediated immunity. However, they are not the major antibody-producing cells. Antibody is produced by the other main group of lymphocytes known as B (bursal) or bone marrow lymphocytes, not influenced by the thymus but requiring some interaction with T cells, and other white cells such as macrophages, in order to produce antibody. So in order to activate fully the immune response an antigen binds to T cells containing a receptor, presumed to be antibody. This stimulates them to divide and then to activate antigen-sensitive B cells which clone to produce large amounts of circulating IgG antibody.

Any population of lymphocytes collected from the circulation will only contain a small population which will respond to a specific antigen stimulus. A method of cloning one population of lymphocytes, while still retaining their ability to be activated by antigen, would greatly help in the elucidation of the role of Ca^{2+} in this physiological response. However, as a model for lymphocyte activation many workers have used plant lectins as non-specific stimuli. These bind to glycoproteins on the cell surface thereby stimulating it to divide, but not necessarily to produce antibody. The plant lectin phytohaemagglutinin (PHA) from *Phaseolus vulgaris*, which as its name implies agglutinates red cells, has been widely used as an activator of lymphocytes (Whitney and Sutherland, 1972; Jensen *et al.*, 1977). It has been particularly popular in investigating the role of intracellular Ca^{2+} in this process. PHA appears to activate directly only T cells and is therefore a model for the delayed hypersensitivity reaction. Its effects can be blocked by wheat germ agglutinin. Other mitogens include concanavalin A and pokeweed mitogen, the latter stimulating cell division from T and B cells, as well as immunoglobulin production.

The activation of T lymphocytes by plant lectins at concentrations of a few μg/ml causes a large number of chemical changes in the cells (Table 8.5), some

Table 8.5 Changes in T lymphocytes induced by plant lectins

Changes detectable within the first 5 min of activation
* Increase in Ca^{2+} uptake
* Increase in glucose uptake
  Increase in cell cyclic AMP and possibly cyclic GMP
* Increased turnover of phospholipids, particularly phosphatidylinositol
  Capping

Changes detectable within 15–30 min of activation
* Increase in RNA synthesis
  Acetylation of arginine-rich histones
* Phosphorylation of non-nuclear proteins
* Increase in Na^+-dependent amino acid uptake

Changes detectable within 2–4 hours of activation
* Increase in protein synthesis
* Increase in glycolysis
  Redistribution of lysosomal hydrolases

Changes detectable within 24–48 hours after activation
  Increase in nucleoside kinase
  Increase in RNA and DNA polymerase
* Increase in DNA synthesis

Changes detectable 48–72 hours after activation
* Mitosis

* Signifies where evidence exists for Ca^{2+} dependency or Ca^{2+} involvement.

detectable within 1 min, others not being observable for several hours. The overall result is that the cells are transformed from G_0 to G_1 phase (see Fig. 8.1), mitosis occurring some 52–72 hours after the initial stimulus. Most workers have assessed transformation by measuring the incorporation of [^{3}H] thymidine into DNA. They are therefore really studying the S phase and not mitosis itself. Much evidence exists that an increase in intracellular free Ca^{2+} caused by the primary stimulus is the internal trigger transforming the cells from G_0 to G_1.

(1) Removal of extracellular Ca^{2+} (Whitney and Sutherland, 1972, 1973; Maino et al., 1974; Jensen et al., 1977). Removal of extracellular Ca^{2+} and addition of EGTA inhibits lymphocyte transformation by PHA or concanavalin A. It prevents the PHA-induced uptake of Ca^{2+}, inhibits the increase in amino acid uptake and phosphatidylinositol turnover, as well as inhibiting DNA synthesis. A Ca^{2+} dose–response relationship can be established over the range of 10 μM – 0.6 mM extracellular Ca^{2+}. Addition of EGTA 16 hours or more after addition of the lectin has no effect, whereas addition at various times up to this progressively reduces inhibition by the Ca^{2+} chelator. This observation would be nicely explained by the unit cell activation hypothesis proposed in Chapter 10.

(2) Increase in $^{45}Ca^{2+}$ uptake. As well as lectins, non-lectin mitogens such as Hg^{2+}, periodate and trypsin also increase Ca^{2+} uptake within a few minutes. The effect of PHA continues for up to 24 hours whereas that of concanavalin A is transient. Evidence for the latter has been obtained by Freedman and Ruff (1975) who showed that no significant increase in $^{45}Ca^{2+}$ uptake could be detected if the radioactive Ca^{2+} was added more than 5 min after the concanavalin A.

(3) Ionophore A23187. A23187 (1–10 μM) increases Ca^{2+} uptake and mimics most of the effects of the plant lectins, including a 5–10-fold stimulation of DNA synthesis 48–72 hours after addition. The effects of A23187 require extracellular Ca^{2+}.

(4) Inhibitors. Both La^{3+} and D600 (50 μM) (see Chapters 2 and 9) inhibit the mitogenic response of lymphocytes to plant lectins.

(5) Measurement of intracellular free Ca^{2+}? The ester of the fluorescent Ca^{2+} indicator quin II (Fig. 2.4) enters lymphocytes. In the cytoplasm of the cells it is hydrolysed by an esterase which converts it back to the acid form. The intracellular concentration of the indicator can be as high as 100–200 μM. Since the apparent dissociation constant for Ca^{2+} is about 0.1 μM, this means that the indicator binds a significant amount of Ca^{2+} relative to the total Ca^{2+} in the cell. Nevertheless, an increase in fluorescence occurs soon after addition of plant lectins to the lymphocytes (Rink and Tsien, 1982; Tsien et al., 1982), indicating that an increase in cytosolic free Ca^{2+} occurs before the onset of DNA synthesis and cell division. Care must be taken with this exciting technique because hydrolysis of the ester produces formaldehyde, in relatively small quantities, but which could potentially harm the cell.

In spite of this fairly convincing array of evidence several problems in interpreting these data have emerged:

(1) In order to produce maximum inhibition with EGTA a 10–30-min preincubation before addition of the lectin is usually required (Greene et al.,

1976). This suggests a possible role for an intracellular store of Ca^{2+}. Ca^{2+} does not appear to be required for lectin binding (Whitney and Sutherland, 1973), though one must be careful to ensure that the Mn^{2+} often present in these proteins is not removed by EGTA.

(2) Hesketh and coworkers (1977) have reported that it is possible to detect effects of concanavalin A on DNA synthesis in the absence of detectable effects on $^{45}Ca^{2+}$ uptake, whereas using A23187 it was possible to produce large uptakes of $^{45}Ca^{2+}$ without initiating DNA synthesis. On the contrary, these workers observed an inhibition of both capping and DNA synthesis by A23187. ('Capping' is the collecting together of receptor–agonist molecules at a particular place on the cell surface.) These experiments might be criticized on the grounds that careful measurements of cell ATP and cell viability were not made. However, these authors point out that only a small movement of Ca^{2+} would cause a significant increase in intracellular free Ca^{2+} and that the increased uptake of $^{45}Ca^{2+}$ may be very difficult to detect.

(3) Both the $(+)$- and $(-)$-isomers of D600 were equally effective in inhibiting lectin-induced DNA synthesis (Jensen et al., 1977) (see Chapters 2 and 9).

(4) The role of cyclic nucleotides is confusing. Lectins induce a transient increase in cyclic AMP and cyclic GMP, and stimulate phosphorylation of high molecular weight non-nuclear proteins (Greene et al., 1976). Furthermore, activation of adenylate cyclase can be demonstrated in membrane preparations. However, long-term addition of cyclic AMP or its dibutyryl derivative, or agents such as isoproterenol (0.1 mM), prostaglandin E_1, theophylline or wheat germ agglutinin which increase intracellular cyclic AMP, are all inhibitory to the response of lymphocytes to lectins. Furthermore, the dibutyryl and bromo derivatives of cyclic GMP can stimulate DNA, RNA and protein synthesis. Since the cell cyclic AMP decreases markedly 1–2 hours after stimulation by lectins it is possible that cyclic AMP may play a role in releasing intracellular Ca^{2+} in the early phase of G_1 but inhibit during the S or G_2 phases. A further possibility is that the physiological changes in cyclic nucleotides are restricted to certain regions within the cell.

(5) Ouabain, a specific inhibitor of the Na^+ pump, can block lymphocyte activation caused by PHA or A23187 (Jensen et al., 1977). How this relates to changes in intracellular Ca^{2+} is not known (see Chapter 4). Nor have the electrical properties of lymphocytes been well characterized.

Similar evidence for Ca^{2+}-dependent activation in other cells involved in the immune system, for example thymocytes (Whitfield et al., 1972, 1974), has been obtained.

But how good a model of the physiological activation of the immune system is the lectin-induced transformation of peripheral lymphocytes? The data of Diamantstein and Odenwald (1974) from spleen cells of immunized and normal mice suggest that, although intracellular Ca^{2+} may be involved in the primary and secondary immune responses, its role may be different from that in PHA-activated lymphocytes. These workers have shown that during the initial 24-hour period after antigenic stimulation the cells do not require extracellular Ca^{2+}, whereas during the following 12 hours, when cell proliferation occurs, they do require Ca^{2+}. This is the opposite of PHA-induced lymphocytes where removal of Ca^{2+} after 16 hours has no effect. Furthermore, in spleen cells removal of Ca^{2+} 48–72 hours after antigenic stimulation actually enhances antibody production. Intracellular Ca^{2+} may therefore play a role in initiating both cell division and cell differentiation to produce specific antibody molecules. Part of the explanation of the apparent discrepancy between these results and those obtained with PHA and peripheral lymphocytes may lie in the fact that spleen cells are mainly B lymphocytes whereas peripheral lymphocytes are mainly T cells.

The controversy of the role of intracellular Ca^{2+} and cyclic nucleotides in the immune response requires answers to the following four questions: (1) What is the time sequence of activation of individual lymphocytes? (2) Can the change in intracellular free Ca^{2+}, demonstrated using a Ca^{2+} indicator, be correlated with the physiological response of each cell? (3) What are the distributional changes in cyclic nucleotides and Ca^{2+} within the cell? (4) What is the molecular basis of the action of Ca^{2+} to stimulate $G_0 \rightarrow G_1$ or mitosis?

8.3.4. Egg fertilization

The process of sexual reproduction is dependent on the ability of the male gamete, the sperm, to incorporate its genetic material into the female gamete, the ovum, forming a zygote which then divides. This phenomenon has been studied in a wide range of invertebrate and vertebrate species. Although specific details vary from species to species a number of generalizations can be made.

(1) The sperm attaches to a site on the membrane of the egg. In some cases only one site of attachment is possible. For example, in eggs of the medaka fish, *Orizias latipes*, the sperm can only penetrate the chorion coat through a hole known as the micropyle (Fig. 8.6).

(2) Almost immediately after sperm attachment an explosive wave of morphological and chemical changes travels down the egg. In some species, such as the sea urchin *Arbacia*, there is a rapid burst in O_2 consumption due to production of O_2^-, which as in phagocytic cells (see Chapter 7 section 7.3.3), is associated with production of chemiluminescence (Foerder et al., 1978). In these eggs responding to sperm there is an explosive reaction of the dense vesicles, called cortical vesicles, in the peripheral layer of the egg, resulting in fusion with the plasma membrane and release of enzymes such as myeloperoxidase (Fig. 8.7). Streaming of the cytoplasm of the egg is also seen in some species. It has been suggested that the release of oxygen radicals and peroxides by this mechanism causes irreversible changes in the surface of the egg, preventing it being fertilized by another sperm.

(3) Injection of the sperm into the egg occurs as a result of the acrosome reaction. This enables penetration of the nucleus and mitochondria to occur from the sperm into the egg. The acrosome is a dense body at the anterior end of the sperm, and breaks down after attachment of the sperm to the egg. The sperm

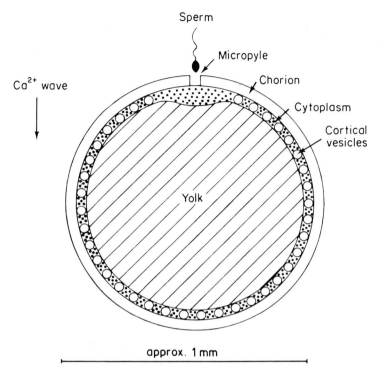

Fig. 8.6. A fish egg fertilized by sperm. From Gilkey et al. (1978). N. B.: Medaka fish egg cytoplasmic thickness is exaggerated

then turns like a screw, thereby injecting itself into the egg. This process may require an increase in intracellular Ca^{2+} (Tilney, 1975).

(4) Before forming the mature diploid zygote, both the pronuclei of the sperm and egg develop and DNA synthesis occurs. The two nuclei then fuse together and a normal mitotic cycle begins, resulting in the cleavage of the fertilized egg into two daughter cells.

(5) In several instances glycogenolysis and glycolysis are enhanced after fertilization.

What is the role of intracellular Ca^{2+} in these events?

For more than 40 years two opposing theories have been expounded by different workers to explain the explosive nature of the changes inside the fertilized egg [see Heilbrunn (1937, 1943) for references]. One proposes that an increase in intracellular Ca^{2+} is the trigger, whereas the other argues that it is a decrease in pH (Lillie, 1941). A release of H^+ does occur in activated eggs; however, the evidence for the Ca^{2+} hypothesis seems more convincing (Table 8.6). This Ca^{2+} hypothesis was first conceived because of the requirement for Ca^{2+} in the parthenogenetic activation of eggs, i.e. activation without sperm (Tyler, 1941).

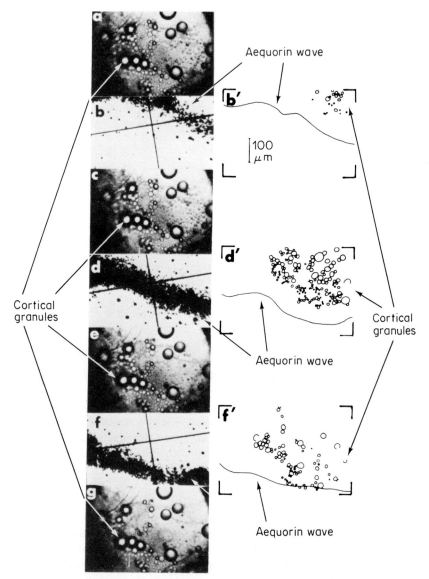

Fig. 8.7. Cortical granule release following 'Ca^{2+} wave'. Fertilized medaka fish egg. (a, c, e, g) Normal photograph, 13 sec between each. (b, d, f) Reverse contrast, showing aequorin wave. (b', d', f') Diagrammatic representation. From Gilkey *et al.* (1978). Reproduced from the *Journal of Cell Biology* by copyright permission from the Rockefeller University Press

Table 8.6. Some systems where evidence exists for intracellular Ca^{2+} activating eggs towards mitosis

Phylum	Genus (common name)	Stimulus	Reference
Annelida	*Nereis* (worm)	Ultraviolet light, $KCl + Ca^{2+}$	Heilbrunn and Wilbur (1937)
Echinodermata	*Arbacia* (sea urchin)	Ca^{2+}	Mazia (1937)
	Paracentrotus (sea urchin)	Ca^{2+}	Hultin (1950)
	Patiria (bat star)	A23187	Steinhardt et al. (1974)
Mollusca	*Acmaea* (limpet)	A23187	Steinhardt et al. (1974)
Protochordata	*Ciona* (sea squirt)	A23187	Steinhardt et al. (1974)
Chordata	*Rana* (frog)	A23187	Schroeder and Strickland (1974)
	Xenopus (toad)	Intracellular Ca^{2+}	Baker and Warner (1972)
	Oryzias (fish)	Sperm, A23187	Gilkey et al. (1978)
	Mesocricietus (hamster)	A23187	Steinhardt et al. (1974)

Early events after sperm attachment

Many of the changes occurring in eggs fertilized by sperm can be mimicked by addition of high Ca^{2+}, or by pricking the egg providing Ca^{2+} is present. Furthermore, extracellular Ca^{2+} is required for the action of many parthenogenetic stimuli (Yamamoto, 1954, 1961). Eggs can also be activated by A23187 (Steinhardt et al., 1974; Foerder et al., 1978) including cortical granule release and chemiluminescence. It has been known for some time that relatively low Ca^{2+} concentrations have effects on egg homogenates similar to those in the intact cell (Hultin, 1950). But perhaps the most convincing evidence for a role for intracellular Ca^{2+} in the fertilization process comes from studies on eggs injected with the Ca^{2+} indicator aequorin (Kiehart et al., 1977; Ridgway et al., 1977; Steinhardt et al., 1977; Gilkey et al., 1978). In sea urchin (*Lytechinus pictus*) eggs the peak in intracellular Ca^{2+} occurs 45–60 sec after activation and lasts 2–3 min, whereas in the freshwater medaka fish *Oryzias latipes*, the time course is longer. In this species there is a 0.8–3-min delay after adding the sperm, probably because time is taken for the sperm to penetrate the micropyle (Fig. 8.6). At this point there is an explosive rise in intracellular free Ca^{2+}, with a time constant of 1–2 sec ($k = 0.5$–1 sec^{-1}). The peak in free Ca^{2+} lasts for 1–2 min and then decays over a period of about 15 min (time constant = 150 sec, $k = 6.7 \times 10^{-3} \text{ sec}^{-1}$).

Steinhardt and coworkers estimated the free Ca^{2+} concentration in sea urchin eggs to be 2.5–4.5 μM, after activation. However, this estimate needs re-examination in view of the image intensification studies by Gilkey et al. (1978). These workers showed that in medaka eggs a wave of free Ca^{2+} of about 30 μM, some 100–1000 times the resting free Ca^{2+} concentration in the cell, starts at the pole of the cell and reaches the other end a few minutes later (Fig. 8.8a). The Ca^{2+} wave takes about 2 min to traverse the egg. This correlates with the occurrence of the explosive cortical granule release, which also traverses the egg over a period of about 2 min. In contrast, A23187 increases both intracellular free Ca^{2+} and cortical granule release at many different sites Fig. 8b.

What is the source of the rise in cytosolic Ca^{2+}?

The activation of eggs from most phyla, either by sperm or A23187, does not require extracellular Ca^{2+}, nor does the increase in free Ca^{2+} in the cell, detected using aequorin, depend on the presence of external Ca^{2+}. It is therefore most likely that the source of the Ca^{2+} increase induced by sperm is an intracellular store. Such a store, requiring MgATP, has been isolated from sea urchin eggs (Kinoshita and Yazaki, 1967; Petzelt and Ledebur-Villiger, 1973). Using fluorescence-labelled antibodies this system has been observed to move to localized areas during different phases of mitosis. A Ca^{2+}-induced release of Ca^{2+}, similar to that in the sarcoplasmic reticulum (see Chapter 5, section 5.3.3), has been proposed as the trigger for the explosive release of Ca^{2+} into the cell cytoplasm (Ridgway *et al.*, 1977; Gilkey *et al.*, 1978). Further evidence for

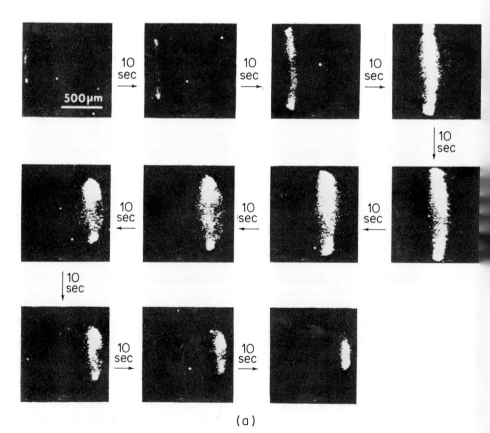

(a)

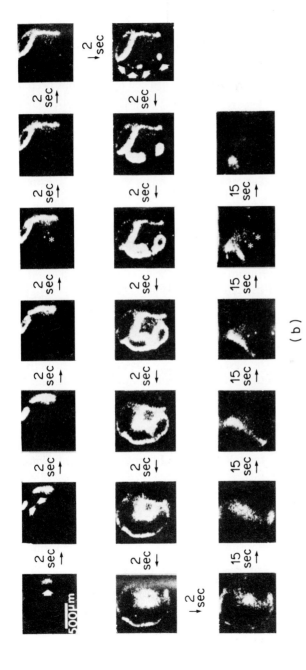

Fig. 8.8. Free 'Ca^{2+} wave' visualized in fish egg using aequorin and image intensification. From Gilkey *et al.* (1978). Reproduced from the *Journal of Cell Biology* by copyright permission of the Rockefeller University Press. (a) Medaka fish egg fertilized by sperm; note single wave moving across egg. (b) A23187 on medaka fish egg produces light from many points simultaneously

internal release of Ca^{2+} comes from the increase in $^{45}Ca^{2+}$ efflux observed from activated eggs.

The cleavage furrow

Cells from many vertebrates and invertebrates in the telophase of mitosis contain a uniform, contractile ring of microfilaments which can be observed from the last few seconds of the anaphase until cell cleavage has been achieved (Schroeder, 1972). Injection of Ca^{2+} chelators at this cleavage furrow site arrests cleavage (Timourian et al., 1972; Schroeder and Strickland, 1974). On the other hand, microinjection of Ca^{2+} or ionophore A23187 induces contractility around the microfilaments (Hollinger and Schuetz, 1976). In a careful study using Ca/EGTA buffers Baker and Warner (1972) found that the threshold for contractility in *Xenopus* eggs was about $0.3-1\,\mu M$ free Ca^{2+}. Ridgway and coworkers (1977) observed a significant rise in aequorin light emission in medaka eggs after the initial wave of light was over, which correlated with cleavage. However, using a Ca^{2+} microelectrode in *Xenopus* eggs no changes in free Ca^{2+} were detected (Rink et al., 1980). In view of the fact that the changes in free Ca^{2+} may be very close to the plasma membrane, it is important to be sure that the method used for measuring intracellular Ca^{2+} is able to detect changes in free Ca^{2+} in different parts of the cell simultaneously.

8.3.5. Conclusions

How convincing is the evidence that intracellular Ca^{2+} regulates different phases of the cell cycle? The best evidence is to be found in the activation of lymphocytes or eggs from their quiescent state, though the molecular basis for this is still lacking. An increase in intracellular Ca^{2+} can cause contraction of the cleavage furrow though this does not necessarily mean that Ca^{2+} is the physiological regulator. For the rest of the cell cycle the evidence for Ca^{2+} is not so good. However, in view of the location of calmodulin in the mitotic apparatus it would be surprising if some role for Ca^{2+} did not exist in controlling mitosis itself.

8.4. Ca^{2+} and meiosis

Complete meiosis consists of two sequential cell divisions in which there is only one duplication of the chromosomes (Fig. 8.3). The result is the formation of a haploid cell, either a mature egg or sperm. Because of the great similarity between meiosis and mitosis the potential role of intracellular Ca^{2+} in each process is very similar (see this chapter, section 8.3.2). It is therefore perhaps not surprising that ionophore A23187 is able to induce meiosis in oocytes (Schuetz, 1972). However, a role for intracellular Ca^{2+} has been discovered in meiosis which is distinct from those already discussed for mitosis.

Oocytes may remain for many months, or even years, in the prophase of the first meiotic division. Maturation of the oocyte into an egg to the metaphase, where it stops after the second meiotic division, requires a hormone (Masui and Clarke, 1979). In amphibians and in other vertebrates this is a steroid, usually progesterone (Fig. 8.9). Steroid hormones also may be the stimuli in invertebrates. However, in some but not all echinoderms, notably starfish, the maturing hormone is 1-methyladenine (Fig. 8.9). This hormone is produced by follicle cells surrounding the egg in response to a neurohormone (Fig. 8.10) and works at a concentration of 0.1–1 μM (Kanatani and Shirai, 1969; Stevens, 1970). In *Xenopus* and other steroid-sensitive animals, progesterone stimulates the production of an intracellular maturation-promoting factor and breakdown of the central germinal vesicle (Fig. 8.10). Such a substance may also exist in starfish oocytes.

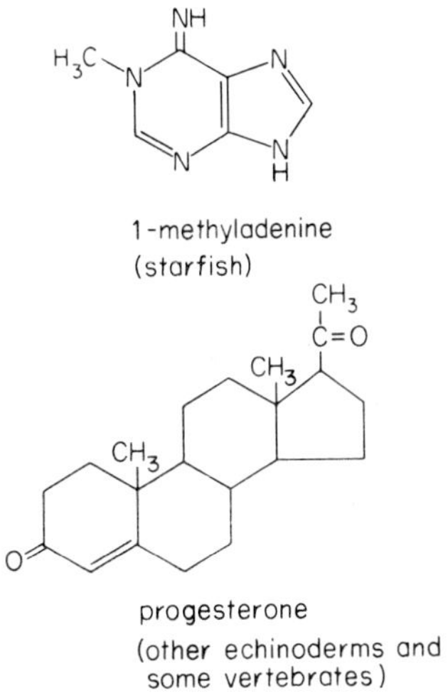

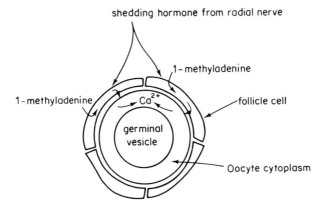

Fig. 8.9. Two oocyte-maturing hormones

Fig. 8.10. Effect of 1-methyladenine on starfish oocyte

In spite of the structural differences between progesterone and 1-methyladenine, their mechanism of action to cause oocyte maturation is very similar. In particular, an increase in intracellular free Ca^{2+} appears to be the internal trigger reactivating meiosis (Table 8.7), presumably via the intracellular maturation-promoting factor (Moreau and Guerrier, 1979; Moreau et al., 1980). Both progesterone and 1-methyladenine seem to bind to receptors on the outside surface of the oocyte, since microinjection of these hormones into the oocytes causes no stimulation. This is particularly surprising for a steroid since much evidence exists in mammalian cells for intracellular receptors.

The hormonal stimuli do not require extracellular Ca^{2+}, though certain pharmacological substances which stimulate meiosis by increasing the permeability of the external membrane to Ca^{2+} do not of course work in the presence of EGTA. Meiosis can be initiated by ionophore A23187 or by microinjection of Ca^{2+} into the oocyte. Oocyte activation, on the other hand, is blocked by Ca^{2+} chelators injected into the oocytes before activation by the hormone or other stimuli. It is therefore proposed that the hormone must cause the release of Ca^{2+} from some internal store. Somewhat surprisingly, therefore, various substances such as D600, verapamil, procaine, La^{3+} and Mn^{2+} which normally block Ca^{2+} entry through the cell membrane (see Chapter 9) also block activation. This may relate to the electrical properties of the cell membrane.

Direct evidence for an increase in intracellular free Ca^{2+} has been obtained using aequorin, and a Ca^{2+} microelectrode, in the starfish *Marthasterias glacialis*, activated by 1-methyladenine (Moreau and Guerrier, 1979) (Fig. 8.11), and in *Xenopus* oocytes activated by progesterone (Moreau et al., 1980). In the starfish oocyte there is a 1–2-sec lag after addition of the hormone, followed by a rapid rise in intracellular free Ca^{2+}, estimated to be from 0.5 to 1.5 μM and lasting for about 30 sec. In the case of *Xenopus*, the Ca^{2+} concentration change is much larger, being from 0.7 to 7 μM, and lasting some 3 to 6 hours. In each case

Table 8.7. Some systems where evidence exists for intracellular Ca^{2+} activating meiosis

Phylum	Genus	Stimulus	Reference
Annelida	*Hydroides* (worm)	Ca^{2+}	Pasteels (1935)
	Nereis (worm)	A23187	Chambers (1974)
	Urechis (worm)	A23187	Paul (1975)
	Chaetopterus (worm)	A23187	Brachet and Denis-Donini (1977)
	Subellaria (worm)	A23187	Peaucellier (1977)
Echinodermata	*Marthesterias* (starfish)	1-Methyladenine, A23187	Moreau et al. (1980); Moreau and Guerrier (1979)
Mollusca	*Barnaea* (rock-boring shellfish)	Ca^{2+}	Pasteels (1938)
	Spirula (cuttlefish)	A23187	Schuetz (1972)
Chordata	*Xenopus* (toad)	Progesterone	Bellé et al. (1977); Beaulieu et al. (1978)
		A23187	Wasserman and Masui (1975)

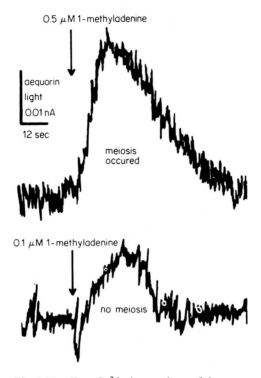

Fig. 8.11. Free Ca^{2+} changes in starfish oocytes induced by 1-methyladenine. From Moreau and Guerrier (1979); reproduced by permission of Elsevier Biomedical Press B.V. Starfish (*Marthasterias glacialis*) oocyte injected with aequorin. Threshold for meiosis $\approx 0.2\,\mu M$ 1-methyladenine

breakdown of the germinal vesicle occurs about 2–3 hours after the peak in intracellular Ca^{2+}. No account has been taken of any possible wave of Ca^{2+} moving down the cell, presumably because the hormone receptors would be expected to be uniformly distributed over the cell surface.

The source of the Ca^{2+}, released internally by hormone action, may be the inner surface of the plasma membrane itself (Doree et al., 1978; Moreau et al., 1980). If this is so it could explain the inhibitory effects of D600, La^{3+}, procaine and other substances which normally act on the cell membrane. An important consequence of release of Ca^{2+} from the plasma membrane is that it is no longer necessary to postulate 'a second messenger' to explain how a hormone acting on the cell surface can release Ca^{2+} from stores within the cell.

Although much needs to be learnt about the molecular mechanism by which Ca^{2+} initiates meiosis, it is significant that a calmodulin-like protein has been isolated from some oocytes (Moreau et al., 1980) which causes reinitiation of meiosis when injected into oocytes.

The role of cyclic AMP in meiosis is not well established. However, Moreau and Guerrier (1979) have reported that dibutyryl cyclic AMP, whilst not triggering meiosis itself, seems to decrease the threshold for 1-methyladenine in starfish oocytes. Furthermore, the regulatory subunit of cyclic AMP protein kinase provokes, and the catalytic subunit inhibits, progesterone-induced meiosis in *Xenopus* eggs (Maller and Krebs, 1977).

8.5. Ca^{2+} and cell transformation

We have seen how an increase in intracellular Ca^{2+} may transform cells to a highly activated metabolic state which leads to cell division, be it mitotic or meiotic. However, examples of cell transformation exist which do not necessarily lead to cell division and where Ca^{2+} also appears to play a role in mediating this transformation. Two particularly fascinating examples are the amoeboid–flagellate transformation of the protozoan *Naegleria gruberii* and the capacitation of sperm.

8.5.1. *Naegleria gruberii*

Naegleria gruberii is a protozoan found all over the world in the soil and in fresh water. It is found most commonly in the amoeboid form, though it can also encyst (Fig. 8.12). Perhaps its most interesting property is its ability, under

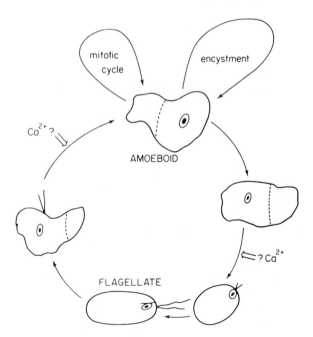

Fig. 8.12. The amoeboid → flagellate transformation of *Naegleria gruberii*. After Fulton (1977). ⇒ = activation

certain nutrient or ionic conditions, to transform to a flagellate form (Fig. 8.12). Willmer (1970) has proposed that the three fundamental eukaryotic cell types are mechanocytes, epitheliocytes and amoebocytes. The ability of a unicellular organism to transform from one type to another is therefore not only of great interest to developmental biologists (Fulton, 1977), but also to those interested in understanding the evolution of cells.

The amoeboid to flagellate conversion is particularly favoured when the organisms, in the amoeboid form, are placed in a medium of low ionic strength. The amoeba attach to a fixed substrate. Gradually the two flagellae appear and the new flagellate begins to rotate around its point of attachment as the flagellae beat. Eventually it swims away, the whole process taking about 30–60 min. This process of 'differentiation' involves both the rearrangement of existing cell components and a change in cell metabolism, as well as the synthesis of new RNA and protein molecules. The result is a conversion of a cell whose movement depends on actin filaments, with little microtubular structure in the cytoplasm, to one with a microtubule cytoskeleton, and no functioning actin microfilament system.

A low molecular weight substance, factor ψ, has been isolated from the flagellate form which when added back to flagellate cells causes changes in their shape. This effect requires extracellular Ca^{2+} and can be mimicked by ionophore A23187. These observations have led Fulton (1977) to propose, as a working hypothesis, that a low intracellular free Ca^{2+} concentration is necessary for the tubulin-based cytoskeleton of the flagellate form and that a release of Ca^{2+} from internal stores causes disassembly and reversion to the amoeboid form, the release of Ca^{2+} being stimulated by factor ψ. In the light of the calmodulin hypothesis (see Chapter 3) it would seem worthwhile investigating whether this protein changes its intracellular location during cell transformation, and in particular whether it is a necessary part of the actin filament system for amoeboid movement.

8.5.2. Capacitation of sperm

Capacitation is the term given to the maturation of sperm, necessary if it is to fertilize an egg. It is associated with changes in membrane structure and permeability, as well as increases in glycolysis and respiration. In the genital tract of mammals capacitation is thought to be induced by a combination of prostaglandins and enzymes. However, *in vitro* a so-called minimal capacitation medium (Singh *et al.*, 1978) containing 110 mM NaCl, HCO_3^-, pyruvate and Ca^{2+} can cause capacitation, which is followed by an increase in motility and an acrosome reaction. The latter only occurs physiologically when the sperm attaches to the egg.

Capacitation does not occur in the absence of external Ca^{2+} (Yanagamachi and Usui, 1974). Furthermore, the minimal capacitation medium involves Ca^{2+} uptake by sperm which occurs in two phases (Singh *et al.*, 1978). The first phase seems to be associated with capacitation, whereas the second phase is associated with the acrosome reaction. This latter process is also activated by A23187. An odd feature of this Ca^{2+} uptake is that it is enhanced by ruthenium red and nupercaine (50 μM), which also speed up the acrosome reaction by about 20–30 min. Both of these substances normally inhibit Ca^{2+} uptake (see Chapters 2 and 9). At higher concentrations (0.4 mM) nupercaine does inhibit motility, and the increase in respiration associated with capacitation.

The possibility exists, therefore, that both sperm capacitation and the acrosome reaction, which is necessary for injection of sperm into the egg, are triggered by an increase in intracellular Ca^{2+}.

8.6. Overall conclusions

We have seen in this chapter that much evidence exists for intracellular Ca^{2+} as the trigger transforming cells from one state to another, both under conditions which involve cell division and those which do not. Each state is defined by certain recognisable morphological and biochemical characteristics. Such transformations provide good examples of Ca^{2+}-dependent 'threshold' activation, the principles of which were outlined in Chapter 2 (section 2.2).

Much still needs to be learned about the free Ca^{2+} concentration in cells before, during and after transformation. This is particularly so in the nucleus where nothing is known of the free Ca^{2+} concentration, and where Ca^{2+} is likely to play an important role in chromosome structure and movement.

CHAPTER 9

The pathology and pharmacology of intra-cellular Ca^{2+}

So far this book has been concerned with the principles governing the ability of intracellular Ca^{2+} to act as a physiological mediator of a wide range of biological phenomena. In view of the large electrochemical gradient of Ca^{2+} across the outer membrane of all cells, one might expect disturbances in intracellular Ca^{2+} to be caused by pathogens, and drugs, which interact with the cell surface. Calcification of injured and necrotic tissue has been known for more than a century (Virchow, 1855) (Table 9.1). During the past 200 years much has been learnt about the aetiology of infectious diseases and of many genetically based diseases. Yet many common diseases remain a mystery. Does intracellular Ca^{2+} play any part in the pathogenesis of these diseases, which include cancer, heart disease, atherosclerosis, rheumatoid arthritis, diabetes, obesity, cystic fibrosis and mental disorders? Are changes in intracellular Ca^{2+} important in the action of any drugs?

Whilst this chapter focuses its attention primarily on such problems relating to man, the considerable social and economic importance of similar considerations of plant and animal diseases should not be overlooked.

9.1. Tissue calcium and the medical problems

The formation of deposits of calcium salts, usually phosphate, carbonate or oxalate, in diseases involving a disturbance in whole-body calcium metabolism and in severe tissue injury has been recognized for many years (Table 9.1). Although textbooks of pathology rarely make it clear whether these deposits originate intra- or extracellularly, the implication is that they are formed mainly outside the cells in the tissue concerned. This type of tissue calcification is classified as either primarily metastatic or dystrophic, though other forms exist (Table 9.1). Extracellular calcification may originate from precipitation of calcium salts (Abraham, 1970), binding of Ca^{2+} to macromolecular deposits for example elastin and collagen in atherosclerosis (Urry, 1971), or it may even be exuded from dying cells in the form of mitochondrial calcium deposits (Lafferty

Table 9.1. Clinical examples of tissue calcification

Type (calcium precipitate)	Tissues affected	Pathology
1. Dystrophic calcification (apatite or $\beta\ Ca_3(PO_4)_2$)	Various soft tissues, many tissues, e.g. myocardial infarct	Scar tissue, fibroids, necrotic tissue
	Blood vessels	Atherosclerotic plaques
	—	Crush syndrome
	—	Paraplegia
	—	Poliomyelitis
	Intervertebral discs	Alkaptonuria
	Vascular and articular	Heredofamilial vascular and articular disease
	Autopsy material	Death
2. Metastatic calcification (calcium phosphate—apatite)	Bone and kidney as well as lungs	Hypercalcaemia, hypoparathyroidism, hyperphosphataemia
	Cornea, arteries, stomach	Sarcoidosis
	Pulmonary vein and left atrium	Hypervitaminosis D, various bone diseases (myelomatosis, leukaemia, Paget's disease, osteomyelitis, metastatic carcinoma)
3. Oxalosis (calcium oxalate)	Arteries, bone marrow, kidney	Hyperoxaluria
4. Chondrocalcinosis or pseudo-gout (calcium pyrophosphate—$Ca_2P_2O_7 \cdot 2H_2O$)	Articular cartilage	Wilson's disease
	Fibrocartilage	Haemochromatosis
	Synovial membrane of joints	Hypercalcaemia
5. Other para-articular calcium deposits (e.g. $CaHPO_4 \cdot 2H_2O$, hydroxyapatite)	Menisci of joints	Calcific polytendinitis, peritendinitis calcanea
6. Calcinosis (cadahlite—$2Ca_3(PO_4)_2 \cdot CaCO_3$)	Connective tissue of skin, panniculus and fascia	Dermatomyositis, scleroderma
7. Tumoral calcinosis ($Ca_3(PO_4)_2$)	Buttocks, trochanters and other bony areas	Tumours

For references see Virchow (1855), Bywater and Glynn (1970), Abraham (1970), Cappell and Anderson (1975), Walter and Israel (1970), Rees and Coles (1969).

et al., 1965). Similarly, the importance of abnormalities in serum Ca^{2+} concentration, and regulators of whole-body calcium metabolism such as parathyroid hormone and vitamin D, in several diseases is well known. During the past 20 years there has been a growing awareness that changes in Ca^{2+} inside a cell can play an important part in cell injury. The main reason for this has been the realization that all normal healthy cells maintain a Ca^{2+} gradient across their outer membrane of some 10 000-fold, and that only a relatively small, but prolonged, change in this gradient would cause both disruption of cell metabolism and Ca^{2+} accumulation. Unfortunately, the problems of studying human tissues, limited particularly by availability of tissue samples as well as variations between individuals, have made studies on the role of intracellular Ca^{2+} in human cell pathology somewhat difficult. However, animal experiments and tissue culture in combination with some studies on human tissue samples now enable us to examine the possible role of intracellular Ca^{2+} in cell injury.

During the lifetime of any individual the cells in each tissue may turn over many thousands of times, apart that is from special cells like nerves. The life span of an average human red cell, for example, is 120 days. Tissue injury can therefore in theory involve either a change in the life span of the cells or a change in the specialized functions of the cells. In some cases it may be necessary to invoke both of these mechanisms to explain pathological changes and impaired function in a tissue. In the dynamic phase of cancer, for example, the cells are multiplying rapidly and no longer perform the functions of their tissue of origin in a controlled fashion. Many toxins have a dramatic effect on the life span of cells by causing a rapid progression to cell death. On the other hand, in diseases like diabetes the clinical manifestations are thought to be due primarily to disturbances in cell metabolism, though effects on the turnover of cells has rarely been thoroughly investigated.

9.1.1. Types of tissue injury

Most of the diseases of man can be grouped into twelve categories based on the primary cause of tissue injury (Table 9.2). The time span of a disease may be only a few days, as in some infectious diseases or toxin-mediated cell injury, or it may

Table 9.2. Types of cell injury

Primary cause	Example
Congenital	
1. Inherited or familial	Phenylketonuria, Down's syndrome, Huntington's chorea
2. Congenital	Spina bifida
Acquired	
1. Physical injury	Heat, trauma, radiation, electric shock
2. Chemical injury	Poisons, insecticides, drugs, bacterial toxins (e.g. tetanus, cholera), fungal toxins (aflatoxin, phalloidin)
3. Inflammations—infections	Bacteria, viruses, protozoa, nematodes
4. Mechanical	Intestinal obstruction, pulmonary collapse
5. Metabolic—intrinsic abnormalities	Gout
nutritional	Vitamin deficiency, starvation, hypervitaminosis, obesity, hypoxia
6. Immunological	Autoimmune disease (e.g. Grave's disease, haemolytic anaemia), complement deficiencies
7. Circulatory	Oedema, thrombosis, infarction
8. Endocrine	Diabetes, Addison's disease, hyperparathyroidism
9. Degenerations and infiltrations	Sarcoid, arthritis
10. Tumours	Primary and secondary

be many years as in diabetes or rheumatoid arthritis. The time taken for the primary mediator of cell injury to cause impairment of tissue function, recognizable clinically, can vary from a few minutes or even seconds in the case of some chemical toxins to many years in other cases. An increase in cell calcium has long been recognized as an early event in many types of cell injury (Cameron, 1956), (Table 9.3), and occurs also in several experimentally induced injuries in animals (Table 9.3). Here again it may take several hours or even days for the change in cell calcium to be observable. Since most of the physiological examples of intracellular Ca^{2+} as a regulator involve acute, transient changes in cell Ca^{2+} (see Chapters 2–8), this raises two crucial questions with respect to the pathology of intracellular Ca^{2+}: (1) How many examples of changes in cell Ca^{2+} are a reflection, rather than a cause, of cell injury? (2) In cases where intracellular Ca^{2+} plays a role in cell injury, how is this compatible with its relatively short-term action in the physiological control of cell function?

9.1.2. Evidence for a role for intracellular Ca^{2+} in cell injury

The experimental approach to the investigation of intracellular Ca^{2+} as a regulator was laid down in Chapter 2, section 2.3. Particular emphasis was placed on the distinction between 'active' and 'passive' Ca^{2+}-binding sites in the cell. Under pathological conditions Ca^{2+}-binding sites normally not involved in physiological cell responses may become significant, particularly if injury to the cell causes an increase in intracellular free Ca^{2+} outside the normal range. Four key questions arise when considering the possible role of intracellular Ca^{2+} in cell injury. (1) Is there a change in intracellular Ca^{2+} in the injured cell, or an alteration in an intracellular Ca^{2+} transient induced by a physiological stimulus? (2) Is this change in intracellular Ca^{2+} directly related to any malfunctions or structural alterations in the injured cell, or is the change in Ca^{2+} simply a result rather than a cause of injury? (3) What is the cause of the change in intracellular Ca^{2+}? (4) What is the molecular basis and time course of the change in intracellular Ca^{2+} and its effect on cell injury?

Because of the problems of studying chemical changes in human cells *in situ*, the evidence for a pathological role for intracellular Ca^{2+} has relied heavily on the use of animal models and cells in tissue culture. The main experimental approaches for investigating the role of intracellular Ca^{2+} in the action of a pathogen can be summarized as follows (see also Chapter 2, section 2.3): (1) Measurement of total cell calcium and intracellular free Ca^{2+}, and correlation with cell injury. (2) The effect of removal of extracellular free Ca^{2+}. (3) The effect of manipulating intracellular Ca^{2+}, e.g. with ionophore A23187, to see if it mimics the cell injury. (4) Localization of cell Ca^{2+}, and in particular intracellular Ca^{2+} stores, by subcellular fractionation or X-ray microprobe analysis. (5) Measurement of $^{45}Ca^{2+}$ uptake and release. (6) Pharmacological intervention: the effect of membrane stabilizers to Ca^{2+}, e.g. local anaesthetics (procaine and dibucaine), phenothiazines (trifluoperazine), and inhibitors of Ca^{2+} channels (verapamil and D600). (7) Identification of changes in enzyme activity, nucleic acids and phospholipids induced by a pathological change in intracellular Ca^{2+}.

There have been many reports of increases in tissue Ca^{2+} occurring under a variety of experimental and pathological conditions (Table 9.3). However, apart from tumours (Bresciani and Auricchio, 1962; Kishi et al., 1937; DeLong et al., 1950), there have been few reports where cell injury results in a decrease in cell Ca^{2+}. The time scale of the increase in tissue Ca^{2+} depends on the cause of the injury and also the method of measuring the Ca^{2+}. For example, an increase in the concentration of intracellular *free* Ca^{2+} can be detected within a few seconds of complement activation (Campbell et al., 1979, 1981), and an increase in $^{45}Ca^{2+}$ uptake produced by leucocidin on white cells can be detected within a few minutes. Both of these changes in Ca^{2+} precede the initial stages of cell malfunction, cell lysis in the case of complement and lysosomal enzyme release in the case of leucocidin. On the other hand, large increases in total cell Ca^{2+} observed on exposure of hepatocytes to toxins such as CCl_4 or thioacetamide, or to hypoxia, are detected only 1-2 hours after the insult and increase some 10-50-fold over the following few hours. In these latter cases the increase in intracellular Ca^{2+} reflects cell necrosis rather than being the primary cause of the disruption of structures and reactions within the cell.

Many pathological increases in intracellular Ca^{2+} are the result of an increase in the permeability of the cell membrane to Ca^{2+}. The large electrochemical gradient of Ca^{2+} causes net movement of Ca^{2+} into the cell thereby increasing cytoplasmic free Ca^{2+}, and increasing Ca^{2+} bound to various ligands and organelles within the cell. Precipitation of calcium phosphate and other salts may occur (see Table 9.1 for examples of extra- and intracellular precipitates). In some cases of increased mitochondrial Ca^{2+} a concomitant increase in phosphate content is not found (Thiers et al., 1960). Other possible mitochondrial calcium salts are calcium oxalate (Carafoli et al., 1971) and Ca-NADH (Vinogradov et al., 1972; Wrogemann et al., 1973). The K_d^{Ca} for the latter is 250 μM and can be measured by the enhanced fluorescence of Ca^{2+}-NADH compared with NADH.

Majno has argued that denatured protein in necrotic tissue is a major site of Ca^{2+} deposition. This is certainly consistent with the high Glu + Asp content of most proteins (Table 3.6) whose low-affinity Ca^{2+}-binding sites would become significant once the gradient of free Ca^{2+} across the cell membrane was lost. However, large increases in intracellular Ca^{2+} appear to involve also mitochondria (Table 9.4) and possibly the nucleus (Maunder-Sewry and Dubowitz, 1979).

Changes in extracellular Ca^{2+} can affect the extent of cell injury (Table 9.5). A requirement for extracellular Ca^{2+}, together with inhibition at high extracellular Ca^{2+}, certainly suggests that an increase in intracellular Ca^{2+} mediates some aspects of the injury to the cell. Drugs such as local anaesthetics (Fig. 9.13) or phenothiazines (Fig. 9.16) often inhibit effects of toxins on cytolysis or on cellular shape changes. Let us examine some particular examples of cell injury in disease to see whether intracellular Ca^{2+} might be involved in the action of the pathogen.

Table 9.3. Raised cell Ca^{2+} in cell injury

Cause	Tissue	Evidence	Reference
1. Anoxia			
permanent anoxia (implant)	Rat liver, lung, muscle, kidney	Total Ca^{2+}	Majno (1964)
transient ischaemia	Rat myocardium	Total Ca^{2+}	Shen and Jennings (1972)
permanent ischaemia	Rat liver and myocardium	Total Ca^{2+}	Chien et al. (1977, 1978, 1979)
ischaemia	Skeletal muscle	Total Ca^{2+}	Siegel et al. (1977)
2. Toxins			
bacteria (leucocidin, streptolysin O)	Leucocytes	Total Ca^{2+}	Woodin and Wineke (1963)
lymphotoxins	Leucocytes	$^{45}Ca^{2+}$ uptake	Woodin (1968)
glucocorticoids	Leucocytes	$^{45}Ca^{2+}$ uptake	Okamoto and Mayer (1977, 1978)
A23187	Leucocytes	$^{45}Ca^{2+}$ uptake	Kaiser and Edelman (1977)
carbon tetrachloride	Rat liver	$^{45}Ca^{2+}$ uptake	Kaiser and Edelman (1977)
thioacetamide	Rat liver	Total Ca^{2+}	Thiers et al. (1960)
galactosamine	Rat liver	Total Ca^{2+}	Thiers et al. (1960)
vitamin D toxicosis	Rat kidney	Total Ca^{2+}	Farber and El-Mofty (1975)
uranyl nitrate	Rat kidney	Total Ca^{2+} and $^{45}Ca^{2+}$ uptake	Empson et al. (1976) Carafoli et al. (1971)
3. Genetic abnormality			
sickle cell anaemia	Human red cells	Total Ca^{2+}	Eaton et al. (1973)

4. Immunological complement activation	Erythrocytes	Intracellular free Ca^{2+}	Campbell et al. (1979, 1981); Campbell and Luzio (1981)
5. Cancer (see Mundy, 1978) carcinogenesis induced by di-methylaminoazobenzene hepatoma	Rat liver	Total Ca^{2+}	Bresciani and Auricchio (1962)
	Rat liver	Total Ca^{2+}	Hickee and Kalant (1967); Everett et al. (1964)
6. ATP depletion	Human erythrocytes	Total Ca^{2+}	Long and Mouat (1973)
7. Muscular dystrophy	Human muscle	Total cell Ca^{2+} (X-ray microprobe analysis)	Bodensteiner and Engel (1978); Maunder-Sewry and Dubowitz (1979); Maunder-Sewry (1980); Maunder et al. (1977)
8. Cystic fibrosis	Leucocytes + cystic serum	Increased $^{45}Ca^{2+}$ uptake	Banschbach (1978)
9. Deficiency disease			
vitamin D	Skeletal muscle	Total Ca^{2+}	Siegel et al. (1972)
vitamin E	Skeletal muscle (mammalian)	Total Ca^{2+}	Grigoryeva and Lytvynenko (1973)
denervation	Skeletal muscle (amphibian)	Total Ca^{2+}	Hines and Knowlton (1933); Kirby et al. (1975)
10. Trauma	Skeletal muscle	Total Ca^{2+}	Meroney et al. (1957).
11. Ca^{2+} uptake causes death	The flagellate *Chilomonas paramecium*	Total Ca^{2+}	Mast and Pace (1939)

Table 9.4. Cell injury leading to elevated mitochondrial Ca^{2+}

Cause of injury	Tissue	Evidence	Reference
Papain	Heart	3	Ruffolo (1964)
Temporary ischaemia	Heart	1, 3	Herdson et al. (1965); Shen and Jennings (1972)
Parathyroid hormone and excess Ca^{2+}	Kidney	3	Caulfield and Schrag (1964)
Excess vitamin D	Kidney		Scarpelli (1965)
CCl_4 poisoning	Liver	1, 3	Thiers et al. (1960)
Excess isoproterenol	Heart	3	Bloom and Cancilla (1969)
Excess isoproterenol	Aortic smooth muscle		Carvallero et al. (1964)
Tetanus toxin	Skeletal muscle		Zacks and Sheff (1974)
Trauma (surgery)	Heart		D'Agostino and Chiga (1970)
Thioacetamide poisoning	Liver	1	Gallagher et al. (1965)
Renal failure	Heart	3	Lafferty et al. (1965)
Carcinogenesis (4-methyl-aminoazobenzene)	Liver	1	Bresciani and Auricchio (1962)
7α-Fluorocortisol and sodium phosphate	Heart and skeletal muscle	3	D'Agostino (1964)
Iodoform	Liver	3	Sell and Reynolds (1969)
Uranium	Kidney	1, 3	Carafoli et al. (1977)
Mg^{2+} deficiency	Heart	3	Heggtweit et al. (1964)
Cystic fibrosis	Fibroblasts	2	Feigal and Shapiro (1979)
Tumour calcinosis	—		Lafferty et al. (1965)
Muscle disease (dystrophy)	Skeletal muscle		Wrogemann and Rose (1976)

Evidence:
1. Subcellular fractionation followed by Ca^{2+} analysis (see Chapter 2).
2. $^{45}Ca^{2+}$ uptake on isolated mitochondria *in vitro*.
3. Electron-opaque deposits observed in the electron microscope.
4. X-ray microprobe analysis (see Chapter 2, section 2.5.3).

9.1.3. Infectious diseases

Many human diseases are caused by bacteria, viruses, protozoa, platyhelminths, or nematodes. Several of these infective agents have to penetrate cells either to replicate or, as in the case of the malaria merozoite, to mature. Divalent cations are known to stabilize several plant, animal and bacterial viruses (Bassel et al., 1971), but there is little direct evidence at present of a role for intracellular Ca^{2+} in these or other types of infection.

In theory, Ca^{2+} could be involved at one or more of four principle stages of the infection: (1) attachment and entry of the pathogen into the cell; (2) modification and maintenance of the pathogen in the cell; (3) replication of the pathogen in the cell; (4) release of infective agent from the cell after maturation or multiplication.

The entry of infectious agents into a cell can occur by one of three mechanisms (Fig. 9.1): (1) endocytosis (pinocytosis and phagocytosis); (2) injection; (3) membrane–membrane fusion. Apart from the bacterial RNA viruses, the bacteriophages, the second of these mechanisms is comparatively rare, although

Table 9.5. Cell injury and the requirement for extracellular Ca^{2+}

Phenomenon	Cause	Reference
Requirement for extracellular Ca^{2+} to produce maximum damage		
1. Lymphocytolysis	Glucocorticoids or A23187	Kaiser and Edelman (1977)
2. Granule (lysosomal) enzyme release in leucocytes	Bacterial toxins (leucocidin, streptolysin), excess vitamin A	Woodin (1968)
3. Maximum rate of cell lysis	Complement activation	Campbell et al. (1981); Campbell and Luzio (1981)
4. Cytolysis of hepatocytes	A23187, lysolecithin, amphotericin B, mellitin, phalloidin, methyl and ethyl methane sulphonate, N-acetoxyacetylaminofluorene, silica, asbestos	Schanne et al. (1979)
5. Spherocytosis in erythrocytes	A23187	White (1974)
6. Target cell	Cytotoxic T-cells	Roitt (1980)
Inhibition by normal extracellular Ca^{2+} (approx. 1 mM)		
1. Cytolysis in various cells	Sendai virus and other fusogens	Kohn (1979)
2. Cytolysis in leucocytes (release of cytoplasmic enzymes)	Bacterial toxins	Woodin (1968)
Inhibition by excess extracellular Ca^{2+} (approx. 2–10 mM)		
1. Lymphocytolysis	Lymphotoxins	Okamoto and Mayer (1978)
2. Haemolysis of erythrocytes	Complement activation	Campbell et al. (1981)
3. Granule enzyme release	Bacterial toxins	Woodin (1968)

the explosive release of the spores of the parasitic protozoa *Microsporidia* through a tubule provides an interesting example of this phenomenon in eukaryotes. No role for Ca^{2+} in this type of infection is known, although removal of Ca^{2+} from the medium surrounding bacteriophages can cause the separation of RNA from its protein coat (Adams, 1949).

The other two mechanisms, endocytosis and fusion, under certain circumstances are thought to depend on an increase in intracellular free Ca^{2+} (see Chapter 7). Uptake of some viruses may occur via cytochalasin B insensitive micropinocytosis and probably does not involve intracellular free Ca^{2+} (Kohn, 1979). However, several other viruses, bacteria and protozoa are taken up by phagocytosis (Fig. 9.1). This type of uptake occurs in cells that are normally phagocytic, for example the protozoan *Leishmania* is taken up by cells of the reticuloendothelial system. It can also occur in cells not usually associated with

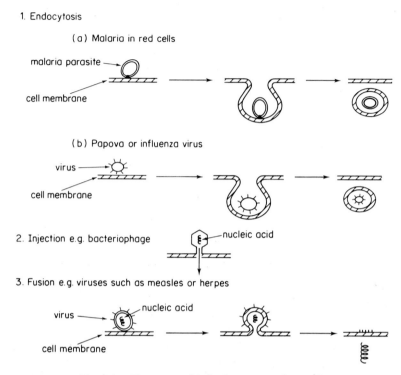

Fig. 9.1. The entry of infectious agents into cells

this phenomenon, for example the uptake of the malaria parasite by red cells (Dvorak and Hyde, 1973). In view of the possible role of intracellular Ca^{2+} in phagocytosis (see Chapter 7, section 7.3), it is necessary to show whether the uptake of infectious agents by this mechanism requires Ca^{2+}.

Several RNA viruses, for example influenza, mumps and rabies, as well as some DNA viruses, for example pox and herpes, are enveloped by a lipid membrane. Some, like influenza virus, are taken up into the cell by endocytosis, others release their genetic material into the cell after fusion of the viral membrane envelope with the outer membrane of the cell (Fig. 9.1). The most well studied of this type are the paramyxoviruses, including Sendai virus, (not infectious to man), measles, and Newcastle disease. Many of these viruses are also able to induce cell–cell fusion (Poste, 1970, 1972; Kohn, 1979), which occurs experimentally but rarely *in vivo* and can occur without infection of the cell by the virus. This is an example of so-called 'fusion from without', in contrast to 'fusion from within', which requires the production of viral proteins within the cell (Poste, 1972).

The role of intracellular Ca^{2+} in this phenomenon is controversial (Pasternak, 1980). Poste has suggested that displacement of Ca^{2+} and ATP from the cell membrane is required for fusion, whereas Lucy (1975) and his group, as well as Volsky and Loyter (1978), propose that the influx of Ca^{2+} which follows

1. Viral site

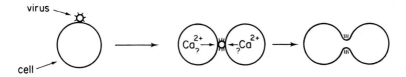

2. Separate site

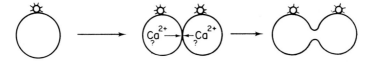

Fig. 9.2. Two possible mechanisms for viral-induced cell fusion

attachment of virus to the cell membrane is essential for cell–cell fusion to occur. In contrast, Pasternak (see Knutton and Pasternak, 1979) believes that the influx of Ca^{2+} is simply the result of cell 'leakiness' induced by the virus and is not required for fusion to occur. Ca^{2+} is not required for attachment of virus to the cell. Unfortunately, investigation of the role of Ca^{2+} in the fusion process is complicated by cytolysis induced by the virus, which is greatly enhanced when Ca^{2+} is removed from the external medium. Volsky and Loyter have reported that cell–cell fusion induced by Sendai virus can be inhibited in the presence of EGTA, whereas virus attachment is not. The inhibition of haemolysis by Ca^{2+} (approx. 1 mM) with enhancement of fusion under these conditions, together with inhibition of fusion by high Ca^{2+} concentrations (2–10 mM) and membrane-stabilizing drugs, for example procaine, chlorpromazine and diphenhydramine, is very reminiscent of the situation of fusion of secretory vesicles with the cell membrane (see Chapter 7).

Sendai virus increases the concentration of intracellular free Ca^{2+} in erythrocyte ghosts containing obelin (Fig. 9.3). This increase occurs before fusion (Fig. 7.9). However, more than 90% fusion of these ghosts with polymorphs can occur without a large consumption of the photoprotein (Hallett and Campbell, 1982a, b) (Fig. 2.12). It would be surprising if the increase in cytosolic free Ca^{2+} induced by the virus did not have some metabolic effects on the cell. If Ca^{2+} is required for cell–cell fusion then its most likely role is in enabling fusion to occur at the maximum rate.

This, however, tells us nothing about the role of Ca^{2+} in fusion of the viral envelope with the cell membrane, which may occur at a separate site from that of cell–cell membrane fusion (Fig. 9.2). Whether Ca^{2+} is required for infectivity of cells by these viruses has not been fully investigated, although Volsky and Loyter (1978) have shown that Ca^{2+} is not required for incorporation of Sendai virus

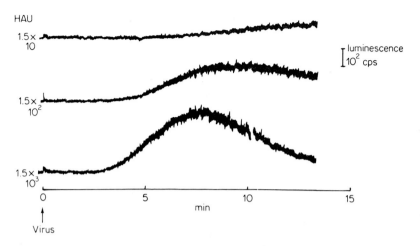

Fig. 9.3. Sendai virus increases intracellular free Ca^{2+}. Pigeon erythrocyte ghosts containing the photoprotein obelin in a balanced salt solution, pH 7.4, containing 1 mM Ca^{2+}, 37°C.↑ = addition of virus (HAU = haemagglutinating units). An increase in obelin luminescence indicates an increase in free Ca^{2+} in the ghosts. The decrease in obelin luminescence occurs because the photoprotein is being consumed. From Hallett and Campbell (1982a)

coat protein into the membrane of erythrocytes. The stimulation by Semliki Forest virus, another fusogen, of mitochondrial Ca^{2+} uptake (Peterhans et al., 1979) suggests that there may be a role for intracellular Ca^{2+} in the infection of cells by this, and other, arbor viruses, as well as in their stimulation of metabolism (glycolysis) and cell–cell fusion.

The means by which endocytosed viruses release their genetic material into the cell cytoplasm from the phagocytic vacuole is still something of a puzzle (Helenius et al., 1980). Some viruses incorporate their nucleic acid into the cell nucleus by fusion of micropinocytotic vesicles with the nuclear membrane. Alternatively, the viral envelope may fuse with the membrane of the secondary lysosome which would then release free genetic material into the cytoplasm. Although little is known of the involvement of Ca^{2+} in these processes, the possible importance of the low cytoplasmic free Ca^{2+} compared with that of the extracellular fluid has been highlighted by the studies of Durham and his colleagues (Durham and Hendry, 1977). Disassembly of many plant (e.g. tobacco mosaic and turnip rosette viruses) and animal viruses, not enveloped by a membrane, into protein subunits and nucleic acid occurs in media where the free Ca^{2+} concentration is less than 10 μM. The binding of Ca^{2+} to these viruses is pH-dependent and may involve both binding to CO_2^- and $>C=O$ moieties in the viral protein. The low intracellular free Ca^{2+} concentration (< 0.1 μM) in most cells would therefore favour virus disassembly. Whilst this idea is attractive there is as yet no direct evidence from cells infected by virus to support it, nor does this hypothesis explain how the new viruses can reassemble in this low Ca^{2+} environment.

Viruses, and other intracellular parasites, often affect cell metabolism and switch off host cell protein and RNA synthesis, a process thought to be dependent on ionic changes in the cell (Carrasco and Smith, 1976). A role for Ca^{2+} in this aspect of virus–cell interaction remains to be established. Infectious influenza virus inhibits the Ca^{2+}–calmodulin regulation of cyclic nucleotide phosphodiesterase, suggesting a possible mechanism for viral-induced changes in cellular cyclic AMP or cyclic GMP (Krizanova et al., 1979; La Porte et al., 1979).

9.1.4. Toxins

During the 1960's Woodin and his colleagues began an investigation into the mechanism of action of certain bacterial toxins. Two toxins attracted their attention in particular. One, called leucocidin, is released by *Staphylococcus aureus*. It consists of two components which act together to cause lysis of white cells. The population of circulating polymorphonuclear leukocytes and macrophages may thereby be reduced during infection by this type of bacterium. The other, called streptolysin O, acts similarly but in this case is produced by *Streptococcus pyogenes*. One of the earliest effects of these toxins on phagocytic cells is the stimulation of the release of lysosomal enzymes and oxygen radicals (see Chapter 7, section 7.3.3), the latter being detectable by chemiluminescence (Allen et al., 1972). These effects depend on the presence of Ca^{2+} in the medium, and an uptake of $^{45}Ca^{2+}$ into the cells can be detected during the first few minutes of exposure to the toxin. Within 15–30 min leucocidin increases the total cell Ca^{2+} from about 2.3 mmoles/kg cell water to 11 mmoles/kg cell water. This is caused by Ca^{2+} accumulation in intracellular vacuoles, and probably mitochondria, and is followed by a decrease in cellular ATP and other nucleotides, not apparently caused by leakage out of the cell, initially at least. Eventually cytolysis and cell death occur.

Ca^{2+}-mediated injury of phagocytic cells caused by these bacterial toxins is in contrast to the non-specific swelling and release of cytoplasmic enzymes which occur when Ca^{2+} is not present in the extracellular fluid. Whether other bacterial toxins such as tetanus and cholera toxins also require Ca^{2+} for their toxic effects is not yet known. Cholera toxin is known to activate adenylate cyclase. Also a considerable number of other toxins, be they biological or chemical in origin, not only disturb the Ca^{2+} balance of cells but may need this disturbance to occur in order to produce the maximum cell damage.

The number of chemical agents known to be toxic to man is very large (Alouf 1976). They include a variety of substances produced by pro- and eukaryotic cells, excess naturally occurring compounds, a wide spectrum of organic and inorganic substances, as well as agents like asbestos which might be classified as environmental toxins. Many, if not all, of these agents cause disturbances in cell calcium (Tables 9.3, 9.4 and 9.6). An increase in $^{45}Ca^{2+}$ uptake, induced by the bacterial toxin leucocidin or excess isoprenaline (isoproteronol), can be detected within a few minutes. It usually takes longer for this to be reflected in a detectable

Table 9.6. Some toxins associated with Ca^{2+}-dependent cell injury

Substance	Nature of substance	Example of injured tissue
1. Biological toxins		
Mellitin	A protein, cytolytic component of bee venom	Liver
Phalloidin	A bicyclic heptapeptide isolated from poisonous toadstool *Amanita phalloides*	Liver (very specific)
Leucocidin	Cytolytic toxin from *Staphylococcus*	Leucocytes
Streptolysin	Cytolytic toxin from *Streptococcus*	Leucocytes
Lymphotoxin	Cytolytic toxin from lymphocytes (45 000 molecular weight)	Leucocytes
Hydroperoxides (oxygen radicals)	O_2^-, $OH\cdot$, H_2O_2 released by leucocytes	Any cell
2. Excess hormone or vitamin		
Fluorocortisol + sodium phosphate	Synthetic glucocorticoid	Muscle
Triamcinolone	Synthetic fluoroglucocorticoid	Thymocytes
Isoprenaline	Synthetic adrenergic agonist	Heart
Adrenaline	Natural hormone	Heart
Vitamin A	Natural vitamin	Leucocytes
Vitamin D	Natural vitamin	Kidney
3. Chemical		
Galactosamine	Sugar	Liver
Plasmocid	Antimalarial drug	Muscle
Carbon tetrachloride	CCl_4	Liver
Thioacetamide	Organic toxin	Liver
Uranyl nitrate	Cation	Kidney
Methyl and ethyl methane sulphonate	Mutagens	Liver
N-Acetoxyacetylaminofluorene	Mutagen and carcinogen	Liver
A23187	Ca^{2+} ionophore (see Fig. 2.2)	Any cell
Fusogens	Polyethylene glycol	Any cell
4. Environmental		
Silica	Dust particles	Liver and macrophages
Asbestos	Dust particles	

See Tables 9.3, 9.4 and 9.5 for references.

increase in total cell Ca^{2+}, or morphologically identifiable Ca^{2+} precipitates in organelles like the mitochondria. A 2–4-fold increase in cell Ca^{2+}, mainly due to deposition of Ca^{2+} salts in the mitochondria and detectable within 1–2 hours, is a common result of the action of many toxins and can be reversible (Tables 9.3 and 9.4). Much larger, but irreversible, increases can be found after several hours or days but these are usually associated with cell necrosis. In the case of CCl_4 or uranium poisoning in experimental systems, cellular Ca^{2+} accumulation can be prevented by mitochondrial inhibitors such as dinitrophenol. Very few measurements have been made of the effect of toxins on intracellular *free* Ca^{2+}.

Most cells will eventually become necrotic if exposed to high enough concentrations of a toxin for long enough. However, under some conditions the effects of a toxin can be reversed, provided that the 'point of no return' for the cell can be identified. Examples of such reversible cell injury can be found in temporary myocardial ischaemia (Shen and Jennings, 1972) and CCl_4 poisoning of the liver (Izutsu and Smuckler 1978). This reversibility is to be distinguished from tissue repair involving replacement of dead cells by new ones.

Does a large, prolonged increase in cytosolic free Ca^{2+} kill the cell?

The proportion of viable cells in a population can be estimated by various biochemical and morphological techniques (Table 9.7). Using the exclusion of the dye trypan blue as a measure of cell viability, it has been shown that several substances toxic to liver and white cells require extracellular Ca^{2+} in order to produce maximum cytolosis and cell death (Table 9.5). The trypan blue exclusion test is notoriously unreliable (Black and Berenbaum, 1964), and in none of the examples quoted has a requirement for Ca^{2+} in the uptake or binding of the toxin by the cells been ruled out. However, the fact that in some cases an uptake of $^{45}Ca^{2+}$ by the cells has been correlated with cytolysis is consistent with intracellular Ca^{2+} mediating some of the cytolytic effects of these toxins. High extracellular Ca^{2+} concentrations are usually inhibitory to cytolysis. The protective action of local anaesthetics, or chlorpromazine and other phenothiazines, against toxins has not been well investigated, though local anaesthetics and cardiac glycosides do seem to interfere with the cytolytic effects of some lymphotoxins (Okamoto and Mayer, 1978). These substances usually require preincubation with the cells before cell injury has been initiated in order to be effective.

Table 9.7. Some tests for cell viability

Test	Time scale
ATP loss	Seconds–minutes
Nucleotide and other metabolite loss	Minutes
K^+ loss	Minutes
Na^+ gain	Minutes
Enzyme release (e.g. lactate dehydrogenase)	Minutes–hours
Decreased metabolism	Minutes–hours
e.g. glucose uptake and metabolism	
hormone sensitivity	
protein synthesis (^{14}C-labelled amino acids)	
RNA synthesis ($[^3H]$uridine)	
DNA synthesis ($[^3H]$thymidine)	
Increased trypan blue uptake	Hours
Electron microscopy	Hours
Light microscopy	Hours–days

9.1.5. Anoxia

Both low and high oxygen tensions can injure and kill cells. High oxygen is toxic because of excess production of superoxide anion, hydroxyl radicals, or H_2O_2. The role of Ca^{2+} in the toxic effects of these highly reactive agents has not been investigated, though an increase in intracellular Ca^{2+} can provoke their production in some cells (Hallett et al., 1981). In contrast, changes in cell Ca^{2+} have been observed in many cells during hypoxia or anoxia, induced experimentally or by ischaemia.

Because of the clinical significance of lack of O_2 supply to the heart, this tissue has been widely investigated in various experimental models. Particular attention has been paid to the changes occurring after O_2 supply to the heart is restored since there is good evidence that it is at this time that irreversible damage to myocardial cells can occur, even in those hearts that have survived the initial transient anoxia (Hearse, 1977; Chien et al., 1979). Transient ischaemia, as one might expect, leads to a decrease in cell ATP and an increase in the permeability of the cell membrane to Ca^{2+} and other ions (Nayler et al., 1979). However, unlike several other experimental systems (Majno, 1964), there is insufficient Ca^{2+} left in the blood vessels of the heart to cause more than a 2-3-fold increase in cell Ca^{2+}. It is perhaps surprising that the mitochondria can take up Ca^{2+} under these conditions (Shen and Jennings, 1972), since the respiratory chain which provides the energy for mitochondrial Ca^{2+} uptake via a pH gradient (see Chapter 4) requires O_2. Once blood flows again, unless the permeability properties of the cell membrane are rapidly restored, the cells can become overloaded with Ca^{2+} (Fleckenstein, 1974, 1977; Fleckenstein et al., 1975) since there will now be plenty of Ca^{2+} available to the cells. This Ca^{2+} overload may damage irreversibly intracellular organelles like the mitochondria, causing eventual necrosis at a time when the cells are adequately oxygenated. It is particularly interesting in the light of this hypothesis that in experimental models local anaesthetics, phenothiazines and Ca^{2+}-channel blockers like D600 or verapamil, can protect against ischaemic cell death in both heart and liver providing that they are administered about 30 min before ischaemia (Fleckenstein, 1974; Chien et al., 1977, 1979). Similar findings have been obtained in adrenaline-induced 'oxygen wastage' in experimental hearts (Opie et al., 1979).

9.1.6. Immune injury

In many diseases interactions of the immune system occur with tissue antigens, involving antibodies, complement and cell-mediated cell injury. Circulating antibodies to constituents on the surface membrane of cells have been identified in several diseases (Table 9.8), and have effects on the tissues primarily responsible for the clinical manifestations of the disease. In diseases like rheumatoid arthritis, immune complexes in the circulation or in the synovial fluid may play an important part in provoking tissue injury. Antibodies may interfere with cell function as a direct result of binding to cell surface receptors (Smith and Hall,

Table 9.8. Diseases where antibodies to plasma membrane antigens have been detected

Disease	Tissue or cell type	Plasma membrane antigen
Haemolytic anaemia	Erythrocyte	Blood group antigens often Rh or I
Myasthenia gravis	Skeletal muscle	Acetylcholine receptor
Graves' disease	Thyroid	Receptor for thyroid-stimulating hormone
Acanthosis nigricans (type B)	Many	Insulin receptor
Allergic rhinitis / Allergic asthma	Lung	β_2-Adrenergic receptor
Idiopathic thrombocytopenic purpura	Platelet	Unidentified
Thyroiditis (including Hashimoto's disease)	Thyroid	Unidentified
Pemphigus	Epidermal cells	Unidentified
Chronic active hepatitis	Liver	Unidentified
Diabetes (juvenile onset)	Pancreatic β-cells	Unidentified

In several diseases cited antibodies to other cell constituents are also observed. For references see Campbell and Luzio (1981).

1974; Flier et al., 1975; Heilbronn and Stalberg, 1978; Venter et al., 1980). Furthermore, Shearer et al. (1976) have reported that antibodies to L cells stimulate $^{45}Ca^{2+}$ uptake.

Antibodies bound to the cell surface activate complement (Fig. 9.4), a pathway which can also be activated by a mechanism independent of antibody. Using the Ca^{2+}-activated photoprotein obelin trapped inside erythrocyte ghosts as a model system, it has been shown that complement activation causes a rise in intracellular free Ca^{2+} within a few seconds and that this effect requires the complete terminal attack complex C5b6789 (Fig. 9.5) (Campbell et al., 1979a, 1981; Campbell and Luzio, 1981). It had previously been thought that Ca^{2+} was only required in complement action for the binding of C1 (C1q, r and s) to the antibody bound to the antigen (Frank et al., 1964, 1965), and that Ca^{2+} was not required further down the proteolytic cascade (Fine et al., 1972; May and Frank, 1973). We (Campbell and Luzio, 1981) have proposed that the earliest intracellular event after formation of the terminal attack complex of complement (C5b6789) is an increase in cytosolic free Ca^{2+}.

Two levels of free Ca^{2+} increase are envisaged:

(1) A rise in cytosolic free Ca^{2+} in the range 0.1–1 μM will cause changes in cell structure and function without causing cell lysis. Under these conditions of mild complement attack, damage to the cell may be reversible, for example by removal of the terminal attack complex from the cell surface or by vesiculation (Hoerl and Scott, 1978; Richardson and Luzio, 1980).

(2) A rise in cytosolic free Ca^{2+} to greater than approx. 10 μM in response to more severe complement attack, will increase the rate of loss of intracellular

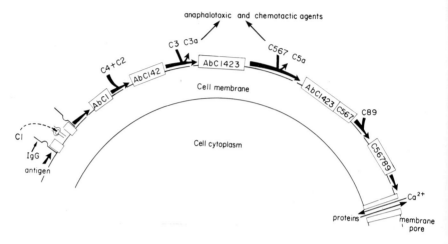

Fig. 9.4. A simplified scheme for complement activation by the 'classical pathway'. This highly simplified scheme is only intended to indicate the main reactions in the pathway. C1 = three proteins, C1q, r, s; binding to IgG or IgM requires Ca^{2+}. C4, C2, C3 and C5 steps all involve proteolytic changes. C5b → 9 involves binding of components to the membrane only. Conventionally cell death is thought to occur, in erythrocytes at least, by colloid-osmotic lysis

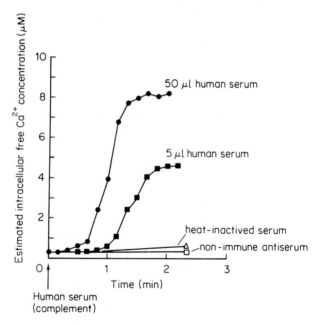

Fig. 9.5. Complement activation increases intracellular free Ca^{2+}. Pigeon erythrocyte ghosts, containing obelin, coated with rabbit anti erythrocyte antibody, 37 °C. From Campbell et al. (1981); reproduced by permission of the Biochemical Society

constituents leading to cell lysis and death, at which time no Ca^{2+} gradient will exist across the plasma membrane. This hypothesis is supported by the complement-induced activation of oxygen-radical production by polymorphs, detected by chemiluminescence (Fig. 9.6). This effect appears to need C9, the terminal component (Fig. 9.7) (Hallett *et al.*, 1981), but occurs in the absence of lysis. The chemiluminescence also occurs after the increase in intracellular free Ca^{2+} (Fig. 9.6).

Conventionally, attack of a cell by the terminal complement complex inevitably leads tö cell lysis. Yet several complement-mediated morphological and chemical changes occur *in vitro* in the absence of lysis. These include shape changes in erythrocytes, reversible membrane depolarization in muscle, release of transmitter from frog motor nerve endings, inhibition of Schwann cell miniature end-plate potentials, inhibition of macromolecular synthesis by bacteria, release of lysosomal enzymes by cartilage cells, release of serotonin by platelets, and stimulation

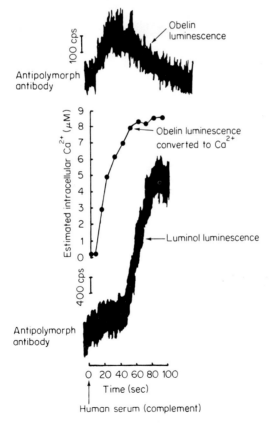

Fig. 9.6. Complement increases free Ca^{2+} in polymorphs before chemiluminescence. From Hallett and Campbell (1982b)

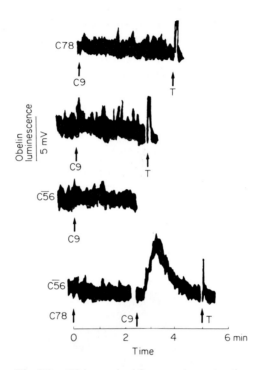

Fig. 9.7. C9 is required for complement activation of obelin luminescence (intracellular Ca^{2+}). Pigeon erythrocyte ghosts containing obelin, 37°C. From Campbell and Luzio (1981); reproduced by permission of Birkhäuser Verlag. T = triton X-100. Analogue luminometer recorded obelin luminescence

of prostaglandin synthesis and bone resorption in an organ culture of foetal rat long bones [see Campbell and Luzio (1981) for references].

The question now arises whether these non-lytic actions of complement are mediated by an increase in intracellular free Ca^{2+}.

The production of plasma membrane vesicles during complement-mediated cell lysis (Richardson and Luzio, 1980), and the possible recovery from complement attack of nucleated cells through cyclic AMP (Boyle et al., 1976) could also involve interactions with intracellular Ca^{2+}. Another example of cell injury which could be dependent on intracellular Ca^{2+} is cell-mediated damage (Grinwich et al., 1978).

An increase in intracellular Ca^{2+} provides a mechanism by which the immune system may alter cell metabolism and morphology, not only in diseases where antibodies to cell-surface components have been found (Table 9.8) but also in diseases such as rheumatoid arthritis, cystic fibrosis (Bogart et al., 1977), muscular dystrophy and multiple sclerosis (Bloom, 1980).

9.1.7 Membrane abnormalities and changes in cell shape

An often ignored characteristic of a cell is its distinctive shape. In a tissue this may be partly imposed on it by its neighbours. However, there is increasing evidence that internal proteins play a major role in determining the shape and polarity of cells as well as that of organelles (e.g. the nucleus). An increase in intracellular Ca^{2+} induced by ionophore A23187 can cause rounding of cells, for example of erythrocytes.

The blood of vertebrates flows through narrow capillaries in the various tissues at an incredible rate, often many hundreds of millilitres per minute. Red cells must therefore be flexible enough to move through the capillaries without damage. Under the microscope red cells from certain groups of animals are immediately recognizable. Birds, for example, have nucleated, ovoid-shaped red cells whereas mammalian red cells are biconcave discs with no nuclei. Changes in red cell shape are associated with several diseases (Fig. 9.8). For example, hereditary disorders like spherocytosis or glucose 6-phosphate deficiency result in roughly spherical cells, whereas haemoglobin S thalassaemias and other genetically abnormal haemoglobin diseases are associated with different shape changes such as sickling. Autoimmune diseases involving attack of red cells by antibodies and complement also tend to cause sphering or rounding of those cells not actually lysed. *In vitro*, too, spherocytes or echinocytes occur in populations of aged cells, particularly in those cells with low ATP and K^+ contents.

Most of these *in vivo* and *in vitro* changes in shape are paralleled by an increase in cell Ca^{2+} and Ca^{2+} permeability (Long and Mouat, 1971, 1973; Eaton *et al.*, 1973; Wolf, 1975; Vaughan and Penniston, 1976; Dreher *et al.*, 1978; Bookchin and Lew, 1980). An increase in the rigidity of the cell membrane also occurs somewhat analogous to the more physiological effect of prostaglandins (Allen

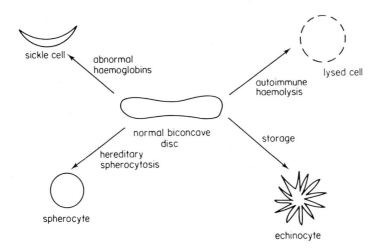

Fig. 9.8. Changes in red cell shape in disease.

and Rasmussen, 1971), which also appears to be mediated by Ca^{2+}. Furthermore, ionophore A23187 induces sphering (White et al., 1975) and vesiculation of red cells (Allan and Michell, 1977) both dependent on Ca^{2+} being present in the extracellular fluid. Chlortetracycline also causes Ca^{2+}-dependent shape changes in erythrocytes (Behn et al., 1977).

The question is, are these changes in shape and membrane flexibility mediated by an increase in intracellular Ca^{2+} and how is it then possible to produce sickling on the one hand and spherocytosis or echinocytosis on the other? A clue to this problem lies in the elaborate and highly organized structure on the inner membrane of the erythrocyte. This involves a protein called spectrin and other protein components, several of which can be cross-linked not by S–S bridges but rather by a peptide link between the εNH_2- of lysine and the $\gamma-CO_2H$ of glutamate (or more correctly the $-CONH_2$ of glutamine). The reaction is catalysed by a transglutaminase (EC 2.3.2.13, glutaminyl-peptide γ-glutamyl transferase) belonging to a group of enzymes also required for cross-linking fibrin in blood clotting. The reaction is

$$\text{protein}^1\text{–N–C glutaminyl peptide} + \text{protein}^2\text{-lys}(NH_2) = NH_3$$
$$\qquad\quad \overset{\|}{O} \qquad\qquad\qquad\qquad\qquad + \text{protein}^1\text{–N–C–glutaminyl-lys protein}^2$$
$$\qquad\qquad\qquad\qquad\qquad\qquad\qquad\qquad\qquad\quad \overset{\|}{O}$$

This is therefore an intermolecular transfer, or endo γ-glutamine : ε-lysine transferase.

The enzyme is known to require Ca^{2+} for significant activity. The K_d^{Ca} seems to be in the mM range, higher than would be expected for an intracellular enzyme regulated by Ca^{2+} under physiological conditions. However, in damaged cells the free Ca^{2+} may be at least as high as 30 μM (Campbell and Luzio, 1981). Furthermore, using gel electrophoresis to analyse components in the red cell membrane, it has been found that ionophore A23187 increases the extent of polymerization, decreasing the amounts of low molecular weight components (Lorand et al., 1976). Ca^{2+} also promotes erythrocyte membrane protein aggregation (Carraway et al., 1975).

It therefore seems likely that at least some of the abnormalities in the diseased red cell membrane are mediated by an increase in intracellular Ca^{2+}, as a result of increased permeability of the membrane. Whether similar structural changes occur in other cells is not known. However, it is significant that ionophore A23187 induces vesiculation of myelin, the membrane sheath of the nerve axon (Schlaepfer, 1977). Vesiculation is a common response to cell injury (Hoerl and Scott, 1978) and may be particularly relevant to diseases such as allergic encephalomyelitis and allergic neuritis, and possibly even the demyelinating disease of multiple sclerosis.

9.1.8 Muscle diseases

In view of the central role played by intracellular Ca^{2+} in the contraction of all types of muscle (see Chapter 5) it is hardly surprising that several authors have invoked a role for Ca^{2+} in muscle disease (Martonosi, 1972; Duncan, 1978). Certainly abnormalities in the Ca^{2+} content of diseased muscle, together with changes in the uptake or release of Ca^{2+} from intracellular stores, have been reported. The question is, do these changes play a primary role in the pathogenesis of the disease or are they simply secondary to cell injury induced by other chemical mechanisms?

Duchenne's muscular dystrophy presents clinically as muscle wasting and weakness. The prognosis is bad. The average life span is 5 to 25 years, and the disease is incurable. It has long been thought that the primary chemical defect is in the muscle cell, though recent studies in tissue culture, where apparently normal muscle cells have been grown from the stem cells of biopsies from patients with the disease, has thrown some doubt on this. Several other types of dystrophy are due to defects in motor nerve innervation. However, in Duchenne's dystrophy identification of the primary cause of injury to the muscle cells lies at the heart of understanding the pathogenesis of this disease. Some muscle fibres in Duchenne's muscular dystrophy can be hypercontracted, indicating a prolonged elevation of intracellular free Ca^{2+}. In the other cells it is possible that an increase in the passive permeability of the cell membrane to Ca^{2+} leads to a prolonged rise in intracellular Ca^{2+}, insufficient to cause contraction but sufficient to cause uptake of Ca^{2+} by mitochondria (Caulfield and Schrag, 1964) and the nucleus (Maunder-Sewry and Dubowitz, 1979) as well as disturbed sarcoplasmic reticulum function (Martonosi, 1972). It has been suggested that Ca^{2+} activation of a neutral protease may also occur under these conditions (Kar and Pearson, 1976). Muscle is a tissue with few lysosomes and this protease could cause significant breakdown of the contractile apparatus, particularly the part associated with the Z line (Fig. 5.11) which seems to be sensitive to this protease. The major problem with this hypothesis is that the neutral protease from muscle requires mM Ca^{2+} concentrations, in the range 0.1–10 mM, to be activated by Ca^{2+} and as yet no convincing effects of calmodulin on this enzyme *in vivo* have been reported, which would allow its activation by μM Ca^{2+} concentrations. The second problem is that of demonstrating directly that muscle cells from dystrophic patients do have an elevated free Ca^{2+} concentration.

Another type of muscle disease is myotonia, which is characterized as far as muscle is concerned by an abnormally delayed relaxation of skeletal muscle. The two main types are myotonia congenita and dystrophica myotonica. Similar clinical symptoms can also be brought on by chronic administration of anticholesterol steroids such as 20,25-diazocholesterol. One might have thought that this would be a good candidate for impaired uptake of Ca^{2+} by the sarcoplasmic reticulum providing the chemical basis for the delayed relaxation. However, studies on human and animal tissue have been unable to demonstrate any significant abnormality in Ca^{2+} uptake by sarcoplasmic reticulum isolated from myotonic muscle. Measurement of intracellular free Ca^{2+} would demonstrate whether the abnormality lay in the regulation of intracellular Ca^{2+}, or in the kinetic constants of troponin C–Ca^{2+} and its interaction with the contractile apparatus.

Ca^{2+} uptake by the sarcoplasmic reticulum has also been shown to be normal in muscle from patients with McArdle's syndrome where the primary abnormality is phosphorylase deficiency. However, in muscle contraction induced by exercise there is evidence for impaired Ca^{2+} uptake by the sarcoplasmic reticulum, whereas in hypokalaemic periodic paralysis the symptoms of the disease may be due to impaired release of Ca^{2+} by the sarcoplasmic reticulum. This latter disease is observed in young people, and results in sudden bouts of flaccid muscular weakness associated with a fall in serum K^+ concentration. The fact that addition of Ca^{2+} to skinned muscle fibres from diseased muscle causes normal contraction shows that the pathogenesis of this disease does not lie in the contractile apparatus itself. Intracellular Ca^{2+} may also play a role in congestive heart disease and cardiac failure, together with the treatment of cardiac disease (see this chapter section 9.4.4).

Before leaving muscle disease it is worth re-emphasizing the importance of appropriate innervation of muscle if normal Ca^{2+}-induced contraction is to be maintained. Denervation of animal muscle results in changes in the capacity of the sarcoplasmic reticulum for Ca^{2+}, whilst increased Ca^{2+} uptake in isolated microsomes has been observed from muscle of patients with amyotrophic lateral sclerosis (Samaha and Gergely, 1965) and the rare Kugelberg–Welander syndrome, both of which involve lesions of motor nerve innervation.

9.2. Possible mechanisms of cell injury induced by a change in intracellular Ca^{2+}

9.2.1. How can a pathological change in cell Ca^{2+} occur?

The consequences of a pathological change in intracellular Ca^{2+} will depend on the extent of the change with respect to both time and concentration. Cytoplasmic free Ca^{2+} concentrations greater than 10 μM are likely to be outside the normal range, whereas local concentrations of about 1 μM, for example close to the cell membrane, maintained for long periods may also be abnormal. Measured changes in intracellular free Ca^{2+} have usually been found to last only a few seconds. This is probably because most of the cells that have been studied are relatively fast, transiently responding cells like nerve or muscle (see Chapters 4 and 5). In fertilized eggs, on the other hand, a Ca^{2+} transient at any point in the egg's cytoplasm only lasts a few seconds since this is sufficient for the cell to pass through a threshold on the way to division (see Chapter 8). Longer-lasting changes in free Ca^{2+} may occur in cells responding to hormones over periods of minutes or hours (see Chapters 5, 6 and 8), for example polymorphs and lymphocytes (Hallett and Campbell, 1982b; Rink and Tsien, 1982). It is likely that an increase in cytoplasmic free Ca^{2+}, in the range 1–10 μM, which lasts for a minimum of several minutes will disrupt intracellular metabolism under pathological circumstances. More severe effects on the cell will occur if the cytoplasmic Ca^{2+} concentration is elevated to higher levels for longer periods. It is possible, in theory, for the cytoplasmic free Ca^{2+} to rise to a higher concentration than that outside the cell, if a membrane potential, negative inside, is maintained across the cell membrane (see Chapter 4 for electrochemical equations). This is unlikely, however, since an increase in the passive permeability of the cell membrane to Ca^{2+} is likely to occur concomitantly with an increase in permeability to other ions with a consequent loss of membrane potential.

An increase in intracellular Ca^{2+} may have several metabolic, electrical and structural consequences (Table 9.9), depending on the extent of the increase. A key question concerns the point at which the cellular changes become irreversible, resulting in the rapid onset of cell death. Decreases in intracellular Ca^{2+} are less common in injured cells, though they have not been looked for so avidly. Some tumours, for example, have a lower total Ca^{2+} content than normal tissue (Table 9.3). Also the possibility exists that decreases in plasma free Ca^{2+}, for example in alkalosis and hypocalcaemia due to parathyroid disease, cause a decrease in cytoplasmic free Ca^{2+}. This could help to explain the hyperexcitability of nerve, resulting in muscle tetany in these conditions, rather than the conventional idea of Ca^{2+} stabilizing the outside surface of excitable cells. Certainly in several experimental models, such as single muscle fibres from the barnacle, a decrease in resting cytoplasmic free Ca^{2+} increases the tendency of the cell to generate action potentials (Hagiwara, 1973). However, the number of cellular changes expected from a decrease in cytoplasmic Ca^{2+} (Table 9.10) are less than those expected from an increase (Table 9.9). Nevertheless, this is an important area requiring further research, since it is likely that several pathological conditions and pharmacological agents act by either decreasing the resting level in the cell or by inhibiting the increase normally induced by a physiological stimulus. This would lead to impairment of cell function without causing cell death.

A change in cytoplasmic free Ca^{2+} can arise by a variety of mechanisms (Table 9.11). Perhaps the commonest cause, or at least the best studied at present, is a change in the permeability of the cell membrane to Ca^{2+} either inwards through passive or specific ion channels, or outwards through active extrusion mechanisms. An increase in the passive permeability of the cell membrane to Ca^{2+} may occur in several diseases (Table 9.12). The precise mechanism of this increase in permeability is not known. Four possibilities can be considered (Fig. 9.9): (1) increased leakage around membrane proteins; (2) vesiculation; (3) micellation; (4) formation of a membrane pore. Vesiculation and micellation both involve severe disruption of the lipid bilayer, locally at least, and the former has been shown to be stimulated by an increasing intracellular Ca^{2+} using ionophore A23187 (Allan et al., 1976a, b; Allan and Michell, 1977a, b). The fourth mechanism is possibly the only one likely to exhibit any significant specificity for Ca^{2+} over other ions. An example might be the action of complement (section 9.1.6). However, even if the permeability change is not specific for Ca^{2+}, because of the enormous membrane gradient of Ca^{2+} compared with other ions, it will be cytoplasmic free Ca^{2+} which exhibits the greatest fractional and earliest change.

Table 9.9. Some possible chemical changes associated with a pathological increase in intracellular Ca^{2+}

Effect	Intracellular location of the effect of Ca^{2+}
1. *Enzyme activation*	
Neutral protease	Cytoplasm
Phospholipase	Cytoplasm and cell membrane
Nuclease	Nucleus and cytoplasm
Ca^{2+} pumps	Various membranes
Transglutaminase	Cell membrane
2. *Enzyme inhibition*	
Adenylate cyclase	Cell membrane
$(Na^+ + K^+)$-MgATPase	Cell membrane
Phosphofructokinase	Cytoplasm
3. *Metabolism*	
Activation of glycogen breakdown by proteolytic cleavage of phosphorylase kinase	Cytoplasm
Activation of lipolysis	Cytoplasm
Decreased ATP formation due to competition with Ca^{2+} uptake	Mitochondria
4. *Electrical*	
Decreased excitability in nerve and muscle	Cell membrane
5. *Activation of Ca^{2+}-dependent mechanisms*	
Exocytosis, e.g. leucocidin on polymorphs	Cell membrane
Shape change, e.g. erythrocytes	Cell membrane
Tetany–hypercontraction, e.g. muscle	myofibrils
6. *Increased organelle permeability*	
NAD	Mitochondria
Ca^{2+}	Microsomes
7. *Cell–cell communication decreased*	Gap junctions
8. *Composition changes*	
Phospholipid content decrease	Microsomes
9. *Precipitation*	
Citrate	Mitochondria
Phosphate	Mitochondria
NADH	Mitochondria
Oxalate	Cytoplasm
Protein	Cytoplasm
Nucleic acid (chromatin condensation)	Nucleus

A steady-state normally exists between the cytoplasmic free Ca^{2+} concentration of 30–300 nM and the bound Ca^{2+} in the cell, approximately 2 mmoles/l cell water. Hence any change in cytoplasmic Ca^{2+} over a significant time period would cause a measurable change in bound, and therefore total, cell Ca^{2+}. The proviso is, of course, that the kinetics of the Ca^{2+}-binding sites are fast enough to respond to the free Ca^{2+} change if this is only transient. Since most of the cell

Table 9.10. Possible consequences of a decrease in intracellular Ca^{2+}

Possible effect	Intracellular location
1. *Response to stimuli*	
Abolished	Cytoplasm, organelles
Concentration of primary stimulus for threshold increased	Cell membrane
2. *Structural changes*	
Disruption of chromatin	Nucleus
3. *Intracellular stores of Ca^{2+} depleted*	
4. *Increased permeability of cell membrane to other ions*	
Hyperexcitability of electrically excitable cells	Cell membrane

Table 9.11. Potential mechanisms for a pathological increase in cytosolic free Ca^{2+}

(A) *Permeability of the cell membrane to Ca^{2+}*
 1. Increase in passive permeability
 2. Sensitization and opening of specific Ca^{2+} channels
 3. Production of a Ca^{2+} ionophore
 4. Decrease in Ca^{2+} efflux, by inhibition of (Ca^{2+})-MgATPase or Na^+–Ca^{2+} exchange

(B) *Intracellular organelles*
 1. Increase in passive pemeability
 2. Provocation of a Ca^{2+} release mechanism
 3. Decrease in intraorganelle binding sites
 4. Decrease in Ca^{2+} uptake by organelles

(C) *Affinity of ligands exposed to the cytoplasm*
 1. Decrease in affinity of ligands for Ca^{2+}
 2. Degradation of surface membrane components which bind Ca^{2+}
 3. Decrease in concentration of soluble Ca^{2+} ligands
 4. Decrease in pH

Table 9.12. Possible diseases involving increase in the permeability of the cell membrane to Ca^{2+}

Disease	Cell	Reference
Muscular dystrophy	Skeletal muscle	Duncan (1978)
Sickle cell anaemia	Erythrocyte	Wolf (1975); Eaton et al. (1973)
Complement-mediated cell injury	Any cell	Campbell and Luzio (1981)
Bacterial toxins	White cells	Woodin (1968)
Intracellular parasite infection	Any cell	Kohn (1979)
Cystic fibrosis	Exocrine cells	Bogart et al. (1977)
Multiple sclerosis	Nerve axon	Schlaepfer (1977)
Rheumatoid arthritis	Polymorph	Hallett et al (1981)

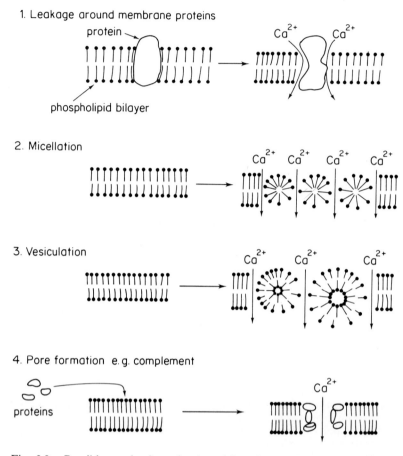

Fig. 9.9. Possible mechanisms for increasing the passive permeability of membranes to Ca^{2+}

Ca^{2+} is bound to relatively low-affinity ligands (pK_d^{Ca} approx. 0.01–1mM), then from Figs. 3.8 and 3.9 it can be seen that to a first approximation a doubling in cytoplasmic free Ca^{2+} (e.g. from 0.1 to 0.2 μM) would cause a doubling in total cell Ca^{2+}. This does not take into account precipitation of calcium salts. A similar change in cell Ca^{2+} in the reverse direction would be expected from a halving in cytoplasmic free Ca^{2+} once a new steady-state is reached. Unfortunately, no mathematical relationship exists as yet to enable the effect of changes in free cytoplasmic Ca^{2+} on total cell Ca^{2+} to be calculated. However, from a knowledge of some of the *in vitro* effects of Ca^{2+} on isolated organelles, proteins and other macromolecules it is possible to explain some of the observed biochemical consequences of a pathological change in cell Ca^{2+}, and even to predict others. Let us therefore examine some of these possibilities in more detail.

9.2.2. Abnormalities in intracellular Ca^{2+} stores

We have seen in Chapters 2, 4 and 5 that mitochondria and endoplasmic reticulum constitute important dynamic intracellular stores of Ca^{2+}. Regulation of the permeability to Ca^{2+} of the membranes of these organelles enables the concentration of cytoplasmic free Ca^{2+} to be changed under physiological conditions. The nucleus is also a major store of internal Ca^{2+}, perhaps as much as 50% of cell Ca^{2+} is found here. Since the amount of Ca^{2+} stored in all of these organelles is determined by a dynamic balance between uptake from and release into the cytoplasm, any abnormal change in cytoplasmic free Ca^{2+} would cause a change in Ca^{2+} stored by these organelles. Mitochondria are known to respond very rapidly to such changes (Bygrave, 1978; Rose and Loewenstein, 1976), which explains why so many instances of mitochondrial Ca^{2+} deposition have been reported (Table 9.4). Presumably Ca^{2+} in the endoplasmic reticulum responds equally rapidly. Unfortunately, this organelle is more difficult to isolate intact after cell fractionation and so data are lacking. Similarly, data for the speed with which the nucleus responds to changes in cytoplasmic free Ca^{2+} are not available.

A pathological change in Ca^{2+} within an intracellular organelle could have two main consequences. Firstly, it will inevitably result in an abnormal response of the cell to a physiological stimulus if the Ca^{2+} store in question is mobilized during the response. Secondly, the uptake of Ca^{2+} itself may cause secondary disturbances in other functions of the organelle. Consider for example a mitochondrion responding to a pathologically elevated cytoplasmic free Ca^{2+}. This could cause: (1) decreased ATP synthesis because of its competition with Ca^{2+} uptake for the proton motive force generated by the respiratory chain; (2) decreased respiration, and hence decreased ATP synthesis, due to mitochondrial precipitation of Ca^{2+}-NADH and NAD leakage (Wrogemann and Nylen, 1978; Wrogemann et al., 1973; Wrogemann and Pena, 1976), which causes the P/O ratio to decrease from 2.8 to 1.7; (3) changes in the activity of Ca^{2+}-regulated enzymes, e.g. pyruvate dehydrogenase; (4) structural damage due to massive Ca^{2+} overloading and Ca^{2+} precipitates.

In the presence of free Ca^{2+} in the physiological range for the resting cell (approx. 30–300 nM), uptake of Ca^{2+} by mitochondria may be as little as 20 nmoles/mg protein. Increasing the free Ca^{2+} to about 10 μM increases the maximum uptake to several hundred nmoles/mg protein. This results in a virtually complete loss of ATP synthesis during the Ca^{2+} uptake phase, and swelling of the organelles. Mitochondria isolated from kidneys injured by uranium (Carafoli et al., 1971) can lose this Ca^{2+} when their capacity for ATP synthesis, measured by P/O ratio, returns to normal. Free Ca^{2+} concentrations in the range 10–200 μM, particularly in the presence of permeant anions such as phosphate, result in massive loading of mitochondria by Ca^{2+}. This can sometimes be as high as 3000 nmoles/mg protein. It causes irreversible injury to the mitochondrion, which eventually releases its Ca^{2+} when the organelle breaks up. Whether factors related to prolonged elevation of cellular Ca^{2+} regulate the

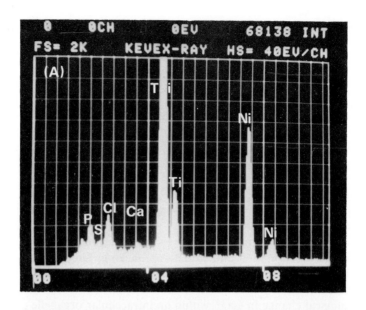

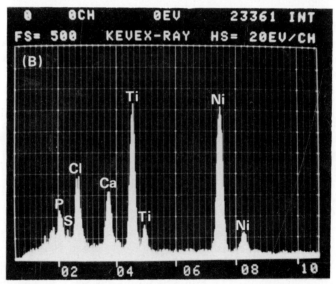

Fig. 9.10. Increased nuclear Ca^{2+} in dystrophic muscle. (A) Normal muscle. (B) Dystrophic muscle. Energy-dispersive spectrum obtained on a partially fixed specimen using X-ray microprobe analysis. By courtesy of Dr. Caroline Sewry

number of mitochondria in the cell or their capacity to take up Ca^{2+}, for example by controlling the amount of the glycoprotein associated with the Ca^{2+} uptake mechanism (see Chapter 4), is not known. It is interesting that in skin fibroblasts grown from patients with cystic fibrosis, or obligate heterozygotes, there is an increased rate of a Ca^{2+} uptake and O_2 consumption by mitochondria isolated from these cells compared with those from control subjects (Feigal and Shapiro, 1979).

Less is known about abnormalities in microsomal or nuclear Ca^{2+}. The physiological and pathological significance of Ca^{2+}-induced release of Ca^{2+} from muscle sarcoplasmic reticulum is still not clear (Endo, 1977; Duncan, 1978). Ca^{2+} uptake by sarcoplasmic reticulum from dystrophic muscle and by endoplasmic reticulum in ischaemic liver is reduced (Chien and Farber, 1977). The latter seems to be due to activation of phospholipid degradation, possibly by a Ca^{2+}-activated phospholipase. The long-term regulation of protein concentration in many eukaryotic cells is often still poorly characterized particularly under pathological conditions. Changes in the concentrations of CaATPases, calsequestrin, and calmodulins over days and months would obviously cause changes in Ca^{2+}-dependent cell regulation. The evidence that changes in nuclear Ca^{2+} can regulate chromatin structure and thus gene expression is still sparse (see Chapter 8). Using X-ray microprobe analysis, an increase in nuclear Ca^{2+} of the order of two-fold compared with controls has been demonstrated in dystrophic muscle (Fig. 9.10) (Maunder-Sewry and Dubowitz, 1979). This technique is open to several criticisms related to specimen preparation (see Chapter 2). However, these observations provide the first evidence for changes in nuclear Ca^{2+} as a result of increased permeability of the cell membrane to Ca^{2+} under pathological conditions.

9.2.3. Abnormalities in the regulation of intracellular free Ca^{2+}

The problems of studying directly, changes in intracellular free Ca^{2+} in mammalian cells have been discussed in Chapter 2. In spite of these difficulties it is possible to outline how possible abnormalities in the regulation of cytoplasmic free Ca^{2+} could be investigated.

A pathological change in cytoplasmic free Ca^{2+} is defined as one occurring in the absence of a physiological stimulus, or as an abnormal response to a cell stimulus. The abnormality may involve an increase or decrease in the free Ca^{2+} concentration, as well as a change in the time course induced by a physiological stimulus. In the absence of such a stimulus the abnormality could arise from any one of three mechanisms: (1) a change in Ca^{2+} bound by organelles or ligands within the cell; (2) a change in the passive permeability to Ca^{2+} of the cell membranes or internal membranes; (3) a change in the activity of Ca^{2+}-efflux mechanisms e.g. Ca^{2+}-activated MgATPase or Na^+-dependent Ca^{2+} efflux.

In addition to these abnormalities, the response of a cell to a primary stimulus would also be modified if the pathological change in the cell modified the response of specific Ca^{2+} channels to the stimulus, or if there was a change in the quantity or affinity of receptors on the cell surface.

In order to define these abnormalities it is essential to be able to measure the concentration of intracellular free Ca^{2+}. It is significant that the earliest detectable intracellular 'event' as a result of complement activation is a large fractional increase in intracellular free Ca^{2+} (Campbell et al., 1979a, 1981) (Figs. 9.6 and 9.7).

9.2.4. Ca^{2+}-dependent regulatory proteins

In view of the ubiquitous occurrence of these proteins (see Chapter 3) it would be surprising if they were not somehow involved in Ca^{2+}-dependent cell injury. Any pathologically induced change in intracellular free Ca^{2+} concentration within the μM range is likely to modify cell responses through a Ca^{2+}-dependent regulatory protein. A role for such a protein has been proposed in the infection of cells by influenza virus I (Krizanova et al., 1979). The stimulation of protein secretion by toxins such as leucocidin, and possibly by activated complement, in diseases such as diabetes and rheumatoid arthritis, as well as cystic fibrosis (Bogart et al., 1977) and hypercontraction in muscle diseases (Duncan, 1978), are other likely candidates for Ca^{2+}-dependent regulatory protein involvement.

It would be interesting to know whether any inherited diseases exist in which these proteins either change their concentration or affinity for Ca^{2+} and the other proteins with which they interact.

The regulation of the concentration of Ca^{2+}-dependent regulatory proteins is poorly understood. Perhaps this, together with internal Ca^{2+} stores such as mitochondria, is dependent on the mean free Ca^{2+} concentration in the cell existing over the cell's resting and activation cycles. In one case, that of the Ca^{2+}-binding protein in the kidney, toxicity induced by excess vitamin D_3 may be mediated by an increase in the intracellular concentration of this protein, thereby increasing Ca^{2+}-binding sites and aiding precipitation of calcium phosphate.

9.2.5. Normally 'passive' or 'latent' Ca^{2+} ligands

Throughout this book a distinction has been drawn between 'active' Ca^{2+} ligands involved in physiological cell responses and 'passive' or 'latent' Ca^{2+} ligands which are not. Few examples of passive Ca^{2+} ligands within the cell have been defined, most exist outside cells, though much of the Ca^{2+} bound to phospholipids and nucleic acids inside cells may fall within this category. In contrast, many proteins within cells contain low-affinity Ca^{2+} sites (Table 3.4), which only become significant when the free Ca^{2+} concentration rises above the physiological range (i.e. $> 10\mu M$) or remains elevated for long enough. In many cases the K_d^{Ca} for these proteins is in the range 0.1–10 mM. Whilst these latent Ca^{2+} ligands are not involved in Ca^{2+}-dependent control of cells under physiological conditions, under pathological conditions they are likely to become significant. The two key questions which must be answered are: (1) What is the magnitude and time course of any changes in intracellular free Ca^{2+}? (2) Can activation or inhibition of the latent Ca^{2+} ligands be demonstrated in the intact cell?

In most cases, at present, the evidence is circumstantial. For example, proteolysis of muscle proteins occurs in dystrophy and other muscle diseases (Duncan, 1978). Muscle cells contain few lysosomes, but do contain a Ca^{2+}-activated neutral protease (Toyo-Oka and Masaki, 1979). This protease can activate phosphorylase kinase and thus phosphorylase, and thereby stimulate glycogen breakdown (Cohen, 1973), a phenomenon commonly associated with cell injury and cell death. The neutral protease (pH optimum 7.5) has a molecular weight of about 85 000 and degrades tropomyosin and troponin I and T, but not the contractile proteins, nor troponin C. It is thought to be responsible for loss of the Z line in injured muscle cells. Unfortunately, the K_m^{Ca} is about 200–300μM. Even quite severely injured cells can maintain a Ca^{2+} gradient across the cell membrane of several hundred-fold (Campbell et al., 1981) (Fig. 9.5). It is therefore difficult to see how this protease could be significantly activated, unless the cell is on a rapid course to death. Similar problems exist in explaining the possible activation by Ca^{2+} of intracellular phospholipases (Chien et al., 1978) and transglutaminases (Lorand et al., 1976). An obvious possibility is that these proteins can also interact with a Ca^{2+}-dependent regulatory protein, thereby increasing their apparent affinity for Ca^{2+}.

A possible example of an important intracellular response of an injured cell to Ca^{2+} concentrations in the μM range is at the gap junction (Deleze and Loewenstein, 1976; Rose and Loewenstein, 1976). No genuine physiological role for Ca^{2+} shutting off this type of cell communication has been established (see Chapter 4, section 4.4.3). However, it could be vital to the survival of a tissue, since it would prevent damage to one cell being transmitted to several others.

Pathological changes caused by these normally latent Ca^{2+} ligands may be rapidly reversible. However, proteolysis or changes in chromatin structure are likely to cause injury to the cell which may remain for some considerable time after the Ca^{2+} levels in the cell have returned to normal.

9.2.6. Precipitation

It has been known for many years that injection of large quantities of Ca^{2+} into cells leads to solubilization of protoplasm, and eventually to denaturation and precipitation of proteins (Heilbrunn, 1943, 1956). In view of the potentially large Glu + Asp content of proteins (Table 3.6), and the possible loss of -NH_2 groups from Glu(NH_2) and Asp(NH_2) side-chains, it is not perhaps surprising that neutralization by cations of the resulting negative charge causes denatured proteins to precipitate. How the proteins denature is not understood. However, this type of precipitation appears to be an important part of the sequence resulting in cell death.

Inorganic precipitates of calcium phosphate, oxalate or even carbonate may occur in cells which have been injured but not irretrievably so. Such precipitates may occur inside organelles like mitochondria (Tables 3.11 and 9.4; Fig. 9.11), though it is unlikely that they could be found in the cytoplasm without the cell being irreversibly injured. The cause of these precipitates is usually an increase in free Ca^{2+} concentration, presumably causing the solubility product to be exceeded. However, in view of the fact that supersaturated solutions of Ca^{2+} salts can often be found in biological fluids (see Chapter 3, section 3.3.4), an additional catalytic factor may be necessary.

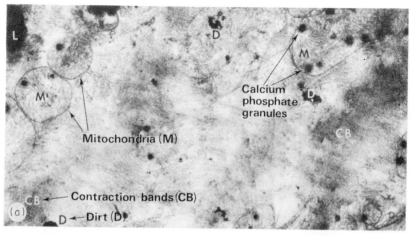

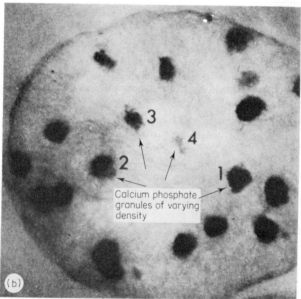

Fig. 9.11. Ca^{2+} precipitates in mitochondria from hypoxic myocardial cells. From Shen and Jennings (1972); reproduced by permission of Lippincott/Harper and Row. (a) Mitochondria in an ischaemic myocardial cell. (b) Mitochondria isolated by centrifugation. L = lipofuscin granules; M = mitochondria; D = dirt; CB = contraction bands

9.3. Intracellular Ca^{2+} and necrosis

To the casual observer the difference between a living organism and something which is dead is obvious. Yet scientists over the past centuries have not found it easy to define life and death precisely at the cellular level, particularly in chemical terms. Apart from erythrocytes and a few other cells, the characteristics of living human cells are respiration, an active metabolism in a continual state of flux with no reactions at true chemical equilibrium, and the ability of the cells to respond to external stimulation, to divide and to carry out certain specialized functions. Pathological conditions cause disruption and impairment of cell structure, metabolism and function. As we have already seen, in many cases these changes can be reversed, allowing the cell to recover provided that the cause of injury is removed. However, sometimes a point of no return is reached where injury to the cell is so severe that within a relatively short space of time, perhaps a few hours, it is morphologically and chemically dead. Some metabolic processes remain, for example proteolytic degradation, but the power supply of the cell, the mitochondria, is functionless, most other reactions have virtually ceased and denaturation of proteins and nucleoproteins has occurred. At this point the cell is recognizable under the microscope as being necrotic [see Abraham (1970) for histological criteria]. Necrosis is observable histologically by gross changes in nuclear structure and loss of stainable cytoplasmic components, both in the light and electron microscope. These morphological changes are often undetectable for several hours after the irreversible chemical changes have occurred in the cells.

Calcium seems to play an important part in the chemical changes associated with cell death (Table 9.13). As has been outlined above, once the free Ca^{2+} concentration in the cytoplasm is in the region of 100 μM, several proteases, nucleases and phospholipases (Chien et al., 1978) may be significantly activated as

Table 9.13. Chemical changes in necrotic cells

Chemical change	Approx. time after cell insult
(A) *Reversible changes*	
1. Initial insult	0
2. Increase in cytosolic free Ca^{2+}	seconds–minutes
3. Loading of mitochondria with Ca^{2+}	minutes
4. Decrease in ATP, increase in ADP, AMP and adenosine	seconds–minutes
5. Activation of glycogen phosphorylase	minutes
6. Increase in intracellular Na^+, decrease in intracellular K^+, loss of membrane potential	minutes
(B) *Irreversible changes*	
1. Irreversible damage to mitochondria	10–30 min
2. Swelling of cell due to H_2O uptake	30–60 min
3. Complete loss of intracellular K^+ and membrane potential	30 min–6 hours
4. Leakage of soluble proteins	30 min–6 hours
5. Activation of proteases, phospholipases and nucleases	1 hour
6. Coagulation of intracellular protein	1–6 hours
7. Loss of RNA	1–6 hours
8. Degradation of protein and DNA	6 hours–several days

For references see Majno (1964), Campbell (1970), Campbell and Hales (1971), Chien et al. (1978), Hearse (1977), Wrogemann and Pena (1976).

well as several ion pumps and key regulatory enzymes inhibited (Table 9.9). The increase in total cell Ca^{2+} can be considered in two phases (Fig. 9.12). The first, about a two- to three-fold increase, is usually reversible and probably involves a cytoplasmic free Ca^{2+} concentration of the order of 10 μM and a large uptake of Ca^{2+} by mitochondria. This mitochondrial Ca^{2+} uptake presumably requires energy, and is detectable morphologically within a few minutes' exposure of cells to toxins like CCl_4 or excess isoproterenol. Under some conditions uncouplers like dinitrophenol added *in vivo* or *in vitro* can release this Ca^{2+} (Carafoli *et al.*, 1971), suggesting that at this stage the mitochondria are not irreversibly damaged. Within 30–60 min the Ca^{2+} accumulation has become irreversible and causes structural injury to the mitochondria. Furthermore, the enormous decrease in cell ATP, which often drops to < 5% of that in the normal cell, causes massive changes in intracellular K^+ and Na^+ and leakage of proteins out of the cell. A

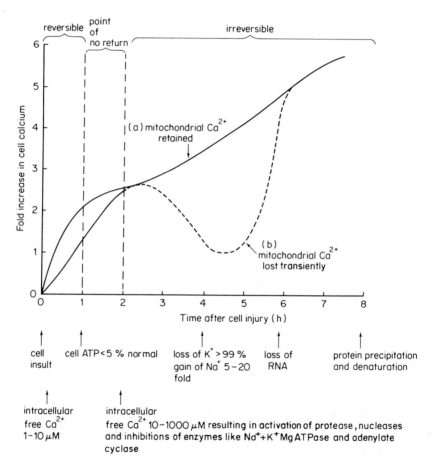

Fig. 9.12. Biphasic changes in mitochondrial Ca^{2+} in necrosis. These figures are only approximate and are intended to illustrate the orders of magnitude involved. Two different situations are illustrated

further increase in intracellular free Ca^{2+} in the range 0.1–1 mM causes further metabolic changes through enzyme inhibition, and aids precipitation and denaturation of many proteins (K_d^{Ca} 0.01–1 mM) remaining in the cell. At this point, usually some 2–8 hours after the initial injury, the Ca^{2+} content may be more than ten times that of the original cell, increasing to 50 times over the next few days. The denatured proteins and nucleic acids with Ca^{2+} bound to them are relatively resistant to hydrolysis since the protein and nucleic acid content of necrotic cells decreases very slowly (Majno, 1964; Campbell and Hales, 1971).

The phenomenon of reversible and irreversible protein coagulation in cells has been studied in some detail by several workers (Heilbrunn, 1956; Majno, 1964). Heilbrunn has shown that it can depend on Ca^{2+} ions being present. The whitish appearance of many dead tissues is the result of this denaturation, and can be measured using dark-field or fluorescence microscopy. Whether this denaturation in necrotic tissue is due to pH and ionic changes, or proteolysis, is not known. It is likely, however, that unfolding of polypeptide chains together with conversion of $Asp(NH_2)$ and $Glu(NH_2)$ residues to Asp and Glu, respectively, by deamidation, producing NH_4^+ ions, will greatly increase the binding capacity of denatured proteins for Ca^{2+}. Inhibition of Ca^{2+} uptake by Ca^{2+}-channel blockers (Fleckenstein, 1974), chlorpromazine (Chien *et al.*, 1977) and local anaesthetics (Hearse, 1977) can increase the time taken for irreversible cell damage to occur.

9.4. The pharmacology of intracellular Ca^{2+}

9.4.1. The problems

Tissue extracts, particularly from plants, have been employed by man for thousands of years in an attempt to treat human ailments, as well as being used for stimulating or repressing mental activity. These pharmacological preparations, as we would now describe them, have multiplied considerably during this century, thanks to the efforts of synthetic organic chemists. The word 'pharmacology' is derived from the Greek *pharmakon* meaning drug or medicine. It was a word which to the Greeks had a somewhat mystical significance. Today, however, much is known of the cellular effects and chemical mechanism underlying many, but by no means all, of the substances used in the treatment of disease. Many substances are also used as experimental tools for investigating physiological control mechanisms. We have already seen in this chapter that substances such as the local anaesthetics or phenothiazines can alleviate certain types of cell injury in which changes in intracellular Ca^{2+} seem to play a key role. Over the past 20 years several other groups of drugs have been identified whose cellular actions are thought to involve disturbances in intracellular Ca^{2+} (Table 9.14). The evidence for this, however, is often poorly documented. Therefore, before examining each of these groups in turn, let us first consider the experimental approach necessary to establish a role for intracellular Ca^{2+} in the action of a pharmacological agent. This involves the characterization of five main aspects of its action: (1) Definition of the effects in the whole organism,

Table 9.14. Some pharmacological substances which may interact directly with cell Ca^{2+}

Clinical use	Example	Specific compound	Reference
1. Anaesthetics	General	Urethane	Baker and Schapira (1980)
	Local	Lignocaine	Lee (1976)
2. Pain relievers	Opiate	Morphine	Sanghvi and Gershon (1977)
3. Antipsychotic drugs	Phenothiazines	Chlorpromazine	Kwant and Seeman (1969)
		Trifluoperazine	Weiss (1978); Weiss and Levin (1978)
4. Cardiovascular agents	Cardiac glycosides	Digoxin (digitalin)	Akern and Brody (1977)
	Ca^{2+}-channel blockers	Verapamil	Fleckenstein (1977); Opie (1980)
5. Antiallergy drugs	Antihistamines		
	Disodium cromoglycate		Foreman and Garland (1976)
7. Stimulants	Methylxanthines	Caffeine	Bianchi (1961)
		LSD and amphetamines	Vaccari et al. (1971)
8. Cations			
	Lanthanum		Weiss (1974)
	Mg^{2+} toxicity		Mordes and Wacker (1978)
9. Miscellaneous	—	Dantrolene sodium	Hainaut and Desmedt (1974, 1979)

See Figs. 9.13–9.22 for structures.

including identification of which cell types are affected and at what dosage. (2) Definition of the electrical and chemical changes induced in cells exposed to different concentrations of the drug. (3) Requirement for extra- or intracellular Ca^{2+}, in the physiological mechanism altered by the drug or in the action of the drug itself. (4) Identification of the site of action within the cell of the drug, including whether it acts on the surface of the cell or internally. In particular, does it interact with one or more of the following molecules: hormone receptors; membrane-bound enzymes, such as adenylate cyclase, or ion pumps; phospholipids; ion channels; intracellular proteins or other macromolecules. (5) The molecular basis of the interaction with Ca^{2+}-dependent processes within the cell.

9.4.2. Anaesthetics

As long ago as 1922, Burridge, studying the action of cocaine, concluded that its effect on beating hearts depended on interference with the tissue's normal interaction with electrolytes, particularly Ca^{2+}. This theme was developed further by Heilbrunn (1943), who advocated a role for internal Ca^{2+} in the action of not only cocaine but also of all 'fat-solvent' anaesthetics, as he called them. His idea was that both cell stimuli and anaesthetics caused release of Ca^{2+} inside the cell from the cell cortex. This cell stimulation resulted in a 'clotting reaction' of proteins in the interior protoplasm of the cell, which was inhibited in the presence of an anaesthetic. Whilst this concept is compatible with the ideas developed in this book concerning the regulatory role of intracellular Ca^{2+}, the validity of the interpretation of many of the experiments quoted by Heilbrunn is uncertain, because no information is available on cell viability after exposure to the anaesthetic. Nevertheless, much evidence about the involvement of cell Ca^{2+} in the action of anaesthetics has accumulated since Heilbrunn wrote his *Outline of General Physiology* in which he expounded these ideas on the role of Ca^{2+} in cells (Table 9.14 and 9.16) [see Halsey *et al.* (1974), Covino and Vassallo (1976) for references]. Only some of these observations, however, are relevant to anaesthetics in clinical use.

Table 9.15. Structure of some general anaesthetics

Compound	Structure
1. *Gaseous*	
Xenon	Xe
Ethylene	C_2H_4
Cyclopropane	C_3H_6
Nitrous oxide	N_2O
2. *Volatile*	
Divinyl ether	$(C_2H_3)O$
Diethyl ether	$(C_2H_5)_2O$
Chloroform	$CHCl_3$
Halothane	$CF_3CH \cdot BrCl$
Urethane	$NH_2COOC_2H_5$
Trichloroethylene	$CHClCCl_2$
Methoxyflurane	$CH_3OCF_2CCl_2H$
3. *Intravenous injection*	
Hexobarbital (a barbiturate)	(structure with ONa, N-CH$_3$, HN, O, CH$_3$, cyclohexenyl)
3α-Hydroxy-5β-pregnan-20-one	(steroid structure with COCH$_3$ and OH)

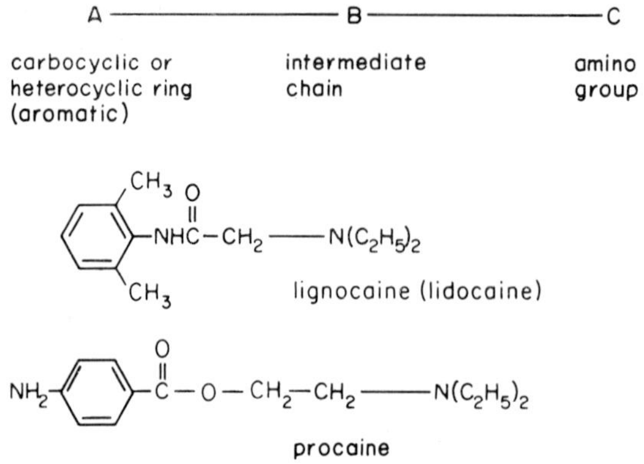

```
    A              B              C
carbocyclic or   intermediate   amino
heterocyclic ring   chain       group
(aromatic)
```

[Structure of lignocaine (lidocaine): 2,6-dimethylphenyl-NHC(O)-CH₂-N(C₂H₅)₂]

lignocaine (lidocaine)

[Structure of procaine: NH₂-C₆H₄-C(O)-O-CH₂-CH₂-N(C₂H₅)₂]

procaine

Fig. 9.13. The essential structure of local anaesthetics

Anaesthetics can be divided into two main classes. (1) General anaesthetics (Table 9.15), which cause reversible depression of the responses of animals to stimuli and loss of consciousness; their effects are mediated almost entirely through the central nervous system. (2) Local anaesthetics (Fig. 9.13), which cause a reversible loss of sensation in the region close to the site of application. The main site of action of both classes of anaesthetic is the nerve synapse. They depress release of transmitter from nerve endings, and inhibit the action of the transmitter post-synaptically. Effects on the conduction of impulses along nerve axons do not seem to play a major role in the clinical action of anaesthetics, though this may be important in the action of local anaesthetics.

General anaesthetics

Halothane and methoxyflurane, two general anaesthetics, inhibit Ca^{2+} uptake by rat liver and brain mitochondria (Rosenberg and Haugaard, 1973). Krnjevic (1971) has proposed that the resulting loss of Ca^{2+} from mitochondria causes prolonged elevation of cytoplasmic free Ca^{2+}. This Ca^{2+} then inhibits evoked transmitter release from the nerve terminal, presumably after an initial stimulation. Ca^{2+} will also inhibit firing of the next nerve by increasing K^+ conductance at the post-synaptic terminal (see Chapter 4, section 4.4.2). Unfortunately, attempts to measure directly the effect of general anaesthetics on intracellular free Ca^{2+} in squid giant axon, using aequorin, have failed to show any increase (Baker and Schapira, 1980). However, an increase in aequorin luminescence was observed due to a direct effect of the general anaesthetic urethane on the protein itself. This has lead to the suggestion that changes in the affinity for Ca^{2+} of Ca^{2+}-dependent regulatory proteins might explain the action of general anaesthetics on cells.

This possibility brings to the fore the two conflicting hypotheses regarding the primary site of action of both general and local anaesthetics (Koblin and Eger, 1979). For some time it has been known that both classes of compounds can affect the permeability to ions of membrane vesicles composed of pure phospholipids. Bangham *et al.* (1967) have proposed that general anaesthetics interact primarily with phospholipids, thereby affecting the fluidity of the membrane and hence its electrical properties. However, it is known that these anaesthetics can have a direct action on cellular proteins (Allison, 1974; Foldes, 1978). In spite of the structural diversity of general anaesthetics (Table 9.15), the alternative hypothesis has been proposed that these anaesthetics interact with particular membrane proteins. In other words, there are protein receptors for anaesthetics, just as there seem to be for hormones and neurotransmitters. These two, apparently conflicting, hypotheses would not be incompatible if the anaesthetic interacted with the specific phospholipid annulus which is often associated with membrane proteins.

Whatever the primary binding site is for general anaesthetics, and whether they affect the membrane fluxes of ions besides Ca^{2+}, it is likely that a direct or indirect interaction with Ca^{2+}-dependent secretory or permeability mechanisms plays an important part in their action. It has also been suggested that they might increase the permeability of the sarcoplasmic reticulum to Ca^{2+}. This could play a role in one of their potentially dangerous side-effects, hyperthermia (Allison, 1974), by increasing the energy consumption in muscle. Any theory concerning the chemical nature of general anaesthesia must be able to explain the very rapid rate at which unconsciousness can be achieved.

Local anaesthetics

Local anaesthetics form a class of compounds more obviously related to each other chemically than are general anaesthetics. The molecule consists of three main portions (Fig. 9.13; compare with Table 9.15): A—an aromatic moiety conferring lipophilic properties on the molecule; B—an intermediate chain linked to A by an ester or amide bond; C—an amine group, often, but not necessarily always, a tertiary amine.

Like so many useful pharmacological substances the first local anaesthetic to be discovered was from a plant. *Erythroxylon coca* is indigenous to South America, where the natives chew the leaves. Cocaine, the active principle from this plant, was isolated by Niermann in 1860. Unfortunately, the compound proved too toxic and was, of course, too addictive to be of clinical use. Non-toxic, and non-addictive, local anaesthetics were first synthesised by Einhorn in 1905. Procaine, the first, was followed by others such as cinchocaine (dibucaine), lignocaine (lidocaine), and amethocaine (tetracaine) which are more potent. These compounds are used widely in medicine and dentistry, the compound chosen depending on the problem (Covino and Vassallo, 1976).

The clinical action of local anaesthetics depends on their ability to inhibit Na^+ conductance, thereby increasing the threshold for excitation of the nerve membrane. This blocks nerve conduction by preventing the development of action potentials. Local anaesthetics also affect many other electrically excitable and non-excitable cells, under experimental conditions at least (Table 9.16). For

Table 9.16. Some effects of local anaesthetics

Effect	Example of anaesthetic used	Reference
Nerve		
1. Inhibition of Na^+ and K^+ conductance in various axons	Procaine	Blaustein and Goldman (1966)
2. Increase of Ca^{2+} efflux from nerve axons	Procaine, amethocaine (tetracaine)	See Covino and Vassallo (1976)
Muscle		
1. Increase in Ca^{2+} efflux from frog sartorius muscle	Amethocaine (tetracaine)	See Covino and Vassallo (1976)
2. Inhibition of smooth muscle contraction stimulated by histamine or carbamylcholine	Cinchocaine (dibucaine) and amethocaine (tetracaine)	Jefferji and Michell (1976)
3. Inhibition of Na^+ and K^+ conductance and action potential propagation in smooth and striated muscle	Procaine and amethocaine (tetracaine)	Feinstein and Paimre (1969)
Secretion		
1. Inhibition of Ca^{2+}-mediated secretion of catecholamines from adrenal medulla (acetylcholine or K^+ stimulation)	Various	Jaanus *et al.* (1967); Douglas (1974)
2. Inhibition of insulin secretion from cells (glucose or Ba^{2+} stimulation)	Cinchocaine* (dibucaine)	Milner and Hales (1968)
3. Inhibition of Ca^{2+}-dependent histamine release from mast cells	Amethocaine (tetracaine)	Foreman *et al.* (1976)
Hormone action		
1. Inhibition of adrenaline-stimulated lipolysis and glucose uptake in rat adipocytes	Cinchocaine (dibucaine)	Siddle and Hales (1974b)
2. Inhibition of cyclic AMP (glucagon) mediated gluconeogenesis and glycogenolysis in rat liver	Amethocaine (tetracaine)	Friedmann and Rasmussen (1970)
3. Increase in glucose 6-phosphate independent glycogen synthesis in rat adipocytes	Procaine	Hope-Gill *et al.* (1976)
Organelles and vesicles		
1. Inhibition of mitochondrial Ca^{2+} transport	Various	Carafoli (1974)
2. Inhibition of Ca^{2+}-induced increase in Na^+ permeability of phospholipid vesicles	Various	Papahadjopoulos (1972)

US names in parenthesis.   * Nupercaine = a trade name used by these authors for cinchocaine.

example, they inhibit the secretion of hormones and neurotransmitters, as well as inhibiting the action of these substances on various cell types. Procaine is one of the least potent compounds, concentrations of 1–10 mM being required for maximum inhibition. In contrast, effects at μM concentrations of more potent local anaesthetics such as cinchocaine (dibucaine) and amethocaine (tetracaine) can be detected, maximum effects being obtained with concentrations of 0.1–0.2 mM.

The action of cocaine on beating hearts depends on its interference with tissue electrolytes, particularly Ca^{2+} (Burridge, 1922). This, together with the well-known stabilizing action of Ca^{2+} on excitable membranes (see Chapter 4) and the fact that the ionised form of the amine seems to be the active form of the local anaesthetic (Fig. 9.15), lead to the suggestion that they act as 'membrane stabilizers'. In spite of this rather vague concept, there is a correlation between their potency as inhibitors of cell activation (Table 9.16) and their ability to displace radioactive Ca^{2+} from biological or pure phospholipid membranes (Fig. 9.14). The sensitivity of cells to low concentrations of local anaesthetics is increased when the extracellular Ca^{2+} concentration is reduced. Increasing the external Ca^{2+} reduces the potency of the anaesthetic. This is to be expected if the R_3N^+H binds to sites on the membrane normally occupied by Ca^{2+}, the hydrophobic moiety of the anaesthetic presumably lying in the phospholipid. The question arises, are the inhibitory effects of local anaesthetics entirely mediated by an effect on membrane structure and conductance or is a direct interaction with the regulation of intracellular free Ca^{2+} required?

Unfortunately, evidence for direct effects of local anaesthetics on intracellular free Ca^{2+} is lacking. However, in several cases their inhibitory effects on cell activation have been shown to be independent of other intracellular regulators such as cyclic AMP (Friedmann and Rasmussen, 1970; Siddle and Hales, 1974b). Local anaesthetics do affect Ca^{2+} efflux, though whether this is an activation (see Covino and Vasallo, 1976; Hope-Gill et al., 1976) or an inhibition (Friedmann and Rasmussen, 1970) depends on the type of cell being studied. Effects on isolated intracellular organelles, such as mitochondria, have been observed (Table 9.16). However, it is not known how much, if any, of the anaesthetic is able to penetrate into the interior of the cell.

The present hypothesis for the role of intracellular and extracellular Ca^{2+} in the action of local anaesthetics can be summarized as follows (Fig. 9.15): (1) Local anaesthetics bind to phospholipid, or perhaps protein receptor sites, on the surface of cells. These binding sites are distinct from voltage-dependent ion channels. Inhibition of passive Na^+ and K^+ permeability follows. The role of membrane proteins is not clear since effects of the anaesthetics can be demonstrated on pure phospholipids (Table 9.16). (2) Ca^{2+} is displaced from the outer surface of the cell as a result of anaesthetic binding. Unlike the bound Ca^{2+}, when the cell is depolarized this bound anaesthetic is not released. Action potentials are therefore more difficult to generate, resulting in a reduction of Ca^{2+} movement into the cell through its voltage-dependent channel. (3) Increases in intracellular free Ca^{2+}, normally caused by physiological agonists

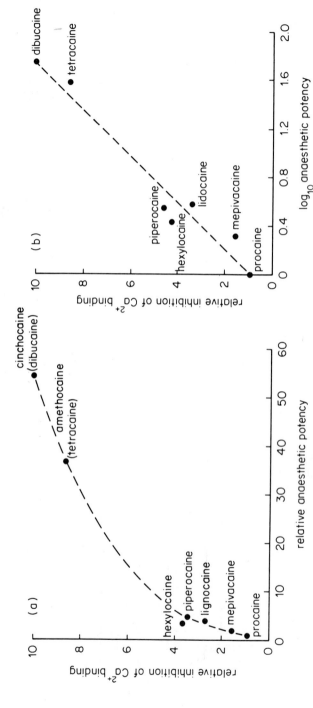

Fig. 9.14. Correlation between local anaesthetic potency and displacement of bound Ca^{2+}. Calculated from data of Blaustein and Goldman (1966). Relative inhibition of Ca^{2+} binding was calculated from the reciprocal of the drug concentrations required to inhibit $^{45}Ca^{2+}$ binding to phosphatidylserine by 20%. Relative anaesthetic potency was calculated from the concentration required to inhibit action potentials in frog sciatic nerve at pH 7.2. Data are plotted on a (a) linear scale, and (b) a semilog scale

increasing the permeability of the cell membrane to Ca^{2+}, are reduced. Cell activation is therefore either inhibited or prevented altogether. (4) Local anaesthetics also inhibit cell activation caused by release of Ca^{2+} from intracellular organelles, for example in smooth and striated muscle and some secretory cells. The precise mechanism of this is not yet clear. (5) Local anaesthetics do not normally inhibit hormone activation of adenylate cyclase. In some cells they elevate cell cyclic AMP when the hormone is present (Siddle and Hales, 1974b).

9.4.3. Phenothiazines

During the 1950's a group of substituted phenothiazines (Fig. 9.16) was discovered which had some unusual neurosedatory effects. The first of these compounds was called chlorpromazine and more than 50 named derivatives have now been synthesised. These so-called neuroleptics have been used to treat psychosis, though many years ago phenothiazine itself was used as a urinary

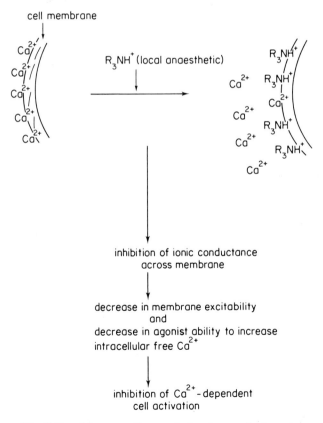

Fig. 9.15. Sequence of events in local anaesthetic action

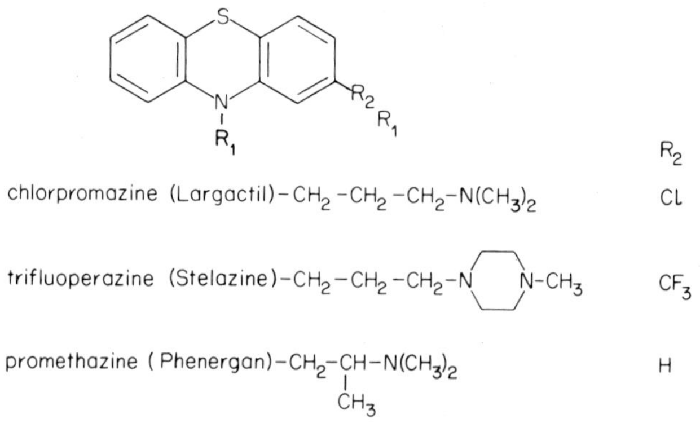

chlorpromazine (Largactil) $-CH_2-CH_2-CH_2-N(CH_3)_2$ R_2 = Cl

trifluoperazine (Stelazine) $-CH_2-CH_2-CH_2-N\diagup\diagdown N-CH_3$ CF_3

promethazine (Phenergan) $-CH_2-CH(CH_3)-N(CH_3)_2$ H

Fig. 9.16. Phenothiazines

antiseptic against nematode infection. The clinical effects of this group of drugs seem to be mediated mainly through actions on the central nervous system, although they do affect many other tissues, particularly under experimental conditions.

The α-adrenergic blocking action of chlorpromazine, together with its interference with neurotransmitter action and electrically excitable cells, lead to the suggestion that phenothiazines act in a similar manner to the local anaesthetics. This hypothesis was strengthened by the observation that chlorpromazine competitively displaced Ca^+ from the membrane of erythrocyte ghosts (Kwant and Seeman, 1969) possibly by binding to triphosphoinositides. However, these drugs can act intracellularly, reducing phospholipid synthesis and causing uncoupling of oxidative phosphorylation in the mitochondria.

Many of the effects of phenothiazines involve interactions with Ca^{2+}-dependent processes in the cell (Table 9.17). Though some of these may be explained by a 'local anaesthetic' type of action, the discovery that they bind to Ca^{2+}-dependent regulatory proteins (Weiss and Levin, 1978) suggested a more direct interaction with the regulation of cells by intracellular Ca^{2+}. Trifluoperazine (TFP), and several other phenothiazines, bind to the Ca^{2+}-dependent regulator of both cyclic AMP phosphodiesterase (Levin and Weiss, 1978) and myosin ATPase (Sheterline, 1980), i.e. calmodulin (see Chapter 3, section 3.4.2). There are two high-affinity sites ($K_d^{TFP} \approx$. 1–3 μM) and a large number, perhaps 50, low-affinity sites ($K_d^{TFP} \approx$ 5 mM). Maximum binding requires saturation of the protein by Ca^{2+} (approx. 10–100 μM), suggesting that the phenothiazine-binding site is not the same as that for Ca^{2+}. Presumably the drug inhibits the action of the regulator protein by preventing it activating the protein to which it normally binds. Troponin C does bind trifluoperazine, though less avidly than calmodulin. Trifluoperazine does not bind to, or affect, the low-affinity Ca^{2+}-binding protein from brain, S-100, or the myosin light chain ATPase in smooth muscle, or the actomyosin ATPase in skeletal muscle. Compounds apparently related structurally to phenothiazines such as W-7

Table 9.17. Some Ca^{2+}-dependent effects of some phenothiazines

Effect	Agent used	Reference
Inhibition of sarcoplasmic reticulum Ca^{2+} pump	Chlorpromazine	Balzer and Makinose (1968)
Inhibition of contraction of mesenteric arterial muscle	Chlorpromazine	Godfraind and Kaba (1969)
Inhibition of myosin light chain kinase of chicken gizzard muscle	Chlorpromazine Trifluoperazine	Hidaka et al. (1979) Sheterline (1980)
Inhibition of Ca^{2+}-activated myosin ATPase in chicken gizzard muscle	Trifluoperazine	Sheterline (1980)
Inhibition of exocytosis in sea urchin eggs	Trifluoperazine	Baker and Whitaker (1979)
Inhibition of Ca^{2+}-activated neutral protease	Trifluoperazine	Sugden et al. (1979)
Inhibition of polymorph chemiluminescence	Trifluoperazine	Hallett et al. (1981)
Inhibition of insulin secretion	Trifluoperazine	Ashcroft et al. (1981)

(Hidaka et al., 1979) (Fig. 9.24), which relax isolated rabbit arterial muscle and inhibit contraction initiated by high K^+, prostaglandin $F_{2\alpha}$, noradrenaline, histamine, 5-hydroxytryptamine or angiotensin II, may also act on Ca^{2+}-calmodulin.

Phenothiazines also have a local anaesthetic action, displacing Ca^{2+} bound to the membrane of cells. Furthermore they can also affect cell viability. Whilst the interaction of phenothiazines with calmodulin is of great experimental interest, it remains to be established whether this is the molecular basis of the clinical effects of this group of drugs.

9.4.4 Cardiovascular agents

The power on the heart of extracts of the foxglove leaf, *Digitalis purpurea*, has been known for centuries. Cardiac glycosides (Fig. 9.17) are currently used in the treatment of congestive heart disease and in the control of ventricular rate in patients with atrial fibrillation. Their positive inotropic action strengthens the contraction of the heart. The other group of drugs used to stimulate the performance of the heart are the β-adrenergic agonists (Fig. 9.18), which not only increase the strength of the heart beat but also its rate, a so-called chronotropic effect (see Chapter 5, section 5.3.5). These agents act by increasing the cyclic AMP concentration in myocardial cells. Cyclic AMP activates a protein kinase which in turn phosphorylates proteins on the inner surface of the cell membrane, and also possibly on the sacroplasmic reticulum and the contractile apparatus. The result is that the voltage-sensitive Ca^{2+} channel is more sensitive to intracellular Ca^{2+}, and the sarcoplasmic reticulum contains more Ca^{2+}. When a myocardial cell is activated by an action potential, adrenaline-stimulated cells produce a greater

Ouabain (G strophanthin)

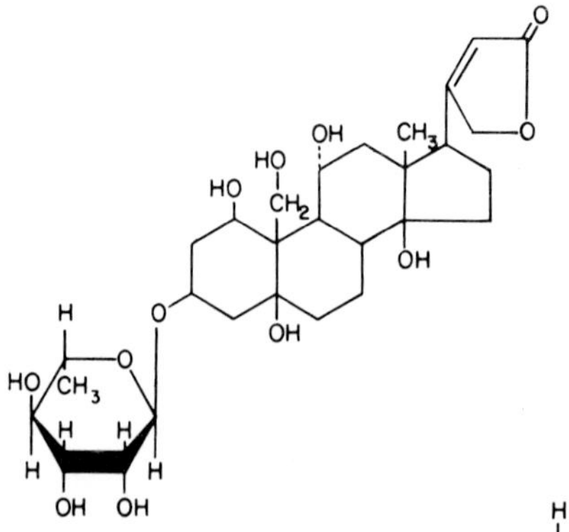

Digitalin

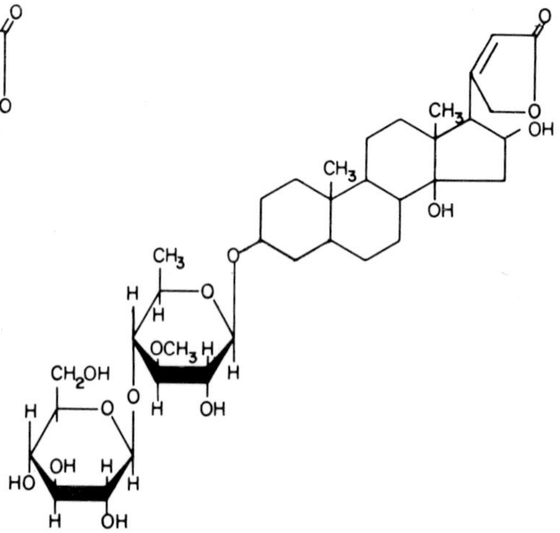

Cassaine

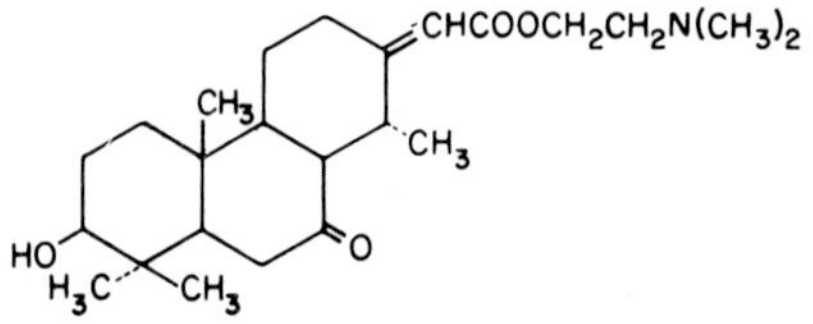

Fig. 9.17. Inotropic agents which inhibit ($Na^+ + K^+$)-MgATPase

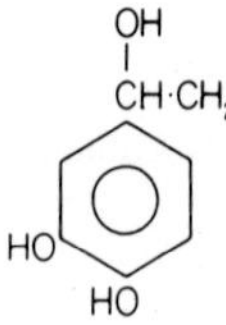

Adrenaline

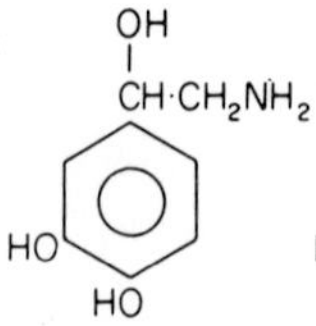

Noradrenaline

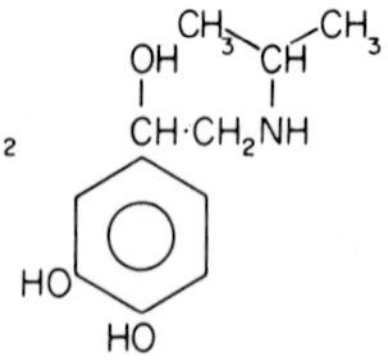

Isoprenaline

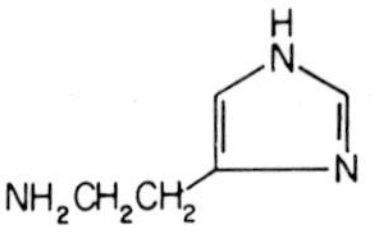

Histamine

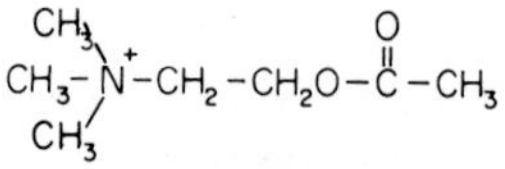

Acetylcholine

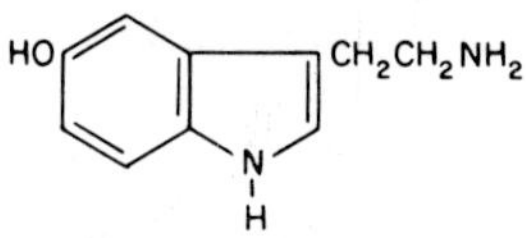

5-Hydroxytryptamine (serotonin)

Fig. 9.18. Receptor agonists

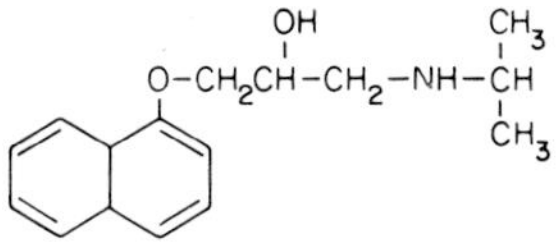

Propranolol (β-adrenergic)

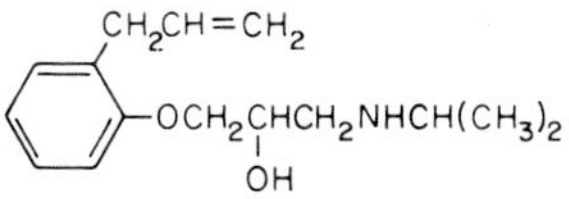

Alprenolol (β-adrenergic)

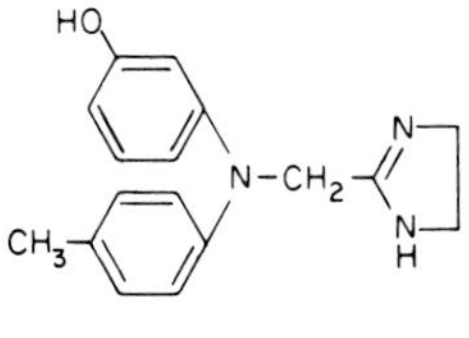

Phentolamine (α-adrenergic)

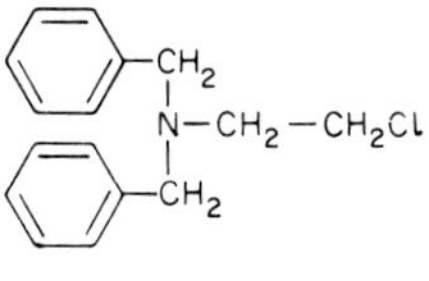

Dibenamine (α-adrenergic)

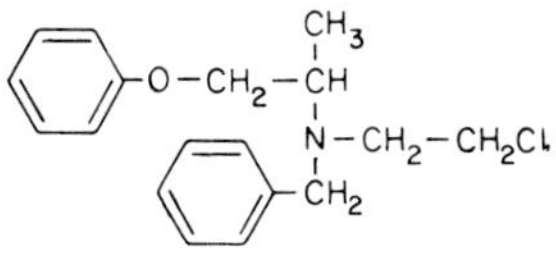

Phenoxybenzamine (α-adrenergic)

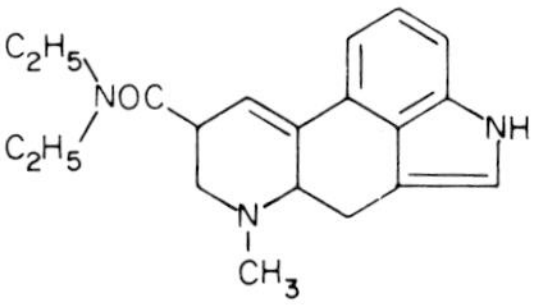

Lysergide (LSD) (5-HT)

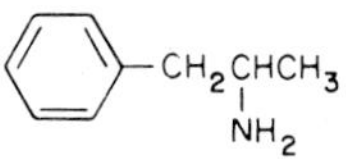

D-amphetamine (dopaminergic)

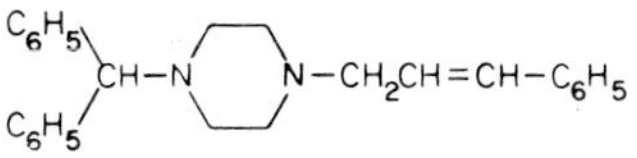

Cinnarizine (histaminergic)

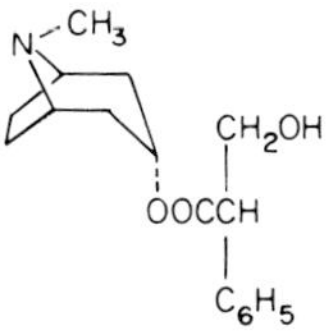

Atropine (cholinergic, muscarinic)

Fig. 9.19. Receptor antagonists (class of receptor that is blocked in parentheses)

increase in intracellular free Ca^{2+} than normal cells (Allen and Blinks, 1978) (Fig. 5.20). The Ca^{2+} transient may also be shorter in the presence of adrenaline.

β-Adrenergic anatagonists (Fig. 9.19) have been used widely in the treatment of high blood pressure and angina pectoris. They act by binding to the β-adrenergic receptor on the outer surface of the myocardial cell, and thus inhibit the action of the natural β-agonists, adrenaline and noradrenaline.

Digoxin

A relationship between Ca^{2+} and the effects of digitalis on the heart was first noted by Loewi in 1917. Since that time there have been many attempts to explain the positive inotropic action of cardiac glycosides by direct effects on intracellular Ca^{2+} stores, or Ca^{2+} influx (Akern and Brody, 1977). Certainly many of their actions on cells, including those on the myocardium, only occur when Ca^{2+} is present in the medium (Schwartz and Lee, 1962; Bourke and Tower, 1966). However, it is now generally agreed that cardiac glycosides act primarily by inhibiting the $(Na^+ + K^+)$-MgATPase, and that most of their cellular effects are secondary to the increase in intracellular Na^+ concentration which occurs as a consequence of this. This increase in intracellular sodium could increase intracellular free Ca^{2+} either through $Na^+ - Ca^{2+}$ exchange, which exists in several excitable membranes including apparently the heart (see Chapter 4, section 4.6.3), or through a release of Ca^{2+} from mitochondria (Crompton and Carafoli, 1979). But cardiac glycosides do not stimulate contraction themselves, nor have any direct measurements on intracellular free Ca^{2+} been made in their presence.

The clinical action of cardiac glycosides is very sensitive to the plasma K^+ concentration. Perhaps rather surprisingly there is an enormous variation in the sensitivity of other tissues to these substances. The liver, for example, responds to mM concentrations of cardiac glycosides like ouabain, whereas effects in the μM range can be obtained on insulin secretion from pancreatic islets, as on the contraction of heart muscle. Nevertheless, many of the non-myocardial effects of cardiac glycosides that can be demonstrated experimentally *in vitro* are Ca^{2+}-dependent (Hales and Milner, 1968). The idea that a small absolute change in intracellular Na^+ could give rise to a large fractional change in free Ca^{2+} in the cell is therefore an attractive one. Under conditions where this rise in intracellular Ca^{2+} is insufficient to activate the cell itself, an increase in endoplasmic reticulum Ca^{2+} would occur, thereby increasing the sensitivity of the cell to Ca^{2+}-dependent stimuli.

Ca^{2+}-channel blockers and Ca^{2+} antagonists

More recently a group of drugs, first introduced in 1964, has been discovered which act on the cardiovascular system by inhibiting Ca^{2+} influx during an action potential (Fleckenstein, 1977; Opie, 1980 (Fig. 9.20). These drugs relax smooth muscle, and their consequent vasodilatory properties are important in the treatment of angina pectoris. They also affect myocardial cells directly, having negative inotropic and chronotropic effects. It is these opposing effects to β-adrenergic agonists that are important in the treatment of supraventricular arrhythmias. Verapamil, nifedipine and indapamide also have hypotensive actions, though nifedipine is used clinically to treat coronary artery spasm.

Although there is little obvious structural relationship between this group of compounds, apart from verapamil and D600 (Fig. 9.20), their mechanism of action seems to be very similar. They all inhibit excitation–contraction coupling in cells where Ca^{2+} is a major current carrier in the action potential (see Chapter 4, section 4.3). They therefore inhibit cardiac and smooth muscle contraction, the latter causing vasodilation of coronary and systemic blood vessels. They can also inhibit the action of neurotransmitters on certain invertebrate muscles. Electrophysiological studies on the heart have shown that verapamil inhibits the plateau of the action potential (Fig. 9.21), which is Ca^{2+}-dependent, but has little effect on the Na^+-mediated rapid phase of the action potential in myocardial cells. Studies on squid giant axons (Baker et al., 1973a, b) have confirmed that verapamil and D600 inhibit the slow voltage-dependent Ca^{2+} channel, but have no effect on the fast Na^+ channel in excitable membranes. The electrophysiology of the other Ca^{2+} antagonistic cardiac-active drugs (Fig. 9.20) has not been so well documented, and the possibility that more than one type of slow Ca^{2+} channel exists has been suggested (Opie, 1980). The fact that their primary site of action is the cell membrane has been confirmed by the lack of any effects, at normally pharmacologically active concentrations, on isolated mitochondria, sarcoplasmic reticulum or myofibrils.

As with cardiac glycosides the concentration required for inhibitory action of Ca^{2+}-channel blockers on excitable cells varies from μM to mM depending on the tissue. Furthermore, it has been found that in vitro preincubation of 5–15 min may be necessary to demonstrate maximum inhibition. The differences in potency of these drugs on cardiac and smooth muscle may have some important clinical implications. It is possible experimentally to use low doses of, for example, verapamil, to inhibit the Ca^{2+} channel in smooth muscle thus causing vasodilation but without significantly affecting cardiac muscle. A further important clinical feature of Ca^{2+} antagonists is that they do not normally inhibit the effect of β-adrenergic agonists on the heart, though they can protect the heart against massive Ca^{2+} overload and necrosis induced experimentally in animals by toxic doses of β-agonists and vitamin D. This means that there is a natural defence mechanism against a moderate overdose, since excessive inhibition of cardiac function will stimulate the release of natural stimulatory neurotransmitters. The considerable variation in potency between these compounds (Table 9.18) is exemplified by nifedipine, which is 50 times more active than verapamil on pig coronary and up to 1000 times more potent in other systems. D600 the methoxy derivative, is also more potent than verapamil and is not safe for clinical use.

Both verapamil and nifedipine inhibit Ca^{2+}-dependent cell activation in non-muscle cells (Table 4.4). For example, they inhibit glucose-stimulated insulin release from the islets of Langerhans (Malaisse et al., 1977a, b). However, several effects of verapamil and D600 have been found, particularly in non-excitable cells such as the liver, which are not explained by their ability to block the voltage-dependent Ca^{2+} channels. In view of their structural resemblance to adrenaline it is perhaps not surprizing that they may act, either as β-adrenergic blockers (Campbell and Siddle, 1978) or as α-adrenergic blockers (Blackmore et al., 1978) at the membrane receptor in a manner independent of Ca^{2+}. They should therefore be used with caution in providing experimental evidence for Ca^{2+}-dependent cell activation, and should satisfy the following criteria: (1) Effects should be dependent on extracellular Ca^{2+} and inhibited by a high concentration of Ca^{2+}. (2) They should inhibit $^{45}Ca^{2+}$ flux across the cell membrane (Godfraind and Kaba, 1969). (3) Effects should not be explicable by competitive inhibition for hormone receptors. (4) In excitable cells they should inhibit the slow, voltage-

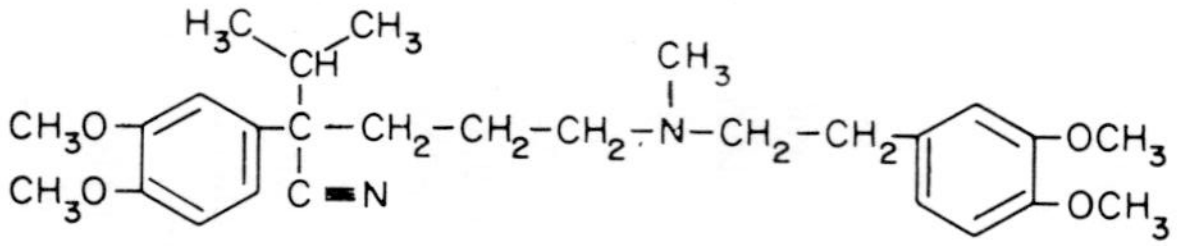

D 600

Prenylamine (Segontin)

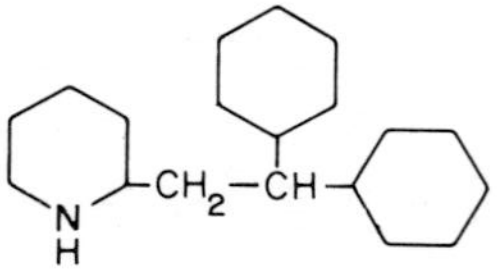

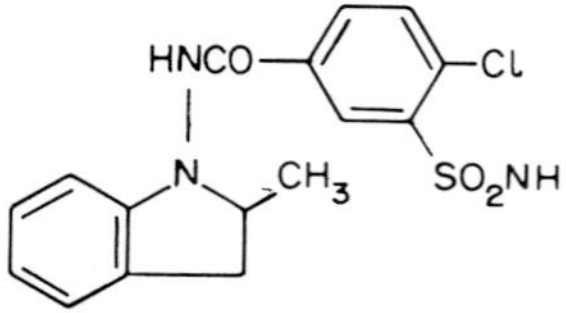

Indapamide

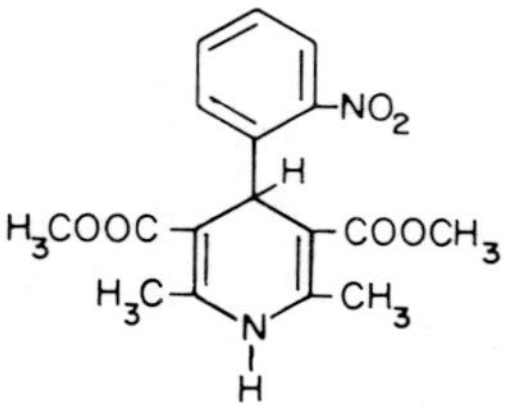

Iproveratril (verapamil)

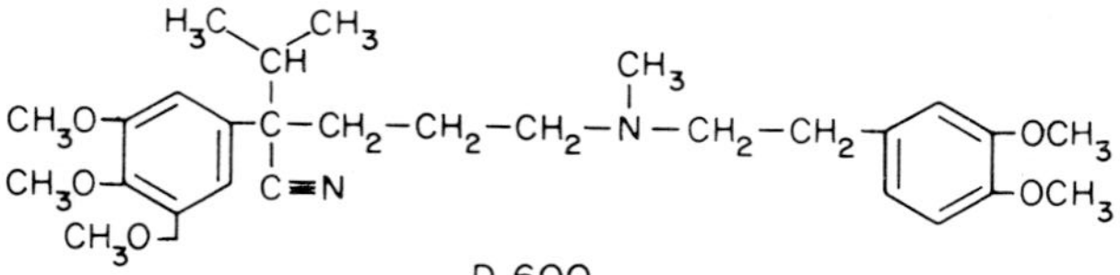

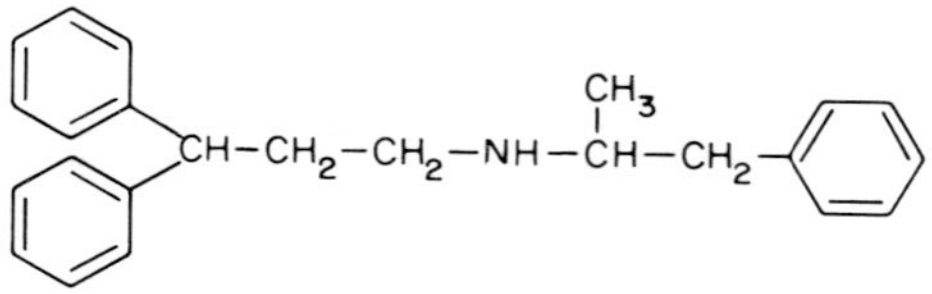

Nifedipine (Adalat, Bay a1040)

Perhexiline (Pexid)

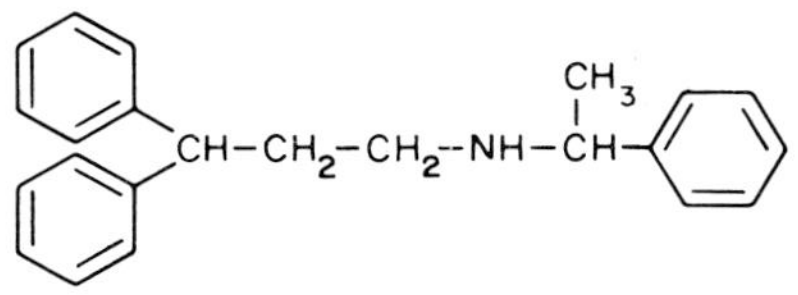

Fendiline (Sensit)

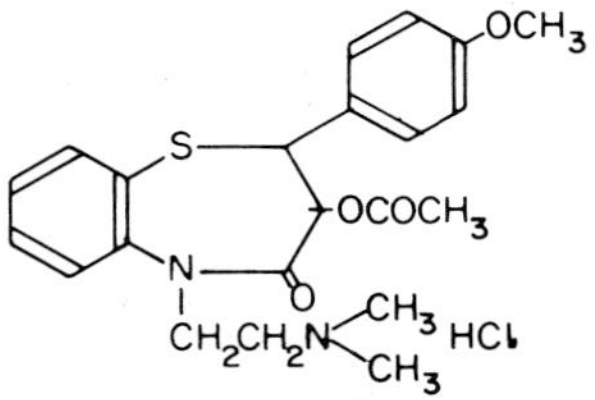

Diltiazem (Herbesser)

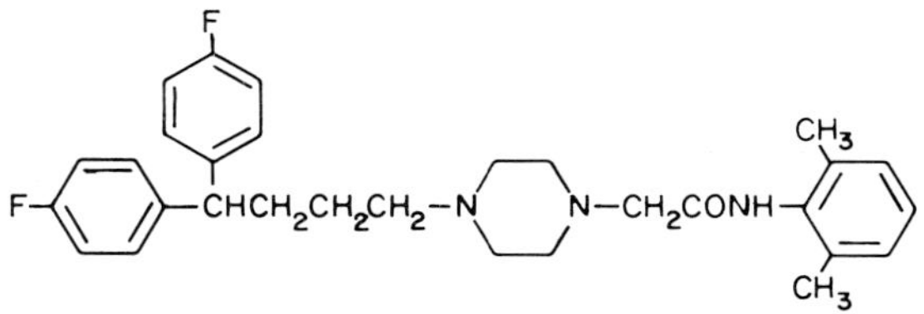

Lidoflazine (Clinium)

Fig. 9.20. Some cardiovascular agents (trade-names in parentheses)

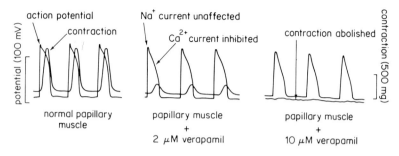

Fig. 9.21. Effect of verapamil on cardiac action potential and contractility. Isolated guinea-pig papillary muscle in Tyrode's solution. Verapamil inhibits plateau phase of action potential, due to Ca^{2+} current, but not the fast phase, due to Na^+ current. This inhibitory effect of verapamil is rapidly overcome by β adrenergic stimuli, e.g. isoproterenol. From Fleckenstein (1977). Reproduced with permission from the *Annual Review of Pharmacology and Toxicology*, vol. 17. 1977 by Annual Reviews Inc.

Table 9.18. Relative potency of Ca^{2+} antagonists

Drug	Approximate concentration for 50% inhibition
Nifedipine	7 nM
Verapamil	0.3 μM
Diltiazem	0.4 μM
Prenylamine	1.6 μM
Fendiline	3 μM
Papaverine	10 μM

Assay: suppression of K^+-induced contraction in strips of pig coronary. Data from Fleckenstein (1977); see also Northover (1977).

dependent Ca^{2+} channels, and electrical events consequent from opening of these channels.

A further note of caution in experimental systems is raised by the possibility that the (+)- and (−)-isomers of verapamil and D600 may act differently (Bayer *et al.*, 1975) (see Chapter 2, section 2.3.3). The (−)-isomer seems to be the only one capable of blocking Ca^{2+} channels. Most reagents are supplied as a racemic mixture.

9.4.5. Disodium cromoglycate and antiallergics

In asthma the binding of reaginic (IgE) antibodies to antigens on the surface of mast cells and basophils causes release of histamine, 5-hydroxytryptamine, slow-reacting substances of anaphylaxis, and prostaglandins. These in turn all cause anaphylaxis. Chromone derivatives such as disodium chromoglycate (Fig. 9.22) are able to give some relief to the many patients who suffer from this type of

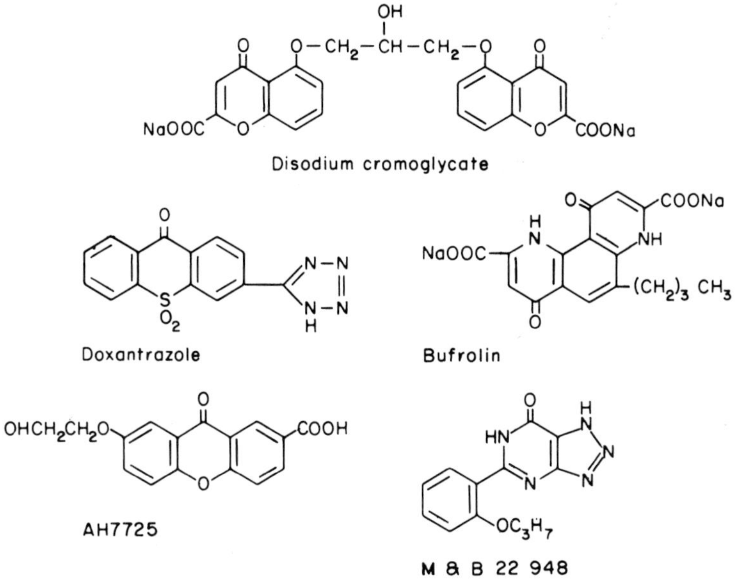

Fig. 9.22. Antiallergy drugs

allergic asthma (Cox, 1967). More recently, Foreman and Garland (1976) showed that several other antiallergy compounds (Fig. 9.22) seem to act in a similar manner. These substances have few general pharmacological actions. They are not, for example, bronchodilators nor are they anti-inflammatory agents, suggesting that they are not able to interact directly with hormone, chemotactic or anaphylactic receptors on the surface of smooth muscle or leukocytes. They are relatively specific for reaginic antibody–antigen effects on mast cells, maximum inhibition being obtained with approx. 0.5 mM disodium cromoglycate (Spataro and Bosmann, 1976).

Although their mechanism of action is not completely understood they do seem to inhibit the antibody–antigen-mediated influx of Ca^{2+} into mast cells, but not that caused by ionophore A23187 which bypasses the receptor-mediated route. It is therefore proposed that they act by inhibiting receptor-mediated Ca^{2+} influx into mast cells and hence inhibit histamine release, since this is dependent on an increase in intracellular free Ca^{2+}. Both cromoglycate and doxantrazole inhibit cyclic AMP phosphodiesterase in broken cells. So it is possible that this might also play a part in their cellular action since increases in cyclic AMP in the cell promote the phosphorylation of a mast cell protein (Theoharides et al., 1980) and inhibit histamine release provoked by IgE.

9.4.6. Xanthines

The word xanthine is derived from the Greek word *xanthos* meaning yellow. Many xanthines occur naturally in plants. For example, tea leaves and coffee beans contain approximately 1–2% of caffeine, whereas cocoa contains 1–3% of theobromine. This means that a strong cup of tea or coffee could contain as much as 0.1–1 mM caffeine. This group of pharmacological compounds (Fig. 9.23) can stimulate all parts of the central nervous system as well as many peripheral tissues. The discovery of cyclic AMP by Sutherland focused attention on the relatively specific inhibition by xanthine derivatives of the enzyme cyclic nucleotide phosphodiesterase. The compounds are normally used experimentally in the concentration range 0.1–10 mM. The relative potency on the enzyme of the three commonest compounds used is: isobutylmethylxanthine > theophylline > caffeine.

An involvement of intracellular Ca^{2+} in the action of xanthines has been established for some time. Caffeine (approx. 10 mM) has long been known to stimulate muscle contraction *in vitro*, in the absence of extracellular Ca^{2+}. Bianchi (1961) therefore proposed that xanthines could stimulate the release of Ca^{2+} from the sarcoplasmic reticulum. This was demonstrated directly, using aequorin as an indicator of free Ca^{2+}, in skinned muscle fibres of barnacle (Ashley et al., 1974). Furthermore, methylxanthines increase Ca^{2+} uptake by heart muscle and stimulate acetylcholine release from motor nerve terminals. They also potentiate several other types of cell activation that are thought to depend on an increase in intracellular free Ca^{2+}. The use of methylxanthines *in vitro* as experimental evidence for cyclic nucleotide mediated cell stimuli should therefore be treated with caution, since release of intracellular Ca^{2+} by methylxanthines may be independent of their action on cyclic AMP.

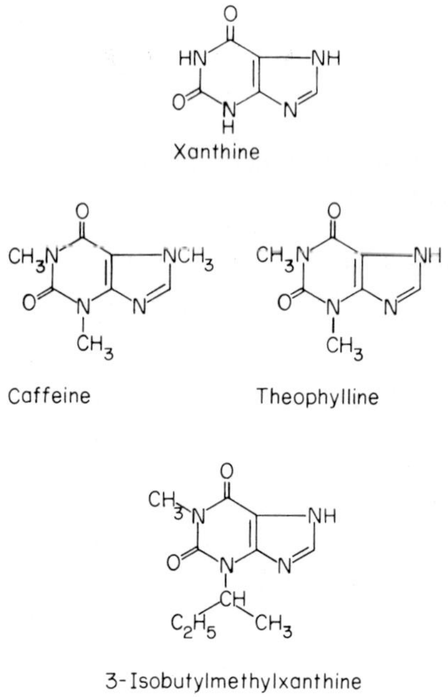

Xanthine

Caffeine

Theophylline

3-Isobutylmethylxanthine

Fig. 9.23. Xanthines

9.4.7. Cations

K^+ has been widely used experimentally as a stimulator of cells. Partial or total replacement of external Na^+ by K^+ depolarizes the cell and provokes action potentials in an electrically excitable cell. This will increase the concentration of intracellular Ca^{2+} by opening the voltage-sensitive Ca^{2+} channels. In contrast, several other cations are inhibtory to Ca^{2+}-dependent processes in cells.

Inhibition of the contraction of frog heart by lanthanum salts was first reported by Mines as long ago as 1910. Since that time many pharmacological effects of La^{3+} in experimental systems have been reported (Weiss, 1974). La^{3+} is usually active at mM concentrations, and inhibits passive and stimulated Ca^{2+} flux in both excitable and non-excitable cells. In cells with a slow voltage-dependent Ca^{2+} channel the inhibitory effect of La^{3+} seems to be relatively specific (Baker et al., 1973a, b). However, in other cells it is not clear whether La^{3+} is a blocker of a specifically activated Ca^{2+}-permeability mechanism, or whether its action is non-specific in a manner analogous to local anaesthetics. Nevertheless, the inhibitory action of La^{3+} has been widely used to provide evidence for Ca^{2+}-mediated cell activation, in spite of the fact that very few direct demonstrations of La^{3+} on intracellular free Ca^{2+} have been made.

Several other cations affect cells *in vivo*, or in experimental systems, by interacting with mechanisms which regulate intracellular Ca^{2+}. Several transition metal cations, for example Mn^{2+}, Co^{2+} or Ni^{2+}, inhibit the slow Ca^{2+} channel (Baker, 1972).

As has already been discussed, high K^+ will depolarize cells, thereby opening any voltage-dependent Ca^{2+} channels. Li^+ replacement of Na^+, on the other hand, leads to an increase in intracellular free Ca^{2+} in cells with a Na^+-dependent Ca^{2+}-efflux mechanism, since under these conditions this will be blocked. Such effects on cytosolic free Ca^{2+} of Na^+ removal have been demonstrated using photoproteins (Ashley et al., 1974; Campbell et al., 1979b).

Li^+ is now widely used in the treatment of manic depression (Schou, 1976) in spite of toxic side-effects on the thyroid and kidneys. During treatment serum Li^+ concentrations are usually maintained within the range 0.5–1.5 mM, a concentration of about 0.7 mM being favoured. The mechanism of action of this clinically important drug is still not known. It may inhibit cation pumps in the cell membrane, but this still is not able to explain its cellular effects. The intracellular concentration of Li^+ after prolonged treatment has not been properly established, so the significance of effects observed experimentally on hormone-activated adenylate cyclase (Campbell and Siddle, 1978) or Ca^{2+}-dependent contraction in muscle (Freer and Smith, 1979), which only occur at very high Li^+ concentrations (for example greater than 50 mM), is not clear. It would seem likely that Li^+ acts either directly or indirectly on certain pre- or post-synaptic nerve terminals to inhibit neurotransmitter release.

Before leaving the pharmacology of cations it is worth noting that competitive interactions between Mg^{2+} and Ca^{2+} may be the basis of the clinical consequences of excess Mg^{2+} (Engbaek, 1952; Mordes and Wacker, 1978), and that interactions of some transition metal cations, e.g. Zn^{2+}, with calmodulin could be significant clinically (Brewer, 1980).

9.4.8. Miscellaneous compounds

Several other compounds not previously discussed (Fig. 9.24) may disturb the normal physiological regulation of intracellular Ca^{2+} (Balzer and Makinose, 1968). Amphetamines and LSD bind to 5-hydroxytryptamine receptors and thus inhibit Ca^{2+} influx, whereas eledoisin may bind to a separate receptor (Vaccari et al., 1971). The analgesic effects of opiates such as morphine may be mediated through inhibition of influx of Ca^{2+} at certain nerve terminals (Sanghvi and Gershon, 1977). This is consistent with the ability of morphine to inhibit the release of acetylcholine from brain slices, and its ability to inhibit Ca^{2+} uptake by binding to membrane gangliosides and other neuronal phospholipids. The existence of a Ca^{2+}-dependent regulator of cyclic nucleotide phosphodiesterase in brain (see Chapter 6, section 6.2.1) may also provide a link between effects of opiates on intracellular Ca^{2+} and cyclic nucleotides. One important problem is to be able to distinguish between the acute and long-term effects of this group of drugs.

9.4.9. Conclusions

Many of the pharmacological substances discussed in this charter act primarily on the cell membrane, usually on the outer surface. The anti-cancer drug doxorubicin (adriamycin) has been shown to inhibit phase transients in pure phospholipid membranes (Tritton et al., 1978). Many other substances, including anaesthetics and phenothiazines, have been shown experimentally to act on phospholipids, causing displacement of Ca^{2+} and inhibition of the permeability of the membrane to Ca^{2+}. However, many workers believe that the clinical and experimental actions of most pharmacological agents are explained by specific receptors on the cell surface. The receptors are presumed to be proteins. It is quite possible that once bound to these proteins, it is the secondary interaction with the lipid which results in changes in Ca^{2+} permeability.

Very few pharmacologically active agents have so far been proposed to affect mitochondrial Ca^{2+}. In contrast, several substances like methylxanthines may stimulate Ca^{2+} release from sarcoplasmic reticulum, or inhibit it like the muscle relaxant dantrolene sodium (Hainaut and Desmedt, 1979) and alkaloids like harmaline or ryanodine (Jones et al., 1979).

There are six ways in which a pharmacological substance can interfere with cell activation in a manner that is dependent on intracellular free Ca^{2+} (see Fig. 9.25):

(1) Competitive or non-competitive inhibitors of the binding of hormones or neurotransmitter to their membrane receptors. For example, α-adrenergic blockers and LSD.

(2) Binding of the substance to specific membrane receptors, or ion channels, which are normally activated by the physiological stimulus. For example,

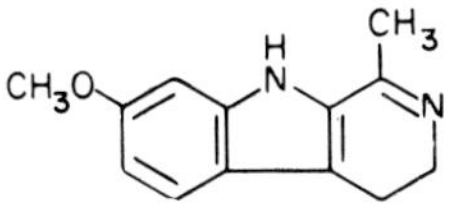

Harmaline

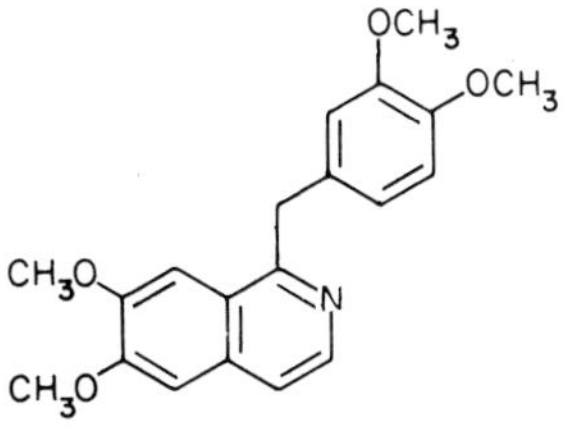

Papaverine

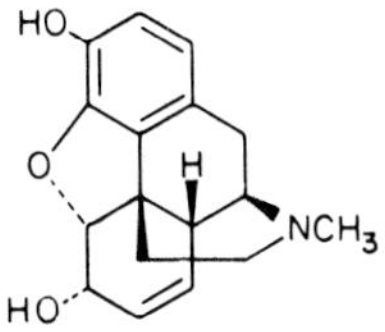

Morphine

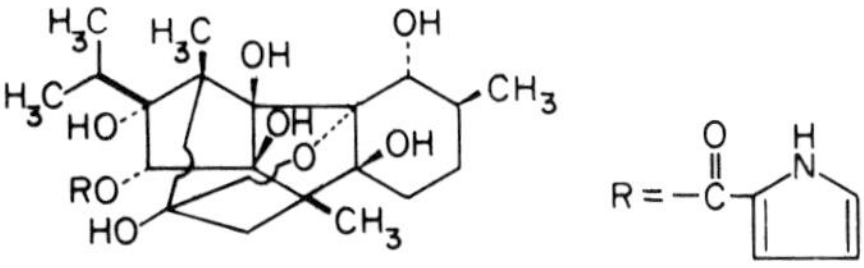

Ryanodine

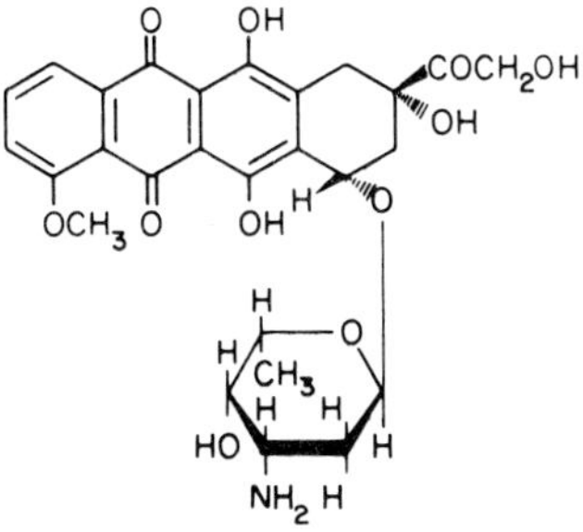

Doxorubicin

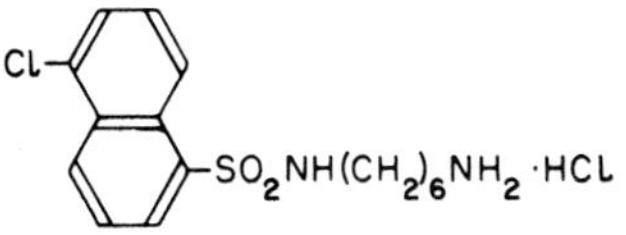

N-(6-aminohexyl)-5-chloro-
-1-naphthalenesulfonamide (W-7)

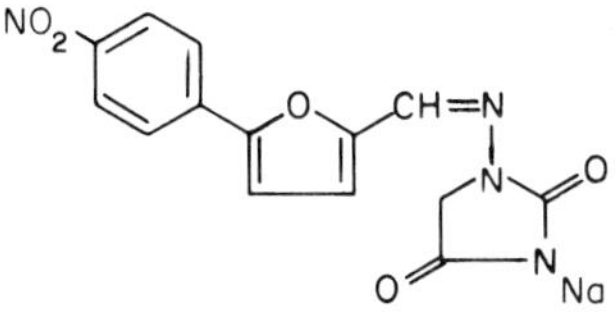

Dantrolene sodium

5-oxo-Pro-Pro-Ser-Lys-Asp-Ala-Phe-Ile-Gly-Leu-Met-NH$_2$

Eledoisin

Fig. 9.24. Miscellaneous compounds

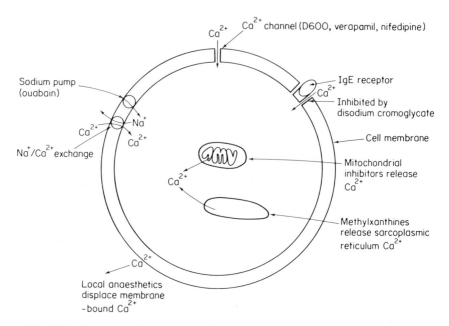

Fig. 9.25. Summary of mechanisms by which drugs can alter intracellular Ca^{2+}

antiallergics such as cromoglycate on histamine secretion; and Ca^{2+}-channel blockers such as verapamil.

(3) Binding of the substance to membrane phospholipids or proteolipids, thereby blocking Ca^{2+} influx and hence the increase in intracellular free Ca^{2+} induced by the cell stimulus. For example, general and local anaesthetics, and possibly phenothiazines.

(4) Changes in intracellular Ca^{2+} mediated through some other substances. For example, cardiac glycosides affecting intracellular Na^+ and Na^+–Ca^{2+} exchange; and cyclic AMP and methylxanthines.

(5) Penetration of the cell by the substance, followed by an interaction with an intracellular Ca^{2+} store. For example, cardiac glycosides, intracellular Na^+ and mitochondria; increase in intracellular free Ca^{2+} from endoplasmic reticulum released by methylxanthines; decrease in intracellular free Ca^{2+} from endoplasmic reticulum by dantrolene sodium.

(6) Inhibition of the action of intracellular Ca^{2+}. For example, inhibition of the calmodulin by phenothiazines like trifluoperazine.

In experimental systems designed to investigate these possibilities care must be taken to relate the concentration of the drug, and the conditions, to those found clinically.

9.5. Overall conclusions

We have seen in this chapter that changes in intracellular Ca^{2+} occur under pathological circumstances, as well as experimentally or clinically in the presence of several drugs. Are these changes the primary cause of the cellular disturbance which leads to the clinical manifestations of the disease or drug in question? How do these pathological and pharmacological effects on intracellular Ca^{2+} relate to the threshold hypothesis for cell activation, outlined in Chapter 2?

So far as the first of these questions is concerned, any significant increase in intracellular Ca^{2+} in disturbed or dying cells will cause further chemical changes in these cells. What remains to be established is whether these Ca^{2+}-dependent changes are directly related to the pathogenesis of a disease, or the clinical actions of drugs.

Many of the cellular effects of pathogens and drugs can be considered to be threshold phenomena (Table 9.19) in that they transform the cell from one state to another. This is seen most clearly with agents that provoke cell death or cell multiplication. Some of these phenomena may also be relevant under physiological circumstances. For instance, plasma glucocorticoid concentration is one of the factors which regulate the number of lymphocytes in the blood (Wyllie *et al.*, 1980). To see whether this concept is relevant to the pathogenesis of a disease, or the action of a drug, three key questions must be answered: (1) Does the pathogen or drug provoke or modify a threshold in the activation of individual cells? (2) Is there a change in cytosolic free Ca^{2+} and total cell Ca^{2+} balance? (3) What is the precise role of Ca^{2+}; is it structural, electrical, or regulatory?

In view of the potential clinic importance of many of the ideas discussed in this chapter, it would perhaps be wise to close with a note of caution to the would-be experimenter. Much work is needed with animal models and tissue-culture systems before the potential benefits of these ideas can be realised in the

Table 9.19. Some examples of pathological and pharmacological threshold phenomena involving intracellular Ca^{2+}

(A) *Pathological*
1. Glucocorticoid-induced killing of lymphocytes (apoptosis*)
2. Toxin- and drug-induced cell injury leading to cell death
3. Complement-mediated cell lysis
4. Complement-mediated cell activation
5. Cellular shape change
6. Viral infection
7. Anoxia
8. Tetany
9. Increase in cell number—cancer
10. Intracellular Ca^{2+} precipitation

(B) *Pharmacological (clinical or experimental)*
1. Ca^{2+} antagonists blocking Ca^{2+}-dependent action potentials
2. Blocking of neuronal transmission by general or local anaesthetics
3. Pharmacological block of cell activation (e.g. disodium cromoglycate on histamine secretion?)
4. Caffeine-induced muscle contraction

* See Wyllie *et al.* (1980).

treatment or prevention of human disease. Whilst several of the pharmacological substances discussed here are used clinically, many others are too toxic to be used in man, and can only be used in experimental systems. In the hope that this chapter has stimulated the clinically minded scientist it cannot be overemphasized that before embarking on any human studies the potential dangers of the drug and its safe dose, if usable at all in humans, should be ascertained.

CHAPTER 10

Synthesis and perspectives

10.1. Synthesis—what is so special about calcium?

The late Marcel Florkin in his work on the history of biochemistry (Florkin, 1975a) made the rather provocative statement that the modern subject had become 'a soulless and mechanistic science'. Much of the present book has been concerned with detailed arguments about the evidence for intracellular Ca^{2+} as a cell regulator, about the mechanisms involved in its action, and the regulation of its concentration in the cell. The time has come, in this the final chapter, to redress the balance, to summarize not only what has been learnt from the preceding chapters but to try to provide an overview to justify the suggestion in the title that 'intracellular Ca^{2+} does play a universal role' as a cell regulator.

10.1.1. Resumé

Of the million or so species representing the 30 major animal phyla (Table 1.11) many examples from arthropods, molluscs, chordates, coelenterates, annelids and echinoderms have been given where intracellular Ca^{2+} seems to be involved in regulating the activity of cells. These six phyla make up about 95 % of the whole animal kingdom. Some examples, though much fewer in number, from the plant kingdom and prokaryotes have also been given.

The main theme of this book has been that primary cell stimuli, be they physical or chemical in origin (Chapter 2, sections 2.1 and 2.2), or secondary regulators which modify their action, alter the concentration of free Ca^{2+} somewhere in the cell. The result is a rapid change in state of the cell. Eukaryotic cells can undergo any one of six changes of state (Table 10.1). They can move, they can secrete, they can take up particles from the surrounding fluid, they can be excited electrically, they can respire and metabolize substrates, and they can divide. Examples have been given (Chapters 4-8) where an increase in intracellular Ca^{2+} appears to be the internal trigger mediating the effect of primary stimuli on these phenomena.

Some phenomena exist which do not apparently require a change in intracellular Ca^{2+}, though one of the other three biological roles for Ca^{2+} (see Chapter 1, section 1.3) could be important. Five such phenomena are: (1) Na^{+}-dependent action potentials generated in some nerves and muscles, for example

Table 10.1. Phenomena that may be regulated by changes in intracellular Ca^{2+}

Phenomenon	Examples of a change of state
1. Cell movement	Muscle contraction
	Ciliate and flagellate movement in eukaryotes
	Chemotaxis in pro- and eukaryotes
2. Excitability of cells capable of generating action potentials	Invertebrate muscle action potential
	Heart muscle beat potential
	Response of the eye and other types of photoreceptor to light
3. Secretion of cell products	Neurotransmitter and hormone secretion
	Fluid secretion from exocrine glands
4. Uptake of particles or soluble substances by vesiculation	Phagocytosis of bacteria
5. Intermediary metabolism and respiration	Glucose production
	Lipolysis
	Prostaglandin synthesis
6. Cell reproduction	Lymphocyte transformation
	Egg fertilization and sperm capacitation
	Egg maturation and meiosis

by the nicotinic acetylcholine receptor; (2) phenomena dependent solely on an increase in cyclic AMP through receptor activation of adenylate cyclase; (3) regulation of intermediary metabolism by substrate supply or changes in regulatory metabolites; (4) long-term regulation of cells by controlling protein concentration, for example steroids or thyroid hormones acting via receptors which interact with mRNA synthesis; (5) internally programmed cell division.

Even in some of these cases intracellular Ca^{2+} may have a role to play. For example, in muscle the increase in internal Ca^{2+} may play a major role in desensitization of the acetylcholine receptors (Miledi, 1980), thereby preventing tetany. Ca^{2+} entering excitable cells through voltage-dependent Ca^{2+} channels may cause an increase in free Ca^{2+} near to the inner surface of the cell membrane which may alter the permeability of the membrane to other ions such as K^+. In the heart adrenaline binds to β-receptors stimulating an increase in cyclic AMP. This may increase Ca^{2+} uptake into the sarcoplasmic reticulum via phosphorylation of a membrane protein (Chapter 5). Thus, even with these apparently Ca^{2+}-independent phenomena, it is difficult to find one, considered as a whole, where intracellular Ca^{2+} has not been proposed to play some role.

On the basis of the wide distribution of organisms and the phenomena involved it is possible to argue that intracellular Ca^{2+} is a universal regulator. The case for this argument depends on the quality of the experimental evidence (Table 10.2). In muscle contraction the evidence is very strong and most of the criteria laid out in Chapter 2 (section 2.3.2) have been fulfilled. In others cases, for example intermediary metabolism, phagocytosis, and cell division, much of the present evidence is circumstantial. Two key experiments are required to strengthen the argument in cases such as these: (1) Direct measurement of

Table 10.2. Important experiments establishing intracellular Ca^{2+} as a universal cell regulator

Date	Experiment	Workers
1808	Discovery of calcium as an element	Davy
1883	Extracellular Ca^{2+} required for frog heart contraction	Ringer
1894	Extracellular Ca^{2+} required for transmission of impulses from nerve to muscle	Locke
1928	Detection of a change in cytoplasmic Ca^{2+} in Amoeba	Pollack
1940	Ca^{2+} action on isolated protoplasm from muscle	Heilbrunn
1940	Ca^{2+} required for acetylcholine release	Harvey and MacIntosh
1942	Ca^{2+} activates myosin ATPase	Bailey, Needham
1943	Ca^{2+} on muscle protoplasm causes contraction	Kâmata and Kinoshita
1947	Ca^{2+} injection into muscle causes contraction	Heilbrunn and Wiercinski
1953	Non-Na^+-dependent action potentials in crustacea	Fatt and Katz
1957	Discovery that cytoplasmic Ca^{2+} is very low ($< 10\,\mu M$)	Hodgkin and Keynes
1961	Ca^{2+} may trigger exocytosis	Douglas and Rubin
1962	ATP-dependent Ca^{2+} uptake system	Ebashi and Lipmann
1963	'Active tropomyosin' leads to the discovery of troponin C	Ebashi
1964	Ca^{2+} spikes in excitable cells	Hagiwara et al.
1967	Direct measurement of a Ca^{2+} transient with aequorin	Ridgway and Ashley
1967	Discovery of calmodulin	Cheung
1974	Visualization of Ca^{2+}-activated photoproteins	Morin and Reynolds
1975	Visualization of localized free Ca^{2+} changes in cytoplasm	Reynolds, Rose and Loewenstein
1981	Free Ca^{2+} measured in a small cell by dye fluorescence	Tsien

intracellular free Ca^{2+}. (2) Identification of the molecular basis of the action of Ca^{2+}, and a demonstration of an effect of Ca^{2+} at concentrations within the physiological range (approx. $0.1-10\,\mu M$) in a reconstituted *in vitro* system, thereby mimicking the *in situ* phenomenon.

All cells have a very low cytosolic free Ca^{2+} concentration, which is some 10 000–100 000 times lower than that in the external medium. Free cytosolic Ca^{2+} in resting cells is less than $0.5\,\mu M$ and is probably in the range 30–100 nM.

Three features enable the resting cell to maintain this enormous gradient across the external membrane in spite of the large membrane potential, negative inside, pulling Ca^{2+} inwards. (1) Biological membranes have a low passive permeability to Ca^{2+}, at least an order of magnitude less than for K^+, Na^+ or Cl^- (Table 4.1). (2) The small passive leak of Ca^{2+} into the cell, down its electrochemical gradient, is compensated for by a Ca^{2+} pump in the cell membrane. This may be a (Ca^{2+})-MgATPase, which need not necessarily require a counterion in that if it were electrogenic its contribution to the resting membrane potential would be very small. Even the sodium pump only contributes a few millivolts. Alternatively, the Ca^{2+}-efflux mechanism may be a Na^+–Ca^{2+} exchange as first proposed by Blaustein, Baker and coworkers from experiments with invertebrate nerve. (3) More than 95% of cell Ca^{2+} is retained in a bound form, mainly within the organelles of the cell.

Activation of the cell increases the cytoplasmic free Ca^{2+} 10–50-fold. However, the total Ca^{2+} required to activate the cell may be up to ten times this since the concentration of the Ca^{2+}-binding protein responsible for mediating the effect of Ca^{2+} can be in the range 10–100 μM. The source of this Ca^{2+} may be external, internal or both. Mitochondria, and other internal buffers, can restrict the increase in intracellular free Ca^{2+} to a particular region of the cytoplasm of the cell. This enables the cell to restrict the increase in Ca^{2+} to the appropriate area within the cell. For example, the sarcoplasmic reticulum releases Ca^{2+} close to the myofibrils responsible for contraction. In many exocrine glands the neurotransmitter or hormone receptors are on the opposite side of the cell from where secretion actually takes place. Here again the source of Ca^{2+} appears to be internal. On the other hand, in some endocrine glands secretion occurs at the same surface as that exposed to the agonist, for example glucose stimulating insulin release from pancreatic β-cells. In this case the source of Ca^{2+} is external. In phagocytes, where both membrane surface events and activation of intermediary metabolism within the cell are required during both phagocytosis and chemotaxis, Ca^{2+} may be released from intracellular stores and come from outside the cell. Even when an internal store is the major source of Ca^{2+} for cell activation an increase in Ca^{2+} influx is often necessary to prevent depletion of Ca^{2+} from the cell, and thereby maintain the cell Ca^{2+} balance.

The main source of internal Ca^{2+} for release is the endoplasmic reticulum. There has been much speculation regarding the role of mitochondrial Ca^{2+}. Some mitochondrial enzymes can be activated by Ca^{2+}. However, the best established role for mitochondria is in restricting free Ca^{2+} changes in the cytoplasm to localized areas and for short durations. Increased entry of Ca^{2+} into the cell occurs through the opening of hypothetical channels, relatively specific for Ca^{2+}. In excitable cells the permeability of these channels can be dependent on the electrical potential across them.

In theory, Ca^{2+} could inhibit or activate proteins in the cell. Many of the inhibitory effects of Ca^{2+} on enzymes have now been discredited (Tables 3.14, 6.1 and 6.8) because they are only significant at mM Ca^{2+}. Most of the physiological effects of intracellular Ca^{2+} as a regulator involve a special class of high-affinity Ca^{2+}-binding proteins. Troponin C in muscle was the first to be discovered and its role in the contraction of certain muscles is well established (Chapter 5, section 5.3.4). A similar protein, leiotonin, has been found in smooth muscle. The ubiquitous occurrence of Ca^{2+}-binding proteins with 4 Ca^{2+} sites per molecule, and named calmodulins, has lead to the hypothesis that they are responsible for mediating the regulatory effects of intracellular Ca^{2+}. In theory, the affinity of Ca^{2+}-binding sites could be regulated, for example by phosphorylation, so as to cause a change in Ca^{2+} saturation without an increase in free

Ca^{2+}. Whilst such a mechanism could be involved in interactions between intracellular Ca^{2+} and cyclic AMP and in secondary regulation (Chapter 2, section 2.1), it is unlikely to be a primary mechanism since there is insufficient Ca^{2+} in the cytoplasm to provide for the 40–400 μM Ca^{2+} sites on the regulatory proteins.

Many proteins are negatively charged at physiological pH, as indicated by the fact that they move towards the positive electrode during electrophoresis. Because of the relatively large Glu + Asp content of most proteins (Table 3.6) it is not surprising that occasionally two of these residues come within about 10 nm and provide a low-affinity Ca^{2+} site, with a mM K_d^{Ca}. The ionic diameter of Ca^{2+} is about 0.2 nm. Ionic interactions obey the inverse-square law (Force α $1/d^2$, where d = distance between the two charges). The weaker van der Waals forces obey $F \alpha\ 1/d^6$ and are virtually ineffective beyond 5–10 nm. Addition of a third or fourth CO_2^- group provides a high-affinity Ca^{2+} site with a K_d^{Ca} in the μM range, and which is selective over Mg^{2+} (cf. calmodulin, Fig. 3.13). Such sites are found on troponin C and calmodulin.

Ca^{2+} has six particular chemical properties which make it ideally suited for its regulatory role. (1) Ca^{2+} is doubly charged, resulting in higher-affinity binding to protein CO_2^- than Na^+ or K^+. (2) Ca^{2+} has a more flexible coordination chemistry than Mg^{2+}. (3) The affinity of proteins for Ca^{2+} is not so high as it can be for transition metals such as Cu^{2+}, Zn^{2+} and Mn^{2+} where kinetically dissociation can be very slow. (4) Ca^{2+} is not *directly* involved in the osmotic balance of the cell or in maintaining the resting membrane potential. A large fold change in Na^+ or K^+ would have disastrous effects on the osmotic and electrical properties of the cell. (5) Unlike Mg^{2+}, Ca^{2+} does not form significant complexes with substrates like ATP at the concentrations of free Ca^{2+} found in the cell. (6) There is no problem with Ca^{2+} having more than one redox state. It only has one.

It is the combination of these properties, together with the enormous electrochemical gradients of Ca^{2+} across the cell membrane, which enables intracellular Ca^{2+} to play its 'special' role as a cell regulator (Fig. 10.1). These are also the reasons why changes in intracellular Ca^{2+} play an important part in the action of several drugs and in cell injury (Campbell and Luzio, 1981) (see Chapter 9). In these cases the increase in free Ca^{2+} may be greater than it is in cells exposed to a physiological stimulus. This means that not only will the physiologically active Ca^{2+} sites have more Ca^{2+} bound to them, but also some normally passive Ca^{2+} sites may become relevant.

10.1.2. Unitary hypothesis

Historical

The ultimate objective of examining in detail the role of intracellular Ca^{2+} as a regulator is to rationalize all of the observations and to formulate a unitary hypothesis. Many attempts have been made to do this over the past one hundred

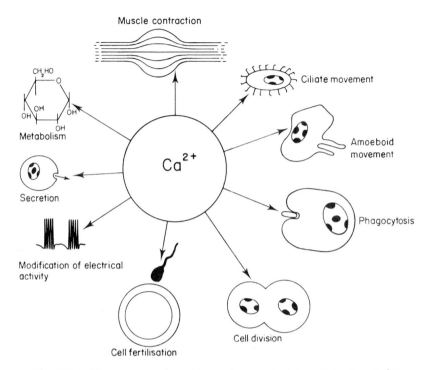

Fig. 10.1. Phenomena activated by an increase in intracellular free Ca^{2+}

years (Table 10.3). Following the experiments of Ringer on the effect of various isotonic salt solutions on cells, many of the early workers made much of the apparent antagonism between different cations. Höber (1945) pointed out that only Sr^{2+} and Ba^{2+} could mimic the actions of Ca^{2+} on cells and that the narcotizing action of Mg^{2+} could be antagonized by Ca^{2+}. This latter phenomenon is well known to biologists wishing to preserve invertebrates in fixatives. For example, to prevent the retraction of the tentacles of hydroids one simply has to immerse them in isotonic $MgCl_2$ (approx. 0.36 M) before adding the formalin fixative. A more complex scheme was proposed by Rubinstein (1927). K^+ and Na^+ were proposed as being antagonistic to each other. Ca^{2+} and Mg^{2+} also antagonized each other but acted synergistically to antagonize Na^+. The problem with these hypotheses was that they failed to distinguish between the different electrical and structural roles of these cations, nor did they have any real chemical basis.

During the 1920's, physiologists such as Loeb, Loewi, Heilbrunn and Pollack came to the important conclusion that Ca^{2+} bound to intracellular protein played a crucial role in the structure and function of the cell. Furthermore, they proposed that certain types of cellular phenomena were initiated by release of Ca^{2+} within the cell. These ideas were synthesized by Heilbrunn (1937, 1943). He separated out some of the external and internal actions of Ca^{2+} by pointing out its

Table 10.3. Unitary hypotheses regarding intracellular Ca^{2+} as a physiological regulator

Date	Hypothesis	Worker
1886	Competitive antagonism or synergism between Na^+, K^+, Ca^{2+} and Mg^{2+}	Ringer
1928	Concept of a change in free Ca^{2+} as a regulator	Pollack
1937	General role of intracellular Ca^{2+} as a cell regulator	Heilbrunn
1961	Stimulus–secretion coupling analogous to excitation-contraction coupling in muscle	Douglas and Rubin
1967	Ca^{2+} acts via activation or inhibition of intracellular enzymes	Bygrave
1970	Interrelationships between Ca^{2+} and cyclic AMP	Rasmussen
1974	Yin-yang hypothesis	Goldberg et al.
1975	Phosphatidylinositol and opening of Ca^{2+} channels	Michell
1976	Mono- and bidirectional regulation of cells	Berridge
1977	Open and closed loops and Ca^{2+}	Goodman and Rasmussen
1979–1980	Universal role of high-affinity Ca^{2+}-binding proteins (e.g. troponin C, calmodulin)	Ebashi; Cheung; Means and Dedman
1980	Phospholipid methylation and opening of Ca^{2+} channels	Hirata and Axelrod
1981	Intracellular Ca^{2+} induces prostaglandins or other such compounds	Barritt
1982	Intracellular Ca^{2+} and threshold phenomena	This book

These dates are only intended to provide a historical guide and are not necessarily the date of the first paper in which the authors introduced their hypothesis.
Expansions of these hypotheses can be found in Chambers (1928), Loeb (1906), Klinke (1928), Locke (1894), Hollingsworth (1941), Hober (1945), Kretsinger (1976a, b) Ashley and Campbell (1979).

ability: (1) to be an essential nutrient of all cells; (2) to form insoluble compounds with inorganic anions and proteins; (3) to bind cells together; (4) to be necessary for the clotting of blood and milk; (5) to act as the internal stimulus for various phenomena including cell movement, muscle contraction and cell division after fertilization. Heilbrunn also realized that the key to understanding the regulatory role of intracellular Ca^{2+} was its interaction with proteins. However, he failed to comprehend the catalytic nature of this interaction, nor did he identify clearly the role of Ca^{2+} in the electrical properties of cells.

In the first chapter of this book (section 1.3) four main biological roles for Ca^{2+} were characterized: (1) structural; (2) electrical; (3) extracellular cofactor; (4) intracellular regulator.

Since Heilbrunn first published his ideas, attempts to rationalize the role of intracellular Ca^{2+} as a regulator can be divided into three categories: (1) the identification and classification of cell stimuli whose effects are mediated through intracellular Ca^{2+}. (2) The characterization and classification of the mechanism by which Ca^{2+} is released into the cytoplasm and how it acts. (3) Rationalization of the interrelationships between Ca^{2+} and other intracellular regulators, such as cyclic nucleotides.

Many attempts have been made to classify the different types of cell stimuli and to correlate this classification with a particular intracellular mechanism. It has been argued, for instance, that certain hormone and neurotransmitter receptors, identifiable pharmacologically, always act through the same mechanism. It is this belief that has led to a search for a single mechanism of action for hormones like insulin (Czech, 1977) and to the proposal that all β-adrenergic receptors activate adenylate cyclase, whereas α-adrenergic receptors act through intracellular Ca^{2+}. Similar examples can be found in muscarinic versus nicotinic acetylcholine receptors, and H_1 and H_2 histamine receptors. It may well be that a particular class of receptor, or a type of physical cell stimulus be it electrical or mechanical, works through the same mechanism regardless of the cell or the phenomenon concerned. But there is no *a priori* reason why this should be so. In this book an attempt has been made to define the physiological problem first and then to see what stimuli and mechanisms have evolved in order to solve it. Such an approach might be open to the criticism that it is too teleological. Yet it is perhaps particularly useful in helping to provide an understanding of the evolutionary forces which have lead to the selection of particular mechanisms for activating cells.

Yin-yang and mono-/bidirectional systems

Since the discovery of hormones by Bayliss and Starling in 1902 there have been several attempts to classify hormones and neurotransmitters on the basis of their structure, physiology and mechanism of action (Starling, 1905; Dale, 1934; Huxley, 1935; Hershko et al., 1971; McMahon, 1974). In 1974 Goldberg and his coworkers proposed that the activation of cells could be divided into 'monodirectional' and 'bidirectional' systems. Their thesis was embodied in the ancient oriental concept known as 'yin-yang', where a dualism exists between opposing forces, sometimes opposing each other and at other times working together. 'Monodirectional' systems were defined as processes where the cells were transformed from a non-functional state to an activated state on addition of the appropriate signal. More than one cell stimulus could exist but no inhibitory factors were known. In contrast 'bidirectional systems' were defined as processes where interaction between stimulatory and inhibitory factors determined the eventual state of the cells. Goldberg and his coworkers further proposed that cyclic AMP and cyclic GMP were the intracellular regulators which formed the axis of the 'yin-yang'. In monodirectional systems such as ACTH stimulation of the adrenal cortex to release cortisol, the two cyclic nucleotides would act cooperatively, whereas in bidirectional systems such as chemotaxis of polymorphonuclear leukocytes, the two cyclic nucleotides had opposing actions. In cases like chemotaxis the system was potentiated by cyclic GMP and inhibited by cyclic AMP, whereas in some other bidirectional systems the converse was true. The hypothesis also included interactions of cyclic nucleotides with electrical phenomena. For example, Greengard (1976) has shown that in some neurones slow inhibitory post-synaptic potentials are mediated by a cyclic AMP induced hyperpolarization, whereas slow excitatory post-synaptic potentials are mediated by a cyclic GMP induced depolarization. Membrane kinases have been discovered which cause phosphorylation of membrane proteins to explain these effects (Schulman and Greengard, 1978).

This mono-/bidirectional concept for cell activation was elegantly developed by Berridge (1976b) who incorporated more fully intracellular Ca^{2+} into the hypothesis. Monodirectional systems included neurosecretion from nerve terminals, acetylcholine-induced release of adrenaline from the adrenal medulla, several types of endo- and exocrine secretion including insulin from pancreatic β-cells, and photoreceptor activation by light. Cell activation was proposed to be caused by intracellular Ca^{2+}. In some cases this could be supported by cyclic AMP, in others cyclic AMP had an independent action distinct from that of Ca^{2+} such as activation of metabolism. In 'bidirectional systems' intracellular Ca^{2+} was again proposed as the internal mediator of the cell response. Agents which inhibited acted to reduce the level of intracellular Ca^{2+} induced by the primary stimulus, for example by a cyclic AMP activated efflux through membrane phosphorylation. Agents which enhanced the response increased the level of cytosolic free Ca^{2+} by increasing Ca^{2+} influx, or by releasing more Ca^{2+} from internal stores. These 'bidirectional systems' included contraction of smooth muscle and heart muscle, platelet aggregation and secretion, histamine secretion from mast cells and the regulation of hepatic glucose metabolism by various hormones and neurotransmitters. For example, intracellular cyclic AMP, induced by prostaglandin E_1, or β-adrenergic agonists (plus an α blocker), inhibits the aggregation–secretion–retraction sequence of platelets provoked by ADP or 5-hydroxytryptamine. Similarly, cyclic AMP inhibits histamine secretion from mast cells.

Attractive as this hypothesis may seem, it does not appear, as it stands, to provide a universal scheme for all types of cell activation dependent on intracellular Ca^{2+}. There are four particular problems: (1) How can one reconcile the effects of electrical and other physical stimuli on individual cells with effects of hormones and neurotransmitters on multicellular tissue like the liver? (2) Several of the examples of mono- and bidirectional systems have been oversimplified. For example, insulin secretion from the pancreatic β-cell (monodirectional) can be stimulated by glucose, certain amino acids and several gut hormones and inhibited by adrenaline. Similarly, photoreceptors (monodirectional) have a mechanism of dark adaptation. (3) It is difficult to decide how to classify cell transformation, egg fertilization and cell division in the light of the mono-/bidirectional hypothesis (Berridge, 1976a, b). (4) The simple yin-yang dualism for cyclic AMP and cyclic GMP does not always hold (Goldberg and Haddox, 1977).

The concept of 'threshold'

I have attempted to incorporate some of the ideas inherent in the mono-/bidirectional hypothesis but to reclassify the cell stimuli and types of cell activation. The latter are divided into threshold and non-threshold phenomena (Chapter 2, section 2.2), a threshold being defined as 'the point at which physiological experience begins'. Cell stimuli are divided into primary and secondary stimuli. The former activate a cell response whether via a threshold or not as the case may be, whereas the latter modify the response. This modification occurs through a change in the level at which the 'threshold' occurs or through an alteration in the time of onset, duration or magnitude of the cell response (Fig. 10.2).

Many possible examples of threshold phenomena have been cited in this book (see Chapter 2 and this chapter, section 10.1.1). They include contraction and other dramatic changes in cell movement, physiological activation of secretion, cell excitability, cell fertilization, cell transformation and division. Phenomena that are *apparently* non-threshold include hormonal activation of metabolism (Chapter 6). It is the hypothesis put forward in the present book that most threshold phenomena, apart from for example the transmission of certain action potentials, are mediated by an increase in cytosolic free Ca^{2+}. Non-threshold phenomena may be mediated by protein phosphorylation induced by cyclic nucleotide dependent protein kinases. Modification of a threshold response by a secondary regulator could occur by preventing the increase of cytosolic Ca^{2+} above a threshold level, by modifying the magnitude or time course of the Ca^{2+} transient once it has passed the threshold level, or by modifying the proteins that are responsible for the phenomenon and its regulation by Ca^{2+}.

Experimentally it may be possible to mimic some of the effects of a physiological stimulus in a continuously graded manner. If, however, the response of the cell *in vivo* is to be explained by a threshold, several criteria must be satisfied (Fig. 10.2). (1) The cell must jump from one state to another when

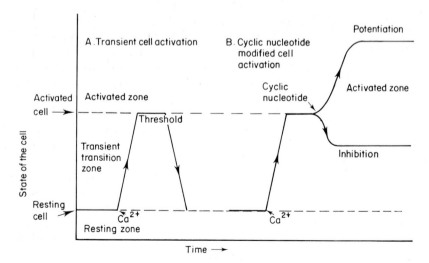

Fig. 10.2. Threshold activation of a cell. Cyclic nucleotides may also alter the level of free Ca^{2+} required to generate a threshold response in the cell

stimulated under normal conditions. (2) Secondary regulators or pharmacological agents must be shown to alter the state of the cell, if and when the threshold response of an individual cell occurs. (3) A molecular explanation for the threshold must be found, for example through cooperative Ca^{2+} binding to calmodulin (Chapter 3, section 3.5, Fig. 3.15), or through cooperative interactions of Ca^{2+}–protein with the structures responsible for the cell response.

The author presents this as an hypothesis in the belief that it may encourage many phenomena of cell activation and cell pathology to be examined in a new light. For example, is chemotaxis a threshold phenomenon, and what of the effect of glucose on insulin secretion or adrenaline on glucose output from the liver? What determines the threshold for cell death, for example glucocorticoids acting on leucocytes or complement-induced cell lysis?

Chemotaxis of polymorphonuclear leukocytes is induced by several peptides, including the complement fragment C5a and bacterial chemotactic factors (see Chapter 5 section 5.4). This phenomenon plays a key role in the inflammatory response in infected wounds and in diseases like rheumatoid arthritis. Chemotaxis can be inhibited, at least *in vitro*, by β-adrenergic agents, PGE_1, PGE_2, histamine (H_2 receptor) or cholera toxin, all of which seem to act through an increase in cyclic AMP. Conversely, the chemotactic response can be enhanced by acetylcholine, α-adrenergic agents or $PGF_{2\alpha}$, which increase cyclic GMP in these cells. The 'unit cell activation' hypothesis described here predicts that there is a threshold for directed cell movement. A further prediction is that the number of cells which give this threshold response is determined by cyclic AMP/cyclic GMP concentrations in the cell. It is further proposed that these substances, through protein phosphorylation, may localize the increase in cytosolic free Ca^{2+} to the appropriate part of the cell.

'Unit cell activation' hypothesis

Put as a question, in its simplest and most extreme form, if a stimulus activates a tissue to 50% of the maximum possible is this explained by 50% of the cells being 'switched on' or by all the cells being activated to give 50% of their individual maximum response?

If the hypothesis is correct then there are three immediate predictions: (1) There should be a relationship between the number of activated cells within the population and the level of stimulus. The stimulus could, for example, vary as the extent or frequency of electrical stimulation, or hormone or neurotransmitter concentration (Fig. 10.3a). (2) There should be a relationship between the mean number of regulatory units occupied per cell (e.g. hormone receptors occupied or calmodulins occupied by Ca^{2+}) and the number of activated cells (Fig. 10.3b). (3) It should be possible to discover a simple mathematical model which describes the number of activated cells as a function of stimulus level or the number of regulatory units occupied (Fig. 10.3c).

The first and second of these predictions can be tested by counting activated cells or by measurements on single cells (see this chapter, section 10.4.1).

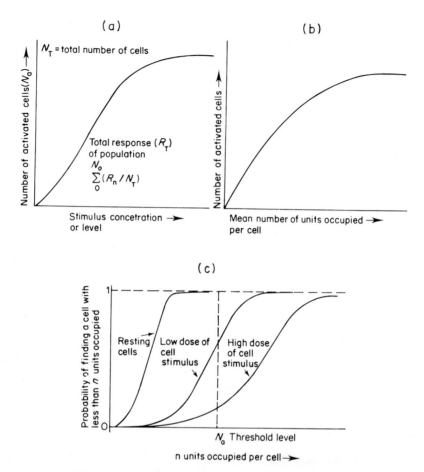

Fig. 10.3. Normal distribution curve for activated cells

Techniques for single-cell analysis are required. Useful possible systems would be the inhibition of complement-induced cell lysis by cyclic AMP, inhibition of polymorph chemotaxis by cyclic AMP, inhibition of histamine secretion from mast cells by cyclic AMP, and the potentiation of glucose-stimulated insulin secretion by cyclic AMP.

For measurements normally made on large numbers of cells (approx. 10^5–10^8) the law of mass action is usually adequate to describe the relationship between hormone (H) concentration and hormone receptor (R) occupancy. For example, when the hormone concentration changes, assuming a simple non-cooperative binding,

$$H + R \rightleftharpoons HR$$

$$K_d = \frac{[H][R]}{[HR]} \tag{10.1}$$

Assuming $[H] = [H_{total}]$ (i.e. tiny fraction of hormone bound for each cell)
then $[R_T] = [HR] + [r_{free}] = r_m + r_{free}$
$K_d = [H_T](R_T - r_m)/r_m$ where r_m = mean number of receptors occupied.

$\therefore$ mean number of receptors occupied per cell $= r_m = V[H_T]R_T/(K_d + [H_T])$

$$\tag{10.2}$$

where V = volume of cell suspension

r_m = receptor occupancy measured by conventional radioligand binding techniques $= \dfrac{\text{total bound hormone}}{\text{total number of cells}}$ (in molecules).

However, when measurements are made on individual cells it may be more appropriate to use a derivation of the Poisson equation (equation 10.3), which predicts the probability of a random, stochastic event occurring.

Consider a cell with r_n receptors occupied on cell 'n'. The probability of finding a cell with r_n receptors occupied $= P_{(r_n)}$

$$P_{(r_n)} = e^{-r_m} r_m^{r_n}/r_n! \tag{10.3}$$

where r_m = mean number of receptors occupied per cell.

This may help to explain why there is an apparent excess of hormone receptors on the surface of cells.

What of the third prediction, the possibility of deriving an equation relating the level of stimulus to the number of activated cells? For a homogeneous population of cells the probability of finding an activated cell (P_a) at a given stimulus is predicted by the binomial distribution

$$P_a = P_A (1 - P_A)^{N_T - N_a} \left(\dfrac{N_T!}{N_a!(N_T - N_a)!} \right) \tag{10.4}$$

P_a = probability of observing N_a number of cells activated from N_T total number of cells sampled.
The probability of not finding an activated cell $= (1 - P_a)$.

P_A = true probability of a cell being activated experimentally for a large sample

$$P_A = \dfrac{N_a}{N_T}.$$

where N_a = number of activated cells, with a variance of var $(N_a) = N_a P_A (1 - P_A)$
Increasing the level of the cell stimulus would increase P_A.

However, this derivation only tests whether a cell population is homogeneous for a given level of stimulus. This is unlikely to be the case. Furthermore, the relationship we want, P_A as a function of stimulus, can only be discovered experimentally until more is known of the mechanisms which generate threshold responses in cells.

An alternative approach to describing the chance of finding an activated cell at a given level of stimulus is to use the Poisson distribution (equation 10.5). For example, the number of quanta of acetylcholine released per stimulus at a nerve terminal, measured by miniature end-plate potentials in the neighbouring muscle, obeys the Poisson distribution (Martin, 1966). This shows that the fusion of secretory granules is a random process and, to a first approximation, the granule population involved is homogeneous.

How can the Poisson distribution be applied to a population of cells? Let us assume that a population of N_T cells responds to a hormone (H), and that inside each cell there are a certain number of regulatory units (U) which become occupied. The chance of finding a cell with n units occupied will be described by the Poisson distribution

$$P_{(n)} = e^{-m} m^n / n! \tag{10.5}$$

where m = mean number of units occupied per cell

$$= \frac{\text{total number of units occupied} \left(\sum_1^{N_T} n \right)}{\text{total number of cells } (N_T)}$$

If the cells undergo a threshold response to the hormone then assume that N_a units must be occupied for any one cell to pass through the threshold. The theory predicts that all cells with more than N_a units occupied (Fig. 10.3c) will be activated.

What could be the molecular basis of these 'regulatory units' in the cell? They could be, for example, calmodulin or the hormone receptor. If the intracellular concentration of calmodulin is 10 μM and the free Ca^{2+} concentration is 1 μM, then in a cell with a volume of 10 pl (i.e. a cell of diameter approx. 27 μm), there will be approx. 60 million molecules of calmodulin and 6 million free calcium ions per cell. Hormone receptors will be even fewer, perhaps only 1000–2000 per cell.

However, the assumption that cell activation is a random process is unlikely to be true and, like the binomial distribution, the Poisson equation cannot be used to predict the number of activated cells as a function of stimulus. To test the third prediction of the 'unit cell hypothesis' therefore requires measurements on single cells and an elucidation of the molecular basis of threshold phenomena.

Suppose one was standing on a station platform and a high-speed train rushed past, blowing its whistle as it does so. How would one describe what was heard? Would one tell someone who had not heard it that the pitch of the whistle had gone down as the train went away, or would one describe it more carefully by saying that what was heard was a decrease in pitch after the train had passed. Without the correct theoretical framework, or at least a perception of our ignorance, it is easy to miss or mistakenly describe phenomena. It is the author's hope that this section will provoke some interesting questions in the mind of the reader, and provide a new framework for understanding some of the puzzles of cell regulation, such as the mechanism of action of insulin.

10.2. Problems

10.2.1. General mechanisms

In spite of an enormous amount of experimental evidence supporting 'intracellular Ca^{2+} as a universal cell regulator', many problems still exist. (1) In many cell phenomena the precise *physiological* function of the primary and secondary regulators has not been fully characterized. (2) In many cells the evidence for intracellular Ca^{2+} is circumstantial because no direct measurements of internal free Ca^{2+} have been made, nor has a Ca^{2+}-dependent response been reconstituted in a broken cell preparation. (3) The precise chemical mechanisms involved in the release and action of intracellular Ca^{2+} remain to be elucidated.

In no cell has a complete calcium balance been obtained. Studies on isolated organelles enable an estimate to be made of the location of bound Ca^{2+} within the cell (Table 2.7, Fig. 2.3). However, this is little more than a guess rather than an accurate assessment of where Ca^{2+} is in the living cell. Cytoplasmic free Ca^{2+} has been measured in relatively few cells (Ashley and Campbell, 1979), although cell-fusion techniques (Hallett and Campbell, 1982a, b) or the synthesis of transiently permeant fluorescent Ca^{2+} indicators (Tsien, 1981) have now opened the way for the study of many types of small cell. A further problem is that the extent of the free Ca^{2+} changes in cells does not always correlate with the magnitude of the cell response. The measurement of free Ca^{2+} within organelles such as the mitochondria or nucleus, is an even more difficult problem in the living cell. The most likely approach will be one based on the fluorescent Ca^{2+} indicators developed by Tsien.

The apparent ubiquitous occurrence of calmodulin in eukaryotic cells has provoked a considerable amount of enthusiasm that it will provide the molecular explanation for most of the regulatory effects of intracellular Ca^{2+}. Certainly this protein can activate many rate-limiting enzymes in the test-tube in the presence of μM Ca^{2+}. For example, phosphorylase kinase when isolated and purified still has calmodulin bound to it. Furthermore, with the use of fluorescent labelled antibodies, calmodulin has been observed at some interesting locations in the cell such as the mitotic spindle (Means and Dedman, 1980). However, in spite of these encouraging observations, experimental evidence for calmodulin regulating phenomena in the living cell is still circumstantial. Several observations sound a cautionary note regarding the precise function of calmodulin in the cell:

(1) Most of the effects of calmodulin *in vitro* have been demonstrated on isolated enzymes. It is now necessary to show that Ca^{2+}–calmodulin can provoke events like membrane fusion, and microfilament or microtubule assembly/disassembly, *in vivo*.

(2) The inhibitory effects of phenothiazines, e.g. trifluoperazine, on live cells may be explained by a 'local anaesthetic' action, rather than by their ability to bind to Ca^{2+}–calmodulin.

(3) Whilst effects can be shown of Ca^{2+} in the μM range on calmodulin-regulated enzymes, maximal activation (> 95%) sometimes requires up to 10–20 μM Ca^{2+}. This is somewhat higher than is to be expected for maximal elevation of cytoplasmic free Ca^{2+} under physiological conditions.

(4) Some of the effects of Ca^{2+}–calmodulin on membrane-bound enzymes, for example (Ca^{2+})-MgATPases in the cell membrane, can be mimicked by certain phospholipids and non-specific detergents. Phosphatidylserine is often active in this way and does not require Ca^{2+}. Furthermore, calmodulin may no longer activate the phospholipid-stimulated enzyme. This raises the question, is there a hydrophobic grouping on such enzymes which interacts with Ca^{2+}–calmodulin *in vitro* but may be occupied by phospholipids *in vivo*, thereby making the calmodulin effect redundant? These doubts are heightened by the observation that Ca^{2+}–calmodulin can activate the adenylate cyclase from bacteria, yet calmodulin has so far been found only in eukaryotes.

Three experimental developments may help to resolve these problems concerning the physiological role of calmodulin. Firstly, highly potent inhibitors of calmodulin are beginning to be synthesized (Fig. 10.4) which may be more specific *in vivo* than trifluoperazine. Secondly, the microinjection of antibodies to calmodulin into living cells, perhaps by cell fusion, should inhibit Ca^{2+}–calmodulin cell responses. Thirdly, it may be possible to produce genetic variants either lacking calmodulin or with a modified protein, with the predictable consequences on cell activation.

Much still needs to be learnt about the molecular mechanisms involved in releasing Ca^{2+} into the cytoplasm when a cell is activated and how it is then removed. What is the molecular basis of Ca^{2+} channels? Are there Ca^{2+} ionophores in the membranes of cells? Does the binding of an agonist to its receptor increase Ca^{2+} permeability by direct modification of phospholipids, for example methylation or phosphatidylinositol release, or does the receptor interact with a protein analogous to the GTP-binding protein required for adenylate cyclase activation? The answers to these questions will require

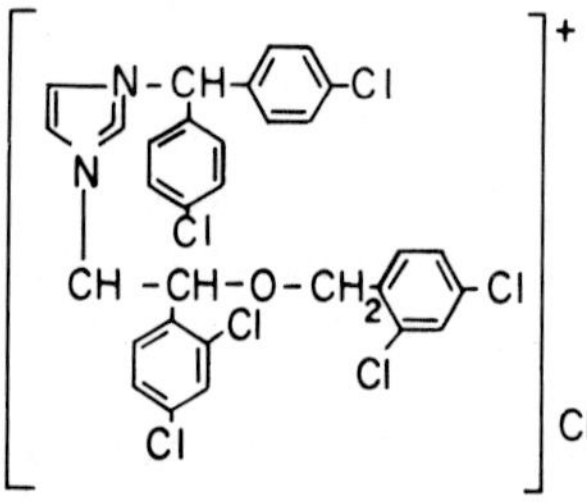

1- [bis (*p*–chlorophenyl) methyl] - 3 - [2,4 - dichloro -
-β- (2,4 - dichlorobenzyloxy) phenethyl] imidazolium chloride
(R24571)

Fig. 10.4. A new, potent inhibitor of calmodulin-activated enzymes given the name R24571. From Van Belle (1981)

advances in the techniques for isolating membrane components and their reconstitution in phospholipid vesicles (liposomes).

A further problem is the physiological relationship between cyclic nucleotides and intracellular Ca^{2+}. Three particular experimental approaches are necessary to resolve these difficulties: (1) The molecular basis of the action of cyclic GMP requires clarification. (2) A direct correlation is required between intracellular free Ca^{2+} and cyclic nucleotide concentration in the intact cell, particularly with respect to the time course of changes in their concentration. (3) Techniques are required to show where and when in the cell changes in intracellular Ca^{2+} and cyclic nucleotide concentation occur.

Most of the processes described in this book are initiated within a few seconds or minutes of exposure of the cell to the stimulus. But what of the more long-term regulatory mechanisms? Do the effects of sex hormones (Bresciani *et al.*, 1973; Liao, 1975), insulin (Jacobs and Krahl, 1973), thyroid hormones (Wallach *et al.*, 1972), and glucocorticoids on protein synthesis, and on the sensitivity of tissues to primary and secondary regulators, involve interactions with intracellular Ca^{2+}? Are the concentrations of proteins concerned with regulating intracellular free Ca^{2+} and mediating its action regulated physiologically through modifications in protein synthesis and degradation? It has been known for many years that denervation of muscle has marked effects on the distribution of acetylcholine receptors. Similar changes in the function of secretory cells are observed after denervation *in vivo*. Are there significant modifications in resting free Ca^{2+} under these experimental conditions? Is the nucleus able to monitor the train of Ca^{2+} transients occurring in excitable cells such as muscle and have a mechanism which responds to it? Finally, vitamin D through its active metabolite 1,25-dihydroxycholecalciferol is known to regulate the concentration of the high-affinity Ca^{2+}-binding protein in the intestine which is required for Ca^{2+} absorption. Abnormalities in muscle contraction have been observed in vitamin D deficiency, which are restored on treatment. These observations raise the possibility that the concentration of high-affinity Ca^{2+}-binding proteins like troponin C and calmodulin may be under long-term hormonal control.

10.2.2. Plants

Ca^{2+} is an essential nutrient for plants (Chapter 1, section 1.2.3) (Burstrom, 1968; Hewitt and Smith, 1974). As in animals it is required for maintaining the integrity and permeability properties of the cell membrane (Wyn-Jones and Lunt, 1967). External Ca^{2+} is also required for several phenomena in plants (Table 10.4). Ca^{2+} therefore plays a structural role in plants, and also has an electrical role (Jaffe *et al.*, 1975; Robinson and Jaffe, 1974; Robinson, 1977), but does Ca^{2+} have a role as an intracellular regulator in plants?

Cell growth and differentiation, metabolism, cell death (e.g. abscission of leaves), fertilization of gametes and cytoplasmic streaming are phenomena found in most plants. They can be provoked or regulated by chemical stimuli or light. Several hormones have been identified in plants (Fig. 10.5). The effects of these

Table 10.4. Some roles for Ca^{2+} in plants

Effect of Ca^{2+}	Reference
Ca^{2+} is transported in aquatic plants	Loewenhaupt (1956)
Ca^{2+} plays a role in serine transport in tobacco	Smith (1978)
Ca^{2+} is required for modulation in leguminous plants	Dixon (1969)
Ca^{2+} is involved in electrical activity of *Nitella*	Weisenseel and Ruppert (1977); van Netten and Belton (1977)
Ca^{2+} is accumulated by chloroplasts	Gross and Hess (1974); Bakker-Grunwald (1974)
Ca^{2+} is involved in differentiation of *Nitella*	Gillet and Lefebvre (1978)
Ca^{2+} is accumulated in pollen tips	Jaffe et al. (1975)
Ca^{2+} is involved in root-tip development	Macklon (1975)
Ca^{2+} is involved in fucoid egg development	Robinson and Jaffe (1974); Nuccitelli and Jaffe (1976); Robinson (1977)
Ca^{2+} inhibits abscission	Poovaiah and Leopold (1973)
Ca^{2+} is involved in auxin-stimulated cell-wall deposition	Morris and Northcote (1977)
Calmodulin activates NAD^+ kinase in the pea, *Pisum*	Anderson and Cormier (1978)
Ca^{2+} induces contraction in a green alga	Salisbury and Floyd (1978)
Ca^{2+} inhibits cytoplasmic motility in *Chara*	Ashley and Williamson (1981); Williamson and Ashley (1982)

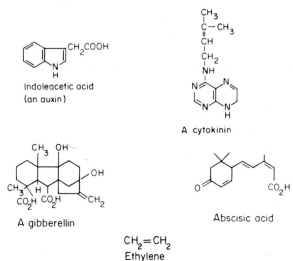

Fig. 10.5. Some plant hormones

substances usually occur over a longer time scale than the acute actions of hormones and neurotransmitters in animals. Nevertheless, there is some evidence that intracellular Ca^{2+} may mediate the increase in cell-wall synthesis induced by auxins (Morris and Northcote, 1977), and the polarization of developing pollen after fertilization. Evidence of the latter has been obtained by studying eggs from the marine brown algae *Fucus* and *Pelvetia* (Robinson and Jaffe, 1974; Jaffe et al., 1975).

Fertilization of *Fucus* eggs by the male gamete stimulates the secretion of an adhesive, rigid wall. Within a few hours the egg is stuck to the substratum. After 12 hours the egg 'germinates', changing from a symmetrical sphere to a pear shape. This polarity is maintained by a current of about 60 pA flowing through the egg from the growing tip to the other end. Thermoelectric equilibrium is maintained by a counter current carried in the other direction by ions in the external medium. Since $^{45}Ca^{2+}$ enters at the growing tip during this polarization phase it has been proposed that the internal current is carried by Ca^{2+} flowing down a concentration gradient. The Ca^{2+} gradient is maintained by differences in permeability of the cell membrane between one end of the egg and the other. The structural polarization would then be produced by the electrical potential and Ca^{2+}. Another interesting example of polarization in one cell is in the giant marine plant cell *Acetabularia*. Action potentials flowing from one end of the cell to another are thought to play a role in developing this polarity, but a role for intracellular Ca^{2+} in this process remains to be established. Evidence for intracellular Ca^{2+} as a regulator in plants is poor relative to animals. Few measurements of intracellular free Ca^{2+} have been made in plants (Ashley and Williamson, 1981; Williamson and Ashley, 1982); nor have the criteria for establishing a role for intracellular Ca^{2+} laid down in Chapter 2, section 2.3 been applied. Calcium oxalate precipitates are found in some plant cells. However, little is known of the physiological role of intracellular stores of Ca^{2+}. In the absence of Mg^{2+} and Mn^{2+} isolated chloroplasts contain two classes of Ca^{2+}-binding site, of $K_d^{Ca} 8 \mu M$ and $51 \mu M$, respectively (Gross and Hess, 1974, Bakker-Grunwald, 1974). Binding of Ca^{2+} to the second site correlates with a change in chlorophyll *a* fluorescence and may be involved in inhibiting the spillover of energy from photosystem II to I during green plant photosynthesis. However, the significance of these *in vitro* findings to ionic conditions within the living cell is not clear, nor is it established whether light can induce uptake of Ca^{2+} by chloroplasts *in vivo*.

The investigation of intracellular Ca^{2+} in plants would seem worth pursuing, especially since a calmodulin which activates NAD^+ kinase to enhance NADP production has been discovered in the pea (Anderson and Cormier, 1978).

10.2.3. Prokaryotes

The role of intracellular Ca^{2+} in bacteria is, as yet, poorly defined. No estimations of free Ca^{2+} have been made inside bacteria. No calmodulin has been found. Prokaryotic cells are much simpler in structure than eukaryotic cells and they are not susceptible to the wide range of primary activators which affect the behaviour of animal cells. The metabolism of nutrients is controlled mainly by repression and by derepression of individual genes and operons. However, as was discussed in Chapter 5, where bacteria respond to a chemotactic stimulus there is a possibility that Ca^{2+} may play a role in controlling the tumbling threshold.

10.3. Guidelines for establishing intracellular Ca^{2+} as a regulator

Suppose a new type of cell activation is discovered, triggered by certain neurotransmitters and modified by a group of hormones. How is it possible to establish whether intracellular Ca^{2+} is the mediator of these primary and secondary cell stimuli?

In order to establish a role for intracellular Ca^{2+} in cell activation four criteria must be satisfied: (1) The phenomenon must be dependent on Ca^{2+}, but not necessarily on external Ca^{2+}. (2) The primary and secondary regulators must alter the concentration of free Ca^{2+} somewhere in the cell. (3) There must be a change, usually an increase, in Ca^{2+} bound to sites responsible for regulating the phenomenon. (4) There must be a change in the flux of Ca^{2+} across either the cell membrane and/or the membrane of one of the organelles within the cell.

The first of these criteria would normally be established by studying the acute and long-term effects of removal of extracellular Ca^{2+}, in the presence of EGTA, and the effect of ionophore A23187 or ionomycin. These experiments will also show whether external or internal Ca^{2+} is required. Care must be taken not to affect cell viability (see Chapter 2). The second criterion is more difficult to satisfy because of the technical problems in incorporating indicators of free Ca^{2+} into living cells, particularly those less than $100 \mu m$ in diameter. If it is possible to measure intracellular free Ca^{2+} directly then an indicator should be selected on the basis of the criteria laid down in Chapter 2, section 2.5.2. Alternatively, it may be possible to develop an indirect method of estimating free Ca^{2+} in the appropriate cell compartment, for example measurement of $^{45}Ca^{2+}$ flux.

In order to satisfy the third criterion the site of regulation by Ca^{2+} must be identified. This may be a high-affinity Ca^{2+} binding like troponin C or calmodulin. However, direct effects of Ca^{2+}, in the μM range, on isolated proteins, ATPases and subcellular structures in the absence of these special Ca^{2+}-binding proteins should also be looked for. Attempts should be made to reconstitute the cellular event in the test-tube and to show that it can be triggered by Ca^{2+} at concentrations and with a time course similar to those found in the intact cell.

Finally, the source of the change in intracellular Ca^{2+} must be identified. If the primary stimulus increases internal Ca^{2+} by increasing the permeability of the cell membrane to Ca^{2+} then removal of external Ca^{2+} should immediately abolish the cell response. Furthermore, an increase in $^{45}Ca^{2+}$ influx will be observed in the presence of external Ca^{2+}. In electrically excitable cells where opening of the slow voltage-dependent Ca^{2+} channel occurs, then the cell response will usually be inhibited by blockers of this channel, such as verapamil, D600, nifedipine, La^{3+} and Co^{2+}. However, the pharmacological properties of these channels can vary considerably, particularly in invertebrates. For example, the voltage-sensitive Ca^{2+} channels in the ciliate protozoan *Paramecium* are not blocked by D600 or transition metal cations. If the source of Ca^{2+} is internal then the two most likely candidates for the internal store are the endoplasmic reticulum and the mitochondria. Incubation of cells with $^{45}Ca^{2+}$, followed by addition of the cell stimulus, will cause a redistribution of Ca^{2+} which can be measured in these organelles after subcellular fractionation. If the mitochondria are thought to be the source of Ca^{2+} for cell activation, then inhibitors such as ruthenium red should have predictable effects on the cell response.

Having established a central role for intracellular Ca^{2+} in the phenomenon being studied, the role of cyclic nucleotides in modifying the cell response can be investigated and the details of the chemical mechanisms can be defined. The aim should be to provide eventually a unified scheme for the regulation of the cells by the various primary and secondary stimuli (Fig. 10.6).

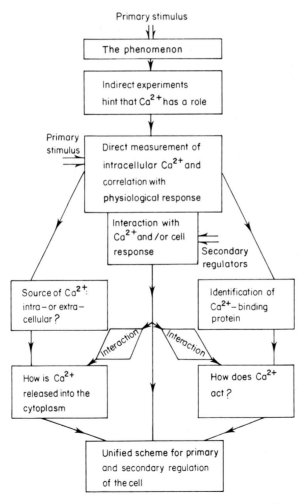

Fig. 10.6. A scheme for investigating intracellular Ca^{2+} as a regulator

10.4. Perspectives

10.4.1. Techniques for chemical studies on intact and single cells

The development of biochemistry in the early years of this century was dominated by the need to break away from the vitalist arguments of the nineteenth century, and to establish the validity of the chemical approach in the elucidation of biological problems (see Needham and Baldwin, 1949; Florkin, 1975a, b). This philosophy was pioneered by Frederick Gowland Hopkins (Fig. 1.5) and his followers, who argued that it was justifiable to study reactions in broken cells provided that the 'event' which occurred in the living cell also occurred in the test-tube (see Needham and Baldwin, 1949). It was not too serious if the magnitude or time scale of this 'event' was altered somewhat in the broken cell system. Alternatively, the event could be generated in the intact cell and then analysed in the broken cell, provided that the original chemical changes that had occurred remained after homogenization. Perhaps the most important role of the biochemist until now has been to isolate and characterize the molecules responsible for these events. It has been this approach which has led to the discovery of the hormones and neurotransmitters which regulate intracellular Ca^{2+}, as well as the MgATPases and Ca^{2+}-binding proteins responsible for the action of Ca^{2+} within the cell.

However, there are dangers in following this approach solely. It has been pointed out throughout this book that many biological molecules will bind Ca^{2+}, and are affected by it, under *in vitro* conditions. This has resulted in many observed effects of Ca^{2+} in broken cell preparations that have no relevance to events within the cell. The event, so far as intracellular Ca^{2+} as a regulator is concerned, is a change in its concentration brought about by a stimulus acting on the cell. In common with many electrical changes in cells this Ca^{2+} event can only be studied while the cell membrane remains intact. This argument is not unique for intracellular Ca^{2+}. For example, the production of oxygen radicals detected by chemiluminescence is abolished in broken cells (Hallett *et al.*, 1981). Changes in regulatory metabolites such as cyclic nucleotides may be restricted to particular locations within the cell.

There is therefore a need for techniques enabling changes in ion concentration, metabolites and enzyme activity to be quantitatively studied in the intact cell. The location of these changes within the cell must also be identified. Such techniques would further circumvent the problems of loss of cell–cell communication, and the difficulties of reconstituting cell functions or the intracellular environment in broken cell preparations. There are three methods currently being developed to tackle this problem: (1) microelectrodes producing an electrical signal in response to a change in metabolite or enzyme activity; (2) luminescent indicators producing a change in light signal; (3) physical techniques, such as nuclear magnetic resonance spectroscopy, which produce signals from endogenous atomic nuclei e.g. ^1H, ^{13}C and ^{31}P.

The latter technique has been used to estimate cytoplasmic free and bound nucleotide concentration (Ackerman *et al.*, 1980), and changes in membrane

Table 10.5. Approximate values for some substances in small cells

Substance	Assumed intracellular concentration	Approx. content per cell
Total K^+	150 mM	600 fmoles (6×10^{-13})
Total Ca^{2+}	2 mM	8 fmoles (8×10^{-15})
Free Ca^{2+}	0.1 μM	400 tmoles* (4×10^{-19})
ATP	10 mM	40 fmoles (4×10^{-14})
AMP	0.2 mM	0.8 fmole (8×10^{-16})
Cyclic AMP	1–10 μM	4–40 amoles (4×10^{-18} to 40×10^{-18})
Cyclic GMP	0.1–1 μM	0.4–4 amoles (0.4×10^{-18} to 4×10^{-18})
Phosphorylase	1 μM	4 amoles (4×10^{-18})
Calmodulin	10 μM	40 amoles (4×10^{-17})
Lactate dehydrogenase	100 μM	400 amoles (4×10^{-16})

These figures are only approximate and intended to give an indication of the relative values for a cell, such as ahepatocyte, 20 μm in diameter (volume approx. 4 pl).
* 1 tipo mole = 10^{-21} moles (*tipyn* is Welsh for small).

structure (De Kruijff et al., 1979). However, the time resolution of this method is of the order of minutes or even hours, and large numbers of cells are required to produce a detectable signal. Giant cells can be injected with absorbing and fluorescent dyes, but these are rarely sensitive enough to detect the minute quantities of enzymes and metabolites which may be present in one small cell (Table 10.5). The analytical method which provides a technique sensitive enough for measurement of enzymes in the pU and nU(1 U = 1 μmole of substrate per min) range or metabolites in the amole to fmole (10^{-18} to 10^{-15}) range is chemiluminescence (Campbell and Simpson, 1979). The high signal-to-noise ratios obtainable provide sufficient sensitivity for the analysis of single small cells and a signal that can be visualized by image intensification (Reynolds, 1979).

10.4.2. The evolutionary significance of intracellular Ca^{2+}

Life began on earth about 3500 million years ago, some 1000–1500 million years after the formation of the earth itself (Table 1.3). Before the existence of recognizable life forms the earth's atmosphere was composed mainly of H_2, NH_3 and CH_4. Energy supplied by ultraviolet light, radioactivity, electric discharges and heat caused reactions between these substrates and H_2O, resulting in the synthesis of the precursors of biological macromolecules, namely sugars, fatty acids, amino acids and nucleotides. These polymerized randomly into proteinoids, polynucleotides and phospholipids. A phase separation of the lipid and protein moeities produced microspheres, the prototypes of the first prokaryotic cells. One of the mysteries of this process still to be resolved is why all living cells became based on only one of the stereoisomers of optically active sugars and amino acids.

About 1500 million years ago the first eukaryotic cell appeared. There is still a considerable degree of controversy as to how this occurred. One hypothesis put forward is that the organelles of the eukaryotic cell developed from involutions of the membrane of a particular prokaryotic cell. However, much of the early microfossil and biochemical data is more consistent with the alternative hypothesis, the symbiotic theory (Margoulis, 1971). In this case it is proposed that the organelles of the eukaryotic cell, including mitochondria, flagella and even the nucleus, originated from cells taken up into the main host cell, forming a symbiotic relation with it until the genetic material eventually became controlled synchronously. Whichever hypothesis is correct the formation of eukaryotic cells eventually lead to multicellular animals and plants evolving through the process of natural selection.

Where does Ca^{2+} fit into this evolutionary scheme?

The discovery of the calcareous remains of blue-green algae more than 1900 million years' old suggests that several primitive cells may have utilized calcium precipitates in a structural role. But what of intracellular Ca^{2+}? At what stage in evolution did cells develop the means of maintaining a resting intracellular free Ca^{2+} concentration in the range 30–300 μM and a large electrochemical gradient of Ca^{2+} across the cell membrane? What were the precursor proteins of those found in contemporary organisms and which are now responsible for mediating the regulatory effects of intracellular Ca^{2+}? What were the environmental and chemical factors that provided the driving force for selecting this type of regulatory mechanism? At a time when the conventional Darwinian-Mendelian evolutionary principles, comprehensible at the level of whole organisms and species, are being questioned particularly at a molecular level, little is known of evolutionary role of ions. Nevertheless, the importance of Ca^{2+} and other ions in controlling the electrical and chemical properties of cells means that any molecular theory of evolution must include this aspect of the chemistry of cells, in addition to that concerned with changes in the genetic material.

The ionic conditions existing in the primordial days before life began are unknown. They were likely to have been very different from those found in contemporary sea or fresh water since the concentrations of ions, particularly Ca^{2+}, have been changed by biological and chemical erosion of rocks during the past 3500 million years (Fig. 1.4). The abundance of calcium in the earth's crust (Fig. 1.2), and the limited solubility of many calcium salts, suggest that the free Ca^{2+} concentration in the primordial soup was likely to have been in the range 10–1000 μM. This conclusion is strengthened by the knowledge of how difficult it is to prepare water in the laboratory with a concentration of Ca^{2+} less than 10 μM. At these concentrations Ca^{2+} would bind to metabolites inside the primitive cell, form precipitates of phosphate or oxalate, inhibit many enzymes, and affect the electrical properties of the cell. These arguments lead to the inevitable conclusion that even the earliest pro- and eukaryotic cells must have had a mechanism for maintaining the free Ca^{2+} concentration at about 0.1 μM. This could have developed in any one of three ways: (1) a membrane potential, positive inside; (2) a Ca^{2+}–cation exchange (antiport) or Ca^{2+}–anion exchange (symport); (3) a Ca^{2+} pump.

All present day cells have a resting membrane potential, negative inside. However, the Gibbs-Donnan equilibrium would enable a Ca^{2+} gradient to be generated across a semipermeable membrane if the concentration of impermeant anions on the outside exceeded those on the inside. Consider two extreme situations A and B (Fig. 10.7), where K^+ and Cl^- are the only permeant ions, X^- is impermeant, and the cell remains at constant volume.
In example A, X^- is inside the cell

At equilibrium, to a first approximation

$$[K_i^+] = [Cl_i^-] + [X^-] \tag{10.6}$$

$$[K_o^+] = [Cl_o^-] \tag{10.7}$$

$$\text{but } [K_i^+] + [Cl_i^-] + [X^-] > [Cl_o^-] + [K_o^+] \tag{10.8}$$

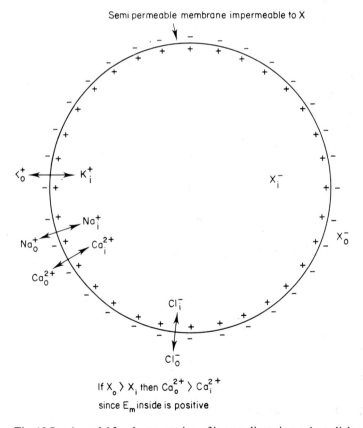

If $X_o > X_i$ then $Ca_o^{2+} > Ca_i^{2+}$
since E_m inside is positive

Fig. 10.7. A model for the generation of ion gradients in a primordial cell

$$E_m = \text{membrane potential} = \frac{RT}{F}\log_e([K_o^+]/[K_i^+]) = \frac{RT}{F}\log_e([Cl_i^-]/[Cl_o^-])$$
(10.9)

$\therefore [K_i^+][Cl_i^-] = [K_o^+][Cl_o^-] = [K_o^+]^2$

$\therefore [K_i^+]([K_i^+] - [X^-]) = [K_o^+]^2$

$\therefore [K_i^+] > [K_o^+]$

$\therefore E_m$ is negative inside.

In example B, the converse is true, i.e. X^- is outside the cell and E_m is positive inside.

In the primordial cell with no energy-requiring pumps, a Ca^{2+} gradient would exist such that $[Ca_i^{2+}] < [Ca_o^{2+}]$ if $[X_o^-] > [X_i^-]$. The other two mechanisms for maintaining a Ca^{2+} gradient across the cell membrane would require a source of energy such as a H^+ or Na^+ gradient, or ATP. Ca^{2+} pumps exist in all present day prokaryotes and eukaryotes. The mechanisms for regulating intracellular Ca^{2+}, and for changing its concentration rapidly inside the cell, must have existed at least 700 million years ago. This was when metazoans began to appear (Table 1.3) enabling Darwinian-Mendelian mechanisms of natural selection to act. What lead to the development of Ca^{2+} pumps and regulatory Ca^{2+}-binding proteins up to this point in evolution is not known. However, the elucidation of the amino acid sequences of various Ca^{2+}-binding proteins is beginning to provide some clues as to the development of regulatory mechanisms, involving intracellular Ca^{2+}, during the most recent stages of evolution.

Comparison of sequences of the same protein from species whose time of divergence from a common ancestor is known from the fossil record, has uncovered a fascinating phenomenon (Wilson et al., 1977). For a given protein the *rate* of change of amino acid residues is approximately constant, at least over the last 100–200 million years where precise knowledge of the fossil record is accurate enough to make the calculations possible. For example, the Ca^{2+}-binding protein parvalbumin took 5 million years to change 1% of its amino acid residues. This compares with 3 million years for albumin and 400 million years for histone H_4. As yet only a few troponin C and calmodulins have been sequenced.

Troponin C shows many homologies with parvalbumin and calmodulin. The sequence homologies between the troponin C from different species seem to be less than those for calmodulins, suggesting that the rate of evolutionary change of calmodulin has been slower than for troponin C.

The appearance of these high-affinity Ca^{2+}-binding proteins depended on the development of at least 2–3 acidic amino acid residues (i.e. Asp or Glu) in close proximity (Fig. 3.12). Examination of the triplet code to see how this might have arisen provides some interesting observations (Fig. 10.8). As with many amino acids the wobble hypothesis provides for redundancy in the third base of the triplet. A mutation here will either retain Asp or Glu or simply exchange one for the other. However, single mutations in either position one or two could lead to a

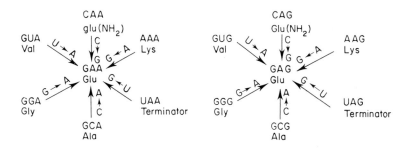
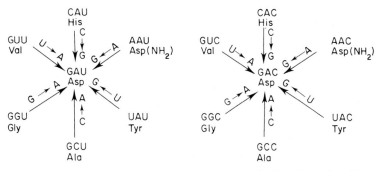

Fig. 10.8. Single base mutations producing the triplet for Glu or Asp. Short arrows represent single base mutation

change from a neutral (Ala, Gly, Val), or even a basic amino acid (Lys), into an acidic one. There is a remarkable similarity in the amino acid changes necessary between Asp and Glu. The mutation of the terminator codons into Glu or Asp would also produce a protein of increased length. The fact that there are 34% homologies between the two halves of the troponin C molecule and between domains 1 and 3 and 2 and 4 of calmodulin (Fig. 3.13) has lead to the suggestion that these proteins have evolved through gene duplication. The process of gene duplication also enables deleterious mutations to occur in nucleotide sequences whilst still retaining the active original sequence. Whether such mutations are inevitably selected for, or are so-called 'neutral' mutations is still a matter of controversy. Even so, changes in amino acid sequence may be advantageous for reasons other than increasing the activity of the protein. It may affect its interaction with another protein or an intracellular structure, it may affect its affinity for ions or metabolites which modify its activity, it may affect the rate of synthesis or storage, and it may affect its susceptibility to proteolysis and thus its turnover rate.

The evolution of multicellular organisms lead to the development of mechanisms of cell activation regulated by both chemical and physical stimuli. How hormones and neurotransmitters evolved is not known. They are often degradation products of amino acids, protein precursors and lipids. Perhaps they and their receptors evolved inside a primitive organelle analogous to the secondary lysosome of present day cells (Hales, 1978). A better understanding of the evolutionary clock of proteins concerned with the regulation of intracellular Ca^{2+} is likely to greatly enhance our comprehension of the evolution of life at the molecular level.

10.4.3. Coda

For a naturalist one of the most exciting things he can encounter is a new animal or plant, or one in an unexpected place doing something unexpected. His love of natural beauty, the sensitization of his intellect, sight, hearing, smell and touch all drive him on, acting as a catalyst to his imagination of what might lie beyond awaiting exploration. Like many areas in modern science the story of intracellular calcium has focused on but one piece of the puzzle confronting the biologist. This story has involved much detailed discussion of experimental method and molecular mechanisms. Yet the 'universal role of intracellular calcium as a regulator' of phenomena in so many different animals and plants also provides for a greater perspective, analogous to that seen by the naturalist.

For the author the study and thought which has gone into this book have been both stimulating and very enjoyable. The exciting consequence has been the realization that, for him, rather than being the beginning of the end of his comprehension, it has turned out to be only the end of the beginning.

References

Aaron, J. (1976). Histology and micro-anatomy of bone. In *Calcium, Phosphate and Magnesium Metabolism*. (Ed. B. E. C. Nordin) pp. 298–356. Churchill Livingstone, Edinburgh.
Abbot, F., Gomez, J. E., Birnhaum, E. T., and Darnall, D. W. (1975). The location of the calcium ion binding site in bovine alpha-trypsin and beta-trypsin using lanthanide ion probes. Biochemistry, **14**, 4935–4943.
Abraham, E. P. (1970). Necrosis, calcification and autolysis. In General Pathology pp. 431–450. (Ed. Florey H. W. Lord). Lloyd-Luke, London.
Ackerman, J. J., Grone, T. H., Wong, G. G., Gadian, D. G., and Radda, G. K., (1980). *Nature (London)*, **283**, 167–170. Mapping of metabolites in whole animals by 31 P NMR using surface coils.
Adams, M. H. (1949). *J. Gen. Physiol.*, **32**, 579—594. The stability of bacterial viruses in solutions of salts.
Adelman, W. J., Jr., and Moore, J. W. (1961). *J. Gen. Physiol.*, **45**, 93–103. Action of external divalent ion reduction on sodium movement in squid giant axon.
Adelstein, R. S., and Eisenberg, E. (1980). *Ann. Rev. Biochem.*, **49**, 921–956. Regulation and kinetics of the actin-myosin-ATP interaction.
Agassi, I. J. (1971). *Faraday as a Natural Philosopher*. University of Chicago Press, Chicago and London.
Ahkong, Q. F., Fisher, D., Tampion, W., and Lucy, J. (1973). *Biochem. J.*, **136**, 147–155. The fusion of erythrocytes by fatty acids, esters, retinol and α-tocopherol.
Ahkong, Q. F., Tampion, W., and Lucy, J. A. (1975). *Nature (London)*, **256**, 208–209. Promotion of cell fusion by divalent cation ionophores.
Aidley, D. J., (1971). *The Physiology of Excitable Cells*. Cambridge University Press, Cambridge.
Akern, T., and Brody, T. M. (1977). *Pharmacol. Rev.*, **29**, 187–220. The role of Na^+, K^+-ATPase in the ionotropic action of digitalis.
Aktar, R. A., and Perry, M. C. (1979). *Biochim. Biophys. Acta*, **585**, 107–116. Insulin action in isolated fat cells. I. Effects of divalent cations on the stimulation by insulin of glucose uptake.
Albertini, D. F., and Anderson, E. (1977). *J. Cell Biol.*, **73**, 111–127. Microtubule and microfilament rearrangements during capping of concanavalin A receptors on cultured ovarian granulosa cells.
Albritton, E. C. (Ed.) (1951). *Standard Values in Blood*. American Institute of Biological Sciences Washington, DC.
Allan, D., and Michell, R. H. (1974). *Biochem. J.* **142**, 599–604. Phosphatidyl inositol cleavage in lymphocytes. Requirement for calcium ions at a low concentration and effects of other cations.
Allan, D., and Michell, R. H. (1977a). *Biochem. J.*, **166**, 495-499. Calcium ion dependent diacylglycerol accumulation in erythrocytes is associated with microvesiculation but not with efflux of potassium ions.
Allan, D., and Michell, R. H. (1977b). *Biochem. Soc. Trans.*, **5**, 55–59. Metabolism of phosphatidate at the plasma membrane.
Allan, D., Billah, M. M., Finean, J. B., and Michell, R. H. (1976a). *Nature (London)*, **261**, 58–60. Release of diacylglycerol-enriched vesicles from erythrocytes with increased intracellular $[Ca^{2+}]$.
Allan, D., Watts, R., and Michell, R. H. (1976b). *Biochem. J.*, **156**, 225–232. Production of 1,2-diacylglycerol and phosphatidate in human erythrocytes treated with calcium ions and ionophore A23187.
Allen, D. G., and Blinks, J. R. (1978). *Nature (London)*, **273**, 509–513. Calcium transients in aequorin-injected frog cardiac muscle.

Allen, J. E., and Rasmussen, H. (1971). *Science*, **174**, 512–514. Human red blood cells; prostaglandin E_2, epinephrine and isoproteronol alter deformability.

Allen, R. C., Stjernholm, R. L., and Steele, R. H. (1972). *Biochem. Biophys. Res. Commun.*, **47**, 679–684. Evidence for the generation of an electronic excitation state(s) in human polymorphonuclear leukocytes and its participation in bactericidal activity.

Allen, R. D. (1974). *Symp. Soc. Exp. Biol.*, **28**, 15–26. Some new insights concerning cytoplasmic streaming.

Allison, A. C. (1974). The effect of inhalational anaesthetics on proteins. In *Molecular Mechanisms in General Anaesthesia* (Eds. M. J. Halsey, P. A. Millar, and J. A. Sutton), pp. 164–182. Churchill Livingstone, Edinburgh.

Allison, A. C., and Davies, P. (1974). *Symp. Soc. Exp. Biol.*, **28**, 419–443. Mechanisms of endocytosis and exocytosis.

Allison, A. C., Davies, P., and De Petris, S. (1971). *Nature (London)*, **232**, 153–155. Role of contractile microfilaments in macrophage movement and endocytosis.

Allison, F., Lancaster, M. G., and Crosthwaite, J. Z. (1963). *Am. J. Pathol.*, **43**, 775–795. Studies on the pathogenesis of acute inflammation. V. An assessment of factors that influence *in vitro* the phagocytic and adhesive properties of leukocytes obtained from rabbit peritoneal exudate.

Alm, B., Efendic, S., and Löw, H. (1970). *Horm. Metab. Res.*, **2**, 142–146. Effect of lipolytic and antilipolytic agents on the uptake of ^{47}Calcium into rat adipose tissue *in vitro*.

Alouf, J. E. (1976). Cell membranes and cytolytic bacterial toxins. In *The Specificity and Action of Animal, Bacterial and Plant Toxins*. (Ed. P. Cuatrecasas), pp. 220–270. Chapman and Hall, London.

Ammann, D., Meier, P. C., and Simon, W. (1979). Design and use of calcium-sensitive microelectrodes. In *Detection and Measurement of Free Ca^{2+} in Cells*. (Eds. C. C. Ashley and A. K. Campbell), pp. 117–129. Elsevier/North-Holland, Amsterdam.

Amos, W. B., Routledge, L. M., and Yew, F. F. (1975) *J. Cell Sci.* **19**, 202–213. Calcium-binding proteins in a vorticellid contractile organelle.

Amos, W. B., Routledge, L. M., Weis-Foch, T. and Yew, F. F. (1976). Soc. Exp. Biol. Symp. **30**, 273–301. The spasmoneme and calcium-dependent contraction in connection with specific calcium binding proteins.

Amos, W. B., Grimstone, A. V., Rothschild, L. J., and Allen, R. D. (1978). *J. Cell Sci.*, **35**, 139–164. Structure, protein composition and birefringence of the costa; a motile flagellar root fibre in the flagellate *Trichomonas*.

Anderson, J. M., and Cormier, M. J. (1978). *Biochem. Biophys. Res. Commun.*, **84**, 595–602. Calcium-dependent regulator of NAD kinase in higher plants.

Anderson, N. C., Ramon, F., and Snyder, A. (1971). *J. Gen. Physiol.*, **58**, 322–339. Studies on calcium and sodium in uterine smooth muscle excitation under current clamp and voltage-clamp conditions.

Apostolov, K., and Poste, G. (1972). *Microbios*, **6**, 247–261. Interaction of Sendai virus with human erythrocytes: a system for the study of membrane fusion.

Argy, W. P., Handler, J. S., and Orloff, J. (1967). *Am. J. Physiol.*, **213**, 803–808. Ca^{2+} and Mg^{2+} effects on toad bladder response to cyclic AMP, theophylline and ADH analogues.

Arthur, E. J., and Sanbon, R. C. (1969). in *Chemical Zoology*, vol. 3, pp 429–464. (Ed. M. Florkin, and B. T. Scheer,) Osmotic and ionic regulation in nematodes

Ashby, J. P., and Speake, R. N. (1975). *Biochem. J.*, **150**, 89–96. Insulin and glucagon secretion from isolated islets of Langerhans. The effect of calcium ionophores.

Ashcroft, S. J., Sugden, M. C., and Williams, I. H. (1980) *Hom. Met. Res. Suppl.* **10**, 1–7 Carbohydrate metabolism and the glucoreceptor mechanism.

Ashley, C. C. (1969). *J. Physiol. (London)*, **203**, 32–33P. Aequorin-monitored calcium transients in single *Maia* muscle fibres.

Ashley, C. C. (1978). *Ann. N. Y. Acad. Sci.*, **307**, 308–329. Calcium ion regulation in barnacle muscle fibres and its relation to force development.

Ashley, C. C., and Campbell, A. K. (1978). *Biochim. Biophys. Acta*, **512**, 429–435. Effect of L-glutamate on aequorin light, tension and membrane responses in barnacle muscle.

Ashley, C. C., and Campbell, A. K. (Eds.) (1979). *Detection and Measurement of Free Ca^{2+} in Cells*. Elsevier/North-Holland, Amsterdam.

Ashley, C. C., and Ellory, J. C. (1972). *J. Physiol. (London)*, **226**, 653–674. The efflux of magnesium from single crustacean muscle fibres.

Ashley, C. C., and Lea, T. J. (1977). Photoproteins and glass scintillators in assessing transport proteins. Ellis, Horwood, Chichester. In *Membranous Elements and Movements of Molecules*, Vol. 6 (Ed. E. Reid), pp. 79–96.
Ashley, C. C., and Lea, T. J. (1978). *J. Physiol. (London)*, **282**, 307–331. Calcium fluxes in single muscle fibres measured with a glass scintillator probe.
Ashley, C. C., and Moisescu, D. G. (1972). *Nature New Biol.*, **237**, 208–211. Model for the action of calcium in muscle.
Ashley, C. C., and Moisescu, D. G. (1977). *J. Physiol. (London)*, **270**, 627–652. Effect of changing the composition of the bathing solutions upon the isometric tension–pCa relationship in bundles of crustacean myofibrils.
Ashley, C. C., and Ridgway, E. B. (1968). *Nature (London)*, **219**, 1168–1169. Simultaneous recording of membrane potential, calcium transient and tension in single barnacle muscle fibres.
Ashley, C. C., and Ridgway, E. B. (1970). *J. Physiol. (London)*, **209**, 105–130. On the relationship between membrane potential, calcium transient and tension in single barnacle muscle fibres.
Ashley, C. C., and Williamson, R. E. (1981). *J. Physiol. (London)*, 319, 103 P. Cytoplasmic free Ca^{2+} and streaming velocity in characean algae; measurements with micro injected aequorin.
Ashley, C. C., Ellory, J. C., and Hainaut, K. (1974). *J. Physiol. (London)*, **242**, 255–272. Calcium movements in single crustacean muscle fibres.
Ashley, C. C., Caldwell, P. C., Campbell, A. K., Lea, T. J., and Moisescu, G. G. (1976). *Symp. Soc. Exp. Biol.*, **30**, 397–422. Calcium movements in muscle.
Ashley, C. C., Ellory, J. C., and Griffiths, P. J. (1977). *J. Physiol. (London)*, **269**, 421–439. Caffeine and the contractility of single muscle fibres from the barnacle *Balanus nubilus*.
Ashley, C. C., Rink, T. J., and Tsein, R. Y. (1978). *J. Physiol. (London)*, **280**, 27 P. Changes in free Ca during muscle contraction, measured with an intracellular Ca-sensitive electrode.
Ashwell, G., and Dische, Z. (1950). *Biochim. Biophys. Acta*, **4**, 276–292. Inhibition of metabolism of nucleated red cells by intracellular ions and its relation to intracellular structural factors.
Assimacopoulos-Jeannet, F. D., Blackmore, P. F., and Exton, J. H. (1977). *J. Biol. Chem.*, **252**, 2662–2669. Studies on α-adrenergic activation of hepatic glucose output.
Atwater, I., and Beigelman, P. M. (1976). *J. Physiol. (Paris)*, **72**, 769–786. Dynamic characteristics of electrical activity in pancreatic β-cells.
Atwater, I., Dawson, C. M., Ribalet, B., and Rojas, E. (1978). *J. Physiol. (London)*, **288**, 575–588. Potassium permeability activated by intracellular Ca^{2+} in the pancreatic β-cell.
Axelrod, A. E., Swingle, K. F., and Elvehjem, C. A. (1941). *J. Biol. Chem.*, **140**, 931–932. The stimulatory effect of calcium upon the succinoxidase activity of fresh rat tissues.
Babcock, D. F., Chen, J.-L. J., Yin, B. P., and Lardy, H. A. (1979). *J. Biol. Chem.*, **254**, 8117–8120. Evidence for mitochondrial localization of the hormone-responsive pool of Ca^{2+} in isolated hepatocytes.
Baddiley, J., Michelson, A. M., and Todd, A. R. (1948). *Nature (London)*, **161**, 761–726. Synthesis of adenosine triphosphate.
Baguet, F. (1975). *Prog. Neurobiol.*, **5**, 97–125. Excitation and control of isolated photophores of luminous fishes.
Bailey, I. A., Gadian, D. G., Matthews, P. M., Radda, G. K., and Seeley, P. J. (1981). *FEBS Lett.*, **123**, 315–318. Studies of metabolism in the isolated perfused rat heart.
Bailey, K. (1942). *Biochem. J.*, **36**, 121–139. Myosin and adenosine triphosphatase.
Bailey, K., and Webb, E. C. (1944). *Biochem. J.*, **38**, 394–398. Purification and properties of yeast pyrophosphatase.
Bailey, K. (1946). *Nature (London)*, **157**, 368–369. Tropomyosin; a new asymmetric protein component of muscle.
Bailey, K. (1948). *Brit. Med. Bull.*, **5**, 338–341. Fibrous proteins as components of biological systems.
Baker, P. F. (1972). *Prog. Biophys. Mol. Biol.*, **24**, 177–223. Transport and metabolism of calcium ions in nerve.
Baker, P. F. (1975). *Recent Adv. Physiol.*, **9**, 51–86. Excitation–secretion coupling.
Baker, P. F. (1976). *Symp. Soc. Exp. Biol.*, **30**, 67–88. The regulation of intracellular Ca^{2+}.
Baker, P. F., and Blaustein, M. P. (1968). *Biochim. Biophys. Acta*, **150**, 167–170. Sodium-dependent uptake of calcium by crab nerve.
Baker, P. F., and Carruthers, A. (1980). *Nature (London)*, **286**, 276–279. Insulin stimulates sugar transport in giant muscle fibres of the barnacle.

Baker, P. F., and Crawford, A. C. (1972). *J. Physiol. (London)*, **227**, 855–874. Mobility and transport of magnesium in squid giant axons.
Baker, P. F., and Crawford, A. C. (1975). *J. Physiol. (London)*, **247**, 209–226. A note on the mechanism by which inhibitors of the sodium pump accelerate spontaneous release of transmitter from motor nerve terminals.
Baker, P. F., and Knight, D. E. (1978). *Nature (London)*, **276**, 620–622. Calcium-dependent exocytosis in bovine adrenal medullary cells with leaky plasma membranes.
Baker, P. F. and Honerjäger, P. (1978). *Nature (London)*, **273**, 160–161. Influence of carbon dioxide on the level of ionized in calcium squid axons.
Baker, P. F., and McNaughton, P. A. (1976a). *J. Physiol. (London)*, **260**, 24P–25P. The effect of membrane potential on the calcium transport process.
Baker, P. F., and McNaughton, P. A. (1976b). *J. Physiol. (London)*, **259**, 103–144. Kinetics and energetics of calcium efflux from intact squid giant axons.
Baker, P. F., and McNaughton, P. A. (1978). *J. Physiol. (London)*, **276**, 127–150. The influence of extracellular calcium binding on the calcium efflux from squid axons.
Baker, P. F., and Reuter, H. (1975). *Calcium in Excitable Cells.* Pergamon Press. Oxford.
Baker, P. F., and Schapira, A. H. V. (1980). *Nature (London)*, **284**, 168–169. Anaesthetics increase light emission from aequorin at constant ionised calcium.
Baker, P. F., and Warner, A. E. (1972). *J. Cell Biol.*, **53**, 579–581. Intracellular calcium and cell cleavage in early embryos of *Xenopus*.
Baker, P. F., and Whitaker, M. J. (1979). *J. Physiol. (London)*, **298**, 55P. Trifluoperazine inhibits exocytosis in sea-urchin eggs.
Baker, P. F., Blaustein, M. P., Hodgkin, A. L., and Steinhardt, R. A. (1969). *J. Physiol. (London)*, **200**, 431–458. The influence of calcium on sodium efflux in squid axons.
Baker, P. F., Hodgkin, A. L., and Ridgway, E. B. (1971). *J. Physiol. (London)*, **218**, 708–755. Depolarization and calcium entry in squid giant axons.
Baker, P. F., Meves, H., and Ridgway, E. B. (1973b). *J. Physiol. (London)*, **231**, 527–548. Calcium entry in response to maintained depolarization of squid axons.
Baker, P. F., Knight, D. E., and Whitaker, M. J. (1980). *Proc. R. Soc. London Ser. B*, **207**, 149–161. The relation between ionized calcium and cortical granule exocytosis in eggs of the sea urchin *Echinus esculentes*.
Bakker-Grunwald, T. (1974). *Biochim. Biophys. Acta*, **347**, 141–143. A Ca^{2+}-stimulated incorporation of phosphate into ATP in chloroplasts; the problem of allotopy.
Balk, S. D. (1971). *Proc. Natl. Acad. Sci. U.S.A.*, **68**, 271–275. Calcium as a regulator of the proliferation of normal, but not of transformed, chicken fibroblasts in a plasma containing medium.
Balk, S. D., Whitfield, J. F., Youdale, T., and Braun, A. C. (1973). *Proc. Natl. Acad. Sci. U.S.A.*, **70**, 675–679. Roles of calcium, serum, plasma and folic acid in the control of proliferation of normal and Rous Sarcoma virus-infected chicken fibroblasts.
Balzer, H., and Makinose, M. (1968). *Arch. Exp. Pathol. Pharmakol.*, **259**, 151–152. The inhibition of the sarcoplasmic calcium pump produced by agents which react with its lipid component: reserpine, prenylamine, chlorpromazine, imipramine.
Bangham, A. D., Standish, M. M., and Watkins, J. C., and Weissmann, G. (1967). *Protoplasma*, **63**, 183–187. The diffusion of ions from a phospholipid model membrane system.
Banschbach, M. W. (1978). *Biochem. Biophys. Res. Commun.*, **84**, 922–927. Cystic fibrosis serum promotes ^{45}Ca uptake by normal human leukocytes.
Baptiste, E. C. D. (1935). *Ann. Bot. (London)*, **49**, 345–366. The effect of some cations on the permeability of cells to water.
Bär, H.-P., and Hechter, O. (1969a). *Biochem. Biophys. Res. Commun.*, **35**, 681–686. Adenyl cyclase and hormone action. III. Calcium requirement for ACTH stimulation of adenyl cyclase.
Bär, H. C., and Hechter, O. (1969b). *Proc. Natl. Acad. Sci. U.S.A.*, **63**, 350–356. Adenyl cyclase and hormone action. I. Effects of adrenocorticotropic hormone, glucagon, and epinephrine on the plasma membrane of rat fat cells.
Barany, M. (1967). *J. Gen. Physiol.*, **50**, 197–218. ATPase of myosin correlated with speed of muscle shortening.
Barany, M., and Barany, K. (1980). *Annu. Rev. Physiol.*, **42**, 275–292. Phosphorylation of the myofibrillar proteins.
Barnes, R. D. (1974). *Invertebrate Zoology.* Saunders, Philadelphia.
Barnett, L. B., and Ball, H. B. (1960). *J. Biophys. Biochem. Cytol.*, **8**, 83–101. Metabolic and ultrastructural changes induced in adipose tissue by insulin.

Barritt, G. J. (1981). *Cell Calcium*, **2**, 53–63. A proposal for the mechanism by which α-adrenergic agonists, vasopressin, angiotensin, and cyclic AMP induce calcium release from intracellular stores in the liver cell: a possible role for the metabolites of arachidonic acid.

Baserga, R. (1976). *Multiplication and Cell Division in Mammalian Cells*. Marcel Dekker, New York and Basel.

Bassel, A., Shaw, M., and Campbell, L. L. (1971). *J. Virol.*, **7**, 661–672. Dissociation by chelating agents and substructure of the thermophilic bacteriophage Te 84.

Bassingthwaite, J. B., and Reuter, H. (1972). Calcium movements and excitation–contraction coupling in cardiac cells. In *Electrical Phenomena of the Heart* (Ed. W. C. De Mello). Academic Press, New York.

Bassot, J.-M., and Bilbaut, A. (1977). *Biol. Cell.*, **28**, 155–162, 163–168. Bioluminescence des elytres d'Acholoë.

Bates, R. G., Staples, B. R., and Robinson, R. A. (1970). *Anal. Chem.*, **42**, 867–871. Ionic hydration and single ion activities in unassociated chlorides at high ionic strengths.

Battacharya, G. (1961). *Biochem. J.*, **79**, 369–377. Effects of metal ions on the utilisation of glucose and on the influence of insulin on it by isolated rat diaphragm.

Batzri, S., and Selinger, Z. (1973). *J. Biol. Chem.*, **248**, 356–360. Enzyme secretion mediated by the epinephrine β receptor in rat parotid slices.

Batzri, S., Selinger, Z., Schramm, M., and Rabinovitch, M. R. (1973). *J. Biol. Chem.*, **248**, 361–368. Potassium release mediated by the epinephrine α-receptor in rat parotid slices. Properties and relation to enzyme secretion.

Baux, G., Simonneau, M., Tauc, L., and Segundo, J. P. (1978). *Proc. Natl. Acad. Sci. U.S.A.*, **75**, 4577–4581. Uncoupling of electronic synapses by calcium.

Bayer, R., Hennekes, R., Kaufmann, R., Kalusche, D., and Mannhold, R. (1975). Naunyn-Schmiedebergs Arch. Pharmacol., **290**, 49–68, 69–80, 81–98. Inotropic and electrophysiological action of verapamil and D-600 on mammalian myocardium.

Bayliss, W. M., and Starling, E. H. (1902). *J. Physiol. (London)*, **28**, 325–353. The mechanism of pancreatic secretion.

Beam, K. G., Nestler, E. J., and Greengard, B. (1977). *Nature (London)*, **267**, 534–536. Increased cyclic GMP levels associated with contraction in muscle fibres of the giant barnacle.

Beaulieu, E.-E., Godeau, F., Schorderet, M., and Schorderet-Slatkine, S. (1978). *Nature (London)*, **275**, 593–598. Steroid-induced meiotic division in *Xenopus laevis* oocytes: surface and calcium.

Beck, M. T. (1970). *Chemistry of Complex Equilibria*. van Nostrand Reinhold, London.

Becker, G. L., and Showell, H. J. (1972). *Zeitschrift. J. Immunitätsforsch., n. Exp. Ther.*, **143**, 466. The effect of Ca and Mg on the chemotactic responsiveness and spontaneous motility of rabbit polymorphonuclear leukocytes.

Beeler, G. W., and Reuter, H. (1970). *J. Physiol. (London)*, **207**, 211–229. The relation between membrane potential, membrane currents and activation of contraction in ventricular myocardium fibres.

Behn, C., Lübbemeier, A., and Weskamp, P. (1977). *Pflügers Arch.*, **372**, 259–268. Chlorotetracycline induces calcium mediated shape changes in human erythrocytes. Is Ca asymmetrically distributed in the red cell membrane?

Beigelman, P. M., Ribalet, B., and Atwater, I. (1977). *J. Physiol. (Paris)*, **73**, 201–217. Electrical activity of mouse pancreatic beta-cells. II. Effects of glucose and arginine.

Bellé, R., Ozon, R., and Stinnakre, J. (1977). *Mol. Cell. Endocrinol.*, **8**, 65–72. Free calcium in full grown *Xenopus laevis* oocyte following treatment with ionophore A23187 or progesterone.

Belocopitow, E., Appleman, M. M., and Torres, H. N. (1965). *J. Biol. Chem.*, **240**, 3473–3478. Factors affecting the activity of muscle glucogen synthetase. II. The regulation by Ca^{2+}.

Bennett, H. S. (1956). *J. Biophys. Biochem. Cytol.*, **2** (Suppl.), 99–103. The concepts of membrane flow and membrane vesiculation as mechanisms for active transport and ion pumping.

Bennett, H. S., and Porter, K. R. (1953). *Am. J. Anat.*, **93**, 61–105. Electron microscope study of sectioned breast muscle of domestic fowl.

Bennett, J. P., Smith, G. A., Houslay, M. D., Hesketh, T. R., Metcalfe, J. C., and Warren, G. B. (1978). *Biochim. Biophys. Acta*, **513**, 310–320. The phospholipid head group specificity of an ATP-dependent calcium pump.

Benzinger, T., Kitzinger, T., Hems, R., and Burton, K. (1959). *Biochem. J.*, **71**, 400–407. Free energy changes of the glutaminase reaction and the hydrolysis of the terminal pyrophosphate bond of adenosine triphosphate.

Benzonana, G., Capony, J.-P., and Pechere, J. F. (1972). *Biochim. Biophys. Acta*, **278**, 110–116. The binding of calcium to muscular parvalbumins.

Benzonana, G., Cox, J., Kohler, L., and Stein, E. A. (1974). *C. R. Acad. Sci.*, **279**, 1491–1493. Caractérisation d'une nouvelle métalloprotéine calcique du myogène de certain crustacés.
Berger, J., Rachlin, A. L., Scott, W. E., Sternback, L. H., and Goldberg, J. (1951). *J. Amer. Chem. Soc.*, **73**, 5295–5298. The isolation of three new crystalline antibiotics from *Streptomyces*.
Berger, W., Dahl, G., and Meissner, H. P. (1975). *Cytobiologie*, **12**, 119–139. Structural and functional alterations in fused membranes of secretory granules during exocytosis in pancreatic islet cells of the mouse.
Berlind, A. (1977). *Int. Rev. Cytol.*, **49**, 172–251. Cellular dynamics in invertebrate neurosecretory systems.
Berlind, A., and Cooke, I. M. (1968). *Gen. Comp. Endocrinol.*, **11**, 458–463. Effect of calcium omission on neurosecretion and electrical activity of crab pericardial organs.
Berliner, K. (1933). *Am. Heart J.*, **8**, 548–562. The action of calcium on the heart—a critical review.
Berneis, K. H., Pletscher, A., and Da Prada, M. (1969). *Nature (London)*, **224**, 281–283. Metal-dependent aggregation of biogenic amines: a hypothesis for their storage and release.
Berneis, K. H., Da Prada, M., and Pletscher, A. (1971). *Experientia*, **27**, 971–972. A possible mechanism for uptake of biogenic amines by storage organelles: incorporation into nucleotide–metal aggregates.
Bernstein, J. (1902). *Pflügers Arch. Gesamte Physiol.*, **92**, 521–567. Untersuchungen zur Thermodynamik der bioelektrischen Ströme.
Berntsson, K.-E., Haglund, B., and Lovtrup, S. (1965). *J. Cell. Comp. Physiol.*, **65**, 101–112. Osmoregulation in the amphibian egg. The influence of calcium.
Berridge, M. J. (1976a). *Symp. Soc. Exp. Biol.*, **30**, 219–231. Calcium, cyclic nucleotides and cell division.
Berridge, M. J. (1976b). *Adv. Cyclic Nucleotide Res.*, **6**, 1–96. The interaction of cyclic nucleotides and calcium in the control of cellular activity.
Berridge, M. J. (1980). *Cell Calcium*, **1**, 217–227. Preliminary measurements of intracellular calcium in an insect salivary gland using a calcium sensitive microelectrode.
Berridge, M. J., Oschman, J. L., and Wall, B. J. (1975). Intracellular calcium reservoirs in *Calliphora* salivary glands. In *Calcium Transport in Contraction and Secretion* (Eds. E. Carafoli, F. Clementi, W. Drabikowsky and A. Margreth), pp. 131–138. North-Holland, Amsterdam.
Betteridge, A. (1980). *Biochem. J.*, **186**, 987–992. Role of Ca^{2+} and cyclic nucleotides in the control of prostaglandin E production in the rat anterior pituitary gland.
Biales, B., Dichter, M. A., and Tischler, A. (1977). *Nature (London)*, **267**, 172-173. Na^+ and Ca^{2+} action potential in pituitary cells.
Bianchi, C. P. (1961). *J. Gen. Physiol.*, **44**, 845–858. The effect of caffeine on radiocalcium movement.
Bianchi, C. P. (1968). *Cell Calcium*. Butterworths, London.
Bianchi, C. P., and Shanes, A. M. (1959). *J. Gen. Physiol.*, **42**, 803–815. Calcium influx in skeletal muscle at rest, during activity, and during potassium contracture.
Bihler, I. (1972). Ionic effect in the regulation of sugar transport in muscle. In *The Role of Membranes in Metabolic Regulation*. (Eds. Mehlmen, M. A. and R. W. Hanson), p. 411. Academic Press, New York.
Bikhazi, A. B., Nadir, N. S., and Hajjar, J. J. (1977). *J. Pharm. Sci.*, **66**, 1308–1312. Calcium–prostaglandin aggregation and its effect on prostaglandin uptake by isolated rabbit intestine.
Billah, M. M., and Michell, R. H. (1979). *Biochem. J.*, **182**, 661–668. Phosphatidylinositol metabolism in rat hepatocytes stimulated by glycogenolytic hormones. Effects of angiotensin, vasopressin, adrenaline, ionophore A23187 and calcium-ion deprivation.
Birch, J. R., and Pirt, S. J. (1971). *J. Cell Sci.*, **8**, 693–700. The quantitative glucose and mineral nutrient requirements of mouse LS (suspension) cells in chemically defined medium.
Birmingham, M. K., Elliott, F. H., and Valère, H. C. P. (1953). *Endocrinology*, **53**, 687–689. The need for the presence of calcium for the stimulation *in vitro* of rat adrenal glands by adrenocorticotrophic hormone.
Birmingham, M. K., Kurlents, E., Lane, R., Muhlstock, B., and Traikov, H. (1960). *Can. J. Biochem. Physiol.*, **38**, 1077–1085. Effects of calcium on the potassium and sodium content of adrenal glands, on the stimulation of steroid production by adenosine 3′,5′-monophosphate and on the response of the adrenal to short contact with ACTH.
Birnbaumer, L., and Rodbell, M. (1969). *J. Biol. Chem.*, **244**, 3477–3482. Adenyl cyclase in fat cells. II. Hormone receptors.
Bitensky, M. W., Miki, N., Keirns, J. J., Keirns, M., Baraban, J. M., Freeman, J., Wheeler, M. A., Lacy, J., and Marcus, F. R. (1975). *Adv. Cyclic Nucleotide Res.*, **5**, 213–240. Activation of photoreceptor disk membrane phosphodiesterase by light and ATP.

Bittar, E. E., Chambers, G., and Schultz, R. (1976). *J. Physiol. (London)*, **257**, 561–579. Mode of stimulation by adenosine 3′:5′-cyclic monophosphate of the sodium efflux in barnacle muscle fibres.
Bjerrum, J. (1941). *Metal Amine Formation in Aqueous Solution*. Hasse, Copenhagen.
Black, L., and Berenbaum, M. C. (1964). *Exp. Cell Res.*, **35**, 9–13. Factors affecting the dye exclusion test for cell viability.
Blackmore, P. F., Brumley, F. T., Marks, J. T., and Exton, J. H. (1978). *J. Biol. Chem.*, **253**, 4851–4858. Studies on α-adrenergic activation of hepatic glucose output.
Blaschko, H. and Welch, A. D. (1953). *Arch. exper. Path. u. Pharmakol.*, **219**, 17–22. Localization of adrenaline in cytoplasmic particles of bovine adrenal medulla.
Blaustein, M. P. (1974). *Rev. Physiol. Biochem. Pharmacol.*, **70**, 33–82. Interrelationships between sodium and calcium fluxes across cell membranes.
Blaustein, M. P., and Goldman, D. E. (1966). *Science*, **153**, 429. Action of anionic and cationic nerve-blocking agents: experiment and interpretation.
Blaustein, M. P., and Hodgkin, A. L. (1969). *J. Physiol. (London)*, **400**, 497–527. The effect of cyanide on the efflux of calcium from squid axons.
Blaustein, M. P., and Weismann, W. P. (1970). *Proc. Natl. Acad. Sci.*, **66**, 664–671. Effect of sodium ions on calcium movements in isolated synaptic terminals.
Blaustein, M. P., Russell, J. M., and Weer, P. (1974). *J. Supramol Struct.*, **2**, 558–581. Calcium efflux from internally dialysed squid axons: the influence of external and internal cations.
Blayney, L., Thomas, H., Muir, J., and Henderson, A. (1977). *Biochim. Biophys. Acta*, **470**, 128–133. Critical re-evaluation of murexide technique in the measurement of calcium transport by cardiac sarcoplasmic reticulum.
Blinks, J. R. (1978). *Photochem. Photobiol.*, **27**, 423–432. Application of calcium-sensitive photoproteins in experimental biology.
Blinks, S. R., Prendegast, F., and Allen, D. G. (1976). *Pharmacol. Rev.*, **28**, 1–93. Photoproteins as biological calcium indicators.
Blinks, J. R., Rüdel, R., and Taylor, S. R. (1978). *J. Physiol. (London)*, **277**, 291–323. Calcium transients in isolated amphibian skeletal muscle fibres: detection with aequorin.
Blioch, Z. L., Glagoleva, I. M., Liberman, E. A., and Nenashev, V. A. (1968). *J. Physiol. (London)*, **199**, 11–35. A study of the mechanism of quantal transmitter release at a chemical synapse.
Blitstein-Willinger, E., and Diamantstein, T. (1978). *Immunology*, **34**, 303–307. Inhibition by Isoptin[R] (a calcium antagonist) of the mitogenic stimulation of lymphocytes prior to the S-phase.
Blitz, A. L., Fine, R. E., and Toselli, Q. A. (1977). *J. Cell Biol.*, **75**, 135–147. Evidence that coated vesicles isolated from brain are calcium-sequestering organelles resembling sarcoplasmic reticulum.
Blondin, G. A., Kessler, R. J., and Green, D. E. (1977). *Proc. Natl. Acad. Sci. U.S.A.*, **74**, 3667–3671. Isolation of an electrogenic K^+/Ca^{2+} ionophore from an ionophore protein of beef heart mitochondria.
Bloom, B. R. (1980). *Nature (Lond.)*, **287**, 275–276. Immunological changes in multiple sclerosis.
Bloom, S., and Cancilla, P. A. (1969). *Am. J. Pathol.*, **54**, 373–391. Myocytolysis and mitochondrial calcification in rat myocardium after low doses of isoproterenol.
Blow, A. M. J. (1978). *FEBS Lett.*, **94**, 305–310. Water and calcium ions in cell fusion induced by poly (ethylene glycol).
Blow, A. M. J., Botham, G. M., and Lucy, J. A. (1979). *Biochem. J.*, **182**, 555–563. Calcium ions and cell fusion.
Blum, J. J. (1967). *Proc. Natl. Acad. Sci. U.S.A.*, **58**, 81–88. An adrenergic control system in *Tetrahymena*.
Blum, R. M., and Hoffman, J. F. (1972). *Biochem. Biophys. Res. Commun.*, **46**, 1146–1152. Ca-induced K transport in human red cells: localisation of the Ca-sensitive site to the inside of the membrane.
Bockaert, J., Roy, C., and Jard, S. (1972). *J. Biol. Chem.*, **247**, 7073–7081. Oxytocin-sensitive adenylate cyclase in frog bladder epithelial cells. Role of calcium, nucleotides, and other factors in hormonal stimulation.
Bodensteiner, J. B., and Engel, A. G. (1978). *Neurology*, **28**, 439–446. Intracellular calcium accumulation in Duchenne dystrophy and other myopathies: a study of 567,000 muscle fibres in 114 biopsies.
Bogart, B. I., Conod, E. J., and Cohover, J. H. (1977). *Pediatr. Res.*, **11**, 131–134. The biologic activities of cystic fibrosis serum. I. The effects of cystic fibrosis sera and calcium ionophore A23187 on rabbit treated explants.

Bolton, T. B. (1979). *Physiol. Rev.*, **59**, 606–718. Mechanism of action of transmitters and other substances on smooth muscle.
Bond, G. H., and Clough, D. C. (1973). *Biochim. Biophys. Acta*, **323**, 592–599. A soluble protein activator of ($Mg^{2+} + Ca^{2+}$)-dependent ATPase in red cell membranes.
Bondi, A. Y. (1978). *J. Pharmacol. Exp. Ther.*, **205**, 49–57. Effects of verapamil on excitation–contraction coupling in frog sartorius muscle.
Bonne, D., Belhadj, O., and Cohen, P. (1977). *Eur. J. Biochem.*, **75**, 101–105. Modulation by calcium of the insulin action and of the insulin-like effect of oxytocin on isolated rat lipocytes.
Bonne, D., Belhadj, O., and Cohen., P. (1978). *Eur. J. Biochem.*, **86**, 261–266. Calcium as a modulator of the hormone-receptors–biological-response coupling system. Effects of Ca^{2+} ions on the insulin activated 2-deoxyglucose transport in fat cells.
Bonner, J. T. (1971). *Annu. Rev. Microbiol.*, **25**, 78–92. Aggregation and differentiation in the cellular slime molds.
Bookchin, R. M., and Lew, V. L. (1980). *Nature (London)*, **284**, 561–563. Progressive inhibition of the Ca pump and Ca:Ca exchange in sickle red cells.
Borle, A. B. (1968). *J. Cell Biol.*, **36**, 567–582. Calcium metabolism in HeLa cells and the effects of parathyroid hormone.
Borle, A. B. (1972). *J. Membrane Biol.*, **10**, 45–66. Kinetic analysis of calcium movements in cell culture. V. Intracellular calcium distribution in kidney cells.
Borle, A. B. (1973). *Fed. Proc.*, **32**, 1944–1950. Calcium movement at the cellular level.
Borle, A. B. (1974). *Annu. Rev. Physiol.*, **36**, 361–390. Calcium and phosphate metabolism.
Borle, A. B. (1975). *Methods Enzymol.*, **39**, 513–573. Methods of assessing hormone effect on calcium fluxes *in vitro*.
Boszormenyi-Nagy, I. (1955). *J. Biol. Chem.*, **212**, 495–499. Formation of phosphopyruvate from phosphoglycerate in haemolyzed human erythrocytes.
Botelho, S. Y., and Dartt, D. A. (1980). *J. Physiol. (London)*, **304**, 397–403. Effect of calcium antagonism or chelation on rabbit lacrimal gland secretion and membrane potentials.
Boucek, M. M., and Snyderman, R. (1976). *Science*, **193**, 905–907. Calcium influx requirement for human neutrophil chemotaxis: inhibition by lanthanum chloride.
Bourke, R. S., and Tower, D. B. (1966). *J. Neurochem.*, **13**, 1071–1097. Fluid compartmentation and electrolytes of cat cerebral cortex *in vitro*. I. Swelling and solute distribution in mature cerebral cortex.
Boveris, A., Cadenas, E., Reiter, R., Filipkowski, M., Nakase, Y., and Chance, B. (1980). *Proc. Natl. Acad. Sci. U.S.A.*, **77**, 347–351. Organ chemiluminescence: non-invasive assay for oxidative radical reactions.
Boyer, P. D., Lardy, H. A., and Phillips, P. H. (1942). *J. Biol. Chem.*, **146**, 673–682. The role of potassium in muscle phosphorylation.
Boyer, P. D., Lardy, H. A., and Phillips, P. H. (1943). *J. Biol. Chem.*, **149**, 529–541. Further studies on the role of potassium and other ions in the phosphorylation of the adenylic system.
Boyle, M. D. P., Ohanian, S. H. and Borsos, T. (1976). *J. Immunol.*, **116**, 1276–1279. Studies on the terminal stages of antibody-complement mediated killing of a tumour cell. II. Inhibition of transformation of T* to dead cells.
Brachet, J., and Denis-Donini, S. (1977). *C. R. Acad. Sci.*, **284**, 1091–1096. Effect de divers agents sur la maturation et la différentiation sans clivage chez l'oeuf de *Chaetoptère*.
Bradford, H. F. (1970). *Brain Res.*, **19**, 239–247. Metabolic response of synaptosomes to electrical stimulation: release of amino acids.
Bradham, L. S. (1972). *Biochim. Biophys. Acta*, **276**, 434–443. Comparison of the effects of Ca^{2+} and Mg^{2+} on the adenyl cyclase of beef brain.
Bradham, L. S., Holt, D. A., and Sims, M. (1970). *Biochim. Biophys. Acta*, **201**, 250–260. The effect of Ca^{2+} on the adenyl cyclase of calf brain.
Brandt, P. W., and Freeman, A. R. (1967). *J. Colloid Interface Sci.*, **25**, 47–56. The role of surface chemistry in the biology of pinocytosis.
Brecht, K., and Gebert, G. (1966). *Experientia*, **22**, 713. Direct action of extracellular calcium ions on skeletal muscle.
Brehm, P., and Eckert, R. (1978). *Science*, **202**, 1203–1206. Calcium entry leads to inactivation of calcium channel in *Paramecium*.
Brehm, P., Morin, J. G., and Reynolds, G. T. (1973). *Biol. Bull. (Woods Hole, Mass.)*, **145**, 426. Bioluminescent characteristics of the ophiuroid *Ophiopsila californica*.
Brehm, P., Dunlap, K., and Eckert, R. (1978). *J. Physiol. (London)*, **274**, 639–654. Calcium-dependent repolarization in *Paramecium*.

Bresciani, F., and Auricchio, F. (1962). *Cancer Res.*, **22**, 1284–1289. Subcellular distribution of some metallic cations in the early stages of liver carcinogenesis.
Bresciani, F., Nola, E., Sica, V., and Puca, G. A. (1973). *Fed. Proc.*, **32**, 2126–2132. Early stages in estrogen control of gene expression and its derangement in cancer.
Brewer, G. J. (1980). *Am. J. Hematol.*, **8**, 231–248. Calmodulin, zinc and calcium in cellular and membrane regulation: an interpretive review.
Brezina, M., and Zuman, P. (1958). *Polarography in Medicine, Biochemistry and Pharmacy.* Interscience, New York.
Brien, P. (1968). The sponges, or Porifera. In *Chemical Zoology*, Vol. 2, (Eds. M. Florkin and B. T. Scheer), pp. 1–30. Academic Press, New York and London.
Brink, F. (1954). *Pharmacol. Rev.*, **6**, 243–298. The role of calcium ions in neural processes.
Brinley, F. J. (1968). *J. Gen. Physiol.*, **51**, 445–447. Sodium and potassium fluxes in barnacle muscle fibres.
Brinley, F. J. (1978). *Annu. Rev. Biophys. Bioeng.*, **7**, 363–392. Calcium buffering in squid axons.
Brinley, F. J., Jr., and Scarpa, A. (1975). *FEBS Lett.*, **50**, 82–85. Ionised magnesium concentration in axoplasm of dialysed squid axons.
Brinley, F. J., Scarpa, A., and Tiffert, T. (1977). *J. Physiol. (London)*, **266**, 545–565. The concentration of ionised magnesium in barnacle muscle fibres.
Brinley, F. J., Tiffert, T., and Scarpa, A. (1979). *J. Gen. Physiol.*, **72**, 101–127. Mitochondria and other calcium buffers of squid axon studied *in situ*.
Brokaw, C. J. (1972). *Science*, **178**, 455–462. Flagellar movement: A sliding filament model.
Brostrom, C. O., Hunkeler, F. L., and Krebs, E. G. (1971). *J. Biol. Chem.*, **246**, 1961–1967. The regulation of skeletal muscle phosphorylase kinase by Ca^{2+}.
Brostrom, C. O., Huang, Y.-C., Breckenridge, B.McL., and Wolff, D. J. (1975). *Proc. Natl. Acad. Sci. U.S.A.*, **72**, 64–68. Identification of calcium-binding protein as a calcium-dependent regulator of brain adenylate cyclase.
Brostrom, M. A., Brostrom, C. O., Breckenridge, B. M., and Wolff, D. J. (1976). *J. Biol. Chem.*, **251**, 4744–4750. Regulation of adenylate cyclase from glial tumour cells by calcium and a calcium binding protein.
Brostrom, M. A., Brostrom, C. O., and Wolff, D. J. (1978). *Arch. Biochem. Biophys.*, **191**, 341–350. Calcium-dependent adenylate cyclase from rat cerebral cortex: activation by guanine nucleotides.
Brown, A. M., Baur, P. S., and Tuley, F. H. (1975). *Science*, **188**, 157–160. Phototransduction in *Aplysia* neurons: calcium release from pigmented granules is essential.
Brown, E. M., Gardner, D. G., and Aurbach, G. D. (1980). *Endocrinology*, **106**, 133–138. Effects of calcium ionophore A-23187 in dispered bovine parathyroid cells.
Brown, H. M., and Brown, A. M. (1972). *Science*, **178**, 755–756. Ionic basis of the photoresponse of *Aplysia* giant neuron: K^+ permeability increase.
Brown, H. M., Pemberton, J. P., and Owen, J. D. (1976). *Anal. Chim. Acta*, **85**, 261–276. A calcium-sensitive microelectrode suitable for intracellular measurement of calcium (II) activity.
Brown, J. E., and Blinks, J. R. (1974). *J. Gen. Physiol.*, **64**, 643–655. Changes in intracellular free calcium concentration during illumination of invertebrate photoreceptors, detection with Aequorin.
Brown, J. E., Coles, J. A., and Pinto, L. H. (1977a). *J. Physiol. (London)*, **269**, 707–722. Effects of injections of calcium and EGTA into the outer segments of retinal rods of *Bufo marinus*.
Brown, J. E., Coles, J. A., and Pinto, L. H. (1977b). *J. Physiol. (London)*, **267**, 299–320. Detection of light-induced changes in intracellular ionised calcium concentration in *Limulus* ventral photoreceptors using arsenazo III.
Bruns, D. E., McDonald, J. M., and Jarett, L. (1976). *J. Biol. Chem.*, **251**, 7191–7197. Energy-dependent calcium transport in endoplasmic reticulum of adipocytes.
Bruns, D. E., McDonald, J. M., and Jarett, L. (1977). *J. Biol. Chem.*, **252**, 927–932. Properties of passive binding of calcium to endoplasmic reticulum from adipocytes.
Budesinsky, B. (1969). Monoarylazo and bis(arylazo) derivatives of chromotropic acid as photometric reagents. In *Chelates in Analytical Chemistry*, Vol. 2 (Eds. A. Flaschka and A. S. Barnard), pp. 1–91. Marcel Dekker, New York.
Bülbring, E., and Tomita, T. (1970). *J. Physiol. (London)*, **210**, 217–232. Effect of Ca removal on the smooth muscle of the guinea-pig taenia coli.
Bülbring, E., and Tomita, T. (1977). *Proc. R. Soc. London Ser. B*, **197**, 271–284. Calcium requirement for the α-action of catecholamines on guinea-pig taenia coli.
Burchfield, J. D. (1975). *Lord Kelvin and the Age of the Earth*. Macmillan, London.

Burgoyne, L. A., Waquar, A. M., and Atkinson, M. R. (1970). *Biochem. Biophys. Res. Commun.*, **39**, 918–922. Initiation of DNA synthesis in rat thymus: correlation of calcium-dependent initiation in thymocytes and in isolated thymus nuclei.
Burnstock, G. (1972). *Pharmacol. Rev.*, **24**, 510–581. Purinergic nerves.
Burnstock, G. (1977). *Fed. Proc.*, **36**, 2434–2438. Cholinergic, adrenergic and purinergic neuromuscular transmission.
Burnstock, G., and Wong, H. (1978). *Br. J. Pharmacol.*, **62**, 293–302. Comparison of the effect of ultraviolet light and purinergic nerve stimulation on the guinea-pig taenia coli.
Burridge, W. (1922). *Arch. Int. Pharmacodyn. Ther.*, **26**, 115–128. Experiments with cocaine.
Burström, H. G. (1968). *Biol. Rev.*, **43** 287–316. Calcium and plant growth.
Burton, R. F. (1973a). *Biol. Rev.*, **48**, 195–231. The significance of ionic concentrations in the internal media of animals.
Burton, R. F. (1973b). *Comp. Biochem. Physiol.*, **44A**, 781–792. The balance of cations in the plasma of vertebrates and its significance in relation to the properties of cell membranes.
Burton, R. F., and London, J. R. (1972). *J. Physiol. (London)*, **220**, 363–381. The antagonistic actions of calcium and magnesium on the superfused ventricle of the snail *Helix pomatia*.
Butcher, F. R. (1975). *Metabolism*, **24**, 409–418. The role of calcium and cyclic nucleotides in α-amylase release from slices of rat parotid: studies with the divalent cation ionophore A23187.
Butcher, R. W. (1933). *J. Ecol.*, **21**, 58–91. Studies on the ecology of rivers. I. On the distribution of macrophytic vegetation in the rivers of Britain.
Butler, J. N. (1968). *Biophys. J.*, **8**, 1426–1433. The thermodynamic activity of calcium ions in sodium chloride electrolytes.
Bygrave, F. L. (1966a). *Biochem. J.*, **101**, 480–487. The effect of calcium ions on the glycolytic activity of Ehrlich ascites tumour cells.
Bygrave, F. L. (1966b). *Biochem. J.*, **101**, 488–494. Studies on the interaction of metal ions with pyruvate kinase from Ehrlich ascites tumour cells and from rabbit muscle.
Bygrave, F. L. (1967). *Nature (London)*, **214**, 667–671. The ionic environment and metabolic control.
Bygrave, F. L. (1977). *Curr. Top. Bioenerg.*, **6**, 260–318. Mitochondrial calcium transport.
Bygrave, F. L. (1978). *Biol. Rev.*, **53**, 43–80. Mitochondria and the control of intracellular calcium.
Bywater, E. G. L., and Glynn, L. E. (1970). Connective tissue disorders. In *Biochemical Disorders in Human Disease* (Eds. R. H. S. Thompson and I. D. P. Wooton), pp. 723–750. J. & A. Churchill, London.
Caldwell, P. C. (1968). *Physiol. Rev.*, **48**, 1–64. Factors governing movement and distribution of inorganic ions in nerve and muscle.
Caldwell, P. C., and Lea, T. J. (1973). *J. Physiol. (London)*, **232**, 4P–5P. Use of intracellular glass scintillator for the continuous measurement of the uptake of ^{14}C-labelled glycine into squid giant axons.
Calkins, G. N. (1933). *The Biology of the Protozoa*. 2nd edn. Baillière, Tindall and Cox, Philadelphia.
Cameron, G. R. (1956). *New Pathways in Cellular Pathology*. Edward Arnold, London.
Campbell, A. K. (1970). Ph.D. Thesis, University of Cambridge.
Campbell, A. K. (1974). *Biochem. J.*, **143**, 411–418. Extraction, partial purification and properties of the Ca^{2+}-activated luminescent protein obelin from the hydroid *Obelia geniculata*.
Campbell, A. K., and Dormer, R. L. (1978). *Biochem. J.*, **176**, 53–66. Inhibition by calcium of cyclic AMP formation in sealed pigeon erythrocyte 'ghosts': a study using the photoprotein obelin.
Campbell, A. K., and Hales, C. N. (1971). *Exp. Cell Res.*, **68**, 33–42. Maintenance of viable cells in an organ culture of mature rat liver.
Campbell, A. K., and Hales, C. N. (1976). Some aspects of intracellular regulation. In *The Cell in Medical Science*, Vol. 4 (Eds. F. Beck and J. B. Lloyd), pp. 105–151. Academic Press, London and New York.
Campbell, A. K., and Luzio, J. P. (1981). *Experienta*, **37**, 1110–1112. Intracellular free calcium as a pathogen in cell damage initiated by the immune system.
Campbell, A. K., and Siddle, K. (1976). *Biochem. J.*, **158**, 211–221. The effect of intracellular calcium ions on adrenaline-stimulated adenosine 3':5'-cyclic monophosphate concentrations in pigeon erythrocytes: studied by using ionophore A23187.
Campbell, A. K., and Siddle, K. (1978). *Mol. Cell. Endocrinol.*, **11**, 79–89. Effect of replacement of extracellular sodium ions and of D-600 on the activation by adrenalin of adenylate cyclase in intact pigeon erythrocytes.

Campbell, A. K., and Simpson, J. S. A. (1979). Chemi- and bio-luminescence as an analytical tool in biology. In *Techniques in Life Sciences, B213* (Ed. C. Pogson), pp. 1–56. Elsevier/North-Holland, Amsterdam.
Campbell, A. K., Daw, R. A., and Luzio, J. P. (1979a). *FEBS Lett.*, **107**, 55–60. Rapid increase in intracellular Ca^{2+} induced by antibody plus complement.
Campbell, A. K., Lea, T. J., and Ashley, C. C. (1979b). Coelenterate photoproteins. In *Detection and Measurement of Free Ca^{2+} in Cells* (Eds. C. C. Ashley and A. K. Campbell), pp. 13–72. Elsevier/North-Holland, Amsterdam.
Campbell, A. K., Daw, R. A., Hallet, M. B., and Luzio, J. P. (1981). *Biochem. J.*, **194**, 551–560. Direct measurement of the increase in intracellular free calcium ion concentration in response to the action of complement.
Campbell, B. J., Woodward, G., and Borberg, B. (1972). *J. Biol. Chem.*, **247**, 6167–6175. Calcium mediated interactions between the antidiuretic hormone and renal plasma membranes.
Cappell, D. F., and Anderson, J. R. (1975). Miscellaneous tissue degradations and deposits. In *Muirs Textbook of Pathology*, pp. 207–209. Edward Arnold, London.
Carafoli, E. (1974). *Biochem. Soc. Symp.*, **39**, 89–110. Mitochondrial uptake of calcium ions and the regulation of cell function.
Carafoli, E., and Crompton, M. (1976). *Symp. Soc. Exp. Biol.*, **30**, 89–115. Calcium ion and mitochondria.
Carafoli, E., Hansford, R. G., Sacktor, B., and Lehninger, A. L. (1971). *J. Biol. Chem.*, **246**, 964–972. Interaction of Ca^{2+} with blowfly flight muscle mitochondria.
Carafoli, E., Crompton, M., and Malmström, K. (1977). *Horm. Cell Regul.*, **1**, 157–166. The reversibility of mitochondrial Ca^{2+} transport and its physiological implications.
Carbonne, E. (1979). Aequorin and fluorescent chelating probes to detect free calcium influx and membrane-associated calcium in excitable cells. In *Detection and Measurement of Free Ca^{2+} in Cells* (Eds. C. C. Ashley and A. K. Campbell), pp. 355–371. Elsevier/North-Holland Amsterdam.
Carchman, R. A., Jaanus, S. A., and Rubin, R. P. (1971). *Mol. Pharmacol.*, **7**, 491–499. The role of adrenocorticotropin and calcium in adenosine cyclic 3′, 5′-phosphate production and steroid release from the isolated, perfused cat adrenal gland.
Carlisle, D. B., and Knowles, F. (1959). *Endocrine control in crustaceans*. Cambridge University Press.
Carrasco, L., and Smith, A. E. (1976). *Nature (London)*, **264**, 807–809. Sodium ions and the shut-off of host cell protein synthesis by picorna viruses.
Carraway, K. L., Triplett, R. B., and Anderson, D. R. (1975). *Biochim. Biophys. Acta*, **379**, 571–581. Calcium-promoted aggregation of erythrocyte membrane proteins.
Carruthers, A., and Simons, A. J. B. (1978). *J. Physiol. (London)*, **284**, 49P. Calcium and ionophore A23187 stimulate sugar transport in pigeon red cells.
Carsten, M. E., and Miller, J. D. (1978). *Arch. Biochem. Biophys.*, **185**, 282–283. Comparison of calcium association constants and ionophoretic properties of some prostaglandins and ionophores.
Case, J. F., and Strause, L. G. (1978). Neurally controlled luminescent systems. In *Bioluminescence in Action* (Ed. P. J. Herring), pp. 331–366. Academic Press, London.
Case, R. M. (1978). *Biol. Rev.*, **53**, 211–354. Synthesis, intracellular transport and discharge of exportable proteins in the pancreatic acinar cell and other cells.
Case, R. M., and Clausen, T. (1973). *J. Physiol. (London)*, **235**, 75–102. The relationship between calcium exchange and enzyme secretion in the isolated rat pancreas.
Cassidy, P., Hoar, P. E., and Kerrick, W. G. C. (1980). *Pflügers Arch.*, **387**, 115–120. Inhibition of Ca^{2+}-activated tension and myosin light chain phosphorylation in skinned smooth muscle strips by the phenothiazines.
Caswell, A. H. (1979). *Int. Rev. Cytol.*, **56**, 145–187. Methods of measuring intracellular calcium.
Caulfield, J. B., and Schrag, B. A. (1964). *Am. J. Pathol.*, **44**, 365–381. Electron microscope study of renal calcification.
Chacko, S., Conti, M. A., and Adelstein, R. S. (1977). *Proc. Natl. Acad. Sci. U.S.A.*, **74**, 129–133. Effect of phosphorylation of smooth muscle myosin on actin activation and on Ca^{2+} regulation.
Chambers, E. L. (1974). *Biol. Bull.*, **147**, 471. Effects of ionophores on marine eggs and cation requirements for activation.
Chambers, R. (1928). *Harvey Lect.*, (1926–1927) 41–58. Nature of living cell as revealed by microdissection.

Chambers, R., and Reznikoff, P. (1926). *J. Gen. Physiol.*, **8**, 396–401. Micrurgical studies in cell physiology. I. The action of the chlorides of Na, K, Ca and Mg on the protoplasm of *Amoeba proteus*.
Chan, T. M., and Exton, J. H. (1977). *J. Biol. Chem.*, **252**, 8645–8651. α-Adrenergic-mediated accumulation of adenosine 3′:5′-monophosphate in calcium-depleted hepatocytes.
Chance, B. (1965). *J. Biol. Chem.*, **240**, 2729–2748. The energy-linked reaction of calcium with mitochondria.
Chance, B., Sies, H., and Boveris, A. (1979). *Physiol. Rev.*, **59**, 527–605. Hydroperoxide metabolism in mammalian organs.
Chandler, J. A. (1977). *X-ray Microanalysis in the Electron Microscope*. North-Holland, Amsterdam.
Chang, K. Y., and Carr, C. W. (1968). *Biochim. Biophys. Acta*, **157**, 127–139. The binding of calcium with deoxyribonucleic acid and deoxyribonucleic acid–protein complexes.
Chapman-Andresen, C. (1977). *Physiol. Rev.*, **57**, 371–385. Endocytosis in freshwater Amoebas.
Cheng, S. C., and Chen, S. S. (1975). *Life Sci.*, **16**, 1711–1716. Stimulation by cyclic nucleotides of calcium efflux in barnacle muscle fibres.
Cheung, W. Y. (1967). *Biochem. Biophys. Res. Commun.*, **29**, 478–482. Cyclic 3′:5′-nucleotide phosphodiesterase: pronounced stimulation by snake venom.
Cheung, W. Y. (1980). *Science*, **207**, 19–27. Calmodulin plays a pivotal role in cell regulation.
Cheung, W. Y., Bradham, L. S., Lynch, T. J., Lin, Y. M., and Tallaut, E. A. (1975). *Biochem. Biophys. Res. Commun.*, **66**, 1055–1062. Protein activator of cyclic 3′:5′-nucleotide phosphodiesterase of bovine or rat brain also activates adenylate cyclase.
Chi, Y., and Francis, D. (1971). *J. Cell Physiol.*, **77**, 169–174. Cyclic AMP and calcium efflux in a cellular slime mould.
Chien, K. R., and Farber, J. L. (1977). *Arch. Biochem. Biophys.*, **180**, 191–198. Microsomal membrane dysfunction in ischemic rat liver cells.
Chien, K. R., Abrams, J., Pfau, R. G., and Farber, J. L. (1977). *Am. J. Pathol.*, **88**, 539–558. Prevention by chlorpromazine of ischemic liver cell death.
Chien, K. R., Abrams, J., Serroni, A., Martin, J. T., and Farber, J. L. (1978). *J. Biol. Chem.*, **253**, 4809–4817. Accelerated phospholipid degradation and associated membrane dysfunction in irreversible, ischemic liver cell injury.
Chien, K. R., Pfau, R. G., and Farber, J. L. (1979). *Am. J. Pathol.*, **97**, 505–530. Ischaemic myocardial cell injury.
Chilinger, G. V., and Bissel, H. J. (1963). *J. Paleontol.*, **37**, 942–943. Note of possible reason for scarcity of calcareous skeletons of invertebrates in pre-Cambrian formations.
Chock, P. B., Rhee, S. G., and Stadtman, E. R. (1980). *Ann. Rev. Biochem.*, **49**, 813–843. Interconvertible enzyme cascades.
Chrisman, T. D., Garbers, D. L., Parks, M. A., and Hardman, J. G. (1975). *J. Biol. Chem.*, **250**, 374–381. Characterisation of particulate and soluble guanylate cyclase from rat lung.
Chubb, I. W., DePotter, W. P., and DeSchaepdryner, A. F. (1972). *Naunyn-Schmiedebergs Arch. Pharmakol.*, **274**, 281–286. Tyramine does not release nor-adrenaline from splenic nerve by exocytosis.
Claret-Berthon, B., Claret, M., and Mazet, J. L. (1977). *J. Physiol. (London)*, **272**, 529–555. Fluxes and distribution of calcium in rat liver cells: kinetic analysis and identification of pools.
Clark, E. P., and Collip, J. B. (1925). *J. Biol. Chem.*, **63**, 461–464. A study of the Tisdall method for the determination of blood serum calcium with a suggested modification.
Clark, G. W. (1921). *J. Biol. Chem.*, **49**, 487–517. The micro determination of calcium in whole blood, plasma and serum by direct precipitation.
Clarke, M., and Spudich, J. A. (1977). *Annu. Rev. Biochem.*, **46**, 797–822. Non-muscle contractile proteins: the role of actin and myosin in cell motility and shape determination.
Claus, T. H., and Pilkis, S. J. (1976). *Biochim. Biophys. Acta*, **421**, 246–262. Regulation by insulin of gluconeogenesis in isolated rat hepatocytes.
Clausen, T. (1975). *Curr. Top. Membr. Transp.*, **6**, 169–226. The effect of insulin on glucose transport in muscle cells.
Clausen, T. (1980). *Cell Calcium*, **1**, 311–325. The role of calcium in the activation of the glucose transport system.
Clausen, T., Elbrink, J., and Dahl-Han, A. B. (1975). *Biochim. Biophys. Acta*, **375**, 292–308. The relationship between the transport of glucose and cations across cell membranes in isolated tissues. IX. The role of cellular calcium in the activation of the glucose transport system in rat soleus muscle.

Clemente, F., and Meldolesi, J. (1975). *J. Cell Biol.*, **65**, 88–102. Calcium and pancreatic secretion. I. Subcellular distribution of calcium and magnesium in the exocrine pancreas of the guinea pig.
Close, R. I. (1972). *Physiol. Rev.*, **52**, 129–197. Dynamic properties of mammalian skeletal muscle.
Clusin, W., Spray, D. C., and Bennett, M. V. L. (1975). *Nature (London)*, **256**, 425–427. Activation of a voltage-insensitive conductance by inward calcium current.
Cobbold, P. H. (1980). *Nature (London)*, **285**, 441–446. Cytoplasmic free calcium and amoeboid movement.
Cochrane, D. E., and Douglas, W. W. (1974). *Proc. Natl. Acad. Sci. U.S.A.*, **71**, 408–412. Calcium induced extrusion of secretory granules (exocytosis) in mast cells exposed to 48/80 or the ionophore A23187 and X537A.
Cockcroft, S., Bennett, J. P., and Gomperts, B. D. (1980). *Nature (London)*, **288**, 275–277. Stimulus–secretion coupling in rabbit neutrophils is not mediated by phosphatidyl inositol breakdown.
Coffino, P., Gray, J. W., and Tomkins, G. M. (1975). *Proc. Natl. Acad. Sci. U.S.A.*, **72**, 878–882. Cyclic AMP, a nonessential regulator of the cell cycle.
Cohen, J. B., Weber, M., and Changeux, J.-P. (1974). *Mol. Pharmacol.*, **10**, 904. Effects of local anaesthetics and calcium on the interaction of cholinergic ligands with the nicotinic receptor protein from *Torpedo marmorata*.
Cohen, P. (1973). *Eur. J. Biochem.*, **34**, 1–14. The subunit structure of rabbit-skeletal-muscle phosphorylase kinase, and the molecular basis of its activation reactions.
Cohen, P. (1974). *Biochem. Soc. Symp.*, **39**, 51–73. The role of phosphorylase kinase in the nervous and hormonal control glycogenolysis in muscle.
Cohen, P., Burchell, A., Foulkes, J. G., and Cohen, P. T. W. (1978). *FEBS Lett.*, **92**, 287–293. Identification of the Ca^{2+}-dependent modulator protein as the fourth subunit of rabbit skeletal muscle phosphorylase kinase.
Cohen, S. M., and Burt, T. M. (1977). *Proc. Natl. Acad. Sci. U.S.A.*, **74**, 4271–4275. ^{31}P nuclear magnetic relaxation studies on phosphocreatine in intact muscle: determination of intracellular free magnesium.
Cohn, M. (1963). *Biochemistry*, **2**, 623–629. Magnetic resonance studies of metal activation of enzymic reactions of nucleotides and other phosphate substrates.
Cohn, M. (1970). *Q. Rev. Biophys.*, **3**, 61–89. Magnetic resonance studies of enzyme–substrate complexes with paramagnetic probes as illustrated by creatine kinase.
Cole, K. S. (1949). *Proc. Natl. Acad. Sci. U.S.A.*, **35**, 558–566. Some physical aspects of bioelectric phenomena.
Cole, K. S., and Curtis, H. J. (1939). *J. Gen. Physiol.*, **22**, 649–670. Electric impedance of the squid giant axon during activity.
Collins, J. H. (1976). *Soc. Symp. Exptl. Biol.*, **30**, 303–334. Structure and evolution of troponin C and related proteins.
Coman, D. R. (1954). *Cancer Res.*, **14**, 519–521. Cellular adhesiveness in relation to invasiveness of cancer: electron microscopy of liver perfused with chelating agent.
Condeelis, J. S., and Taylor, D. L. (1977). *J. Cell Biol.*, **74**, 901–921. The contractile basis of amoeboid movement. V. The control of gelation, solation, and contraction in extracts from *Dictyostelium discoideum*.
Cone, R. A. (1972). *Nature (London)*, **236**, 39–43. Rotational diffusion of rhodopsin in the visual receptor membrane.
Constantin, L. L. (1968). *J. Physiol. (London)*, **195**, 119–132. The effect of calcium on contraction and conductance thresholds in frog skeletal muscle.
Constantin, L. L., Franzini-Armstrong, C., and Podolsky, R. J. (1965). *Science*, **147**, 158–159. Localisation of calcium-accumulating structures in striated muscle fibres.
Conway, E. J. (1943). *Proc. R. Irish Acad.*, **48B**, 161–212. The chemical evolution of the ocean.
Conway, E. J. (1945). *Biol. Rev.*, **20**, 56–72. The physiological significance of inorganic levels in the internal medium of animals.
Cook, B. J., Holman, G. M., and Marks, E. P. (1975). *J. Insect Physiol.*, **21**, 1807–1814. Calcium and cyclic AMP as possible mediators of neurohormone action in the hindgut of the cockroach *Leucophaea maderae*.
Cooper, R. H., Randle, P. J., and Denton, R. M. (1974). *Biochem. J.*, **742**, 625–641. Regulation of heart muscle pyruvate dehydrogenase kinase.
Copp, D. H. (1969). The ultimobranchial glands and calcium regulation. In Fish Physiology, Vol. II (Eds. W. S. Hoar and D. J. Randall), pp. 377–398. Academic Press, New York and London.

Copp, D. H. (1970). *Annu. Rev. Physiol.*, **32**, 61–86. Endocrine regulation of calcium metabolism.
Copp, D. H. (1976). Comparative endocrinology of calcitonin. In *Handbook of Physiology*, Vol. VII (Ed. G. D. Aurbach), pp. 431–442. American Physiological Society, Washington, DC.
Cormier, M. J. (1978). In *Bioluminescence in Action* (Ed. P. J. Herring), pp. 75–108. Academic Press, London. Comparative biochemistry of animal systems.
Cosmos, E. (1962). *Anal. Biochem.*, **3**, 90–94. Autoradiography of Ca^{45} in ashed sections of frog skeletal muscle.
Covino, B. G., and Vassallo, H. G. (1976). *Local Anaesthetics—Mechanism of Action and Clinical Use*. Grune and Stratton, New York.
Cox, J. S. G. (1967). *Nature (London)*, **216**, 1328–1329. Disodium cromoglycate (FPL 670) (Intal): a specific inhibitor of reaginic antibody–antigen mechanisms.
Cranefield, P. F., Aronson, R. S., and Wit, A. L. (1974). *Circ. Res.*, **34**, 204–213. Effect of verapamil on the normal action potential and on a calcium-dependent slow response of canine cardiac Purkinje fibres.
Crompton, M., and Carafoli, E. (1979). The application of calcium-selective electrodes and inhibitor-stop techniques to the resolution of calcium fluxes in mitochondria. In *Detection and Measurement of Free Ca^{2+} in Cells* (Eds. C. C. Ashley and A. K. Campbell), pp. 373–382. Elsevier/North-Holland, Amsterdam.
Crompton, M., Capano, M., and Carafoli, E. (1976). *Eur. J. Biochem.*, **69**, 453–462. The sodium-induced efflux of calcium from heart mitochondria.
Curry, D. L., Bennett, L. L., and Grodsky, G. M. (1968). *Am. J. Physiol.*, **214**, 174–178. Requirement for calcium ion in insulin secretion by the perfusing rat pancreas.
Curtis, A. S. G. (1962). *Biol. Rev.*, **37**, 82–129. Cell contact and cell adhesion.
Czech, M. P. (1977). *Annu. Rev. Biochem.*, **46**, 359–384. Molecular basis of insulin action.
Dabrowska, R., and Hartshorne, D. J. (1978). *Biochem. Biophys. Res. Commun.*, **85**, 1352–1359. A Ca^{2+}- and modulator-dependent myosin light chain kinase from non-muscle cells.
Dabrowska, R., Aromatorio, D., Sherry, J. M. F., and Hartshorne, D. J. (1977). *Biochem. Biophys. Res. Commun.*, **78**, 1267–1272. Composition of the myosin light chain kinase from chicken gizzard.
Dagley, S., and Dawes, E. A. (1955). *Biochim. Biophys. Acta*, **17**, 177–184. Citric desmolase: its properties and mode of action.
D'Agostino, A. N. (1964). *Am. J. Pathol.*, **45**, 633–644. An electron microscopic study of cardiac necrosis produced by 9-alpha-fluorocortisol and sodium phosphate.
D'Agostino, A. N., and Chiga, M. (1970). *Am. J. Clin. Pathol.*, **53**, 820–824. Mitochondrial mineralization in human myocardium.
Dahl, G., and Gratzl, M. (1976). *Cytobiologie*, **12**, 344–355. Calcium induced fusion of isolated secretory vesicles from the islet of Langerhans.
Dahlstrom, A. (1971). *Philos. Trans. R. Soc. London Ser. B*, **261**, 325–358. Axoplasmic transport (with particular respect to adrenergic neurones).
Daimon, T., Mizuhira, V., and Uchida, K. (1978). *Histochem.* **55**, 271–279. Ultrastructural localisation of calcium around the membrane of the surface connected system in the human platelet.
Dalcq, A. (1925). *Arch. Biol. (Paris)*, **34**, 507–574. Récherches experimentales et cytologiques sur la maturation et l'activation de l'oeuf d'*Asterias glacialis*.
Dale, H. (1934). *Br. Med. J.*, **i**, 835–841. Chemical transmission of the effects of nerve impulses.
Dally, D. K., and Gray, W. D. (1974). *Can. J. Microbiol.*, **20**, 935–936. Inorganic ion content of fungal mycelia and spores.
Dalquist, R. (1974). *Acta Pharmacol. Toxicol.*, **35**, 1–10. Determination of ATP-induced ^{45}calcium uptake in rat mast cells.
Dambach, N. G., and Friedmann, N. (1974). *Biochim. Biophys. Acta*, **332**, 374–386. The effect of varying ionic composition of the perfusate on liver membrane potential, gluconeogenesis and cyclic AMP.
Dan, K. (1947). *Biol. Bull.*, **93**, 267–286. Electrokinetic studies of marine ova. V. Effects of pH changes on the surface potentials of sea-urchin eggs.
Danielli, J. F. (1943). *J. Exp. Biol.*, **20**, 167–176. The biological action of ions and concentration of ions at surfaces.
Davis, E. (1931). *J. Physiol. (London)*, **71**, 431–441. Relations between the actions of adrenaline, acetylcholine and ions on the perfused heart.
Davis, J. O., and Freeman, R. H. (1976). *Physiol. Rev.*, **56**, 1–56. Mechanisms regulating renin release.

Davy, H. (1808a). *Philos. Trans. R. Soc. London*, **98**, 1–44. On some new phenomena of chemical changes produced by electricity, particularly the decomposition of the fixed alkalies and the exhibition of the new substances which constitute their bases.

Davy, H. (1808b). *Philos. Soc. Trans. R. Soc. London*, **98**, 333–370. Electro-chemical researches, on the decomposition of the earths; with observations on the metals obtained from the alkaline earths, and on the amalgam from ammonium.

Davy, H. (1810). *Philos. Trans. R. Soc. London*, **100**, 16–74. On some new electrochemical researches on various objects, particularly metallic bodies, from the alkalies, and earths, and on some combinations of hydrogene.

Dawes, C. M. (1975). *Biol. Rev.*, **50**, 351–371. Acid–base relationships within the avian egg.

Dawson, R. M. C., and Hauser, H. (1970). Ca^{2+} binding to phospholipids. In *Calcium and Cellular Function* (Ed. A. W. Cuthbert), pp. 17–41. Cambridge University Press, Cambridge.

Day, F. H. (1963). *The Chemical Elements in Nature*. George Harrup, London.

Dean, P. M. (1975). *J. Theor. Biol.*, **54**, 289–308. Exocytosis modelling: an electrostatic function for calcium in stimulus–secretion coupling.

Dean, P. M., and Matthews, E. K. (1970a). *J. Physiol. (London)*, **210**, 255–264. Glucose-induced electrical activity in pancreatic islet cells.

Dean, P. M., and Matthews, E. K. (1970b). *J. Physiol. (London)*, **210**, 265–272. Electrical activity in pancreatic islet cells: effect of ions.

Dean, P. M., and Matthews, E. K. (1972). *J. Physiol. (London)*, **225**, 1–13. Pancreatic acinar cells: measurement of membrane potential and immature depolarisation potentials.

Dedman, J. R., Potter, J. D., and Means, A. R. (1977). *J. Biol. Chem.*, **252**, 2435–2440. Biological cross-reactivity of rat testis phosphodiesterase activator protein and rabbit skeletal muscle troponin-C.

Dedman, J. R., Welsh, M. J., and Means, A. R. (1978). *J. Biol. Chem.*, **253**, 7515–7521. Ca^{2+}-dependent regulator. Production and characterisation of a monospecific antibody.

De Duve, C. (1963). *Ciba Found. Symp.*, 126. Definition of exocytosis. Eds. A. V. S. de Reuck and M. P. Cameron.

Deery, D. J. (1974). *Biochem. Biophys. Res. Commun.*, **60**, 326–333. A calcium supported adenylyl cyclase activity in the pars distalis of the dogfish pituitary.

De Kruijff, B., Verkely, A. J., van Echted, C. J., Gerritsen, W. J., Mombers, C., Noordan, P. C., and De Girer, J. (1979). *Biochim. Biophys. Acta*, **555**, 200–209. The occurrence of lipidic particles in lipid bilayers as seen by ^{31}P NMR and freeze fracture electron-microscopy.

Del Castillo, J., and Katz, B. (1956). *Prog. Biophys.*, **6**, 121–170. Biophysical aspects of neuromuscular transmission.

Del Castillo, J., and Stark, L. (1952). *J. Physiol. (London)*, **116**, 507–515. The effect of calcium on the motor end-plate potentials.

Deleze, J., and Loewenstein, W. R. (1976). *J. Membr. Biol.*, **28**, 71–86. Permeability of a cell junction during intracellular injection of divalent cations.

deLong, R. P., Coman, D. R., and Zeidman, I. (1950). *Cancer*, **3**, 718–721. Significance of low calcium and high potassium content in neoplastic tissue.

De Lorenzo, R. J., Freedman, S. D., Yoke, W. B., and Maurer, S. C. (1979). *Proc. Natl. Acad. Sci. U.S.A.*, **76**, 1838–1842. Stimulation of Ca^{2+}-dependent neurotransmitter release and presynaptic nerve terminal protein phosphorylation by calmodulin and a calmodulin-like protein isolated from synaptic vesicles.

DeLuca, H., Gran, F. C., and Steenbock, H. (1957). *J. Biol. Chem.*, **224**, 201–208. Vitamin D and citrate oxidation.

DeLuca, H., Engstrom, G., and Rasmussen, H. (1962). *Proc. Natl. Acad. Sci. U.S.A.*, **48**, 1604–1609. The action of vitamin D and parathyroid hormone *in vitro* on calcium uptake and release by kidney mitochondria.

DeMello, W. (1974). *Fed. Proc.*, **33**, 445. Intracellular Ca injection and cell communication in heart.

DeMello, W. (1975). *Experientia*, **31**, 460–461. Uncoupling of heart cells produced by intracellular sodium injection.

Denton, R. M. (1977). *Horm. Cell Regul.*, **1**, 167–170. Pools of free and bound calcium in mitochondria.

Denton, R. M., Richards, D. A., and Chin, J. G. (1978). *Biochem. J.*, **176**, 899–906. Calcium ions and the regulation of NAD^+-linked isocitrate dehydrogenase from the mitochondria of rat heart and other tissues.

Denton, R. M., McCormack, J. G., and Edyell, N. J. (1980). *Biochem. J.*, **190**, 107–117. Role of calcium ions in the regulation of intramitochondrial metabolism.

Derksen, A., and Cohen, P. (1975). *J. Biol. Chem.*, **250**, 9342–9347. Patterns of fatty acid release from endogenous substrates by human platelet homogenates and membranes.
DeRubertis, F. R., and Craven, P. A. (1976). *J. Biol. Chem.*, **251**, 4651–4658. Properties of the guanylate cyclase–guanosine 3′:5′:monophosphate system of rat renal cortex. Activation of guanylate cyclase and calcium-independent modulation of guanosine 3′:5′-monophosphate by sodium azide.
Devis, G., Somers, G., VanObberghen, E., and Malaisse, W. J. (1975). *Diabetes*, **24**, 547–551. Calcium antagonists and islet cell function. I. Inhibition of insulin release by verapamil.
Devore, D. I., and Nastuk, W. L. (1977). *Nature (London)*, **270**, 441–443. Ionophore-mediated calcium influx effects on the post-synaptic muscle fibre membrane.
Diamantstein, T., and Odenwald, M. V. (1974). *Immunology*, **27**, 531–541. Control of the immune response *in vitro* by calcium ions.
Diamond, J. (1973). *Am. J. Physiol.*, **225**, 930–937. Phosphorylase, calcium and cyclic AMP in smooth-muscle contraction.
Dickens, F., and Greville, G. D. (1935). *Biochem. J.*, **29**, 1468–1483. Metabolism of normal and tumour tissue; neutral salt effects.
Dicker, S. E. (1966). *J. Physiol. (London)*, **185**, 429–444. Release of vasopressin and oxytocin from isolated pituitary glands of adult and new born rats.
Diegel, J., Cunningham, M., and Coburn, R. E. (1980). *Biochim. Biophys. Acta*, **619**, 482–493. Calcium dependence of prostaglandin release from guinea pig taenia coli.
Dikstein, S., Weller, C. P., and Sulman, F. G. (1963). *Nature (London)*, **200**, 1106. Effect of calcium ions on melanophore dispersal.
DiPolo, R. (1973). *Biochim. Biophys. Acta*, **298**, 279–283. Sodium-dependent calcium influx in dialysed barnacle muscle fibres.
DiPolo, R. (1978). *Nature (London)*, **274**, 340. Ca pump driven by ATP in squid axons?
Dixon, J. K., Weith, A. J., Argyle, A. A., and Salley, D. J. (1949). *Nature (London)*, **163**, 845. Measurement of the adsorption of surface-active agent at a solution–air interface by a radiotracer method.
Dixon, M., and Webb, E. C. (1964). *The Enzymes.* Longmans, London.
Dixon, R. O. D. (1969). *Annu. Rev. Microbiol.*, **23**, 137–158. Rhizobia.
Dodge, F. A., and Rahamimoff, R. (1967). *J. Physiol. (London)*, **193**, 419–432. Co-operative action of calcium ions in transmitter release at the neuromuscular junction.
Donnellan, J. F., and Beechey, R. B. (1969). *J. Insect Physiol.*, **15**, 367–372. Factors affecting the oxidation of glycerol-1-phosphate by insect flight muscle mitochondria.
Doree, M., Moreau, M., and Guerrier, P. (1978). *Expl. Cell Res.*, **275**, 251–260. Hormonal control of meiosis: *in vitro* induced release of calcium ions from the plasma membrane in starfish oocytes.
Dorman, D. M., Barritt, G. J., and Bygrave, F. L. (1975). *Biochem. J.*, **150**, 389–395. Stimulation of hepatic mitochondrial calcium transport by elevated plasma insulin concentrations.
Dormer, R. L., and Ashcroft, S. J. H. (1974). *Biochem. J.*, **144**, 543–550. Studies on the role of calcium ions in the stimulation by adrenaline of amylase release from rat parotid.
Dormer, R. L., Newman, G. R., and Campbell, A. K. (1978). *Biochim. Biophys. Acta*, **538**, 87–105. Preparation and characterization of liposomes containing the calcium activated photoprotein obelin.
Dormer, R. L., Poulsen, J. H., Licko, V., and Williams, J. A. (1981). *Am. J. Physiol*, **240**, G38–G49. Calcium fluxes in isolated pancreatic acini: effects of secretagogues.
Douglas, W. W. (1974). *Biochem. Soc. Symp.*, **39**, 1–28. Involvement of calcium in exocytosis and the exocytosis–vesiculation sequence.
Douglas, W. W., and Poisner, A. M. (1963). *J. Physiol. (London)*, **165**, 528–541. The influence of calcium on the secretory response of the submaxillary gland to acetyl choline or to noradrenaline.
Douglas, W. W., and Poisner, A. M. (1964). *J. Physiol. (London)*, **172**, 1–18. Stimulus–secretion coupling in a neurosecretory organ: the role of calcium in the release of vasopressin from the neurohypophesis.
Douglas, W. W., and Rubin, R. P. (1961). *J. Physiol. (London)*, **159**, 40–57. The role of calcium in the secretory response of the adrenal medulla to acetylcholine.
Douglas, W. W., and Rubin, R. P. (1963). *J. Physiol. (London)*, **167**, 288–310. The mechanism of catecholamine release from the adrenal medulla and the role of calcium in stimulus–secretion coupling.
Douglas, W. W., Kanno, T., and Sampson, S. R. (1961). *J. Physiol. (London)*, **191**, 107–121. Effects of acetyl choline and other medullary secretagogues and antagonists on the membrane potential of adrenal chromaffin cells: an analysis employing techniques of tissue culture.

Drahota, Z., Carafoli, E., Rossi, C. S., Gamble, R. L., and Lehninger, A. L. (1965). *J. Biol. Chem.*, **240**, 2712–2720. The steady state maintenance of accumulated Ca^{2+} in rat liver mitochondria.
Draper, M. H., and Weidmann, S. (1951). *J. Physiol. (London)*, **115**, 74–94. Cardiac resting and action potentials recorded with an intracellular electrode.
Dreher, K. L., Eaton, J. W., Kuettner, J. F., Breslawec, K. P., Blackshear, P. L., and White, J. G. (1978). *Am. J. Pathol.*, **92**, 215–216. Retention of water and potassium by erythrocytes prevents calcium-induced membrane rigidity.
Dreifuss, J. J., Kalnins, I., Kelly, J. S., and Ruf, K. B. (1971). *J. Physiol. (London)*, **215**, 805–807. Action potentials and release of neurohypophyseal hormones *in vitro*.
Drummond, G. I., and Duncan, L. (1966). *J. Biol. Chem.*, **241**, 3097–3103. The action of calcium ion on cardiac phosphorylase *b* kinase.
Duane, M. (1974). Interactions of metal ions with nucleic acid. In *Metal Ions in Biological Systems*, Vol. 3 (Ed. H. Sigel), pp. 2–45. Marcel Dekker, New York.
DuBois, K. P., and Potter, V. R. (1943). *J. Biol. Chem.*, **150**, 185–195. The assay of animal tissues for respiratory enzymes. III. Adenosine triphosphatase.
Duffus, J. H., and Patternson, L. J. (1974). *Nature (London)*, **251**, 626–627. Control of cell division in yeast using ionophore A23187 with calcium and magnesium.
Dugal, B. S., and Göpel, G. (1975). *Biochim. Biophys. Acta*, **410**, 407–413. Effect of calcium ions on pyruvate carboxylase from pigeon liver.
Dulbecco, R. (1975). *Proc. R. Soc. London Ser. B*, **189**, 1–14. The control of cell growth by regulation by tumour-inducing viruses: a challenging problem.
Dulbecco, R., and Elkington, J. (1975). *Proc. Natl. Acad. Sci. U.S.A.*, **72**, 1584–1588. Induction of growth in resting fibroblastic cell cultures by Ca^{2+}.
Dumont, J. E., and Lamy, F. (1980). The regulation of thyroid cell metabolism, function growth and differentiation. In *Comprehensive Endocrinology: the Thyroid Gland* (Ed. M. De Visscher), pp. 153–167. Raven Press, New York.
Dumont, J. E., and Mockel, J. (1977). *Biochem. Soc. Trans.*, **5**, 883–884. Calcium ions, cyclic guanosine monophosphate and the regulation of thyroid membrane function.
Dumont, J. E., Boeynaems, J. M., Van Saude, J., Erneux, C., Decoster, C., Van Cauter, E., and Mockel, J. (1977). *Horm. Cell Regul.*, **1**, 171–194. Cyclic AMP, calcium and cyclic GMP in the regulation of thyroid function.
Duncan, C. J. (Ed.) (1976). *Symp. Soc. Exp. Biol.*, **30**. Calcium in biological systems.
Duncan, C. J. (1978). *Experientia*, **34**, 1531–1535. Role of intracellular calcium in promoting muscle damage: a strategy for controlling the dystrophic condition.
Dunham, E. T., and Glynn, I. M. (1961). *J. Physiol. (London)*, **156**, 274–293. Adenosine-triphosphatase activity and the active movements of alkali metal ions.
Durham, A. C. H. (1974). *Cell*, **2**, 123–136. A unified theory of the control of actin and myosin in non-muscle movements.
Durham, A. C. H., and Hendry, D. A. (1977). *Virology*, **77**, 510–533. Cation binding by tobacco mosaic virus.
Durham, A. C. H., and Ridgway, E. B. (1976). *J. Cell Biol.*, **69**, 218-223. Control of chemotaxis in *Physarum polycephalum*.
Dvorak, J. A., and Hyde, T. P. (1973). *Exp. Parasitol.*, **34**, 268–283. *Trypanosoma curzi* interaction with vertebrate cells *in vitro*. I. Individual interactions at the cellular and subcellular level.
Eagle, H. (1956). *Arch. Biochem. Biophys.*, **61**, 356–366. Nutrition needs of mammalian cells in tissue culture.
Eaton, J. W., Skelton, T. D., Swofford, H. S., Kolpin, C. E., and Jacob, H. S. (1973). *Nature (London)*, **246**, 105–106. Elevated erythrocyte calcium in sickle cell disease.
Ebashi, S. (1960). *J. Biochem.*, **48**, 150–151. Calcium binding and relaxation in actomyosin system.
Ebashi, S. (1961). *J. Biochem.*, **50**, 236–244. Calcium binding activity of vesicular relaxing factor.
Ebashi, S. (1963). *Nature (London)*, **200**, 1010. Third component participating in the superprecipitation of 'natural actomyosin'.
Ebashi, S. (1974). *Essays Biochem.*, **10**, 1–36. Regulatory mechanism of muscle contraction with special reference to the Ca-troponin–tropomyosin system.
Ebashi, S. (1976) *Annu. Rev. Physiol.*, **38**, 293–314. Excitation–contraction coupling.
Ebashi, S. (1980). *Proc. R. Soc. London Ser. B.* **207**, 259–286. Regulation of muscle contraction.
Ebashi, S., and Lipmann, F. (1962). *J. Cell Biol.*, **14**, 389–400. Adenosine triphosphate-linked concentration of calcium ions in a particular fraction of rabbit muscle.
Eckert, R., and Brehm, P. (1979). *Annu. Rev. Biophys. Bioeng.*, **8**, 353–383. Ionic mechanisms of excitation in Paramecium.

Eckert, R., Naitohy, Y., and Machemer, H. (1976). *Symp. Soc. Exp. Biol.*, **30**, 233–255. Calcium in the bioelectric and motor functions of *Paramecium*.

Eichhorn, G. L., Clark, P., and Tarien, E. (1969). *J. Biol. Chem.*, **244**, 937–942. The interaction of metal ions with polynucleotides and related compounds.

Einstein, A. (1906). *Ann. Phys.*, **19**, 371–381. Zür Theorie der Brownschen Bewegung.

Elkeles, T. S., Lazarus, J. H., Siddle, K., and Campbell, A. K. (1975). *Clin. Sci. Mol. Med.*, **48**, 27–31. Plasma adenosine 3':5'-cyclic monophosphate response to glucagon in thyroid disease.

Ellar, D. J., Eaton, M. W., and Postgate, J. (1974). *Biochem. Soc. Trans.*, **2**, 947–948. Calcium release and germination of bacterial spores.

Elliot, D. A., and Rizack, M. A. (1974). *J. Biol. Chem.*, **249**, 3985–3990. Epinephrine and adrenocorticotropic hormone-stimulated magnesium accumulation in adipocytes and their plasma membranes.

Empson, R. N., Bucci, T. J., Morrissey, R. L., and Chandler, J. S. (1976). *Am. J. Pathol.*, **82**, 55a–56a. Restriction of renal lesions induced by vitamin D_3 to tubules containing calcium-binding protein.

Endo, M. (1972). *Nature New Biol.*, **237**, 211–213. Stretch-induced increase in activation of skinned muscle fibres by calcium.

Endo, M. (1977). *Physiol. Rev.*, **57**, 71–108. Calcium release from the sarcoplasmic reticulum.

Enfield, D. L., Ericsson, L. H., Blum, H. F., Fischer, E. H., and Neurath, H. (1975). *Proc. Natl. Acad. Sci. U. S. A.*, **72**, 1309–1313. Amino-acid sequence of parvalbumin from rabbit skeletal muscle.

Engbaek, L. (1952). *Pharmacol. Rev.*, **4**, 396–414. The pharmacological actions of magnesium ions with particular reference to the neuromuscular and cardiovascular system.

Engelhardt, W. A., and Ljabimova, M. N. (1939). *Nature (London)*, **144**, 668–669. Myosin and adenosine triphosphatase.

Erlanger, B. F. (1976). *Annu. Rev. Biochem.*, **45**, 267–283. Photoregulation of biologically active macromolecules.

Esmon, C. T., and Jackson, C. M. (1974). *J. Biol. Chem.*, **249**, 7791–7797. The conversion of prothrombin to thrombin. IV. The function of the fragment of 2 region during activation in the presence of factor V.

Eto, S., Wood, J. M., Hutchins, M., and Fleischer, N. (1974). *Am. J. Physiol.*, **226**, 1315–1320. Pituitary $^{45}Ca^{2+}$ uptake and release of ACTH, GH and TSH: effect of verapamil.

Ettienne, E. M. (1970). *J. Gen. Physol.*, **56**, 168–179. Control of contractility in *Spirostomum* by dissociated calcium.

Ettienne, E. (1972). *J. Cell Biol.*, **54**, 179–184. Subcellular localisation of calcium repositories in plasmodia of the acellular slime mould *Physarum polycephalum*.

Eubesi, F., Miledi, R., and Takahasi, T. (1980). *Nature (London)*, **284**, 560–561. Calcium transients in mammalian muscles.

Everett, J. L., Day, C. L., and Bergel, F. (1964). *J. Pharm. Pharmacol.*, **16**, 85–90. Analysis of August rat liver for calcium, copper, iron, magnesium, manganese, molybdenum, potassium, sodium and zinc.

Fabiato, A., and Fabiato, F. (1977). *Circ. Res.*, **40**, 119–129. Calcium release from the sarcoplasmic reticulum.

Fain, J. N., and Berridge, M. J. (1979). *Biochem. J.*, **180**, 665–661. Relationship between phosphatidyl inositol synthesis and recovery of 5-hydroxytryptamine-responsive Ca^{2+} flux in blowfly salivary glands.

Fain, J. N., Tolbert, M. E. M., Pointer, R. H., Butcher, F. R., and Arnold, A. (1975). *Metabolism*, **24**, 395–407. Cyclic nucleotides and gluconeogenesis by rat liver cells.

Fairbridge, R. W. (1972). The Encyclopedia of Geochemistry and Environmental Sciences: Encyclopedia of Earth Sciences Series Vol. IV A. Van Nostrand Reinhold Co., New York.

Fanburg, B., and Gergely, J. (1965). *J. Biol. Chem.*, **240**, 2721–2728. Studies on adenosine triphosphate-supported calcium accumulation by cardiac subcellular particles.

Farber, J. L., and El-Mofty, S. K. (1975). *Am. J. Pathol.*, **81**, 237–250. The biochemical pathology of liver cell necrosis.

Farese, R. V. (1971a). *Endocrinology*, **89**, 1057–1063. On the requirement for calcium during the steroidogenic effect of ACTH.

Farese, R. V. (1971b). *Science*, **173**, 447–450. Calcium as a mediator of adrenocorticotropic hormone action on adrenal protein synthesis.

Fatt, P., and Katz, B. (1953). *J. Physiol. (London)*, **120**, 171–204. The electrical properties of crustacean muscle fibres.

Fawcett, D. W., Long, J. A., and Jones, A. L. (1969). *Rec. Prog. Horm. Res.*, **25**, 315–380. The ultrastructure of endocrine glands.

Fay, F. S., Shelvin, H. H., Granger, W. C., and Taylor, S. R. (1979). *Nature (London)*, **280**, 506–508. Aequorin luminescence during activation of single isolated smooth muscle cells.
Fearon, D. T., and Austen, K. F. (1976). *Essays Med. Biochem.*, **2**, 1–35. The human complement system: biochemistry, biology and pathobiology.
Fedelesova, M., and Ziegelhöffer, A. (1975). *Experientia*, **31**, 518–520. Enhanced calcium accumulation related to increased protein phosphorylation in cardiac sarcoplasmic reticulum induced by cyclic $3',5'$-AMP or isoproterenol.
Feigal, R. J., and Shapiro, B. L. (1979). *Nature (London)*, **278**, 276–277. Mitochondrial calcium uptake and oxygen consumption in cystic fibrosis.
Feiman, R. D., and Detwiler, T. C. (1974). *Nature (London)*, **249**, 172–173. Platelet secretion induced by divalent cation ionophores.
Feinstein, M. B., and Fraser, C. (1974). *J. Gen. Physiol.*, **66**, 561–581. Human platelet secretion and aggregation induced by calcium ionophores. Inhibition by PGE_1 and dibutyryl cyclic AMP.
Feinstein, M. B., and Paimre, M. (1969). *Fed. Proc.*, **28**, 1643–1648. Pharmacological action of local anaesthetics on excitation–contraction coupling in striated and smooth muscle.
Feldman, H., Rodbard, D., and Slevine, D. (1972). *Anal. Biochem.*, **45**, 530–556. Mathematical theory of cross-reactive radioimmunoassay and ligand-binding systems at equilibrium.
Fine, D. P., Marney, S. R., Colley, D. G., Sergent, J. S., and Desprez, R. M. (1972). *J. Immunol.*, **109**, 807–809. C3 shunt activation in human serum chelated with EGTA.
Fischer, E. H., and Krebs, E. G. (1955). *J. Biol. Chem.*, **216**, 121–132. Conversion of phosphorylase *b* to phosphorylase *a* in muscle extracts.
Fisher, G., Kaneshiro, E. S., and Peters, P. D. (1976). *J. Cell Biol.*, **69**, 429–442. Divalent cation affinity sites in *Paramecium aurelia*.
Fiske, C. H., and Subbarow, Y. (1929). *Science*, **70**, 381–382. Phosphorus compounds of muscle and liver.
Flatman, P., and Lew, V. L. (1977). *Nature (London)*, **267**, 360–362. Use of ionophore A23187 to measure and to control free and bound cytoplasmic Mg in intact red cells.
Flatt, P. R., Boquist, L., and Hellman, B. (1980). *Biochem. J.*, **190**, 361–372. Calcium and pancreatic β-cell function. The mechanism of insulin secretion studied with the aid of lanthanum.
Fleckenstein, A. (1974). *Adv. Cardiol.*, **12**, 183–197. Drug-induced changes in cardiac energy.
Fleckenstein, A. (1977). *Annu. Rev. Pharmacol. Toxicol.*, **17**, 149–166. Specific pharmacology of calcium in myocardium, cardiac pacemakers, and vascular smooth muscle.
Fleischer, N., Zimmerman, G., Schindler, W., and Hutchins, M. (1972). *Endocrinology*, **91**,1436–1441. Stimulation of adrenocorticotropin (ACTH) and growth hormone (GH) release by ouabain, relationship to calcium.
Flier, J. S., Kahn, C. R., Roth, J., and Bar, R. S. (1975). *Science*, **190**, 63–65. Antibodies that impart insulin receptor binding in an unusual diabetic syndrome with severe insulin resistance.
Fletcher, J. M., Greenfield, B. F., Hardy, C. J., Scargill, D., and Woodhead, J. L. *J. Chem. Soc.*, **1961**, 2000–2006. Ruthenium red.
Florkin, M. (1975a). *Comprehensive Biochemistry*, Vol. 33.
Florkin, M. (1975b). The discovery of cell free fermentation. In *Comprehensive Biochemistry*, Vol. 33 (Eds. M. Florkin, and E. H. Stotz), p.33. Elsevier, Amsterdam.
Foden, S., and Randle, P. J. (1978). *Biochem. J.*, **170**, 615–625. Calcium metabolism in rat hepatocytes.
Foerder, C. A., Klebanoff, S. J., and Shapiro, B. M. (1978). *Proc. Natl. Acad. Sci. U.S.A.*, **75**, 3183–3187. Hydrogen peroxide production of chemiluminescence and the respiratory burst of fertilisation: intracellular events in early sea urchin development.
Fogg, G. E., Stewart, W. D. P., Fay, P., and Walsby, A. E. (1973). *The Blue-Green Algae*. Academic Press, London and New York.
Foldes, F. F. (Ed.) (1978). *Enzymes in Anesthesiology*. Springer Verlag, New York.
Ford, L. E., and Podolsky, R. J. (1970). *Science*, **167**, 58–59. Regenerative calcium release within muscle cells.
Foreman, J. C., and Garland, L. G. (1976). *Br. Med. J.*, **i**, 820–821. Cromoglycate and other antiallergic drugs: a possible mechanism of action.
Foreman, J. C., Garland, L. G., and Mongar, J. L. (1976). *Symp. Soc. Exp. Biol.*, **30**, 193–218. The role of calcium in secretory processes: model studies in mast cells.
Foreman, J. C., Hallett, M. B., and Mongar, J. L. (1977). *J. Physiol. (London)*, **271**, 193–214. The relationship between histamine secretion and 45-calcium uptake by mast cells.
Forskål, P. (1775). *Fauna Arabica*, pp.110–11. Heineck and Faber, London.

Frank, G. B. (1960). *J. Physiol. (London)*, **151**, 518–538. Effects of changes in extracellular calcium concentration on the potassium-induced contracture of frog's skeletal muscle.

Frank, M. M., Rapp, H. J., and Borsos, T. (1964). *J. Immunol.*, **93**, 409–413. Studies on the terminal steps of immune hemolysis. I. Inhibition by trisodium ethylene-diamine tetracetate (EDTA).

Frank, M. M., Rapp, H. T., and Borsos, T. (1965). *J. Immunol.*, **94**, 295–300. Studies on the terminal steps of immune hemolysis. II. Resolution of the E transformation reaction into multiple steps.

Frankenhauser, B. (1957). *J. Physiol. (London)*, **137**, 245–260. The effect of calcium on myelinated nerve fibre.

Frankenhauser, B., and Hodgkin, A. L. (1957). *J. Physiol. (London)*, **137**, 218–244. The action of calcium on the electrical properties of squid axons.

Freedman, M. H., and Ruff, M. C. (1975). *Nature (London)*, **255**, 378–382. Induction of increased calcium uptake in mouse T lymphocytes by concanavalin A and its modulation by cyclic nucleotides.

Freer, R. J., and Smith, A. B. (1979). *Am. J. Physiol.*, **236**, C171–C176. Lithium dissociation of calcium- and angiotensin-induced contractions in depolarised rat uterus.

Frei, E., and Zahler, P. (1979). *Biochim. Biophys. Acta*, **550**, 450–463. Phospholipase A_2 from sheep erythrocyte membranes. Ca^{2+} dependence and localization.

Friedmann, N. (1974). *Biochim. Biophys. Acta*, **274**, 214–225. Effects of glucagon and cyclic AMP on ion fluxes in the perfused liver.

Friedmann, N., and Park, C. R. (1968). *Proc. Natl. Acad. Sci. U.S.A.*, **61**, 505–508. Early effects of 3′,5′-adenosine monophosphate on the fluxes of calcium and potassium in the perfused liver of normal and adrenalectomised rats.

Friedmann, N., and Rasmussen, H. (1970). *Biochim. Biophys. Acta*, **222**, 41–50. Calcium, magnesium and hepatic gluconeogenesis.

Friesen, A. J. D., Oliver, N., and Allen, G. (1969). *Am. J. Physiol.*, **217**, 445–450. Activation of cardiac glycogen phosphorylase by calcium.

Fuchs, F. (1971). *Biochim. Biophys. Acta*, **245**, 221–229. Ion exchange properties of the calcium receptor site of troponin.

Fuchs, P., Gruber, E., Gitelman, J., and Kohn, A. (1980). *J. Cell. Physiol.*, **103**, 271–278. Nature of permeability changes in membranes HeLa cells adsorbing Sendai virus.

Fukuda, T. R. (1935). *J. Cell. Comp. Physiol.*, **7**, 301–311. Ionic antagonism in the water permeability of sea urchin eggs.

Fuller, G. M., Ellison, J., McGill, M., and Brinkley, B. R. (1975). *J. Cell Biol.*, **67**, 126a. The involvement of mitochondria and calcium in regulating microtubule assembly *in vitro*.

Fulton, C. (1963). *J. Cell. Comp. Physiol.*, **61**, 39. Rhythmic movements in *Cordylophora*.

Fulton, C. (1977). *Annu. Rev. Microbiol.*, **31**, 597–629. Cell differentiation in *Naegleria gruberii*.

Fyfe, W. S. (1974). *Geochemistry*. Clarendon Press, Oxford.

Gabbiani, G., and Montaudon, D. (1977). *Int. Rev.Cytol.*, **48**, 187–219. Reparative processes in mammalian wound healing: the role of contractile phenomena.

Gabella, G., and North, R. A. (1974). *J. Physiol. (London)*, **240**, 28P–30P. Proceedings: intracellular recording and electron microscopy of the same myenteric plexus neurone.

Gagliardino, J. J., Harrison, D. E., Christie, M. R., Gagliardino, E. E., and Ashcroft, S. J. H. (1980). *Biochem. J.*, **192**, 919–927. Evidence for the participation of calmodulin in stimulus–secretion coupling in the pancreatic β-cell.

Gail, M. H., Boore, C. W., and Thompson, C. S. (1973). *Exp. Cell Res.*, **79**, 386–390. A calcium requirement for fibroblast mobility and proliferation.

Gallagher, C. H., Koch, J. H., and Mann, D. M. (1965). *Biochem. Pharmacol.*, **14**, 799–811. The effect of phenothiazine on the metabolism of liver mitochondria.

Gallin, J. J., and Quie, P. G. (Eds.) (1978). *Leukocyte Chemotaxis: Methods, Physiology and Clinical Implications*. Raven Press, New York.

Gallin, J. I., and Rosenthal, A. S. (1973). *Fed. Proc.*, **32**, 819. Divalent cation requirements and calcium fluxes during human granulocyte chemotaxis.

Gallin, J. I., and Rosenthal, A. S. (1974). *J. Cell Biol.*, **62**, 594–609. The regulatory role of divalent cations in human granulocyte chemotaxis. Evidence for an association between calcium exchanges and microtubule assembly.

Galvani, L. (1791). *Commentary on the Effect of Electricity on Muscular Motion* (translated by R. M. Green, 1953). E. Licht, Cambridge, Massachusetts.

Gambetti, P., Erulkar, S. E., Somlyo, A. P., and Gonotas, N. K. (1975). *J. Cell Biol.*, **64**, 322–330. Calcium containing structures in vertebrate glial cells. Ultrastructural and microprobe analysis.

Garcia, A. G., Kirpekar, S. M., and Prat, A. (1975). *J. Physiol. (London)*, **244**, 253–262. A Ca^{2+}-ionophore stimulating the secretion of catecholamine from cat adrenal.
Garcia-Sainz, J. A., and Fain, J. W. (1980). *Biochem. J.*, **186**, 781–789. Effect of insulin, catecholamines and calcium ions on phospholipid metabolism in isolated white fat-cells.
Gardos, G. (1958). *Biochim. Biophys. Acta*, **30**, 653–654. The function of calcium in the potassium permeability of human erythrocytes
Garrison, J. C. (1978), *J. Biol. Chem.*, **253**, 7091–7100. The effects of glucagon, catecholamines, and the calcium ionophore A23187 on the phosphorylation of rat hepatocyte cytosolic proteins.
Gaskin, F., Kramer, S. B., Cantor, C. R., Adelstein, R. and Shelanski, M. L. (1974). *FEBS Lett.*, **40**, 281–286. A dynein-like protein associated with neurotubules.
Gautvik, K. M., and Tashjian, A. (1973). *Endocrinology*, **93**, 793–799. Effects of cations and colchicine on the release of prolactin and growth hormone by functional pituitary tumour cells in culture.
Geiger, A. (1940). *Biochem. J.*, **34**, 465–482. Glycolysis in cell-free extracts of brain.
Gemsa, D. (1979). *Exp. Cell Res.*, **118**, 55–62. Ionophore A23187 raises cyclic AMP levels in macrophages by stimulating prostaglandin E formation.
Gerschenfeld, H. M. (1973). *Physiol. Rev.*, **53**, 1–119. Chemical transmission in invertebrate central nervous systems and neuromuscular junctions.
Geschwind, I. I. (1970). Mechanism of action of hypothalamic adenohypophysiotropic factors. In *Hypophysiotropic Hormones of the Hypothalamus* (Ed. J. Meites), p. 298. Williams and Wilkins, Baltimore.
Gevers, W., and Krebs, H. A. (1966). *Biochem. J.*, **98**, 720–735. The effects of adenine nucleotides on carbohydrate metabolism in pigeon-liver homogenates.
Gibbons, B. H., and Gibbons, I. R. (1974). *J. Cell Biol.*, **63**, 970–985. Properties of flagellar 'rigor waves' formed by abrupt removal of adenosine triphosphate from actively swimming sea urchin sperm.
Gilbert, I. G. F. (1972). *Eur. J. Cancer*, **8**, 99–105. The effect of divalent cations on the ionic permeability of cell membranes in normal and tumour tissue.
Gilbert, D. L., and Ehrenstein, G. (1969). *Biophys. J.*, **9**, 447–463. Effect of divalent cations on potassium conductance of squid axons: determination of surface charge.
Gilbert, D. S (1975). *J. Physiol. (London)*, **253**, 303–319. Axoplasm chemical composition in *Myxicola* and solubility properties of its structural proteins.
Gilbert, D. S. (1977). *J. Physiol. (London)*, **266**, 81–83P. Neurofilament rings from giant axons.
Gilkey, J. C., Jaffe, L. F., Ridgway, E. B., and Reynolds, G. T. (1978). *J. Cell Biol.*, **76**, 448–466. A free calcium wave traverses the activating egg of the medaka, *Oryzias latipes*.
Gillespie, E. (1975). *Ann. N. Y. Acad. Sci.*, **253**, 771–779. Microtubules, cyclic AMP, calcium and secretion.
Gillespie, M., and Thornton, J. W. (1932). *J. Pharm. Exp. Ther.*, **45**, 419–426. Effect of calcium on response of isolated bronchi to histamine.
Gillet, C., and Lefebvre, J. (1978). *J. Exp. Bot.*, **29**, 1155–1159. Ionic diffusion through the *Nitella* cell wall in the presence of calcium.
Gilula, N. B., and Epstein, M. L. (1976). *Symp. Soc. Exp. Biol.*, **30**, 257–272. Cell-to-cell communication, gap junctions and calcium.
Gingell, D., Garrad, D. R., and Palmer, J. F. (1970). Divalent cations and cell adhesion. In *Calcium and Cellular Function* (Ed. A. W. Cuthbert), pp. 59–71. Macmillan, London.
Ginsburg, M., Jayasena, K., and Thomas, P. J. (1966). *J. Physiol. (London)*, **184**, 387–401. The preparation and properties of porcine neurophysin and the influence of calcium on the hormone-neurophysin complex.
Glaessner, M. F. (1962). *Biol. Rev.*, **37**, 467–494. Pre-cambrian fossils.
Glitsch, H. G., Reuter, H., and Scholz H. (1969). *Naunyn-Schmiedebergs Arch. Pharmacol.*, **264**, 236–237. Influence of intracellular sodium concentrations on calcium influx isolated guinea-pig auricles.
Godfraind-De Becker, A., and Godfraind, T. (1980). *Int. Rev. Cytol.*, **67**, 141–170. Calcium transport systems: a comparative study in different cells.
Godfraind, T., and Kaba, A. (1969). *Brit. J. Pharmacol.*, **36**, 549–560. Blockade or reversal of the contraction induced by calcium and adrenaline in depolarized arterial smooth muscle.
Goldberg, N. D., and Haddox, M. K. (1977). *Annu. Rev. Biochem.*, **48**, 823–896. Cyclic GMP metabolism and involvement in biological regulation.

Goldberg, N. D., Haddox, M. K., Dunham, E., Lopez, C., and Haddox, J. W. (1974). In the Cold Spring Harbor Symposium on the Regulation of Proliferation of Animal cells pp 609–675. Ed. Clarkson, B., and Baserga, R. Cold Spring Harbor Laboratory, New York. The Ying-Yang hypothesis of biological control-opposing influences of cyclic GMP and cyclic AMP in the regulation of cell proliferation and other biological processes. Cited in Goldberg, N. D., Haddox, M. K., Nicol. S. E., Glass, D. R., Sanford, C. H., Kuehe, FA. Jr., and Estensen. R. (1975), Adv. cyclic Nucl. Res., **5**, 307–330.

Goldman, D. E. (1943). *J. Gen Physiol.*, **27**, 37–60. Potential, impedance and rectification in membranes.

Gomez-Puyou, A., and Gomez-Lojero, C. (1977). *Curr. Top. Bioenerg.*, **6**, 222–259. The use of ionophore and channel formers in the study of the function of biological membranes.

Good, N. E., Winget, G. D., Winter, W., Connolly, T. N., Izauro, S., and Singh, R. M. M. (1966). *Biochemistry*, **5**, 467–477. Hydrogen ion buffers for biological research.

Goodenough, D. A. (1975). *Am. J. Clin. Pathol.*, **63**, 636–645. The structure of cell membranes involved in intercellular communication.

Goodford, P. J., (1967). *J. Physiol. (London)*, **792**, 145–157. The calcium content of the smooth muscle of guinea-pig taenia coli.

Gopinath, R. M., and Vincenzi, F. F. (1977). *Biochem. Biophys. Res. Commun.*, **77**, 1203–1209. Phosphodiesterase protein activator mimics red blood cell cytoplasmic activator of $(Ca^{2+} - Mg^{2+})$ ATPase.

Gordon, E. E., and Ferris, R. K. (1971). *Biochem. Pharmacol.*, **26**, 1089–1091. Stimulation of renal gluconeogenesis by verapamil and D-600.

Gordon, S., and Cohn, Z. A. (1973). *Int. Rev. Cytol.*, **36**, 171–214. The macrophage.

Goreau, T. J. (1977). *Proc. R. Soc. London Ser. B*, **196**, 291–315. Coral skeletal chemistry: physiological and environmental regulation of stable isotopes and trace metals in *Montrastrea annularis*.

Gorman, A. L. F., and Thomas, M. V. (1978). *J. Physiol. (London)*, **275**, 357–376. Changes in intracellular concentration of free calcium ions in a pace-maker neurone, measured with metallochromic indicator dye arsenazo III.

Goswami, A., and Rosenberg, I. N. (1978). *Endocrinology*, **103**, 2223–2233. Thyroid hormone modulation of epinephrine-induced lipolysis in rat adipocytes—possible role of calcium.

Granit, R. (1933). *J. Physiol. (London)*, **77**, 207–240. The components of the retinal action potential and their relation to the discharge in the optic nerve.

Gratzer, W. B., and Beaven, G. H. (1977). *Anal. Biochem.*, **81**, 118–129. Use of the metal-ion indicator, arsenazo III in the measurement of calcium binding.

Gratzl, M., and Dahl, G. (1976). *FEBS Lett.*, **62**, 142–145. Ca^{2+} induced fusion of Golgi derived secretory vesicles isolated from rat liver.

Gray, E. G. (1975). *Proc. R. Soc. London*, **190**, 367–372. Pre-synaptic microtubules and the association with synaptic vesicles.

Gray, J. (1922). *Proc. R. Soc. London Ser. B*, **93**, 122–131. The mechanism of ciliary movement. II. The effect of ions on the cell membrane.

Gray, J. (1924). *Proc. R. Soc. London Ser. B*, **96**, 95–114. The mechanism of ciliary movement. IV. The relation of ciliary activity to oxygen consumption.

Greaser, M. L., and Gergeley, J. (1971). *J. Biol. Chem.*, **246**, 4226–4233. Reconstitution of troponin activity from three protein components.

Green, H. S., and Casley-Smith, J. R. (1972). *J. Theor. Biol.*, **35**, 103–111. Calculations on the passage of small vesicles across endothelial cells by Brownian motion.

Greene, W. C., Parker, C. M., and Parker, C. W. (1976). *Cell. Immunol.*, **25**, 74–89. Calcium and lymphocyte activation.

Greengard, P. (1976). *Nature (London)*, **260**, 101–108. Possible role for cyclic nucleotides and phosphorylation of membrane proteins in post synaptic actions of neurotransmitters.

Greenwald, I. (1938) *J. Biol. Chem.*, **124**, 437–452. The dissociation of some calcium salts.

Greenwald, I. (1941). *J. Biol. Chem.*, **141**, 789–796. The dissociation of calcium and magnesium carbonates and bicarbonates.

Greenwald, I., Redish, J., and Kibrick, A. C. (1940). *J. Biol. Chem.*, **135**, 65–76. The dissociation of calcium and magnesium phosphates.

Grigoryeva, V. A., and Lytvynenko, O. O. (1973). *Ukr. Biokhem. Zh.*, **45**, 28 (cited in Wrogemann and Pena, 1976).

Grinwich, K. D., Alexander, T. S., and Cemy, J. (1978). *Cell. Immunol.*, **37**, 285–297. Evidence for a role for calcium in immunosuppression by tumour cells *in vitro*.

Grodsky, G. M., and Bennet, L. C. (1966). *Diabetes*, **15**, 910–913. Cation requirement for insulin secretion in the isolated perfused pancreas.
Gross, E. L., and Hess, S. C. (1974). *Biochim. Biophys. Acta*, **339**, 334–346. Correlation between calcium ion binding to chloroplast membranes and divalent cation-induced structural changes and changes in chlorophyll and fluorescence.
Guazzi, M., Olivari, M. T., Polese, A., Fiorentini, C., Magrini, F., and Moruzzi, P. (1977). *Clin. Pharmacol. Ther.*, **22**, 528–532. Nifedipine, a new anti-hypertensive with rapid action.
Guggenheim, E. A. (1935). *Philos. Mag.*,**19** (7th series), 588–643. The specific thermodynamic properties of aqueous solutions of strong electrolytes.
Güntelberg, E. (1926). *Z. Phys. Chem.*, **123**, 199–247. Untersuchungen über Ioneninteraktion.
Gupta, B. L., and Hall, T. A. (1978). *Ann. N. Y. Acad. Sci.*, **307**, 28–51. Electron microprobe X-ray analysis of calcium.
Habener, J. F. (1976). *Ciba Found. Symp.*, **41**, 197–224. New concepts in the formation, regulation of release and metabolism of parathyroid hormone.
Habener, J., Kemper, B. W. Rich, A., and Potts, J. T. (1977). *Rec. Prog. Horm. Res.*, **33**, 249–308. Biosynthesis of parathyroid hormone.
Haga, T., Abe, T., and Kurokawa, M. (1974). *FEBS Lett.*, **39**, 291–295. Polymerization and depolymerization of microtubules *in vitro* as studied by flow birefringence.
Hagins, W. A. (1972). *Annu. Rev. Biophys. Bioeng.*, **1**, 131–177. The visual process: excitatory mechanisms in the primary receptor cells.
Hagins, W. A., and Yoshikami, S. (1974). *Exp. Eye. Res.*, **18**, 299–305. A role for Ca^{2+} in excitation of retinal rods and cones.
Hagiwara, S. (1973). *Adv. Biophys.*, **4**, 71–102. Ca spike.
Hagiwara, S., Naka, K. I. and Chichibu, S. (1964). *Science*, **143**, 1446–1448. Membrane properties of barnacle muscle.
Hainaut, K., and Desmedt, J. E. (1974). *Nature (London)*, **252**, 728–730. Effect of dantrolene sodium on calcium movements in single muscle fibres.
Hainaut, K., and Desmedt, J. E. (1979). Steady and phasic changes of intracellular free calcium detected with aequorin in single muscle cells. In *Detection and Measurement of Free Ca^{2+} in Cells* (Eds. C. C. Ashley and A. K. Campbell), pp. 133–152. Elsevier/North-Holland, Amsterdam.
Haldane, J. B. S. (1930). *Enzymes*. Longmans, London.
Hales, C. N. (1968). The glucose–fatty acid cycle and diabetes mellitus. In *Biological Basis of Medicine*, Vol. 2, (Eds. E. E.Bittar and N. Bittar), pp. 309–338. Academic Press, New York.
Hales, C. N. (1978). *FEBS Lett.*, **94**, 10–16. Proteolysis and the evolutionary origin of polypeptide hormones.
Hales, C. N., and Milner, R. D. (1968). *J. Physiol. (London)*, **194**, 725–743. The role of sodium and potassium in insulin secretion from rabbit pancreas.
Hales, C. N., Luzio, J. P., Chandler, J. A., and Herman, L. (1974). *J. Cell Sci.*, **15**, 1–15. Localization of calcium in the smooth endoplasmic reticulum of rat isolated fat cells.
Hales, C. N., Campbell, A. K., Luzio, J. P., and Siddle, K. (1977). *Biochem. Soc. Trans.*, **5**, 866–872. Calcium as a mediator of hormone action.
Hales, C. N., Luzio, J. P., and Siddle, K. (1979). *Biochem. Soc. Symp.*, **43**, 97–135. Hormonal control of adipose tissue lipolysis.
Haljamäe, H., and Wood, D. C. (1971). *Anal. Biochem.*, **42**, 155–170. Analysis of picomole quantities of Ca, Mg, K, Na, and Cl in biological samples by ultramicro flame spectrometry.
Hallett, M. B., and Campbell, A. K. (1980). *Biochem. J.*, **192**, 587–596. Uptake of liposomes containing the photoprotein obelin by rat isolated adipocytes. Adhesion, endocytosis or fusion?
Hallett, M. B. and Campbell, A. K. (1982a). Applications of coelenterate luminescent proteins. In *Biomedical Applications of Chemiluminescence* (Eds. L. Kriska and T. N. Carter). Marcel Dekker, New York.
Hallett, M. B., and Campbell, A. K. (1982b): *Nature (London)*, **295**, 155–158. Measurement of changes in cytoplasmic free Ca^{2+} in fused cell hybrids.
Hallett, M. B., Luzio, J. P., and Campbell, A. K. (1981). *Immunology*, **44**, 569–576. Stimulation of Ca^{2+}-dependent chemiluminescence in rat polymorphonuclear leucocytes by polystyrene beads and the non-lytic action of complement.
Halsey, M. J., Millar, R. A., and Sutton, J. A. (1974). *Molecular Mechanisms in General Anaesthesia*. Churchill Livingstone, Edinburgh.
Halverson, H. O. (1963). *Soc. Gen. Micro Symp.*,**13**, 343–368. Sequential expression of biochemical events during intracellular differentiation.

Ham, A. W. (1965). *Histology*, 5th Edition. Pitman Medical Publishing, London, and J. P. Lippincott, Philadelphia.
Hamman, J. P., and Seliger, H. H. (1972). *J. Cell. Physiol.*, **80**, 397–408. The mechanical triggering of bioluminescence in marine dinoflagellates: chemical basis.
Hammerschlag, R., Bakhit, C., Chiu, A. Y., and Dravid, A. R. (1977). *J. Neurobiol.*, **8**, 439–451. Role of calcium in the initiation of fast axonal transport of protein: effects of divalent cations.
Hamon, M., Bourgoin, S., Artaud, F., and Hery, F. (1977). *J. Neurochem.*, **28**, 811–818. Rat brain stem tryptophan hydroxylase: mechanism of activation by calcium.
Hamon, M., Bourgoin, S., Hery, F., and Simonnet, G. (1978). *Mol. Pharmacol.*, **14**, 99–110. Activation of tryptophan hydroxylase by adenosine triphosphate, magnesium and calcium.
Hanlon, D. P., and Westhead, E. W. (1965). *Biochim. Biophys. Acta*, **96**, 537–540. Conformation changes of yeast phosphopyruvate hydratase (enolase) induced by activating and inhibiting metal ions.
Hansford, R. G., and Chappell, J. B. (1967). *Biochem. Biophys. Res. Commun.*, **27**, 686–692. The energy-dependent accumulation of phosphate by blowfly mitochondria and its effect on the rate of pyruvate oxidation.
Hanski, E., Sevilla, N., and Levitzki, A. (1977). *Eur. J. Biochem.*, **76**, 513–520. The allosteric inhibition by calcium of soluble and partially purified adenylate cyclase from turkey erythrocytes.
Hanson, J., and Lowy, J. (1960). Structure and function of the contractile apparatus in the muscles of invertebrate animals. In *Structure and Function of Muscle*, Vol. 1 (Ed. G. Bourne), pp. 263–365. Academic Press, New York.
Hanström B. (1939) *Hormones in invertebrates*. Clarendon Press, Oxford.
Harary, I., Renaud, J. F., Sato, E., and Wallace, G. A. (1976). *Nature (London)*, **261**, 60–61. Calcium ions regulate cyclic AMP and beating in cultured heart cells.
Harfield, D., and Tenenhouse, A. (1973). *Can. J. Physiol. Pharmacol.*, **51**, 997–1001. Effect of EGTA on protein release and cyclic AMP accumulation in rat parotid gland.
Harling, W. B. (1967). *The Fossil Record*. Geological Society, London.
Harris, E. J., and Berent, C. (1969). *Biochem. J.*, **115**, 645–652. Calcium ion-induced uptakes and transformations of substrates in liver mitochondria.
Harris, J. R., and Milne, J. F. (1975). *J. Physiol. (London)*, **251**, 23P–24P. The behaviour of isolated cell nuclei when divalent cations are removed.
Harrison, H. E., and Harrison, H. C. (1974). *Biomembranes*, **4B**, 743–846. Calcium.
Harrison, G. E., and Raymond, W. H. A. (1953). *Analyst*, **78**, 528–531. The determination of microgram amounts of calcium.
Hart, A. D., Fisher, T., Hallinan, T., and Lucy, J. A. (1976). *Biochem. J.*, **158**, 141–145. Effects of calcium ions and the bivalent cation ionophore A23187 on the agglutination and fusion of chicken erythrocytes.
Hartshorne, D. J. and Boucher, L. J. in *Calcium Binding Proteins*. pp. 29–49. Ed. Drabikowski, W., Strzelecka-Golaszewska, H. and Carafoli, E. PWN-Polish scientific publishers, Warszawa and Elsevier, Amsterdam. Ion binding by troponin.
Hartshorne, D. J., and Mueller, H. (1968). *Biochem. Biophys. Res. Commun.*, **31**, 647–653. Fractionation of troponin into two distinct proteins.
Hartshorne, D. J., and Pyun, H. Y. (1971). *Biochim. Biophys. Acta*, **229**, 698–711. Calcium binding by the troponin complex, and the purification and properties of troponin A.
Harvey, A. M., and MacIntosh, F. C. (1940). *J. Physiol. (London)*, **97**, 408–416. Calcium and synaptic transmission in a sympathetic ganglion.
Harvey, E. N. (1952). *Bioluminescence*. Academic Press, New York.
Haslam, J. M., and Griffiths, D. E. (1968). *Biochem. J.*, **109**, 921–928. Factors affecting the translocation of oxaloacetate and L-malate into rat liver mitochondria.
Haslam, R. J. and Siay, A. (1975). *Fed. Proc.*, **34**, 231. Effects of acetylcholine and carbachol on cyclic AMP levels in dog blood platelets in relation to platelet function: role of extracellular Ca^{2+} ions.
Hasselbach, W. (1966). *Ann. N. Y. Acad. Sci.*, **137**, 1041–1048. Structural and enzymatic properties of the calcium transporting membranes of the sarcoplasmic reticulum.
Hasselbach, W. (1978). *Biochim. Biophys. Acta*, **515**, 23–53. The reversibility of the sarcoplasmic calcium pump.
Hasselbach, W., and Makinose, M. (1963). *Biochem. Z.*, **339**, 94–111. Über den Mechanismus des Calcium Transportes durch die Membranen des Sarkoplasmatischen Reticulums.
Hastings, A. B., Murray, C. D., and Sendroy, J., Jr. (1926). *J. Biol. Chem.*, **71**, 723–781. Studies on the solubility of calcium salts. I. The solubility of calcium carbonate in salt solutions and biological fluids.

Hastings, A. B., Teng, C. T., Nesbett, F. B., and Sinex, F. M. (1952). *J. Biol. Chem.*, **194**, 69–81. Studies on carbohydrate metabolism in rat liver slices. I. The effect of cations in the media.
Hatano, S. (1970). *Exp. Cell Res.*, **61**, 199–203. Specific effects of Ca^{2+} on movement of plasmodial fragment obtained by caffeine treatment.
Hathaway, D. R., and Adelstein, R. S. (1979). *Proc Natl. Acad. Sci. U. S. A.*, **76**, 1653–1657. Human platelet myosin light chain kinase requires the calcium-binding protein calmodulin for activity.
Hauser, H., and Dawson, R. M. C. (1968). *Biochem. J.*, **109**, 909–916. Displacement of calcium ions from phospholipid monolayers by pharmacologically active and other organic bases.
Hauser, D. C. R., Petrylak, D., Singer, G., and Levandowsky, M. (1978). *Nature (London)*, **273**, 230–231. Calcium-dependent sensory motor response of a marine dinoflagellate.
Hausmann, K. (1978). *Int. Rev. Cytol.*, **52**, 197–276. Extrusive organelles in protists.
Hearse, D. J. (1977). *J. Mol. Cell. Cardiol.*, **9**, 605–616. Reperfusion of the ischemic myocardium.
Heaton, G. M., and Nicholls, D. G. (1976). *Biochem. J.*, **156**, 635–646. The calcium conductance of the inner membrane of rat liver mitochondria and the determination of the calcium electrochemical gradient.
Hechter, O. (1958). *Vit. Horm.*, **13**, 293–346. Concerning possible mechanisms of hormone action.
Heggtveit, A. H. Herman, L., and Mishra, R. R. (1964). *Am. J. Pathol.* **45**, 757–782. Cardiac necrosis and calcification in experimental magnesium deficiency. A light and electron microscopic study.
Heilenius, E., and Stalberg, E. (1978). *J. Neurochem.*, **31**, 5–11. The pathogenesis of myasthenia gravis.
Heilbrunn, L. V. (1923). *Am. J. Physiol.*, **64**, 481–502. The colloid chemistry of protoplasm. I. General considerations. II. The electrical charges of protoplasm.
Heilbrunn, L. V. (1927). *Arch. f. exper. Zellforsch.*, **4**, 246–263. Colloid chemistry of protplasm. V. A preliminary study of the surface precipitation reaction of living cells.
Heilbrunn, L. V. (1928). *The Colloid Chemistry of Protoplasm.* Borntraeger, Berlin.
Heilbrunn, L. V. (1937, 1st Edition; 1943, 2nd Edition). *An Outline of General Physiology.* Saunders, Philadelphia, Pennsylvania.
Heilbrunn, L. V. (1940). *Physiol. Zool.*, **13**, 88–94. The action of calcium on muscle protoplasm.
Heilbrunn, L. V. (1956). *The Dynamics of Living Protoplasm.* Academic Press, New York.
Heilbrunn, L. V. (1958). *Protoplasmatologia*, **2** (C1), 1–109. The viscosity of protoplasm.
Heilbrunn, L. V., and Wiercinski, F. J. (1947). *J. Cell Comp. Physiol.*, **29**, 15–32. The action of various cations on muscle protoplasm.
Heilbrunn, L. V., and Wilbur, K. M. (1937). *Biol. Bull.*, **73**, 557–564. Stimulation and nuclear breakdown in the *Nereis* egg.
Heisler, S., Fast, D., and Tenenhouse, A. (1972). *Biochim. Biophys. Acta*, **279**, 561–572. Role of Ca^{2+} and cyclic AMP in protein secretion from rat exocrine pancreas.
Helenius, A., Marsh, M., and White, J. (1980). *TIBS*, **5**, 104–106. The entry of viruses into animal cells.
Hellam, D. C., and Podolsky, R. J. (1969). *J. Physiol. (London)*, **200**, 807–819. Force measurements in skinned muscle fibres.
Hemminki, K. (1975). *Acta Physiol. Scand.*, **95**, 117–125. Accumulation of calcium by retinal outer segments.
Hendricks, Th., Daemen, F. J. M., and Bonting, S. L. (1974). *Biochim. Biophys. Acta*, **345**, 468–473. Biochemical aspects of the visual process XXV. Light-induced calcium movements in isolated frog rod outer-segments.
Henkart, M., Reese, T. H., and Brinley, F. J., Jr. (1978). *Biophys. J.*, **24**, 187a. Oxalate produces precipitates in endoplasmic reticulum of Ca-loaded axons.
Henry, J. P., and Ninio, M. (1978). *Biochim. Biophys. Acta*, **504**, 40–59. Control of the Ca^{2+}-triggered bioluminescence of *Veretillum cynomoium* lumisomes.
Henry, R. J. (1964). *Clinical Chemistry. Principles and Techniques*, pp. 345–421. Harper-Row, New York.
Herbst, C. (1900). *Arch. Entwicklungsmech.*, **9**, 424–463. Über das Anseinandergehen von Furchungs-und Gewebezellen in kalkfreiem Mediuon.
Herd, P. M. (1978). *Arch. Biochem. Biophys.*, **188**, 220–225. Thyroid hormone–divalent cation interactions. Effect of thyroid hormone on mitochondrial calcium metabolism.
Herdson, P. B., Sommers, H. M., and Jennings, R. B. (1965). *Am. J. Pathol.*, **46**, 367–386. A comparative study of the fine structure of normal and ischemic dog myocardium with special reference to early changes following temporary occlusion of a coronary artery.
Herman, L., Sato, T., and Hales, C. N. (1973). *J. Ultrastruct. Res.* **42**, 298–311. The electron microscopic localisation of cations to pancreatic islets of Langerhans and their possible role in insulin secretion.

Hermann, S. (1932). *Naunyn-Schmiedebergs Arch. Exp. Pathol. Pharmakol.*, **167**, 82–84. Die Beeinflussbarkeit der Zustands Form des Calciums in Organismus durch Adrenalin.

Herrera, A. A. (1977). *J. Cell Biol.*, **75**, 113a. Ca ion couple membrane excitation to intracellular luminescence.

Herring, P. J. (Ed.) (1978). *Bioluminescence In Action.* Academic Press, London.

Herrman-Erlee, M. P. M., Gaillard, P. J., Hekkelman, J. W., and Nijweide, P. J. (1977). *Eur. J. Pharmacol.*, **48**, 51–58. The effect of verapamil on the action of parathyroid hormone on embryonic bone *in vitro*.

Hershko, A. P., Mamont, R., Shields, R., and Tomkins, G. M. (1971). *Nature (London)*, **232**, 206–211. Pleiotypic response.

Hesketh, T. R., Smith, G. A., Houslay, M. D., Warren, G. B., and Metcalfe, G. B. (1977). *Nature (London)*, **267**, 490–494. Is an early calcium flux necessary to stimulate lymphocytes?

Hewitt, E. J., and Smith, T. A. (1975). *Plant Mineral Nutrition.* The English Universities Press.

Heyer, C. B., and Lux, H. D. (1976). *J. Physiol. (London)*, **262**, 349–382. Control of the delayed outward potassium currents in bursting pacemaker neurons of the snail *Helix pomatia*.

Hickee, R. A., and Kalant, H. (1967). *Cancer Res.*, **27**, 1053–1057. Calcium and magnesium content of rat liver and Morris hepatoma.

Hidaka, H., Naka, M., and Yamaki, T. (1979). *Biochem. Biophys. Res. Commun.*, **90**, 694–699. Effect of novel specific myosin light chain kinase inhibitors on Ca^{2+}-activated Mg^{2+}-ATPase of chicken gizzard actomyosin.

Highnam, K. C., and Hill, L. (1977). *The Comparative Endocrinology of the Invertebrates.* Edward Arnold, London.

Hill, A. V. (1948). *Proc. R. Soc. London Ser. B.*, **135**, 446–453. On the time required for diffusion and its relation to processes in muscle.

Hillarp, N. A. (1958). *Acta Physiol. Scand.*, **43**, 242–302. The release of catecholamines from the amine containing granules of the adrenal medulla.

Hille, B. (1972). *J. Gen. Physiol.*, **59**, 637–658. The permeability of the sodium to metal cations in myelinated nerve.

Hillman, D. E., and Llinas, R. (1961). *J. Cell Biol.*, **61**, 146–155. Calcium-containing electron-dense structures in the axons of the squid giant synapse.

Hines, H. M., and Knowlton, G. C. (1933). *Am. J. Physiol.*, **104**, 379–391. Changes in skeletal muscle following denervation.

Hirata, F., and Axelrod, J. (1980). *Science*, **209**, 1082–1090. Phospholipid methylation and biological signal transmission.

Hirata, M., Mikawa, T., Nonomura, Y., and Ebashi, S. (1980). *J. Biochem.*, **87**, 369–378. Ca^{2+} regulation in vascular smooth muscle.

Hitchcock, S. F. (1977). *J. Cell Biol.*, **74**, 1–15. Regulation of motility in non-muscle cells.

Höber, R. (1945). *Physical Chemistry of Cells and Tissues.* J. & A. Churchill, London.

Hodgkin, A. L. (1951). *Biol. Rev.*, **26**, 339–409. The ionic basis of electrical activity in nerve and muscle.

Hodgkin, A. L., and Huxley, A. F. (1945). *J. Physiol. (London)*, **104**, 176–195. Resting and action potentials in single nerve fibres.

Hodgkin, A. L., and Katz, B. (1949a). *J. Exp. Biol.*, **26**, 292–294. The effect of calcium on the axoplasm of giant nerve fibres.

Hodgkin, A. L., and Katz, B. (1949b). *J. Physiol. (London)*, **108**, 37–77. The effect of sodium ions on the electrical activity of the giant axon of the squid.

Hodgkin, A. L., and Keynes, R. D. (1957). *J. Physiol. (London)*, **138**, 253–281. Movements of labelled calcium in squid giant axons.

Hoerl, B. J., and Scott, L. E. (1978). *Virchows Arch. B*, **27**, 335–341. Plasma membrane vesiculation: a cellular response to injury.

Hoffstein, S., Goldstein, I. M., and Weismann, G. (1977). *J. Cell Biol.*, **73**, 242–256. Role of microtubule assembly in lysosomal enzyme secretion from human polymorphonuclear lymphocytes.

Hokin, L. E. (1966). *Biochim. Biophys. Acta*, **115**, 219–221. Effect of calcium omission on acetylcholine-stimulated amylase secretion and phospholipid synthesis in pigeon pancreas slices.

Holland, W. C., and Porter, M. T. (1969). *Fed. Proc.*, **28**, 1663–1669. Pharmacological effects of drugs on excitation–contraction coupling in cardiac muscle.

Hollinger, T. G., and Schuetz, A. A. (1976). *J. Cell Biol.*, **71**, 395–401. 'Cleavage' and cortical granule breakdown in *Rana pipiens* oocytes induced by direct microinjection of calcium.

Hollingsworth, J. (1941). *Biol. Bull.*, **81**, 261–276. Activation of *Cumingia* and *Arbacia* eggs by divalent cations.
Holloszy, J. O., and Narahara, H. T. (1965). *J. Biol. Chem.*, **240**, 3493–3500. Studies of tissue permeability X. Changes in permeability to 3-methyl-glucose associated with contraction of isolated frog muscle.
Holloszy, J. O., and Narahara, H. T. (1967). *Science*, **155**, 573–575. Nitrate ions: potentiation of increased permeability to sugar associated with muscle contraction.
Hölm-Hansen, O. (1968). *Annu. Rev. Microbiol.*, **22**, 47–70. Ecology, physiology, and biochemistry of blue-green algae.
Holmsen, H. (1972). The platelet: its membrane physiology, and biochemistry. In *Clinics in Haematology*, Vol. I. Ed. O'Brien, J. R., pp. 236–266. Saunders W. B. London, Philadelphia.
Holt, C., Parker, T. G., and Dalgleish, D. G. (1975). *Biochim. Biophys. Acta*, **379**, 638–644. The thermochemistry of reactions between α_{s1}-casein and calcium chloride.
Holwill, M. E. J., and McGregor, J. L. (1975). *Nature (London)*, **255**, 157–158. Control of flagellar wave movement in *Crithidia oncopelti*.
Hope-Gill, H. F., Kissebah, A. H., Clarke, P., Vydelingum, N., Tulloch, B., and Fraser, T. R. (1976). *Horm. Metab. Res.*, **8**, 184–140. Effects of insulin and procaine hydrochloride on glycogen synthetase activation and adipocyte calcium flux: evidence for a role of calcium in insulin activation of glycogen synthetase.
Hopkins, F. G. (1936). *London Naturalist*, pp. 40. The naturalist in the laboratory. (address of the Honorary President, London Natural History Society) see Needham, J. and Baldwin, E. pp. 280.
Horton, E. W. (1969). *Physiol. Rev.*, **49**, 122–161. Hypotheses on physiological role of prostaglandins.
Horwitz, B. A., Horwitz, J. M., and Smith, R. E. (1969). *Proc. Natl. Acad. Sci. U.S.A.*, **64**, 113–120. Norepinephrine-induced depolarisation of brown fat cells.
Houssay, B. A., and Molinelli, E. A. (1928). *C.R. Soc. Biol.*, **99**, 172–174. Excitabilité des fibres adrénalino-sécrétories du nerf grand splanchnique. Fréquences, seuil et optimum des stimulus. Rôle de l'ion calcium.
Hovi, T., Williams, S. C., and Allison, A. C. (1975). *Nature (London)*, **256**, 70–72. Divalent cation ionophore A23187 forms lipid soluble complexes with leucine and other amino acids.
Howard, A. and Pelc, S. R. (1951). *Exptl Cell Res.*, **2**, 178–187. Nuclear incorporation of 32P as demonstrated by autoradiographs.
Howell, S. L. (1977). *Biochem. Soc. Trans.*, **5**, 875–879. Intracellular localisation of calcium in pancreatic β-cells.
Hsiu, J., Fischer, E. H., and Stein, E. A. (1964). *Biochemistry*, **3**, 61–66. Alpha-amylases as calcium-metalloezymes-II. Calcium and the catalytic activity.
Hughes, B. P., and Barritt, G. J. (1978). *Biochem. J.*, **176**, 295–304. Effects of glucagon and $N^6, O^{2'}$-dibutyryl adenosine 3':5'-cyclic monophosphate on calcium transport in isolated rat liver mitochondria.
Hughes, H. S., and Barritt, G. J. (1971). *Proc. Aust. Biochem. Soc.*, **10**, 36. The effects of glucagon, adrenalin and dibutyryl adenosine 3'5'-monophosphate on calcium transport in rat liver mitochondria.
Hukovic, A., and Muscholl, E. (1962). *Naunyn-Schmiedebergs Arch. Pharmakol.*, **244**, 81–96. Die Noradrenalin: Abgabe aus demisolierten Kaninchenherzen bei sympathischen Nervenreizung und ihre pharmakologische Beeinflussung.
Hultin, T. (1950). *Expl. Cell Res.*, **1**, 159–168. On the oxygen uptake of *Paracentrotus lividus* egg homogenates after the addition of calcium.
Huston, R. B., and Krebs, E. G. (1968). *Biochemistry*, **7**, 2116–2122. Activation of skeletal muscle phosphorylase kinase by Ca^{2+}. II. Identification of the kinase activating factor.
Hutner, S. A. (1972). *Annu. Rev. Microbiol.*, **26**, 311–346. Inorganic nutrition.
Hutton, J. C., Penn, E. J., Jackson, P., and Hales, C. N. (1981). *Biochem. J.*, **193**, 875–855. Isolation and characterization of calmodulin from an insulin-secreting tumour.
Huxley, A. F., and Niedergerke, R. (1954). *Nature (London)*, **173**, 971–973. Structural changes in muscle during contraction.
Huxley, A. F., and Taylor, R. E. (1958). *J. Physiol. (London)*, **144**, 426–447. Local activation of striated muscle fibres.
Huxley, J. S. (1935). *Biol. Rev.*, **31**, 427–441. Chemical regulation and the hormone concept.
Hynie, S., and Sharp, G. W. G. (1971). *Biochim. Biophys. Acta*, **230**, 40–51. Adenyl cyclase in the toad bladder.

Ichikawa, A., (1965). *J. Cell Biol.*, **24**, 369–385. Fine structural changes in response to hormonal stimulation of the perfused canine pancreas.
Ignarro, L. J., and Colombo, C. (1973). *Science*, **180**, 1181–1183. Enzyme release from polymorphonuclear leukocyte lysosomes: regulation by autonomic drugs and cyclic nucleotides.
Impraim, C. C., Micklem, K. J., and Pasternak, C. A. (1979). *Biochem. Pharmacol.*, **28**, 1963–1969. Calcium, cells and virus—alterations caused by paramyxoviruses.
Iqbal, Z., and Ochs, S. (1978). *J. Neurochem.*, **31**, 409–418. Fast axoplasmic transport of a calcium-binding protein in mammalian nerve.
Irving, E. A., and Watts, P. S. (1961). *Biochem. J.*, **79**, 429–432. Estimation of calcium and magnesium in blood serum by the cathode-ray polarograph.
Irving, H., and Williams, R. J. P. *J. Chem. Soc.*, **1953**, 3192–3210. The stability of transition-metal complexes.
Isenberg, G. (1975). *Nature (London)*, **253**, 273–274. Is potassium conductance of cardiac Purkinjé fibres controlled by $[Ca_i^{2+}]_i$?
Ishida, R., Akiyoshi, H., and Takahashi, T. (1974). *Biochem. Biophys. Res. Commun.*, **56**, 703–710. Isolation and purification of calcium and magnesium dependent endonuclease from rat liver nuclei.
Israel, M., Dunant, Y., and Manaranche, R. (1979). *Prog. Neurobiol.*, **13**, 237–275. The present status of the vesicular hypothesis.
Israel, M., Manaranche, R., Marsal, J., Meunier, F. M., Morel, N. Frachon, P., and Lesbats, B. (1980). *J. Membr. Biol.*, **54**, 115–126. ATP-dependent calcium uptake by cholinergic synaptic vesicles isolated from *Torpedo* electric organ.
Itaya, K. (1978). *J. Pharm. Pharmacol.*, **30**, 315–317. Need for calcium ions in the lipolytic action of 5-hydroxytryptamine in rat brown adipose tissue.
Iversen, L. C., Iversen, S. D., Bloom, F., Douglas, C., Brown, M., and Vale, W. (1978). *Nature (London)*, **273**, 161–163. Calcium-dependent release of somatostasin and neurotensin from rat brain *in vitro*.
Iwasaki, S., and Sato, Y. (1971). *J. Gen. Physiol.*, **57**, 216–238. Sodium- and calcium-dependent spike potentials in the secretory neuron soma of the X-organ of the crayfish.
Iwatsuki, N., and Petersen, O. H. (1977). *Nature (London)*, **268**, 147–149. Acetyl choline like effects of intracellular calcium injection in pancreatic acinar cells.
Izutsu, K. T., and Smuckler, E. A. (1978). *Am. J. Pathol.*, **90**, 145–158. Effects of carbon tetrachloride on rat liver plasmalemmal calcium adenosine triphosphatase.
Jaanus, S. D., Miele, E., and Rubin, R. P. (1967). *Br. J. Pharmacol. Chemother.*, **31**, 319–330. The analysis of the inhibitory effect of local anaesthetics and propranolol on adreno-medullary secretion evoked by calcium or acetylcholine.
Jacobs, B. O., and Krahl, M. E. (1973). *Biochim. Biophys. Acta*, **319**, 410–415. The effects of divalent cations and insulin on protein synthesis in adipose cells.
Jacobson, A., Schwarts, M., and Rehm, W. S. (1975). *Am. J. Physiol.*, **209**, 134–140. Effects of removal of calcium from bathing media on frog stomach.
Jaffe, L. A., Weisenseel, M. H., and Jaffe, L. F. (1975). *J. Cell Biol.*, **67**, 488–492. Calcium accumulations within the growing tips of pollen tubes.
Jafferji, S. S., and Michell, R. H. (1976). *Biochem. J.*, **160**, 163–169. Effects of calcium-antagonistic drugs on the stimulation by carbamoylcholine and histamine of phosphatidylinositol turnover in longitudinal smooth muscle of guinea-pig ileum.
Jahn, T. L., and Bovee, E. C. (1969). *Physiol. Rev.*, **49**, 793–862. Protoplasmic movements within cells.
Jahn, T. L., and Rinaldi, R. A. (1959). *Biol. Bull. (Woods Hole Mass.)*, **117**, 100–118. Protoplasmic movement in the foraminiferan, *Allogromia laticellaris*, and a theory of its mechanism.
Jamieson, G. A., Jr., Vanaman, T. C., and Blum, J. J. (1979). *Proc. Natl. Acad. Sci. U.S.A.*, **76**, 6471–6475. Presence of calmodulin in *Tetrahymena*.
Jamieson, J. D., and Palade, G. E. (1967). *J. Cell Biol.*, **34**, 597–615. Intracellular transport of secretory proteins in the pancreatic exocrine cell. II. Transport to condensing vacuoles and zymogen granules.
Janakidevi, K., Dewey, V. C., and Kidder, G. W. (1966). *J. Biol. Chem.*, **241**, 2576–2578. The biosynthesis of catecholamines in two genera of protozoa.
Jarrett, H. W., and Penniston, J. T. (1977a). *Biochem. Biophys. Res. Commun.*, **77**, 1210–1216. Partial purification of the Ca^{2+} Mg^{2+} ATPase activator from human erythrocytes: its similarity to the activator of 3′:5′-cyclic nucleotide phosphodiesterase.

Jarrett, H. W., and Penniston, J. T. (1977b). *J. Biol. Chem.*, **253**, 4676–4682. Purification of the Ca^{2+}-stimulated ATPase activator from human erythrocytes. Its membership in the class of Ca^{2+}-binding modulator proteins.

Jenden, D. J., and Fairhurst, A. S. (1969). *Pharmacol. Rev.*, **21**, 1–25. The pharmacology of ryanodine.

Jensen, P., Winger, L., Rasmussen, H., and Nowell, P. (1977). *Biochim. Biophys. Acta*, **496**, 374–383. The mitogenic effect of A23187 in human peripheral lymphocytes.

Jetley, M., and Weston, A. H. (1976). *Br. J. Pharmacol.*, **58**, 287P–288P. Some effects of D600, nifedipine and sodium nitroprusside on electrical and mechanical activity in rat portal vein.

Jöbsis, F. F., and O'Connor, M. J. (1966). *Biochem. Biophys. Res. Commun.*, **25**, 246–252. Calcium release and reabsorption in the sartorius muscle of the toad.

Johnson, J. D., Collins, J. H., and Potter, J. D. (1978). *J. Biol. Chem.*, **253**, 6451–6458. Dansylaziridine-labelled troponin C. A fluorescent probe of Ca^{2+} binding to the Ca^{2+}-specific regulatory sites.

Johnson, J. D., Collins, J. H., Robertson, J. P., and Potter, J. D. (1980). *J. Biol. Chem.*, **255**, 9635–9640. A fluorescent probe study of Ca^{2+} binding to the Ca^{2+}-specific sites of cardiac troponin and troponin C.

Jones, H. P., Lenz, R. W., Palevitz, B. A., and Cormier, M. J. (1980). *Proc. Natl. Acad. Sci.*, **77**, 2772–2776. Calmodulin localization in mammalian spematozoa.

Jones, H. P., Matthews, J. C., and Cormier, M. J. (1974). *Biochemistry*, **18**, 55–60. Isolation and characterisation of Ca^{2+}-dependent modulator protein from the marine invertebrate *Renilla reniformis*.

Jones, L. M., and Michell, R. H. (1978). *Biochem. Soc. Trans.*, **6**, 22–38. Stimulus–response coupling at α-adrenergic receptors.

Jones, L. R., Besch, H. R., Jr., Sutko, J. L., and Willerson, J. T. (1979). *J. Pharmacol. Exp. Ther.*, **209**, 48–55. Ryanodine-induced stimulation of net Ca^{++} uptake by cardiac sarcoplasmic reticulum vesicles.

Joos, R. W., and Carr, C. W. (1967). *Proc. Soc. Exp. Biol. Med.*, **124**, 1268–1272. The binding of calcium in mixtures of phospholipids.

Judah, J. D., and Ahmed, K. (1963). *Biochim. Biophys. Acta*, **71**, 34–44. Role of phosphoproteins in ion transport: interactions of sodium with calcium and potassium in liver slices.

Judah, J. D., and Ahmed, K. (1964). *Biol. Rev.*, **39**, 160–193. The biochemistry of sodium transport.

Junge, D. (1967). *Nature (London)*, **215**, 546–548. Multi-ionic potentials in molluscan giant neurones.

Jutisz, M., and de la Llosa, M. P. (1970). *Endocrinology*, **86**, 761–768. Requirement of Ca^{2+} and Mg^{2+} ions for the *in vitro* release of follicle-stimulating hormone from rat pituitary glands and its subsequent biosynthesis.

Juzu, H. A., and Holdsworth, E. S. (1980). *J. Membr. Biol.*, **52**, 185–186. New evidence for the role of cyclic AMP in the release of mitochondrial calcium.

Kachmar, J. F., and Boyer, P. D. (1953). *J. Biol. Chem.*, **200**, 669–682. Kinetic analysis of enzyme reactions. II. The potassium activation and calcium inhibition of pyruvic phosphoferase.

Kahl, R. C., Zawalich, W. S., Ferrendelli, J. A., and Matschinsky, F. M. (1975). *J. Biol. Chem.*, **250**, 4575–4579. The role of Ca^{2+} and cyclic adenosine 3′:5′-monophosphate in insulin release induced *in vitro* by the divalent cation ionophore A23187.

Kaiser, N., and Edelman, I. S. (1977). *Proc. Natl. Acad. Sci. U.S.A.*, **74**, 638–642. Calcium dependence of glucocorticoid-induced lymphocytolysis.

Kakiuchi, S., Sobue, K., Yamazuki, R., Nagao, S., Umeki, S., Nozawa, Y., Yazawa, M., and Yagi, K. (1981). *J. Biol. Chem.*, **256**, 19–22. Ca^{2+}-dependent modulator proteins from *Tetrahymena pyriformis*, sea anemone and scallop and guanylate cyclase activation.

Kalix, P. (1977). *Pflügers Arch.*, **326**, 1–14. Uptake and release of calcium in rabbit vagus nerve.

Kamada, T. (1940). *Proc. Imp. Acad. (Tokyo)*, **16**, 241. Ciliary reversal of Paramecium.

Kamata, T., and Kinoshita, H. (1943). *Jpn. J. Zool.*, **10**, 469–493. Disturbances initiated from naked surface of muscle protoplasm.

Kamiya, N. (1959). *Protoplasmatologia*, **8** (3a), 1–198. Protoplasmic streaming.

Kamiya, N., and Kuroda, K. (1956). *Bot. Mag. (Tokyo)*, **69**, 544–544. Velocity distribution of protoplasmic streaming in *Nitella* cells.

Kanagasuntheram, P., and Randle, P. J. (1976). *Biochem. J.*, **160**, 547–564. Calcium metabolism and amylase release in rat parotid acinar cells.

Kanatani, H., and Shirai, H. (1969). *Biol. Bull.*, **137**, 297–311. Mechanism of starfish spawning II. Some aspects of action of a neural substance obtained from radial nerve.

Kanno, T.(1972). *J. Physiol. (London)*, **226**, 353–371. Calcium-dependent amylase release and electrophysiological measurements in cells of the pancreas.

Kanno, T., Cochrane, D. E., and Douglas, W. W. (1973). *Can. J. Physiol. Pharmacol.*, **51**, 1001–1004. Exocytosis (secretory granule extrusion) induced by injection of calcium in mast cells.

Kaplan, J. (1981). *Science*, **212**, 14–20. Polypeptide-binding membrane receptors: analysis and classification, membrane receptors analysis and classification.

Kar, N. C., and Pearson, C. M. (1976). *Clin. Chim. Acta*, **73**, 293–297. A calcium-activated neutral protease in normal and dystrophic human muscle.

Katsumi, W., Kamberi, I. A., and McCann, S. M. (1969). *Endocrinology*, **85**, 1046–1056. In vitro response of the rat pituitary to gonadotrophin-releasing factors and to ions.

Katz, A. M., Tada, M., and Kirchberger, M. (1975). *Adv. Cyclic Nucleotide Res.*, **5**, 453–472. Control of calcium transport in the myocardium by cyclic AMP-protein kinase system.

Katz, B., and Miledi, R. (1965a). *Proc. R. Soc. London Ser. B*, **161**, 483–495. The measurement of synaptic delay and the time course of acetyl choline release at the neuromuscular junction.

Katz, B., and Miledi, R. (1965b). *Natural (London)*, **207**, 1097–1098. Release of acetylcholine from a nerve terminal by electric pulses of variable strength and duration.

Katz, B., and Miledi, R. (1967). *Proc. R. Soc. London Ser. B*, **167**, 1–7. Modification of transmitter release by electrical interference with motor nerve endings.

Katz, B., and Miledi, R. (1969). *J. Physiol. (London)*, **203**, 459–487. Tetrodotoxin-resistant electric activity in presynaptic terminals.

Katz, R. I., and Kopin, I. I. (1969). *Biochem. Pharmacol.*, **18**, 1835–1839. Release of norepinephrine-^{3}H and serotonin-^{3}H evoked from brain slices by electrical-field stimulation—Calcium dependency and the effects of lithium, ouabain and tetrodotoxin.

Kaupp, U. B., and Junge, W. (1977). *FEBS Lett.*, **81**, 229–232. Rapid calcium release by passively loaded retinal discs of photoexcitation.

Kaupp, U. B., Schnetkamp, P. P. M., and Junge, W. (1979). Flash-spectrometry with arsenazo III invertebrate photoreceptor cells. In *Detection and Measurement of Free Ca^{2+} in Cells* (Eds. C. C. Ashley and A. K. Campbell). pp. 287–308. Elsevier/North-Holland, Amsterdam.

Kaye, G. W. C., and Layby T. H. (1959). *Tables of Physical and Chemical Constants*. Longmans, London.

Kendrick, N. C. (1976). *Anal. Biochem.*, **76**, 487–501. Purification of arsenazo III, a Ca^{2+}-sensitive dye.

Kendrick-Jones, J., Lehman, W., and Szent-Györgyi, A. G. (1970). *J. Mol. Biol.*, **54**, 313–326. Regulation in molluscan muscles.

Keppens, S., Vandenheede, J. R., and DeWulf, H. (1977). *Biochim. Biophys. Acta*, **496**, 448–457. On the role of calcium as a second messenger in liver for the hormonally induced activation of glycogen phosphorylase.

Keynes, R. D., and Lewis, P. R. (1956). *J. Physiol. (London)*, **134**, 399–407. The intracellular calcium contents of some invertebrate nerves.

Kiehart, D. P., Reynolds, G. T., and Eisen, A. (1977). *Biol. Bull.*, **153**, 432. Calcium transients during the early development in echinoderms and teleosts.

Kimball, J. W. (1974). *Biology*, 3rd Edition. Addison-Wesley, Reading, Massachusetts.

Kimberg, D. V., and Goldstein, S. A. (1966). *J. Biol. Chem.*, **241**, 95–103. Binding of calcium by liver mitochondria of rats treated with steroid hormones.

Kimizuka, H., and Koketsu, K. (1962). *Nature (London)*, **196**, 995–996. Binding of calcium ion to lecithin film.

Kimmich, G. A., and Rasmussen, H. (1969). *J. Biol. Chem.*, **244**, 190–199. Regulation of pyruvate carboxylase activity by calcium in intact rat liver mitochondria.

Kimura, H., and Murad, F. (1975). *J. Biol. Chem.*, **249**, 6910–6916. Evidence for two different forms of guanylate cyclase in rat heart.

Kimura, S., and Rasmussen, H. (1977). *J. Biol. Chem.*, **252**, 1217–1255. Adrenal glucocorticoids, adenine nucleotide translocation and mitochondrial calcium accumulation.

Kinoshita, S., and Yazaki, I. (1967). *Exp. Cell Res.*, **47**, 449–458. The behaviour and localisation of intracellular relaxing system during cleavage in the sea urchin egg.

Kirby, A. C., Lindley, B. D., and Picken, J. R. (1975). *J. Physiol. (Lond.)*, **253**, 37–52. Calcium content and exchange in frog skeletal muscle.

Kirtland, S. J., and Baum, H. (1972). *Nature (London)*, **236**, 47–49. Prostaglandin E_1 may act as a 'calcium ionophore'.

Kishi, S., Fujiwara, T., and Nakahara, W. (1937). *Gann*, **31**, 1–11. Comparison of chemical composition between hepatoma and normal liver tissues; sodium, potassium, calcium, magnesium, iron, iodine and chloride, including sodium chloride.
Kissebah, A. H., Vydelingum, N., Tulloch, B. R., Hope-Gill, H. F., and Fraser, T. R. (1974a). *Horm. Metab. Res.*, **6**, 247–255. The role of calcium in insulin action: I. Purification and properties of enzyme regulating lipolysis in human adipose-tissue: effects of cyclic AMP and calcium ions.
Kissebah, A. H., Tulloch, B. R., Vydelingum, N., Hope-Gill, H., Clarke, P., and Fraser, T. R. (1974b). *Horm. Metab. Res.*, **6**, 357–364. The role of calcium in insulin action: II. Effect of insulin and procaine-hydrochloride on lipolysis.
Kissebah, A. H., Clarke, P., Vydelingum, N., Hope-Gill, H., Tulloch, B., and Fraser, T. R. (1975). *Horm. Metab. Res.*, **7**, 194–196. The role of calcium in insulin action: III. Calcium distribution in fat cells; its kinetics and the effects of adrenaline, insulin and procaine-HCl.
Klebanoff, S. J., and Clark, R. A. (1978). *The Neutrophil: Function and Clinical Disorders.* North-Holland, Amsterdam.
Klee, C. B., Crouch, T. H., and Richman, P. G. (1980). *Annu. Rev. Biochem.*, **49**, 489–515. Calmodulin.
Klein, R. L., Horton, C. R., and Thureson-Klein, A. (1970). *Am. J. Cardiol.*, **25**, 300–310. Studies on nuclear amino acid transport and cation content in embryonic myocardium of the chick.
Klinke, K. (1928). *Ergeb. Physiol.*, **26**, 235–319. Neuere Ergebnisse der Calciumforschung.
Knauf, P. A., Proverbio, F., and Hoffmann, J. F. (1974). *J. Gen. Physiol.*, **63**, 324–336. Electrophoretic separation of different phosphoproteins associated with Ca-ATPase and Na, K-ATPase in human red cells.
Kneer, N. M., Bosch, A. L., Clark, M. G., and Lardy, H. A. (1974). *Proc. Natl. Acad. Sci. U.S.A.*, **71**, 4523–4527. Glucose inhibition of epinephrine stimulation of hepatic gluconeogenesis by blockade of the α-receptor junction.
Knutton, S., and Pasternak, C. A. (1979). *TIBS*, **4**, 220–223. The mechanism of cell–cell fusion.
Koblin, D. D., and Eger, E. I. (1979). *N. Engl. J. Med.*, **301**, 1222–1224. Theories of narcosis.
Kohn, A. (1979). *Adv. Virus Res.*, **24**, 223–276. Early interactions of viruses with cellular membranes.
Komnick, H. (1962). *Protoplasma*, **55**, 414–418. Elektronenmikroskopische Lokalisation von Na^{2+} und Cl^- in Zelhon und Geweben.
Kornberg, A., Spudich, J. A., Nelson, D. L., and Deutscher, M. P. (1975). *Annu. Rev. Biochem.*, **37**, 51–78. Origin of proteins in sporulation.
Koshland, D. E. (1979). *Physiol. Rev.*, **59**, 811–862. A model regulatory system: bacterial chemotaxis.
Krahl, M. E. (1966). *Fed. Proc.*, **25**, 832–834. Insulin-like and anti-insulin effects of chelating agents on adipose tissue.
Kramer, B., and Tisdall, F. F. (1921). *J. Biol. Chem.*, **48**, 1–12. Methods for direct qualitative determination of sodium, potassium, calcium, magnesium in urine and stools.
Krebs, E. G., and Beavo, J. A. (1979). *Annu. Rev. Biochem.*, **48**, 923–959. Phosphorylation–dephosphorylation of enzymes.
Krebs, E. G., and Fischer, E. H. (1956). *Biochim. Biophys. Acta*, **20**, 150–157. The phosphorylase *b* to a converting enzyme of rabbit skeletal muscle.
Krebs, H. A. (1971). *Curr. Top. Cell Regul.*, **1**, 45–55. The role of equilibria in the regulation of metabolism.
Krebs, H. A., Bennett, D. A. H., DeGasquet, P., Gascoyne, T., and Yoshida, T. (1963). *Biochem. J.*, **86**, 22–27. Renal gluconeogenesis: the effect of diet on the gluconeogenic capacity of rat kidney cortex slices.
Kretsinger, R. H. (1976a). *Int. Rev. Cytol.*, **46**, 323–393. Evolution and function of calcium-binding proteins.
Kretsinger, R. H. (1976b). *Annu. Rev. Biochem.*, **45**, 239–268. Calcium binding proteins.
Kretsinger, R. H., and Nelson, D. J. (1977). *Coord. Chem. Rev.*, **18**, 29–124. Calcium in biological systems.
Krieger, N. S., and Tashjian, A. H., Jr. (1980). *Nature (London)*, **287**, 843–845. Parathyroid hormone stimulates bone resorption via a Na–Ca exchange mechanism.
Krizanova, O., Solarikova, L., and Hana, L. (1979). *Acta Virol.*, **23**, 295–302. Role of calcium-dependent regulator protein (CDR) in inhibition of 3′,5′-c′AMP-phosphodiesterase by influenza virus. I. Isolation and purification of CDR and CDR-dependent 3′,5′-c′AMP-phosphodiesterase from chick embryos.
Krnjevic, K. (1971). *Anesthesiology*, **34**, 215–217. The mechanism of general anesthesia.

Krnjevic, K. (1974). *Physiol. Rev.*, **54**, 418–540. Chemical nature of synaptic transmission in vertebrates.
Krnjevic, K., and Lisiewicz, A. J. (1972). *J. Physiol. (London)*, **225**, 363–390. Injection of calcium ions into spinal motoneurones.
Kruegar, B. K., Forn, J., and Greengard, P. (1977). *J. Biol. Chem.*, **252**, 2764–2773. Depolarization-induced phosphorylation of specific proteins, mediated by calcium ion influx, in rat brain synaptosomes.
Kumagai, H., Ebashi, S., and Takeda, F. (1955). *Nature (London)*, **176**, 166. Essential relaxing factor in muscle other than myokinase and creatine phosphokinase.
Kuntziger, H., Antonetti, A., Couette, S., Coureau, C., and Amiel, C. (1974). *Anal. Biochem.*, **60**, 449–454. Ultramicro (nanoliter range) determination of calcium concentration (10^{-3} M) by atomic absorption.
Kuo, I. C. Y., and Coffee, C. J. (1976). *J. Biol. Chem.*, **251**, 1603–1609. Purification and characterization of a troponin-C-like protein from bovine adrenal medulla.
Kuroda, Y. (1978). *J. Physiol. (Paris)*, **74**, 463–470. Physiological roles of adenosine derivatives which are released during neurotransmission in mammalian brain.
Kurokawa, K., and Rasmussen, H. (1973). *Biochim. Biophys. Acta*, **313**, 17–31. Ionic control of renal gluconeogenesis: I. The interrelated effect of calcium and hydrogen ions.
Kurokawa, K., Ohno, T., and Rasmussen, H. (1973). *Biochim. Biophys. Acta*, **313**, 32–41. Ionic control of renal gluconeogenesis. II. Effects of Ca^{2+} and H^+ upon the response to parathyroid hormone and cyclic AMP.
Kusano, K., Miledi, R., and Stinnakre, J. (1975). *Proc. R. Soc. London*, **189**, 49–56. Post-synaptic entry of calcium induced by transmitter action.
Kwant, W. O., and Seeman, P. (1969). *Biochim. Biophys. Acta*, **193**, 338–349. The displacement of membrane calcium by a local anaesthetic (chlorpromazine).
Labeyrie, E., and Koechlin, Y. (1979). *J. Neurosci. Methods*, **1**, 35–39. Photoelectrode with a very short time-constant for recording intracerebrally Ca^{2+} transients at a cellular level.
Lacy, P. E., and Malaisse, W. J. (1973). *Rec. Prog. Horm. Res.*, **29**, 199–228. Microtubules and Beta cell secretion.
Lafferty, F. W., Reynolds, E. S., and Pearson, O. H. (1965). *Am. J. Med.*, **38**, 105–117. Tumoral calcinosis: a metabolic disease of obscure etiology.
La Ganga, T. S. (1974). Calculi. In *Clinical Chemistry: Principles and Techniques* (Eds. D. C. Cannon and J. W. Windelman), pp. 1569–1583. Harper and Row, New York.
Lagnado, J., Tan, L. P., and Reddington, M. (1975). *Ann. N.Y. Acad. Sci.*, **253**, 577–597. The in situ phosphorylation of microtubular protein in brain cortex slices and related studies on the phosphorylation of isolated brain tubulin.
Lagunoff, D. (1973). *J. Cell Biol*, **57**, 232–250. Membrane fusion during mast cell secretion.
Laidlaw, P. P., and Payne, W. W. (1922). A method for the estimation of small quantities of calcium. *Biochem. J.*, **16**, 494–498.
Lamb, J. F., and Linsay, R. (1971). *J. Physiol. (London)*, **218**, 691–708. Effect of Na, metabolic inhibitors and ATP on calcium movements in L-cells.
Lane, C. E. (1968). Coelenterata: chemical aspects of ecology: Pharmacology and toxicology. In *Chemical Zoology*, Vol. 2 (Eds. M. Florkin and B. T. Scheer), pp. 263–284. Academic Press, New York and London.
Langer, G. A., Sato, M., and Seraydarian, M. (1969). *Circ. Res.*, **24**, 589–597. Calcium exchange in a single layer of rat cardiac cells studied by direct counting of cellular activity of labelled calcium.
Langford, G. M. (1977). *Exp. Cell Res.*, **111**, 139–151. In vitro assembly of dogfish brain tubulin and the induction of coiled ribbon polymers by calcium.
Lansdowne, D., and Ritchie, J. M. (1971). *J. Physiol. (London)*, **212**, 503–517. On the control of glycogenolysis in mammalian nervous tissue by calcium.
La Porte, D. C., Gidwitz, S., Wener, M. J., and Storm, D. R. (1979). *Biochem. Biophys. Res. Commun.*, **86**, 1169–1177. Relationship between changes in calcium dependent regulatory protein and adenylate cyclase during viral transformation.
Lauter, F., and Erne, D. (1980). *Anal. Chem.*, **52**, 2400–2403. Neutral carrier based ion selective electrode for intracellular magnesium activity studies.
Lawaczek, H. (1928). *Dtsch. Arch. Klin. Med.*, **160**, 302–309. Ueber das Verhalten des Kalziums unter Adrenalin.
Laycock, S. G., Warner, W., and Rubin, R. P. (1977). *Endocrinology*, **100**, 74–81. Further studies on the mechanisms controlling prostaglandin biosynthesis in the cat adrenal cortex: the role of calcium and cyclic AMP.
Lazarus, N. R., and Davis, B. (1977). *Biochem. Soc. Trans.*, **5**, 884–886. Insulin β-granule–plasma membrane interactions: effects of glucose and glucose 6-phosphate.

Lea, T. J., and Ashley, C. C. (1978). *Nature (London)*, **275**, 236–238. Increase in free Ca^{2+} in muscle after exposure to CO_2.
Leavis, P. C., and Kraft, E. L. (1978). *Arch. Biochem. Biophys.*, **186**, 411–415. Calcium binding to cardiac troponin C.
Lee, A. G. (1976). *Nature (London)*, **262**, 545–548. Model for action of local anaesthetics.
Lee, C. O., Taylor, A., and Windhager, E. C. (1980). *Nature (London)*, **287**, 859–861. Cytosolic calcium ion activity in epithelial cells of *Necturus* kidney.
Lefkowitz, R. J., Roth, J., and Pastan, I. (1970). *Nature (London)*, **228**, 864–866. Effects of calcium on ACTH stimulation of the adrenal: separation of hormone binding from adenyl cyclase activation.
Lehman, W. (1976). *Int. Rev. Cytol.*, **44**, 55–92. Phylogenetic diversity of the proteins regulating muscular contraction.
Lehman, W. (1977). *Biochem. J.*, **163**, 291–296. Calcium ion-dependent myosin from decapod-crustacean muscles.
Lehman, W., Jones, J. K., and Szent-Györgi, A. (1972). *Cold Spring Harbor Symp. Quant. Biol.*, **37**, 319–330. Myosin-linked regulatory systems: comparative studies.
Lehninger, A. L. (1950). *Physiol. Rev.*, **30**, 393–429. Role of metal ions in enzyme systems.
Lehninger, A. L. (1962). *Physiol. Rev.*, **42**, 467–517. Water uptake and extrusion by mitochondria in relation to oxidative phosphorylation.
Lehninger, A. L. (1970). *Biochem. J.*, **119**, 129–138. Mitochondria and calcium ion transport.
Le John, H. B., Jackson, S. G., Klassen, G. R., and Sawula, R. V. (1969). *J. Biol. Chem.*, **244**, 5345–5356. Regulation of mitochondrial glutamic dehydrogenase by divalent metals, nucleotides, and α-keto-glutarate.
Le Marchand, Y., and Jeanrenaud, B. (1977). *Horm. Cell Regul.*, **1**, 77–90. Role of microtubules in hepatic secretory processes, and metabolic consequences of colchicine administration *in vivo*.
Lenhoff, H. M., and Bovaird, J. (1959). *Science*, **130**, 1474–1476. Requirement of bound calcium for the action of surface chemoreceptors.
Lenhoff, H. M., and Loomis, W. F. (Eds.) (1961). *The Biology of Hydra and Some Other Coelenterates*. University of Miami Press, Florida.
Leoni, S., Spaguolo, S., and Panzali, A. (1978). *FEBS Lett.*, **92**, 63–67. Rat liver adenylate cyclase and phosphodiesterase dependence on Ca^{2+} and on cytoplasmic factors during liver regeneration.
Le Peuch, C. J., Haiech, J., and Demaille, J. G. (1979). *Biochemistry*, **18**, 5150–5157. Concerted regulation of cardiac sarcoplasmic reticulum calcium transport by cyclic adenosine monophosphate dependent and calcium-calmodulin-dependent phosphorylations.
Levandowsky, M., and Hauser, D. C. R. (1978). *Int. Rev. Cytol.*, **53**, 145–210. Chemosensory responses of swimming algae and protozoa.
Levin, R. M., and Weiss, B. (1976). *Mol. Pharmacol.*, **12**, 581–589. Mechanism by which psychotropic drugs inhibit adenosine cyclic 3′,5′-monophosphate phosphodiesterase of brain.
Levin, R. M., and Weiss, B. (1978). *Biochim. Biophys. Acta*, **540**, 197–204. Specificity of the binding of trifluoperazine to the calcium-dependent activator of phosphodiesterase and to a series of other calcium-binding proteins.
Levine, B. A., Coffman, D. D., and Thornton, J. M. (1977). *J. Mol. Biol.*, **115**, 743–760. Calcium binding by troponin-C. A proton magnetic resonance study.
Levine, B. A., Thornton, J. M., Fernandes, R., Kelly, C. M., and Mercola, D. (1978). *Biochim. Biophys. Acta*, **535**, 11–24. Comparison of the calcium and magnesium induced structural changes in troponin-C. A magnetic resonance study.
Levitzki, A., and Reuben, J. (1973). *Biochemistry*, **12**, 41–44. Abortive complexes of α-amylase with lanthanides.
Lewis, B. A., Freyssinet, J. M., Holbrook, J. J. (1978). *Biochem. J.*, **169**, 397–402, An equilibrium study of metal ion binding to human plasma coagulation factor XIII.
Li, H. J., Hu, A. W., Maciewicz, R. A., Cohen, P., Santella, R. M., and Chang, P. (1977). *Nucleic Acid Res.*, **4**, 3834–3854. Structural transition in chromatin induced by ions in solution.
Liao, S. (1975). *Int. Rev. Cytol.*, **41**, 87–172. Cellular receptors and mechanisms of action of steroid hormones.
Lichtman, M. A., Jackson, A. H., Peck, W. A. (1972). *J. Cell. Physiol.*, **80**, 383–396. Lymphocyte cation metabolism: cell volume, cation content and cation transport.
Lillie, R. S. (1936). *Biol. Rev.*, **11**, 181–201. The passive iron wire mould of protoplasmic and nervous transmission and its physiological analogues.
Lillie, R. S. (1941). *Physiol. Zool.*, **14**, 239–267. Further experiments on artificial parthenogenesis in starfish eggs, with a review.

Limas, C. J., and Cohn, J. N. (1973). *Proc. Soc. Exp. Biol. Med.*, **142**, 1230–1234. Regulation of myocardial prostaglandin dehydrogenase activity. The role of cyclic 3′,5′-AMP and calcium ions.
Ling, R. C. (1965). *Bull. Math. Biophys.*, **26**, 291–294. On the stimulation and excitation of a system.
Lipicky, J., and Bryant, S. H. (1963). *Am. J. Physiol.*, **208**, 480–482. Potassium efflux of squid giant axons in low external divalent cation.
Lipton, S. A., Rasmussen, H., and Dowling, J. E. (1977). *J. Gen. Physiol.*, **70**, 771–797. Electrical and adaptive properties of rod photoreceptors in *Bufo marinus*. II. Effects of cyclic nucleotides and prostaglandins.
Liu, C.-M., and Hermann, T. E. (1978). *J. Biol. Chem.*, **253**, 5892–5894. Characterization of ionomycin as a calcium ionophore.
Llinas, R., and Nicholson, N. C. (1975). *Proc. Natl. Acad. Sci. U.S.A.*, **72**, 187–190. Calcium role in depolarisation-secretion coupling: an aequorin study in squid giant synapse.
Locke, F. S. (1894). *Zentralbl. Physiol.*, **8**, 166–167. Notiz über den Einfluss, physiologischer Kochsalzlösung anf die Erregbarkeit von Muskel und Nerv.
Loeb, J. (1906). *The Dynamics of Living Matter*. The Columbia University Press, New York.
Loeb, J. (1922). *Proteins and the Theory of Colloidal Behaviour*. New York.
Loew, D. (1892). *Flora*, **75**, 368–394. Nober die Physiologischen Functionen der Calcium-und Magnesiumsalze in Pflanzen organismus.
Loewenhaupt, B. (1956). *Biol. Rev.*, **31**, 371–395. The transport of calcium and other cations in submerged aquatic plants.
Loewenstein, W. R. (1966). *J. Colloid Interface Sci.*, **25**, 34–46. Cell surface membranes in close contact. Role of calcium and magnesium ions.
Loewenstein, W. R. (1975). *Cold Spring Harbor Symp. Quant. Biol.*, **40**, 49–63. Permeable junctions.
Loewenstein, W. R. (1979). *Biochim. Biophys. Acta*, **560**, 1–65. Junctional intercellular communication and the control of cell growth.
Loewenstein, W. R., and Penn, R. D. (1967). *J. Cell Biol.* **33**, 235–242. Intercellular communication and tissue growth II. Tissue regeneration.
Loewenstein, W. R., and Rose, B. (1978). *Ann. N.Y. Acad. Sci.*, **307**, 285–307. Calcium in (junctional) intercellular communication and a thought on its behaviour in intracellular (junctional) communication.
Loewi, O. (1917). *Naunyn-Schmiedebergs Arch. Pharmacol.*, **82**, 131–158. Über den Zusammonhang zwischen Digitalis und Calcium-wirkung.
Loewi, O. (1918). *Arch. Exp. Pathol. Pharmakol.*, **82**, 131–158. Über den zusammenhang zwischen Digitalis- und Kalziumwirkung.
Loewi, O. (1921). *Pflügers Arch. Gesamte Physiol.*, **189**, 239–242. Über humorale Ubertragbarkeit de Herz-nervenwirkung.
Loewy, A. R. (1952). *J. Cell. Comp. Physiol.*, **40**, 127–156. An actomyosin-like substance from the plasmodium of a myxomycete.
Lohmann, K. (1929). *Naturwissenschaften*, **17**, 624. Nachweise von Atomtrümmern aus Aluminium mit dem Hoffmannischen Elektrometer.
Long, C., and Mouat, B. (1971). *Biochem. J.*, **123**, 829–836. The binding of calcium ions by erythrocytes and 'ghost'-cell membranes.
Long, C., and Mouat, B. (1973). *Biochem. J.*, **132**, 559–570. The influx of calcium ions to human erythrocytes during cold storage.
Loomis, W. F. (1954). *J. Exp. Zool.*, **126**, 223–234. Environmental factors controlling growth in *Hydra*.
Lopatin, R. N., and Gardner, J. D (1978). *Biochim. Biophys. Acta*, **543**, 465–475. Effects of calcium and chelating agents on the ability of various agonists to increase cyclic GMP in pancreatic acinar cells.
Lorand, L., Weissmann, L. B., Epel, D. L., and Brumer-Lorand, J. (1976). *Proc. Natl. Acad. Sci. U.S.A.*, **73**, 4479–4481. Role of intrinsic transglutaminase in the Ca^{2+}-mediated cross-linking of erythrocytes proteins.
Lucké, B., and McCutcheon, M. (1932). *Physiol. Rev.*, **12**, 68–139. Living cell as osmotic system and its permeability to water.
Lucy, J. A. (1970) *Nature (London)* **227**, 815. Refusion of biological membranes.
Lucy, J. A. (1975). *Biochem. Soc. Trans.*, **3**, 611–613. Lipids and cell fusion.
Lukas, R. J., Morimoto, H., and Bennett, E. L. (1979). *Biochemistry*, **18**, 2384–2395. Effects of thio-group modification and Ca^{2+} on agonist–specific state transitions of a central nicotinic acetylcholine receptor.
Lundberg, A. (1958). *Physiol. Rev.*, **38**, 21–40. Electrophysiology of salivary glands.

Lüscher, E. F., Probst, E., and Bettex-Galland, M. (1972). *Ann. N.Y. Acad. Sci.*, **201**, 122–130. Thrombostenin: structure and function.
Luther, R. Z. (1898). *Phys. Chem.*, **27**, 364.
Lüttgau, H. C. (1963). *J. Physiol. (London)*, **168**, 679–697. The action of calcium ions on potassium contractures of single muscle fibres.
Lux, H. D., and Hofmeier, G. (1979). Effects of calcium currents and intracellular free calcium in *Helix* neurones. In *Detection and Measurement of Free Ca^{2+} in Cells* (Eds. C. C. Ashley and A. K. Campbell), pp. 409–422. Elsevier/North-Holland, Amsterdam.
Lynne, W. S., and Mukherjee, C. (1978). *Am. J. Pathol.*, **91**, 581–593. Motility of human PMN's.
Macan, T. T. (1963). *Freshwater Ecology*. Longmans, London.
Macan, T. T., and Worthington, E. B. (1974). *Life in Lakes and Rivers*. New Naturalist, Collins, London.
Macklon, A. E. S. (1975). *Planta (Berlin)*, **122**, 131–141. Cortical cell fluxes and transport to the stele in excised root segments of *Allium cepa* L. II. Calcium.
MacLennan, D. H., and Campbell, K. P. (1979). *TIBS*, **4**, 148–151. Structure, function and biosynthesis of sarcoplasmic reticulum proteins.
MacLennan, D. H., and Wong, P. T. S. (1971). *Proc. Natl. Acad. Sci. U.S.A.*, **68**, 1231–1235. Isolation of a calcium-sequestering protein from sarcoplasmic reticulum.
MacLennan, D. H., Yip, C. C., Iles, G. H., and Seeman, P. (1972). *Cold Spring Harbor Symp. Quant. Biol.*, **37**, 469–477. Isolation of sarcoplasmic reticulum proteins.
MacLeod, R. A. (1965). *Bact. Rev.*, **29**, 9–23. The question of existence of specific marine bacteria.
MacLeod, R. A. and Matula, T. I. (1961). *Nature (London)* **192**, 1209–1210. Solute requirements for preventing lysis of some marine bacteria.
Maggio, B., Akhong, Q. F., and Lucy, J. A. (1976). *Biochem. J.*, **158**, 647–650. Polyethylene glycol, surface potential and cell fusion.
Maino, V. C., Green, N. W., and Crumpton, M. J. (1974). *Nature (London)*, **251**, 324–327. The role of calcium ions in initiating transformation of lymphocytes.
Majno, G. (1964). Death of liver tissue. In *The Liver*, vol. II (Ed. Ch. Rouiller), pp. 267–313. Academic Press, New York.
Malaisse, W. (1977). *Biochem. Soc. Trans.*, **5**, 872–875. Calcium ion fluxes and insulin release in pancreatic islets.
Malaisse, W. J., and Couturier, E. (1978). *Nature (London)*, **275**, 664–665. Ionophoretic model for Na–Ca counter transport.
Malaisse, W. J., Devis, G., Pipeleers, D. G., and Somers, G. (1976). *Diabetologia*, **12**, 77–81. Calcium-antagonists and islet function: VI. Effect of D600.
Malaisse, W. J., Devis, G., and Somers, G. (1977a). *Experientia*, **33**, 1035–1036. Inhibition by verapamil of ionophore-mediated calcium translocation.
Malaisse, W. J., Herchuelz, A., Levy, J., and Sener, A. (1977b). *Biochem. Pharmacol.*, **26**, 735–740. Calcium antagonists and islet function. III. The possible site of action of verapamil.
Maller, J. L., and Krebs, E. G. (1977). *J. Biol. Chem.*, **252**, 1712–1718. Progesterone stimulated meiotic cell division in *Xenopus* oocytes. Induction by regulatory subunit and inhibition by catalytic subunit of adenosine 3′ : 5′-monophosphate dependent protein kinase.
Malmström, K., and Carafoli, E. (1975). *Arch. Biochem. Biophys.*, **171**, 418–423. Effects of prostaglandins on the interaction of Ca^{2+} with mitochondria.
Manery, J. F. (1954). *Physiol. Rev.*, **34**, 334–417. Water and electrolyte metabolism.
Manery, J. F. (1961). Minerals in non-osseous connective tissues (including the blood, lens and cornea). In *Mineral Metabolism*, Vol. I, Part 1B (Eds. C. L. Comer and F. Bronner), pp. 551–608. Academic Press, New York and London.
Manery, J. F. (1966). *Fed. Proc.*, **25**, 1804–1810. Effects of Ca ions on membranes.
Manery, J. F. (1969). Calcium and membranes. In *Mineral Metabolism*, Vol. 3 (Eds. C. L. Comer and F. Bronner), pp. 405–452. Academic Press, New York and London.
Mannherz, H. G., and Goody, R. S. (1976). *Ann. Rev. Biochem.*, **45**, 427–466. Proteins of contractile systems.
March, B. B. (1951). *Nature (London)*, **167**, 1065–1066. A factor modifying muscle fibre syneresis.
Marchbanks, R. M. (1975). The origin and some possible mechanisms of the release of acetylcholine at synapses. In *Metabolic Compartmentation and Neurotransmission*. (Eds. S. Berl *et al.*), Plenum Press, New York.
Marcum, J. M., Dedman, J. R., Brinkley, B. R., and Means, A. R. (1978). *Proc. Natl. Acad. Sci. U.S.A.*, **75**, 3771–3775. Control of microtubule assembly–disassembly by calcium-dependent regulator protein.

Margoulis, L. (1973). *Int. Rev. Cytol.*, **34**, 333–361. Colchicine-sensitive microtubules.
Margoulis, L. (1971). *Origin of Eukaryotic Cells.*, Yale University Press, New Haven.
Margreth, A., Cantani, C., and Schiaffino, S. (1967). *Biochem. J.* **102**, 35C–37C. Isolation of microsome-bound phosphofructokinase from frog skeletal muscle and its inhibition by calcium ions.
Martin, A. R. (1966). *Physiol. Rev.*, **46**, 51–66. Quantal nature of synaptic transmission.
Martin, D. R., and Williams, R. J. P. (1975). *Biochem. Soc. Trans.*, **3**, 166–167. The nature and function of alamethicin.
Martin, R. B., and Richardson, F. S. (1979). *Q. Rev. Biophys.*, **12**, 181–209. Lanthanides as probes for calcium in biological systems.
Martonosi, A. (1972). *Curr. Top. Membr. Transp.*, **3**, 83–197. Biochemical and clinical aspects of sarcoplasmic reticulum function.
Martonosi, A. N., Chyn, T. L., and Schibeci, A. (1978). *Ann. N. Y. Acad. Sci.*, **307**, 148–159. The calcium transport of sarcoplasmic reticulum.
Mason, M. (1974). *Biochem. Biophys. Res. Commun.*, **60**, 64–69. Effects of calcium ions and quinolinic acid on rat kidney mitochondrial kynurenine aminotransferase.
Mason, W. T., Fager, R. S., and Abrahamson, E. W. (1974). *Nature (London)*, **247**, 562–563. Ion fluxes in disk membranes of retinal rod outer segments.
Massini, P., and Luscher, E. F. (1974). *Biochim. Biophys. Acta*, **372**, 109–121. Some effects of ionophores for divalent cations on blood platelets. Comparison with effects of thrombin.
Mast, S. O., and Pace, D. M. (1939). *J. Cell. Comp. Physiol.*, **14**, 261–279. The effects of calcium and magnesium on metabolic processes in *Chilonomas* and *Paramecium*.
Masui, Y., and Clarke, H. J. (1979). *Int. Rev. Cytol.*, **57**, 185–282. Oocyte maturation.
Matthews, E. K. (1970). Calcium and hormone release. In *Calcium and Cell Regulation* (Ed. A. W. Cuthbert). Macmillan, London.
Matthews, E. K. (1975). Calcium and stimulus–secretion coupling in pancreatic islets cells. In *Calcium Transport in Contraction and Secretion* (Eds. E. Carafoli, F. Clementi, W. Drabikowsky and A. Margreth), pp. 203–210. North-Holland, Amsterdam.
Matthews, E. K. (1977). *Horm. Cell Regul.*, **1**, 57–76. Insulin secretion.
Matthews, E. K., and Saffran, M. (1973). *J. Physiol. (London)*, **234**, 43–64. Ionic dependence of adrenal steroidogenesis and ACTH-induced changes in the membrane potential of adrenocortical cells.
Maunder, C. A., Yarom, R., and Dubowitz, V. (1977). *J. Neurol. Sci.*, **33**, 323–334. Electron-microscopic X-ray microanalysis of normal and diseased human muscle.
Maunder-Sewry, C. A. (1980). Ph.D. Thesis, University of London.
Maunder-Sewry, C. A., and Dubowitz, V. (1979). *J. Neurol. Sci.*, **42**, 337–347. Myonuclear calcium in carriers of Duchenne muscular dystrophy: an X-ray microanalysis study.
May, J. E., and Frank, M. M. (1973). *Proc. Natl. Acad. Sci. U.S.A.*, **70**, 649–652. A new complement-mediated cytolytic mechanism—the C1-by-pass activation pathway.
Mazia, D. (1937). *J. Cell. Comp. Physiol.*, **10**, 291–304. The release of calcium in *Arbacia* eggs upon fertilisation.
McCall, D. (1976). *J. Gen. Physiol.*, **68**, 537–549. Effect of verapamil and of extracellular Ca and Na on contraction frequency of cultured heart cells.
McDonald, J. M., Bruns, D. E., and Jarett, L. (1976). *Proc. Natl. Acad. Sci. U.S.A.*, **73**, 1542–1546. Ability of insulin to increase calcium binding by adipocyte plasma membranes.
McDonald, J. M., Bruns, D. E., and Jarett, L. (1978). *J. Biol. Chem.*, **253**, 3504–3508. Ability of insulin to increase calcium uptake by adipocyte endoplasmic reticulum.
McDonald, K., Pickett-Heaps, J. D., MeIntosh, J. R., and Tippitt, D. H. (1977). *J. Cell Biol.*, **74**, 377–388. On the mechanism of anaphase spindle elongation in *Diatoma vulgare*.
McDowell, J. H., and Kühn, H. (1977). *Biochemistry*, **16**, 4054–4060. Light-induced phosphorylation of rhodopsin cattle photoreceptor membranes: substrate activation and inactivation.
McIlwain, H. (1952). *Biochem. J.*, **52**, 289–295. Phosphates of brain during *in vitro* metabolism; effects of oxygen, glucose, glutamate, glutamine, and calcium and potassium salts.
McKinley, D., and Meissner, G. (1978). *J. Membr. Biol.*, **44**, 159–186. Evidence for a K^+, Na^+ permeable channel in sarcoplasmic reticulum.
McLean, F. C., and Hastings, A. B. (1934). *J. Biol. Chem.*, **107**, 337–350. A biological method for the estimation of calcium ion concentration.
McLean, F. C., and Hastings, A. B. (1935). *J. Biol. Chem.*, **108**, 285–322. The state of calcium in the fluids of the body. I. Conditions affecting the ionisation of calcium.
McMahon, D. (1974). *Science*, **185**, 1012–1021. Chemical messengers in development: a hypothesis.

Means, A. R., and Dedman, J. R. (1980). *Nature (London)*, **285**, 73–77. Calmodulin—an intracellular calcium receptor.
Means, A. R., Tash, J. S., and Chafouleas, J. G. (1982). *Physiol. Rev.*, **62**, 1–39. Physiological implications of the presence, distribution and regulation of calmodulin in eukaryotic cells.
Meech, R. W. (1976). *Symp. Soc. Exp. Biol.*, **30**, 161–191. Intracellular calcium and the control of membrane permeability.
Meech, R. W. (1979). *Annu. Rev. Biophys. Bioeng.*, **7**, 1–18. Calcium-dependent potassium activation in nervous tissue.
Meech, R. W., and Standen, N. B. (1975). *J. Physiol. (London)*, **249**, 211–239. Potassium activation in *Helix aspersa* neurones under voltage clamp: a component mediated by calcium influx.
Meeves, H. (1968). *Pflügers Arch.*, **304**, 215–241. The ionic requirements for the production of action potentials in *Helix pomatia* neurones.
Meissner, H. P., and Schmelz, H. (1974). *Arch. Gesamte Physiol.*, **351**, 195–206. Membrane potential of beta-cells in pancreatic islets.
Mela, L. (1968). *Fed. Proc.*, **27**, 828. La^{3+} as a specific inhibitor of divalent cation-induced mitochondrial electron transfer.
Mellanby, J., and Williamson, D. A. (1963). *Biochem. J.*, **88**, 440–444. The effect of calcium ions on ketone-body production by rat liver slices.
Mennes, P. A., Yates, J., and Klahr, S. (1978). *Proc. Soc. Exp. Biol. Med.*, **157**, 168–174. Effects of ionophore A23187 and external calcium concentrations on renal gluconeogenesis.
Meroney, W. H., Arney, G. K., Segar, W. E., and Balch, H. H. (1957). *J. Clin. Invest.*, **36**, 825–832. The acute calcification of traumatized muscle, with particular reference to renal insufficiency.
Metchnikoff, E. (1905). *Immunity in Infective Diseases*. Cambridge University Press, Cambridge.
Meyer, W. L., Fischer, E. H., and Krebs, E. G. (1964). *Biochemistry*, **3**, 1033–1039. Activation of skeletal muscle phosphorylase *b* kinase by Ca^{2+}.
Michell, R. H. (1975). *Biochim. Biophys. Acta*, **415**, 81–147. Inositol phospholipids and cell surface receptor function.
Michell, R. H., Jones, L. M., and Jafferji, S. S. (1977). *Biochem. Soc. Trans.*, **5**, 77–81. A possible role for phosphatidyl inositol breakdown in muscarinic cholinergic stimulus–response coupling.
Mikawa, T., Toy-Ota, T., Nonomura, Y., and Ebashi, S. (1977). *J. Biochem.*, **81**, 273–275. Essential factor of gizzard 'troponin'. A new type of regulatory protein.
Mikawa, T., Nonomura, Y., Hirata M., Ebashi, S., and Kakinchi, S. (1978). *J. Biochem.*, **84**, 1633–1636. Leiotonin, a new Ca^{2+} binding protein from smooth muscle.
Mildvan, A. S., and Cohn, M. (1970). *Adv. Enzymol.*, **33**, 1–69. Aspects of enzyme mechanisms studied by nuclear spin relaxation induced by paramagnetic probes.
Miledi, R. (1971). *Nature (London)*, **229**, 410–411. Lanthanum ions abolish the 'calcium response' of nerve terminals.
Miledi, R. (1973). *Proc. R. Soc. London Ser. B*, **183**, 421–425. Transmitter release induced by injection of calcium ions into nerve terminals.
Miledi, R. (1980). *Proc. R. Soc. London Ser. B*, **209**, 447–452. Intracellular calcium and desensitization of acetyl choline receptors.
Miledi, R., and Slater, C. R. (1966). *J. Physiol. (London)*, **184**, 473–498. The action of calcium on neuronal synapses in the squid.
Miller, C. (1978). *J. Membr. Biol.*, **40**, 1–23. Voltage-gated cation conductance channel from fragmented sarcoplasmic reticulum: steady-state electrical properties.
Miller, D. J., and Moisescu, D. G. (1976). *J. Physiol. (London)*, **259**, 283–308. The effects of very low external calcium and sodium concentrations on cardiac contractile strength and calcium–sodium antagonism.
Miller, R. L., and Brokaw, C. J. (1970). *J. Exp. Biol.*, **52**, 699–706. Chemotactic turning behaviour of *Tubularia* spermatozoa.
Milligan, J. V., and Kraicer, J. (1974). *Endocrinology*, **94**, 435–443. Physical characteristics of the Ca^{2+} compartments associated with *in vitro* ACTH release.
Milner, R. G., and Hales, C. N. (1968). *Biochim. Biophys. Acta*, **150**, 165–167. Cations and the secretion of insulin.
Milutinovic, S., Argent, B., Schulz, I., Haase, W., and Sachs, G. (1977). *J. Membr. Biol.*, **36**, 281–295. Studies on isolated subcellular components of cat pancreas. II. A Ca^{2+} dependent interaction between membranes and zymogen granules of the cat pancreas.
Mimura, N., and Asano, A. (1978). *Nature (London)*, **272**, 273–276. Actin-related gelation of Ehrlich tumour cell extracts is reversibily inhibited by low concentrations of Ca^{2+}.

Minchin, E. A. (1898). *Q. J. Microsc. Sci.*, **40**, 469. Calcareous spicules in sponges.
Mines, G. R. (1910). *J. Physiol. (London)*, **40**, 327–346. The action of beryllium, lanthanum, yttrium and cerium on the frog's heart.
Mines, G. R. (1911). *J. Physiol. (London)*, **42**, 251–266. On the replacement of calcium in certain neuro-muscular mechanisms by allied substances.
Mines, G. R. (1913). *J. Physiol. (London)*, **46**, 188–235. On functional analysis by the action of electrolytes.
Mitchell, P. (1966). *Biol. Rev.*, **41**, 445–502. Chemiosmotic coupling in oxidative and photosynthetic phosphorylation.
Mitchell, P. (1968). *Chemiosmotic Coupling and Energy Transduction.* Glynn Research Ltd., Bodmin.
Moe, O. A., and Butler, L. G. (1972). *J. Biol. Chem.*, **247**, 7315–7319. Yeast inorganic pyrophosphatase. III. Kinetics of Ca^{2+} inhibition.
Moisescu, D. G., and Pusch, H. (1975). *Pflugers Arch. Gesamte Physiol.*, **355**, suppl. 1 R 122. A pH-metric method for determining the concentration relationship Ca/EGTA.
Mongar, J. L., and Schild, H. O. (1958). *J. Physiol. (London)*, **140**, 272–284. The effect of calcium and pH on the anaphyactic reaction.
Montague, W., Green, I. C., and Howell, S. L. (1976). Insulin secretion—the role and mode of action of cyclic AMP. In *Eukaryotic Cell Function and Growth* (Eds. J. E. Dumont, B. L. Brown and M. J. Marshall), pp. 609–631. Plenum Press, New York.
Moore, C. L. (1971). *Biochem. Biophys. Res. Commun.*, **42**, 298–305. Specific inhibition of mitochondrial Ca^{2+} transport of ruthenium red.
Moore, C., and Pressman, B. C. (1964). *Biochem. Biophys. Res. Commun.*, **15**, 562–567. Mechanism of action of valinomycin on mitochondria.
Moore, E. W. (1970). *J. Clin. Invest.*, **49**, 318–334. Ionised calcium in normal serum, ultrafiltrates, and whole blood determined by ion-exchange electrodes.
Moore, L., and Landon, E. J. (1979). *Life Sci.*, **25**, 1029–1034. Reversible inhibition of renal microsome calcium pump by furosemide.
Moore, L., Fitzpatrick, D. F., Chen, T. S., and Landon, E. J. (1974). *Biochim. Biophys. Acta*, **345**, 405–418. Calcium pump activity of the renal plasma membrane and renal microsomes.
Moore, L., Chen, T., Knapp, H. R., Jr., and Landon, E. J. (1975). *J. Biol. Chem.*, **250**, 4562–4568. Energy-dependent calcium sequestration activity in rat liver microsomes.
Moore, L., Davenport, G. R., and Landon, E. J. (1976). *J. Biol. Chem.*, **251**, 1197–1201. Calcium uptake of a rat liver microsomal subcellular fraction in response to *in vivo* administration of carbon tetrachloride.
Mordes, J. P., and Wacker, W. E. C. (1978). *Pharmacol. Rev.*, **29**, 273–300. Excess magnesium.
Moreau, M., and Guerrier, P. (1979). Free calcium changes associated with hormone action in oocytes. In *Detection and Measurement of Free Ca^{2+} in Cells* (Eds. C. C. Ashley and A. K. Campbell), pp. 219–226. Elsevier/North-Holand, Amsterdam.
Moreau, M., Villain, J. P., and Guerrier, P. (1980). *Dev. Biol.*, **78**, 201–214. Free calcium changes associated with hormone action in amphibian oocytes.
Morin, J., and Reynolds, G. T. (1974). *Biol. Bull.*, **147**, 397–410. The cellular origin of bioluminescence in the colonial hydroid *Obelia*.
Morris, M. R., and Northcote, D. H. (1977). *Biochem. J.*, **166**, 603–618. Influence of cations at the plasma membrane in controlling polysaccharide secretion from sycamore suspension cells.
Moyle, J., and Mitchell, P. (1977). *FEBS Lett.*, **73**, 131–136. Electric charge stoichiometry of calcium translocation in rat liver mitochondria.
Mueller, P., and Rudin, D. O. (1968). *Nature (London)*, **217**, 713–719. Action potentials induced in bimolecular lipid membranes.
Murphy, E., Coll, K., Rich, T. L., and Williamson, J. R. (1980). *J. Biol. Chem.*, **255**, 6600–6608. Hormonal effects on calcium homeostasis in isolated hepatocytes.
Murphy, R. C., Hammarström, S., and Samuelsson, B. (1979). *Proc. Natl. Acad. Sci. U.S.A.*, **76**, 4275–4279. A slow-reacting substance from murine mastocytoma cells.
Mustard, J. F., and Packham, M. A. (1968). *Series Haematol.*, **1**, 168–184. Platelet phagocytosis.
Naccache, P. H., Showell, H. J., Becker, E. L., and Sha'afi, R. I. (1979a). *Biochem. Biophys. Res. Commun.*, **89**, 1224–1230. Pharmacological differentiation between the chemotactic factor induced intracellular calcium redistribution and transmembrane calcium influx in rabbit neutrophils.
Naccache, P. H., Volpi, M., Showell, H. J., Becker, E. L., and Sha'afi, R. I. (1979b). *Science*, **203**, 461–463. Chemotactic factor-induced release of membrane calcium.
Nagao, S., Suzuki, Y., Watanabe, Y., and Nozawa, Y. (1979). *Biochem. Biophys. Res. Commun.*, **90**, 261–268. Activation by a calcium-binding protein of guanylate cyclase in *Tetrahymena pyriformis*.

Nagata, N., and Rasmussen, H. (1968). *Biochemistry*, **7**, 3728–3733. Parathyroid hormone and renal cell metabolism.
Nagata, N., and Rasmussen, H. (1970). *Biochim. Biophys. Acta*, **215**, 1–16. Renal gluconeogenesis: effects of Ca^{2+} and H^+.
Nairn, A. C., and Perry, S. V. (1979). *Biochem. J.*, **179**, 89–97. Calmodulin and myosin light chain kinase of rabbit fast skeletal muscle.
Naitoh, Y., and Kaneko, H. (1972). *Science*, **176**, 523–524. Reactivated triton-extracted models of *Paramecium*: modification of ciliary movement by calcium ions.
Nakagawa, G., and Tanaka, M. (1962). *Talenta*, **9**, 847–853. Quantitative treatment of exchange equilibriums involving complexes. I. Polarographic determination of calcium in the presence of magnesium.
Nakamaru, Y., and Schwartz, A. (1971). *Arch. Biochem. Biophys.*, **144**, 16–29. Adenosine triphosphate dependent calcium binding vesicles, Mg^{2+}, Ca^{2+} adenosine triphosphatase and Na^+, K^+ adenosine triphosphatase: distributions in dog brain.
Nakos, M., Higashino, S., and Loewenstein, W. R. (1966). *Science*, **151**, Uncoupling of an epithelial cell membrane junction by calcium ion removal.
Namm, D. H. (1971). *J. Pharmacol. Exp. Ther.*, **178**, 299–310. The activation of glycogen phosphorylase in arterial smooth muscle.
Namm, D. H., Mayer, S. E., and Maltbie, M. (1968). *Mol. Pharmacol.*, **4**, 522–530. The role of potassium and calcium ions in the effect of epinephrine on cardiac adenosine 3′:5′-monophosphate, phosphorylase kinase and phosphorylase.
Nanninga, L. B. (1961a). *Biochim. Biophys. Acta*, **54**, 330–338. The association constant of the complexes of adenosine triphosphate with magnesium, calcium, strontium and barium ions.
Nanninga, L. B. (1961b). *Biochim. Biophys. Acta*, **54**, 338–344. Calculation of free magnesium, calcium and potassium in muscle.
Naora, H., Naora, H., Mirsky, A. E., and Allfrey, V. G. (1961). *J. Gen. Physiol.*, **44**, 713–742. Magnesium and calcium in isolated nuclei.
Naora, H., Naora, H., Izawa, M., Allfrey, V., and Mirsky, A. E. (1962). *Proc. Natl. Acad. Sci. U.S.A.*, **48**, 853–854. Some observations on differences in composition between the nucleus and cytoplasm of the frog oocyte.
Nastuk, W. L., and Liu, J. H. (1966). *Science*, **154**, 266–267. Muscle post junctional membrane-changes in chemosensitivity produced by calcium.
Nayler, W. G., Poole-Wilson, P. A., and Williams, A. (1979). *J. Mol. Cell. Cardiol.*, **7**, 683–706. Hypoxia and calcium.
Needham, D. M. (1942). *Biochem. J.*, **36**, 113–120. The adenosine triphosphatase activity of myosin preparations.
Needham, D. M. (1971). *Machina Carnis: The Biochemistry of Muscular Contraction in its Historical Development*. Cambridge University Press, Cambridge.
Needham, J., and Baldwin, E. Ed (1949). Hopkins and Biochemistry 1861–1947. Heffer and Sons, Cambridge.
Needham, J., and Needham, D. M. (1925–1926). *Proc. R. Soc. London Ser. B*, **99**, 173–199. The hydrogen ion concentration and oxidation–reduction potential of the cell interior before and after fertilisation and cleavage: A micro-injection study on marine eggs.
Nering, I. R., and Morgan, K. G. (1981). *Nature (London)*, **288**, 585–587. Use of aequorin to study excitation-contraction coupling in mammalian smooth muscle.
Neufeld, A. H., Miller, W. H., and Bitensky, M. W. (1972). *Biochim. Biophys. Acta*, **266**, 67–71. Calcium binding to retinal rod disk membranes.
Newsholme, E. A., and Crabtree, B. (1976). *Biochem. Soc. Symp.*, **41**, 61–109. Substrate cycles in metabolic regulation.
Newsholme, E. A., and Gevers, W. (1967). *Vit. Horm.*, **25**, 1–87. Control of glycolysis and gluconeogenesis in liver and kidney cortex.
Newsholme, E. A., and Start, C. (1973). *Regulation in Metabolism*. Wiley, Chichester.
Nichols, D. G. (1978). *Biochem. J.*, **170**, 511–522. Calcium transport and proton electrochemical potential gradient in mitochondria from guinea-pig cerebral cortex and rat heart.
Nichols, K. M., and Rikmenspoel, R. (1978). *Exp. Cell Res.*, **116**, 333–340. Control of flagellar motility in *Euglena* and *Chlamydomonas*. Microinjection of EDTA, EGTA, Mn(2)+ and Zn(2)+.
Niedergerke, R. (1963). *J. Physiol. (London)*, **167**, 551–580. Movements of Ca in beating ventricles of the frog.
Nimmo, H. G., and Cohen, P. (1977). *Adv. Cyclic Nucleotide Res.*, **8**, 145–266. Hormonal control of protein phosphorylation.
Njus, D., and Radda, G. K. (1978). *Biochim. Biophys. Acta*, **463**, 219–244. Bioenergetic processes in chromaffin granules: a new prospective on some old problems.

Noble, D. (1962). *J. Physiol. (London)*, **160**, 317–352. A modification of the Hodgkin-Huxley equations applicable to Purkinje fibre action pace-maker potentials.
Nordmann, J. J., and Currell, G. A. (1975). *Nature (London)*, **253**, 646–647. The mechanism of calcium ionophore-induced secretion from the rat neurohypophysis.
Nordmann, J. J., and Dyball, R. E. J. (1975). *Nature (London)*, **255**, 414–415. New calcium-mobilising agent.
Northover, B. J. (1977). *Gen. Pharmacol.*, **8**, 293–296. Indomethacin—a calcium antagonist.
Nuccitelli, R., and Jaffe, L. F. (1976). The ionic components of the current pulses generated by developing fucoid eggs. *Dev. Biol.*, **49**, 518–531.
Ochs, S., Worth, R. M., and Chan, S.-Y. (1977). Calcium requirement for axoplasmic transport in mammalian nerve. *Nature (London)*, **270**, 748–750.
O'Dea, R. F. and Zatz, M. (1976). *Proc. Natl. Acad. Sci. U.S.*, **73**, 3398–3402. Catecholamine-stimulated cyclic GMP accumulation in the rat pineal: apparent presynaptic site of action.
Ogata, E., Kondo, K., Kimura, S., and Yoshitoshy, Y. (1972). *Biochem. Biophys. Res. Commun.*, **46**, 640–645. Dependency on Ca^{2+} of ATP-stimulated uncoupled oxidation of succinate in rat liver mitochondria.
Ohnishi, S. T. (1978). *Anal. Biochem.*, **85**, 165–179. Characterisation of the murexide method: dual wavelength spectrophotometry of cations under physiological conditions.
Okada, Y., and Murayama, F. (1966). *Exp. Cell Res.*, **44**, 527–551. Requirement of calcium ions for the cell fusion reaction of animal cells by HVJ.
Okamoto, M., and Mayer, M. M. (1977). *J. Immunol.*, **120**, 272–285. Studies on mechanism of action of guinea-pig lymphotoxin. I. Membrane active substances prevent target cell lysis by lymphotoxin.
Okamoto, M., and Mayer, M. M. (1978). *J. Immunol.*, **120**, 279–285. Studies on the mechanism of action of guinea-pig lymphotoxin. II. Increase of calcium uptake rate in LT-damaged target cells.
Oliveira-Castro, G. M., and Barcinski, M. A. (1974). *Biochim. Biophys. Acta*, **352**, 338–343. Calcium-induced uncoupling in communicating human lymphocytes.
Oliveira-Castro, G. M., and Loewenstein, W. R. (1971). *J. Membr. Biol.*, **5**, 51. Junctional membrane permeability: effects of divalent cations.
Olmsted, J. B., and Borisy, G. G. (1975). *Biochemistry*, **14**, 2996–3005. Ionic and nucleotide requirements for microtubule polymerisation *in vitro*.
Opie, L. H. (1980). *Lancet*, **i**, 806–810. Calcium antagonists and the heart.
Opie, L. H., Thandroyen, F. T., Muller, C., and Bricknell, O. L. (1979). *J. Mol. Cell. Cardiol.*, **11**, 1073–1094. Adrenaline induced 'oxygen wastage' and enzyme release from working heart. Effects of calcium antagonism, β-blockade, nicotinic acid and coronary artery ligation.
Orci, L., Blondel, B., Malaisse-Lagae, F., Ravazzola, M., Wollheim, C., Malaisse, W. J., and Renold, A. E. (1974). *Diabetologia*, **10**, 382. Cell motility and insulin release in monolayer cultures of endocrine pancreas.
Ordal, G. W. (1977). *Nature (London)*, **270**, 66–67. Calcium ion regulates chemotactic behaviour in bacteria.
Oschman, J. L., and Wall, B. J. (1972). *J. Cell Biol.*, **55**, 58–73. Calcium binding to intestinal membranes.
Ostlund, P. E. (1977). *J. Cell Biol.*, **73**, 78–87. Muscle actin filaments bind pituitary secretory granules *in vitro*.
Ostrom, K. M., and Simpson, T. L. (1978). *Dev. Biol.*, **64**, 332–338. Calcium and the release from dormancy of freshwater sponge gemmules.
Ostwald, T. J., and Heller, J. (1972). *Biochemistry*, **11**, 4679–4686. Properties of magnesium- or calcium-dependent adenosine triphosphatase from frog photoreceptor outer segment disks and its inhibition by illumination.
Overton, E. (1902). *Pflügers Arch. Gesamte Physiol.*, **92**, 346–386. Beiträge zur allgemeinen Muskel- und Nerven-physiologie. II. Ueber die Unentbehrlichkeit von Natrium (oder Lithium-) Ionen für den Contractions act des Muskels.
Overton, E. (1904). *Pflügers Arch. Gesamte Physiol.*, **105**, 176–290. Beitrage zur allgemeinen Muskel- und Nerven physiologie. III Mittheilung. Studien über die Wirkung der Alkali- und Erdalkali-salze auf Skelettmuskeln und Nerven.
Owen, J. D. (1976). *Biochim. Biophys. Acta*, **451**, 321–325. The determination of the stability constant for calcium-EGTA.
Owen, J. D., and Brown, H. M. (1979). Comparison between free calcium-selective micro-electrodes and arsenazo III. In *Detection and Measurement of Free Ca^{2+} in Cells*. (Eds. C. C. Ashley, and A. K. Campbell), pp. 395–408. Elsevier/North-Holland, Amsterdam.

Owen, J. D., Brown. H. M., and Pemberton, J. P. (1977). *Anal. Chim. Acta*, **90**, 241–244. Neurophysiological applications of a calcium-sensitive microelectrode.
Owens, O. V. H., Gey, M. K., and Gey, G. O. (1956). *Fed. Proc.*, **15**, 140. Mineral requirement of mammalian cells grown in aqueous fluid media.
Owens, O. V. H., Gey, M. K., and Gey, G. O. (1958). *Cancer Res.*, **18**, 968–973. Effect of Ca and Mg on the growth and morphology of mouse lymphoblasts
Ozawa, E., Hosoi, K., and Ebashi, S. (1967). *J. Biochem. (Tokyo)*, **61**, 531–533. Reversible stimulation of muscle phosphorylase *b* kinase by low concentration of calcium ions.
Page, R. C., Davies, P., and Allison, A. C. (1978). *Int. Rev. Cytol.*, **52**, 119–157. The macrophage as a secretory cell.
Pannbacker, R. G. (1973). *Science*, **182**, 1138–1140. Control of guanylate cyclase activity in rod outer segment.
Pantin, C. F. A. (1926). *Br. J. Exp. Biol.*, **3**, 275. On the physiology of amoeboid movement. III. The action of calcium.
Papahadjopoulos, D. (1972). *Biochim. Biophys. Acta*, **265**, 169–186. Studies on the mechanism of action of local anesthetics with phospholipid model membranes.
Pappano, A. J. (1970). *Circ. Res.*, **27**, 379–390. Calcium-dependent action potentials produced by catecholamines in guinea-pig atrial muscle fibres depolarised by potassium.
Parello, J., Lilja, H., Cave, A., and Lindman, B. (1978). *FEBS Lett.*, **87**, 191. A ^{43}Ca NMR study of the binding of calcium to parvalbumins.
Parfili, E., Sandri, G., Sottocasa, G. L., Lunazzi, G., Liut, G., and Graziosi, G. (1976). *Nature (London)*, **264**, 185–186. Specific inhibition of mitochondrial Ca^{2+} transport by antibodies directed to Ca^{2+}-binding glycoproteins.
Park, C. S., and Malvin, R. L. (1978). *Am J. Physiol.*, **235**, F22–F25. Calcium in the control of renin release.
Parker, I. (1979). Use of arsenazo III for recording calcium transients in frog skeletal muscle. In *Detection and Measurement of Free Ca^{2+} in Cells*. (Eds. C. C. Ashley and A. K. Campbell), pp. 269–286. Elsevier/North-Holland, Amsterdam.
Parks, R. E., Ben-Gershom, E., and Lardy, H. A. (1957). *J. Biol. Chem.*, **227**, 231–242. Liver fructokinase.
Pasteels, J. J. (1935). *Arch. Biol.*, **46**, 229–262. Récherches sur le déterminisme de l'entrée en maturation de l'oeuf chez divers Invertébrés marins.
Pasteels, J. (1938). *Trav. Stn. Zool. Wimereux*, **13**, 515–530. Le rôle de calcium dans l'activation de l'oeuf de pholade.
Pasternak, C. A. (1980). *Biochem. Soc. Trans.*, **8**, 700–702. Virally mediated changes in cellular permeability.
Paul, J. (1961, 1972, 1975). *Cell and Tissue Culture*. Churchill-Livingstone, Edinburgh.
Paul, M. (1975). *Dev. Biol.*, **43**, 299–312. Release of acid and changes in light scattering properties following fertilisation of *Urechis caupo* eggs.
Pautard, F. G. E. (1970). Calcification in unicellular organisms. In *Biological Calcification, Cellular and Molecular Aspects*, (Ed. H. Schraer). Meredith Corp., New York.
Peach, M. J. (1972). *Proc. Natl. Acad. Sci. U.S.A.*, **69**, 834–836. Stimulation of release of adrenal catecholamine by adenosine 3′ : 5′-cyclic monophosphate and theophylline in the absence of extracellular Ca^{2+}.
Peachey, L. D. (1964). *J. Cell Biol.*, **20**, 95–109. Electron microscopic observations on the accumulation of divalent cations in intra-mitochondrial granules.
Pease, D. C., Jenden, D. J., and Howell, J. N. (1965). *J. Cell. Comp. Physiol.*, **65**, 141–154. Calcium uptake in glycerol-extracted rabbit psoas muscle fibres. II. Electron microscopic localisation of uptake sites.
Peaucellier, G. (1977). *Exp. Cell Res.*, **106**, 1–14. Inhibition of meiotic maturation by specific protease in oocytes of the polychaete annelid *Sabellaria alveolata*.
Pelc, S. R., and Howard, A. (1956). *Exp. Cell Res.*, **11**, 128–134. A difference between spermatogonia and somatic tissues of mice in the incorporation of [8-^{14}C]-adenine into deoxyribonucleic acid.
Penn, R. D. (1966). *J. Cell Biol.*, **29**, 171–174. Ionic communication between liver cells.
Perris, A. D., Whitfield, J. F., and Tölg, P. K. (1968). *Nature (London)*, **219**, 527–529. Role of calcium in control of growth and cell division.
Perry, S. V. (1974). *Biochem. Soc. Symp.*, **39**, 115–132. Calcium ions and the function of contractile proteins of muscle.
Pershadsingh, H. A., and McDonald, J. M. (1979). *Nature (London)*, **281**, 495–497. Direct addition of insulin inhibits a high affinity Ca^{2+}-ATPase in isolated adipocyte plasma membranes.
Peterhans, E., Haenggeli, E., Wild, P., and Wyler, R. (1979). *J. Virol.*, **29**, 143–152. Mitochondrial calcium uptake during infection of chicken embryo cells with Semliki Forest virus.

Petersen, O. H. (1976). *Physiol. Rev.*, **56**, 535–577. Electrophysiology of mammalian gland cells.
Petersen, O. H., and Matthews, E. K. (1972). *Experientia*, **28**, 1037–1038. The effect of pancreozymin and acetylcholine on the membrane potential of the pancreatic acinar cells.
Pett, L. B., and Wynne, A. M. (1933). *Biochem. J.*, **27**, 1660–1671. Studies on bacterial phosphatase. II. The phosphatases of *Clostridium acetobutylicum* Weitzmann and *Propioni bacterium jensenii* Van Niel.
Petzelt, C., and Ledebur-Villiger, M. (1973). *Exp. Cell Res.*, **81**, 87–94. Ca^{2+}-stimulated ATPase during the early development of parthenogenetically activated eggs of the sea urchin *Paracentrotus lividus*.
Pfeiffer, D. R., Reed, P. W., and Lardy, H. A. (1974). *Biochemistry*, **13**, 4007–4013. Ultraviolet and fluorescent spectral properties of the divalent cation ionophore A23187 and its metal ion complexes.
Pfeiffer, D. R., Taylor, R. W., and Lardy, H. A. (1978). *Ann. N. Y. Acad. Sci.*, **307**, 402–423. Ionophore A23187: cation binding and transport properties.
Pfeiffer, D. R., Kauffmann, R. F., and Lardy, H. A. (1978). *J. Biol. Chem.*, **253**, 4163–4171. Effects of N-ethylmaleimide on limited uptake of Ca^{2+}, Mn^{2+} and Sr^{2+} by rat liver mitochondria.
Pickett, W. C., Jesse, R. L., and Cohen, P. (1976). *Biochim. Biophys. Acta*, **486**, 209–213. Initiation of phospholipase A2 activity in human platelets by the calcium ion ionophore A23187.
Picton, C., Klee, C. B., and Cohen, P. (1981). *Cell Calcium*, **2**, 281–294. The regulation of muscle phosphorylase kinase by calcium ions, calmodulin and troponin C.
Piddington, R. W. (1976). *Symp. Soc. Exp. Biol.*, **30**, 55–66. Laser measurement of biological movement.
Pipeleers, D. G., Pipeleers-Marichal, M. A., and Kipnis, D. M. (1976). *Science*, **191**, 88–90. Microtubule assembly and the intracellular transport of secretory granules in pancreatic islets.
Plattner, H. (1974). *Nature (London)*, **252**, 722–724. Intramembranous changes on cationophore triggered exocytosis in *Paramecium*.
Pointer, R. H., Butcher, F. R., and Fain, J. N. (1976). *J. Biol. Chem.*, **251**, 2987–2992. Studies on the role of cyclic guanosine $3':5'$ monophosphate and extracellular Ca^{2+} in the regulation of glycogenolysis in rat liver cells.
Poisner, A. M., and Cooke, P. (1975). *Ann. N. Y. Acad. Sci.*, **253**, 653–669. Microtubules and the adrenal medulla.
Poisner, A. M., and Douglas, W. W. (1966). *Proc. Soc. Exp. Biol. Med.*, **123**, 62–64. The need for calcium in adrenomedullary secretion evoked by biogenic amines, polypeptides, and muscarinic agents.
Pollack, H. (1928). *J. Gen. Physiol.*, **11**, 539–545. Micrurgical studies in cell physiology. VI. Calcium ions in living protoplasm.
Pollard, T. D., and Weihing, R. R. (1974). *Crit. Rev. Biochem.*, **2**, 1–65. Actin and myosin and cell movement.
Poo, M. M., and Cone, R. A. (1973). *Exp. Eye Res.*, **17**, 503–510. Lateral diffusion of rhodopsin in *Necturus* rods.
Poovaiah, B. W., and Leopold, A. C. (1973). *Plant Physiol.*, **51**, 848–851. Inhibition of abscission by calcium.
Popescu, L. M., and Diculescu, I. (1975). *J. Cell Biol.*, **67**, 911–918. Calcium in smooth muscle sarcoplasmic reticulum *in situ*.
Portzehl, H., Caldwell, P. C., and Rüegg, J. C. (1964). *Biochim. Biophys. Acta*, **79**, 581–591. The dependence of contraction and relaxation of muscle fibres from the crab *Maia squinado* on the internal concentration of free calcium ions.
Posner, A. S. (1969). *Physiol. Rev.*, **49**, 760–792. Crystal chemistry of bone mineral.
Poste, G. (1970). *Adv. Virus Res.*, **16**, 303–356. Virus-induced polykaryocytosis and the mechanism of cell fusion.
Poste, G. (1972). *Int. Rev. Cytol.*, **33**, 157–252. Mechanisms of virus-induced cell fusion.
Poste, G., and Allison, A. C. (1973). *Biochim. Biophys. Acta*, **300**, 421–465. Membrane fusion.
Poste, G., and Papahadjopoulos, D. (1976). *Methods in Cell Biol.*, **14**, 23–32. Fusion of mammalian cells by lipid vesicles.
Potter, J. D., and Gergely, J. (1975). *J. Biol. Chem.*, **250**, 4628–4633. The calcium and magnesium binding sites of troponin and their role in the regulation of myofibrillar adenosine triphosphatase.
Potter, J. D., Seidal, J. C., Leavis, P. C., Lehner, S. S., and Gergeley, (1974). in *Calcium Binding Proteins*, pp. 129–152. Eds. W., Drabikowski, H. Strzelecka-Golaszewska, and E. Carafoli, PWN-Polish Scientific Publishers, Warszawa; Elsevier Scientific Publishing Co., Amsterdam.

Potter, V. R., Siekevitz, P., and Simonson, H. C. (1953). *J. Biol. Chem.*, **205**, 893–908. Latent adenosinetriphosphatase activity in resting rat liver mitochondria.
Potts. W. T. W. (1954). *J. Exp. Biol.*, **31**, 376–385. The inorganic composition of the blood of *Mytilus edulis* and *Anodonta cygnea*.
Pressman, B. C. (1976). *Ann. Rev. Biochem.*, 501–530. Biological applications of ionophores.
Price, H. L. (1974). *Anesthesiology*, **41**, 576–579. Calcium reverses myocardial depression caused by halothane.
Price, W. J. (1972). *Analytical Atomic Absorption Spectrometry*. Heyden, London.
Prince, W. T., Rasmussen, H., and Berridge, M. (1973). *Biochim. Biophys. Acta*, **329**, 98–107. The role of calcium in fly salivary gland secretion analysed with the ionophore A23187.
Quie, P. G., Mills, E. L., and Holmes, B. (1971). *Prog. Hematol.*, **10**, 193–210. Molecular events during phagocytosis by human neutrophils.
Raaflaub, J. (1960). *Methods Biochem. Anal.*, **3**, 301–325. Applications of metal buffers and metal indicators in biochemistry.
Rabinovitch, M., and De Stefano, M. J. (1971). *Exp. Cell Res.*, **64**, 275–284. Phagocytosis of erythrocytes by *Acanthamoeba* sp.
Rahamimoff, R., Erulkar, S. D., Alnaes, E., Meiri, H., Rotshenker, S., and Rahaminoff, H. (1975). *Cold Spring Harbor Symp. Quant. Biol.*, **40**, 107–116. Modulation of transmitter release by calcium ions and nerve impulse.
Raisz, L. G., Sandberg, A. L., Goodson, M. J., Simmons, H. A., and Mergenhayer, S. E. (1974). *Science*, 789–791. Complement dependent stimulation of prostaglandin synthesis and bone reabsorption.
Raminez-Munoz, J. (1968). *Atomic-Absorption Spectroscopy*. Elsevier, Amsterdam.
Ramwell, P. W., and Shaw, J. E. (1970). *Recent Prog. Horm. Res.*, **26**, 138–187. Biological significance of the prostaglandins.
Randle, P. J., Garland, P. B., Hales, C. N., Newsholme, E. A., Denton, R. M., and Pogson, C. I. (1966). *Recent Prog. Horm. Res.*, **22**, 1–48. Interactions of metabolism and the physiological role of insulin.
Rasmussen, H. (1970). *Science*, **170**, 404–412. Cell communication, calcium ion, and cyclic adenosine monophosphate.
Rasmussen, H., and Goodman, D. B. P. (1977). *Physiol. Rev.*, **57**, 421–509. Relationship between calcium and cyclic nucleotides in cell activation.
Rasmussen, H., Goodman, D. B. P., and Tenenhouse, A. (1972). *Crit. Rev. Biochem.*, **1**, 95–148. The role of cyclic AMP and calcium in cell activation.
Rasmussen, H., Goodman, D. B. P., Friedmann, N., Allen, J. E., and Kurokawa, K. (1976). Ionic control of metabolism., In *Handbook of Physiology*, Vol. 7, Section 7, pp. 225–264. Eds. R. O. Geep, and E. B. Astwood. American Physiological Society, Washington.
Ray, T. K., Tomasi, V., and Marinetti, G. V. (1970). *Biochim. Biophys. Acta*, **211**, 20–30. Hormone action at the membrane level. I. Properties of adenyl cyclase in isolated plasma membrane of rat liver.
Rebhun, L. (1972). *Int. Rev. Cytol.*, **32**, 93–137. Polarised intracellular particle transport: saltatory movements and cytoplasmic streaming.
Rebhun, L. I. (1977). *Int. Rev. Cyt.*, **49**, 1–54. Cyclic nucleotides, calcium and cell division.
Reed, P. W., and Lardy, H. A. (1972). *J. Biol. Chem.*, **247**, 6970–6977. A23187: a divalent cation ionophore.
Rees, J. K. H., and Coles, H. A. (1969). *Br. Med. J.*, **ii**, 670–671. Calciphylaxis in man.
Reeves, J. P. (1975). *J. Biol. Chem.*, **250**, 9428–9430. Calcium-dependent stimulation of 3-*O*-methyl glucose uptake in rat thymocytes by the divalent cation ionophore A23187.
Reeves, J. P. (1977). *Arch. Biochem. Biophys.*, **183**, 298–305. Heat-induced stimulation of rat thymocyte 3-*O*-methyl glucose transport by mitogenic stimuli.
Reid, G. K. (1961). *Ecology of Inland Waters and Estuaries*. Van Nostrand Reinhold, New York.
Reinhold, M., and Stockem, W. (1972). *Cytobiologie*, **6**, 182. Darstellung eines ATP-sensitiven Membransystems mit Ca^{2+}-Transportierenderfuktion bei Amöben.
Reuter, H. (1973). *Prog. Biophys. Mol. Biol.*, **26**, 1–43. Divalent cations as charge carriers in excitable membranes.
Reuter, H. (1974). *J. Physiol. (London)*, **242**, 429–451. Localisation of beta adrenergic receptors, and effects of nor-adrenaline and cyclic nucleotides on action potentials, ionic currents and tension in mammalian cardiac muscle.
Reuter, H., Blaustein, M. P., and Haeusler, G. (1973). *Philos. Trans. R. Soc. London Ser. B*, **265**, 87–94. Na–Ca exchange and tension development in arterial smooth muscle.

Reynolds, G. T. (1978). *Photochem. Photobiol.*, **27**, 405–421. Application of photosensitive devices to bioluminescence studies.

Reynolds, G. T. (1979). Localization of free ionized calcium in cells by means of image intensification. In *Detection and Measurement of Free Ca^{2+} in Cells* (Eds. C. C. Ashley and A. K. Campbell), pp. 227–244. Elsevier/North-Holland, Amsterdam.

Reznikoff, P., and Chambers, R. (1926–1927). *J. Gen. Physiol.*, **10**, 731. Micrurgical studies in cell physiology. III. The action of CO_2 and some salts of Na, Ca and K on the protoplasm of *Amoeba dubia*.

Richardson, P. J., and Luzio, J. P. (1980). *Biochem. J.*, **186**, 897–906. Complement-mediated production of plasma-membrane vesicles from rat cells.

Ridgway, E. B., and Ashley, C. C. (1967). *Biochem. Biophys. Res. Commun.*, **29**, 229–234. Calcium transients in single muscle fibres.

Ridgway, E. B., and Durham, A. C. H. (1976). *J. Cell Biol.*, **69**, 223–226. Oscillations of calcium ion concentrations in *Physarum polycephalum*.

Ridgway, E. B., Gilkey, J. C., and Jaffe, L. F. (1977). *Proc. Natl. Acad. Sci. U.S.A.*, **74**, 623–627. Free calcium increases explosively in activating meduka eggs.

Riemann, F., Zimmermann, U., and Pilwat, G. (1975). *Biochim. Biophys. Acta*, **394**, 449–462. Release and uptake of haemoglobin and ions in red blood cells induced by dielectric breakdown.

Ringer, S. (1882). *J. Physiol. (London)*, **3**, 380–393. Concerning the influence exerted by each of the constituents of the blood on the contraction of the ventricle.

Ringer, S. (1883a). *Practitioner*, **31**, 81–93. An investigation regarding the action of strontium and barium salts compared with the action of lime on the ventricle of the frog's heart.

Ringer, S. (1883b). *J. Physiol. (London)*, **4**, 29–43. A further contribution regarding the influence of different constituents of the blood on the contractions of the heart.

Ringer, S. (1883c). *J. Physiol. (London)*, **4**, 222–225. A third contribution regarding the influence of the inorganic constituents of the blood on centricular contraction.

Ringer, S. (1886). *J. Physiol. (London)*, **7**, 291–308. Further experiments regarding the influence of small quantities of lime, potassium and other salts on muscular tissue.

Ringer, S. (1890). *J. Physiol. (London)*, **11**, 79–84. Concerning experiments to test the influence of lime, sodium and potassium salts on the development of ova and growth of tadpoles.

Ringer, S., and Sainsbury, H. (1894). *J. Physiol. (London)*, **16**, 1–9. The action of potassium, sodium and calcium salts on *Tubifex rivulorum*.

Rink, T. J., and Tsien, R. Y. (1982). *Biochem. Soc. Trans.*, **10**, 209. Free calcium measurements in very small cells.

Rink, T. J., Tsien, R. Y., and Warner, A. E. (1980). *Nature (London)*, **283**, 658–660. Free calcium in *Xenopus* embryos measured with ion-selective electrodes.

Rizack, M. (1964). *J. Biol. Chem.*, **239**, 392–395. Activation of an epinephrine-sensitive lipolytic activity from adipose tissue by adenosine 3′,5′ cyclic phosphate.

Robblee, L. S., Shepro, D., and Belamarich, F. A. (1973). *J. Gen. Physiol.*, **61**, 462–481. Calcium uptake and associated adenosine triphosphatase activity of isolated platelet membranes.

Robertson, J. D. (1941). *Biol. Rev.*, **16**, 106–133. The function and metabolism of calcium in the Invertebrata.

Robertson, J. D. (1949). *J. Exp. Biol.*, **26**, 182–200. Ionic regulation in some marine invertebrates.

Robertson, J. D. (1953). *J. Exp. Biol.*, **30**, 277–296. Further studies on ionic regulation in marine invertebrates.

Robinson, B. H., and Chappell, J. B. (1967). *Biochem. J.*, **105**, 18P. An apparent energy requirement for oxaloacetate penetration into mitochondria provided by permeant cations.

Robinson, K. R. (1977). *Planta*, **136**, 153–158. Reduced external calcium or sodium stimulates calcium influx in *Pelvetia* eggs.

Robinson, K. R., and Jaffe, L. F. (1974). *Science*, **187**, 70–72. Polarising fucoid eggs drive a calcium current through themselves.

Robinson, R. B., and Sleator, W. W. (1977). *Am. J. Physiol.*, **233**, H203–210. Effects of Ca^{2+} and catecholamines on the guinea-pig atrium action potential plateau.

Robinson, R. A., and Stokes, R. H. (1968). *Electrolyte Solutions*. Butterworths, London.

Robison, G. A., Butcher, R. W., and Sutherland, E. W. (1971). *Cyclic AMP*. Academic Press, New York.

Roitt, I. M. (1980). *Essential Immunology*. 4th Edition, pp. 79–80. Blackwell, Oxford.

Romero, P. J., and Whittam, R. (1971). *J. Physiol. (London)*, **214**, 481–507. The control of internal calcium of membrane permeability to sodium and potassium.

Roobol, A., and Alleyne, G. A. O. (1973). *Biochem. J.*, **134**, 157–165. Regulation of renal

gluconeogenesis by calcium ions, hormones and adenosine 3′:5′-cyclic monophosphate.
Roos, A., and Boron, W. F. (1981). *Physiol. Rev.*, **61**, 296–434. 'Intracellular pH.
Rose, B., and Loewenstein, W. R. (1971). *J. Membr. Biol.*, **5**, 20–50. Junctional membrane permeability. Depression by substitution of Li^+ for extracellular Na and by long-term lack of Ca and Mg; restoration by cell repolarisation.
Rose, B., and Loewenstein, W. R. (1975). *Nature (London)*, **254**, 250–252. Permeability of cell junction depends on local cytoplasmic calcium activity.
Rose, B., and Loewenstein, W. R. (1976). *J. Membr. Biol.*, **28**, 87–119. Permeability of a cell junction and the local cytoplasmic free ionised calcium concentration: a study with aequorin.
Rose, I. A. (1968). *Proc. Natl. Acad. Sci. U.S.A.*, **61**, 1079–1086. The state of magnesium in cells as estimated from the adenylate kinase equilibrium.
Rosen, B. L., and McClees, J. S. (1974). *Proc. Natl. Acad. Sci. U.S.A.*, **71**, 5042–5046. Active transport of calcium in inverted vesicles of *Escherichia coli*.
Rosenberg, H., and Haugaard, N. (1973). *Anesthesiology*, **39**, 44–53. The effects of halothane on metabolism and calcium uptake in mitochondria of the rat liver and brain.
Rosenfeld, A. C. (1976). *FEBS Lett.*, **65**, 144–147. Magnesium stimulation of calcium binding to tubulin and calcium induced depolarisation of microtubules.
Ross, E. M., and Gilman, A. G. (1980). *Ann. Rev. Biochem.*, **49**, 533–564. Biochemical properties of hormone-sensitive adenylate cyclase.
Ross, J. W. (1967). *Science*, **156**, 1378–1379. Calcium-selective electrode with liquid ion exchanger.
Rosselin, G., Freychet, P., and Bataille (1974). *Isr. J. Med. Sci.*, **10**, 1313–1323. Polypeptide hormone-receptor interactions. A new approach to the study of pancreatic and gut glucagons.
Rossi, E. C. (1976). *J. Chron. Dis.*, **29**, 215–219. Platelets and thrombosis.
Rouf, M. A. (1964). *J. Bactiol.*, **88**, 1545–1549. Spectrochemical analysis of inorganic elements in bacteria.
Routledge, L. B. (1978). *J. Cell Biol.*, **77**, 358–370. Calcium binding proteins in the vorticellid spasmoneme: extraction and characterization by gel electrophoresis.
Rubin, R. P. (1970). *Pharmacol. Rev.*, **22**, 389–423. The role of calcium in the release of neurotransmitter substances and hormones.
Rubin, H., and Koide, T. (1976). *Proc. Natl. Acad. Sci. U.S.A.*, **73**, 168–172. Mutual potentiation by magnesium and calcium of growth in animal cells.
Rubin, R. P. (1974). *Calcium and Secretion*. Plenum Press, New York.
Rubino, A. (1936) *Boll. Oculist.*, **15**, 279–292. Comportamento del cristallino in soluzioni varianti in contenuto Ca/K.
Rubinstein, D. L. (1927). *Biochem. Z.*, **182**, 50–64. Ueber die wirkung physiologisch aquilibrierter Salzlösungen.
Rudel, R. (1979). Detection of free calcium transients in single frog muscle fibres using aequorin. In *Detection and Measurement of Free Ca^{2+} in Cells*. (Eds. C. C. Ashley and A. K. Campbell), pp. 153–158. Elsevier/North-Holland, Amsterdam.
Ruffolo, P. R. (1964). *Am. J. Pathol.*, **45**, 741–749. The pathogenesis of necrosis. I. Correlated light and electron microscopic observations of the myocardial necrosis induced by the intravenous injection of papain.
Russell, F. S. (1953, 1969). *The Medusae of the British Isles*, Vols. 1 and 2. Cambridge University Press, Cambridge.
Russell, J. M., and Blaustein, M. B. (1974). *J. Gen. Physiol.*, **63**, 144–167. Calcium efflux from barnacle muscle fibres. Dependence on external cations.
Ruzicka, J., Hansen, E. H., and Tjell, J. C. (1973). *Anal. Chim. Acta*, **67**, 155–178. Selectrode—the universal ion-selective electrode. Part IV. The Calcium (II) selectrode employing a new ion exchanger in a non-porous membrane and solid state reference system.
Sabbatani, L. (1901). *Arch. Ital. Biol.*, **36**, 397; cited by Hasting, A. B., McLean, F. C., Eichelberger, L., Hall, J. L., and Da Costa, E. (1934). *J. Biol. Chem.*, **107**, 351–370.
Sacktor, B., Wu, N.-C., Lescure, O., and Reed, W. D. (1974). *Biochem. J.*, **137**, 535–542. Regulation of muscle phosphorylase kinase activity by inorganic phosphate and calcium ions.
Salisbury, J. C., and Floyd, G. L. (1978). *Science*, **202**, 975–977. Calcium-induced contraction of the rhizoplast of a quadriflagellate green alga.
Samaha, F. J., and Gergely, J. (1965). *J. Clin. Invest.*, **44**, 1425–1431. Ca^{++} uptake and ATPase of human sarcoplasmic reticulum.
Samli, M. H., and Gershwind, I. I. (1968). *Endocrinology*, **82**, 225–231. Some effects of energy-transfer inhibitors and of Ca^{2+} free or K^+-enhanced media on the release of luteinizing hormone (LH) from the rat pituitary gland *in vitro*.

Samuelsson, B. (1978). *Recent Prog. Horm. Res.*, **34**, 239–258. Prostaglandins and thromboxanes.
Sanghvi, I. S., and Gershon, S. (1977). *Biochem. Pharmacol.*, **26**, 1183–1185. Brain calcium and morphine action.
Satir, P. (1975). *Science*, **190**, 586–588. Ionophore-mediated calcium entry induces mussel gill ciliary arrest.
Scarpa, A. (1972). *Methods Enzymol.*, **24**, 343–351. Spectrophotometric measurement of calcium by murexide.
Scarpa, A., and Graziotti, P. (1973). *J. Gen. Physiol.*, **62**, 756–772. Mechanisms for intracellular calcium regulation in heart. I. Stopped flow measurements of Ca^{++} uptake by cardiac mitochondria.
Scarpa, A. (1974). *Biochemistry*, **13**, 2789–2794. Indicators of free magnesium in biological systems.
Scarpa, A. (1979). Measurement of calcium ion concentrations with metallochromic indicators. In *Detection and Measurement of Free Ca^{2+} in Cells* (Eds. C. C. Ashley and A. K. Campbell), pp. 85–115. Elsevier/North-Holland, Amsterdam.
Scarpa, A., Brinley, F. J., Tiffert, T., and Dubyak, G. R. (1978). *Ann. N. Y. Acad. Sci.*, **307**, 86–112. Metallochromic indicators of ionized calcium.
Scarpa, A., Malmstrom, K., Chiesi, M., and Carafoli, E. (1976). *J. Membr. Biol.*, **29**, 205–208. On the problems of the release of mitochondrial calcium by cyclic AMP.
Scarpa, A., Brinley, F. J., and Dubyak, G. (1979). *Biochemistry*, **17**, 1378–1386. Antipyrylazo III, a 'middle range' Ca^{2+} metallochromic indicator.
Scarpelli, D. G. (1965). *Lab. Invest.*, **14**, 123–141. Experimental nephrocalcionosis. A biochemical and morphologic study.
Schachter, D., Kowarski, S., and Reid, P. (1970). Active transport of calcium by intestine: studies with a calcium activity electrode. In *Calcium and Cellular Function* (Ed. A. N. Cuthbert). Macmillan, London.
Schadt, Ch., and Pelks, D. (1975). *FEBS Lett.*, **59**, 36–39. Influence of the ionophore A23187 on myogenic cell fusion.
Schanne, F. A. X., Kane, A. B., Young, E. E., and Farber, J. L. (1979). *Science*, **206**, 700–702. Calcium dependence of toxic cell death: a final common pathway.
Schatzmann, H. J. (1966). *Experientia*, **22**, 364. ATP-dependent Ca^{2+} extrusion from human red cells.
Schatzmann, H. J. (1973). *J. Physiol. (London)*, **235**, 551–569. Dependence on calcium concentration and stoichiometry of the calcium pump in human red cells.
Schatzmann, H. J. (1975). *Curr. Top. Membr. Trans.*, **6**, 125–168. Active calcium transport and Ca^{2+}-activated ATPase in human red cells.
Schell-Frederick, E. (1974). *FEBS Lett.*, **48**, 37–40. Stimulation of the oxidative metabolism of polymorphonuclear leukocytes by the calcium ionophore A23187.
Schimmel, R. J. (1973). *Biochim. Biophys. Acta*, **326**, 272–278. The influence of extracellular calcium ion on hormone-activated lipolysis.
Schimmel, R. J. (1976). *Horm. Metab. Res.*, **8**, 195–201. The role of calcium ion in epinephrine activation of lipolysis.
Schlaepfer, W. W. (1977). *Nature (London)*, **265**, 734–736. Vesicular disruption of myelin simulated by exposure of nerve to calcium ionophore.
Schlaepfer, W. W., and Bunge, R. P. (1973). *J. Cell Biol.*, **59**, 456–470. Effect of calcium ion concentration on the degeneration of amputated axons in tissue culture.
Schliwa, M. (1976). *J. Cell Biol.*, **70**, 527–540. The role of divalent cations in the regulation of microtubule assembly. *In vivo* studies on microtubules of the heliozoan axopodium using the ionophore A23187.
Schneider, F. H. (1972). *J. Pharmacol. Exp. Ther.*, **183**, 80–89. Amphetamine-induced exocytosis of catecholamines from the cow adrenal medulla.
Schofield, J. G., and Bicknell, R. J. (1978). *Mol. Cell. Endocrinol.*, **9**, 255–268. Effects of somatostatin and verapamil on growth hormone release and 45 Ca fluxes.
Scholey, J. M., Taylor, K. A., and Kendrick-Jones, J. (1980). *Nature (London)*, **287**, 233–235. Regulation of non-muscle myosin assembly by calmodulin-dependent light chain kinase.
Schopf, J. W. (1970). *Biol. Rev.*, **45**, 319–352. Precambrian micro-organisms and evolutionary events prior to the origin of vascular plants.
Schopf, J. W. (1975). *Endeavour*, **122**, 51–58. The age of microscopic life.
Schou, M. (1976). *Ann. Rev. Pharmacol. Toxicol.*, **16**, 231–243. Pharmacology and toxicology of lithium.
Schroeder, T. E. (1972). *J. Cell Biol.*, **53**, 419–434. The contractile ring. II. Determining its brief existence, volumetric changes, and its vital role in cleaving *Arbacia* eggs.

Schroeder, T. W., and Strickland, O. C. (1974). *Exp. Cell Res.*, **83**, 139–142. Ionophore A23187, calcium and contractility in frog eggs.
Schudt, C., Gaertner, U., and Pette, D. (1976). *Eur. J. Biochem.*, **68**, 103–111. Insulin action on glucose transport and calcium fluxes in developing muscle cell *in vitro*.
Schuetz, A. (1972). *J. Exp. Zool.*, **191**, 433–440. Induction of nuclear breakdown and meiosis in *Spirula solidissima* oocytes by calcium ionophore.
Schulman, H., and Greengard, P. (1978). *Proc. Natl. Acad. Sci. U.S.A.*, **75**, 5432–5436. Ca^{2+}-dependent protein phosphorylation system in membranes from various tissues, and its activation by calcium-dependent regulator.
Schultz, G., and Hardman, J. G. (1975). *Adv. Cyclic Nucl. Res.*, **5**, 339–351. Regulation of cyclic GMP levels in ductus deferens of the rat.
Schultz, G., Hardman, J. G., Schultz, K., Baird, C. E., and Sutherland, E. W. (1973). *Proc. Natl. Acad. Sci. U.S.A.*, **70**, 3889–3993. The importance of calcium ions for the regulation of guanosine 3': 5'-cyclic monophosphate levels.
Schultz, G., Schultz, K., and Hardman, G. (1975). *Metabolism*, **24**, 429–437. Effects of norepinephrine on cyclic nucleotide levels in ductus deferens of the rat.
Schwartz, A., and Lee, K. S. (1962). *Circ. Res.*, **10**, 321–332. Study of heart mitochondria and glycolytic metabolism in experimentally induced cardiac failure.
Schwarzenbach, G. (1954) in Chemical Specificity in Biological Interactions. pp. 164. Ed. F. N. Gurd, Academic Press, New York.
Schwarzenbach, G., and Anderegg, G. (1957). *Helv. Chim. Acta*, **40**, 1229–1231. Die Erdalkalikomplexe von Adenosintetraphosphat, Glycerophosphat und Fructose phosphat.
Schwarzenbach, G., Senn, H., and Anderegg, G. (1957). *Helv. Chim. Acta*, **40**, 1886–1900. Komplexone. XXIX. Eingrosser Chelateffekt besonderer Art.
Selinger, Z., and Naim, E. (1970). *Biochim. Biophys. Acta*, **203**, 335–337. The effect of calcium on amylase secretion by rat parotid slices.
Selinger, Z., Naim, E., and Lasser, M. (1970). *Biochim. Biophys. Acta*, **203**, 326–334. ATP-dependent calcium uptake by microsomal preparations from rat parotid and submaxillary glands.
Selinger, Z., Eimerl, S., and Schramm, M. (1974). *Proc. Natl. Acad. Sci. U.S.A.*, **71**, 128–131. A calcium ionophore stimulating the action of epinephrine on the α-adrenergic receptor.
Sell, D. E., and Reynolds, E. S. (1969). *J. Cell Biol.*, **41**, 736–752. Liver parenchymal cell injury. 8. Lesions of membranous cellular components following iodoform.
Serck-Hanssen, G., and Christiansen, T. (1973) *Biochim. Biophys. Acta*, **307**, 404–414. Uptake of calcium in chromaffin granules of bovine adrenal medulla stimulated *in vitro*.
Severson, D., Drummond, G., and Sulakhe, P. (1972). *J. Biol. Chem.*, **247**, 2949–2958. Adenylate cyclase in skeletal muscle.
Severson, D. L., Denton, R. M., Pask, H. T., and Randle, E. A. J. (1974). *Biochem. J.*, **140**, 225–237. Calcium and magnesium ions as effectors of adipose-tissue pyruvate dehydrogenase phosphatase.
Severson, D. L., Denton, R. M., Bridges, B. J., and Randle, P. J. (1976). *Biochem. J.*, **154**, 209–223. Exchangeable and total calcium pools in mitochondria of rat epididymal fat pads and isolated fat-cells. Role in the regulation of pyruvate dehydrogenase activity.
Shainberg, A., Yagil, G., and Yaffe, D. (1969). *Exp. Cell Res.*, **58**, 163–167. Control of myogenesis *in vitro* by Ca^{2+} concentration in nutritional medium.
Shamoo, A. E. (Ed.) (1975). *Ann. N. Y. Acad. Sci.*, **264**, Carriers and Channels in Biological systems.
Shamoo, A. E., and Goldstein, D. A. (1977). *Biochim. Biophys. Acta*, **472**, 13–53. Isolation of ionophores from ion transport systems and their role in energy translocation.
Shanes, A. M. (1958). *Pharmacol. Rev.*, **10**, 59–164, 165–273. Electrochemical aspects of physiological and pharmacological action in excitable cells. I. The resting cell and its alteration by extrinsic factors. II. The action potential and excitation.
Shaw, J. (1955). *J. Exp. Biol.*, **32**, 383–396. Ionic regulation in the muscle fibres of *Carcinus maenas*.
Shaw, T. I., and Newby, B. J. (1972). *Biochim. Biophys. Acta*, **255**, 411–412. Movement in a ganglion.
Shearer, W. T., Atkinson, J. P., and Parker, C. W. (1976). *J. Immunol.*, **117**, 973–980. Humoral immunostimulation. VI. Increased calcium uptake by cells treated with antibody and complement.
Sheldrake, A. R. (1974). *Nature (London)*, **250**, 381–385. The ageing, growth and death of cells.
Shen, A. C., and Jennings, R. B. (1972). *Am. J. Pathol.*, **67**, 417–452. Myocardial calcium and magnesium in acute ischaemic injury.
Sherwood, L. M., Herrmann, I., and Bassett, C. A. (1970). *Nature (London)*, **225**, 1056–1058. Parathyroid hormone secretion *in vitro*: regulation by calcium and magnesium ions.
Sheterline, P. (1980). *Biochem. Biophys. Res. Commun.*, **93**, 194–200. Trifluoperazine can distinguish between myosin light chain kinase-linked and troponin C-linked control of actomyosin interaction by Ca^{2+}.

Shimomura, O., and Johnson, F. H. (1973). *Biochem. Biophys. Res. Commun.*, **53**, 490–494. Further data on the specificity of aequorin luminescence.
Shimomura, O., Johnson, F. H., and Saiga, Y. (1962). *J. Cell. Comp. Physiol.*, **59**, 223–239. Extraction, purification and properties of aequorin, a bioluminescent protein from the luminous hydromedusan *Aequorea*.
Shimomura, O., Johnson, F. H., and Saiga, Y. (1963). *J. Cell. Comp. Physiol.*, **62**, 1–8. Further data on the bioluminescent protein, aequorin.
Shooter, R. A., and Grey, G. O. (1952). *Br. J. Exp. Pathol.*, **33**, 98–103. Studies of mineral requirements of mammalian cells.
Shooter, R. A., and Wyatt, H. V. (1955). *Br. J. Exp. Pathol.*, **36**, 341–350. Mineral requirements for growth of *Staphylococcus pyogenes*. Effect of magnesium and calcium ions.
Siddle, K., and Hales, C. N. (1974a), *Biochem. J.*, **142**, 97–103. The relationship between the concentration of adenosine 3′: 5′-cyclic monophosphate and the anti-lipolytic action of insulin in isolated rat fat-cells.
Siddle, K., and Hales, C. N. (1974b). *Biochem. J.*, **142**, 345–351. The action of local anaesthetics on lipolysis and on adenosine 3′:5′-cyclic monophosphate content in isolated rat fat-cells.
Siddle, K., and Hales, C. N. (1980). *Horm. Metab. Res.*, **12**, 489–562. The effect of bivalent cation chelating agents on some actions of adrenalin and insulin in rat isolated fat cells.
Siegel, B. A., Engel, W. K., and Derrer, E. C. (1977). *Neurology*, **27**, 230–238. Localization of technetium-99 m diphosphonate in acutely injured muscle. Relationship to calcium deposition.
Siegel, P. J. (1978). *Biophys. J.*, **22**, 341–346. The effects of Ca^{2+} and Mg^{2+} on the electrophoretic mobility of chromaffin granules measured by electrophoretic light scattering.
Sieghart, W., Theoharides, T. C., Alper, S. L., Douglas, W. W., and Greengard, P. (1978). *Nature (London)*, **275**, 329–331. Calcium-dependent protein phosphorylation during secretion by exocytosis in the mast cell.
Siekevitz, P., and Potter, V. R. (1953a). *J. Biol. Chem.*, **200**, 187–196. The adenylate kinase of rat liver mitochondria.
Siekevitz, P., and Potter, van R. (1953b). *J. Biol. Chem.*, **201**, 1–13. Intramitochondrial regulation of oxidative rate.
Sillen, L. G., and Martell, D. E. (1964). *Stability Constants of Metal Ion Complexes*. The Chemical Society, Special Publication No. 17, London.
Silverman, M. (1976). *Biochim. Biophys. Acta*, **457**, 303–351. Glucose transport in the kidney.
Simkiss, K. (1961). *Biol. Rev.*, **36**, 321–367. Calcium metabolism and avian reproduction.
Simkiss, K. (1964). *Biol. Rev.*, **39**, 487–505. Phosphates as crystal poisons of calcification.
Simkiss, K. (1974). *Endeavour*, **120**, 119–123. Calcium translocation by cells.
Simons, T. J. (1975). *J. Physiol. (London)*, **256**, 227–244. Calcium-dependent potassium exchange in human red cell ghosts.
Simpson, L. L. (1968). *J. Pharm. Pharmacol.*, **20**, 889–910. The role of calcium in neurohumoral and neurohormonal extrusion processes.
Singh, J. P., Babcock, D. F., and Lardy, H. A. (1978). *Biochem. J.*, **172**, 549–556. Evidence that increased influx of calcium ions is a component of the capacitation of guinea-pig spermatozoa.
Skankar, K., and Bard, R. C. (1952). *J. Bacteriol.*, **63**, 279–290. Effect of metallic ions on growth and morphology of *Clostridium perfringens*.
Skoryna, S. C., ed. (1981) *Handbook of Stable Strontium Compounds*. Plenum Press, New York and London.
Sleigh, M. (1979). *Nature (London)*, **277**, 263–264. Contractility of the roots of flagella and cilia.
Smith, A. D., DePotter, W., Merman, E. J. and DeSchaepnyuer, A. F. (1970). *Tissue Cell*, **2**, 547–568. Release of dopamine β-hydroxylase and chromogranin A upon stimulation of the splenic nerve.
Smith, B. R., and Hall, R. (1974). *Lancet 2*, 427–430. Thyroid-stimulating immunoglobulins in Graves' disease.
Smith, H. G., Fager, R. S., and Litman, B. J. (1977). *Biochemistry*, **16**, 1399–1405. Light-activated calcium release from sonicated bovine retinal rod out segment disks.
Smith, I. K. (1978). *Plant Physiol.*, **62**, 941–948, Role of calcium in serine transport in tobacco cells.
Smith, L. H., and Williams, H. E. (1971). Kidney stones. In *Diseases of the Kidney*, Vol. II (Eds. M. B. Strauss and L. G. Welt), pp. 973–996. Little, Brown and Co., Boston.
Smith, M. W., and Thorn, N. A. (1965). *J. Endocrinol.*, **32**, 141–151. The effects of calcium on protein-binding and metabolism of arginine vasopressin in rats.

Smith, R. J., and Ignarro, L. J. (1975). *Proc. Natl. Acad. Sci. U.S.A.*, **72**, 108–112. Bioregulation of lysosomal enzyme secretion from human neutrophils: roles of guanosine 3′: 5′-monophosphate and calcium in stimulus-secretion coupling.
Sneddon, J. M. (1972). *Nature (London)*, **236**, 103–104. Divalent cations and the blood platelet release reaction.
Soifer, D. (Ed.) (1975). *Ann. N. Y. Acad. Sci.*, **253**. The biology of cytoplasmic microtubules.
Solomon, F. (1977). *Biochemistry*, **16**, 358–363. Binding sites for calcium on tubulin.
Somers, G., Blondel, B., Malaisse, W. J., and Orci, L. (1976). *Diabetologia*, **12**, 420. Motion analysis in pancreatic endocrine cells: saltatory movements of secretory granules along oriented pathways.
Sorkin, E., Stecher, V. J., and Borel, J. F. (1970). *Ser. Haematol.*, **III**, 131–162. Chemotaxis of leukocytes and inflammation.
Spangeberg, D. B., and Beck, C. W. (1968). *Trans. Am. Microsc. Soc.*, **87**, 329–335. Calcium sulphate dihydrate in *Aurelia*.
Spataro, A. C., and Bosmann, H. B. (1976). *Biochem. Pharmacol.*, **25**, 505–510. Mechanism of action of disodium cromoglycate-mast cell calcium ion influx after a histamine-releasing stimulus.
Srivastava, A. K., Waisman, D. M., Brostrom, C. O., and Soderling, T. R. (1979). *J. Biol. Chem.*, **254**, 583–586. Stimulation of glycogen synthase phosphorylation by calmodulin-dependent regulator protein.
Stahl, W. L., and Swanson, P. D. (1972). *J. Neurochem.*, **19**, 2395–2407. Calcium movements in brain slices in low sodium or calcium media.
Stanier, R. Y., Doudoroff, M., and Adelberg, E. A. (1964). *General Microbiology*. Macmillan, London.
Starling, E. H. (1905). *Lancet*, **ii**, 339–341. The chemical correlation of the functions of the body.
Statland, B. E., Heagan, B. M., and White, J. G. (1969). *Nature (London)*, **233**, 521–522. Uptake of calcium by platelet releasing factor.
Steinberg, M. S. (1958). *Am. Nat.*, **92**, 65–81. On the chemical bonds between animal cells. A mechanism for type-specific association.
Steiner, M. (1978). *Nature (London)*, **272**, 834–835. 3′,5′-Cyclic AMP binds to and promotes polymerisation on platelet tubulin.
Steinhardt, R. A., Epel, D., Carroll, E. J., and Yanagimachi, R. (1974). *Nature (London)*, **252**, 41–43. Is calcium ionophore a universal activator for unfertilised eggs?
Steinhardt, R. A., Zucker, R., and Schatten, G. (1977). *Dev. Biol.*, **58**, 185–196. Intracellular calcium release at fertilisation in the sea urchin egg.
Stephens, R. E., and Edds, K. T. (1976). *Physiol. Rev.*, **56**, 709–777. Microtubules: structure, chemistry and function.
Stevens, M. (1970). *Exp. Cell Res.*, **59**, 482–484. Procedures for induction of spawning and meiotic maturation of starfish oocytes by treatment with 1-methyladenine.
Stinnakre, J., and Tauc, L. (1973). *Nature (London)*, **242**, 113–115. Calcium influx in active *Aplysia* neurones detected by injected aequorin.
Stossel, T. P. (1973). *J. Cell Biol.*, **58**, 346–356. Quantitative studies of phagocytosis kinetic effects of cations and heat labile opsonin.
Stossel, T. P. (1974). *N. Engl. J. Med.*, **290**, 717–723, 774–780, 833–839. Phagocytosis.
Stossel, T. P., and Hartwig, J. H. (1976). *J. Cell Biol.*, **68**, 602–619. Interactions of actin, myosin, and a new actin-binding protein of rabbit pulmonary macrophages. II. Role in cytoplasmic movement and phagocytosis.
Straub, W. (1912). *Verh. Ges. Dtsch. Naturforsch. Ärtze*, **84**, 194–214. Bedeutung der Zellmembran.
Stroobant, P., Dame, J. B., and Scarborough, G. A. (1980). *Fed. Proc.*, **39**, 2437–2441. The *Neurospora* plasma membrane Ca^{2+} pump.
Stubbs, M., Kirk, C. J., and Hems, D. A. (1976). *FEBS Lett.*, **69**, 199–202. Role of extracellular calcium in the action of vasopressin on hepatic glycogenolysis.
Suelter, C. H. (1970). *Science*, **168**, 789–795. Enzymes activated by monovalent cations.
Sugden, M. C., Christie, M. R., and Ashcroft, S. J. H. (1979). *FEBS Lett.*, **105**, 95–100. Presence and possible role of calcium-dependent regulator (calmodulin) in rat islets of Langerhans.
Sulakhe, P. V., Sulakhe, S. J., Leung, N. L.-K., St. Louis, P. J., and Hicke, R. A. (1976a). *Biochem. J.*, **157**, 705–712. Guanylate cyclase: subcellular distribution in cardiac muscle, skeletal muscle, cerebral cortex and liver.
Sulakhe, S. J., Leung, N. L.-K., and Sulakhe, P. V. (1976b). *Biochem. J.*, **157**, 713–719. Properties of particulate, membrane-associated and soluble guanylate cyclase from cardiac muscle, skeletal muscle, cerebral cortex and liver.

Sumner, J. B. (1926). *J. Biol. Chem.*, **69**, 435–441. Isolation and crystallization of enzyme urease: preliminary paper.
Sun, S.-T., Day, E. P., and Ho, J. T. (1978). *Proc. Natl. Acad. Sci. U.S.A.*, **75**, 4325–4328. Temperature dependence of calcium-induced fusion of sonicated phosphatidylserine vesicles.
Sutherland, E. W. (1961–1962). *Harvey Lect.*, **57**, 17–33. The biological role of adenosine-3'-5'-phosphate.
Suttie, J. W., and Jackson, C. M. (1977). *Physiol. Rev.*, **57**, 1–70. Prothrombin structure, activation and biosynthesis.
Swinnerton, H. H. (1970). *Fossils*. New Naturalist Series, Collins, London.
Szuts, E. Z., and Cone, R. A. (1974). *Fed. Proc.*, **33**, 1471. Rhodopsin: light activated release of calcium.
Tada, M., Yamomoto, T., and Tonomura, Y. (1978). *Physiol. Rev.*, **58**, 1–79. Molecular mechanism of active calcium transport by sarcoplasmic reticulum.
Tait, J. F., Tait, S. A. S., and Bell, J. B. G. (1980). *Essays Biochem.*, **16**, 99–174. Steroid hormone production by mammalian adrenocortical dispered cells.
Takagi, A., Yonemoto, K., and Sugita, H. (1978). *Neurology*, **28**, 497–499. Single-skinned human muscle fibres: activation by calcium and strontium.
Taljedahl, I. B. (1974). *Biochim. Biophys. Acta*, **372**, 154–161. Interaction of Na^+ and Mg^{2+} with Ca^{2+} in pancreatic islets as visualized by chlortetracycline fluorescence.
Taunton, O. D., Roth, J., and Pastan, I. (1969). *J. Biol. Chem.*, **244**, 247–253. Studies on the adrenocorticotropic hormone-activated adenyl cyclase of a functional adrenal tumour.
Taylor, D. L., and Condeelis, J. S. (1979). *Int. Rev. Cytol.*, **56**, 57–144. Cytoplasmic structure and contractility in amoeboid cells.
Taylor, D. L., Condeelis, J. S., Moore, P. L., and Allen, R. D. (1973). *J. Cell Biol.*, **59**, 378–394. The contractile basis of amoeboid movement. I. The chemical control of motility in isolated cytoplasm.
Taylor, D. L., Reynolds, G. T., and Allen, R. D. (1975). *Biol. Bull.*, **149**, 448. Detection of free calcium ions in amoebae by aequorin bioluminescence.
Taylor, E. W. (1958). *The Examination of Waters and Water Supplies* (Thresh, Beale and Suckling). pp. viii and 841. Churchill, London.
Telfer, A., Barber, J., and Nicholson, J. (1975). *Biochim. Biophys. Acta*, **396**, 301–309. Availability of monovalent and divalent cations within intact chloroplasts for the action of ionophores nigericin and A23187.
Tempest, D. W., Dicks, J. W., and Hunter, J. R. (1966). *J. Gen. Microbiol.*, **45**, 135–146. The interrelationship between potassium, magnesium and phosphorus in potassium-limited chemostat cultures of *Aerobacter aerogenes*.
Terepka, A. R., Coleman, J. R., Armbrecht, H. J., and Gunter, T. E. (1976). *Symp. Soc. Exp. Biol.*, **30**, 117–139. Transcellular transport of calcium.
Theoharides, T. C., Sieghart, W., Greengard, P., and Douglas, W. W. (1980). *Science*, **207**, 80–82. Antiallergic drug cromolyn may inhibit histamine secretion by regulating phosphorylation of a mast cell protein.
Thiers, R. E., and Vallee, B. L. (1957). *J. Biol. Chem.*, **226**, 911–920. Distribution of metals in subcellular fractions of rat liver.
Thiers, R. E., Reynolds, E. S., and Vallee, B. L. (1960). *J. Biol. Chem.*, **235**, 2130–2133. The effect of carbon tetrachloride poisoning on subcellular metal ion distribution in rat liver.
Thomas, R. C. (1978). *Ion-sensitive Intracellular Microelectrodes: How to Make and Use Them*. Academic Press, London.
Thomson, D. L., and Collip, J. B. (1932). *Physiol. Rev.*, **12**, 309–383. The parathyroid glands.
Thomson, M. P., and Williamson, D. H. (1976). *FEBS Lett.*, **62**, 208–211. The ability of the calcium ionophore A23187 to mimic some of the effects of adrenaline of the metabolism of rat submaxillary gland.
Tilney, L. G. (1975). *J. Cell Biol.*, **64**, 289–310. Actin filaments in the acrosomal reaction of *Limulus* sperm.
Tilney, L. G., and Gibbins, J. R. (1969). *J. Cell Biol.*, **41**, 227–250. Microtubules in the formation and development of the primary mesenchyme in *Arbacia punctulata*. II. An experimental analysis of their role in development and maintenance of cell shape.
Timourian, H., Clothier, G., and Watchmaker, G. (1972). *Exp. Cell Res.*, **75**, 296–298. Cleavage furrow: calcium as determinant site.
Tischler, A. S., Dichter, M. A., Biales, B., DeLellis, R. A., and Wolfe, H. (1976). *Science*, **192**, 902–904. Neural properties of cultured human endocrine tumour cells of proposed neural crest origin.

Tolbert, M. E. M., and Fain, J. N. (1974). *J. Biol. Chem.*, **249**, 1162–1166. Studies on the regulation of gluconeogenesis in isolated rat liver cells by epinephrine and glucagon.
Tomita, T. (1970). *Q. Rev. Biophys.*, **3**, 179–222. Electrical activity of vertebrate photoreceptors.
Toyo-Oka, T., and Masaki, T. (1979). *J. Mol. Cell. Cardiol.*, **8**, 769–786. Ca^{2+}-activated neutral protease from bovine ventricular muscle: isolation and some of its properties.
Tritton, T. R., Murphree, S. A., and Sartorelli, A. C. (1978). *Biochem. Biophys. Res. Commun.*, **84**, 800–808. Adriamycin: a proposal on the specificity of drug action.
Trutter, M. R. (1975). *Philos. Trans. R. Soc. London Ser. B*, **272**, 29–41. Recognition of metal cations by biological systems.
Trutter, M. R. (1976). *Symp. Soc. Exp. Biol.*, **30**, 19–40. Chemistry of the calcium ionophores.
Tsien, R. Y. (1980). *Biochemistry*, **19**, 2396–2404. New calcium indicators and buffers with selectivity against magnesium and protons: Design, synthesis and properties of prototype structures.
Tsien, R. Y. (1981). *Nature (London)*, **290**, 527–528. A non-disruptive technique for loading calcium buffers and indicators into cells.
Tsien, R. Y., Pozzan, T., and Rink, T. J. (1982). *Nature (London)*, **295**, 68–71. T-cell mitogens cause early changes in cytoplasmic free Ca^{2+} and membrane potential in lymphocytes.
Turvey, J. R., and Simpson, P. R. (1966). Polysaccharides from *Corallina officinalis*. In *Proceedings of the Fifth International Seaweed Symposium*. (Eds. E. Young and J. L. McLachlan), pp. 323–327. Pergamon Press, Oxford.
Tyler, A. (1941). *Biol. Rev.*, **16**, 291–336. Artificial parthenogenesis.
Umen, M. J., and Scarpa, A. (1978). *J. Med. Chem.*, **21**, 505–506. New synthetic calcium selective ionophores. Design, synthesis and transport properties.
Urry, D. W. (1971). *Proc. Natl. Acad. Sci. U.S.A.*, **68**, 810–814. Neutral sites for calcium binding to elastin and collagen.
Urry, D. W., Cunningham, W. D., and Ohnishi, T. (1973). *Biochim. Biophys. Acta*, **292**, 853–857. A neutral polypeptide–calcium ion complex.
Utter, M. F., and Werkman, C. H. (1942). *J. Biol. Chem.*, **146**, 289–300. Effect of metal ions on the reactions of phosphopyruvate by *Escherichia coli*.
Vaccari, A., Vertua, R., and Furlani, A. (1971). *Biochem. Pharmacol.*, **20**, 2603–2612. Decreased calcium uptake by rat fundal strips after pretreatment with neuraminidase of LSD *in vitro*. Effect of serotonin, D-amphetamine and eledoisin on this uptake.
Vale, W., Burgus, R., and Guillemin, R. (1967). *Experientia*, **23**, 853–855. Presence of calcium ions as a requisite for the *in vitro* stimulation of TSH-release by hypothalamic TRF.
Vallee, B. L. (1955). *Adv. Protein Chem.*, **10**, 317–384. Zinc and metallo enzymes.
Vallee, B. L., and Wacker, W. E. C. (1970). *The Proteins*, Vol. 5. Academic Press, New York and London.
Vallee, B. L., Stein, E. A., Sumerwell, W. N., and Fischer, E. H. (1959). *J. Biol. Chem.*, **234**, 2901–2905. Metal content of α-amylases of various origins.
Van Bellé, H. *Biochem. Soc. Trans.*, **9**, 133P (S2-19)
Van Breeman, C., Farinas, B. R., Casteels, R., Gerba, P., Wuytack, F., and Deth, R. (1973). *Philos. Trans. R. Soc. London Ser. B*, **265**, 57–71. Factors controlling cytoplasmic Ca^{2+} concentration.
Van Eerd, J.-P., and Takahashi, K. (1975). *Biochem. Biophys. Res. Commun.*, **64**, 122–128. The amino acid sequence of bovine cardiac troponin-C. Comparison with rabbit skeletal troponin-C.
Van Netten, C., and Belton, P. (1977). *Can. J. Physiol. Pharmacol.*, **55**, 1023–1027. The effects of calcium deficiency on the electrical activity of *Nitella flexilis*.
Van Oss, C. J., and Stinson, M. W. (1970). *J. Reticul. Endothel. Soc.*, **8**, 397–406. Immunoglobulins as a specific opsonins.
Vasington, F. D., and Murphy, J. V. (1961). *J. Biol. Chem.*, **237**, 2670–2677. Ca^{++} uptake by rat kidney motochondria and its dependence on respiration and phosphorylation.
Vaughan, H., and Newsholme, E. A. (1969). *FEBS Lett.*, **5**, 124–126. The effects of Ca^{2+} and ADP on the activity of NAD-linked isocitrate dehydrogenase of muscle.
Vaughan, H., Thornton, S. D., and Newsholme, E. A. (1973). *Biochem. J.*, **132**, 527–535. The effects of calcium ions on the activities of trehalase, hexokinase, phosphofructokinase, fructose diphosphatase, and pyruvate kinase from various muscles.
Vaughan, L., and Penniston, J. T. (1976). *Biochem. Biophys. Res. Commun.*, **73**, 200–205. Cation control of erythrocyte membrane shape: Ca^{++} reversal of discocyte to echinocyte transition caused by Mg^{2+} and other cations.
Vedeckis, W. V., Bolhenbacher, W. E., and Gilbert, L. I. (1976). *Mol. Cell. Endocrinol.*, **5**, 81–88. Insect prothoracic glands: a role for cyclic AMP in the stimulation of α-ecdysone secretion.
Veloso, D., Guynn, R., Oskarsson, M., and Veech, R. L. (1973). *J. Biol. Chem.*, **248**, 4811–4819. The concentration of free and bound magnesium in rat tissues.

Venter, J. C., Fraser, C. M., and Harrison, L. C. (1980). *Science*, **207**, 1361–1363. Autoantibodies to β_2-adrenergic receptors: a possible cause of adrenergic hyporesponsiveness in allergic rhinitis and asthma.
Verati, E. (1902). *Arch. Ital. Biol.*, **37**, 449–454. Ricerche sulla fine struttura della fibra muscolane striata.
Vianna, A. L. (1975). *Biochim. Biophys. Acta*, **410**, 389–406. Interaction of calcium and magnesium in activating and inhibiting the nucleoside triphosphatase of sarcoplasmic reticulum.
Vinogradov, A., Scarpa, A., and Chance, B. (1972). *Arch. Biochem. Biophys.*, **152**, 646–654. Calcium and pyridine nucleotide interaction in mitochondrial membranes.
Virchow, R. (1855) cited in Abraham, E. P. (1970). pp 435.
Volsky, D. J., and Loyter, A. (1978). *J. Cell Biol.*, **78**, 465–479. Role of Ca^{2+} in virus-induced membrane fusion. Ca^{2+} accumulation and ultrastructural changes induced by Sendai virus in chicken erythrocytes.
von der Wense (1934). *Naunyn-Schmiedeberys Arch.*, **176**. Untersuchungen Über die Einwirkungen von Adrenalin auf Paramecium Ein Beitrag zur Frage der kolloid-Chemischen Wirkung des Adrenalins auf des Protoplasma cited in Hanström B. (1939) pp. 13.
Voorheis, P., and Martin, B. R. (1981). *Eur. J. Biochem.*, **138**, 1–5. 'Swell dialysis' demonstrates the adenylate cyclase in *Trypanosoma brucei* is regulated by calcium ions.
Wacker, W. E. C., and Williams, R. J. P. (1968). *J. Theor. Biol.*, **20**, 65–78. Magnesium/calcium balances of biological systems.
Waisman, D. M., Steven, F. C., and Wang, J. H. (1978). *J. Biol. Chem.*, **253**, 1106–1113. Purification and characterisation of a Ca^{2+}-binding protein in *Lumbricus terrestris*.
Walker, J.-L., and Brown, H. M. (1977). *Physiol. Rev.*, **57**, 729–778. Intracellular ionic activity measurements in nerve and muscle.
Wallach, D., and Pastan, I. (1976). *Biochem. Biophys. Res. Commun.*, **72**, 859–864. Stimulation of membranous guanylate cyclase by concentrations of calcium that are in the physiological range.
Wallach, S., Bellavia, J. V., Camponia, P. J., and Bristrim, P. (1972). *J. Clin. Invest.*, **51**, 1572–1577. Thyroxine-induced stimulation of hepatic cell transport of calcium and magnesium.
Walser, M. (1970). *Am. J. Physiol.*, **218**, 582–589. Calcium transport in the toad bladder: permeability to calcium ions.
Walter, J. B., and Israel, M. S. (1970). *General Pathology*. J. and A. Churchill, London.
Walter, M. F., and Satir, P. (1979). *Nature (London)*, **278**, 69–70. Calcium does not inhibit active sliding of microtubules from mussel gill cilia.
Ward, W. W., and Seliger, H. H. (1974). *Biochemistry*, **13**, 1491–1499. Extraction and purification of calcium-activated photoproteins from the ctenophores. *Mnemiopsis* Sp. and *Beroë ovata*.
Warner, W., and Carchman, R. (1978). *Biochim. Biophys. Acta*, **528**, 409–415. Effect of ruthenium red, A23187 and D-600 on steroidogenesis in Y-1 cells.
Wasserman, J., and Masui, Y. (1975). *J. Exp. Zool.*, **193**, 369–375. Initiation of meiotic maturation in *Xenopus laevis* oocytes by the combination of divalent cations and ionophore A23187.
Watanabe, A., Tasaki, I., Singer, I., and Lerman, L. (1967). *Science*, **155**, 95–97. Effect of tetrodotoxin on excitability of squid giant axons in sodium-free media.
Watterson, D. M., van Eldik, J. J., Smith, R. E., and Vanaman, T. C. (1976). *Proc. Natl. Acad. Sci. U.S.A.*, **73**, 2711–2715. Calcium dependent regulatory protein of cyclic nucleotide metabolism in normal and transformed chicken embryo fibroblasts.
Watterson, D. M., Iverson, D. B., and van Eldik, L. J. (1980). *Biochemistry*, **19**, 5762–5768. Spinach calmodulin: isolation, characterization and comparison with vertebrate calmodulin.
Watterson, D. M., Sharief, F., and Vanaman, T. C. (1980). *J. Biol. Chem.*, **255**, 962–975. The complete amino acid sequence of the Ca^{2+}-dependent modulator protein (calmodulin) of bovine brain.
Weast, R. C., and Atle, M. J. (Ed.) (1980). *CRC Handbook of Chemistry and Physics*. CRC Press, Florida.
Weber, A., Herz, R., and Reiss, I. (1964). *Proc. R. Soc. London Ser. B*, **160**, 489–501. The regulation of myofibrillar activity by calcium.
Weber, A., Herz, R., and Reiss, I. (1966). *Biochem. Z.*, **345**, 329–369. Study of the kinetics of calcium transport by isolated fragmented sarcoplasmic reticulum.
Weeds, A. (1982). *Nature (London)*, **296**, 811–816. Actin-binding protein-regulators of cell architecture and motility.
Weir, E. G., and Hastings, A. B. (1936). *J. Biol. Chem.*, **114**, 397–406. The ionisation constants of calcium proteinate determined by the solubility of calcium carbonate.

Weise, E. (1934). *Arch. Exp. Pathol. Pharmakol.*, **176**, 367–377. Untersuchungen zur Frage der Verteilung und der Bindangsart des Calcium ion Muskel.
Weisenberg, R. C. (1972). *Science*, **177**, 1104–1105. Microtubule formation *in vitro* in solutions containing low calcium concentrations.
Weisenseel, M. H., and Ruppert, H. K. (1977). *Planta (Berlin)*, **137**, 225–229. Phytochrome and calcium ions are involved in light-induced membrane depolarisation in *Nitella*.
Weiss, G. B. (1974). *Ann. Rev. Pharmacol.*, **14**, 343–354. Cellular pharmacology of lanthanum.
Weiss, G. B. (Ed.) (1978). *Calcium and Drug Action*. Plenum Press, New York.
Weiss, G. B., and Levin, R. M. (1978). *Adv. Cyclic Nucleotide Res.*, **9**, 285–304. Mechanism for selectively inhibiting the activation of cyclic nucleotide phosphodiesterase and adenylate cyclase by antipsychotic agents.
Weissmann, G., Smolen, J. E. and Korchak, H. M. (1980). *New Eng. J. Med.*, **303**, 27–34. Release of inflammatory mediators from stimulated neutrophils.
Weissmann, N., and Pileggi, V. J. (1974). Inorganic ions. In *Clinical Chemistry: Principles and Techniques* (Eds. J. D. Henry, D. C. Cannon and J. W. Winkleman), pp. 639–754. Harper and Row, New York.
Weller, M., and Rodnight, R. (1974). *Biochem. J.*, **142**, 605. Protein kinase activity stimulated by adenosine 3′ : 5′-cyclic monophosphate in synaptic-membrane fragments from ox brain. Inhibition of intrinsic activity by free and membrane-bound calcium ions.
Welsh, M. J., Dedman, J. R., Brinkley, B. R., and Means, A. R. (1979). *J. Cell Biol.*, **81**, 624–634. Tubulin and calmodulin. Effects of microtubule and microfilament inhibitors on localization in the mitotic apparatus.
White, J. G. (1974). *Am. J. Pathol.*, **77**, 507–518. Effects of an ionophore, A23187, on the surface morphology of normal erythrocytes.
Whitfield, J. F., MacManus, J. P., Franks, D. J., Braceland, B. M., and Gillan, D. J. (1972). *J. Cell. Physiol.*, **80**, 315–328. Calcium-mediated effects of calcitonin on cyclic AMP formation and lymphoblast proliferation in thymocyte populations exposed to prostaglandin E.
Whitfield, J. F., Rixon, R. H. MacManus, J. P., and Balk, S. D. (1973). *In Vitro*, **8**, 257–278. Caclium, cyclic adenosine 3′, 5′-monophosphate and the control of cell proliferation: a review.
Whitfield, J. F., MacManus, J. P., Boynton, A. L., Gillan, D. J., and Isaacs, R. J. (1974). *J. Cell. Physiol.*, **84**, 445–458. Concanavalin A and the initiation of thymic DNA synthesis and proliferation by a calcium-dependent increase in cyclic GMP level.
Whitfield, M. (1975). Sea water as an electrolyte solution. In *Chemical Oceanography*, Vol. 1 (Eds. J. P., Riley, and G., Skirrow) pp. 44–171. Academic Press, New York.
Whitney, R. B., and Sutherland, R. M. (1972). *J. Cell. Physiol.*, **80**, 329–337. Requirement for calcium ions in lymphocyte transformation stimulated by phytohaemagglutinin.
Whitney, R. B., and Sutherland, R. M. (1973). *Biochim. Biophys. Acta*, **298**, 790–797. Effects of chelating agents on the initial interaction of phytohemagglutinin with lymphocytes and the subsequent stimulation of amino acid uptake.
Whittam, R. (1961). *Nature (London)*, **191**, 603–604. Active cation transport as a pace-maker of respiration.
Whitton, P. D., Rodrigues, L. M., and Hems, D. A. (1977). *Biochem. Soc. Trans.*, **5**, 992–994. Influence of extracellular calcium ions on hormonal stimulation of glycogen breakdown in hepatocyte suspensions.
Widdowson, E. M., and Dickerson, J. W. T. (1964). Chemical composition of the body. In *Mineral Metabolism*, Vol. 2 A (Eds. C. L. Comar and F. Bronner), pp. 1–247. Academic Press, New York.
Wilbur, K. M., and Simkiss, K. (1968). Calcified shells. In *Comprehensive Biochemistry*, Vol. 26 (Eds. M. Florkin and E. H. Stotz), pp. 229–295. Elsevier, Amsterdam.
Williams, J. A. (1972). *Endocrinology*, **90**, 1459–1463. Effects of Ca^{2+} and Mg^{2+} on secretion *in vitro* by mouse thyroid glands.
Williams, J. A. (1980). *Am. J. Physiol.*, **238**, G269–G279. Regulation of pancreatic acinar cell function by intracellular calcium.
Williams, L. P. (1965). *Michael Faraday: A Biography*. Chapman, London and New York.
Williams, R. J. P. (1970). *Q. Rev. Chem. Soc.*, **24**, 331–365. The biochemistry of sodium, potassium magnesium and calcium.
Williams, R. J. P. (1974). *Biochem. Soc. Symp.*, **39**, 133–138. Calcium ions: their ligands and their functions.
Williams, R. J. P. (1976). *Symp. Soc. Exp. Biol.*, **30**, 1–17. Calcium chemistry and its relation to biological function.

Williams, R. J. P., and Wacker, W. E. C. (1967). *J. Am. Med. Assoc.*, **210**, 18–22. Cation balance in biological systems.
Williamson, M. B., and Gulick, A. (1944). *J. Cell. Comp. Physiol.*, **23**, 77–82. The calcium and magnesium content of mammalian cell nuclei.
Williamson, R. E. and Ashley, C. C. (1982). *Nature (London)*, **296**, 647–651. Free Ca^{2+} and cytoplasmic streaming in the alga *Chara*.
Willis, J. B. (1963). *Methods Biochem. Anal.*, **11**, 1–67. Analysis of biological materials by atomic absorption spectroscopy.
Willis, J. B. (1965). *Clin. Chem.*, **11**, (Suppl. 2), 251–258. The analysis of biological materials by atomic-absorption spectroscopy.
Willmer, E. N. (Ed.) (1965). *Cells and Tissues in Culture*, Vols. I–III. Academic Press, New York.
Willmer, E. N. (1970). *Cytology and Evolution*. Academic Press, New York.
Willmer, E. N. (1974). *Biol. Rev.*, **49**, 321–363. Nemertines as possible ancestors of the vertebrates.
Willmer, E. N. (1977). Cytology and Evolution, 2nd Edition. Academic Press, New York.
Wilson, A. C., Carlson, S. S., and White, T. J. (1977). *Annu. Rev. Biochem.*, **46**, 573–639. Biochemical evolution.
Wilson, M. E., Trush, M. A., van Dyke, K., and Neal, W. (1978). *FEBS Lett.*, **94**, 387–390. Induction of chemiluminescence in human polymorphonuclear leukocytes by the calcium ionophore A23187.
Wimhurst, J. M., and Manchester, K. L. (1970). *FEBS Lett.*, **10**, 33–37. Role of bivalent cations in control of enzymes involved in gluconeogenesis.
Wimhurst, J. M., and Manchester, K. L. (1972). *FEBS Lett.*, **27**, 321–326. Comparison of ability of Mg and Na to activate the key enzymes of glycolysis.
Winegrad, S. (1961). *Circulation*, **24**, 523–529. The possible role of calcium in excitation–contraction coupling of heart muscle.
Winegrad, S. (1965). *J. Gen. Physiol.*, **48**, 455–479. Autoradiographic studies of intracellular calcium in frog skeletal muscle.
Winegrad, S. (1968). *J. Gen. Physiol.*, **51**, 65–83. Intracellular calcium movements of frog skeletal muscle during recovery from tetanus.
Winkler, H. (1976). *Neuroscience*, **1**, 65–80. The composition of adrenal chromaffin granules: an assessment of controversial results.
Wolf, P. L. (1975). *Am. J. Pathol.*, **78**, 15a. Enhancement of sickling secondary to increased calcium erythrocyte influx and ATP depletion.
Wollheim, C. B., Blondel, B., Trueheart, P. A., Renold, A. E., and Sharp, G. W. G. (1975). *J. Biol. Chem.*, **250**, 1354–1360. Calcium-induced insulin release in monolayer culture of the endocrine pancreas. Studies with ionophore A23187.
Wong, P. Y.-K., and Cheung, W. Y. (1979). *Biochem. Biophys. Res. Commun.*, **90**, 473–480. Calmodulin stimulates human platelet phospholipase A_2.
Wood, E. H., Heppner, R. L., and Weidmann, S. (1969). *Circ. Res.*, **24**, 409–445. Ionotropic effects of electric currents. I. Positive and negative effects of constant electric currents or current pulses, applied during cardiac action potentials. II. Hypothesis: Calcium movements, excitation–contraction coupling and ionotropic effects.
Woodhead, J. S. (1982). Regulation of whole body calcium in man. In *Biochemical Aspects of Human Disease*. (Ed. R. S. Elkeles). Blackwell, Oxford.
Woodin, S. M. (1968). The basis of leucocidin action. In *The Biological Basis of Medicine*, Vol. 2. (Ed. E. E. Bittar), pp. 373–396. Academic Press, New York.
Woodin, A. M., and Wieneke, A. A. (1963). *Biochem. J.*, **87**, 487–495. The accumulation of caclium by polymorphonuclear leucocytes treated with staphylococcal leucocidin.
Woodward, A. A. (1949). *Biol. Bull. (Woods Hole, mass.)*, **97**, 264. The release of radioactive Ca^{45} from muscle during stimulation.
Wray, H. L., Gray, R. R., and Olsen, R. A. (1973). *J. Biol. Chem.*, **248**, 1496–1498. Cyclic adenosine 3′, 5′-monophosphate-stimulated protein kinase and a substrate associated with cardiac sarcoplasmic reticulum.
Wrogemann, K., and Nylen, E. G. (1978). *J. Mol. Cell. Cardiol.*, **10**, 185–196. Mitochondrial calcium overloading in cardiomyopathic hamsters.
Wrogemann, K., and Pena, S. D. J. (1976). *Lancet*, **i**, 672–674. Mitochondrial calcium overload: a general mechanism for cell-necrosis in muscle diseases.
Wrogemann, K., Jacobson, B. E., and Blanchaer, M. C. (1973). *Arch. Biochem. Biophys.*, **159**, 267–278. On the mechanism of a calcium-associated defect of oxidative phosphorylation in progressive muscular dystrophy.
Wyatt, H. V. (1961). *Exp. Cell Res.*, **23**, 97–107. Cation requirements of He La cells in tissue culture.

Wyatt, H. V. (1964). *J. Theor. Biol.*, **6**, 441–470. Cations, enzymes and control of cell metabolism.
Wyatt, H. V., Reed, G. W., and Smith, A. H. (1962). *Nature (London)*, **195**, 100–101. Calcium requirement for growth of *Staphylococcus pyogenes*.
Wyllie, A. H., Kerr, J. F. R., and Currie, A. R. (1980). *Int. Rev. Cytol.*, **68**, 251–306. Cell death: The significance of apoptosis.
Wyn-Jones, R. G., and Lunt, O. R. (1967). *Bot. Rev.*, **33**, 407–426. The function of calcium in plants.
Yagi, K., Yazawa, M., Kakiuchi, S., Ohshima, M., and Uenishi, K. (1978). *J. Biol. Chem.*, **253**, 1338–1340. Identification of an activator protein for myosin light chain kinase as the Ca^{2+} modulator protein.
Yamamoto, T. (1954). *Exp. Cell Res.*, **6**, 56–68. The role of calcium ions in activation of *Oryzias* eggs.
Yamamoto, Y. (1961). *Int. Rev. Cytol.*, **12**, 361–405. Physiology of fertilization in fish eggs.
Yamauchi, T., and Fujisawa, H. (1979). *Biochem. Biophys. Res. Commun.*, **90**, 28–35. Activation of tryptophan 5-monooxygenase by calcium dependent regulator protein.
Yamazaki, R. K. (1975). *J. Biol. Chem.*, **250**, 7924–7930. Glucagon stimulation of mitochondrial respiration.
Yanagamachi, R., and Usui, N. (1974). *Exp. Cell Res.*, **89**, 161–174. Calcium dependence of the acrosome reaction and activation of guinea-pig spermatozoa.
Yanovsky, A., and Loyter, A. (1972). *J. Biol. Chem.*, **247**, 4021–4028. The mechanism of cell fusion: energy requirements for virus-induced fusion of Ehrlich ascites tumour cells.
Yerna, M.-J., Hartshorne, D. J., and Goldman, R. D. (1979). *Biochemistry*, **18**, 673–678. Isolation and characterisation of baby hamster kidney (BHK-21) cell modulator protein.
Yoshikami, S., and Hagins, W. A. (1977). Control of the dark current in vertebrate rods and cones. In *Biochemistry and Physiology of Visual Pigments* (Ed. H. Langer), pp. 245–255. Springer Verlag, New York.
Young, C. M. (1940). *Scientific Report of the Great Barrier Reef Expedition.*, London, **1**, 353.
Zacks, S. I., and Sheff, M. F. (1964). *J. Neuropathol. Exp. Neurol.*, **23**, 306–323. Studies on tetanus toxin. I. Formation of intramitochondrial dense particles in mice acutely poisoned with tetanus toxin.
Zanella, J., and Rall, T. W. (1973). *J. Pharmacol. Exp. Ther.*, **186**, 241–252. Evaluation of electrical pulses and elevated levels of potassium ions as stimulants of adenosine 3′,5′-monophosphate (cyclic AMP) accumulation in guinea-pig brain.
Zierler, K. L. (1972) in *Handbook of Physiology*, Section 7, vol. I. Eds. R. O. Greep., and E. P. Astwood, pp. 347–368. Insulin, ions and membrane potential. American Physiological Society.
Zenser, T. V., and Davis, B. B. (1978). *Am. J. Physiol.*, **235**, F213. Effects of calcium on prostaglandin E_2 synthesis by rat inner medullary slices.
Ziance, R. J., Azzaro, A. J., and Rutledge, C. O. A. (1972). *J. Pharmacol. Exp. Ther.*, **182**, 284–294. Characteristics of amphetamine-induced release of norepinephrine from rat cerebral cortex *in vitro*.

Species Index

Acanthamoeba, 349
Acetabularia, 473
Acholoe, 184
Acmaea, 383
Actinosphaericum eichorni, 215, 216
Aequorea forskalea, 49, 51, 52, 53, 54, 55, 56, 182, 184
Allogromia, 208, 222
Amanita phalloides, 406
Amoeba, 62, 207, 222, 457
Amoeba proteus, 19, 222
Amphipholis, 185
Aplysia, 62, 63, 136, 144, 146, 147, 150, 156, 157, 163, 164
Arbacia, 215, 383
Ascaris, 9
Azotobacter, 198

Bacillus, 42, 121, 198, 372
Bacillus cereus, 372
Bacillus subtilis, 16, 208
Bacillus thermoproteolyticus, 16
Balanus, 62, 136, 207
Balanus nubilus, 237
Barnaea, 388
Beroe, 53
Blastocladiella, 260

Calliphora, 62, 354, 356
Carchesium, 254
Chaetopterus, 181, 388
Chaos, 222, 223
Chara, 222, 472
Chilomonas paramecium, 399
Chironomus, 31, 62, 165, 167
Ciona, 383
Clostridium, 121, 261, 372
Clostridium histolyticum, 16
Colombia, 62
Corallina officionalis, 7
Crithidia oncopelti, 219

Crypthecodinium, 252
Cypridina hilgendorfi, 305

Dictyostelium, 207, 222, 223, 252, 254
Digitalis purpurea, 439
Diplocardia, 182
Drosophila, 370

Electroplax, 126
Elliptio, 219
Erythroxylon coca, 433
Escherichia coli, 123, 198, 260
Euglena, 42, 207

Fucus, 473
Fundulus, 19

Gammarus pulex, 8

Harmothoë, 184
Helix, 122, 150, 163
Homo sapiens, 62, 207
Hydra, 252
Hydra littoralis, 12
Hydroides, 388

Latia, 182
Leishmania, 401
Leptospira, 207
Limulus, 30, 172, 174
Loligo forbesi, 31, 62, 148
Loligo peallii, 66
Lumbricus, 126
Lytechinus pictus, 383

Maia squinado, 49, 150, 234
Marthasterias glacialis, 62, 388, 389
Mesocricietus, 383.
Microsporidia, 401
Mnemiopsis, 184
Mytilus edulis, 19, 223

Naegleria gruberii, 29, 209, 370, 390
Necturus, 62
Nereis limbata, 363, 383, 388
Neurospora, 198
Nitella, 208, 212, 472
Noctiluca, 184

Obelia geniculata, 52, 53, 154, 183, 184, 199
Obelia lucifera, 52, 53, 181
Ophiopsila, 184
Oryzias latipes, 62, 380

Paracentrotus, 383
Paramecium, 19, 35, 66, 150, 152, 157, 158, 159, 160, 162, 164, 199, 207, 217, 219, 252, 309, 316, 327, 474
Patiria, 383
Pelvetia, 473
Phasoelus vulgaris, 377
Pholas dactylus, 179, 181
Photinus, 182, 184
Physalia, 9
Physarum polycephalum, 62, 208, 222, 223, 252
Pionosyllis, 181
Pisum, 472
Porichtys, 184
Pyrophorus, 179

Rana pipens, 62, 236, 383
Renilla, 126, 182, 184
Rhizobia, 12

Sphaerotilus, 207
Spirostomum, 29, 62
Spirula, 388
Spongilla lacustris, 372
Sporosarcina, 121, 372
Staphylococcus, 405, 406
Streptococcus pyrogenes, 405, 406
Streptomyces chartreusensis, 34
Streptomyces conglobatus, 34
Subellaria, 388

Thalassicola, 53, 181, 184
Torpedo, 313
Trichinella spiralis, 7
Trichoderma viride, 91
Trypanosoma, 207
Tubularia, 252

Urechis, 388

Vortićella, 124, 207, 230, 254, 255

Xenopus, 62, 383, 386, 387, 388, 390

Zoothamnium, 254

Subject Index

A23187 (*see also* Ca^{2+} ionophores), 34, 138, 161, 169, 177, 182, 194, 195, 209, 215, 216, 217, 219, 223, 254, 271, 272, 279, 280, 283, 289, 295, 300, 311, 337, 339, 340, 350, 360, 378, 379, 383, 384, 385, 386, 388, 391, 392, 396, 398, 401, 406, 414, 447
Abscission, 363, 472
Acetoxy methyl esters, 31, 59
Acetyl choline, 24, 26, 28, 35, 91, 135, 146, 147, 161, 197, 233, 249, 264, 272, 284, 299, 306, 307, 308, 309, 311, 312, 319, 321, 330, 333, 344, 354, 355, 356, 357, 440
Acidic amino acids, *see* Aspartate, Glutamate
Acrosome reaction, 220, 380, 392
ACTH, 24, 29, 268, 271, 300, 321, 354, 358–361
Actin, 209, 210, 211, 218, 219–231, 232, 245, 246, 254, 257, 336, 337, 391
α-actinin, 223, 225, 227
β-actinin, 221
Actinogelin, 221
Actomyosin, 210, 218, 225–231
Action potentials, 135, 136, 147, 148, 149, 155, 161, 170, 172, 207, 242, 245, 250, 277, 280, 283, 284, 308, 316
  calcium, 16, 63, 135, 136, 150, 151, 152, 153, 154, 155, 156, 157, 158, 159, 164
  sodium, 15, 16, 135, 136, 150, 151
  'sodium theory', 136, 148
Activity coefficient, 100–104
  Ca^{2+}, 59, 102
  calculations, 102
  equations, 101, 102, 103
  K$^+$, 102
Adenosine, 28, 202, 233, 264, 268, 317, 345
Adenylate cyclase, 24, 91, 123, 124, 154, 250, 267, 269, 274, 289, 292, 355, 359, 360, 405, 449

Adipose tissue, 24, 27, 29, 161, 166, 188, 197, 258, 260, 268, 271, 272, 273, 279, 283, 285, 286, 291–296, 298, 299, 301, 348, 434
ADP, 90, 112, 191, 200, 240, 264, 268, 286, 321
Adrenaline, 11, 19, 26, 27, 28, 29, 30, 35, 146, 152, 154, 194, 197, 215, 249–252, 268, 269, 271, 273, 277, 280, 284, 285, 288, 289, 290, 291, 292, 295, 296, 298, 299, 300, 308, 312, 313, 314, 315, 316, 317, 321, 326, 327, 328, 344, 345, 354, 356, 357, 406, 407, 442
Adrenal cortex, 24, 154, 265, 268, 271, 273, 291, 300, 303, 330, 354–361
Adrenal medulla, 72, 126, 199, 215, 257, 258, 265, 277, 308, 312, 316, 320, 321, 326, 327, 328, 330, 333, 337, 344, 434
α- and β-adrenergic effects and receptors, 356, 358, 439, 440, 443
α-adrenergic blockers, 154, 155, 288, 344, 438, 441, 443
β-adrenergic blockers, 156, 288, 344, 441, 443
Aequorin, 45, 50, 51, 53, 61, 62, 63, 65, 66, 70, 71, 83, 94, 124, 153, 155, 156, 163, 164, 168, 169, 171, 174, 182, 184, 188, 191, 196, 203, 235, 236, 237, 250, 382–386, 388, 389, 432
Alamethecin, 91, 92, 107
Albumin, 94, 122
Aldosterone, 354, 360
Algae, 6
Alizarin, 46, 57
Alizarin sulphonate, 19, 46, 49, 50, 57, 66
Alkali metals, *see also* Potassium, Sodium
  discovery, 3
  roles, 13, 14
Alkaline earths, *see also* Barium, Calcium, Magnesium, Radium, Strontium
  discovery, 1–3

541

Amino acids (*see also* Aspartate, Glutamate), 186, 308
Ammonia, 87
Ammonium purpurate, *see* Murexide
Amoeba, *see* Protozoa
Amoeboid-flagellate transformation (*see also* Intracellular Ca^{2+}), 206, 209, 211, 390, 391
Amoeboid movement, 206, 207, 208, 209, 210, 220, 222, 224, 253
AMP, 90, 112, 267, 281, 283, 477
AMP PCP, 204
AMP PNP, 204
Amphetamines, 430
α-amylase, 16, 91, 122, 309, 325
Anaesthetics (*see also* General anaesthetics, Local anaesthetics), 20
Anaphylatoxins, 410, 446
Angiotensin, 283, 288, 299, 354, 360, 439
Annelida, 22, 233, 383, 388
Anoxia, *see* Intracellular calcium
Antiallergy drugs, 430, 446, 447
Antibodies, 61, 193, 282, 317, 342, 376–380, 409–411
Antidiuretic hormone, *see* Vasopressin
Antigen, 28, 308, 311
Antimycin A, 33, 191
Antipyrylazo III, 46, 50, 56–58
Aplysia, *see* Molluscan neurones
Arbor viruses, 340, 341, 404
Arsenazo III, 45, 46, 50, 56–59, 62, 63, 65, 153, 156, 163, 164, 174, 175, 196, 236
Arthropeda, 6, 8, 9, 22, 137, 150, 166, 184, 191, 198, 199, 210, 230, 233, 234, 264, 265, 355, 356, 357, 372
Ascites, 260, 339
Aspartate, 94, 124, 127, 130, 246, 254, 425, 429, 480, 481
Association constant (*see also* Dissociation constant), 99
Atomic absorption spectrometry, 45, 48, 49, 191
ATP, 14, 72, 78, 82, 89, 90, 107, 110, 112, 120, 121, 200, 204, 206, 214, 291, 306, 314, 344, 399, 402, 405, 408, 413, 417
    ATPMg, 21, 121, 138, 158, 201, 221, 240, 241, 250, 254, 267, 269, 270, 272, 276, 280, 283, 286, 292, 293, 313, 384
    discovery, 17
    requirement for muscle contraction, 20, 21, 234
    synthesis (including chemiosmotic theory), 188, 189, 191, 192, 258, 421

ATPase, *see* Ca^{2+}-activated MgATPase
Auxin, 309, 472
Avogadro's number, 101, 195
Axoneme, 218
Axoplasm, 42
Axoplasmic flow, 208, 222, 336
Axopodia, 215, 216
Azide, 189

Bacteria, 12, 13, 171, 182, 198, 206–209, 253, 257, 264, 274, 277, 312, 347, 349, 363, 398, 405, 406, 411, 419
Bacterial Ca^{2+} chemotaxis, 29, 35, 206, 207
    content, 13, 42
    requirement, 12
    spores, 7, 12, 42, 121, 363, 372, 373
Bacteriophage, 15, 401, 402
Balance point, 190
Barium, 37, 81, 85, 97, 98, 236, 254, 319, 320, 321, 326, 434, 460
Barnacle, *see* Muscle
Bathmotropy, 249
Bayliss and Starling, 18, 261
Bernstein, 15
Berzelius, 1
Bicarbonate (*see also* Sodium bicarbonate), 116
Biomineralization, 6, 7, 8, 13, 14
Bladder, 268, 273, 354, 358
Blepharismin, 89
Blue-green algae (calcareous), 1, 5, 6, 206, 363
Boltzmann distribution 335
Bone, 1, 6, 15, 117, 154, 199, 268, 303, 412
Boyle, Robert, 179
Bradykinin, 272, 299
Brain, 42, 83, 126, 186, 188, 193, 199, 215, 258, 268, 269, 271, 272, 273, 274, 277, 282, 308, 340, 432
Brown fat, 161, 193
Brownian motion, 335, 336, 337

Caffeine (*see also* Xanthines), 170, 187, 430, 448
Calcification, 7, 394
Calcineurin, 282
^{43}calcium (*see also* Nuclear magnetic resonance), 3, 107
^{45}calcium
    decay reaction, 73
    half-life, 3, 73
    use, 49, 66, 72–76, 107, 133, 151, 154, 175, 191, 199, 201, 203, 239, 242, 253, 254, 279, 286, 289,

298, 328, 329, 335, 378, 379,
 397, 398, 399, 405, 443, 473
[47]calcium
 decay reaction, 73
 half-life, 3, 73
 use, 73
Calcium
 activation of nuclease, 91, 427
 activation of phospholipases, 16, 122,
  126, 293, 294, 298, 303, 343,
  423, 425, 427
 activation of protease, 23, 121, 171,
  261, 415, 424, 427
 activation of transglutaminase, 16, 23,
  91, 107, 122, 414
 active versus passive binding, 23, 24,
  37, 38, 120–130, 144
 activity and activity coefficient, 99,
  100–104
 atomic properties, 85
 bacteria, 12
 binding calculations, 104–114
 binding to inorganic ligands, 88
 binding to NADH, 397
 binding to nucleic acids, 15, 92, 93, 105,
  107, 122, 186, 375, 429
 binding to phospholipids, 15, 92, 93,
  105, 107, 122, 124, 186, 272,
  434, 435, 436, 437, 450, 452
 binding to proteins, 14, 15, 104, 105,
  185, 186, 404, 429
 binding to small organic ligands, 89, 90,
  186
 biogeochemistry, 13, 14
 biological roles, 13–17
 blood clotting, 16, 23, 25, 29, 107
 bound in cell (and to ligands), 13, 112
 cell and tissue content, 13, 41, 42, 43,
  333, 393, 394, 405, 413, 477
 cell division, 362–392
 chemical potential, 98, 99
 chemistry, 85–134
 complement action, 16, 28, 29, 122,
  409–412
 coordination chemistry, 96–98,
  117–120
 cell growth, 11, 362–364
 cofactor, 14, 16
 concentration in body fluids and
  balanced salt solutions, 8, 9, 10,
  11, 25, 116
 concentration in water, 8, 9, 11
 discovery as an element, 1–3, 85, 457
 effect of removal, 10, 11, 12, 148, 166,
  167, 169, 202, 215, 217, 239,
  254, 271, 272, 288, 295, 306,
  307, 322, 323, 324, 348, 352,
  354, 360, 378, 379, 391, 401,
  403, 407
 electrical role, 14, 15, 16, 135–205,
  329–332
 enzyme activation, 24, 91, 107, 122,
  123, 124
 enzyme inhibition, 91, 107, 122, 123,
  124
 extraction, 45, 46
 extracellular (free), 9, 10, 142
 extracellular (total), 8, 9, 349
 free Ca^{2+}, 49–66
 geology, 4–6
 historical, 1, 2, 17–21, 123, 136, 234,
  235
 hydration energy, 98
 inorganic ligands, 88
 ionic radius (and hydrated), 85, 97
 isotopes (natural and radioactive), 3
 ionophores, see Calcium ionophores
 kinetics of binding, 113, 114, 121, 127
 lack of requirement for, 29, 30, 310,
  321
 measurement of binding to ligands, 106,
  107
 medical problems, 393–395
 methods for measuring total, 45
 microfilaments, 126, 209, 210,
  219–225, 374, 375
 microtubules, 126, 209, 210, 212–219,
  375
 minerals, 4, 5
 mobility, see Diffusion
 natural history, 1–22
 natural occurrence, 1–8
 need to measure free see Intracellular
  Ca^{2+}
 need to measure total, 39–43
 nuclear content, 40, 41, 93, 186, 333,
  397, 415, 421, 422
 origin of the word (*calx*), 2
 passive, 23, 24, 144, 424, 425
 pathological deposits of salts, 7, 8, 394,
  406
 pharmacology, see Pharmacology
 plants, 6, 7, 12, 471–473
 polymeric ligands, 90–95
 precipitation of salts, 1, 5, 6, 47, 66,
  394, 406, 425, 426
 reaction with nitrogen and oxygen, 2
 requirement for action of complement,
  29
 requirement for life and growth, 8–13
 requirement of phenomena for
  extracellular, 19, 28, 29, 30
 role as an intracellular regulator, 14, 15,
  17, 23–84

Calcium (contd.)
  role in blood clotting, 16, 23, 25, 29
  role in extracellular enzymes, 14, 15, 16
  small organic ligands, 89, 90, 272
  solvation and solvation energy, 96, 97, 98
  solubility of salts, 114–117
  statocysts, 7
  stereochemistry, 87, 88
  structural role, 14, 15
  thermodynamics of binding, 96–120
  whole body metabolism, 9
Calcium-activated MgATPase (see also Sarcoplasmic reticulum), 24, 91, 124, 126, 138, 143, 185–205, 226, 240, 275, 296, 298, 333, 341, 346
  reversible, 202
Calcium-activated K^+ current, see K^+ conductance
Calcium binding
  Kd(Ka), 25, 39, 201, 202
  stereochemistry, 87, 88
Calcium binding proteins (see also Calmodulin, Calsequestrin, Gelsolin, Leiotonin)
  amino acid composition, 94
  concentration in cells, 238
  how they work, 95, 131, 132
  mitochondrial Ca^{2+} binding glycoprotein, 193
  myosin, 229–231, 235
  non-physiological, 109, 123, 259, 260, 273, 274, 289, 296, 303, 414
  pathology, 424
  physiological, 93, 94, 109, 121, 123, 124, 182, 186, 229–231, 260, 273, 274, 289, 303
  properties, 124, 125, 126, 201, 202, 203, 205
Calcium buffers, 11, 20, 23, 28, 30, 44, 76–82, 87, 107, 185
  BAPTA, 78
  EDTA, 78, 82
  EGTA, 20, 78, 79, 80, 82, 234, 282, 286, 295, 320, 356, 378, 403
  problems, 77, 79, 80, 81
  structures, 78
  use in cells, 20, 30, 31, 160, 163, 164, 167, 223, 234, 235, 280
Calcium–calcium exchange, 204
Calcium carbonate, 6, 15, 45, 47, 85, 88, 114, 116, 394, 425
Calcium channels, 293, 301, 449
  blockers (see D-6000, Verapamil, Nifedipine), 32, 33, 37, 148, 151, 152, 153, 328, 331, 360, 388, 408, 429, 430, 442–446, 452

  cells with, 150, 154, 242
  comparison with Na^+ and K^+, 153
  evidence for and properties, 148, 149, 150–160
  structures of blockers, 33, 148, 444
Calcium channel blockers, see Calcium channels, D-600, Verapamil, Nifedipine
Calcium chelators, see Calcium buffers
Calcium citrate, see Citrate
Calcium conductance, 158
Calcium currents (see also Calcium channels, Membrane potential), 24, 149–160
Calcium dipicolinate, 7, 89, 90, 118, 372, 373
Calcium efflux, 41, 190, 198–204, 239
Calcium electrodes, 4, 9, 50, 59–61, 62, 63, 65, 103, 107, 191, 236, 357, 388
Calcium equilibrium potential, see Membrane potential
Calcium flux
  analysis, 72–76
  ^{45}Ca versus ^{47}Ca, 73
  interpretation of data (problems), 75
  measurement, 72
  passive across membranes, 138
  pools, 72
Calcium-'free' media and calcium contamination, 11, 28, 30
Calcium granules, 186
Calcium-H^+ transport, 138, 192, 198
Calcium ionophores (see also A23187, Ionomycin, X537A), 32–37, 43, 44, 50, 89, 91, 98, 107, 138, 193, 300, 301, 303, 326, 327
Calcium molybdate, 45, 47
Calcium NADH, 397, 421
Calcium nitride, 2, 3
Calcium oxide, 2, 3
Calcium oxalate, 7, 45, 47, 66, 67, 72, 114, 116, 117, 187, 188, 189, 240, 333, 394, 397, 425
Calcium phosphate, 6, 15, 72, 88, 89, 98, 114, 115, 116, 117, 118, 138, 191, 192, 193, 394, 425, 426
Calcium picrolonate, 47
Calcium pumps, see Ca^{2+}-MgATPase, Endoplasmic reticulum, Sarcoplasmic reticulum
Calcium pyroantimonate, 66, 67, 187, 297, 333, 334
Calcium stearate, 85
Calcium sulphate, 7, 88, 114, 116
Calcium thymidylate, 118
Calculi, 8

Calmodulin
    action, 126, 127, 202, 212, 217, 218,
        224, 254, 261, 267, 274, 275,
        276, 281, 282, 283, 284, 289,
        296, 343, 344, 376, 405
    activated kinase, 221, 224, 229–231,
        248, 337
    affinity for Ca^{2+}, 24, 72, 93, 105, 132,
        272, 282
    amino acid composition, 94, 124
    amino acid sequence, 125
    antibodies and immuno localization,
        127, 130, 282, 374–376
    Ca^{2+} binding, 107, 110, 112, 273
    concentration, 112, 468, 477
    criteria for establishing role, 127
    discovery, 123, 235, 272, 457
    enzyme activation, 126, 286
    inhibitor R24571, 470
    molecular weight, 94, 124, 226
    phenothiazine inhibition, 127, 328, 438,
        439, 452
    properties, 93
    sources, 123, 210, 214, 221, 226, 328,
        389, 472
    trifluoperazine affinity constant, 438
    Zn^{2+}, 450
Calsequestrin, 91, 124, 187, 226, 235, 241,
    242
Campbell, A. K., 54
Capacitation (*see also* Spermatozoa), 364,
    391, 392
Cancer, 398, 400
Carbonate, 116
Cardiac glycosides (*see also* Ouabain),
    140, 197, 407, 430, 439–442
Cardiac muscle, *see* Muscle
Cardiovascular agents, *see* Intracellular
    Ca^{2+}
Cartilage, 411
Casein, 122
Catecholamines (*see also* Adrenaline,
    Isoprenaline, Nor-adrenaline), 233
Cations
    biological roles, 14, 21
CDTA, *see* Ca^{2+} buffers
Cell adhesion, 11, 15
Cell aggregation, 29, 35
Cell communication (*see also* Gap junctions), 15, 17, 35, 165–171
Cell cycle, 362, 364–368
Cell death and necrosis, 365, 393, 407,
    411, 412, 417, 425, 426–429, 443
Cell differentiation, 26
Cell division, 17, 26, 208, 277, 299,
    362–392
Cell dormancy, 372, 373

Cell fusion, *see* Membrane fusion
Cell membrane, *see* Membranes
Cell motility
    eukaryote, 25
    prokaryote, 25
Cell movement, 17, 25, 28, 29, 35,
    157–160, 206–256
Cell polarization, 208, 210
Cell shape, 35, 208, 210, 401, 413, 414
Cell stimuli classification, 25–28
Cell surface area, 133, 140
Cell transformation, 17, 26, 29, 30, 35,
    370, 390, 391
Cell viability (criteria), 29, 30, 407
Cell volume, 133, 140, 155
Centroacinar cells (exocrine pancreas),
    353–356, 358
Chaetognatha, 230
Chambers, 19
Chelex 100, *see* Calcium-free media
Chemiluminescence, *see* Luminescence
Chemoreceptors, 178
Chemotaxis, 17, 25, 28, 29, 207, 220, 222,
    252–254, 277, 299, 318, 349
Chemotactic peptides (C3a, C5a, N
    formylmet-leu-phe), 252–254, 277,
    298, 299, 301, 347, 349, 410
Chicago sky blue, 167
Choride, 9, 16, 139, 140, 141, 355, 357
Chloroplasts, 35, 472
Chlorpromazine (*see also* Phenothiazines),
    429, 430, 437, 438, 439
Chlortetracycline, 46, 58, 59, 62, 253, 329,
    414
Cholecystokinin, 322
Cholera toxin, 405
Choline, 149, 154, 333
Chordata (vertebrates), 5, 6, 9, 22, 184,
    230, 233, 264, 265, 286, 383, 388
Chorioallantoic membrane, 199, 205
Chromaffin granules, 72, 186, 312, 313,
    314, 326
Chromatin, 15, 375
Chromone, 446
Chromosomes, 215
Chronotropy, 249, 250
Cilia, 19, 40, 156–160, 161, 206, 207, 208,
    209, 210, 217, 218, 219
Citrate, 23, 78, 79, 82, 89, 90, 112, 186,
    362
Citrate desmolase, 260
Clathrin, 347
Cleavage furrow, 31, 220, 374, 375, 376,
    386
Clot retraction, 220
Co^{2+} 32, 33, 151, 152, 153, 159, 331,
    449

545

Coelenterata (hydroids, jelly fish, corals, sea anemones, sea combs), 6, 9, 12, 22, 29, 35, 49–53, 154, 166, 178, 179, 180, 181, 182, 183, 184, 207, 233, 309, 363
Colchicine, 213, 337
Cole, 15, 136
Collagenase, 16, 122
Complement (including C3a, C5a, C3b), 28, 122, 178, 207, 252, 253, 303, 342, 348, 349, 351, 397, 401, 409–412, 423
Concanavilin A, 298, 299, 347
Contractile vacuoles, 208
Coordination number, 87, 118, 119
Cortical granules, 327, 380–386
Covalent modification of proteins (see also Phosphorylation), 259, 261, 276
Crab, see Arthropoda, Maia
Crustacea, 6
CTP, 291, 293
Curtis, 15, 136
Cyanide (CN⁻), 32, 140, 167, 191, 203, 327, 351
Cyclic AMP, 27, 38, 154, 161, 165, 171, 172, 176, 177, 197, 204, 207, 212, 214, 218, 221, 249, 251, 252, 253, 254, 263, 266, 267–276, 277, 288, 289, 290, 292, 295, 300, 301, 317, 318, 321, 344, 346, 349, 354, 355, 357, 358, 359, 360, 362, 372, 379, 390, 405–412, 435, 437, 439, 462, 463, 477
Cyclic AMP protein kinase, 212, 214, 218, 221, 224, 248, 251, 256, 355, 359, 390, 439
Cyclic CMP, 266
Cyclic GMP, 35, 161, 171, 176, 177, 249, 254, 263, 266, 267–276, 288, 289, 290, 295, 300, 318, 344, 347, 362, 377, 379, 405, 462, 463, 477
Cyclic GMP protein kinase, 274, 275
Cyclic nucleotide phosphodiesterase, see Phosphodiesterase
Cystic fibrosis, 398, 400, 412, 419
Cytochalasin B, 205, 213, 337
Cytochrome (respiratory) chain, 136
Cytoplasmic streaming, 208, 210, 220, 222, 223, 472
Cytoskeleton, see Microfilaments

D_2O, 214
D-600, 32, 33, 37, 148, 151, 152, 153, 154, 155, 159, 163, 289, 300, 378, 389, 444
D-600 isomers, 153, 155, 379, 446

Dantrolene sodium, 33, 430
Dark adaptation, 176–177
Davson and Danielli, 137
Davy, Humphry, 1, 2, 85, 457
Debye–Huckel law, 100, 101, 103
Deficiency diseases, see Intracellular Ca^{2+}
Deuterostomia, 230
Diabetes, 424
3-Diazo SIBA, 298
Dibutyryl cyclic AMP, 197, 346
Dichlorophonazo III, 46, 57
Dielectric constant, 101, 103, 104
Diffusion, 20, 49, 65, 132, 133
Digitalis, 11
Digoxin (see also Cardiac glycosides), 430, 441
Diltiazem, 33
Dinitrophenol, 32, 33, 167, 189, 191, 327
Dinoflagellates, 178, 184, 252
Discs, 173, 174, 175, 188
Disodium cromoglycate, 430, 446, 447, 452
Dissociation (and Association) constant, 24, 57, 82, 99, 109, 110, 112
Divalent cations, 21, 85
DNA (see also Calcium, Chromatin), 364, 365, 374, 376, 379, 381, 402
DNAase, 16, 24, 91, 121, 124
Donnan equilibrium, 139, 140, 141, 479
Dopamine, 233, 264
Dromotropy, 249
Dubois, R., 51, 179
Ductus deferens, 272, 274
Duodenal juice, 24
Dynein, 213, 214, 215, 218, 374

Ebashi, 12, 93, 123, 235, 457, 461
Ecdysone, 264, 354, 361
Echinodermata (sea urchins, starfish), 7, 19, 22, 29, 30, 62, 166, 184, 188, 230, 264, 327, 383, 386–390, 439
EDTA (ethylenediamine tetraacetic acid), see Calcium buffers
EF hand, 119, 120
EGTA (ethyleneglycol-bis-(-amino-ethyl ether) tetraacetic acid), see Calcium buffers
Egg fertilization, 11, 17, 19, 26, 28, 30, 31, 35, 62, 166, 188, 215, 217, 222, 309, 321, 326, 327, 328, 329, 339, 362, 364, 374, 380–386, 472, 473
Einstein equation, 336
Electron spin resonance, 107, 137, 301
Electroreceptors, 162
Elements in the earth's crust, 4
Embryonic growth and tissue development, 371

Emiocytosis (*see also* Exocytosis), 305
Endocrine pancreas, *see* Pancreatic β cells
Endocytosis, 26, 208, 222, 305–307, 315, 347–353
Endoperoxides, 249, 264, 301, 302, 303, 347
Endoplasmic reticulum Ca^{2+} store (*see also* Muscle), 41, 186, 187, 188, 215, 289, 296, 297
Endoplasmic reticulum effectors, 32, 33
Energetics, 200
Epidermal cells, 409
Epithelia, 166, 207, 219
Equilibrium dialysis, 106
Eriochrome Blue, 83
Erythrocytes, 13, 24, 35, 42, 83, 126, 154, 162, 163, 198, 201, 202, 204, 258, 260, 266, 271, 272, 273, 277, 279, 284, 339, 398, 401, 409, 411, 413, 414, 419
Erythrocyte 'ghosts', 61, 64, 163, 201, 202, 339, 340, 341, 342, 410, 411, 412, 438
ETH 1001, 60
Eukaryte motility, 209, 210
Evolution, 5, 477–482
Excitable membranes, 12, 14, 15, 19, 25, 28, 135–165, 330
Excitation–contraction coupling (*see also* Muscle), 154, 319
Excitation–secretion coupling, 319
Exocrine pancreas, 147, 161, 197, 272, 273, 279, 300, 309, 312, 316, 322, 325, 326, 327, 329, 300, 332, 346, 353–358, 419
Exocytosis, 26, 220, 305–347, 310
Eye, *see* Photoreceptors

Faraday, Michael, 3
Fe receptors, 348, 349
Fibrin, *see* Calcium—blood clotting
Fibroblasts, 29, 166, 207, 252, 273, 299, 337, 347, 363, 400
Firefly, 182, 184
Fish eggs, *see* Egg fertilization
Flagella and cilia, 206, 207
Flagellin, 206
Flame photometry, 45, 47, 48
Fluid secretion, 356–358
Fluorescein, 167, 169, 170
Fluorescence, 107
Fluorescent Ca^{2+} indicators, *see* Chlortetracycline and Quin II
Fluorescent membrane probes, 137
Fluoride, 163
Foraminifera, 6, 208, 222

Forskal, 49˙
Fossils, 5
Fragmin, 221
Frog heart, 11
Frog muscle, *see* Muscle
Fructose diphosphatase, 105, 109, 123, 258, 260, 287, 288
Fugacity, 101

GABA (γ-amino butyric acid), 233, 249, 340
Galvani, 136
Gap junctions, 15, 29, 31, 35, 71, 165–171, 332, 425
Gastrin, 354, 356
GDP, 269, 270
Gelation (gel–sol), 223, 224
Gelsolin, 126, 221, 223, 225
General anaesthetics, 430–433
Genetic code, 127, 130, 480, 481
Germinal cell layers, 371
Glass scintillator probes, 73, 203
Glucocorticoids, 197, 262, 265, 279, 295, 354, 358–361, 398, 401
Glucagon, 28, 29, 35, 197, 263, 264, 265, 268, 271, 273, 277, 280, 283, 285, 287, 288, 289, 290, 300
Gluconeogenesis, 27, 154, 194, 277, 286–291
Glucose, 28, 154, 215, 257, 258, 265, 277–291, 299, 304, 330, 332, 337
Glucose–fatty acid cycle, 259
Glucose transport, 27, 278–280
Glutamate, 89, 94, 110, 112, 118, 124, 127, 130, 137, 147, 154, 233, 237, 246, 249, 254, 299, 425, 429, 480, 481
Glutamate dehydrogenase, 260
Glutathione, 252
α-Glycerophosphate dehydrogenase, 194, 260
Glycocalyx, 72
Glycolysis, 194, 260, 277, 280, 284–286, 287, 304, 349, 391, 404
Glycogen (*see also* Phosphorylase, Phosphorylase kinase), 257, 258, 265, 276, 277, 278, 280–284, 299, 304, 349
Glycogen synthetase, 261, 276, 280–284
Glyoxylate cycle, 7
GMP, 267
Goldman constant field equation (Goldman–Hodgkin–Katz equation), 140, 141, 146, 174
Gouy–Chapman theory, 338
Growth hormone, 154, 279, 295

GTP, 212, 214, 267, 269, 272
GTP binding protein (G protein), 250, 269, 270, 292
GTPase, 212, 269, 270
GTP binding protein, 269, 270
Guanylate cyclase, 126, 176, 267, 269–274, 357
Gut, 198, 205, 269, 279, 303, 344, 354, 356

H^+, 191, 192, 313, 381
H^+–Ca^{2+} exchange, see Ca^{2+}–H^+
HEDTA, see Calcium buffers
Haemolymph, 10, 257
Haemolysin, 16
Harmaline, 451
Heart muscle (see also Muscle), 10, 11, 19
Heilbrunn, L. V., Frontispiece, 19, 31, 234, 235, 236, 336, 362, 425, 457, 460, 461
HeLa cells, 198
Hemichordata, 230
Hepatocytes, see liver
Heptopancreas, 333
Hexokinase, 260, 284
High-voltage pulse holes in membranes, 31, 326, 327
Hill, A. V., 20, 133, 254
Histamine (see also Mast cell), 233, 264, 268, 269, 272, 298, 299, 308, 311, 321, 322, 325, 344, 439, 440, 446
Hober, 19
Hodgkin, A. L., 15, 136, 141, 457
Hopkins, Frederick Gowland, 1, 18, 476
Hormones (+ receptors), 16, 18, 26, 28, 29, 30, 35, 154, 257–304, 368, 437, 466, 482
Huxley, A., 15, 136
Hydroids, see Coelenterata
5-Hydroxytryptamine (serotonin), 29, 72, 170, 233, 262, 264, 268, 299, 321, 344, 354, 356, 357, 360, 439, 440, 446

IgE, 299, 308, 311, 337, 446, 452
Image intensification, 70–72, 166, 169, 184
Immune cell injury, see Intracellular calcium
Infectious diseases, 400–405
Inotropic effect, 249, 250, 283, 442
Insects, see Arthropoda
Insulin, 27, 28, 30, 35, 154, 161, 194, 197, 215, 263, 264, 265, 266, 268, 271, 272, 277, 279, 283, 285, 286, 287, 291, 295, 296, 299, 304, 308, 314, 321, 325, 330, 344, 348, 439

Intact cells, 303
Intercellular pathways, 257, 258
Intermediary metabolism, 17, 19, 257–304
Intracellular Ca^{2+}
  anaesthetics, 430–437
  anoxia, 398, 400, 408
  antiallergy drugs, 446–447
  bioluminescence, 178–184
  bound to ligands, 112, 184
  cancer, 398, 400
  cardiovascular agents, 439–446
  cations, 449
  cell–cell communication and gap junctions, 165–171, 425
  cell change associated with pathological increase, 398, 399
  cell death (necrosis), 365, 398, 399
  cell movement, 157–160, 206–256
  cell shape, 398, 411, 413, 414, 419
  cell transformation 206, 390, 391, 398, 411, 413, 414, 419
  chemistry, 85–134
  chloroplasts, see Chloroplasts
  complement (see also Complement), 398
  criteria for establishing role, 28–39
  criteria for good indicator, 49–51, 61–65
  cystic fibrosis, 399, 400
  cytoplasmic concentration, see Intracellular free Ca^{2+}
  deficiency disease, 399
  diffusion coefficient, 49
  distribution, 65–72
  effects on the electrical properties of cells, 144, 160–177
  egg fertilization, 380–386
  electrical activity of cells, 135–205
  erythrocytes and erythrocyte 'ghosts', see Intracellular free Ca^{2+}
  evidence in cell injury, 396–400
  evolutionary significance, 477–482
  experimental approach to investigation, 27–39, 396, 457
  free concentration, 20, 21, 25, 142
  gluconeogenesis, see Gluconeogenesis
  glucose transport, 279, 280
  glycolysis (see also Glycolysis), 278
  guidelines for establishing as a regulator, 474, 475
  historical, 459–462
  historical experiments, 19, 20, 21, 457
  immune cell injury, 397, 398, 401, 408–412, 419
  incorporation of indicators into cells, 19, 61, 64

infectious diseases, 400–405
injection into cells, 20, 30, 31, 160, 174, 234, 236, 326, 327, 375, 383
intermediary metabolism, 257–304
internal stores, 173, 185–205, 254, 255, 389, 421–423
investigation, 23–84
local anaesthetics, *see* Local anaesthetics
long-term regulatory mechanisms, 471
lymphocyte activation, 376–380
manipulation of, 20, 30–37
measurement, 20
mechanisms of cell injury, 416–425
meiosis, 386–390
methods for measuring distribution, *see* Image intensification, X-ray microprobe analysis
methods for measuring free, 49–60
miscellaneous drugs, 450, 451
mitochondrial enzymes, 91, 121, 122, 123, 194
mitosis, 373–380
molecular action, 38, 39, 335
muscle contraction (*see also* Muscle), 20, 25
muscle disease, 399, 400
need to measure free, 43, 44
pathology, 39, 42, 43, 393–429, 453
perspectives, 476–482
phagocytosis, 348, 349
pharmacology, 39, 429–452, 453
phenomena regulated by, 17, 456, 460
phenothiazines, *see* Phenothiazines
pools, 20, 329
problems, 20, 469–473
regulation by cyclic nucleotides, 268, 275
release from intracellular stores, 242–244, 333–335, 360, 382–386
resumé, 455–459
sites bound in cells, 41, 186
sperm capacitation, 391, 392
synthesis, 455–468
threshold phenomena, *see* Threshold phenomena
toxins, 398, 405–407
trauma, 399
unitary hypotheses, 459–468
unit cell activation hypothesis, 465–468
xanthines (*see also* Xanthines, 447–448
Intracellular pH
as a regulator, *see* H$^+$
mitochondria, 408
value, 79, 282

Intracellular free Ca^{2+}
abnormalities in regulation of, 423
absolute concentration, 20, 21, 46, 71, 116, 142, 143, 163, 164, 169, 174, 189, 190, 223, 235, 238, 280, 284, 378, 410, 414, 477
cells in which measured, 13, 43, 62, 135
concentration and changes in
erythrocyte ghosts, 342, 404, 410, 411, 412
fertilized eggs, 382–386
kidney, 62
lymphocytes, 59, 62, 378
muscle, 62, 63, 235–238, 250, 280, 284
nerve cells, 62, 142, 156, 157, 163, 164, 319, 329
oocytes, 319, 329, 388, 389
phagocytes, 62, 253, 254, 329, 342
photoreceptors, 174
platelets, 62
protozoa and amoebae, 62, 157–160, 223
red cells, 62
secretory cells, 62, 168, 169, 329, 357
distribution in cell, 65
effect of Ca^{2+} ionophores, 385
effect of cell antibody-complement, 61, 62, 64, 409–412
effect of CO$_2$, 252
effect of L-glutamate, 63, 237
effect of membrane potential (action potential, hyper- or depolarization and voltage clamp), 147, 148, 155–160
effect of 1-methyl adenine, 389
effect of Sendai virus, 404
effect of sperm, 381–385
effect on Ca^{2+} conductance, 160
effect on cilia, 158
effect on C$^-$ conductance, 162
effect on excitable cells, 161
effect on gap junctions, *see* Gap junctions
effect on K$^+$ conductance, 160–165
effect on membrane potential, ion permeability and membrane conductance, *see* Membrane potential
effect on Na$^+$ conductance, 162, 171–177
effect on photoreceptors, 161
effect on secretion, 161
estimation using ^{45}Ca, 49
estimation using various indirect methods, 49, 50

Intracellular free Ca^{2+} (contd.)
  image intensification and Ca^{2+} distribution (see also Image intensification), 65, 70–72, 168, 169, 382–386
  measurement, see Ca^{2+} electrodes, Photoproteins, Arsenazo III, Quin II
  release from intracellular stores, 135, 335, 388, 389
  resting, 135
Intracellular pH, 166, 171, 196, 252
Intracellular volume, 40
Inulin, 40
Invertebrate hormones, 264, 265
Invertebrate muscle, see Muscle
Iodoacetate, 163, 264
Ionomycin (see also Ca^{2+} ionophores), 32, 34, 301
Ionophoresis, 174
Iproverati and derivatives, see Ca^{2+} channel blockers, Verapamil, D-600
Islets of Langerhans, see Pancreatic β cells
Isoprenaline (Isoproteronal), 250, 288, 405, 406, 440

K^+, see Potassium
Kallikrein, 252, 253
Katz, 136, 137, 141, 457
Kelvin, Lord (Thomson, William), 17, 96
Ketone bodies, 296
Kidney (see also Renal cortex, Renal medulla), 7, 62, 154, 184, 197, 199, 205, 260, 271, 272, 277, 285, 286, 287, 288, 289, 308, 399, 400, 424, 449
Kynurenine aminotransferase, 260

Lacrimal gland, 154
Lactate, 258
Lactate dehydrogenase, 477
Lanthanum (La^{3+}), 33, 35, 48, 85, 151, 152, 153, 159, 192, 193, 203, 300, 328, 331, 360, 378, 388, 389, 430, 449
Latent Ca^{2+} ligands, 424, 425
L-cells, 198, 363
Leiotonin, 24, 126, 229–231, 235, 246, 248
Leucocidin, 309, 341, 398, 401, 405, 406, 424
Leucocytes (see also Lymphocytes, Macrophages, Monocytes, Polymorphs), 398, 399, 401, 406, 419
Leukotrienes, 300, 303, 347
Li^+, 32, 154, 333, 449
Ligand, 86, 87, 98, 104

Lipid synthesis (Lipogenesis), 27, 259, 265, 277, 290, 291, 296, 304
Lipolysis, 257, 258, 265, 291–296, 304
Liposomes, 61, 62, 137, 164, 339, 434, 436
Liver, 27, 29, 42, 83, 138, 154, 161, 166, 188, 193, 197, 258, 260, 266, 268, 269, 271, 272, 273, 274, 277, 279, 280, 283, 284, 285, 286–291, 299, 300, 309, 312, 318, 330, 397, 398, 399, 400, 401, 406, 407, 408, 409, 432, 434
Local anaesthetics, 32, 33, 243, 288, 289, 300, 328, 392, 397, 403, 407, 429, 430, 432–437, 439
LSD, 430, 450
Luciferin–luciferase, 51, 179, 182
Luminescence
  apparatus and measurement of, 54, 55
  bio-, 25, 26, 29, 35, 154, 161, 178–184, 199
  chemi-, 29, 30, 35, 58, 64, 179, 301, 306, 342, 350–353, 380, 383, 411
  classification, 179
Luminol, 58, 64, 342, 360, 411
Luminous coelenterates, see Coelenterata
Lung, 268, 274, 398, 409
Luteinizing hormone (LH), 268, 269
Lymphocyte, 28, 29, 35, 59, 62, 154, 170, 210, 279, 284, 298, 299, 363, 364, 374, 376–380, 401, 406, 416
Lymphokines, 253
Lymphotoxins, 398, 406, 407

Macrophage, 29, 207, 222, 223, 252, 253, 277, 303, 306, 310, 347, 349, 351, 406
Magnesium, 151, 153, 170, 240, 306, 314, 326, 331, 332, 348, 400, 430, 449, 460
  atomic properties, 85
  mgATP, see ATPMg
  MgATPase, 213
  compared with Ca^{2+}, 98
  fluid concentration, 8, 9
  intracellular Mg^{2+} concentration, 83, 84, 282
  ionic radius, 85, 97
  ligand binding, 110, 114
  measurement of free, 83
  role, 14, 21, 219
  total cell, 13, 93
Mammalian neurones, see Vertebrate neurones
Mast cell, 31, 138, 298, 299, 308, 311, 316, 317, 320, 321, 322, 325, 326, 327,

330, 337, 339, 340, 344, 346, 434, 446
Mechanisms of cell injury, 395–429
Medaka fish (*Oryzias latipes*), 380–386
Meiosis, 5, 35, 208, 364, 368–370, 386–390
Melanocyte stimulating hormone (MSH), 268
Melanophores, 268, 336
Melatonin, 268
Mellitin, 253, 401, 406
Membranes
  as an electrical circuit, 145
  Ca^{2+}, 41, 93, 105, 135, 402, 438
  Ca^{2+} gradient, 43, 201
  capacitance, 145, 151, 195
  electrical properties, 135, 136, 274, 472, 473
  low permeability to cations, 274
  non-excitable, 146
  passive movement of Ca^{2+}, 138, 413, 417, 420
  pathology, 413, 420
  permeability coefficients, 138, 141, 146, 174
  resistance, 136
  resting potential (*see also* Membrane potential), 117
  structure, 137, 215, 272, 274, 413
  vesiculation, 412, 414
  water permeability (*see also* Osmotic balance), 14, 21
Membrane fluidity, 15
Membrane fusion, 29, 35, 61, 64, 65, 120, 220, 232, 329, 331, 337–344, 402–404
Membrane potential, *see also* Calcium channels, Calcium currents
  action potential, *see* Action potentials
  Ca^{2+} equilibrium potential, 141, 142, 143, 157, 158, 195
  effect of complement, 411
  effect of cyclic nucleotides, 161
  effect of extracellular ions, 135, 148, 149, 150, 153, 154
  effect of hormones, 161, 249, 330, 331, 349, 360
  effect of intracellular Ca^{2+}, 144, 145, 147, 148, 160–165, 168, 169, 171–177
  effect of transmitters, 135, 147, 161, 237, 249, 330, 331
  electronic potentials, 145, 146, 172, 330
  evolution, 478, 479, 480
  hyperpolarization, 145, 161, 174, 203, 331, 349, 357
  Mitochondrial, 191–195
  resting, 117, 137, 139–144
  voltage clamp, *see* Voltage clamp
Metabolic poisons, 163
Metallochromic indicator dyes, *see* Arsenazo III, Antipyrulazo III, Chlorophonazo III, Murexide
1-Methyl adenine, 368, 387–390
N-methyl histidine, 219, 220
Microfilaments, 126, 209–213, 215, 219–225, 254, 336, 337, 375, 376
Microinjection of cells (*see also* Calcium, 19, 20, 61, 64, 71
Microsomes, *see* Endoplasmic reticulum
Microtubules, 126, 209–219, 254, 336, 337, 374, 375, 376
Microtubule-associated proteins, 214, 217
Midge (*see also* Chironomus), 165, 167
Mitochondria, 41, 96, 117, 201, 258, 278, 289, 291, 408
Mitochondria Ca^{2+}, 41, 43, 96, 114, 167, 169, 170, 186, 187, 188–198, 205, 244, 275, 285, 286, 287, 289, 393, 397, 400, 406, 415, 421, 425, 426, 428, 429, 432, 434
Mitochondrial effectors, 32, 33, 192, 452
Mitochondrial volume, 96, 195
Mitosis, 208, 210, 211, 217, 363, 364–368, 371–380, 381
Mn^{2+}, 32, 81, 85, 134, 151, 152, 153, 159, 182, 272, 286, 328, 331, 388, 449
Mollusca (squid, Aplysia), 6, 8, 9, 19, 22, 181, 219, 230, 233, 264, 383, 388
Molluscan neurones, 24, 31, 61, 63, 136, 138, 142, 143, 144, 146, 147, 150, 152, 154, 155, 156, 157, 158, 159, 160, 161, 162, 163, 164, 188, 198, 199, 203, 223, 443
Mono- and bi-directional systems, 317, 460, 462, 463
Monocytes, 327
Monovalent cations, 21, 82
Morphine, 450, 451
Multiple sclerosis, 412, 419
Murexide, 45, 46, 47, 50, 56–58, 107, 191, 234
Muscle, 339
  barnacle, 35, 62, 63, 65, 83, 136, 147, 150, 152, 154, 160, 199, 232, 233, 235, 236, 237, 252, 271, 272, 279, 280
  Ca^{2+} mechanism, 95, 185, 191, 245–249, 250, 251, 254
  contraction, 17, 23, 25, 27, 29, 35, 161, 206, 207, 221, 231–252, 275, 277, 279, 299

Muscle (*contd.*)
  correlation of MgATPase with force, 238
  crab, 137, 144, 150, 162, 199, 232, 236
  decapod, 229
  end plate, 210, 234, 244, 249, 306, 307, 310, 319, 322, 325
  evidence for intracellular Ca^{2+}, 17–21, 161, 234–238
  heart, 19, 28, 29, 42, 62, 135, 138, 147, 150, 151, 152, 154, 157, 161, 162, 166, 170, 193, 197, 199, 207, 229, 231, 232, 233, 236, 249–252, 260, 268, 269, 271, 273, 274, 277, 279, 282, 283, 285, 303, 400, 406, 407, 408, 416, 426, 439–446
  historical, 19, 20, 21, 136
  mechanism of intracellular Ca^{2+} release, 242–245
  metabolism, 268, 269, 279, 280–286
  molluscan, 229, 230
  pathology, 398, 399, 400, 406, 407, 409, 411, 412, 414–416
  primary regulators, 207, 233
  proteins and enzymes, 260, 261, 279
  secondary regulators, 233, 249–252
  problems, 231–234
  regulation, 238–244
  relaxation, 244, 251
  sarcoplasmic reticulum, 13, 20, 24, 91, 274
  size, 232
  skeletal, 19, 24, 29, 62, 83, 126, 135, 136, 150, 154, 159, 161, 162, 186, 187, 199, 207, 229, 232, 233, 236, 245, 260, 273, 279, 282, 283, 400, 409, 419, 434, 438
  smooth, 19, 24, 29, 62, 126, 135, 150, 154, 162, 198, 226, 229, 232, 233, 236, 245, 249, 269, 272, 279, 283, 299, 300, 301, 303, 400, 434, 438, 439
  types, 232, 233
  ultrastructure, 68–70, 225, 227
Muscle disease, 69, 70
Myofibrils, 225
Myosin (and Myosin MgATPase), 126, 210, 220–231, 232, 234, 235, 245, 246, 257
Myosin light chain kinase, 221, 226, 228, 229–231

Na^+–Ca^{2+} exchange, see Sodium dependent Ca^{2+} efflux
Na^+–K^+ MgATPase (*see also* Sodium pump), 123, 440, 442

NAD 189, 421
NADH, 192, 194, 196, 397, 421
NAD kinase, 126, 472
NAD-linked isocitrate dehydrogenase, 260
NADPH oxidase, 350, 351, 352
Needham, 18, 19, 20
Nematocyst, 309, 316
Nematoda, 9, 12, 22, 233
Nernertini, 264
Nernst equation, 59, 139
Nerve terminals, 150, 186, 199, 202, 244, 312, 324, 325, 327, 329, 330, 344, 345, 411, 432, 449
Neurophysin, 344
Neurohormones, 387
Neuromuscular junction, 135, 457
Neurones (*see also* Mollusca, Vertebrate), 166
Neurosecretion (*see also* Secretion), 239
Neurotransmitters (and receptors), 16, 18, 24, 26, 28, 207, 233, 239, 308
Nexin, 218
Ni^{2+}, 32, 33, 151, 152, 153, 328, 331, 449
Nifedipine, 33, 151, 445, 446
Non-vesicular secretions, 312, 353–361
Noradrenaline, 135, 161, 249, 264, 265, 268, 272, 277, 285, 287, 291, 299, 308, 321, 355, 439, 440
NTA (nitrilotriacetic acetic acid), see Calcium buffers
Nuclear Ca^{2+}, see Calcium
Nuclear magnetic resonance, 83, 107, 312, 313, 476
$5'$-nucleotidase, 342
Nucleic acids (*see also* Calcium), 92, 93

Obelin, 45, 50, 51, 53, 54, 55, 56, 61, 62, 64, 124, 342, 403, 404, 409, 410, 411, 412
Oestradiol, 197
Ohm's law, 233
Oligomycin, 32, 33, 189
Orcyte maturation, 364, 386–390
Opiates, see Morphine
Origin of life, 1, 5
Origin of the earth, 1, 5
Osmotic balance, 14, 15, 21, 161
Ouabain, 140, 202, 241, 308, 333, 379, 440, 442
Ovary, 268, 269
Oxalate, 116
Oxygen, 179, 191, 252
Oxygen radicals, 306, 318, 347, 349–353, 380, 406, 411
α-Oxoglutarate dehydrogenase, 194, 286
oxytocin, 233, 283, 288, 308, 344

Pancreas, 265, 272
Pancreozymin, see Cholecystokinin
Pancreatic β cells, 28, 62, 126, 147, 150, 154, 188, 199, 215, 306, 308, 314, 316, 318, 321, 322, 325, 326, 327, 328, 330, 332, 333, 334, 337, 339, 344, 345, 346, 409, 434, 443
Paramyxoviruses, 338, 339, 402
Paramyosin, 228
Parathyroid hormone, 7, 9, 12, 24, 154, 197, 205, 321, 363, 394
Parotid, 272, 300, 309, 316, 328, 330, 345, 346, 353, 357, 358
Parthenogenesis, 19, 362, 364, 381
Parvalbumin, 94, 107, 118, 119, 124, 128, 186, 226
Passive Ca^{2+}, see also Calcium, 37, 38, 424, 425
Pathology, 7, 114, 393–429
Permeability coefficients, see Membranes
Phaetohaemagglutinin (PHA), 154, 377–380
Phalloidin, 406
Phagocytosis, 26, 27, 29, 62, 220, 305, 347, 348, 349, 350, 404
Pharmacology, 429–452
Phenol Red, 72
Phenothiazines, 33, 248, 403, 407, 430, 437–439
Pheremones, 178
Phonoreceptors, 178
Phosphatase, 221, 224, 231, 276, 286
Phosphate (see also Calcium phosphate), 116, 191, 240
Phosphatidyl inositides (PI effect), 89, 297–301, 334, 357, 438
Phosphodiesterase, 93, 94, 105, 109, 123, 126, 176, 177, 231, 438, 447, 448
Phosphoenolpyruvate carboxykinase, 287, 289, 293, 299, 300, 301
Phosphofructokinase, 91
Phosphoglycerate kinase, 107
Phospholamban, 250, 251
Phospholipids and metabolism, 89, 92, 93, 137, 240, 241, 343, 377, 420, 438
Phospholipid methylation, 297–301, 334
Phospholipases, see Calcium
Phosphorylase, 226, 261, 276, 278, 280–284, 416, 477
  Ca^{2+}, see Phosphorylase kinase
  cyclic AMP, see Phosphorylase kinase
Phosphorylase kinase, 226
  Ca^{2+}, 24, 91, 124, 260, 261, 275, 276, 280–284
  Calmodulin, (δ subunit), 126
  proteolysis, 123, 260, 425
  structure, 281

Phosphorylation, 120, 171, 176, 177, 212, 221, 222, 224, 229, 230, 231, 241, 248, 249, 250, 251, 261, 274, 275, 276, 281, 282, 283, 284, 286, 289, 295, 344, 354, 355, 359, 379, 463
Photomultiplier, 55, 172
Photoproteins, see Aequorin, Obelin
Photoreceptors, 28, 30, 35, 62, 161, 162, 166, 171–177, 178, 186, 199
Phyla, 22
Pineal gland, 272, 299
Pinocytosis, 26, 29, 220, 305, 347, 348
Pituitary, 154, 199, 273, 303, 308, 316, 321, 330, 337, 340, 344, 346
Plants, 6, 12, 21, 22, 126, 171, 177, 191, 207, 210, 222, 257, 264, 265, 267, 277, 306, 309, 312, 363, 374, 404, 471–473
Plant hormones, 261, 472
Plant lectins, 368, 377, 378, 379, 380
Plant mitochondria, 186, 194
Plasmodium, 223
Platelets, 35, 62, 72, 126, 188, 220, 268, 272, 299, 301, 308, 312, 316, 321, 326, 327, 330, 333, 345, 346, 347, 409, 411
Platyhelmintha (tapeworms), 6, 22
Pleiotypic response, 364, 370
Poisson distribution and equation, 100, 101, 467, 468
Polarography, 50
Pollack, 19, 49, 457, 461
Polyamine, 48, 180, 321, 337
Polychaetes, 180, 181, 184, 265
Polyethylene glycol, 339, 340, 406
Polymorphs, 259, 272, 277, 330, 340, 342, 403, 405, 416, 419
  chemiluminescence, 29, 65, 179, 306, 350–353, 439
  chemotaxis, 29, 207, 222, 252, 253, 254, 277
  phagocytosis, 310, 347, 348, 349, 350
  secretion, 306, 309, 312, 315, 316, 327, 329, 337, 341, 345, 346, 347
Pontin, 2
Porifera (sponges), 6, 7, 72, 264, 372
Potassium, 138, 139, 140, 144, 307, 355, 357, 439, 449
  cell and tissue content, 13, 413, 477
  conductance, 50, 160, 162–165
  discovery, 3
  fluid concentration, 8, 9, 10
  intracellular free concentration, 13
  ^{42}K, 195
Potassium channels, 148, 152, 153, 242
Potassium 'gun', 178

554

Primary regulators, 27, 28, 38, 233, 265, 266, 291, 315, 316, 317, 318, 319, 345, 354, 368, 370
Procain (*see also* Local anaesthetics), 388, 389
Progesterone, 387, 388, 390
Prokaryote motility, 207–209
Prokaryotes (*see also* Bacteria, Blue-green algae), 5, 21, 22, 473
Prostaglandins, 28, 197, 233, 249, 253, 262, 264, 268, 292, 295, 298, 299, 300, 301–303, 315, 334, 347, 391, 412, 439, 446
Protein kinase, 123, 126, 176, 274, 276, 281, 282, 283, 284, 288, 295, 296
Protein phosphorylation, see Phosphorylation
Protein synthesis, 198, 359, 379, 405
Proteolysis, 261, 265
Prothoracic gland, 354
Prothrombin, *see* Thrombin
Protochordata (sea squirts), 383
Protoplasm, 18, 19, 21, 208, 425
Protoplasmic flow or streaming, 19, 222
Protostomia, 230
Protozoa (includes Amoeba), 6, 19, 22, 29, 35, 51, 62, 126, 150, 152, 157–160, 161, 162, 184–188, 199, 207, 208, 217, 219, 222, 252, 264, 265, 306, 310, 316, 347, 349, 370, 399, 401, 402
Pseudopod formation, 19, 208, 222
Purinergic nerves, 319
Purkinje fibres, 150, 163, 170, 233
Pyruvate carboxylase, 91, 123, 194, 260, 285, 287
Pyruvate dehydrogenase, 91, 122, 194, 260, 261, 278, 284, 285, 286, 296
Pyruvate kinase, 91, 123, 260, 284, 285, 289

Quin II, 46, 50, 59, 62, 198, 378

Radiolarians (*see also* Protozoa, Thalassicola), 179, 180, 182, 184
Radium, 85
^{81}Rb, 195
Receptor antagonists, 441
Redox reactions, 14
Regulators, 23
Regulatory Ca^{2+} binding proteins, *see* Calcium, Calcium binding proteins
Relaxation time, 114
Renal cortex, 268, 284, 287, 289
Renal medulla, 268, 273, 277
Renin, 308

Reticulocytes, 299
Retina, 269, 316, 345
Retinal, 172, 173
Rheumatoid arthritis, 349, 412, 419, 421
Rhodopsin, 94, 172, 173, 175, 176, 177
Ringer, Sidney, 10, 11, 17, 18, 19, 234, 457, 461
RNA (*see also* Calcium—nucleic acids), 261, 376, 401, 402, 405
Rotifera, 22
Rubenstein, 19
Ruthenium red, 32, 33, 37, 167, 189, 191, 192, 193, 286, 327, 360, 392
Ryanodine, 32, 33

S-100, 94, 186
Saliva, 24
Salivary gland, 30, 31, 42, 71, 162, 167–171, 188, 197, 268, 271, 299, 300, 310, 312, 327, 353, 354, 355, 356, 357
Sarcoplasmic reticulum (*see also* Muscle), 13, 20, 24, 91, 135, 161, 186, 187, 189, 226, 239–245, 250, 251, 433, 448, 450
  Ca^{2+} induced Ca^{2+} release, 189, 423
  discovery, 20, 235, 239, 457
  electrical induced Ca^{2+} release, 189
  pathology, 415, 416
  proteins, 226, 239–242
  structure, 240
Scandium, 73
Schwann, 17
Sea firefly, *see* Cypridina hilgendorfi
Sea hare, *see* Aplysia, Molluscan neurones
Sea urchin, *see* Echinodermata
Second messengers (*see also* Intracellular Ca^{2+}, Cyclic AMP, Cyclic GMP, 261, 266
Secondary regulators, 27, 28, 38, 233, 265, 266, 291, 315–319, 344–347, 354
Secretin, 300, 354, 355
Secretion, 17, 95, 208, 268, 271, 277, 299
  classification, 26, 307–315
  effect of removal of external Ca^{2+}, 29, 30, 306, 307, 308, 309, 310, 321, 322, 325, 326
  electrophysiology, 329–332
  evidence for intracellular Ca^{2+}, 35, 36, 93, 319–329
  non-vesicular, 310, 312, 353–361
  pathway, 315
  phenothiazine inhibition, 439
  stimuli, 28, 146, 316, 344, 411
  vesicular, 307–347

Secretory granules (vesicles), 186, 208, 215, 312, 335–344
Sendai virus, 29, 35, 338, 339, 340, 341, 342, 353, 401–405
Semliki Forest virus, 341, 404
Sensory mechanisms (and receptors), 27, 29, 178, 210
Sexual reproduction, 368–370
Shells, 1, 6, 186, 205
Sialic acid, 72, 92
Siemen, 159
Single cell analysis, 303, 466, 476, 477
Singlet oxygen ($^1O_2$)
Skeletal muscle, see Muscle
Skin, 42
Slim mould, 29, 62, 207, 208, 222
Slow reacting substance, 321, 446
Smooth muscle, see Muscle
Soap, 8, 85
Sodium, 8, 9, 10, 136, 138, 139, 140, 144, 170, 172
  fluid concentration, 8, 9, 10, 279, 280
  intracellular concentration, 13
  tissue content, 13
Sodium bicarbonate ($NaHCO_3$), 353–356, 358
Sodium channels, 148, 150, 151, 152, 153, 155, 158, 159, 173, 177, 244
Sodium dependent Ca^{2+} efflux, 32, 124, 138, 170, 198–201, 203, 204, 449, 452
Sodium pump (see also Na^+–K^+MgATPase), 139, 140, 201, 452
Solubility product, 114, 115, 116
Somatostatin, 308
Spasmin, 254
Spasmoneme, 99, 124, 207, 254, 255
Spermatozoa, 28, 30, 35, 126, 207, 218, 220, 252, 254, 309, 321, 339, 362, 363, 364, 368, 380–386, 390–392
Squid, see Loligo, Molluscan neurones
Staircase phenomenon, 157
Starfish, see Echinodermata, Marthasterias
Statocysts, 7
Statoreceptors, 178
Steroids (see also Glucocorticoids), 197, 264, 304, 312, 358–361
Steroidogenesis, 154, 358–361
Steroid secretion, 312, 358–361
Streptolysin O, 341, 398, 401, 405, 406
Stretch receptors, 178
Strontium, 37, 81, 85, 97, 98, 170, 254, 319, 326, 460
Substrate cycles, 193, 287
Succinate dehydrogenase, 260

Succinate oxidase, 194
Sulphate, 116
Synaptosomes, 199, 328
Sweat, 354

Taeniae coli, 231
Tango receptors, 178
Tau factor, 214
T and B lymphocytes, see Lymphocytes
TEA (tetraethylammonium), 149, 152, 153, 164
Teeth, 7, 15
Tetany, 12, 156, 271, 272
Tetrodotoxin, 148, 151, 152, 153, 159, 182, 328, 331
Theophylline (see also Xanthines), 320, 321, 346, 448
Thermoreceptors, 178
Thrombin (and prothrombin), 23, 24, 91, 109, 122, 299, 308, 321
Thromboxanes, 249, 264, 300, 301–303, 347
Threshold phenomena, 25, 26, 27, 130, 131, 132, 171, 177, 205, 210, 234, 254, 255, 266, 303, 304, 317, 348, 364, 370, 392, 453, 463–468
Thymocyte, 279, 364
Thyroglobulin, 301, 344, 347, 348
Thyroid, 268, 272, 300, 309, 310, 316, 317, 321, 344, 345, 346, 347, 409, 449
Thyroid stimulating hormone (TSH), 268, 301, 309, 321, 348
Thyroxine (and T_3), 197, 262, 264, 265, 300, 304, 317
Tipomole, 195, 196, 477
Tissue calcification, 6–8, 393–395
Tolbutamide, 331
Tonofilaments, 210
Toxins, 397, 398, 401, 405–407, 419
Trancellular Ca^{2+} transport, 205
Transglutaminase, see Calcium
Transition metals, 13, 14, 21, 134
Trichocyst, 309, 316, 327
Trifluoperazine (see also Phenothiazines), 127, 352, 430, 438, 439
Triglyceride, see Lipolysis
Trimethyl lysine, 125, 127
Triton X-100 on cells (and other detergents), 158, 218
Tropomyosin, 220, 221, 224, 226–231, 235, 245, 246, 247, 251
Troponin (and troponin I and T), 93
  discovery, 93, 1123
  other names, 229

Troponin (*contd.*)
  role in muscle contraction, 225–231, 245, 247, 248, 251
Troponin-C, 210, 217, 221, 272, 281, 282, 284, 415
  amino acid sequence, 94, 129
  Ca^{2+} binding, 24, 86, 91, 93, 107, 109, 113, 124, 126, 132, 245–249, 251, 282
  Cardiac, 246
  concentration in muscle, 126, 238, 282
  discovery, 93, 123, 235
  kinetics, 248
  Mg^{2+} binding, 246, 248
  phenothiazines, 438
  role in muscle contraction, 220, 229–231
  stereochemistry, 118, 119, 120, 247
Trypanosomes, 273
Trypsin(ogen), 17, 24, 91, 94, 109, 122, 218
Tryptophan hydrolase, 124
Tryptophan monooxygenase, 126
Tubulin, 211–219, 254, 336, 337, 375, 391
Tumours, 268, 273
Tunicate eggs, 160
Types of tissue injury, 395, 396
Tyramine, 317, 320, 321

Unitary hypothesis, 19, 459–468
Unit cell activation hypothesis, 465–468

Vasopressin (antidiuretic hormone), 29, 35, 268, 271, 273, 280, 283, 287, 288, 299, 308, 344, 354
Verapamil (*see also* Calcium channels), 35, 37, 151, 153, 154, 430, 444, 446

Veratride, 333
Vertebrates, *see* Chordata
Vertebrate neurones, 62, 150, 162, 222, 283, 306, 307, 316, 434
Vesicular versus non-vesicular secretion, 307, 312
Villin, 221
Vinblastine, 213, 215, 337
Vincristine, 215
Vinculin, 221, 225
Viruses (*see also* Semliki Forest, Sendai), 15, 29, 364, 368, 402–405, 424
Vision, 17, 28, 29, 30
Vitalism, 18
Vitamin A, 172, 341, 401, 406
Vitamin D, 7, 9, 25, 94, 197, 205, 394, 398, 406, 424, 443, 471
Voltage clamp (*see also* Membrane potential), 151, 156, 157, 170, 172

Water (hard and soft), 8, 9, 13, 85

Xanthines (*see also* Caffeine, Theophylline), 267, 327, 430, 447, 448, 452
X537A (*see also* Ca^{2+} ionphores), 32–37, 204
X-ray microprobe analysis, 66–70, 187, 188, 205, 254, 255, 333, 334, 397, 399, 422, 423

Yeast, 42, 191, 260
Yin–Yang hypothesis, 461, 462, 463

Zeta potential, 103
Zn^{a+}, 314, 450

SUFFOLK UNIVERSITY
MILDRED F. SAWYER LIBRARY
8 ASHBURTON PLACE
BOSTON, MA 02108